Biotechnological Applications of Biomass

Edited by Thalita Peixoto Basso, Thiago Olitta Basso and Luiz Carlos Basso

Published in London, United Kingdom

IntechOpen

Supporting open minds since 2005

Biotechnological Applications of Biomass
http://dx.doi.org/10.5772/intechopen.89320
Edited by Thalita Peixoto Basso, Thiago Olitta Basso and Luiz Carlos Basso

Contributors
Djamila Kerrouche, Nabila Rouba, Vincent Chaleix, Haresh Manyar, Gunjan Deshmukh, Jiang Wei, Abdul Basit, Kashif Rahim, Mateus S. Souza Amaral, Arnaldo M.R. Prata, Messias B. Silva, Carla C.A. Loures, Fabiano L. Naves, Gisella L. Samanamud, Abdelilah Meddich, Abderrahim Boutasknit, Mohamed Anli, Meriame Ait Ahmed, Abdelilah El Abbassi, Hanane Boutaj, Boumezzough Ali, Łukasz Wajda, Magdalena Januszek, Tariq Umar, Rini Setiati, Septoratno Siregar, Deana Wahyuningrum, Robson Evaldo Gehlen Bohrer, Danni Maisa da Silva, Divanilde Guerra, Daniela Mueller De Lara, Diego Armando Amaro Da Silva, Eduardo Lorensi De Souza, Marciel Redin, Maiara Figueiredo Ramires, Ramiro Pereira Bisognin, Mastrângello Enivar Lanzanova, Chitra Murugan, Bruna Chagas, Adriano Rangel, Izabel Pereira, Cintia Castro, Bruno De Moura, Roberto Sassi, Stela Urbano, Fabiana Camara, Dinh Quan Nguyen, Thanh Phong Mai, Takuyuki Yoshioka, Temitope Olumide Olumide Olugbade, Anthony Anukam, Jonas Berghel, Alfred Mitema, Naser Aliye Feto, Richard Sayre, Meghna Rajvanshi, Donald Tyoker Kukwa, Maggie Manimagalay Chetty, Lucielen Oliveira Santos, Pedro Garcia Pereira Silva, Sharlene Silva Costa, Taiele Blumberg Machado, Brenno S. Santos Leite, Sibele A. F. Leite, Daniel J.O. Ferreira, Vanessa F. C. Lins, Cintia de Faria Ferreira Carraro, Carla Cristina Almeida Loures, André Celestino Celestino Martins, Ana Carolina da Silva Faria, Safiya Noorzai, Casparus J.R. Johannes Verbeek, Adeolu Adesoji Adediran, Joan Nyika, Adeniyi Olayanju, Olanrewaju Seun Adesina, Francis Odikpo Odikpo Edoziuno, Flora Itzel Beltran Ramirez, Elizabeth Quintana Rodríguez, Domancar Orona Tamayo, José Nicacio Gonzalez-Cervantes, María Alejandra Rivera Trasgallo, Adriana Berenice Espinoza Martinez, Abdurrahman Garba, Emmanuel Galiwango, Ali H. Al-Marzouqi, Edward Kwaku Armah, Jeremiah Adebisi Adedeji, Rita Maria De Brito Alves, Ana Luíza Freitas Ferreira, Alisson Dias Da Silva Ruy, Antônio Ésio Bresciani, Luiz Antônio Magalhães Pontes, Nusrat Iqbal, Amrish Agrawal, Saurabh Dubey, Jitendra Kumar, Cid Marcos G. Andrade, Mauro Antônio da Silva Sá Ravagnani, Eugênia Leandro Almeida, Sérgio Inácio Gomes, Emmanuel Chukwuma, Faith C. Okey-Onyesolu, Daniel C. Anizoba, Joseph C. Ubah, Didiana Galvez-Lopez, Raymundo Rosas-Quijano, Abraham Ontiveros-Cisneros, Alfredo Vazquez-Ovando, Noé Montes-García, Arturo Diaz-Franco

First published in London, United Kingdom, 2021 by IntechOpen
IntechOpen is the global imprint of INTECHOPEN LIMITED, registered in England and Wales, registration number: 11086078, 5 Princes Gate Court, London, SW7 2QJ, United Kingdom
Printed in Croatia

British Library Cataloguing-in-Publication Data
A catalogue record for this book is available from the British Library

Additional hard and PDF copies can be obtained from orders@intechopen.com

Biotechnological Applications of Biomass
Edited by Thalita Peixoto Basso, Thiago Olitta Basso and Luiz Carlos Basso
p. cm.
Print ISBN 978-1-83881-180-8
Online ISBN 978-1-83881-182-2
eBook (PDF) ISBN 978-1-83881-183-9

For EU product safety concerns:
IN TECH d.o.o., Prolaz Marije Krucifikse Kozulić 3, 51000 Rijeka, Croatia, info@intechopen.com or visit our website at intechopen.com.

Meet the editors

Thalita Peixoto Basso received her bachelor's degree in Agriculture Engineering. During this period, she studied the fermentation characteristics of *Saccharomyces cerevisiae* isolated from ethanol industrial processes. She obtained her master's degree from the University of Sao Paulo (ESALQ/USP), Brazil. During this time, she isolated and selected fungi with high cellulose activity for enzymatic hydrolysis of sugarcane bagasse. She received her Ph.D. from the Agricultural Microbiology Program, ESALQ/USP, with a period as a visiting scholar at the University of California Berkeley and the Energy Bioscience Institute. Meanwhile, she worked on the improvement of *S. cerevisiae* by hybridization for increased tolerance toward inhibitors from second-generation ethanol substrates. Currently, she is a collaborating professor and postdoctorate working with metabolomics and proteomics of fermentation processes in the Genetics Department, ESALQ/USP.

Thiago Olitta Basso is an assistant professor in the Chemical Engineering Department, University of Sao Paulo (USP), Brazil. Currently, he is involved with teaching and researching microbial physiology, microbial interactions, and strategies of metabolic and evolutionary engineering for the biotechnology sector. He worked at Novozymes as a senior scientist in its R&D Department, acting as a project leader in yeast physiology and fermentation optimization for traditional (1G) and advanced (2G) biofuels. He holds a BS in Pharmacy and Biochemistry from USP and an MSc in Biotechnology from the University of Abertay, Scotland, UK. He obtained his Ph.D. in Biotechnology from USP, under Andreas Gombert´s supervision. Part of his Ph.D. was done at the Delft University of Technology, in the group of Prof. Jack Pronk.

Luiz Carlos Basso holds a bachelor's degree in Agriculture Engineering from the University of Sao Paulo (ESALQ/USP) in 1969, a master's degree in Soil and Plant Nutrition from ESALQ/USP in 1973, Ph.D. in Biological Science from São Paulo State University (UNESP), and postdoctorate from Institut des Produits de la Vigne, Montpellier and Superior Technical Institute, Lisbon, in 1989. Since 1980, he has been involved with yeast biochemistry and physiology, aiming to increase ethanol yield using the fed-batch industrial process. During the last fifteen years, he has conducted a yeast selection program resulting in the most widely used *Saccharomyces cerevisiae* strains (PE-2 and CAT-1) in the Brazilian ethanol industry. Currently, he is a senior professor at ESALQ/USP teaching Biochemistry (for undergraduate students), Biochemistry and Physiology of Yeast Fermentation (for graduate students), and selecting tolerant strains for lignocellulosic inhibitors.

Contents

Preface

Using biomass to cogenerate energy and produce biological and biochemical products as well as bioproducts with economic importance is a great alternative to using fossil fuels, which have adverse effects on the environment. Biomass, as a sustainable and renewable energy source, has considerable importance in the production of several molecules with vast application in industries such as agriculture and pharmaceuticals. Additionally, alternative biomass sources and the development of thermochemical processes for efficient conversion of several biomass substrates contribute to the development of biofuel and bioenergy production.

The present book, *Biotechnological Applications of Biomass*, provides a comprehensive overview of the current state of the art of biomass utilization in agriculture and pharmaceuticals. It contains thirty-two chapters over eight sections: "Biomass (Pre)Treatments or Conversion," "Biomass for (Bio)Energy," "Alternative Biomass Sources," "Biochemical Products," "Microalgae Biomass," "Biotechnological Products," "Thermochemical Processes," and "Kinetic Models and Tools for Biomass Measurement".

This book will be interesting for researchers around the world involved with biomass utilization. It also contributes to the diverse research areas of biomass utilization and production of bioproducts.

Thalita Peixoto Basso,
Luiz de Queiroz College of Agriculture,
University of São Paulo,
Piracicaba, Brazil

Thiago Olitta Basso and Luiz Carlos Basso,
University of Sao Paulo,
Piracicaba, Brazil

Section 1

Biomass (Pre)Treatments or Conversion

Chapter 1

Biomass Conversion Technologies for Bioenergy Generation: An Introduction

Abdurrahman Garba

Abstract

Over the last century, there has been increasing debate concerning the use of biomass for different purposes such as foods, feeds, energy fuels, heating, cooling and most importantly biorefinery feedstock. The biorefinery products were aimed to replace fossil fuels and chemicals as they are renewable form of energy. Biomass is a biodegradable product from agricultural wastes and residues, forestry and aquaculture. Biomass could be sourced from a variety of raw materials such as wood and wood processing by-products, manure, fractions of organic waste products and agricultural crops. As a form of renewable energy, they have the advantages of easy storage, transportation, flexible load utilization and versatile applications. The aim of this study is to provide an overview for thermochemical and biochemical biomass conversion technologies that were employed currently. Attention was also paid to manufacture of biofuels because of their potentials as key market for large-scale green sustainable biomass product.

Keywords: biomass, bioenergy, biochemicals, conversion routes, green chemistry

1. Introduction

Biomass can be regarded as any organic material that originated from plants or animals. Thus, the United Nations Framework Convention on Climate Change UNFCCC in 2005 [1], defined it as follows;

"A non-fossilized and biodegradable organic material originating from plants, animals and micro-organisms. This shall also include products, by-products, residues and waste from agriculture, forestry and related industries as well as the non-fossilized and biodegradable organic fractions of industrial and municipal wastes".

The world's energy demand has been steadily increasing in the last several decades. This is due to rapid increase in industrialization, population and the quest for improvement of the living standards for societies. On the other hand, this has caused an irreversible damage to the environment which leads to global warming, and climate change. These issues have been the topic of discussion among scientist and policy makers at national and international levels on how to mitigate the problem. The modern society is emphasizing on shift from non-renewable to renewable energy (such as wind, solar, tide, geothermal and biomass) in their search for energy source. Before the discovery of fossil fuels such as petroleum products, coal, and natural gas, biomass was the main source of energy for heating and cooking [2]. Biomass is the term used to describe all materials that contain carbon in an organic form. This organic form of carbon can be transformed into inorganic through photosynthesis by forming bonds with other elements such as hydrogen, and oxygen using solar energy. The demolishing of these bonds (molecules) through physical or biological means, causes a closure in the cycle and making CO_2 to be regenerated. During the regeneration process, energy is released which can be converted into other forms of energy. Therefore, as long as these equilibrium is maintained between use and regeneration, biomass is a renewable or inexhaustible source of energy [3]. Biomass is expected to be the leading form of energy with a significant global energy load of about 10–15%. However, biomass has a share of about 90% of total energy requirements for remote and rural areas of the developing countries. Therefore, it is likely to remain the future leading source of energy feedstock for the developing countries since about 90% of the world population is expected to live in the developing world by 2050 [4–6].

Biomass accumulates chemical energy in form of carbohydrates through combination of solar power and carbon dioxide during the process of photosynthesis. This has made it to be a potential energy source since the carbon dioxide captured during photosynthesis could be released when it burns. It is cheap and available in all forms such as forest and agricultural residues, wood, by-products of biological materials, organic components of municipal and sludge wastes, etc. There are several ways to convert biomass into useful products which largely depends on biomass characteristics and the end product [7]. The technologies applied in the conversion of biomass are mainly categorized under thermochemical or biological methods.

2. Types of biomass conversion technologies

Biomass can be converted into several useful products for energy generation and chemicals. There are some factors that influence the choice of a conversion technology to be applied on the biomass. These factors include quality and quantity of the biomass feedstock, availability, choice of end-products, process economics and environmental issues (**Figure 1**) [9].

2.1 Thermochemical methods

The major options within thermochemical biomass conversion processes include combustion, gasification, pyrolysis, and liquefaction (**Figure 2**). The most practiced thermochemical conversion of biomass industrially is combustion process, which is used for heat and electricity generation. Most of biomass thermochemical conversions were carried out with or without the use of catalysts, though the use of catalyst has distinct effects on the end-products [10].

2.1.1 Gasification

The process of biomass gasification was discovered independently in France and England by the year 1798. The technology did not come into its limelight until 60 years later. The gasification process continued flourishing until 30 years later when natural gas from oil fields was discovered. Until 1970, the use of natural gas for cooking and lighting was substituted with liquid fuels due to discovery of oil. Generally, biomass gasification is an endothermic thermochemical conversion of solid biomass fuel using gasifying agents such as air, steam or CO_2 to form a mixture of combustible gases which may include H_2, CH_4, CO and CO_2. The process is carried out at temperatures between 800 and 1300°C. Nowadays, flexibility of the

DOI: http://dx.doi.org/10.5772/intechopen.93669

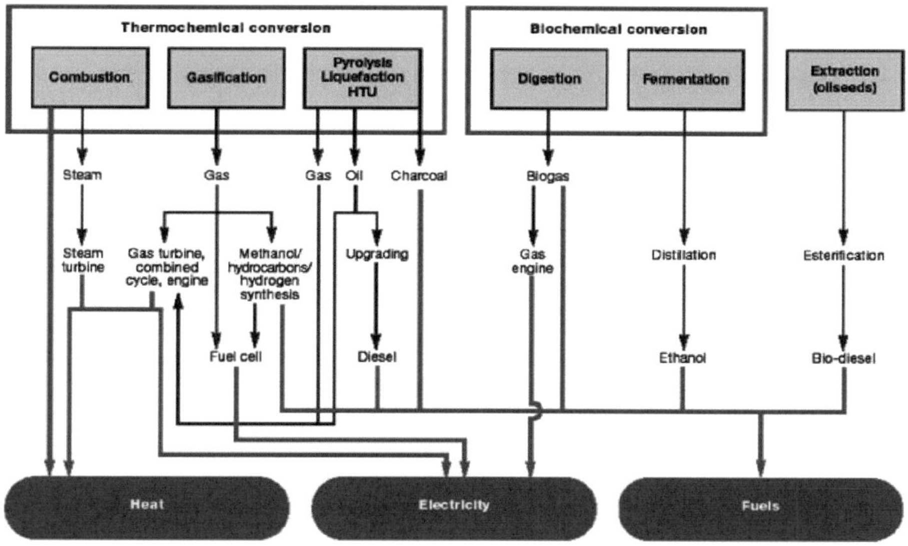

Figure 1.
Main biomass conversion routes [8].

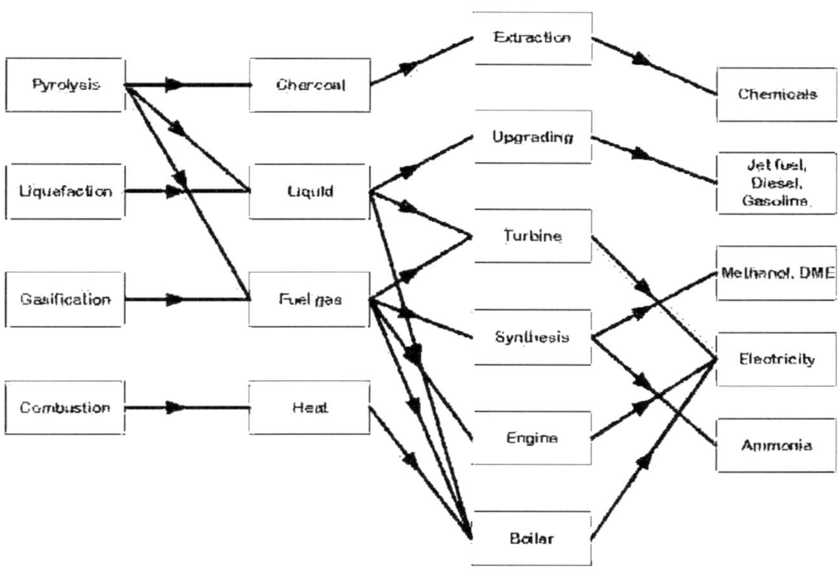

Figure 2.
Thermochemical conversion processes and end products [10].

gasification technology coupled with the different uses of the produced syngas, allows for the integration of biomass gasification with many industrial processes and as well with power generation systems [7].

Biomass feedstock characteristics such as particle size, moisture content, shape, heating value, carbon content and ash content significantly affect the gasifier performance. However, knowledge on feedstock parameters such as volatility, elemental analysis, heat content and biomass potential for fouling or slagging is essential for evaluation of gasification process [11]. Therefore, feedstock with low volatile contents are preferred for partial oxidation gasification, while those with high volatile content are more suitable for indirect gasification process [12].

Feedstocks for biomass gasification exists in different forms with each type having peculiar issues. Therefore, it is vital to predict suitable type of biomass for a specific gasifier type under defined conditions. Although, characteristics within specific biomass feedstock species is identical, the shape and size of the feedstock particles are useful in determining the difficulties that may arise during movement, delivery and as well as the feedstock behavior in the gasifier. The size and size distribution of the feedstock affect the gasification zone thickness, pressure drop in the bed and the maximum hearth load. To overcome some of this problems, biomass feedstock of uniform size were utilized [7].

Gasifier operation depends on moisture content of the biomass feed used. The use of feedstock with high moisture content reduces biomass conversion efficiency and as well the production rate. This is because the process discharges more fuel or heat in order to vapourize the excess moisture to the temperature of the syngas [13]. During the pyrolysis/gasification process, water need about 2.3 MJ/kg to vapourize and as well 1.5 MJ/kg to raise it to 700°C. Also, high moisture content in a biomass reduces the achieved temperature in the oxidation zone which results in incomplete cracking of the products released in the pyrolysis zone. Consequently, high moisture content in the biomass feedstock affect the syngas composition or quality due to production of CO_2 from reaction between the moisture. Furthermore, using feedstock that has high moisture content results in the production of syngas with high moisture, which subsequently course additional stress on downstream cooling and filtering equipment [14].

2.1.1.1 Hydrothermal gasification

Hydrothermal gasification is a biomass treatment that involves the use of water at high temperatures and pressures. Products formed during this process is as a result of different reactions that takes place in the biomass which mainly depends on factors like temperature, pressure, and time of treatment. To understand the process, behavior of water at high temperature and pressure must be known. **Figure 3** indicated the phase diagram of water, where at 273.15 K and atmospheric pressure (0.101325 MPa), ice melts to liquid water, while at 373.15 K liquid water boils and vapourized to steam. However, boiling point of water is affected by pressure and this means at high pressure the boiling point decreases, while at low pressure it increases. Likewise, pressure has

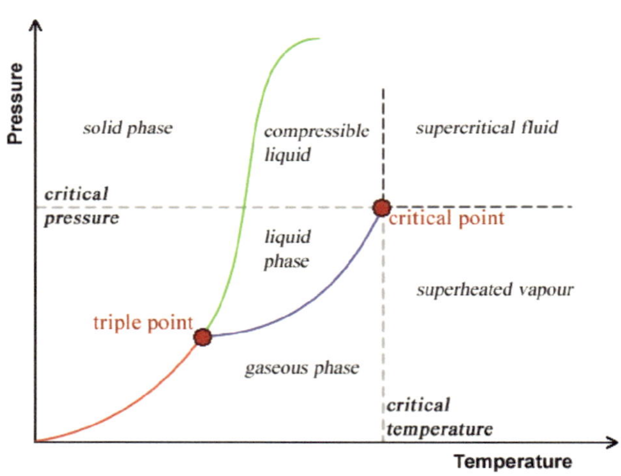

Figure 3.
Phase diagram of water [9].

effect on volume of water when it changes to steam. The volume of water increases greatly when it changes to steam. This change in volume is as high as 1600 times under atmospheric pressure.

At increased pressure, the volume of liquid water is not affected when compared to steam volume. Therefore, under increased pressure, the increase in volume associated with the phase change becomes smaller (**Figure 4**). The volumes for both water and steam were found to be equal at 22.1 MPa at the phase change. Also, when the pressure is higher than this value, no noticeable phase change is seen. At this point, the pressure is called the critical pressure of water, while the temperature is called critical temperature of water which corresponds to 647 K. This point on the phase diagram, is called the critical point. If the temperature and pressure are above these critical values, the water is called supercritical water, while when the values are below the critical values, the water is called subcritical [9].

Hydrothermal treatment of biomass can be carried out in either supercritical or subcritical water. That is when the temperature and pressure of the water is high. The process employs low temperatures ranging between 150 and 250°C. Under these conditions, the polymeric components of the biomass such as hemicellulose and lignin are dissolved together with small fraction of cellulose [15]. This process is mainly physical and requires harsh reaction conditions since the decomposition of the polymeric substances is limited. The process is often employed for saccharification of cellulose (**Figure 5**) or for an increased biomethanation of lignocellulosic biomass [16–18].

2.1.2 Pyrolysis

The term pyrolysis is defined as the thermal depolymerization of organic matter in the presence of nitrogen or absence of oxygen. Pyrolysis is an exothermic reaction with heat requirements that ranges between 207 and 434 kJ/kg of which many wood based and agricultural biomass were heated in an inert atmosphere to produce vapours and a carbon rich residue. The vapours composed of fragments from cellulose, hemicellulose and lignin polymers. These vapours can be condensed into free flowing organic liquid known as the bio-oil. On the other hand, the remaining carbon residues is left as bio-char (**Figure 6**) [20].

The polymeric substances distribution in bio-oil largely depends on the lignocellulosic contents of the biomass feed [21]. Many researchers investigated the individual pyrolysis characteristics of cellulose, hemicellulose and as well lignin. Hemicellulose was observed to decomposes at 220-315°C, cellulose decomposes between the range of 314-400°C, while lignin decomposition takes place from 160 to 900°C and it

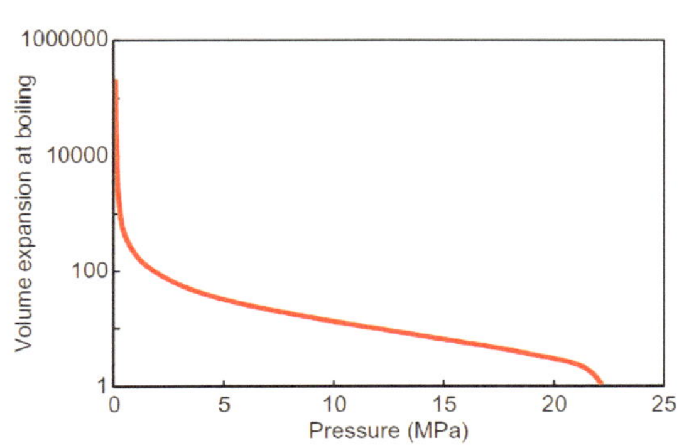

Figure 4.
Pressure effect on volume change when water changes into steam [9].

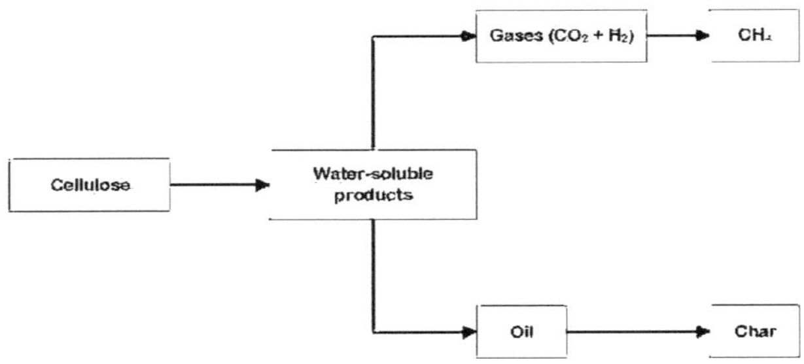

Figure 5.
Reaction network for hydrothermal gasification of cellulose [9].

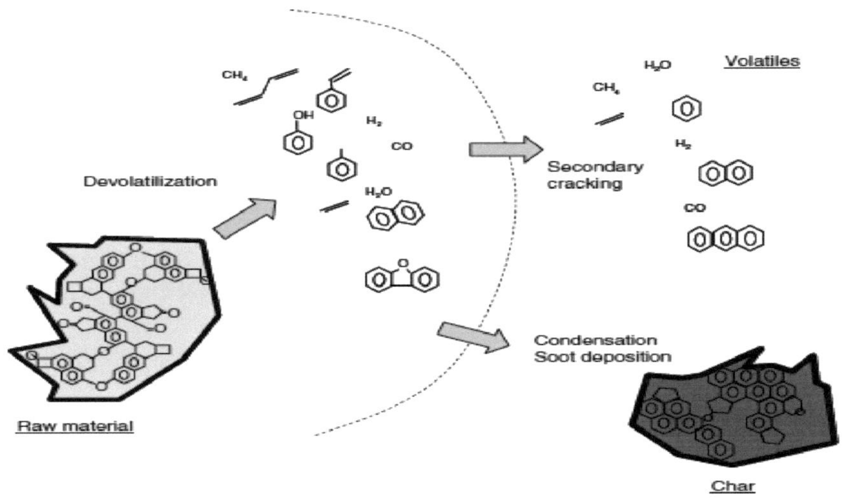

Figure 6.
Carbonization reaction scheme of a carbonaceous material [19].

generates a solid residue with highest percentage of about 40% [22]. From energy view point, cellulose pyrolysis was observed to be an endothermic reaction, while the reactions of hemicellulose and lignin is an exothermic. The gaseous products obtained from pyrolysis of these three components were similar and mainly comprises of CO_2, CO, CH_4 and other organic gases. Micro-GC was employed to analyzed the releasing behaviour of the H_2 and total gases released when the three gases were pyrolyzed in a packed bed. Hemicellulose was observed to have higher yield for CO_2, cellulose gives higher yield for CO with high presence of aromatic ring and methoxyl, while the lignin cracking and deformation yields higher H_2 and CH_4. Cellulose pyrolysis involves the cleavage of glycosidic groups via dehydration which is followed by the breakdown of anhydroglucose units. The dehydration and breakdown of sugar molecules at lower temperatures, results in the formation of char. Shafizadeh and Fu [23] reported char yield of 34.2% for the pyrolysis of pure cellulose in the absence of air and at 300°C. At high temperatures, there is enough energy to initiate the rapid cleavage of glycosidic bonds and evaporation of gaseous products was favoured. However, the distribution of cellulose, hemicellulose and lignin in a bio-oil is predominantly determined by the interactions between these components rather than just their quantities. Rowell [24] suggested that hemicellulose and cellulose were bonded through hydrogen bond, while hemicellulose and lignin were covalently bonded via ester bonds. The bonds that exist between these polymeric substances influence the pyrolytic behaviour of the biomass which may bring about a difference in products distribution when compared to a sample prepared synthetically by physical mixing. Couhert *et al.* [25] pyrolyzed two mixtures at 950°C containing the three components. One of the mixtures was prepared by simple mixing, while the other was prepared by intimate mixing. He discovered that, the yield for CO_2 increases with an increase in intimacy of the mixture. Hence, the effect of components interaction may differ in a physical mixture in comparison with the actual biomass sample, because the structure of the biomass can affect pyrolysis outcome which alter selectivity for certain products [26].

The necessary conditions for pyrolysis are temperature, pressure, heating rate, residence time, environment, catalyst, etc. This conditions greatly determines the nature of the products formed after pyrolysis [27]. Therefore, the pyrolysis conditions can be adjusted to obtain a desired product. It is well known from literatures that high temperature and short residence time favours formation of condensable fractions, high temperatures and longer residence time favours non-condensable gaseous products, and as well solids fractions are only favoured at low temperatures [28]. Depending on the pyrolysis conditions, the process can be classified as follows;

2.1.2.1 Fast pyrolysis

Recently, fast pyrolysis which is an advanced technology is gaining attention because of an increasing need for the production of fuel oil from biomass. As a continuous process, fast pyrolysis is aimed to prevent further cracking of the pyrolytic fractions to non-condensable compounds. During the process, the parameters that give high oil yield were carefully controlled in which the primary parameter is high rates of heat transfer. This parameter could be achieved by grinding the biomass feed finely. The finely ground biomass feed is heated rapidly at high temperatures between 450–600°C for a very short residence time of typically less than 2 seconds. The liquid yield for wood fast pyrolysis was reported to be as high as 75% [29, 30]. Since the process takes place in a very short period, not only chemical kinetics, but rate of heat and mass transfer, and as well transition phenomena plays an important role in determining the chemistry of the end products. Tailored products could be obtained by setting the necessary parameters at optimum [29].

2.1.2.2 Intermediate pyrolysis

In comparison with fast pyrolysis, intermediate pyrolysis is operated at optimum temperature range of 300–500°C. The liquid products obtained during the process is less viscous and contains low tar. However, the chemical reactions taking place during intermediate pyrolysis are more controlled and thus the process offers a wide range of parameter variations for process optimization. Although low yield for liquids of up to 55% were obtained during this operation, large sizes for biomass feed are acceptable that may be coarse, chopped, shredded or ground [31].

2.1.2.3 Slow pyrolysis

Slow pyrolysis is the carbonization of a biomass feed without condensing the pyrolysis products. The process is carried out in batches at low temperatures, slow heating rate and for a long residence time. Though, most of the literatures present about the process were based on its use to produce solid fuels such as charcoal and bio-char, but it can also be used to produce liquid fuels and bio-gas [32]. Temperatures as low as 0.1–2°C were reported by literatures. Slow pyrolysis is the oldest technique used for biomass conversion when the desired end product is charcoal or biochar. The vapours produced during the process were not condensed usually, but they could be used in the process to directly or indirectly provide heating. Moisture of about

15–20% were reported and it affects the properties of the solid fuels produced during the process [20]. The biomass feed sizes can vary from ground to a whole log.

2.1.3 Torrefaction

Torrefaction is a slow and mild pyrolysis process that is usually carried out at low temperatures between 225°C-300°C. The process is aimed at increasing the biomass energy density and as well its fuel properties [33]. This is achieved by removal of biomass moisture content and other superfluous volatiles. During the process, the biopolymeric substances such as cellulose, hemicellulose and lignin were partly decomposed to release organic volatiles. The product obtained at the end of the process is a dry and black residual solid regarded as torrified biomass. The torrified biomass is hydrophobic and soft which can easily be crush, grind or pulverized [20, 33].

2.1.4 Combustion

The process of combustion is a widely applied biomass conversion technology that was functional to a sizeable portion of human race since the advent of human civilization. It is widely applied even today for burning of wood and agricultural residues to make pot fires and stoves in order to provide heat and light energy for cooking and heating. Combustion process is frequently used for the conversion of lignin-rich biomass. The process could be applied in two broad ways, that is either by direct conversion of the whole biomass feedstock or by biochemical conversion in which some portions of the biomass remained. Compared with the other biomass conversion technologies, the process is largely non-selective in terms of the biomass feedstock. During the process, biomass feedstock is converted to CO_2 and water including smaller amount of other species which depends on the composition of the biomass and the process parameters. However, combustion of biomass largely depends on energy content of the feedstock. The amount of heat energy released during the process depends on feedstock energy content and as well as the conversion efficiency of the reaction. The fact that biomass feedstock composition plays a vital role in the combustion process was well established by many researchers worldwide in various reports [34–36]. The major share of energy in the biomass is formed by the assembly of organic matter during photosynthesis and respiration in plants. However, the inorganic fractions in the biomass are important in design and operation of the combustion system, especially when using the fluidized bed reactor. The amount of volatile matter in biomass feedstock is higher when compared with its fossil counterpart in which it is around 70–80%. The presence of this high volatile matter, greatly influence the thermal decomposition of the biomass feedstock and as well as the combustion performance of the solid fuels. This is because, large portion of the biomass feedstock has to be vapourized before the homogeneous combustion reaction takes place and the remaining char will then undergo heterogeneous combustion reaction.

The main elements that constitutes the biomass feedstock are C, H, and O, while herbaceous feedstock such as agricultural waste and grasses contain higher amounts of ash forming minerals [37, 38]. Biomass is more oxygenated compared to the conventional fossil fuel. This is due to the biomass carbohydrate structure and its dry mass usually contains about 30–40% oxygen [37]. During the combustion process, part of the oxygen required is supplied by the organically bonded oxygen from the biomass, while the rest is supplied through air injection into the system. The primary constituent of a biomass is carbon which made up about 30–60% by weight of dry matter depending on its ash content. The carbon present in biomass feedstock is in partly oxidized form and this justifies the low gross calorific value of biomass feedstock when compared to coal. Of the biomass organic components, hydrogen is the third most important constituent that made up of about 5–6% of the dry matter. Other elements that are found in smaller quantities in the biomass (less than 1%) are Nitrogen, Sulfur and Chlorine, with the exception of agricultural residues where their figures are sometimes above 1% [39, 40]. The presence of high amount of such inorganic elements in a biomass feedstock leads to serious operational problems such as agglomeration, deposition, fouling, sintering and corrosion or erosion. Combustion process, unlike biochemical and other thermochemical conversion technologies, is largely nonselective in terms of biomass feedstock selection and the process aims to reduce the entire fuel to simple products. However, this shows that the complex nature of the biomass has substantial influence on its combustion performance. Inorganic elements such as Si, K, S, Cl, P, Ca, Mg and Fe are associated with reactions that leads to ash fouling and slagging (**Figure 7**) [36].

2.2 Biochemical methods

Biochemical biomass conversion technologies refer to conversion of biomass through biological pre-treatments. These pre-treatments were aimed to turn the biomass into a number of products and intermediates through selection of different microorganisms or enzymes. The process provides a platform to obtain fuels and chemicals such as biogas, hydrogen, ethanol, butanol, acetone and a wide range of organic acids [42]. However, this process was aimed at producing products that could replace petroleum-based products and as well as those obtained from the grains. Biomass biochemical conversion technologies are clean, pure, and efficient when compared with the other conversion technologies [43].

2.2.1 Digestion

Anaerobic digestion (AD) is one of the most sustainable and cost-effective technology for lignocellulosic and other form of waste treatment for energy recovery in form of biofuels. This process does not only minimize the amount of waste, but also transforms such waste into bioenergy. Also, the digestates produced during the process are rich in nutrients, which can serve as fertilizer for agricultural purposes [44].

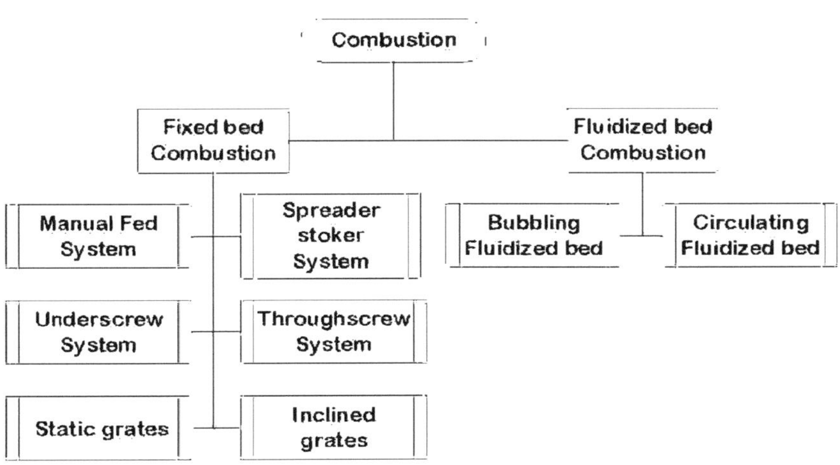

Figure 7.
Various reactors for combustion process [41].

The digestion of lignocellulosic biomass anaerobically produces energy rich methane (CH_4). The CH_4 yield per unit area is usually employed for the determination of energy output of an individual feedstock which significantly varies between species and as well with maturity, location and inputs (such as fertilizer, water etc.) within the same variety (Yang et al., 2013). The Biochemical methane potential (BMP) test is commonly used to evaluate the anaerobic digestibility of a biomass substrate. The biomass yield and CH_4 production potentials of some selected feedstocks were presented in **Table 1** [45].

Anaerobic digestion is a process used to produce biogas through biological treatment of biomass. It is performed at temperature ranges between 30 and 35°C, or 50 and 55°C using two stages. The first stage is the breaking down of the complex organics in the biomass by acid-forming bacteria into simpler compounds such as acetic and propionic acids along with volatiles. The second stage is conversion of such acids into CO_2 and CH_4 commonly called biogas through the use of methane producing bacteria. Usually, both stages of biogas production are performed in a single tank. The produced biogas contains about 60% CH_4, 35% CO_2, and a mixture of other gases such as H_2, NH_3, CO, and H_2S which account for about 5%. The biogas has a heating value of about 22,350 KJ/m^3 for a mixture that contains a ratio (CH_4:CO_2:inerts) of 60: 35: 5 (**Figure 8**) [46].

2.2.2 Fermentation

Fermentation is a biological process that is commonly facilitated by secretion of enzymes sourced from microorganisms which converts simple sugars to low molecular weight structures such as alcohols and acids. The fermentation of two most common sugars follow the two reactions below:

$$\text{Glucose}: C_6H1_2O_6 \rightarrow 2C_2H_6O + 2CO_2 \quad (1)$$

$$\text{Xylose}: 3C_5H_{10}O_5 \rightarrow 5C_2H_6O + 5CO_2 \quad (2)$$

During fermentation, biomass could be converted into alcohols through biochemical pathways. These pathways involved several schemes in which hydrolysis

Biomass	Biomass yield (ton wet weight/ha)	CH_4 potential (Nm^3 CH_4/tonVS)
Sugar beet	40–70	387–408
Fodder beet	80–120	398–424
Maize	40–60	291–338
Wheat	30–50	351–378
Triticale	28–33	319–335
Sorghum	40–80	286–319
Grass	22–31	286–324
Red clover	17–25	297–347
Sunflower	31–42	231–297
Wheat grain	06–10	371–398

Table 1.
The biomass yield and methane potential of some selected lignocellulosic biomass [45].

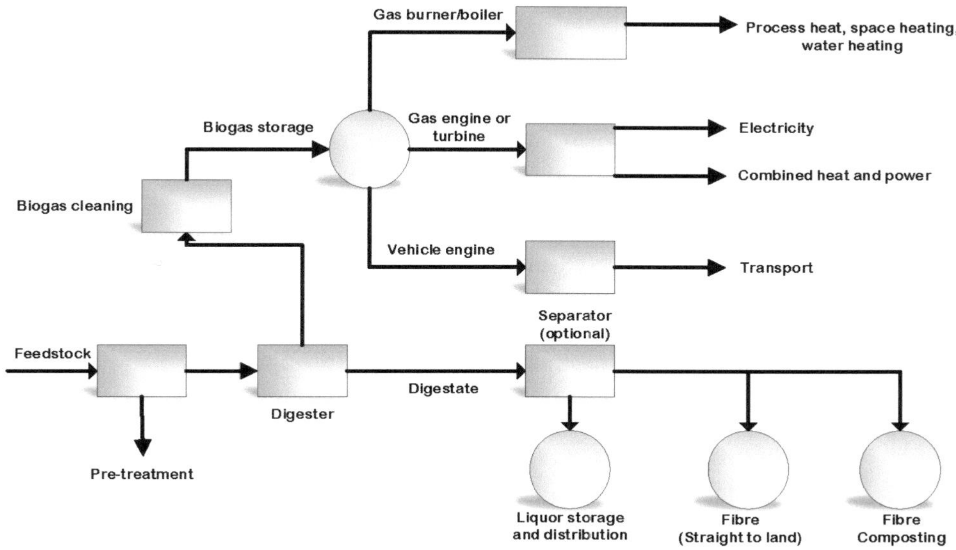

Figure 8.
Anaerobic digestion process [46].

Process	Substrate	Pre-treatment	Ethanol Conc., g/L	Ethanol Pro., g/L/h
SHF	*Parthenium hysterophorus* L.	Acids/alkali	13.6	—
	Arundo donax	Steam explosion	20.6	0.21
	Wheat straw	Steam explosion	—	0.313
SSF	*Miscanthus sacchariflorus*	CHEMET with NaOH	69.2	1.24
	Reed	phosphoric acid-acetone	55.5	0.57
	Reed	Liquid hot water	39.4	0.66
	Liriodendron tulipifera	Acid-free organosolv	29.9	0.42
	Corn stover	Steam explosion	25.7	0.36
	Miscanthus giganteus	Dilute oxalic acid	12.1	0.13
	Industrial hemp	Steam explosion	21.3	0.30
SSCF	Wood chips	Steam explosion	32.9	0.34
	Wheat straw	Steam explosion	—	0.7
CBP	Corn stover	Acid hydrolysis	—	0.27

SHF = *Separate hydrolysis and fermentation.*
SSF = *Simultaneous saccharification and fermentation.*
SSCF = *Simultaneous saccharification and co-fermentation.*
CBP = *Consolidated Bioprocessing.*

Table 2.
Processes in bio-ethanol production [47].

and fermentation process are carried out either concurrently in the same reactor or separately [47]. The different processes involved for alcohols production are presented in **Table 2**.

Conversion of biomass feedstocks through fermentation process is a vital issue because it allows for the production of wide range of substances under mild conditions. The extent of fermentation on organic substances largely depends on composition and structure of the biomass feedstock. Only feedstocks that are not competing with the food items in terms of demand should be selected for biofuel production. Consequently, residues and waste materials from agriculture and forestry were considered as the most interesting sources of biomass.

High hydrolysis ratio is also an important requirement for the effective utilization of monosugars present in lignocellulosic structures. From biochemical perspective, organic substances present in the hydrolyzed solution can be categorized into several groups such as simple and complex carbohydrates, lipids, proteins, and heteropolymers. The potentials for biogas and biohydrogen generation from lignocellulosic biomass is huge due to utilization of different microorganisms in the conversion of cellulose and hemicellulosic fractions of the agricultural and forestry residues [47]. However, a major setback is usually encountered during biofuels production which is the conversion ratio of the polymeric substances into fermentable sugars like hexoses and pentoses due to production of inhibitors along with the desired products. To minimize such inhibitors and maximize hexoses and pentoses production, microbial metabolism in the degradation and saccharification of the biomass cell wall were considered [48, 49].

3. Conclusion

Currently, the use of lignocellulosic biomass as raw material for the generation of bioenergy has received a considerable attention for the development of sustainable ways for production of energy. Most of the researches conducted for biomass conversion technologies heads towards discovery of advanced ways to produce energy fuels so as to tackle its shortage that the world is facing. Also, the studies are aimed towards reduction of greenhouse gases and other harmful effects posed by fossil fuels to the environment.

From above, it can be concluded that biomass is a green source of energy in recent times. The study also indicated that thermochemical and biochemical technologies for the conversion of biomass into different energy products was started several decades ago, but it slowed down due to the discovery of fossil fuels. The biomass conversion technologies gained momentum recently due the fact that it is clean, sustainable and renewable source of energy.

Author details

Abdurrahman Garba
Department of Chemistry, Al-Qalam University Katsina, Katsina, Katsina State, Nigeria

*Address all correspondence to: agarba2008@yahoo.com

References

[1] UNFCCC. Clarifications of definitions of biomass and consideration of changes in carbon pools due to a CDM project activity. Framework convention on climate change—Secretariat. CDM-EB-20, Appendix 8, July 8, 2005

[2] Overend RP. Direct combustion of biomass. In: Sphilrain EE, editor. Renewable Energy Sources Charged with Energy from the Sun and Originated from Earth-Moon Interaction. UNESCO: Encyclopedia of life systems; 2009. p. 1

[3] Fiala M, Nonini L. Biomass and biofuels. In: Joint EPS-SIF International School on Energy; Course 4: Advances in Basic Energy Issues. Édition Diffusion Presse Sciences (EDP Sciences), Springer Science+ Business Media, Società Italiana di Fisica (SIF). EPJ Web of Conferences; 2018

[4] Bhavanam A, Sastry R. Biomass gasification processes in Downd raft fixed bed reactors: A review. International Journal of Chemical Engineering and Applications. 2011;**2**(6):425

[5] Küçük M, Demirbaş A. Biomass conversion processes. Energy Conversion and Management. 1997;**38**(2):151-165

[6] Pathak B, Chaudhari S, Fulekar M. Biomass-resource for sustainable development. International Journal of Advanced Research in Technology. 2013;**2**(6):271-287

[7] Sansaniwal SK et al. Recent advances in the development of biomass gasification technology: A comprehensive review. Renewable and Sustainable Energy Reviews. 2017;**72**:363-384

[8] Faaij A. Modern biomass conversion technologies. Mitigation and Adaptation Strategies for Global Change. 2006;**11**(2):343-375

[9] Matsumura Y. Hydrothermal gasification of biomass. In: Recent Advances in Thermo-Chemical Conversion of Biomass. Elsevier; 2015. pp. 251-267

[10] Kataki R et al. Feedstock suitability for thermochemical processes. In: Recent Advances in Thermo-Chemical Conversion of Biomass. Elsevier; 2015. pp. 31-74

[11] Ammendola P, Scala F. Attrition of lignite char during fluidized bed gasification. Experimental Thermal and Fluid Science. 2012;**43**:9-12

[12] Ramzan N et al. Simulation of circulating fluidized bed gasification for characteristic study of pakistani coal. Polish Journal of Chemical Technology. 2015;**17**(1):66-78

[13] Kore S et al. Steam gasification of coffee husk in bubbling fluidized bed gasifier. In: Proceedings of the Fourth Internation Conference on Bioenvironment, Biodiversity and Renewable Energies. Bionature: Citeseer; 2013

[14] Richardson Y, Blin J, Julbe A. A short overview on purification and conditioning of syngas produced by biomass gasification: Catalytic strategies, process intensification and new concepts. Progress in Energy and Combustion Science. 2012;**38**(6):765-781

[15] Thomsen MH, Thygesen A, Thomsen AB. Hydrothermal treatment of wheat straw at pilot plant scale using a three-step reactor system aiming at high hemicellulose recovery, high cellulose digestibility and low lignin hydrolysis. Bioresource Technology. 2008;**99**(10):4221-4228

[16] Petchpradab P et al. Hydrothermal pretreatment of rubber wood for the saccharification process. Industrial & Engineering Chemistry Research. 2009;**48**(9):4587-4591

[17] Sasaki M et al. Cellulose hydrolysis in subcritical and supercritical water. The Journal of Supercritical Fluids. 1998;**13**(1):261-268

[18] Yoshida T, Nonaka H, Matsumura Y. Hydrothermal treatment of cellulose as a pretreatment for ethanol fermentation: Cellulose hydrolysis experiments. Journal of the Japan Institute of Energy. 2005;**84**(7):544-548

[19] Bandosz TJ. Activated Carbon Surfaces in Environmental Remediation. Vol. 7. Academic Press; 2006

[20] Dhyani V, Bhaskar T. A comprehensive review on the pyrolysis of lignocellulosic biomass. Renewable Energy. 2018;**129**:695-716

[21] Mythili R et al. Characterization of bioresidues for biooil production through pyrolysis. Bioresource Technology. 2013;**138**:71-78

[22] Yang H et al. Characteristics of hemicellulose, cellulose and lignin pyrolysis. Fuel. 2007;**86**(12-13):1781-1788

[23] Shafizadeh F, Fu Y. Pyrolysis of cellulose. Carbohydrate Research. 1973;**29**(1):113-122

[24] Williams R. Handbook of Wood Chemistry and Wood Composites. Boca Raton, FL, USA: CRC Press; 2005. pp. 139-185

[25] Couhert C, Commandre J-M, Salvador S. Is it possible to predict gas yields of any biomass after rapid pyrolysis at high temperature from its composition in cellulose, hemicellulose and lignin? Fuel. 2009;**88**(3):408-417

[26] Wu S et al. Cellulose-hemicellulose interactions during fast pyrolysis with different temperatures and mixing methods. Biomass and Bioenergy. 2016;**95**:55-63

[27] Varma AK, Mondal P. Pyrolysis of sugarcane bagasse in semi batch reactor: Effects of process parameters on product yields and characterization of products. Industrial Crops and Products. 2017;**95**:704-717

[28] Bridgwater AV. Review of fast pyrolysis of biomass and product upgrading. Biomass and Bioenergy. 2012;**38**:68-94

[29] Bridgwater A, Hofbauer H, Van Loo S. Thermal Biomass Conversion. Wheelers Books: CPL Press; 2009

[30] Hornung A. Transformation of Biomass: Theory to Practice. John Wiley & Sons; 2014

[31] Hornung A, Apfelbacher A, Sagi S. Intermediate pyrolysis: A sustainable biomass-to-energy concept-biothermal valorisation of biomass (BtVB) process. Journal of Scientific and Industrial Research. 2011;**70**:664-667

[32] Stamatov V, Honnery D, Soria J. Combustion properties of slow pyrolysis bio-oil produced from indigenous Australian species. Renewable Energy. 2006;**31**(13):2108-2121

[33] Prins MJ, Ptasinski KJ, Janssen FJ. Torrefaction of wood: Part 1. Weight loss kinetics. Journal of Analytical and Applied Pyrolysis. 2006;**77**(1):28-34

[34] Jenkins B et al. Combustion properties of biomass. Fuel Processing Technology. 1998;**54**(1-3):17-46

[35] Vassilev SV et al. An overview of the organic and inorganic phase composition of biomass. Fuel. 2012;**94**:1-33

[36] Vassilev SV, Baxter D, Vassileva CG. An overview of the behaviour of biomass during combustion: Part I. Phase-mineral transformations of organic and inorganic matter. Fuel. 2013;**112**:391-449

[37] McKendry P. Energy production from biomass (part 1): Overview of biomass. Bioresource Technology. 2002;**83**(1):37-46

[38] Saxena R, Adhikari D, Goyal H. Biomass-based energy fuel through biochemical routes: A review. Renewable and Sustainable Energy Reviews. 2009;**13**(1):167-178

[39] Arvelakis S, Koukios E. Physicochemical upgrading of agroresidues as feedstocks for energy production via thermochemical conversion methods. Biomass and Bioenergy. 2002;**22**(5):331-348

[40] Vassilev SV et al. An overview of the chemical composition of biomass. Fuel. 2010;**89**(5):913-933

[41] Bhaskar T, Pandey A. Advances in thermochemical conversion of biomass—Introduction. In: Recent Advances in Thermo-Chemical Conversion of Biomass. Elsevier; 2015. pp. 3-30

[42] Chen H, Qiu W. Key technologies for bioethanol production from lignocellulose. Biotechnology Advances. 2010;**28**(5):556-562

[43] Chen H, Wang L. Technologies for Biochemical Conversion of Biomass. Academic Press; 2016

[44] Li Y, Chen Y, Wu J. Enhancement of methane production in anaerobic digestion process: A review. Applied Energy. 2019;**240**:120-137

[45] Sawatdeenarunat C, Surendra K, Takara D, Oechsner H, Khanal SK. Anaerobic digestion of lignocellulosic biomass: Challenges and opportunities. Bioresource Technology. 2015;**178**:178-186

[46] Balat M. Biomass energy and biochemical conversion processing for fuels and chemicals. Energy Sources, Part A: Recovery, Utilization, and Environmental Effects. 2006;**28**(6):517-525

[47] Abo BO, Gao M, Wang Y, Wu C, Ma H, Wang Q. Lignocellulosic biomass for bioethanol: An overview on pretreatment, hydrolysis and fermentation processes. Reviews on Environmental Health. 2019;**34**(1):57-68

[48] Zhang CM et al. Effective ethanol production by reutilizing waste distillage anaerobic digestion effluent in an integrated fermentation process coupled with both ethanol and methane fermentations. Bioprocess and Biosystems Engineering. 2010;**33**(9):1067-1075

[49] Cheng X-Y, Li Q, Liu C-Z. Coproduction of hydrogen and methane via anaerobic fermentation of cornstalk waste in continuous stirred tank reactor integrated with up-flow anaerobic sludge bed. Bioresource Technology. 2012;**114**:327-333

Chapter 2

Biomass Pretreatment and Characterization: A Review

Anthony Anukam and Jonas Berghel

Abstract

Biomass has the potential to replace conventional fuels in a number of applications, particularly in biofuel production. It is an abundantly available renewable material with great potential as a feedstock for bioconversion processes for the production of energy, fuels and a variety of chemicals. Due to its biogenic origin, the carbon dioxide released from its combustion process does not impact atmospheric carbon dioxide. Despite these merits, a major problem hindering its widespread use has always been its recalcitrant nature, in terms of its inherent characteristics, which are unfavorable to its use in bioconversion and bio refinery processes. This makes it necessary for biomass to be pretreated before use in any conversion process for maximum product recovery. However, a major issue with regards to biomass pretreatment is the lack of rapid, high throughput and reliable tools for assessing and tracing biopolymer components of biomass relevant to the energy production potential of the biomass. This chapter therefore presents an overview of the pretreatment and characterization of biomass relevant to energy, fuels and chemicals production. The information provided will bequeath readers with the basic knowledge necessary for finding an auspicious solution to pretreatment problems and the production of energy from pretreated biomass.

Keywords: biomass, bioconversion, bio refinery, energy systems, energy production

1. Introduction

The definition of biomass may vary due to a number of factors such as the heterogeneity of biomass, its application and origin [1]. However, any organic material directly or indirectly derived from the process of photosynthesis is considered biomass. Generally, biomass is a biological material that is present in various forms such as *wood*, *sugarcane bagasse*, *rice husk*, *rice straw*, *corn cob*, *paper waste*, *municipal solid waste*, *animal manure*, *sewage*, *algae*, *etc*. Traditionally, the use of biomass resources is becoming increasingly vital because of their substantial annual volumes and great economic potential. The main application of biomass is as a source of energy, and it has been reported that global biomass land and aquatic reserves stand at around 1.8 trillion tons for biomass, and about 4 billion tons for aquatic reserves, respectively [1]. In spite of these huge reserves, biomass remains under exploited, accounting for only about 14% of global energy with ca. 56 million TJ/year [1]. This is because the application of biomass as a source of energy has not shown a homogeneous distribution across the globe. In some developed countries, energy production from biomass averages about 11% of the total energy produced; while in developing countries this can be up to 50% [2, 3]. For example, Europe generates about 3.5% of its energy from biomass, which is ca. 40 Mtoe/year; while countries like Austria, Finland and Sweden have about 13, 18 and 17% of their total energy produced from biomass resources; the United States on the other hand generates 3% of its energy from biomass [4]. In general, there is a huge potential for the exploitation of biomass as a source of energy because some countries in central and northern Europe have actually established large plants for heat and power production from biomass. However, there are two main routes by which biomass can be converted to energy and bio products. These are biochemical and thermo-chemical conversion routes. While the former involves breaking down biomass into gaseous and/or liquid fuels (such as biogas and bioethanol) through the use of bacteria, microorganisms and enzymes; the latter involves the use of heat to disrupt the complex chemical structure of biomass (particularly lignocellulosic biomass such as wood) into various products that includes heat, power, fuels, bio oil, biochar and chemicals [5, 6].

It is evident from the above that the application of biomass as a source of energy continues to attract global attention even in the midst of its gross underutilization. The low quality of biomass, often defined in terms of its inherent characteristics (high moisture content, low energy density, low bulk density, irregular size and shape) has been one of the major reasons for its under exploitation [7]. In other words, biomass in its natural form is difficult to use for the purpose of energy production hence pretreatment is often required to overcome its recalcitrant nature and make the biomass amenable for conversion through either of the conversion routes previously mentioned for maximum product recovery; the pretreatment leads to physical, chemical and structural changes to the biomass plays a vital role in the commercial viability of the energy production process of biomass [7–9]. Thus, bioconversion and bio refinery interests define the type of pretreatment measures for biomass [7, 10]. However, there are different categories of pretreatment viz; physical, chemical and biological pretreatments. These three classes of pretreatment are described in greater detail in Section 3.

Another critical step to the optimization of biomass conversion and bio refinery processes is related to the characterization of biomass to determine its suitability for the aforementioned conversion processes. This means that the effectiveness and impact of pretreatment on biomass can be determined through the use of a variety of high-quality analytical techniques able to provide information on quintessential biomass characteristics that can be used to maximize product recovery. Depending on the end application of biomass, some of the useful characterization techniques include *atomic force microscopy (AFM)*, *Fourier transform infrared spectroscopy (FT-IR)*, *scanning electron microscopy (SEM)*, *transmission electron microscopy (TEM)*, *X-ray diffraction (XRD)*, *solid state nuclear magnetic resonance (SSNMR)*, to name a few. An overview of these analytical tools is presented in subsequent sections. For a full understanding of the characteristics of biomass, its value and information for the design and operation of the energy conversion systems using the biomass as feedstock, it is vital to undertake biomass characterization before and after pretreatment. Due to the complex nature of biomass, the study of the mechanisms involved in its conversion process to energy and bio products is quite challenging. The lack of rapid, high throughput and reliable tools for assessing and tracing biomass components relevant not just to energy production but also to other value added products remains a major bottleneck in studying the impact of biomass pretreatment and process parameters. This chapter therefore presents a critical review of biomass pretreatment and characterization and discusses the applications of state-of-the-art analytical techniques commonly used to understand the features of both pretreated and non-pretreated biomass relevant to the production of fuels and chemicals. It is believed that improving the fundamental knowledge of biomass pretreatment and characterization will lead to significant advances in the field of sustainable energy and chemicals production from biomass.

2. Overview of biomass composition

The composition of biomass is largely diverse and dependent upon its origin and species. Besides plant biomass, which are commonly referred to as lignocellulosic biomass (LCB) due to their three major cross-linked polysaccharide constituents (cellulose, hemicellulose and lignin), there are other biomass materials whose primary components include *lipids*, *proteins*, *starch*, *inorganics and minerals*. These types of biomass materials are regarded as non-lignocellulosic biomass (NLCB) and include *sewage sludge*, *animal manure*, *algae*, etc. The major constituents of both the LCB and NLCB are organic in nature and determine the characteristics of the entire biomass [7]. In comparison to LCB however, NLCB pose a greater threat to the environment because of higher content of heavy metals and heteroatoms such as nitrogen (N) and phosphorus (P); [11–13]. The heavy metals can pollute water systems, accumulate in food chains and cause serious health issues [14, 15]. Even though the N and P composition of NLCB can serve as a source of nutrients for plants, excessive amounts of these elements can lead to eutrophication of a water body, a condition linked to the deterioration of water quality via excessive growth of algae and other aquatic plants, resulting in oxygen depletion of the water body, leading to the death of aquatic animals [12].

While the structural unit of NLCB is such that various atoms are arranged in an orderly manner, LCB is characterized by complex internal structure with main components that equally displays structural multiplicities. **Figure 1** shows the internal structure of lignocellulosic biomass and how its three primary components are distributed.

The internal structure of lignocellulosic biomass reveals a crystalline fibrous structure of cellulose, which forms the core of the complex structure of plant biomass. Positioned between the micro- and macrofibrils of the cellulose matrix is hemicellulose; while lignin plays a structural role that encapsulates both cellulose and hemicellulose.

The composition of biomass varies significantly depending on the source of the biomass. In addition to the three primary components of LCB (**Figure 1**), there are other minor components such as *extractives*, *proteins*, *water* and *inorganic components*

Figure 1.
A structural representation of lignocellulosic biomass. Reproduced with permission from [1].

Type of lignocellulosic biomass	Cellulose (%)	Hemicellulose (%)	Lignin (%)
Hardwood	40–55	24–40	18–25
Softwood	45–50	25–35	25–35
Grasses	25–40	35–50	10–30
Leaves	15–20	80–85	–
Sugarcane bagasse	40–45	30–35	20–30
Wheat straw	33–40	20–25	15–20
Sweet sorghum bagasse	45	27	21

Table 1.
The composition of the organic fractions of various lignocellulosic biomass materials (dry basis) [17–20].

such as *silicon (Si)*, *sodium (Na)*, *potassium (K)*, *calcium (Ca)*, *magnesium (Mg)* and *aluminum (Al)*; these minor constituents do not markedly contribute to the formation of the total structure of the biomass [16]. The organic components of various LCB materials are summarized in **Table 1**.

The composition of NLCB, on the other hand, vary from material to material and contains more multifarious elements (such as those previously mentioned) that are embedded in its structural unit; in contrast to LCB, the different compositions of NLCB leads to different thermochemical conversion behaviors when these materials (NLCB) are used as feedstock in the mentioned bioenergy conversion routes [21].

3. Biomass pretreatment

Pretreatment is a necessary process step for both biochemical and thermochemical conversion of biomass and involves structural alteration aimed at overcoming the recalcitrant nature of biomass. It is required to improve biomass characteristics in order to enhance the energy utilization efficiency of the biomass [7, 22]. In pretreatment processes requiring heat, the degradation ability of LCB is controlled by its polymeric and aromatic constituents (cellulose, hemicellulose and lignin), while the heteroatoms and inorganic elemental components of NLCB act as catalysts to facilitate decomposition, leading to the formation of a product with a carbon framework and structural changes that increases the performance of the pretreated material in bioconversion processes [23–25]. The most important barriers facing current pretreatment technologies are high costs and how to obtain a pretreated product with minimal degradation of vital components. These issues are yet to be convincingly tackled by past and present research and development hence extensive studies aimed at the development of technologies that will further exploit the physical, chemical and biological pretreatment approaches are required. The pretreatment methods must be specifically tailored toward biomass origin and its application in bioconversion and bio refinery processes. **Figure 2** shows a schematic layout of the classes and types of pretreatment processes required for the two main conversion routes (biochemical and thermochemical) for biomass.

The following subsections present a further description of the main classes of pretreatment.

3.1 Physical pretreatment

Physical pretreatment of biomass is intended to reduce particle size by mechanical comminution in order to increase surface area and pore size. For LCB materials,

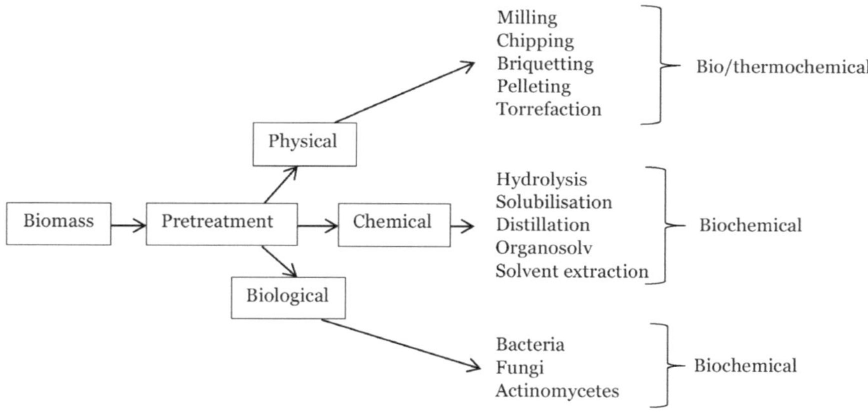

Figure 2.
A layout showing different pretreatment methods for biomass and their corresponding energy conversion routes.

physical pretreatment decreases cellulose crystallinity and degree of polymerization. It is a vital step prior to both the biochemical and thermochemical conversion of biomass [7, 26]. However, there is limited information about the mode of action of physical pretreatment processes, particularly with respect to how the chemical composition of biomass is modified or how its structure is affected. The application of biomass determines the type of physical pretreatment method to be applied. For instance, biochemical conversion of LCB will require size reduction through milling in order to improve enzymatic digestibility of lignocellulosic components. For thermochemical conversion applications of biomass, milling is required for densification, pelletization and even torrefaction prior to thermochemical conversion. In both conversion pathways (biochemical and thermochemical), prior size reduction is necessary in order to eliminate mass and heat transfer limitations. Chipping is also another physical pretreatment technique commonly employed when the biomass is to be used as feedstock in thermochemical conversion processes. This is because most thermochemical conversion systems are size specific hence require feedstock size in the range 50 mm by 50 mm in diameter [7].

As previously stated, densification, pelletization and torrefaction are all considered physical pretreatment methods for biomass intended as feedstock for thermochemical conversion processes [7]. These pretreatment techniques use heat to initiate changes that leads to improved biomass characteristics. A major drawback of the physical pretreatment technique however is its lack of ability to get rid of the lignin content of LCB materials, which renders the cellulose content of the material inaccessible. Other shortcomings include high energy consumption and the prohibitive cost of scale-up for commercial purposes. Studies [26, 27] have shown that the process of lignin removal from LCB materials could be one of the major reasons for the high energy demand of physical pretreatment techniques hence the overall energy efficiency of a bio refinery process may be ultimately affected by this pretreatment method [26].

3.2 Chemical pretreatment

Chemical pretreatment of biomass involves the use of organic or inorganic compounds to bring about disruption of biomass structure through interaction with intra- and inter polymer bonds of primary organic components. Biomass, particularly LCB materials are resistant to chemical breakdown hence biomass is regarded as being recalcitrant in nature. A host of factors are responsible for the recalcitrance nature of biomass including the structural complexity and heterogeneity of biomass, the crystalline nature of its cellulose content, and the extent of lignification [7, 28]. Throughout the chemical pretreatment process, the structural recalcitrance of LCB is disrupted, resulting in the reduction of cellulose crystallinity and depolymerization as well as the degradation of cellulose and the breakdown of lignin [29, 30]. For biochemical conversion of biomass, particularly LCB, chemical pretreatment is commonly undertaken in order to isolate the respective biopolymeric constituents of the material. **Figure 3** shows the effect of chemical pretreatment on LCB.

Examples of compounds that have been used for the chemical pretreatment of biomass and which had significant effect on its structure include acids, alkali, organic solvents, and ionic liquids [32, 33].

3.3 Biological pretreatment

Biological pretreatment of biomass is mostly linked to the action of enzyme-producing fungi that are able to degrade, depolymerize and cleave the cellulose, hemicellulose and the lignin contents of biomass. This pretreatment method has several advantages over other pretreatment methods including its limited or no generation of toxic substances, high yield of needed products, low energy requirements and specificity of substrate and process reaction [34, 35]. However, its major disadvantages include the fact that the process is too slow and needs careful control of conditions of fungi growth as well as the large space required to carry out the process [36]. According to Agbor et al., 2011 [37], the residence time required for biological pretreatment processes is in the range 10 and 14 days. In addition, the organic components of biomass are consumed by the microorganisms' hence biological pretreatment processes faces techno-economic challenges and are considered commercially less attractive when compared to other pretreatment methods [38–40]. The types of fungi commonly used in biological pretreatment of biomass include *brown-*, *white-* and *soft-rot fungi*, *actinomycetes* and *bacteria*. These fungi are particularly known for their removal of hemicellulose and lignin as well as for their mild effect on cellulose. Nevertheless, *white-* and *brown-rot fungi* have a variety of ways to access and degrade LCB materials such as wood, and their very strong metabolism has been applied in industrial operations. For instance, *white-* and

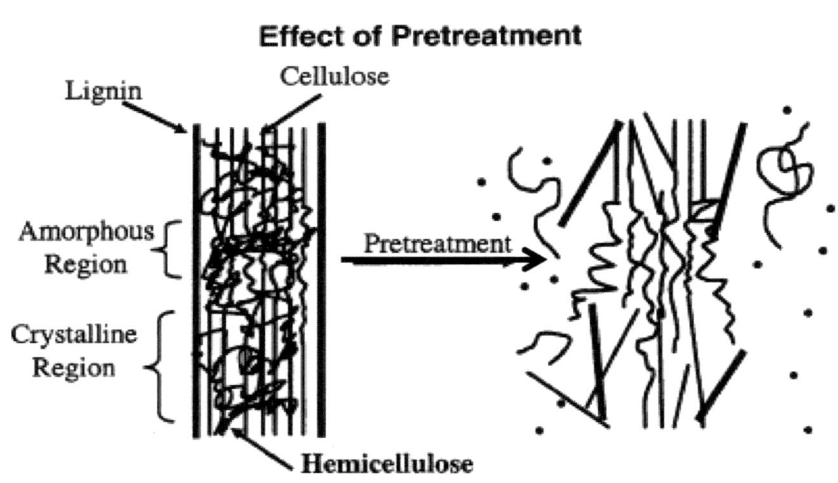

Figure 3.
A schematic representation of the impact of pretreatment on the surface and internal structure of lignocellulosic biomass. Adapted from [31].

Pretreatment methods	Advantages	Disadvantages	Reference
Physical pretreatment	• Simple and easy operation • Very large volumes of biomass can be handled • Increases surface area and pore size • Increases bulk and energy densities • Reduces cellulose crystallinity • Does not involve the use of chemicals	• Power consumption quite high • Require additional pretreatment steps	[8, 25, 43, 44]
Chemical pretreatment	• Increases accessibility to cellulose • Alters the structure of lignin • Hydrolyzes hemicellulose into various sugar fractions	• High cost of chemicals • Corrosion related issues with equipment • Liable to form inhibitory substances • Requires long residence time (time consuming)	[45–47]
Biological pretreatment	• Simple equipment • Efficiently degrades both cellulose and hemicellulose • Suitable for both high and low biomass moisture contents • Low energy requirements	• Slightly degrades lignin • Very slow rate of hydrolysis • The process requires a large space • Require long residence times	[36, 37, 48]

Table 2.
A summary of the advantages and disadvantages of the classes of pretreatment.

brown-rot fungi are known to have a brightening effect on kraft pulp obtained from hardwoods, with cost reductions in bleaching chemicals and potentially decreasing the environmental impact on paper mill operations [41].

In spite of the fact that many pretreatment methods have been investigated, while others are still in their developmental stages, it is quite onerous to assess and compare pretreatment technologies. This is because most pretreatment technologies involve upstream and downstream costs associated with processing, capital investment as well as complicated chemical recycling and waste treatment systems [42].

Table 2 presents a summary of the advantages and disadvantages of the different classes of biomass pretreatment.

4. Biomass characterization

Heterogeneity remains an inherent characteristic of biomass. The feasibility and viability of products recovery from biomass depends upon its properties. The two main conversion pathways earlier mentioned are basically used to recover products of value from biomass. The choice of the conversion route also depends on the features of biomass hence characterization is essential to better understand quintessential physicochemical properties of biomass that will determine how suitable the material is for conversion; these properties are keys to the efficient utilization of biomass in bioconversion processes [1, 7]. However, the characteristics of biomass are largely swayed by its primary organic components (cellulose, hemicellulose and lignin), which vary depending on biomass source, species, climatic conditions, etc. Depending on the end use of biomass, characterization of biomass is commonly determined and reported in terms of proximate and ultimate analysis using a variety of analytical tools some of which are described in subsequent sections of this review. This provides vital information for evaluating various application potential of biomass, particularly its energy production potential, which also takes into

Type of biomass	Proximate analysis (wt.%)				Ultimate analysis (wt.%)				
LCB	MC	VM	FC	A	C	H	O	S	N
Wood & woody biomass	5–63	30–80	6–26	1–8	49–57	5–10	32–45	<1–1	<1–1
Herbaceous biomass	4–48	41–77	9–35	1–19	42–58	3–9	34–49	<1–1	<1–3
NLCB									
Animal & human waste biomass	3–9	43–62	12–13	23–34	57–61	7–8	21–25	1–2	6–12
Aquatic biomass	8–14	42–53	22–33	11–38	27–43	4–6	34–46	1–3	1–3

Abbreviations: Moisture content (MC), Volatile matter (VM), Fixed carbon (FC), Ash (A), Carbon (C), Hydrogen (H), Oxygen (O), Sulfur (S), Nitrogen (N).

Table 3.
The properties of various lignocellulosic and non-lignocellulosic biomass materials [49].

account heating value when the biomass is used as feedstock in thermochemical conversion processes such as gasification [25]. **Table 3** presents the most important characteristics of various lignocellulosic and non-lignocellulosic biomass materials.

The proximate analysis estimates the physical properties of biomass with direct influence on the combustion performance of biomass, while ultimate analysis provides a qualitative and quantitative estimation of chemical properties in terms of the weight fractions of elemental components (such as C, H and O) and determines the heating value of biomass, a vital property of biomass materials intended as feedstock for thermochemical conversion processes; the amounts of other elements such as N, S and Si can help determine the environmental impact of using biomass as a fuel [25, 50, 51]. The proximate and ultimate analyses are discussed further in the following section.

4.1 Analytical techniques for biomass characterization

As earlier alluded, the determination of biomass characteristics often requires the use of a wide variety of state-of-the-art analytical techniques able to provide not just compositional information, but also pretreatment process validation. However, the type of analytical technique to be used is defined by the application choice of the biomass so that analyses results are interpreted in relation to the specific application. A few of the analytical techniques commonly used in characterization studies involving biomass materials are discussed in the following subsections.

4.1.1 Fourier transform infra-red spectroscopic analyzer

The Fourier transform infra-red spectroscopic (FTIR) is a sensitive technique for the qualitative and quantitative analysis of organic materials such as biomass. It identifies chemical bonds by generating a range of infrared (IR) retention in the form of spectra that represents sample profile. This analytical tool is particularly useful for tracing and tracking changes in biomass molecular structure caused by pretreatment and can distinguish between functional groups [51]. It is a useful technique for the characterization of biomass materials intended for a whole range of applications including biochemical and thermochemical conversion applications. For example, the FTIR can be used to investigate the possibilities of the removal of hemicellulose and lignin from a chemically pretreated biomass in a pulping process. It can also be used to understand the most reactive components of biomass materials and how these components affect the thermochemical conversion process of biomass.

The FTIR technique relies on the fact that most organic materials absorb light within the IR region of the electromagnetic spectrum. The frequency of absorption of light is measured in wave numbers that is typically in the range 4000–600 cm^{-1}.

4.1.2 X-ray diffraction analyzer

To the best of the author's knowledge, the X-ray diffraction (XRD) is the only analytical instrument able to reveal comprehensive structural information of materials. Structural information that can be obtained includes *chemical composition*, *deformation*, *crystal structure*, *crystal size* and *orientation* as well as *layer thickness*. This instrument can be used to analyze a wide range of materials including nanomaterials. The XRD is equally a valuable analytical tool for studies involving biomass characterization for various applications. In XRD analysis, the extent of crystallinity is calculated based on an equation: (Eq. (1)) [52, 53]:

$$CrI\ (\%) = \left(\frac{I_{002} - I_{am}}{I_{002}}\right) \times 100\% \tag{1}$$

where CrI is the crystallinity index, while I_{002} represents the overall peak intensity and I_{am} the baseline intensity.

XRD is especially useful for the determination of the efficiency of hydrolysis for chemically pretreated biomass materials intended for the production of various chemical compounds such as sugar molecules and oligosaccharides, which are products of a fermentation process for the production of bioethanol.

4.1.3 Solid state nuclear magnetic resonance spectroscopy

For the analysis of biomass materials considered for the purpose of biofuels, biochar or chemicals production, the solid state nuclear magnetic resonance spectroscopy (SSNMR) is the ideal technique. This technique allows detailed structural elucidation of major constituents of biomass, particularly lignocellulosic biomass such as wood. It complements the XRD technique since the degree of cellulose crystallinity can also be determined. Nonetheless, the XRD is based on the proposition that X-ray scattering can be divided into two structural components that are amorphous and crystalline [54], while SSNMR is characterized by orientation-dependent interactions that are observed in a very broad spectrum that provides detailed information on material chemistry, structure and dynamics in the solid state. The production of biofuels and chemicals from lignocellulosic biomass require an unfettered access to cellulose and hemicellulose, thus SSNMR can be used to comprehend bioconversion of biomass as a function of process conditions [55].

The chemical barriers resulting from lignin renders the hydrolysis process of biomass into fermentable sugars quite complicated. Therefore, advances in spectroscopic techniques, especially spectroscopic methods, have enabled researchers to elucidate the structural characteristics of biomass in relation to specific applications.

4.1.4 Thermo-gravimetric analysis analyzer

When there is a need to investigate the combustion behavior of biomass materials, the thermo-gravimetric analysis (TGA) is particularly useful and valuable for determining thermal parameters relevant to the thermochemical conversion of biomass. Proximate analysis data can be obtained from TGA. In this technique, the sample is combusted at desired heating rates in a chemically inactive atmosphere of nitrogen or argon such that the mass of the sample is monitored as temperature increases. The change in mass of the sample is usually plotted as a function of time or temperature. The TGA is a high temperature analytical instrument that adequately mimics the conditions existing in a typical thermal energy production system [56]. For studies involving the need to determine the kinetics of thermal decomposition of biomass, TGA is equally very helpful as it provides qualitative information that can be used to understand process conditions and design parameters of thermochemical conversion systems [57]. This requires that TGA be conducted at different heating rates and its derivative (DTG) used to simplify the reading of the characteristic peaks obtained from the thermogram of change in mass versus temperature.

4.1.5 Differential scanning calorimetry

The differential scanning calorimetry (DSC) is a thermoanalytical tool used to directly assess the heat energy uptake that occurs in a sample within a controlled increase or decrease in temperature. The instrument monitors phase transitions that lead to heat flow between crucibles since the process involves the heating of two crucibles (one which contains the sample to be analyzed, and the other serving as a reference without a sample). In this analysis, heat flow is measured as a function of temperature so that combustion profiles that will help determine the series of stages that characterizes the thermal performance of a material can be evaluated. In some instances, the DSC can be used as a complementary analytical tool to the TGA, particularly when monitoring softening or glass transition temperature range [8]. The DSC is very valuable for the analysis of biomass materials intended as feedstock for thermochemical conversion processes and the data from DSC can help improve the understanding of the effect of activation energy on the rate of biomass conversion in the aforementioned thermal energy systems [58]. The reactivity of biomass materials can also be determined using DSC and the volatiles from the analysis can be identified using gas analyzers.

4.1.6 CHNS analyzer

The CHNS analyzer is the carbon (C), hydrogen (H), nitrogen (N) and sulfur (S) analyzer used to measure the weight percentages of these elements in a given material. In CHNS analysis, the weight fraction of oxygen is usually determined by difference with respect to a generally accepted equation (Eq. (2)).

$$O = 100 - (C + H + N + S) \tag{2}$$

The CHNS analyzer is an elemental analyzer whose principle of operation is based on combustion that allows the quantitative determination of the above elements without the need for time consuming sample preparation steps or the use of toxic chemicals. Elemental composition is one of the most important features for biomass utilization [25, 59]. In addition to facilitating the determination of the environmental compatibility of using biomass as a fuel in thermal energy systems, CHNS analysis can be used to obtain information about calorific value and establish the combustion performance of biomass, if the weight proportions of primary elemental components such as C, H and O are known. One of the simplest ways to calculate the calorific value of biomass without the need for analytical tools is from an equation developed by Sheng and Azevedo, 2005 [60]:

$$CV(MJ/kg) = -1.3675 + 0.3137C + 0.7009H + 0.0318O \quad (3)$$

where CV is the calorific value of biomass.

Calorific value is an important property of biomass for design calculations or numerical simulation of thermochemical conversion systems using biomass as feedstock [25, 60].

4.1.7 Pyrolysis-gas chromatography/mass spectrometry (Py-GC/MS)

The Py-GC/MS is a technique used to identify non-volatile compounds. It involves high temperature heating of a sample to decomposition into smaller molecules that are separated by gas chromatography and identified by mass spectrometry. This technique is particularly suited for the analysis of biomass materials intended as feedstock in pyrolysis or hydrothermal liquefaction (HTL) processes for the production of charcoals and bio-oils as the mechanisms involved in these two thermochemical processes (pyrolysis and hydrothermal liquefaction) can be conveniently investigated. For example, pyrolysis of biomass is a relatively complex process that involves both simultaneous and successive chemical reactions which occurs when the biomass is heated in an unreactive environment. Due to the compositional and structural variability of biomass, major constituents degrade under non-identical mechanisms at different temperature ranges and at different rates. Therefore, to explore the complexity of this process, cutting-edge analytical tools such as the Py-GC/MS are required.

4.1.8 Scanning electron microscope

The scanning electron microscope (SEM) is a type of electron microscope that produces the image of a sample by scanning the surface of the sample with a focused beam of electrons that interact with the sample to produce a variety of signals used to obtain information about surface composition and topography. The macroscopic nature of biomass requires that some form of pretreatment, such as size reduction, be performed in order to reveal properties of interest for any microscopic and nanoscopic analyses. Thus, by employing imaging techniques such as the SEM, it is possible to study the physical and chemical underpinnings of the prodigy of biomass recalcitrance to breakdown. The SEM can be used to investigate the morphological properties of biomass relevant to the specific application of the biomass. The information obtained can then be used to hone biomass pretreatment methods that will enhance biomass susceptibility to biochemical or thermochemical conversion. It is however worthy to mention that the moisture content of biomass can be very problematic to some microscopic techniques (such as the SEM) since analyses using these techniques are usually performed on dry samples. As such, samples with reduced moisture content are often required before analysis to avoid the introduction of structural artifacts that may interfere with the SEM images of the sample.

5. Future prospects

The large-scale substitution of fossil fuel with biomass resources is a topical issue not just for the production of energy but also for the production of chemicals, bio products and materials. Moreover, due to the large availability of biomass throughout the world, the production of the high value-added products from biomass can be achieved under any geographical conditions and the feasibility and viability of the production of the value-added products depends on biomass characteristics and the pretreatment method employed. However, biomass complexity and the high capital and operation costs associated with biomass pretreatment as well as the mechanisms involved in the conversion process of biomass are some of the challenges associated with the use of biomass for the production of energy, chemicals and fuels. Therefore, efforts should be geared toward the design of more easy-to-use and cost effective technologies at all levels so as to encourage the widespread application of biomass and attract investment in this field. In addition, not much is known about the optimal biomass pretreatment conditions because they are seldom reported. Consequently, for the efficient and feasible utilization of biomass in bioconversion processes, information about the optimum conditions of pretreatment is vital and efforts made to report such information. It has been reported [61] that researchers and policy makers are in need of useful information that may lead to the much needed improvements in this field of research. So, efforts made to report optimal pretreatment conditions for biomass will create further awareness on the advantages of the exploitation of biomass resources for the production of renewable energy and other bio products.

On the other hand, characterization of biomass also faces significant challenges ranging from the nature of the biomass and the multiphase bioconversion processes using the biomass as feedstock as well as the lack of experimental validation of the cutting-edge analytical techniques used for biomass characterization. Efforts made to address these barriers through continued research will equally lead to optimization of bioconversion and bio refinery processes. Needless to say that, since a single analytical technique cannot provide all of the needed information simultaneously with optimal resolution and high sensitivity, complementary techniques are often required to achieve full understanding of the physical and chemical underpinnings of the prodigy of biomass recalcitrance as it undergoes bioconversion processing. This may ease the challenges associated with experimental validation. Nevertheless, it is vital to mention that each type of characterization technique has its own merits and demerits under a particular set of circumstances and that the shortcomings of one technique may be compensated for by the merits of the other.

6. Conclusions

Pretreatment and characterization of biomass are key steps for the efficient utilization of biomass materials in bioconversion processes. A determination of the best pretreatment method and parameters requires an evaluation of its effects on biomass using cutting-edge analytical tools able to provide information that will facilitate better understanding of the origins of biomass recalcitrance and the mechanism and impact of pretreatment relevant to the optimization of different bioconversion pathways.

Acknowledgements

The author would like to thank the Department of Engineering and Chemical Sciences of Karlstad University for providing a conducive environment and an opportunity to conduct this synopsis.

Conflict of interest

The authors declare no conflict of interest.

Notes/Thanks/Other declarations

The first author deems it a great pleasure to place on record his sincere gratitude to his dear wife, Tuliwe-Ndudula Anukam, who stood by him and kept a peaceful home environment in the midst of numerous challenges.

Abbreviations

LCB	biomass
NLCB	Non lignocellulosic biomass
FTIR	Fourier transform infra-red spectroscopic
IR	infrared
XRD	X-ray diffraction
SSNMR	solid state nuclear magnetic resonance spectroscopy
TGA	thermo-gravimetric analysis
DTG	derivative thermogravimetry
DSC	differential scanning calorimetry
SEM	scanning electron microscope

Author details

Anthony Anukam* and Jonas Berghel
Environmental and Energy Systems, Department of Engineering and Chemical Sciences, Karlstad University, Karlstad, Sweden

*Address all correspondence to: anthonyanukam16@gmail.com

References

[1] Tursi A. A review on biomass: Importance, chemistry, classification, and conversion. Biofuel Research Journal. 2019;**22**:962-979

[2] Chum H, Faaij A, Moreira J, Berndes G, Dhamija P, Dong H, et al. Bioenergy. In: Edenhofer O, Pichs-Madruga R, Sokona Y, Seyboth K, Matschoss P, Kadner S, et al., editors. IPCC Special Report on Renewable Energy Sources and Climate Change Mitigation. Cambridge: Cambridge University Press; 2011

[3] World Bioenergy Association. WBA Global bioenergy statistics. Summary Report 2018. Available from: https://www.worldenergy.org/ [Accessed: 06 July 2020]

[4] Habert G, Bouzidi Y, Chen C, Jullien A. Development of a depletion indicator for natural resources used in concrete. Resources, Conservation and Recycling. 2010;**54**(6):364-376

[5] Zafar S. Biochemical conversion of biomass. Available from: https://www.bioenergyconsult.com/biochemical-conversion-technologies/ [Accessed: 14 July 2020]

[6] Zhang J, Zhang X. The thermochemical conversion of biomass into biofuels. In: Verma D, Fortunati E, Jain S, Zhang X, editors. Biomass, Biopolymer-Based Materials, and Bioenergy. 1st ed. Cambridge: Elsevier; 2019. pp. 519-532

[7] Anukam A, Mamphweli S, Reddy P, Meyer E, Okoh O. Pre-processing of sugarcane bagasse for gasification in a downdraft biomass gasifier system: A comprehensive review. Renewable and Sustainable Energy Reviews. 2016;**66**: 775-801

[8] Anukam AI, Berghel J, Frodeson S, Famewo EB, Nyamukamba P. Characterization of pure and blended pellets made from Norway spruce and pea starch: A comparative study of bonding mechanism relevant to quality. Energies. 2019;**12**(23):1-22

[9] Yang B, Wyman CE. Pretreatment: The key to unlocking low-cost cellulosic ethanol. Biofuels, Bioproducts and Biorefining. 2008;**2**:26-40

[10] Meng X, Pu Y, Yoo CG, Li M, Bali G, Park DY, et al. An in-depth understanding of biomass recalcitrance using natural poplar variants as the feedstock. ChemSusChem. 2017;**10**(1): 139-150

[11] Shi W, Liu C, Ding D, Lei Z, Yang Y, Feng C, et al. Immobilization of heavy metals in sewage sludge by using subcritical water technology. Bioresource Technology. 2013;**137**:18-24

[12] Qian T-T, Jiang H. Migration of phosphorus in sewage sludge during different thermal treatment processes. ACS Sustainable Chemistry & Engineering. 2014;**2**(6):1411-1419

[13] Tian K, Liu WJ, Qian TT, Jiang H, Yu HQ. Investigation on the evolution of N-containing organic compounds during pyrolysis of sewage sludge. Environmental Science and Technology. 2014;**48**(18):10888-10896

[14] Shi W, Liu C, Shu Y, Feng C, Lei Z, Zhang Z. Synergistic effect of rice husk addition on hydrothermal treatment of sewage sludge: Fate and environmental risk of heavy metals. Bioresource Technology. 2013;**149**:496-502

[15] Wang X, Li C, Zhang B, Lin J, Chi Q, Wang Y. Migration and risk assessment of heavy metals in sewage sludge during hydrothermal treatment combined with pyrolysis. Bioresource Technology. 2016;**221**:560-567

[16] Raven PH, Evert RF, Eichhorn SE. Biology of Plants. 6th ed. New York: Worth Publishers; 1992. p. 115

[17] Sun Y, Cheng J. Hydrolysis of lignocellulosic materials for ethanol production: A review. Bioresource Technology. 2002;**83**(1):1-11

[18] Cardona CA, Quintero JA, Paz IC. Production of bioethanol from sugarcane bagasse: Status and perspectives. Bioresource Technology. 2010;**101**:4754-4766

[19] Talebnia F, Karakashev D, Angelidaki I. Production of bioethanol from wheat straw: An overview on pretreatment, hydrolysis and fermentation. Bioresource Technology. 2010;**101**:4744-4753

[20] Kim M, Day DF. Composition of sugar cane, energy cane, and sweet sorghum suitable for ethanol production at Louisiana sugar mills. Journal of Industrial Microbiology and Biotechnology. 2011;**38**(7):803-807

[21] Li D-C, Jiang H. The thermochemical conversion of non-lignocellulosic biomass to form biochar: A review on characterizations and mechanism elucidation. Bioresource Technology. 2017;**246**:57-68

[22] Chiang K-Y, Chien K-L, Lu C-H. Characterization and comparison of biomass produced from various sources: Suggestions for selection of pretreatment technologies in biomass-to-energy. Applied Energy. 2012;**100**: 164-171

[23] Yoshida T, Antal MJ. Sewage sludge carbonization for terra preta applications. Energy and Fuels. 2009; **23**(11):5454-5459

[24] Liu W-J, Jiang H, Yu H-Q. Development of biochar-based functional materials: Toward a sustainable platform carbon material. Chemical Reviews. 2015;**115**(22): 12251-12285

[25] Anukam A, Mamphweli S, Okoh O, Reddy P. Influence of torrefaction on the conversion efficiency of the gasification process of sugarcane bagasse. Bioengineering. 2017;**4**:22-23

[26] Aslanzadeh S, Ishola MM, Richards T, Taherzadeh MJ. Biorefineries: Integrated Biochemical Processes for Liquid Biofuels. In: Qureshi N, Hodge D, Vertes AA, editors. An overview of existing individual unit operations. 1st ed. Cambridge: Elsevier; 2014. pp. 3-36

[27] Qureshi N, Singh V. Biorefineries integrated biochemical processes for liquid biofuels. In: Qureshi N, Hodge DB, Vertès AA, editors. Process Economics of Renewable Bio Refineries: Butanol and Ethanol Production in Integrated Bioprocesses from Lignocellulosics and Other Industrial by-Products. 1st ed. Cambridge: Elsevier; 2014. pp. 237-254

[28] Guerriero G, Hausman JF, Strauss J, Ertan H, Siddiqui KS. Lignocellulosic biomass: Biosynthesis, degradation, and industrial utilization. Engineering in Life Sciences. 2015;**16**:1-16

[29] Loow YL, Wu TY, Tan KA, Lim YS, Siow LF, Jahim J, et al. Recent advances in the application of inorganic salt pretreatment for transforming lignocellulosic biomass into reducing sugars. Journal of Agricultural and Food Chemistry. 2015;**63**(38):8349-8363

[30] Chen H, Liu J, Chang X, Chen D, Xue Y, Liu P, et al. A review on the pretreatment of lignocellulose for high-value chemicals. Fuel Processing Technology. 2017;**160**:196-206

[31] Harmsen PFH, Huijgen W, Bermudez L, Bakker R. Literature Review of Physical and Chemical Pretreatment Processes for Lignocellulosic Biomass. A Technical

Report of Biosynergy. Wageningen: Wageningen UR - Food & Biobased Research; 2010

[32] Fengel D, Wegener G. Wood Chemistry, Ultrastructure, Reactions. Berlin, New York: Walter de Gruyter GmbH; 1984

[33] Swatloski RP, Spear SK, Holbrey JD, Rogers RD. Ionic liquids: New solvents for nonderivitized cellulose dissolution. Abstracts of Papers of the American Chemical Society. 2002;**224**:U622

[34] Ria-Millati I, Syamsiah S, Niklasson C, Cahyanto MN, Lundquist K, Taherzadeh MJ. Biological pretreatment of lignocelluloses with white-rot fungi and its applications: A review. BioResources. 2011;**6**:1-36

[35] Saritha M, Arora A, Nain L. Biological pretreatment of lignocellulosic substrate for enhanced delignification and enzymatic digestibility. Indian Journal of Microbiology. 2012;**52**(2):122-130

[36] Chandra R, Bura R, Mabee W, Berlin A, Pan X, Saddler J. Substrate pretreatment: The key to effective enzymatic hydrolysis of lignocellulosics. Advances in Biochemical Engineering/ Biotechnology. 2007;**108**:67-93

[37] Agbor VB, Cicek N, Sparling R, Berlin A, Levin DB. Biomass pretreatment: Fundamentals toward application. Biotechnology Advances. 2011;**29**:675-685

[38] Singh P, Suman A, Tiwari P, Arya N, Gaur A, Shrivastava AK. Biological pretreatment of sugarcane trash for its conversion to fermentable sugars. World Journal of Microbiology and Biotechnology. 2008;**24**(5):667-673

[39] Kuhar S, Nair LM, Kuhad RC. Pretreatment of lignocellulosic material with fungi capable of higher lignin degradation and lower carbohydrate degradation improves substrate acid hydrolysis and the eventual conversion to ethanol. Canadian Journal of Microbiology. 2008;**54**(4):305-313

[40] Behera S, Arora R, Nandhagopal N, Kumar S. Importance of chemical pretreatment for bioconversion of lignocellulosic biomass. Renewable and Sustainable Energy Reviews. 2014;**36**: 91-106

[41] Kirk TK, Yang HH. Partial delignification of unbleached Kraft pulp with ligninolytic fungi. Biotechnology Letters. 1979;**1**(9): 347-352

[42] Jeoh T, Ishizawa CI, Davis MF, Himmel ME, Adney WS, Johnson DK. Cellulase digestibility of pretreated biomass is limited by cellulose accessibility. Biotechnology and Bioengineering. 2007;**98**(1):112-122

[43] Prasad S, Singh A, Joshi HC. Ethanol as an alternative fuel from agricultural, industrial and urban residues. Resources, Conservation and Recycling. 2007;**50**:1-39

[44] Chen J, Adjallé K, Lai TT, Barnabé S, Perrier M, Paris J. Effect of mechanical pretreatment for enzymatic hydrolysis of woody residues, corn Stover and alfalfa. Waste and Biomass Valorization. 2019;**12**:1-10

[45] Bak JS, Ko JK, Han YH, Lee BC, Choi IG, Kim KH. Improved enzymatic hydrolysis yield of rice straw using electron beam irradiation pretreatment. Bioresource Technology. 2009;**100**: 1285-1290

[46] Zaldivar J, Martinez A, Ingram LO. Effect of selected aldehydes on the growth and fermentation of ethanologenic Escherichia coli. Biotechnology and Bioengineering. 1999;**65**(1):24-33

[47] Mosier NS, Wyman C, Dale B, Elander R, Lee YY, Holtzapple M, et al. Features of promising technologies for pretreatment of lignocellulosic biomass. Bioresource Technology. 2005;**96**: 673-686

[48] Duff SJB, Murray WD. Bioconversion of forest products industry waste cellulosics to fuel ethanol: A review. Bioresource Technology. 1996;**55**:1-33

[49] Vassilev SV, Baxter D, Andersen LK, Vassileva CG. An overview of the chemical composition of biomass. Fuel. 2010;**89**(5):913-933

[50] Saidur R, Abdelaziz EA, Demirbas A, Hossain MS, Mekhilef S. A review on biomass as a fuel for boilers. Renewable and Sustainable Energy Reviews. 2011;**15**(5):2262-2289

[51] Anukam AI, Goso BP, Okoh OO, Mamphweli SN. Studies on characterization of corn cob for application in a gasification process for energy production. Journal of Chemistry. 2017;**2017**:1-9

[52] Liu Z, Fei B. Characteristics of moso bamboo with chemical pretreatment. In: Chandel AK, da Silva S, editors. Sustainable Degradation of Lignocellulosic Biomass– Techniques, Applications and Commercialization. Rijeka: Intechopen; 2013. pp. 1-14

[53] Segal L, Creely JJ, Martin AE Jr, Conrad CM. An empirical method for estimating the degree of crystallinity of native cellulose using the X-ray diffractometer. Textile Research Journal. 1959;**29**:786-794

[54] Zhang YP, Lynd LR. Toward an aggregated understanding of enzymatic hydrolysis of cellulose: Noncomplexed cellulase systems. Biotechnology and Bioengineering. 2004;**88**:797-824

[55] Le Brech Y, Delmotte L, Raya J, Brosse N, Gadiou R, Dufour A. High resolution solid state 2D NMR analysis of biomass and biochar. Analytical Chemistry. 2015;**87**:843-847

[56] Anukam A, Mamphweli S, Reddy P, Okoh O, Meyer E. An investigation into the impact of reaction temperature on various parameters during torrefaction of sugarcane bagasse relevant to gasification. Journal of Chemistry. 2015;**2015**:1-12

[57] Font R. Decomposition of organic wastes: Thermal analysis and evolution of volatiles. In: Vyazovkin S, Koga N, Schick C, editors. Handbook of Thermal Analysis and Calorimetry. 2nd ed. New York: Elsevier; 2018. pp. 339-397

[58] Fink JK. Epoxy resins. In: Fink J, editor. Reactive Polymers Fundamentals and Applications. A Concise Guide to Industrial Polymers. 2nd ed. New York: Elsevier; 2013. pp. 95-153

[59] Shen J, Zhu S, Liu X, Zhang H, Tan J. The prediction of elemental composition of biomass based on proximate analysis. Energy Conversion and Management. 2010;**51**(5):983-987

[60] Sheng C, Azevedo JLT. Estimating the higher heating value of biomass from basic analysis data. Biomass and Bioenergy. 2005;**28**(5):499-507

[61] Amin FR, Khalid H, Zhang H, Rahman SU, Zhang R, Liu G, et al. Pretreatment methods of lignocellulosic biomass for anaerobic digestion. AMB Express. 2017;**7**:1-12

Chapter 3

Getting Environmentally Friendly and High Added-Value Products from Lignocellulosic Waste

Elizabeth Quintana Rodríguez, Domancar Orona Tamayo, José Nicacio González Cervantes, Flora Itzel Beltrán Ramirez, María Alejandra Rivera Trasgallo and Adriana Berenice Espinoza Martínez

Abstract

In recent years, alternatives have been sought for the reuse of lignocellulosic waste generated by agricultural and other industries because it is biodegradable and renewable. Lignocellulosic waste can be used for a wide variety of applications, depending on their composition and physical properties. In this chapter, we focus on the different treatments that are used for the extraction of natural cellulose fibers (chemical, physical, biological methods) for more sophisticated applications such as reinforcement in biocomposites. Due to the different morphologies that the cellulose can present, depending from sources, it is possible to obtain cellulose nanocrystals (CNCs), micro- nanofibrillated cellulose (MFC/NFC), and bacterial nanocellulose (BNC) with different applications in the industry. Among the different cellulose nanomaterials highlighted characteristics, we can find improved barrier properties for sound and moisture, the fact that they are environmentally friendly, increased tensile strength and decreased weight. These materials have the ability to replace metallic components, petroleum products, and nonrenewable materials. Potential applications of cellulose nanomaterials are present in the automotive, construction, aerospace industries, etc. Also, this chapter exhibits global market predictions of these new materials or products. In summary, lignocellulosic residues are a rich source of cellulose that can be extracted to obtain products with high value-added and eco-friendly characteristics.

Keywords: lignocellulosic waste, cellulose, lignin, surface treatments, nanomaterials

1. Introduction

The comprehensive use of lignocellulosic waste coincides with the concept of circular economy because these wastes are renewable, abundant in nature, and generated in large volumes. In addition, they are a main source of natural fibers, chemical compounds, and other industrial products. Lignocellulosic residues are used in various applications depending on their composition and physical properties. Generally, lignocellulosic residues are constituted of cellulose, hemicellulose, lignin, pectin, waxes, and ash [1]. One of its main applications is the production of biofuels, where cellulose is subjected to various physical (mechanical, ozonolysis, pyrolysis), chemical (acid, alkali, organosolv), and biological (commonly used white-rot fungi) pretreatments. However, this review is focused on the different treatments used on the surface of natural fibers in order to improve their compatibility with a polymeric matrix and thus obtain materials with ecological, lightweight, and excellent mechanical properties, called biocomposites. It is important to mention that when carrying out some of these treatments, residues are generated, which can be processed to recover some high value-added compounds (antioxidants, sugars, bioactive phenols, organic acids, polysaccharides, and polyphenolics). Furthermore, the different types of biomaterials that can be obtained from cellulose (MCF, NFC, CNC, BNC) are described. Finally, an investigation of the market size of some of the products derived from lignocellulosic residues was carried out.

2. Biocomposites

The biocomposites are materials formed by a polymer matrix and natural fibers, which act as reinforcements. Among their main advantages, we can highlight the following: low density, low cost, high resistance, and they are eco-friendly as well. However, they have a disadvantage, incompatibility between polymer matrix and natural fibers, because polymers are hydrophobic and natural fibers have a hydrophilic nature. This is reflected in the mechanical performance of biocomposites. Because of this, chemical and physical treatments have been developed to promote interfacial adhesion between polymer and natural fibers, in addition, to improve dimensional stability and water absorption capacity of biocomposites [2]. In comparison to chemical and physical treatments, biological treatments are considered efficient and environmentally friendly processes. In nature, a great variety of microorganisms capable of degrading lignin, cellulose, and hemicellulose are found [3]. Inside these microorganisms, we can find out fungi that have the enzymatic structure necessary to degrade this type of polymers [4]. The main applications of biocomposites are automotive parts (door panel/inserts, seatbacks, spare tyre covers, interior panels, etc.), circuit boards, aerospace industry, building materials, etc.

2.1 Chemical treatment

As mentioned above, the main objective of the chemical treatment is to improve the adhesion between the natural fibers and the polymer matrix, in addition, it is possible to reduce the absorption of moisture, therefore the mechanical properties are improved. Chemical treatments including alkali, silane, acetylation, benzoylation, acrylation, maleated coupling agents, isocyanates, and others are commonly used. *Alkali treatment.* This method changes the surface morphology of the fibers, due to the breaking of the hydrogen bonds causing a roughness surface. The aqueous sodium hydroxide (NaOH) applied to natural fibers promotes the ionization of the hydroxyl group to the alkoxide [5]. It also removes certain quantity of lignin, oils, and waxes from the fiber surface; this treatment depolymerizes the cellulose in such a way that the cellulose crystals are left exposed on the fiber surface, increasing the reaction sites. This type of treatment is widely used with natural fibers that act as a reinforcement in either thermoplastic or thermosets polymers. Vinayaka et al. [6] found that biocomposites containing alkali-treated castor plant fibers have better mechanical properties than those with untreated castor plant fibers. Finally, alkaline processing directly influences the cellulosic fibril, the degree of polymerization, and the extraction of lignin and hemicellulosic compounds [7]. Several studies have been focused on the accurate concentration of NaOH, the temperature, and the time of the treatment over the fibers surface, in order to obtain biocomposites able to present satisfactory mechanical properties [8]. *Silane treatment.* According to Xie et al. [9], to effectively couple the natural fibers and polymer matrices, the silane molecule should have bifunctional groups, which may respectively react with the two phases thereby forming a bridge in between them. The general chemical structure of silane coupling agents consists of $R_{(4-n)}$–Si–(R'X)$_n$ (n = 1, 2), where *R* is alkoxy, *X* represents an organofuncionality, and *R'* is an alkyl bridge connecting the silicon atom and the organofuncionality. Silanes can be dissolved in organic solvents or in a water/solvent mixture; this solution can be sprayed on the surface of natural fibers. Silane coupling agents have been found to be efficient improving the compatibility between natural fibers and the polymeric matrix by increasing the tensile strength of the biocomposite. Nishitani et al. [10] studied the effects of silane coupling agents on surface of hemp fiber, and they found that the tribological properties of the biocomposites were improved with the surface treatment by the silane coupling agent. *Acetylation.* Acetylation is a reaction that introduces an acetyl functional group into an organic compound. In natural fibers, the acetyl group reacts with the hydroxyl groups of the fiber and an esterification is generated, which reduces its hydrophilic nature. The advantages of using this method is that it increases the thermal stability as well as the dispersion of the fibers in a polymeric matrix [11]. *Benzoylation.* Benzoylation is an important transformation in organic synthesis [12]. Benzoyl chloride is most often used in fiber treatment. Benzoylation of fiber improves fiber-matrix adhesion, thereby considerably increasing the strength of the composite, decreasing its water absorption, and improving its thermal stability [13]. *Maleated coupling agents.* These agents are mainly used to increase the compatibility between the polymeric matrix and the natural fiber. Generally, maleic anhydride is applied to modify the fiber's surface, and the polypropylene (MAAP) enhances the interfacial bonds, as a result of that the mechanical properties increase (Impact strength, young's modulus, flexural modulus, and hardness) [14]. *Permanganate.* Most of the permanganate treatments are conducted by using potassium permanganate ($KMnO_4$) solution (with acetone) in different concentrations with a soaking duration from 1 to 3 min after alkaline pretreatment [15, 16]. Paul et al. [17] studied the electrical properties of short-sisal fiber-reinforced low-density polyethylene composites using different surface treatments. As a result of permanganate treatment, the hydrophilic nature of the sisal fibers is reduced, and therefore, the water absorption decreases. At higher concentrations of $KMnO_4$, there are possibilities to lead to the degradation of cellulosic fiber by the formation of polar groups. The dielectric constant values increase as the concentration of $KMnO_4$ increases. *Peroxide.* Organic peroxides tend to decompose easily to form free radicals and attack the most available hydrogen in the polymer matrix and natural fibers. Benzoyl peroxide (BP) and dicumyl peroxide (DCP) are used in natural fiber surface modifications [11]. As a result of peroxide treatment, the hydrophilicity of the fiber decreases [17] and the tensile properties increase. *Isocyanate.* The isocyanate functional group (–N=C=O) is highly susceptible to react with the hydroxyl groups of cellulose and lignin in fibers. Joseph and Thomas [18] studied the chemical treatment of the cardanol derivative of toluene diisocyanate (CTDIC) in sisal fiber-LDPE composites. It was demonstrated that CTDIC composites show superior tensile properties than other chemically treated sisal fiber composites due to their better compatibility between sisal fibers and LDPE.

2.2 Physical treatment

There are different types of physical treatments used to modify only the surface of natural fibers without changing their chemical composition. Physical treatments promote the separation of the fiber bundle into individual fibrils and thus increase the surface area of the fibers and the compatibility with the polymer matrix. According to Ahmed et al. [11], these physical treatments can be classified as follows: mechanical treatment (stretching, calendaring, or rolling), solvent extraction treatment, and electric discharge (plasma treatment, corona treatment, ionized air treatment, thermal treatment, steam explosion, electron radiation, dielectric barrier, and ultraviolet). The *mechanical treatments* promote the interactions between the natural fibers and the polymeric matrix by increasing the surface area of the fibers and decreasing the density and stiffness; therefore, a better distribution of the fibers in the polymer matrix is achieved [19]. *Solvent extraction* can increase the surface area and remove soluble impurities for natural fibers and fillers. Hence, fibers with high cellulose content are obtained. However, this treatment is not widely used because it generates dangerous stems that pollute the environment [20]. *Electric discharge* improve the compatibility between the hydrophilic fiber and the polymer matrix through roughness of the natural fiber surface and structure [21]. Plasma treatment does not need the use of chemicals, which makes it environmentally friendly and cheaper as well. Fazeli et al. [22] modified cellulose fibers by using plasma treatment for the development of biocomposites using a thermoplastic starch matrix (TPS), obtaining a biocomposite with acceptable mechanical properties due to a good interfacial interaction between cellulose fibers and TPS, verified by scanning electron microscope (STEM). Corona treatment changes the surface of natural fibers (surface energy can decrease or increase and free radicals can be produced) by using different types of gases and cold plasma [23]. The steam explosion and alkaline extraction treatments are the most efficient for the removal of hemicellulose fibers. Ultraviolet rays treatment oxidizes the surface of the natural fibers and improves the mechanical properties due to a good interfacial adhesion between natural fibers and the polymer matrix [24].

2.3 Biological pretreatment

Biological pretreatment is based in the extracellular enzymes released by microorganisms in which enzymes degrade the noncellulosic components of the fiber surface. Biological pretreatment of fiber offers relevant advantages, such as low chemical and energy use that make it eco-friendly [25]. A great variety of microorganisms exists in nature, they are able to hydrolyze lignin, being the fungi the most studied [3]. Basidiomycetes white-rot fungi are responsible for lignin degradation in nature; they can break down not only lignin but also hemicellulose and cellulose. It has been reported that these microorganisms degrade lignin in a selective way that is able to offer potential biotechnological application [26]. However, recent studies have shown that many bacteria are able to break down lignin [27]. Likewise, enzymes have an enormous potential to be used for lignin valorization.

2.3.1 Fungal lignin degradation

The breaking down of lignin by fungi has been reported mainly for white-rot fungi due to their highly efficient enzymatic system. White-rot fungi are able to degrade lignin in such an efficiently and selectively way that gives them utility in the industry. These fungi have been applied by different industries such as paper, biofuels, and biorefinery for delignifying biomass [28]. According to the selected strain, it is possible to obtain 20–100% for lignin removal. Black liquor from a pulp and paper mill, treated with the fungi *Pleurotus ostreatus,* reduced 70% its lignin content [29]. Sugarcane bagasse treated with *Lentinula edodes* and *P. ostreatus* presented, after the treatment, 87 and 85% of lignin, respectively [30]. Biological pretreatment of bamboo culms with *Punctularia* sp. Strain TUFC20056 showed more than 50% on lignin degradation [31]. High ligninolytic capabilities have been found in the fungi *Polyporus brumalis* using wheat straw as substrate [32]. The fungal lignin degradation is based in an oxidative system. The oxidative and ligninolytic system is based in extracellular enzymes, which break down lignins and open phenyl rings; these enzymes are divided into two families: polyphenol oxidases (laccases) and lignin-modifying heme-containing peroxidases (LMPs); this second family comprises: lignin peroxidases (LiP), manganese peroxidases (MnP), and versatile peroxidases (VP) [33].

Laccases use molecular oxygen to oxidize aromatic and nonaromatic compounds, such as phenols, arylamines, anilines, thiols, and lignins [34]. The oxidation leads to the constitution of free radicals that act as intermediate for the enzymatic reactions. Likewise, these mediators can react with others high redox potential compounds and mediate nonenzymatic reactions [26]. White-Rot fungi are mainly reported to produce laccases such as, *Phlebia radiata*, *P. ostreatus,* and *Trametes versicolor* [35]. Although this enzyme is generally found in fungi, it has been found in bacteria as well, such as *Streptomyces lavendulae*, *S. cyaneus*, and *Marinomonas mediterranea* [36]. Laccases present an enormous potential because they work efficiently on a broad range of substrates with applications on paper industries, biosensors (identifying morphine or codeine), food industries, textile industries, soil bioremediation, and in the production of polymers [37].

LMPs belong to class II peroxidases, named plant, and fungal peroxidases, which contain protoporphyrin IX as a prosthetic group [38]. LiP enzymes oxidize different phenolic aromatic compounds and nonphenolic lignin compounds due to the fact that they are not very specific to their substrates [39]. LiP enzymes have been found only in a few white-rot fungi such as the genera: *Bjerkandera*, *Phanerochaete*, *Phlebia*, and *Trametes* [40–42]. The most common peroxidases found in white-rot fungi and other litter-decomposing fungi are the glycoproteins MnP [43]. The MnP glycoproteins catalyze the oxidation of Mn (II) to Mn (III), which is released in complex with oxalate or others chelators [44]. MnP enzymes are found in white-rot wood and litter-decomposing fungi such as *Dichomitus squalens*, *Agaricus bisporus*, and *Agrocybe praecox* [45]. VP enzymes present molecular similarities to LiP and MnP, oxidizing substrates as LiP and Mn^{2+} with a similar catalytic site to MnP [38]. VP enzyme has been found in white-rot fungal in the genera *Pleurotus* and *Bjerkandera* [46]. A variety of low molecular weight aromatic compounds are obtained from fungal lignin degradation, such as, guaiacol, coniferyl alcohol, p-coumarate, ferulate, protocatechuate, p-hydroxybenzoate, and vanillate [47]. The resulting liquor can be used by bacteria that can metabolize lignin-derived aromatics compounds [48].

2.3.2 Bacterial lignin degradation

It has been reported that bacteria are able to degrade lignin through a complex of enzymes, such as extracellular peroxidases, Dye-decolorizing peroxidases (DyPs), and laccases. Among the reported bacterial genus, we found *Rhodococcus*, *Pseudomonas*, *Streptomyces, Novosphingobium*, and *Bacillus* [49]. The bacteria *S. viridosporus* and *N. autotrophica* were able to degrade lignin through extracellular peroxidases, whereas *P. putida*, *Rhodococcus RHA1*, and *Rhodococcus* sp. were active in hydrogen peroxide absence suggesting the presence of extracellular laccases [50]. DyP peroxidases are able to oxidize lignin, aromatic dye, and other phenolic compounds [51]. In spite of finding at first the DyP peroxidases in fungi, recent studies have shown that these enzymes are prominent in bacteria [52]. Bioinformatic analysis showed that *R. jostii* sp. presents two peroxidases members of the DyP peroxidase family, and the deletion mutant gene assay in these genes showed reduced lignin degradation [53]. Bacterial laccases have showed high tolerance to temperature, salt, and acid/alkaline conditions, which make them valuable in the industry, being the first bacterial laccases identified in *Azospirillum lipoferum* [54]. Many soil bacteria, actinobacteria, and α-, β-, and γ-proteobacteria have shown bacterial laccase genes [55]. A higher laccase production was reached by the bacteria *Streptomyces* sp. KS1025A compared with white-rot fungi in reduced time [56].

2.3.3 Lignin-derived aromatic compounds breaking down by microorganisms

Low molecular weight aromatic compounds are obtained after fungal lignin depolymerization, such as guaiacol, coniferyl alcohol, p-coumarate, ferulate, protocatechuate, p-hydroxybenzoate, and vanillate [57]. Bacteria have the enzymatic machinery to metabolize-derived aromatic compounds that could allow the generation of value-added products such as flavors, polymer building blocks, and energy storage compounds (**Figure 1**). *R. opacus* DSM 1068 and PD630 strains were able to convert lignin into triacylglycerols under nitrogen-limiting conditions [58].

P. paucimobilis is able to metabolize β-aryl ether lignin dimer compounds to yield vanillic acid [59]. In the catecholic compounds production, *O*-demethylation is an essential process with ring cleavage catalyzed by dioxygenase [60]. *Sphingobium* is a bacterial genus characterized for the catabolism of lignin-derived aromatic compounds sp. being able to produce protocatechuate/gallate and 3-*O*-methylgallate [61]. While *Ralstonia eutropha* strain H16 was able to synthesize the biopolyester

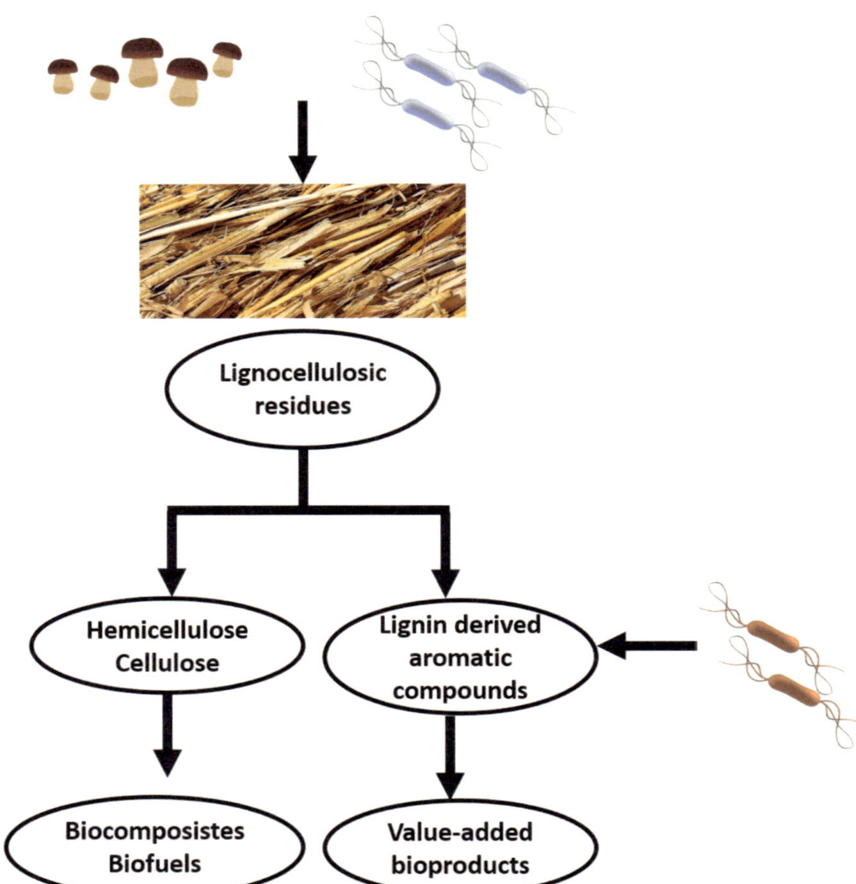

Figure 1.
Biorefinery based on microbial pretreatment in lignocellulosic residues.

polyhydroxyalkanoate (PHA) from lignin derivates [62], *Pandorea* sp. ISTKB converts lignin and its derivates into a value-added product PHA [63].

Fungal and bacterial lignin degraders (BLD) depolymerize the lignocellulosic residues, thus obtaining hemicellulose and cellulose that can be used to produce biocomposites or biofuels and lignin-derived aromatic compounds which can be transformed by bacteria to value-added bioproducts.

2.3.4 Challenges in microbial lignin degradation

Biological lignin degradation process does not involve high temperatures and pressures and does not generate any undesirable products. However, it is a time-consuming process, and there is not an accurate control on it [64]. Long time is necessary to achieve microbial lignin degradation that can range from 10 to 100 days, which is not suitable for commercial applications [28]. Several efforts have been made to engineer microorganisms in order to be more efficient to metabolize lignin-derived compounds with remarkable biotechnological applications, such as pretreatment of lignocellulosics, pulping and bleaching in the paper industry, and decolorization in the textile industry [49]. *Yarrowia lipolytica* was transformed with laccases genes from *Pycnoporus cinnabarinus* offering an efficient model for the engineering of laccases with industrial applications [65]. A dye-decolorizing peroxidase from *P. putida* MET94 strain was engineered to enhance 100-fold the catalytic efficiency when oxidizing phenolic lignin model substrates [66]. On the other hand, multi-copy recombinant *Pichia pastoris* strain expressed lignin peroxidase from *P. chrysosporium* reaching a maximum activity after 12 h induction [67]. Systems among ligninolytic microorganisms and enzymes demonstrate an enormous potential to enhance the lignin degradation [68].

2.3.5 Purified enzymes

The application of enzymes is an attractive alternative due to its shortened time, improved yield, and simple processing [69]. The most common enzymes used to break down lignin are peroxidase and laccase, catalyzing lignin oxidation. Among the most studied peroxidases are lignin peroxidases and manganese-dependent peroxidases. These enzymes degrade lignin randomly converting the phenolic group to free radicals, which lead to lignin depolymerization [70]. Fungal peroxidase from *P. ostreatus* shown lignin degradation at 30°C and pH 4 yielding 2,6-dimethoxy-1,4-benzoquinone, benzoic acid, butyl phthalate, and bis(2-ethylhexyl) phthalate [71]. Laccase can be isolated from fungi and bacteria; it is able to oxidize phenolic compounds; however, it can cooperate with mediators (small molecules able to transfer an electron) to degrade nonphenolic compounds [72]. In spite fungal laccases are selected, not only bacterial laccases have higher thermostability and an extended pH range of use but also they represent a good alternative to lignin depolymerization [73]. The company MetGen Oy has designed the enzyme MetZyme® LIGNO™, a genetically laccase of bacteria origin that can perform its activity in extremely alkaline pH and at elevated temperatures [74]. The enzyme immobilization has also been attempted to improve product separation and catalyzation because the enzymes can be made reusable through techniques such as cross-linking of enzymes, immobilization onto nanomaterials, or entrapping on beads [75]. Laccases from *Fomes fomentarius* and *T. versicolor* were cross-linked showing higher catalytic efficiency, stabilities, and high reusability compared with the free laccase [76]. Other efforts have been made to design multienzyme biocatalysts to improve stability and efficiency of lignin degradation. Co-immobilization of laccase and horseradish peroxidase by cross-linking maintains their activity and improves enzyme stability [77].

3. New biomaterials

Cellulose is the most abundant polymer in the world. It is a linear polymer of β-D-glucose molecules linked by β(1 → 4) bonds. Due to this bond, each molecule has the ability to rotate 180° with regard to the previous one, forming long linear chains that are stabilized by the presence of hydrogen bonds and join chains to others. The cellulose micelle is made up approximately from 60 to 70 cellulose chains, and the union of 20 or 30 cellulose micelles achieves a semicrystalline packing and the formation of microfibrils. However, the morphology, size, and other characteristics depend on the cellulose origin, and according to the above, cellulose microfibrils (MFC)/nanofibrils (NFC), cellulose nanocrystals (CNCs), and bacterial nanocellulose (BNC) can be obtained [78].

3.1 Micro/nanofibrillated cellulose (MFC/NFC)

Microfibrillated cellulose (MFC) is obtained with the longitudinally disintegration of cellulose fibers by multiple mechanical shearing actions; in this way, a three-dimensional network of cellulose microfibrils (10–100 nm) is achieved, which has a higher surface area than conventional cellulose fibers. Due to its structure, MFC has the ability to form gels. Different mechanical treatment procedures have been reported to obtain MFC (high-pressure homogenization and grinding for example) and various pretreatments to facilitate the mechanical treatment (enzymatic, acid hydrolysis, mechanical cutting pretreatments, etc.) [79]. The mechanical properties of MFCs are higher compared to lignocellulosic fibers because they have a more homogeneous structure. The main application of MFCs is in the packaging industry due to its excellent mechanical and barrier properties, which are required in this sector [80]. Adel et al. [81] obtained micro/nanofibrillated cellulose from lignocellulosic residues (rice straw, sugarcane bagasse, cotton stalk) and botnia softwood Kraft pulp. First, the lignocellulosic residues were subjected to an alkaline pretreatment to eliminate the lignin, and later, the mechanical treatment was applied to them using a mill. According to their results, the crystallinity index of MFC increased and the length of the fibers that correspond to lignocellulosic residues decreased compared to the fibers of the pulp. And they concluded that the MFC obtained have optimal mechanical and optical properties; therefore, they can be used as reinforcement in the paper-making industry. Nanofibrillated cellulose (NFC) is obtained by delamination of wood pulp (wood, sugar beet, potato tuber, hemp, flax, etc.) by mechanical pressure before and/or after chemical enzymatic treatment with a diameter between 5 and 60 nm and its length in several micrometers. It exhibits amorphous and crystalline domains and high specific surface area. Nanofibrillated cellulose (NFC)/polyvinyl alcohol (PVA) nanocomposites are prepared by dispersion of nanofibers obtained from several biomass sources, normally at low contents (1–10%), into PVA aqueous solutions typically followed by solvent casting. Frone et al. [82] also used cellulose nanofibers obtained from microcrystalline cellulose by ultrasonic treatment as reinforcement (at lower 1–5 wt%) dispersed in PVA. In summary, these materials exhibit a high aspect ratio and specific surface area, excellent flexibility and strength, low thermal expansion, high optical transparency, and barrier properties. Consequently, they can be used to form strong transparent films and aerogels, as a rheology modifier and strength additive in the paper-making industry, like a constituent of food packaging and in different biomedical applications (drug delivery) [79].

3.2 Bacterial cellulose (BC)

Bacterial cellulose is produced by bacteria such as *Acetobacter xylinus* or *Gluconacetobacter xylinus* [78]. Its structure is similar to the original cellulose but with an ultrafine three-dimensional network of nanofibers with an average diameter 100 times thinner than that of common plant fibers [79]. BC has high water retention due to the fact of being very hydrophilic and having high crystallinity, is relatively inexpensive to produce, and is widely used in biomedical applications (carriers for drug delivery, artificial skin and blood vessels, tissue engineering, etc.); hence, it promotes physical interaction with microorganisms and other active compounds because of its high porosity and surface area [83]. Azeredo et al. [84] explored the possibility of using BC as a raw material in the food and packaging industry applications, and they concluded that the use of this material is increasing and therefore its production cost is decreasing. However, research in this area continues to develop.

3.3 Cellulose nanocrystals (CNCs)

Cellulose nanocrystals are obtained by enzymatic hydrolysis and have the following characteristics: elongated, less flexible, cylindrical, and rod-like nanoparticles with 4–70 nm in width, 100–6000 nm in length, and 54–88% crystallinity index [85]. Gopi et al. [86] used hydrochloric acid to carry out the hydrolysis of cellulose and obtained an improvement in the thermal stability of the CNCs but with a significant agglomeration of the crystals. Park et al. [87] demonstrated a facile and green method of CNC extraction that uses only an high-pressure homogenization (HPH). The obtained CNCs presented rod-like shapes with a size distribution of 4–14 nm for width and 60–20 nm for length. Nanocrystalline cellulose (CNC) was dispersed in an alginate matrix for film application by Huq et al. [88]. They observed that with a small amount of CNC (approximately 5% wt), the mechanical and barrier properties of the films made were improved by comparing with an alginate film. According to the results obtained by infrared spectroscopy (FTIR), they concluded that there was a molecular interaction between the CNC and the alginate through hydrogen bonds. In summary, the morphology and size of cellulose nanocrystals vary according to the kind of lignocellulosic biomass, extraction method, and manufacturing conditions. Nanocellulosic materials can be characterized by employing a variety of techniques [89]: X-ray diffraction (XRD), X-ray photoelectron spectroscopy (XPS), helium pycnometer, differential scanning calorimetry (DSC), thermogravimetric analysis, transmission electron microscopy (TEM), field emission scanning electron microscopy (FESEM), and atomic force (AFM) among others. On the other hand, cellulose nanocrystals not only consist of primary reactive sites (i.e., hydroxyl groups) but also they possess higher surface area to volume ratio, making CNC highly reactive and easy to be functionalized. The most common surface modifications of CNCs are sulfonation, TEMPO-mediated oxidation, esterification, etherification, silylation, urethanization, amidation, polymer grafting, etc. The applications having the greatest potential due to the high available amount of volume on cellulose nanomaterials are placed in the following industries: automotive (body components, interiors), construction (air and water filtration, insulation, and soundproofing), packaging (fiber/plastic replacement, filler, coating, film), paper (filler, coatings), personal care (cosmetics), textiles (clothing), aerogels, aerospace (structural, interiors), industrial (viscosity modifiers, water purification), paint, sensors (medical, environmental, and industrial), electronics, photonic structures, etc. [90].

4. Recovery of chemical compounds of industrial interest

Diverse processes can be used to release lignin as the main product for the revaluation of different biomasses with high-value applications. Each process uses respective chemical agents to extract and obtain different materials from lignocellulosic biomass and produces other materials with different compositions and properties. There are distinct chemical processes of biomass hydrolysis, which use acids, bases, or enzymatic hydrolysis and others (other processes can be used, but their description would come out of the focus of this chapter) whose choice mainly depends on the material structure and characteristics desired for the products to be recovered. However, various sources of lignocellulosic materials need to be considered separately since they have different compositions of cellulose, hemicellulose, and lignin. Against all odds, the depolymerization process of the lignocellulosic biomass is a common goal for all different feedstocks for the production of all types of chemicals [91]. In particular, polyphenolic acids are a group of chemical compounds that are widely distributed in plant biomasses. Those compounds are important antioxidants that efficiently interact with biomolecules such as DNA, RNA, lipids, proteins, enzymes, and other cellular molecules to produce desired results. Due to the benefic effects, that can be useful for preventing the oxidation in foods, and therapeutic human disorders [92], all of them can be used with potential applications in the pharmacy, food, cosmetic, and nutraceutical industries.

4.1 Chemicals derived from alkaline-based methods

Alkaline pretreatment is one of the most intensively studied technologies for biomass delignification [93], and the application of alkaline liquid with NaOH into the bagasse to obtain a black liquor that contains value-added chemicals has been investigated. This procedure is useful for the releasing of chemical compounds in different biomasses; particularly, this method has been commonly used for the processing of the switchgrass (*Panicum virgatum*), corn stover (*Zea mays*), and forestry biomasses. Due to their abundance and availability, the use of the process can produce a different number of high-value fine chemicals such as sugars, vanillin, isoeugenol, guaiacylpropanol, guaiacylethanol, ferulic, p-coumaric, and syringic acids [94, 95]. For example, different woody species of *Quercus* and *Robinia* were subjected to alkaline hydrolysis, and liquors were analyzed by GC-MS. The authors recovered and identified specific bioactive phenols for each woody species such as gallic acid, coniferyl alcohol, vanillic acid, syringaldehyde, and traces of epicatechin and catechin [96]. For the optimum alkali treatment concentration in sweet sorghum bagasse, different types of phenolic species were determined with the use of alkali treatments between concentrations of 3.0 and 6.0 M NaOH, resulting in high concentrations of phenol, 4-ethylphenol, and guaiacol [97].

4.2 Chemicals derived from acid-based methods

The acidic pretreatment is a contemporaneous method for the processing of different cereal straws. Nowadays, acidic and alkaline methods are used especially with other methods such as enzymatic hydrolysis for the production of fermentable sugar and polyphenols. Dilute sulfuric acid pretreatment was used on corn stover feedstock and storage for 3 months, resulting in nonobservable microbial infestation. The cellulose content was stable while the hemicellulose content exhibited a slight decrease in furfural and oligomers, and the concentration of chemical compounds such as *O*-glucose and *O*-xylose was also constant [98]. In recent years, the focus has been on the use of other types of biomasses of fruit, for example, apple pomace, citrus, bananas, and mango among others [99]. In that aspect, different solutions of sulfuric acids were used for the valorization of apple pomace and the production of fermentable sugars and organic acids; the hemicellulose of the biomass was hydrolyzed, and the obtained liquor contained different concentrations of sugars such as glucose, xylose, arabinose, rhamnose, and galacturonic acid [100]. Those are new examples of the use of the acidic digestion of new biomasses with new co-products with a high application mainly in food industries.

4.3 Chemicals derived from hot water methods

Hot water, also known as autohydrolysis, hydrolyzes hemicellulose to release acetyl chemical groups and diverse polyphenols and removes lignin, making cellulose fibers more accessible [101]. The hot water method is very extreme, due to the fact that this method uses water at high temperatures usually between 170 and 230°C [102]. The resulting liquor contains different concentrations of sugars and chemical constituents such as polyphenols. Polyphenol compounds are covalently attached to the cell wall constituents such as cellulose, hemicelluloses, lignin, pectin, and structural proteins [103]. For example, hydroxycinnamic and hydroxybenzoic acids form ether linkages with lignin through their hydroxyl groups in the aromatic ring and ester linkages with structural carbohydrates and proteins through their carboxylic group [104]. Therefore, the recovery of the polyphenols can be made by selective extraction with ethyl acetate, purified and cleaned with resins to obtain a high yield of polyphenols with a direct use in food industries [105]. Ares-Péon et al. characterized phenolic compounds from liquors of stems maize (*Zea mays*) and *Eucalyptus globulus* with the use of hot water. Those authors found high recoveries of different polyphenols such as vanillin, ferulic, coumaric, sinapinic, hydroxybenzoic acids, guaiacol, and others. In addition, strong antioxidant activities have been reported in oligosaccharides esterified with polyphenols compounds derived from cell wall of diverse biomasses subjected to hot water methods. For example, in heteroxylans, such as arabinoxylan or glucuronoxylan, the main and predominant component is the hemicellulosic chain polymer, found in hardwoods, brans, and other softwoods [106], which can link some esterified phenolic acids to the oligosaccharides chain. In that sense, Rivas et al. [107] analyzed, by autohydrolysis, samples of liquor from rice husks, *Eucalyptus globulus* wood, and *Pinus pinaster* and found high amounts of hemicellulose-derived saccharides with esterified polyphenols. The samples displayed higher antiradical activities against strong antioxidants such as DPPH, ABTS, and ferric-reducing power; in addition, the polyphenol samples exerted high antioxidant protection to β-carotene-linoleic emulsions. The authors concluded that those antioxidant activities were mainly due to the esterification of polyphenols such as ferulic, syringic, and vanillic acids found in these polymers. Autohydrolysis has been used to extract polysaccharides and polyphenolic compounds from different biomass sources such as coffee, *Eucalyptus* and hazelnut shells among others, with high antioxidant activities [108, 109].

4.4 Chemicals derived from enzymatic-based methods

Different enzymes have been involved in the lignin break down in order to release value-added chemical compounds, with different uses in the food industries. It is important to note that alkaline and acidic methods can support the delignification of the biomasses residues to support the use of enzymatic digestion and obtain mainly sugars, polyphenols, and organic acids. Biomasses such as sugarcane, maize, agave, and sweet sorghum bagasse are widely used for the sugar and phenol extractions [110]. There are other nonconventional biomasses that can use this type of acidic or alkaline pretreatments for the degradation of hemicellulose and therefore obtain fermentable sugars and release antioxidant molecules. For example, biomasses such as corn cobs, orange, and pomegranate peels produced high yields of glucose and reduced sugars employing alkaline and enzymatic treatments [111]. Pomegranate biomass contains a high concentration of fermentable sugars that can be used in ethanol production and secondary polyphenols derived from the chemical hydrolysis. Pomegranate biomass contains a high concentration of fermentable sugars that can be used in ethanol production and secondary polyphenols derived from the chemical hydrolysis, due to this fact, pomegranate peels were subjected to acidic hydrolysis, and after an enzymatic process with cellulase there were released different fermentable sugars, moreover, bioethanol in presence of ethanol-producing microorganisms was produced. High concentrations of different sugars were released, with acid hydrolysis, such as glucose, xylose, cellobiose, arabinose, and fructose, with a range of ethanol production between 4.2 and 14.3 g/L [112]. Similarly, Talekar et al. [113] incorporated hydrothermal processing in combination with acid and enzymatic hydrolysis in pomegranate peels to recover pectin, phenols, and bioethanol. They recovered pectin ranges of 19–21% and phenolic compounds between 10.6 and 11.8%.

5. Pellets elaboration

Pellets are a type of biomass fuel, that is made from different agroindustrial biomasses; as an example, pellets are a derivative of forest biomass such as wood, sawdust, fruit shells, and kernels as well as agricultural remains derived from straw, corn stove, rice husk, and additionally from plant species with energetic potential such as *Jatropha* and *Ricinus communis* [114], which serve as a source of energy; therefore, it is a good way to use and recycle agricultural surpluses. However, the pellet production is not only focused on using them in the energy industries as solid fuel and thus avoid the use of nonrenewable energy resources such as coal, natural gas, nuclear energy, and oil [115]. Nowadays, the high cost of fossil fuels has led to a high consumption of energy pellets, mainly, since some biomasses are capable of producing a similar calorific index than the oil. Hence, the use of biomass as a heating fuel had an increase in the last decade [116]. Besides, biomass is considered as a carbon-neutral fuel due to the fact that there are no additional carbon dioxide concentrations like fossil energies [117]. However, for the pellets to be used in restaurant kitchens and home kitchens, the biomass must be treated to avoid toxic pollutants for health. For example, it is known that after the consumption of biomass pellets, these produce ashes, which in their contents have high concentrations of chlorides, sulfides, carbonates, and silica among others that can be toxic to the health [118]. Different authors have pretreated the biomass with methodologies such as alkaline hydrolysis and heat treatment to obtain liquors rich in ashes, sugars, and other chemicals. In that sense, Retsina and Pylkkanen (2014) [119] used different treatments of the feedstock to produce an extract liquor that contained different chemicals such as soluble ash, hemicellulosic oligomers, acetic acid, dissolved lignin, and cellulose; the authors produced low-ash biomass ready to be transformed into energetic pellets. One of the most important parameters in the pellet production is its durability and is given by the pellet durability index (PDI). In order to achieve those parameters of PDI, strategies have been implemented to remove lignocellulose and sugars efficiently with the use of alkaline hydrolysis. Those molecules influence the final PDI of the pellet and its energetic capacity. For example, Tang et al. (2018), evaluated the release of lignin, soluble sugars, and whole particle size on the PDI of the untreated and treated Poplar (*Populus* spp.) wood sawdust, with a combination of alkaline and acid pretreatments and steam. The authors presented that PDI increased with those treatments, more specifically, with acidic pretreatment.

6. Market of eco-friendly and high added-value products derived from lignocellulosic wastes

In recent years, a great number of studies have focused on the use of lignocellulosic waste due to the high volume generated by the agroindustrial sector and the need to manufacture new eco-friendly materials. Through a specialized search in the innovation platform "Lens" and using the keywords "cellulose," "hemicellulose," "lignin," "nanocellulose," and "novel" between 2006 and 2020, an increase is shown in the production of research papers regarding cellulose, lignin, and nanocellulose. On the other hand, **Table 1** shows the estimated market size of some of the major high value-added products from lignocellulosic waste before the

Lignocellulosic waste	Estimated market size before COVID-19	Applications	Negative impact	Opportunities	References
Cellulose	$48.37 billion USD by 2025	Textile, paper, fiber-reinforced, and starch foams	Stranded supply chains, breach of contracts, supply chain shortage, and temporary closure of department stores	Increased the digital market, strengthening of the local supply chain, new buying and selling cycle, personal hygiene and protections equipment, made of corrugated paper, demand for toilet paper and sanitizing wipes, and medical materials packaging	[120–123]
Hemicellulose	$1.3 billion USD by 2007	Ethanol and fermentation products	Fuel ethanol consumption decreased	Opportunities in disinfection of medical materials and equipment	[124, 125]
Lignin	Lignin market size worth $1.12 billion USD by 2027	Adhesives and binders	Temporary business closure, automotive supply chain, and automotive adhesives	Packaging adhesives and adhesives for medical applications	[126, 127]
Nanocellulose	$0.78 billion USD by 2025	Biomedical, personal care, oil gas, paint, coatings, food, paper processing, and composites	Disruption in production and supply chains	Development antimicrobial surfaces and packaging	[128, 129]
Biocomposites	$46.30 billion USD by 2025	Transport, construction, and electronics	Temporary closure of assembly plants	Medical applications	[92]

Table 1.
Estimated market of products from lignocellulosic waste.

COVID-19 pandemic as well as the negative impacts and area of opportunity caused by COVID-19. Based on the report by Global Market Insight [130], the market size for nanocellulose was close to 146.7 million USD in 2019 and is expected to grow to 418.2 million USD in 2026 because the global nanocellulose market indicates an increase in demand for certain applications by 2026, like paper processing, food and beverage packaging, paint and coatings, among others. It is important to mention that the term “nanocellulose” used in this report includes micro/nanofibrillated cellulose, cellulose nanocrystals, and bacterial nanocellulose. Among the main nanocellulose manufacturing companies [128], we can mention: Fiberlan technologies (UK), Borregard (Norway), Nippon Paper Industries (Japan), Celluforce (Canada), etc. Due to the COVID-19 pandemic, demand also increased in the pulp and paper industry, mainly in personal hygiene paper products, food packaging products, corrugates packaging materials, and medical specialty papers [120]. Based on the above, we can conclude that the materials obtained from lignocellulosic residues have a wide field of application and have been successfully positioning themselves in the market before and after COVID-19.

7. Conclusions

The use of lignocellulosic waste is an alternative to generate environmentally friendly products with high added value. There is a variety of methods to modify the surface of cellulose fibers both to obtain biofuels and to improve their compatibility with a polymeric matrix and in this way, develop biocomposites with high mechanical performance to be used mainly in the automotive and packaging sectors. Likewise, from the chemical treatment waste, the black liquor is generated, and it can be reused for the generation of high added-value compounds. On the other hand, lignocellulosic residues have had a high growth potential in the market in a wide variety of applications; however, the COVID-19 pandemic has increased the use of some of these products mainly in medical applications and in the packaging industry.

Acknowledgements

We want to thank to Alejandro Carreón for kind suggestions in the early version of this manuscript. The authors gratefully acknowledge the financial support of CONACYT through project CB A1-31735.

Conflicts of interest

The authors of this chapter do not have potential conflicts of interest.

Author details

Elizabeth Quintana Rodríguez[1], Domancar Orona Tamayo[1],
José Nicacio González Cervantes[1], Flora Itzel Beltrán Ramirez[1]*,
María Alejandra Rivera Trasgallo[1] and Adriana Berenice Espinoza Martínez[2]*

1 Center for Applied Innovation in Competitive Technologies (CIATEC), León, Guanajuato, México

2 Center for Research in Applied Chemistry (CIQA), Saltillo, Coahuila, México

*Address all correspondence to: fbeltran@ciatec.mx and adriana.espinoza@ciqa.edu.mx

References

[1] van Maris AJA, Abbott DA, Bellissimi E, van den Brink J, Kuyper M, Luttik MAH, et al. Alcoholic fermentation of carbon sources in biomass hydrolysates by *Saccharomyces cerevisiae*: Current status. Antonie van Leeuwenhoek/International Journal of Molecular Microbiology. 2006;**90**(4):391-418

[2] Xie Y, Xiao Z, Gruneberg T, Militz H, Hill CAS, Steuernagel L, et al. Effects of chemical modification of wood particles with glutaraldehyde and 1,3-dimethylol-4,5-dihydroxyethyleneurea on properties of the resulting polypropylene composites. Composites Science and Technology. 2010;**70**(13):2003-2011

[3] Vats S, Maurya D, Shaimoon M, Agarwal A, Negi S. Development of a microbial consortium for production of blend of enzymesfor hydrolysis of agricultural wastes into sugars. Journal of Scientific and Industrial Research (India). 2013;**72**:585-590

[4] Sánchez C. Lignocellulosic residues: Biodegradation and bioconversion by fungi. Biotechnology Advances. 2009;**27**(2):185-194

[5] Agrawal R, Saxena N, Sharma K, Thomas S, Sreekala M. Activation energy and crystallization kinetics of untreated and treated oil palm fibre reinforced phenol formaldehyde composites. Materials Science and Engineering A [Internet]. 2000;**277** (1-2):77-82. Available from: https://linkinghub.elsevier.com/retrieve/pii/S0921509399005560

[6] Vinayaka DL, Guna V, Madhavi D, Arpitha M, Reddy N. Ricinus communis plant residues as a source for natural cellulose fibers potentially exploitable in polymer composites. Industrial Crops and Products [Internet]. 2017;**100**:126-131. Available from: http://www.sciencedirect.com/science/article/pii/S092666901730122X

[7] Li X, Tabil LG, Panigrahi S. Chemical treatments of natural fiber for use in natural fiber-reinforced composites: A review. Journal of Polymers and the Environment [Internet]. 2007;**15**(1):25-33. Available from: http://link.springer.com/10.1007/s10924-006-0042-3

[8] Fiore V, Di Bella G, Valenza A. The effect of alkaline treatment on mechanical properties of kenaf fibers and their epoxy composites. Composites. Part B, Engineering [Internet]. 2015 Jan;**68**:14-21. Available from: https://linkinghub.elsevier.com/retrieve/pii/S1359836814003576

[9] Xie Y, Hill CAS, Xiao Z, Militz H, Mai C. Silane coupling agents used for natural fiber/polymer composites: A review. Composites. Part A, Applied Science and Manufacturing [Internet]. 2010;**41**(7):806-819. Available from: https://linkinghub.elsevier.com/retrieve/pii/S1359835X10000850

[10] Nishitani Y, Kajiyama T, Yamanaka T. Effect of silane coupling agent on tribological properties of hemp fiber-reinforced plant-derived polyamide 1010 biomass composites. Materials (Basel) [Internet]. 2017;**10**(9):1040. Available from: http://www.mdpi.com/1996-1944/10/9/1040

[11] Ahmad R, Hamid R, Osman SA. Physical and chemical modifications of plant fibres for reinforcement in cementitious composites. Advances in Civil Engineering [Internet]. 2019;**2019**:1-18. Available from: https://www.hindawi.com/journals/ace/2019/5185806/

[12] Paul S, Nanda P, Gupta R. PhCOCl-Py/basic alumina as a versatile reagent for benzoylation in solvent-free conditions. Molecules [Internet].

2003;**8**(4):374-380. Available from: http://www.mdpi.com/1420-3049/8/4/374

[13] Chand N, Fahim M. Natural fibers and their composites. In: Chand N, Fahim M, editors. Tribology of Natural Fiber Polymer Composites [Internet]. Elsevier; 2008. p. 1-58. Available from: https://linkinghub.elsevier.com/retrieve/pii/B9781845693930500014

[14] Sanjay MR, Madhu P, Jawaid M, Senthamaraikannan P, Senthil S, Pradeep S, et al. Biofibers from agricultural byproducts for industrial applications. Industrial Crops and Products [Internet]. 2018;**67**(1):387-395. Available from: https://ac.els-cdn.com/S0960852409009638/1-s2.0-S0960852409009638-main.pdf?_tid=433a2b00-b469-11e7-8291-00000aacb362&acdnat=1508375166_2b803526c852a7db86de40cf1f3ba9a9

[15] Joseph K, James B, Thomas S, Carvalho LH De. A review on sisal fiber reinforced polymer. Revista Brasileira de Engenharia Agrícola e Ambiental 1999;3(083):367-379.

[16] Sreekala M, Kumaran MG, Seena J, Jacob M, Thomas S. Oil palm fibre reinforced phenol formaldehyde composites: Influence of fibre surface modifications on the mechanical performance. Applied Composite Materials. 2000;7:295-329

[17] Paul A, Joseph K, Thomas S. Effect of surface treatments on the electrical properties of low-density polyethylene composites reinforced with short sisal fibers. Composites Science and Technology [Internet]. 1997;**57**(1):67-79. Available from: https://linkinghub.elsevier.com/retrieve/pii/S0266353896001091

[18] Joseph K, Thomas S, Pavithran C. Effect of chemical treatment on the tensile properties of short sisal fibre-reinforced polyethylene composites. Polymer (Guildf) [Internet]. 1996;**37**(23):5139-5149. Available from: https://linkinghub.elsevier.com/retrieve/pii/0032386196001449

[19] Costa VLD, Costa AP, Amaral ME, Oliveira C, Gama M, Dourado F, et al. Effect of hot calendering on physical properties and water vapor transfer resistance of bacterial cellulose films. Journal of Materials Science [Internet]. 2016;**51**(21):9562-9572. Available from: http://link.springer.com/10.1007/s10853-016-0112-4

[20] Fuqua MA, Huo S, Ulven CA. Natural fiber reinforced composites. Polymer Reviews [Internet]. 2012;**52**(3):259-320. Available from: http://www.tandfonline.com/doi/abs/10.1080/15583724.2012.705409

[21] Pickering KLL, Efendy MGA, Le TMM. A review of recent developments in natural fibre composites and their mechanical performance. Composites. Part A, Applied Science and Manufacturing [Internet]. 2016;**83**:98-112. Available from: https://linkinghub.elsevier.com/retrieve/pii/S1359835X15003115

[22] Fazeli M, Florez JP, Simão RA. Improvement in adhesion of cellulose fibers to the thermoplastic starch matrix by plasma treatment modification. Composites. Part B, Engineering. 2019;**163**(August 2018):207-216

[23] Adekunle KF. Surface treatments of natural fibres—A review: Part 1. Open Journal of Polymer Chemistry [Internet]. 2015;**05**(03):41-46. Available from: http://www.scirp.org/journal/doi.aspx?DOI=10.4236/ojpchem.2015.53005

[24] Azwa ZN, Yousif BF, Manalo AC, Karunasena W. A review on the degradability of polymeric composites based on natural fibres. Materials and Design [Internet]. 2013;**47**:424-442. Available from:

https://linkinghub.elsevier.com/retrieve/pii/S0261306912007832

[25] FitzPatrick M, Champagne P, Cunningham MF, Whitney RA. A biorefinery processing perspective: Treatment of lignocellulosic materials for the production of value-added products. Bioresource Technology. Elsevier. 2010;**101**:8915-8922

[26] Dashtban M, Schraft H, Syed TA, Qin W. Fungal biodegradation and enzymatic modification of lignin. International Journal of Biochemistry and Molecular Biology. e-Century Publishing Corporation. 2010;**1**:36-50

[27] Machuca A, Ferraz A. Hydrolytic and oxidative enzymes produced by white- and brown-rot fungi during *Eucalyptus grandis* decay in solid medium. Enzyme and Microbial Technology. 2001;**29**(6-7):386-391

[28] Chen Z, Wan C. Biological valorization strategies for converting lignin into fuels and chemicals. Renewable and Sustainable Energy Reviews. 2017;**73**:610-621

[29] Wu J, Xiao YZ, Yu HQ. Degradation of lignin in pulp mill wastewaters by white-rot fungi on biofilm. Bioresource Technology. 2005;**96**(12):1357-1363

[30] Dong XQ, Yang JS, Zhu N, Wang ET, Yuan HL. Sugarcane bagasse degradation and characterization of three white-rot fungi. Bioresource Technology. 2013;**131**:443-451

[31] Suhara H, Kodama S, Kamei I, Maekawa N, Meguro S. Screening of selective lignin-degrading basidiomycetes and biological pretreatment for enzymatic hydrolysis of bamboo culms. International Biodeterioration and Biodegradation. 2012;**75**:176-180

[32] Herpoel Gimbert dIsabelle. Efficient biomass pre-treatment using the white-rot fungus *Polyporus brumalis*. Fungal Genomics and Biology [Internet]. 2017;**07**(01):3. Available from: https://www.omicsgroup.org/journals/efficient-biomass-pretreatment-using-the-whiterot-fungus-polyporus-brumalis-2165-8056-1000150.php?aid=87333

[33] Xiao-fei T, Fang Z, Guo F. Impact and prospective of fungal pre-treatment of lignocellulosic biomass for enzymatic hydrolysis. Biofuels, Bioproducts and Biorefining. 2012;**6**(3):335-350

[34] Claus H. Laccases: Structure, reactions, distribution. Micron [Internet]. 2004;**35**(1-2):93-96. Available from: https://linkinghub.elsevier.com/retrieve/pii/S0968432803001616

[35] Desai SS, Nityanand C. Microbial laccases and their applications: A review. Asian Journal of Biotechnology. 2011;**3**(2):98-124

[36] Jimenez-Juarez N, Roman-Miranda R, Baeza A, Sánchez-Amat A, Vazquez-Duhalt R, Valderrama B. Alkali and halide-resistant catalysis by the multipotent oxidase from Marinomonas mediterranea. Journal of Biotechnology. 2005;**117**(1):73-82

[37] Mahapatra A. New functions for an old enzyme. ACS Chemical Biology. Elsevier. 2007;**2**:762

[38] Janusz G, Pawlik A, Sulej J, Świderska-Burek U, Jarosz-Wilkolazka A, Paszczyński A. Lignin degradation: Microorganisms, enzymes involved, genomes analysis and evolution. FEMS Microbiology Reviews. Oxford University Press. 2017;**41**:941-962

[39] Wong DWS. Structure and action mechanism of ligninolytic enzymes. Applied Biochemistry and Biotechnology. 2009;**157**(2):174-209

[40] Riley R, Salamov AA, Brown DW, Nagy LG, Floudas D, Held BW, et al. Extensive sampling of basidiomycete genomes demonstrates inadequacy of the white-rot/brown-rot paradigm for wood decay fungi. Proceedings of the National Academy of Sciences of the United States of America. 2014;**111**(27):9923-9928

[41] Floudas D, Binder M, Riley R, Barry K, Blanchette RA, Henrissat B, et al. The paleozoic origin of enzymatic lignin decomposition reconstructed from 31 fungal genomes. Science. 2012;**336**(6089):1715-1719

[42] Mäkelä MR, Hildén KS, de Vries RP. Degradation and modification of plant biomass by fungi. In: Fungal Genomics [Internet]. Berlin, Heidelberg: Springer Berlin Heidelberg; 2014. p. 175-208. Available from: http://link.springer.com/10.1007/978-3-642-45218-5_8

[43] Wesenberg D. White-rot fungi and their enzymes for the treatment of industrial dye effluents. Biotechnology Advances [Internet]. 2003 Dec;22(1-2): 161-87. Available from: https://linkinghub.elsevier.com/retrieve/pii/S0734975003001344

[44] Hatakka A, Hammel KE. Fungal biodegradation of lignocelluloses. In: Industrial Applications [Internet]. Berlin, Heidelberg: Springer Berlin Heidelberg; 2011. p. 319-40. Available from: http://link.springer.com/10.1007/978-3-642-11458-8_15

[45] Chowdhary P, Shukla G, Raj G, Ferreira LFR, Bharagava RN. Microbial manganese peroxidase: A ligninolytic enzyme and its ample opportunities in research. SN Applied Sciences. 2019;**1**(1):45

[46] Pogni R, Baratto MC, Giansanti S, Teutloff C, Verdin J, Valderrama B, et al. Tryptophan-based radical in the catalytic mechanism of versatile peroxidase from Bjerkandera adusta. Biochemistry. 2005;**44**(11):4267-4274

[47] Bandounas L, Wierckx NJP, de Winde JH, Ruijssenaars HJ. Isolation and characterization of novel bacterial strains exhibiting ligninolytic potential. BMC Biotechnology. 2011;**11**(1):94

[48] Ajith S. Bacterial degradation of lignin: A prospective for lignocellulosic biofuels. International Journal for Innovative Research in Science & Technology. 2019;**4**:11

[49] Li X, Zheng Y. Biotransformation of lignin: Mechanisms, applications and future work. Biotechnology Progress. Wiley Online Library. 2020;**36**:e2922

[50] Ahmad M, Taylor CR, Pink D, Burton K, Eastwood D, Bending GD, et al. Development of novel assays for lignin degradation: Comparative analysis of bacterial and fungal lignin degraders. Molecular BioSystems. 2010;**6**(5):815-821

[51] Lambertz C, Ece S, Fischer R, Commandeur U. Progress and obstacles in the production and application of recombinant lignin-degrading peroxidases. Bioengineered. 2016;7(3):145-154

[52] Colpa DI, Fraaije MW, Van Bloois E. DyP-type peroxidases: A promising and versatile class of enzymes. Journal of Industrial Microbiology and Biotechnology. Springer. 2014;**41**:1-7

[53] Ahmad M, Roberts JN, Hardiman EM, Singh R, Eltis LD, Bugg TDH. Identification of DypB from rhodococcus jostii RHA1 as a lignin peroxidase. Biochemistry. 2011;**50**(23):5096-5107

[54] Xu R, Zhang K, Liu P, Han H, Zhao S, Kakade A, et al. Lignin depolymerization and utilization by bacteria. Bioresource Technology. 2018;**269**:557-566

[55] Fernandes TAR, Silveira WB da, Passos FML, Zucchi TD. Laccases from *Actinobacteria*—What we have and what to expect. Advances in Microbiology 2014;04(06):285-296.

[56] Aoyama A, Yamada K, Suzuki Y, Kato Y, Nagai K, Kurane R. Newly-isolated Laccase high productivity *Streptomyces* sp. grown in cedar powder as the sole carbon source. International Journal of Waste Resources. 2014;**4**(2):1-5

[57] Masai E, Katayama Y, Fukuda M. Genetic and biochemical investigations on bacterial catabolic pathways for lignin-derived aromatic compounds. Bioscience, Biotechnology, and Biochemistry. 2007;**71**:1-15

[58] Kosa M, Ragauskas AJ. Bioconversion of lignin model compounds with oleaginous Rhodococci. Applied Microbiology and Biotechnology. 2012;**93**(2):891-900

[59] Masai E, Kubota S, Kawai S, Morohoshi N, Katayama Y, Yamasaki M. Characterization of the Cα-dehydrogenase gene involved in the cleavage of β-aryl ether by *Pseudomonas paucimobilis*. Bioscience, Biotechnology, and Biochemistry. 1993;**57**(10):1655-1659

[60] Kamimura N, Sakamoto S, Mitsuda N, Masai E, Kajita S. Advances in microbial lignin degradation and its applications. Current Opinion in Biotechnology. 2019;**56**:179-186

[61] Abe T, Masai E, Miyauchi K, Katayama Y, Fukuda M. A tetrahydrofolate-dependent O-demethylase, LigM, is crucial for catabolism of vanillate and syringate in *Sphingomonas paucimobilis* SYK-6. Journal of Bacteriology. 2005;**187**(6):2030-2037

[62] Tomizawa S, Chuah JA, Matsumoto K, Doi Y, Numata K. Understanding the limitations in the biosynthesis of polyhydroxyalkanoate (PHA) from lignin derivatives. ACS Sustainable Chemistry & Engineering. 2014;**2**(5):1106-1113

[63] Kumar M, Singhal A, Verma PK, Thakur IS. Production and characterization of Polyhydroxyalkanoate from lignin derivatives by *Pandoraea* sp. ISTKB. ACS Omega. 2017;**2**(12):9156-9163

[64] Ravindran R, Jaiswal AK. A comprehensive review on pre-treatment strategy for lignocellulosic food industry waste: Challenges and opportunities. Bioresource Technology. 2016;**199**:92-102

[65] Madzak C, Otterbein L, Chamkha M, Moukha S, Asther M, Gaillardin C, et al. Heterologous production of a laccase from the basidiomycete *Pycnoporus cinnabarinus* in the dimorphic yeast *Yarrowia lipolytica*. In: FEMS Yeast Research. Oxford, UK: Blackwell Publishing Ltd; 2005. pp. 635-646

[66] Brissos V, Tavares D, Sousa AC, Robalo MP, Martins LO. Engineering a bacterial DyP-type peroxidase for enhanced oxidation of lignin-related phenolics at alkaline pH. ACS Catalysis. 2017;**7**(5):3454-3465

[67] WANG W, WEN X. Expression of lignin peroxidase H2 from *Phanerochaete chrysosporium* by multi-copy recombinant *Pichia* strain. Journal of Environmental Sciences. 2009;**21**(2):218-222

[68] Liu ZH, Le RK, Kosa M, Yang B, Yuan J, Ragauskas AJ. Identifying and creating pathways to improve biological lignin valorization. Renewable and Sustainable Energy Reviews. Elsevier. 2019;**105**:349-362

[69] Chen S, Zhang X, Singh D, Yu H, Yang X. Biological pretreatment of lignocellulosics: Potential, progress and

challenges. Biofuels. Taylor & Francis. 2010;**1**:177-199

[70] Chio C, Sain M, Qin W. Lignin utilization: A review of lignin depolymerization from various aspects. Renewable and Sustainable Energy Reviews. Elsevier. 2019;**107**:232-249

[71] Shin KS, Lee YJ. Depolymerisation of lignosulfonate by peroxidase of the white-rot basidiomycete, *Pleurotus ostreatus*. Biotechnology Letters. 1999;**21**(7):585-588

[72] Kinnunen A, Maijala P, Jarvinen P, Hatakka A. Improved efficiency in screening for lignin-modifying peroxidases and laccases of Basidiomycetes. Current Biotechnology. 2016;**6**(2):105-115

[73] Santhanam N, Vivanco JM, Decker SR, Reardon KF. Expression of industrially relevant laccases: Prokaryotic style. Trends in Biotechnology. Elsevier. 2011;**29**:480-489

[74] Hämäläinen V, Grönroos T, Suonpää A, Heikkilä MW, Romein B, Ihalainen P, et al. Enzymatic processes to unlock the lignin value. Frontiers in Bioengineering and Biotechnology. 2018;**6**(MAR):20

[75] Kirupa Sankar M, Ravikumar R, Naresh Kumar M, Sivakumar U. Development of co-immobilized tri-enzyme biocatalytic system for one-pot pretreatment of four different perennial lignocellulosic biomass and evaluation of their bioethanol production potential. Bioresource Technology. 2018;**269**:227-236

[76] Vršanská M, Voběrková S, Jiménez Jiménez AM, Strmiska V, Adam V. Preparation and optimisation of cross-linked enzyme aggregates using native isolate white rot fungi *Trametes versicolor* and *Fomes fomentarius* for the decolourisation of synthetic dyes. International Journal of Environmental Research and Public Health. 2018;**15**(1):23

[77] Crestini C, Melone F, Saladino R. Novel multienzyme oxidative biocatalyst for lignin bioprocessing. Bioorganic & Medicinal Chemistry. 2011;**19**(16):5071-5078

[78] Habibi Y. Key advances in the chemical modification of nanocelluloses. Chemical Society Reviews. 2014;**43**(5):1519-1542

[79] Vilela C, Pinto RJB, Figueiredo ARP, Neto CP, Silvestre AJD, Freire CSR. Development and applications of cellulose nanofibres based polymer nanocomposites. In: Bafekrpour E, editor. Advanced Composite Materials: Properties and Applications [Internet]. De Gruyter Open Poland; 2017. p. 1-65. Available from: https://www.degruyter.com/view/book/9783110574432/10.1515/9783110574432-001.xml

[80] Osong SH, Norgren S, Engstrand P. Processing of wood-based microfibrillated cellulose and nanofibrillated cellulose, and applications relating to papermaking: A review. Cellulose [Internet]. 2016;**23**(1):93-123. Available from: http://link.springer.com/10.1007/s10570-015-0798-5

[81] Adel AM, El-Gendy AA, Diab MA, Abou-Zeid RE, El-Zawawy WK, Dufresne A. Microfibrillated cellulose from agricultural residues. Part I: Papermaking application. Industrial Crops and Products [Internet]. 2016;**93**(April):161-174. DOI: 10.1016/j.indcrop.2016.04.043

[82] Frone AN, Panaitescu DM, Donescu D, Spataru CI, Radovici C, Trusca R, et al. Cellulose nanofibers, PVA films. BioResources. 2011;**6**(1):487-512

[83] Rajwade JM, Paknikar KM, Kumbhar JV. Applications of bacterial cellulose and its composites in biomedicine. Applied Microbiology and Biotechnology [Internet]. 2015;**99**(6):2491-2511. Available from: http://link.springer.com/10.1007/s00253-015-6426-3

[84] Azeredo HMC, Barud H, Farinas CS, Vasconcellos VM, Claro AM. Bacterial cellulose as a raw material for food and food packaging applications. Frontiers in Sustainable Food Systems [Internet]. 2019;**3**(February):7. Available from: https://www.frontiersin.org/article/10.3389/fsufs.2019.00007/full

[85] Naz S, Ali JS, Zia M. Nanocellulose isolation characterization and applications: A journey from non-remedial to biomedical claims. Bio-Design and Manufacturing [Internet]. 2019;**2**(3):187-212. Available from: http://link.springer.com/10.1007/s42242-019-00049-4

[86] Gopi S, Balakrishnan P, Chandradhara D, Poovathankandy D, Thomas S. General scenarios of cellulose and its use in the biomedical field. Materials Today Chemistry [Internet]. 2019;**13**:59-78. Available from: https://linkinghub.elsevier.com/retrieve/pii/S2468519419300072

[87] Park N-M, Choi S, Oh JE, Hwang DY. Facile extraction of cellulose nanocrystals. Carbohydrate Polymers [Internet]. 2019;**223**(July):115114. DOI: 10.1016/j.carbpol.2019.115114

[88] Huq T, Salmieri S, Khan A, Khan RA, Le C, Riedl B, et al. Nanocrystalline cellulose (NCC) reinforced alginate based biodegradable nanocomposite film. Carbohydr Polym [Internet]. 2012;**90**:1757-1763. Available from: https://linkinghub.elsevier.com/retrieve/pii/S0144861712007382

[89] Trache D, Tarchoun AF, Derradji M, Hamidon TS, Masruchin N, Brosse N, et al. Nanocellulose: From fundamentals to advanced applications. Frontiers in Chemistry [Internet]. 2020;**8**:392. Available from: https://www.frontiersin.org/article/10.3389/fchem.2020.00392/full

[90] Shatkin JA, Wegner TH, Bilek EM, Cowie J. Market projections of cellulose nanomaterial-enabled products—Part 1: Applications. Tappi Journal. 2014;**13**(5):9-16

[91] Isikgor FH, Becer CR. Lignocellulosic biomass: A sustainable platform for the production of bio-based chemicals and polymers. Polymer Chemistry. 2015;**6**(25):4497-4559

[92] Beltrán-Ramírez F, Orona-Tamayo D, Cornejo-Corona I, Luz Nicacio González-Cervantes J, de Jesús E-CJ, Quintana-Rodríguez E. Agro-Industrial Waste Revalorization: The Growing Biorefinery. In: Biomass for Bioenergy - Recent Trends and Future Challenges [Internet]. IntechOpen; 2019. p. 13. Available from: https://www.intechopen.com/books/biomass-for-bioenergy-recent-trends-and-future-challenges/agro-industrial-waste-revalorization-the-growing-biorefinery

[93] Xu J, Zhang X, Cheng JJ. Pretreatment of corn Stover for sugar production with switchgrass-derived black liquor. Bioresource Technology. 2012;**111**:255-260

[94] Sun Z, Fridrich B, De Santi A, Elangovan S, Barta K. Bright side of lignin depolymerization: Toward new platform chemicals. Chemical Reviews. ACS Publications. 2018;**118**:614-678

[95] Al Arni S, Zilli M, Converti A. Solubilization of lignin components of food concern from sugarcane bagasse by alkaline hydrolysis. Ciência e Tecnologia de Alimentos. 2007;**5**(4):271-277

[96] Ianni F, Segoloni E, Blasi F, Di Maria F. Low-molecular-weight phenols

recovery by eco-friendly extraction from *Quercus* spp. wastes: An analytical and biomass-sustainability evaluation. PRO. 2020;**8**(4):387

[97] Sehume TZ, Strydom CA, Bunt JR, Schobert HH. Bio-oil production from sweet Sorghum bagasse via liquefaction using alkaline solutions and identification of phenolic products. Waste and Biomass Valorization. 2020;**11**(7):3593-3607

[98] Zhang J, Shao S, Bao J. Long term storage of dilute acid pretreated corn Stover feedstock and ethanol fermentability evaluation. Bioresource Technology. 2016;**201**:355-359

[99] Durán D, Figueroa A, Gualdrón MA, Sierra R. Potential of Tropical Fruit Waste for Sustainable Biorefineries. In: 26th European Biomass Conference and Exhibition [Internet]. Copenhagen; 2018. p. 166-174. Available from: http://www.etaflorence.it/proceedings/?detail=15627

[100] Luo J, Ma Y, Xu Y. Valorization of apple pomace using a two-step slightly acidic processing strategy. Renewable Energy. 2020;**152**:793-798

[101] Baruah J, Nath BK, Sharma R, Kumar S, Deka RC, Baruah DC, et al. Recent trends in the pretreatment of lignocellulosic biomass for value-added products. Frontiers in Energy Research. Frontiers. 2018;**6**:141

[102] Zhuang X, Wang W, Yu Q, Qi W, Wang Q, Tan X, et al. Liquid hot water pretreatment of lignocellulosic biomass for bioethanol production accompanying with high valuable products. Bioresource Technology. Elsevier. 2016;**199**:68-75

[103] Wong WH, Lee WX, Ramanan RN, Tee LH, Kong KW, Galanakis CM, et al. Two level half factorial design for the extraction of phenolics, flavonoids and antioxidants recovery from palm kernel by-product. Industrial Crops and Products. 2015;**63**:238-248

[104] Ares-Peón IA, Garrote G, Domínguez H, Parajó JC. Phenolics production from alkaline hydrolysis of autohydrolysis liquors. CyTA Journal of Food. 2016;**14**(2):255-265

[105] Farzaneh V, Carvalho IS. A review of the health benefit potentials of herbal plant infusions and their mechanism of actions. Industrial Crops and Products. Elsevier. 2015;**65**:247-258

[106] Heinze T. Polysaccharides I Structure, Characterisation and Use [Internet]. Heinze T, ed. (Advances in Polymer Science; Vol. 186). Berlin, Heidelberg: Springer Berlin Heidelberg; 2005. 105 p. DOI: 10.1007/b136812

[107] Rivas S, Conde E, Moure A, Domínguez H, Parajó JC. Characterization, refining and antioxidant activity of saccharides derived from hemicelluloses of wood and rice husks. Food Chemistry. 2013;**141**(1):495-502

[108] Ballesteros LF, Teixeira JA, Mussatto SI. Extraction of polysaccharides by autohydrolysis of spent coffee grounds and evaluation of their antioxidant activity. Carbohydrate Polymers. 2017;**157**:258-266

[109] Rivas S, Moure A, Parajó JC. Pretreatment of hazelnut shells as a key strategy for the solubilization and valorization of hemicelluloses into bioactive compounds. Agronomy. 2020;**10**(6):760

[110] Hernández C, Escamilla-Alvarado C, Sánchez A, Alarcón E, Ziarelli F, Musule R, et al. Wheat straw, corn Stover, sugarcane, and Agave biomasses: Chemical properties, availability, and cellulosic-bioethanol production potential in Mexico. Biofuels, Bioproducts and Biorefining. 2019;**13**(5):1143-1159

[111] Pocan P, Bahcegul E, Oztop MH, Hamamci H. Enzymatic hydrolysis of fruit peels and other lignocellulosic biomass as a source of sugar. Waste and Biomass Valorization. 2018;**9**(6):929-937

[112] Demiray E, Karatay SE, Dönmez G. Improvement of bioethanol production from pomegranate peels via acidic pretreatment and enzymatic hydrolysis. Environmental Science and Pollution Research. 2019;**26**(28):29366-29378

[113] Talekar S, Patti AF, Vijayraghavan R, Arora A. An integrated green biorefinery approach towards simultaneous recovery of pectin and polyphenols coupled with bioethanol production from waste pomegranate peels. Bioresource Technology. 2018;**266**:322-334

[114] Greinert A, Mrówczynska M, Szefner W. Study on the possibilities of natural use of ash granulate obtained from the combustion of pellets from plant biomass. Energies. 2019;**12**(13):2569

[115] Kargbo H, Harris JS, Phan AN. "Drop-in" fuel production from biomass: Critical review on techno-economic feasibility and sustainability. Renewable and Sustainable Energy Reviews. Elsevier. 2021;**135**:110168

[116] Bajwa DS, Peterson T, Sharma N, Shojaeiarani J, Bajwa SG. A review of densified solid biomass for energy production. Renewable and Sustainable Energy Reviews. Elsevier. 2018;**96**:296-305

[117] Pradhan P, Mahajani SM, Arora A. Production and utilization of fuel pellets from biomass: A review. Fuel Processing Technology. 2018;**181**:215-232

[118] Zając G, Szyszlak-Bargłowicz J, Gołębiowski W, Szczepanik M. Chemical characteristics of biomass ashes. Energies. 2018;**11**(11):2885

[119] Retsina T, Pylkkanen V. Processes for producing fermentable sugars and low-ash biomass for combustion or pellets. Google Patents; 2014

[120] Liu K, Wang H, Liu H, Nie S, Du H, Si C. COVID-19: Challenges and perspectives for the pulp and paper industry worldwide. BioResources. 2020;**15**(3):4638-4641

[121] Grand View Research. Cellulose Fibers Market Size to Reach $29.6 Billion by 2020 [Internet]. 2015. Available from: https://www.grandviewresearch.com/press-release/global-cellulose-fibers-market

[122] Russell M. Does the Clothing Supply Chain Need to Push the Reset Button? [Internet]. Apparel Industry Analysis. 2020. Available from: https://www.just-style.com/analysis/does-the-clothing-supply-chain-need-to-push-the-reset-button_id139218.aspx

[123] Barrantes R. La Industria de la Moda en México ante el COVID-19 [Internet]. 2020. Available from: https://enriqueortegaburgos.com/industria-moda-en-mexico-ante-covid-19/

[124] Wysokińska Z. Market for starch, hemicellulose, cellulose, alginate, its salts and esters, and natural polymers, including chitin and chitosan: Analysis results. Fibres & Textiles in Eastern Europe. 2010;**83**(6):7-13

[125] Healy B. Post-Pandemic Opportunities Emerge for Ethanol [Internet]. Ethanol Producer Magazine. 2020. Available from: http://www.ethanolproducer.com/articles/17224/post-pandemic-opportunities-emerge-for-ethanol

[126] Grand View Research. Lignin Market Size Worth $1.12 Billion by

2027 | CAGR: 2.0% [Internet]. Grand View Research; 2020. Available from: www.grandviewresearch.com/press-release/global-lignin-market

[127] Grand View Research. Film Adhesive Market Size, Share & Trends Analysis Report By Application (Electronics & Electrical, Medical, Automotive), By Region (North America, Europe, APAC, CSA, MEA), And Segment Forecasts, 2020-2027 [Internet]. Grand View Research. 2020. p. 1-120. Available from: https://www.grandviewresearch.com/industry-analysis/film-adhesive-market/request/rs1

[128] Nanocellulose Market by Type (MFC & NFC, CNC/NCC, and Others), Application (Pulp&Paper, Composites, Biomedical & Pharmaceutical, Electronics & Sensors, and Others), Region (Europe, North America, APAC, and Rest of World)—Global Forecast to 2025 [Internet]. Markets and Markets. 2020. Available from: https://www.marketsandmarkets.com/Market-Reports/nano-cellulose-market-56392090.html

[129] Azeredo HMC, Rosa MF, Mattoso LHC. Nanocellulose in bio-based food packaging applications. Industrial Crops and Products. [Internet]. 2017;**97**:664-671. DOI: 10.1016/j.indcrop.2016.03.013

[130] Pulidindi K, Pandey H. Nanocellulose Market Size by Product (Nano Fibrillated Cellulose, Nanocrystalline Cellulose), by Application (Composites, Paper Processing, Food & Beverages, Paints & Coatings, Oil & Gas, Personal Care), Industry Analysis Report, Regional Outlook, Growth [Internet]. 2020. Available from: https://www.gminsights.com/industry-analysis/nanocellulose-market

Chapter 4

Laboratory Optimization Study of Sulfonation Reaction toward Lignin Isolated from Bagasse

Rini Setiati, Septoratno Siregar and Deana Wahyuningrum

Abstract

Bagasse is scientifically defined as waste from the extraction of sugarcane liquid after the grinding process. Bagasse is biomass which is used as raw material to be processed into surfactants. Bagasse fiber cannot be dissolved in water because it consists mostly of cellulose, pentosane and lignin. The optimum conditions for obtaining the highest yield and the best conversion of bagasse to lignin were achieved when used 80 mesh bagasse and 3 M NaOH as a hydrolysis agent. Then lignin is reacted with 0.25 sodium bisulfite to the surfactant sodium lignosulfonate. Lignin and sodium lignosulfonate were further characterized using a FTIR spectrophotometer to determine the components contained therein. The lignin component consists of phenolic functional group elements, aliphatic and aromatic groups, ketone groups, aren functional groups, amine groups and alkyl groups along with standard lignin components. Likewise with lignosulfonates, with indicator components consisting of C=C alkenes, Sulfate S=O, C=O carboxylic acids and S-OR esters. The NMR test was resulted the monomer structure of SLS surfactant bagasse. The results indicate that the lignin isolation process from bagasse has been successfully. Likewise, the sulfonation of lignin to lignosulfonate is also successful.

Keywords: bagasse, FTIR, isolation, lignin, lignosulfonate, sulfonation

1. Introduction

In specific, bagasse is scientifically defined as a waste of sugarcane liquid extraction after milling process and is in a fibrous form. Bagasse is one of the biomass resources that is widely used as a boiler fuel in sugar factory, source of animal feed, material of paper, cement and brick reinforcement material [1, 2]. The amount of bagasse production each year is abundant, easily obtained, and economical. Based on the data from Indonesian Sugar Farm Research Center (P3GI) [3], bagasse amounts to approximately 32% of milled sugarcane weight or about 10.2 million ton/year or mill/season all around Indonesia. Furthermore, bagasse contains 48–52% water, sugar (approximately3.3% in average), and fiber at an average of 47.7% [4, 5]. Bagasse fiber is unable to be dissolved in water because mostly it consists of cellulose, pentosane, and lignin [6]. Bagasse waste could be used as a raw material in producing surfactant due to its high lignin content, which is estimated to be approximately 25% [7]. Lignin can be separated

IntechOpen

from bagasse waste by lignin isolation method and hydrolysis process using sodium hydroxide (NaOH) solution [8, 9]. The process also depends on creating bagasse surface enlargement by minimizing the size of its fiber in order to have the better yields of the isolated product. Lignosulfonate is a derivate of lignin that can be produced by reacting lignin with sodium bisulfite ($NaHSO_3$) at certain reaction conditions via the electrophilic addition reaction [10, 11]. The presence of double bonds within the lignin structure has made lignin to be available for the addition reaction using various electrophilic substances, for instance, the hydrogen sulfite ($-HSO_3$) group of sodium bisulfite [12]. Thus, the product is categorized as sodium lignosulfonate (SLS) surfactant [13]. In addition, lignosulfonate is one of the variants of anionic surfactant that is often utilized in a chemical injection process of enhanced oil recovery (EOR) in the oil industry [14]. Therefore, the high lignin content in bagasse have made bagasse to be an eligible candidate to produce surfactant and became the aim of this research, which is to produce the lignosulfonate surfactant via sulfonation reaction of lignin previously isolated from bagasse. Based on the observations and search for existing patents, what have been found are patent Nos. 2,837,435 and 4,304,361 regarding the use of bagasse as a raw material for building needs, cutting of bagasse fiber for growing media needs, methods for producing bio-aromatic-based chemicals, bio-based aromatic fuels, and lignin residues [15, 16]. Whereas, the No. 8529731 was found to contain the process of fractionation of bagasse into cellulose, hemicellulose (xylene), and lignin with high-purity α-cellulose, which is a useful raw material for the manufacture of cellulose esters such as cellulose triacetate and cellulose plastics [17]. Amri [18] has shown research on sodium lignosulfonate surfactant which has characteristics of water solubility, hygroscopic, and color properties as well as the polydispersity properties of sample SLS which are generally in accordance with commercial SLS.

The lignin isolation method (hydrolysis) can excite lignin with acid, resulting in acid lignin as shown in **Figure 1**.

Isolation of lignin is generally carried out using sulfuric acid or hydrochloric acid. Under acidic conditions, the charged lignin will become neutral. Lignin will not dissolve in water and will settle. The resulting solid can be separated by filtering. To change the nature of water-insoluble lignin, lignin can be modified through the sulfonation process to become lignosulfonate [20]. Sulfonation is intended to change the hydrophilic nature of the less polar lignin into a more polar/water-soluble lignosulfonate salt by inserting the sulfonate group and its salt into the lignin hydroxyl group so that the lignosulfonate salt has a structure as a surface-active agent or surfactant [19]. The sulfonate group in the lignosulfonate is a hydrophilic group that causes the lignosulfonate to have an amphipathic structure (surfactant). **Figure 2** shows the structure of the lignosulfonate.

The existence of the sulfonate group can be determined by the general formula $R\text{-}SO_3Na$ which is a simplification of the sulfate $R\text{-}O\text{-}SO_3Na$ [21]. The R group is a group of C_8-C_{22} aromatic carbon atoms which is a hydrophilic group, while the hydrophobic group consists of carboxylates, sulfonates, phosphates, or other organic acids. The sulfonation process is the core process for producing

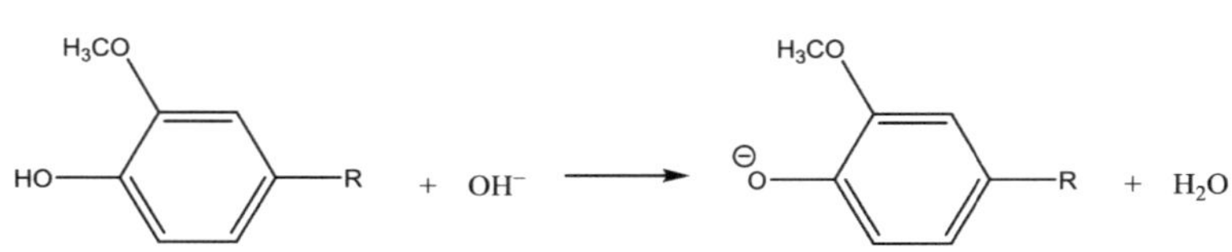

Figure 1.
The reaction of lignin and NaOH in the delignification process [19].

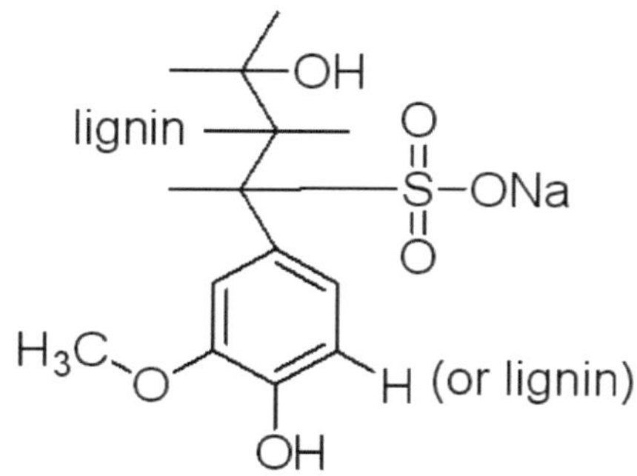

Figure 2.
Lignosulfonate structure.

lignosulfonate salts. The reaction occurs between lignin and sulfite salts. There are several types of sulfite salts that can be used in this process, including using sodium bisulfite ($NaHSO_3$) in addition to other ingredients such as Na_2SO_3, NaOH + CH_2 (OH) SO_3Na, HCHO + NaOH, C_2C_{14} + C_lSO_3H, or SO_3^{2-} + CH_2O [22].

Several studies on the manufacture of sodium lignosulfonate that have been tried include raw materials for oil palm empty bunches [23], oil palm shells [24], palm frond biomass [18], and bagasse [25, 26]. The results of this study were limited to the manufacture of sodium lignosulfonate products which were correlated with the size of bagasse powder and the concentration of sodium bisulfite. Lignosulfonates, as a result of lignin sulfonation, are currently widely used as emulsifiers in iron ore processing, oil field chemicals, and pesticide formulas [27] as well as dust emission control and stabilizer for the fertilizer industry, animal feed industry, gypsum agent wallboard dispersant, oil well drilling mud additive, brick reinforcement, cement, and mortar [28].

2. Materials and methods

In this study of sulfonation reaction toward lignin isolated from bagasse, this study used bagasse as the main raw material, with chemical reagents being sodium hydroxide, sulfuric acid, sodium bisulfite, and distilled water. Range and specifications are used in the bagasse lignin isolation process consisting of bagasse size 40, 60, and 80 mesh; sodium hydroxide concentration 0.6, 2, 3, 6, 8, and 10 M; and sodium bisulfate concentration 0.25 M. The equipment used in the process of lignin isolation and surfactant sulfonation consists of a sieve shaker; hot plate magnetic stirrer; two- or three-neck flask; condenser; beaker glass 200, 500, and 1000 mL; measuring cup 250 mL; thermometer; rod mixer; burette; gloves; glasses; mask; fume hood; pH meter paper; Buchner funnel; Whatman paper; watch glass; oven; digital balance; 250 and 500 mL reagent bottles; 10-mL vial bottle; and desiccator. The mechanism process of the lignosulfonate surfactant occurs through two reactions, namely, hydrolysis and sulfonation [29]. Hydrolysis is a reaction to break down lignin molecules into smaller molecules so that they can dissolve in water. Sulfonation is a reaction between bisulfite ions and lignin molecules. Previous research results reported that the surfactant methyl ester sulfonate (MES) could be synthesized from the direct sulfonation of palm kernel oil methyl ester using sodium bisulfite solution. The important from this previous research is the sulfuric acid concentration factor which affects the value of the decrease in surface tension, the decrease in interface tension, the stability of the emulsion, and the color of the surfactant [30].

The method of processing bagasse into lignosulfonate is carried out through two processes, namely, the isolation process of lignin from bagasse and the sulfonation process of lignin into sulfonates. The bagasse from the sugar factory was previously sifted coarsely and then to oven to dry completely. Then the oven bagasse is sieved again with a sieve shaker to obtain a particle size of bagasse with a certain mesh, namely, 40 mesh, 60 mesh, 80 mesh, and 100 mesh. **Figure 3** shows the bagasse that has been dried and then sieved using a sieve shaker to become a fine powder (**Figure 4**) [31].

The method used in this study is a development from previous researchers who modified lignosulfonate from lignin. In his research, lignin isolation was carried out using NaOH reaction by heating at a temperature of 60–100°C for 3–10 hours [20]. In this research, the lignin isolation process begins by inserting the bagasse that has been sieved with a sieve shaker into the reaction flask and reflux directly in sodium hydroxide solution at a various concentration for 5 hours at a temperature of 90–100°C. The result of reflux of NaOH is then filtered, diluted, and neutralized by adding dropwise concentrated sulfuric acid (H_2SO_4) to pH = 2 and allowed to stand for at least 8 hours until a precipitate appears, then filtered, and dried in an oven at 70°C. In this filtering process, it is accompanied by rinsing with distilled water because lignin does not dissolve in water and this rinsing with distilled water will dissolve the remaining glucose that may still be present in the results of the lignin isolation. The precipitate obtained is lignin isolated from bagasse and after drying using a vacuum oven, it becomes a brown powder.

The lignin isolation process starts with 5 gram of dry bagasse powder of each mesh size which is put into a three-neck flask, then NaOH is added until the bagasse is submerged and heated for 5 hours using a hot plate magnetic stirrer at a

Figure 3.
Bagasse.

Figure 4.
Mesh of bagasse [31].

temperature of 90–100°C. The reflux filtrate which still contains NaOH is taken and diluted with water at a volume ratio of 1:1. The solution is then added dropwise to H_2SO until it reaches pH = 2, then this solution is left to stand to get a precipitate for at least 8 hours. The precipitate that is formed is filtered and then dried in an oven. The structure of isolated lignin product was determined through FTIR spectrophotometric measurements which were then compared with the standard lignin FTIR spectrum. In the lignin isolation process, optimization was also carried out using the concentration of NaOH used, namely, with a concentration range of 2, 3, 6, 8, and 10 M. Each NaOH concentration is used in the lignin isolation process by varying the size of the bagasse mesh.

The synthesis of bagasse into sodium lignosulfonate begins with the preparation of bagasse powder which will be isolated to separate the lignin from the bagasse. After lignin is formed, a Fourier transform infrared (FTIR) [32, 33] test must be carried out to ensure the presence of lignin-forming components. The standard lignin used is commercial lignin from the lignin product of Aldrich and Kraft. If the component has not been formed, it must return to the isolation process again with changes to the variables used. There are three components of the main functional groups as indicators of lignin formation, namely, the phenolic O—H functional groups, the aliphatic and aromatic —CH— stretching groups, and the C=C aromatic functional groups. In the lignin isolation process, the variables used are NaOH concentration, duration of the isolation process, and temperature in the isolation process. This looping process is carried out continuously until the lignin component is obtained that is in accordance with the existing commercial lignin standards. If the lignin formed meets the component requirements, it can be continued to the sulfonation process. The result of this sulfonation process is a brown powder of sodium lignosulfonate (SLS) surfactant. This product must also perform component characterization using the FTIR test. If the FTIR test results do not show any lignosulfonate-forming components, then a looping process is carried out until the sulfonation process produces a lignosulfonate component that matches the standard lignosulfonate. The standard lignosulfonates used are Patricia and Aldrich standards [34].

The components of the lignosulfonate that must be present include the stretching vibration of the alkene functional group —C=C—aromatic, the stretching vibration of the sulfonate functional group S=O, the bending vibration of the C=O functional group carboxylate group, and the bending vibration of the S-OR ester functional group. At this stage, it can be said that the synthesis process is complete, as illustrated in **Figure 5**. The process of synthesis of bagasse into sodium lignosulfonate surfactant as a whole can be seen in **Figure 5**.

The sulfonation process is a procedure in the form of adaptation and modification from research conducted by Ari [25] and Furi [26]. A total of 8 gram of isolated bagasse lignin was put into a three-neck flask, then sodium bisulfite solution was added, and then heated (refluxed) at 150°C for 5 hours. The reaction product is cooled and precipitated and further dried in a vacuum oven. From this sulfonation process, it produces a surfactant called sodium lignosulfonate (SLS). The structure of SLS surfactant was determined through FTIR, LCMS, and NMR spectrophotometric measurements [36]. The FTIR test results were then compared with the main components of the commonly used commercial lignosulfonate [34]. If it is in accordance with the components forming the SLS surfactant, this product can be said to have been successfully obtained. If it is not accordance with the standard components that should be present in the lignosulfonate, the sulfonation process is repeated with different parameters.

Furthermore, the lignosulfonate monomer structure test was carried out using gas chromatograph mass spectrum (GCMS) and nuclear magnetic resonance

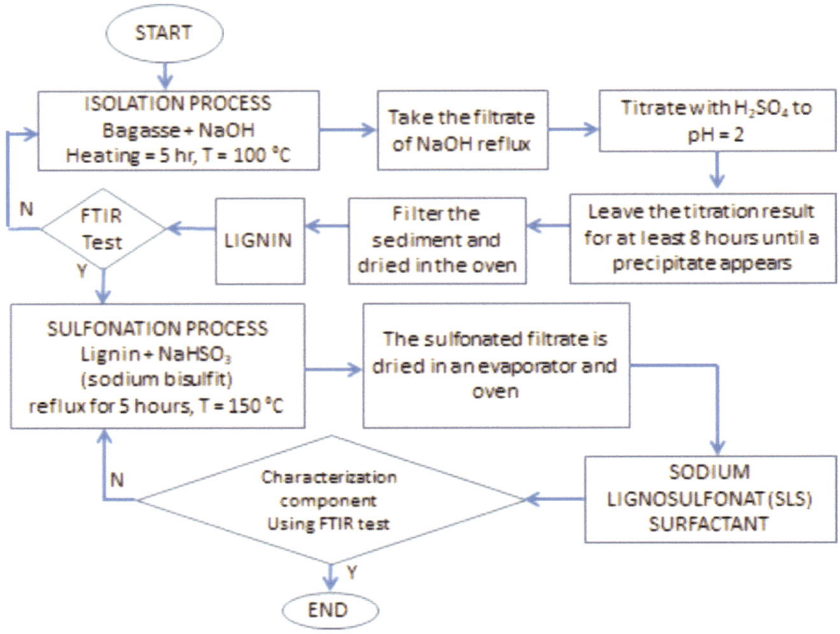

Figure 5.
Schematic synthesis of bagasse into sodium lignosulfonate [35].

(NMR). The structure of the lignosulfonate monomer is needed in order to help see the suitability of the use of the surfactant lignosulfonate against the crude oil to be injected by the lignosulfonate.

3. Results and discussion

The lignin isolation process has been carried out several times with variations in the concentration of NaOH and the size of the mesh bagasse. Variations in bagasse mesh sizes used were 40 mesh, 60 mesh, 80 mesh, and 100 mesh, and the concentration of NaOH was 2, 3, 6, 8, and 10 M. **Figure 6** shows the results of lignin isolation in the form of a dark brown powder.

From the experiment as many as 15 variations, only four variations met the requirements, namely, lignin results above 60% and they had lignin-forming components, namely, lignin (80–3), lignin (60–8), lignin (40–10), and lignin (80–10). The results of lignin recovery can be seen in **Table 1**. In this table, it can be seen that

Figure 6.
Lignin from bagasse isolation.

No.	Concentration of NaOH (M)	Lignin (%)			
		mesh 40	mesh 60	mesh 80	mesh 100
1	2	22.46	20.66	48.60	18.00
2	3	63.36	34.36	61.80	22.22
3	6	32.26	13.07	24.43	35.80
4	8	66.80	75.73	38.36	24.30
5	10	62.85	51.80	63.79	26.10

Table 1.
Results of lignin isolation at variations in bagasse size and NaOH concentrations.

the highest percentage of lignin recovery occurs in the lignin isolation process with a concentration of 3 M NaOH—40 mesh size of 63.36%, 8 M NaOH—60 mesh size of 75.73%, and 10 M NaOH—80 mesh size 63.79%.

Based on the results of the percentage lignin obtained and the results of the lignin functional group absorption test, it turns out that not all research variations have three indicators of the lignin-forming functional groups. The lignin results were compared by looking at the percentage transmittance value; the best lignin results were lignin (80–3), which is bagasse lignin processed with 80 mesh size variations using NaOH 3 M. Lignin (80–3) is then compared with lignin commercial standards which are lignin of Aldrich and Kraft. **Figure 7** shows the FTIR test results on the sample result of isolated and sample of standard lignin.

Figure 7 shows the combined FTIR results for the four most lignin isolation processes, which produce lignin yields of more than 60%. The four variations of lignin isolation are represented as curve a, curve b, curve c, and curve e. This FTIR graphic overlay is then combined with the standard lignin FTIR results, namely, curve "d" at this figure (**Table 2**).

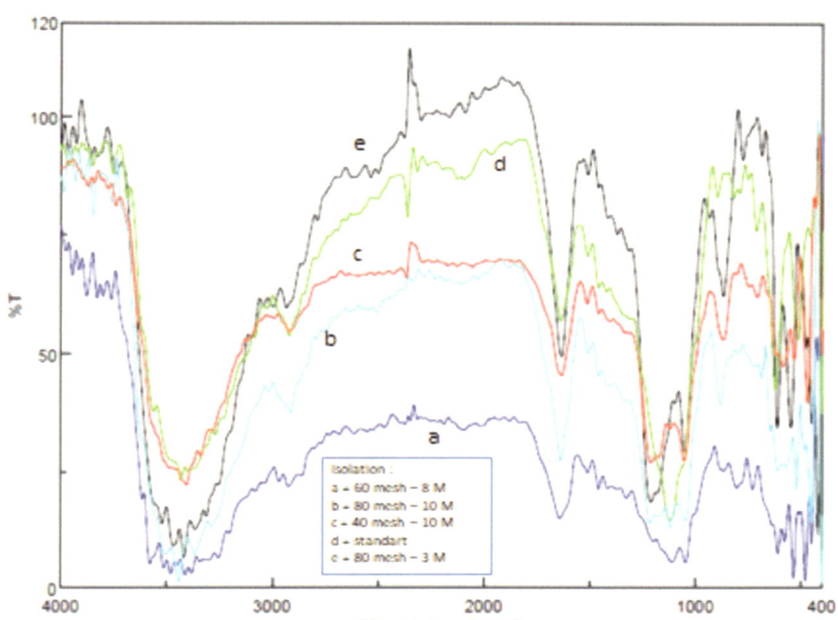

Figure 7.
FTIR test results on lignin isolation.

No.	Typical functional group vibrations in lignin structure	Wave number (cm^{-1})			
		Standard	Bagasse lignin (80–3)	Aldrich lignin	Kraft lignin
1.	Stretch the phenolic O-H	3200–3550	3405.67	3436.62	3414
2	Aliphatic and aromatic stretch groups –CH–	2900	2919.70	2930.17	2926.01
3.	Stretch the arena–C=C	1500–1600	1511.92	1599.14	1614.42
4.	Amine C–N	1000–1250	1100		
5.	Alkyl C–H	600–700	650		

Table 2.
Comparison of the typical absorption peak wave numbers of bagasse lignin with commercial standard lignin FTIR spectrum by Aldrich and Kraft [35].

Based on the reference, standard lignin consists of five main components, namely, phenolic O–H functional groups at wave number 3200–3550 cm^{-1}, aliphatic and aromatic –CH– stretching groups at wave number 2900 cm^{-1}, the C=C aromatic functional groups at wave number 1500–1600 cm^{-1}, amine C–N, and alkyl C–H [37]. There are three main components that are the same as Aldrich lignin and Kraft lignin, namely, phenolic, aliphatic aromatic, and arenas.

In **Figure 7**, for the four curves that have a shape similar to the standard curve, curve "e" (colored black) shows peaks at phenolic, aliphatic, and aromatic wavelengths. So that based on the overlay of the FTIR results, it can be said that the most similar to the standard conditions is the "e" (black) curve which is the result of 80 mesh lignin isolation with 3-M NaOH reagent.

The selected lignin was then continued for the sulfonation process with several variations in the concentration of sodium bisulfite. The sulfonation process has been done with various variations in the concentration of sodium bisulfite and sulfonation time. The best results were achieved in the sulfonation process with a concentration of 0.25 M sodium bisulfite and a sulfonation time of 5 hours. Sulfonation process repeated three times and compare to find spectrum that compose lignosulfonate. The final result of the sulfonation process is lignosulfonate in the form of a light brown powder, as shown in the figure below **Figure 8** (**Table 3**).

From the result of FTIR test, lignosulfonate has been formed, indicated by difference a wavelength spectrum of lignosulfonates and a wavelength spectrum of lignin. The sulfonation process was done in 3 repetitions and the results were tested again by FTIR. With 3 repetitions of the process, the results are almost the same, so you

Figure 8.
Sodium lignosulfonate surfactant from bagasse.

No.	Functional groups in the structure of lignosulfonates	Wave numbers (cm^{-1})		
		SLS standard (Patricia)	SLS standard (Aldrich)	SLS bagasse
1.	Stretch alkene =C=C	1630–1680	1608.34	1635.34
2	Stretch Sulfonate S=O	1350	1365	1384.64
3.	Carboxylate C=O	1000–1300	1187.94	1114.64
4.	Ester S-OR	500–540	499.83	462.83

Table 3.
Comparison of the FTIR spectrum of SLS surfactant-synthesized bagasse and the FTIR spectrum of SLS standard Patricia and Aldrich.

can say the process is correct. To ensure the perfect result of the sulfonation process, a comparison was made with other lignosulfonates [34]. The standard lignosulfonate used for comparison were SLS Aldrich and SLS Patricia. From the FTIR results, the spectrum of SLS surfactant synthesized bagasse and sodium lignosulfonate standard spectrum, the absorption peak and its wave number in the FTIR spectrum of SLS surfactant synthesized from bagasse showed conformity with the spectrum of FTIR standard. This shows that the sulfonation process of lignin to lignosulfonate has been successfully.

In **Figure 9**, it is clear that there is a difference between the FTIR results of lignin and surfactant, where on the blue curve line, as in the surfactant FTIR curve, there is a shift in the absorption peak that occurs, especially at a wavelength of 1635.34 cm^{-1} as a function of the alkene group, at a wavelength of 1384.64 cm^{-1} as a function of the sulfate group, at a wavelength of 1114.65 cm^{-1} as a function of the carbolic acids group, and at a wavelength of 462.832 cm^{-1} as the ester functional group.

Some of the peaks read on FTIR showed lignin and lignosulfonate bagasse components. The lignin component consists of phenolic functional group elements OH, aliphatic and aromatic groups **–**CH**–**, C=O ketone groups, arena functional groups **–**C=C**–**, CN amine groups, and CH alkyl groups with similarity values for standard spectrum wavelengths, such as those shown in **Table 4** (**Figure 10**).

Likewise for lignosulfonates, with indicator components consisting of C=C alkenes, sulfate S=O, C=O carboxylic acids, and S-OR esters, with spectrum

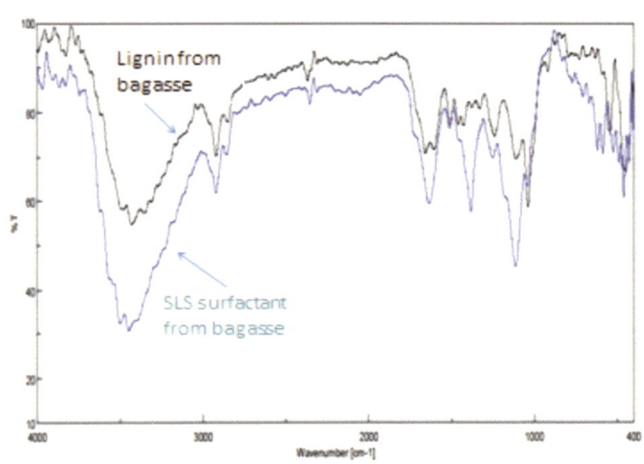

Figure 9.
Overlay of FTIR surfactant—Lignin from bagasse.

Indicator	Component	Wavelength (cm^{-1})
Lignin	Phenolic O-H	3400
	Aliphatic and aromatic –CH–	2910
	Ketone C=O	1450
	Arena –C=C	—
	Amine C–N	1100
	Alkyl C–H	650
Lignosulfonates	Alkene C=C	1635.34
	Sulfate S=O	1384.64
	Carboxylic acids C=O	1114.65
	Ester S-OR	462.832

Table 4.
FTIR of lignin and lignosulfonate bagasse.

Figure 10.
Sugarcane becomes lignosulfonate [35].

wavelengths close to the standard spectrum wavelength values. Lignin from bagasse can be completely synthesized into sodium lignosulfonate surfactant completely with lignosulfonate components consisting of alkene, sulfonate, carboxylate, and ester.

Furthermore, from the results of the NMR test, the components form the lignosulfonate. In the HMQC data, it can be seen that the proton nuclei are directly correlated with carbon-13 (13C) or have one bond (1JC, H) so that their own pairs can be known with certainty. The broad singlet signal on the δ H 6.64 ppm chemical shift (2H, bs, H^{-3}, and H-5) correlates directly with carbon at δ C 102.2 ppm (C-3 and C-5). In addition, the HMQC spectrum also indicates the presence of methylene protons bound to C-9, methane bound to oxygen, and sulfate bound to C-8 and C-7, respectively.

From the HMBC spectrum, it can be seen that there is a correlation between protons and carbon with a distance of two bonds (2 J) to three bonds (3 J), which can be seen in **Figure 3**. From the HMBC data, it can be seen that there is a correlation between H-3 and H-5 with C-5/C-3, C-1, and C-7; H-7 correlates with C-8 and H-9 correlates with C-8 and C-7. These data support the existence of phenyl propanoid compounds as the basis for lignosulfonates [38]. The correlation between HMQC and HMBC can be seen in **Figure 11**. With the results that look like this, it shows that the isolation process of lignin from bagasse has been successful. Likewise, the sulfonation of lignin to lignosulfonate has also been successful.

DOI: http://dx.doi.org/10.5772/intechopen.93662

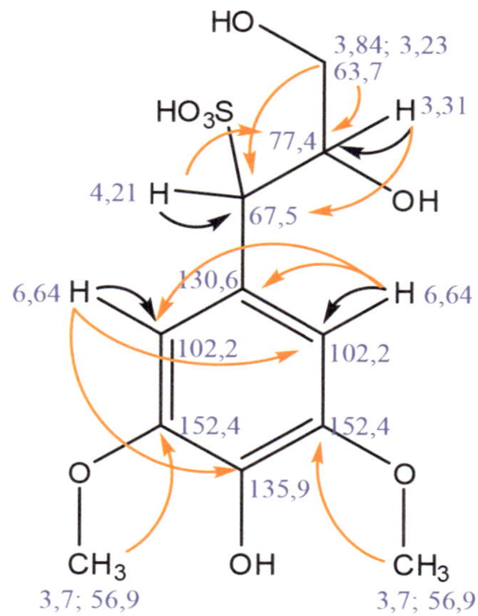

Figure 11.
NMR test results—HSQC and HMBC correlation of bagasse lignosulfonate H_4S_4 isolates.

4. Conclusions

Based on the results of the lignin sulfonation process on lignin sulfonation optimization, several conclusions can be drawn, namely:

1. Bagasse as biomass is a raw material that can be processed into lignosulfonate surfactants. The lignosulfonate obtained from bagasse is processed in two stages, namely, the lignin isolation process using sodium hydroxide and the sulfonation process using sodium bisulfite.

2. Based on the FTIR test, the lignin-forming components were shown by the presence of phenolic functional groups O–H, aliphatic –CH– and aromatic stretching groups, and C=O ketone functional groups, while the lignosulfonate-forming components were indicated by the presence of alkene groups, sulfate groups, and carbocyclic acids and ester functional groups, each with a spectrum wavelength corresponding to the standard spectrum.

3. Based on the results of the NMR test, the presence of phenyl propanoid compounds as the basis of the lignosulfonate compounds indicates that the sulfonation process has reached the expected target, namely, the formation of lignosulfonates completely.

Author details

Rini Setiati[1*], Septoratno Siregar[2] and Deana Wahyuningrum[3]

1 Petroleum Engineering, FTKE, Universitas Trisakti, Jakarta, Indonesia

2 Petroleum Engineering, FTTM, Institut Teknologi Bandung, Bandung, Indonesia

3 Chemistry, FMIPA, Institut Teknologi Bandung, Bandung, Indonesia

*Address all correspondence to: rinisetiati@trisakti.ac.id

References

[1] Teixeira SR, Arenales A, de Souza AE, Magalhães R d S, Peña AFV, Aquino D, et al. Sugarcane bagasse: Applications for energy production and ceramic materials. Journal of Solid Waste Technology and Management. 2016;**41(3)**:229. DOI: 10.5276/JSWTM

[2] Hallersbo M, Onoszko E. An Investigation of new Markets for the Bagasse in Cuban Sugar Mills, Bachelor of Science Thesis EGI-2015. 2015

[3] Taslim S. Sugar Development Policy in Indonesia, Food and Fertilizer Tecnology for the Asian and Pacific Region. 2014. Available at: https:// ap.fftc.org.tw/article/701

[4] Lois-Correa J, Flores-Vela A, Ortega-Grimaldo D, Berman-Delgado J. Experimental evaluation of sugar cane bagasse storage in bales system. Journal of Applied Research and Technology. 2010;**8**(3):365-377

[5] Anwar SI. Determination of moisture content of bagasse of Jaggery unit using microwave oven. Journal of Engineering Science and Technology. 2010;**5**(4):472-478

[6] Varma AJ. Process for Fractionating Sugarcane Bagasse into High A-Cellulose Pulp, Xylan and Lignin, Patent Application Publication, No.: US 2010/0276093 A1. 2010

[7] Hermiati E, Mangunwidjaja D, Sunarti TC, Suparno O, Prasetya B. Pemanfaatan Biomassa Lignoselulosa Ampas Tebu untuk Produksi Bioetanol. Jurnal Litbang Pertanian. 2010;**29**(4):121-130. (In Indonesian)

[8] Abdel-Halim ES. Chemical modification of cellulose extracted from sugarcane bagasse: Preparation of hydroxyethyl cellulose. Arabian Journal of Chemistry. 2014;**7**:363-371

[9] Akpan EI, Odeosun SO. Sustainable Lignin for Carbon Fibers: Principles, Techniques, and Applications. Nature Switzerland AG: Springer; 2019

[10] Kumar P, Barrett DM, Delwiche MJ, Stroeve P. Methods for pretreatment of lignocellulosic biomass for efficient hydrolysis and biofuel production. Industrial and Engineering Chemistry Research. 2009;**48**:3713-3729. DOI: 10.1021/ie801542g

[11] Yao L, Yong-Chao L, Hong-Qin H, Xie F-J, Wei X-Y, Fan X. Structural characterization of lignin and its degradation products with spectroscopic methods. Journal of Spectroscopy. 2017;**2017**:15. DOI: 10.1155/2017/8951658

[12] Krawczyk N, Moller A, Geigle P, Larionov E, Hartwig J. Process for the Production of Lignin Derived Low Molecular Products, Patent No. WO2018/146343Al. 2018

[13] Sreela Pal MM, Banat F, Al Sumaiti AM. Review of surfactant-assisted chemical enhanced oil recovery for carbonate reservoirs: Challenges and future perspectives. Petroleum Science. 2018;**15**:77-102

[14] Ranjan P, Subrata BG, Pankaj T. Enhanced oil recovery by alkaline-surfactant-alternated-gas/CO_2 flooding. Journal of Petroleum Exploration and Production Technology. 2019;**9**:247-260

[15] Miller AC. Bagasse Concrete, Paten No. 2837435, California. 1958

[16] Campbell GE. Bagasse Process and Product, Paten No. 4304361, Fairway. 1981

[17] Varma AJ. Process for Fractionating Sugarcane Bagasse into High Celloluse Pulp, Xylan and Lignin, Patent No. 8529731 B2, New Delhi. 2013

[18] Amri A, Zulfansyah, Iwan M, Suryani R. Pembuatan Sodium Lignosulfonat Dengan Metode Sulfonasi Langsung Biomasa Pelepah Sawit, JurusanTeknik Kimia, FT Universitas Riau. Jurnal Sains dan Teknologi. 2008;**8**(2):61-69

[19] Heradewi. Isolasi Lignin lindi Hitam dari Pemasakan Organosolv Serat Tandan Kosong Kelapa Sawit, Tesis, Fakultas Teknologi Pertanian. Institut Pertanian Bogor; 2007

[20] Yanhua J, Weihong Q, Zongshi L, Lubai C. A study on the modified lignosulfonate from lignin. Energy Sources. 2010;**26**(4):409-414. DOI: 10.1080/00908310490281528

[21] Fujimoto T. New Introduction to Surface Active Agent. Kyoto: Sanyo Chemical Industry; 1985

[22] Aziz MM, Rachmadi H, Wintoko J, Yuliansyah AT, Hasokowati W, Purwono S, et al. On the use of sodium lignosulphonate for enhanced oil recovery. In: International Conference on Biomass: Technology, Application, and Sustainable Development IOP Publishing, IOP Conf. Series: Earth and Environmental Science. 2017. DOI: 10.1088/1755-1315/65/1/012030

[23] Ismiyati SM. Pembuatan natrium lignosulfonat berbahan dasar lignin isolat tandan Kosong Kelapa sawit: Identifikasi dan uji kenerjanya sebagai bahan pendispersi. Jurnal Teknologi Industri Pertanian. 2008;**19**(1):25-29

[24] Kurniati E. Pemanfaatan Cangkang Kelapa Sawit Sebagai Arang Aktif. Jurnal Penelitian Ilmu Teknik. 2008;**8** (2, 2008):96-103

[25] Ari PH. Studi Awal Mengenai Pembuatan Surfaktan dari Ampas Tebu. Jurusan Teknik Kimia. Universitas Diponegoro. Semarang; 2009

[26] Furi TA, Coniwanti P. Pengaruh Perbedaan Ukuran Partikel Dari Ampas Tebu Dan Konsentrasi Natrium Bisulfit (NaHSO3) Pada Proses Pembuatan Surfaktan. Jurnal Teknik Kimia Unsri. 2012;**4**(18):49-58

[27] Rivai M. Analisis Kinerja Surfaktan Metil Ester Sulfonat (MES) Ramah Lingkungan Dari CPO, JPO dan CNO. IPB, Pusat Penelitian Surfaktan dan Bioenergi. Bogor: IPB; 2008

[28] Chen G, Gao J, Chen W, So S, Peng Z. Method for Preparing Concrete Water Reducer by Grafting of Lignosulfonate with Carbonyl Aliphatics, Patent Application Publication, No.: US 2011/0124847 A1. 2011

[29] Setiati R, Prakoso S, Siregar S, Marhaendrajana T, Wahyuningrum D, Fajriah S. Improvement of bagasse become lignosulfonate surfactant for oil industry. IOP Conf Series: Earth and Environmental Science. 2017;**106**(2017):012105. DOI: 10.1088/1755-1315/106/1/012105

[30] Suryani A, Putra ND. Kajian Pengaruh Konsentrasi H2SO4 Pada Proses Produksi Surfaktan Metil Ester Sulfonat (MES) Dengan Metoda Sulfonasi. Journal of Agroindustrial Technology. 2004;**14**(2):67-73. Available from: http://journal.ipb.ac.id/index.php/jurnaltin/article/view/4388/2955

[31] Setiati R, Wahyuningrum D, Siregar S, Marhaendrajana T. Optimasi Pemisahan Lignin Ampas Tebu Dengan Menggunakan Natrium Hidroksida, Prosiding SNaPP. Sains dan Teknologi Unisba; 2015

[32] Naumann A, Peddireddi S, Kües U, Polle A. Fourier Transform Infrared Microscopy in Wood Analysis. 2007

[33] Merlin N, Nogueira AB, de Lima VA, dos Santos LM. Application of Fourier transform infrared spectroscopy, chemical and chemometrics analyses

to the characterization of agro-industrial waste. Quimica Nova. 2014;**37**(10):1584-1588

[34] Patricia RJ. Relationship between the structure of Fe-Lignosulfonate complexes determined by FTIR spectroscopy and their reduction by the leaf Fe reductase. In: The Proceedings of the International Plant Nutrition Colloquium XVI. Davis: University of California; 2009

[35] Setiati R. Synthesis and Characterization of Sodium Lignosulfonate from Bagasse: The Effects of Concentration and Salinity toward the Performance of Oil Injection in Core [dissertation]. Bandung, Indonesia: Bandung Institute of Technology (ITB); 2017

[36] Holladay JE. Result of Screening for Potential Candidates from Biorefinery Lignin. University of Tennesse, Departement of Energy. Springfield, VA: US Departement of Energy; 2007

[37] Areskogh D. Structural Modification of Lignosulfonate. Stockholm: KTH Royal Institut of Technology, School of Chemical Science and Engineering; 2011

[38] Lutnaes BF, Myrvold BO, Lauten RA, Endeshaw MM. 1Hand 13C NMR Data of Benzylsulfonic Acids – Model Compounds for Lignosulfonate. Magnetic Resonance in Chemistry Willey Interscience; 2007

Section 2

Biomass for (Bio)Energy

Chapter 5

Agroenergy from Residual Biomass: Energy Perspective

Cintia de Faria Ferreira Carraro, André Celestino Martins, Ana Carolina da Silva Faria and Carla Cristina Almeida Loures

Abstract

The search for energy alternatives from renewable and clean sources has been gaining prominence at the international level, due to the increased demand for energy and the future depletion of fossil fuels, coupled with the concern with environmental issues. The generation of electricity distributed from the use of biomass can contribute to the conservation of the environment, the diversification of the energy matrix, the national economic development, the generation of jobs in the agro-industry and in the distribution of clean energy, as a sustainable alternative. This chapter aims to present information related to the use of different residual biomass as an energy alternative for Brazil, with a focus on electricity generation, based on a bibliographic survey, where it is highlighted as the best sources of biomass for electricity generation in the country, observing the profitability and viability for logistics and national economy.

Keywords: biomass residues, power generation, energy efficiency, environment, sustainable

1. Introduction

The constant growth of the world population and the economy and social development are the main drivers of the increase global energy demand that is currently supported by fossil fuels. The global energy market depends heavily on fossil fuel energy sources such as coal, oil, and natural gas. Currently, oil is the main source of energy, although its reserves are considered finite, since it takes millions of years for these fuels to be formed on earth, they are soon subject to depletion as they are consumed. In addition to the production processes in the oil fields, they are recognized as polluting and harmful to the environment and the climate. Many efforts have been made to find alternative ways of obtaining energy through cleaner and more sustainable processes. The only natural and renewable resource based on carbon that is vast enough to be used as a replacement for fossil fuels it is biomass [1–5].

Biomass is all organic matter, of vegetable or animal origin, used in the production of energy. It is obtained through the decomposition of a variety of renewable resources, such as plants, wood, food scraps, excrement, garbage, and agricultural waste. The advantages of biomass energy over other energy sources may explain the growing interest in its consumption. First, biomass energy can be used for many different purposes, such as cooking, heating, electricity generation, and transportation. Among the types of renewable energy, biomass conversion process can generate solid, liquid or gaseous fuels, the biomass energy being the only one that can be converted into liquid fuel. Second, biomass energy is renewable energy, abundant, and easily produced source. The use of biomass energy will help countries to reduce their dependence on fossil energy resources and ensure national energy security. Third, biomass energy production contributes to creating more job opportunities, thereby increasing income and reducing poverty among the rural labor force. Finally, and most importantly, biomass energy is a "carbon neutral" source. Compared with fossil energy, biomass energy is less polluting and environmentally safer. Using biomass energy can help to mitigate greenhouse gas emissions and tackle climate change [6–9].

In Brazil, different forms of agro-industrial production work are in parallel with agricultural production. Most processing is directly conditioned to the generation of products and, consequently, the generation large amounts of waste. The production of waste from agro-industrial works is originally derived from the processing of sugar and alcohol industries, biodiesel, cassava, citrus, beers, pulp, and paper, participate expressively in the production of waste. Inadequate disposal of residual biomass can cause soil contamination, compromise the quality of water resources, and cause environmental disturbance among species. Several other factors are related to the disposal of biomass, including many associated with public health problems. Brazil has vast reserves of residual biomass energy from agricultural activities, such as sugarcane bagasse, cassava, and soybeans, which has been gaining interest as a source of energy resources, due to the energetic potential. It is possible to verify that these species together have in Brazil an energy potential of 2615,360 GWh/year [10–15].

2. Biomass agricultural origin

Biomass is one of the most environmentally friendly fuels, since bagasse offers the advantage of being a cheap, abundant, and low-polluting fuel [16]. Biomass has been considered a promising and "environmentally friendly" energy source about energy production. One reason for this renewed interest is due to the way they spread, their sustainable character, and their potential to reduce global emissions of greenhouse gases [17]. World biomass amount is estimated at 1.8 Tt on a dry basis, with a potential thermal yield close to 138 EJ [18–20].

Biomass has been widely recognized as a source of renewable energy with increasing potential to replace conventional fossil fuels in the energy market. Furthermore, using biomass for energy production, another part of a problem is solved, which is waste disposal.

2.1 Bagasse sugar cane

As Brazil is the largest producer of sugarcane, accounting for 36% of global production [21–23], great importance to mention this biomass. In addition, the cultivation of sugarcane has the potential to increase environmental benefits, increasing carbon sequestration, optimizing the agricultural production chain, and thus moderating local environmental impacts [21].

When compared to other agricultural residues, the bagasse has a high yield in terms of the solar energy reservoir and the capture of chemical energy. The sugarcane combustion/gasification produces the same amount of CO_2 that it consumes during its growth; therefore it has a carbon neutral [24].

Edreis et al. [16] studied the effect of the gasification heating rate and the thermal kinetic behavior of sugarcane bagasse coals prepared at 500, 800, and

900°C during CO_2 gasification, and they found that the sugar cane coal gasification occurred in a one stage and that the maximum mass loss rate and its corresponding temperature are directly proportional to the high pyrolysis temperature and the gasification heating rate. In thermal analysis, the activation energy mainly affects the temperature sensitivity of the reaction rate.

2.2 Cassava

The cassava starch produces a significant quantity of residues, which must be rationally used for minimizing the environmental impact of the agricultural activities.

Cassava is widely grown in the tropical and subtropical regions of Asia, Africa, and South America. Brazil occupies a prominent position in world cassava production, alongside only Nigeria and Thailand [25]. Serious attempts have been carried out by the industrial and agro-industrial sectors aiming at the use of this waste profitably [26], but still further opportunities are yet to be developed mainly due to the variety of biomasses and bi-products obtained during the processing steps. The cassava bagasse can be considered as the remaining fraction of the processing of cassava for starch production and consists of 75% of starch, on average, on a dry basis [25].

2.3 Corn stalk

The behavior of corn stalk pyrolysis was studied by Sun et al. [27]. Their research showed that hydrogen-rich gas could be generated by decomposing of the pyrolysis gas at a higher temperature. They concluded that the residual charcoal produced—consisting of fixed carbon and ash—is a good fuel with higher activity and heat value [28].

Corn stands out among agricultural species with the potential to provide biomass for energy production, as it has a large planted area of approximately 177 million acres worldwide [29] and grain production of almost 900 million hectares. Tons [30] resulting in approximately the same amount of residual biomass [31]. This biomass has a high calorific value, ranging from 15.6 to 18.3 MJ Kg^{-1}, like the values of species cultivated exclusively for energy production, such as Eucalyptus sp. [32–33]. Due to the different energy content and amounts of biomass produced by different parts of the corn plant, its potential for energy generation varies significantly [34–37].

3. Process of energy conversion of biomass

High moisture content biomass, such as the herbaceous plant sugarcane, lends itself to a "wet/aqueous" conversion process, involving biologically mediated reactions, such as fermentation, while a "dry" biomass such as cassava and corn stalks, is more economically suited to gasification, pyrolustion. Aqueous processing is used when the moisture content of the material is such that the energy required for drying would be inordinately large compared to the energy content of the product formed. It is the inherent properties of the biomass source that determines both the choice of conversion process and any subsequent processing difficulties that may arise. Equally, the choice of biomass source is influenced by the form in which the energy is required, and it is the interplay between these two aspects that enables flexibility to be introduced into the use of biomass as an energy source.

The World Energy Council defines bioenergy to include traditional biomass (example forestry and agricultural residues), modern biomass and biofuels [38].

The typical biomass materials used for power generation are bagasse, cotton stalk, straw, rice husk, soya husk, saw dust, de-oiled cakes, coconut shells, coffee waste, groundnut shells, Neem, *Jatropha curcas*, Mahua, and Jute wastes, [39]. These biomass materials are being converted into energy via two major energy conversion routes, that is, thermochemical and biochemical. The possible ways in thermochemical route are illustrated in **Figure 1** [40].

The main material properties of interest, during subsequent processing as an energy source, relate to:

- moisture content (intrinsic and extrinsic),
- calorific value,
- proportions of fixed carbon and volatiles,
- ash/residue content,
- alkali metal content,
- cellulose/lignin ratio.

For dry biomass conversion processes, the first five properties are of interest, while for wet biomass conversion processes, the first and last properties are of prime concern.

In biomass the moisture content is presented as intrinsic (without the influence of climate effects) and extrinsic (the influence of climate in the moisture content).

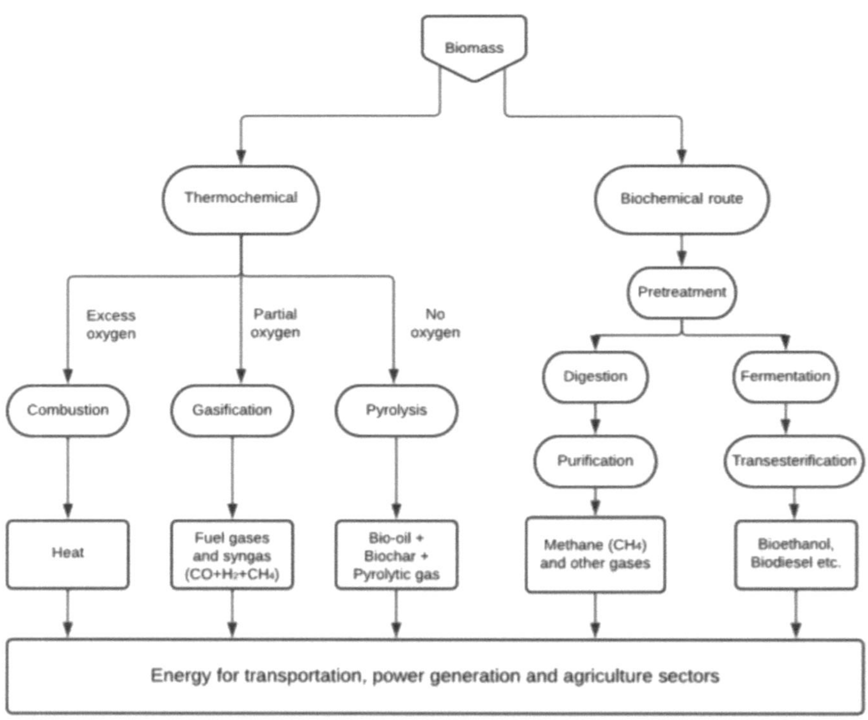

Figure 1.
Thermochemical and biochemical routes for conversion of biomass to energy [40].

In practical terms, it is only concerned with the extrinsic moisture content because the intrinsic moisture content is usually only achieved, or applicable, in a laboratory setting.

The calorific value (CV) of a material is an expression of the energy content, or heat value, released when burnt in air. The CV is usually measured in terms of the energy content per unit mass, or volume; hence MJ/kg for solids, MJ/l for liquids, or MJ/Nm^3 for gases. The CV of a fuel can be expressed in two forms, the gross CV (GCV), or higher heating value (HHV) and the net CV (NCV), or lower heating value (LHV). In practical terms, the latent heat contained in the water vapor cannot be used effectively and therefore, the LHV is the appropriate value to use for the energy available for subsequent use. In **Table 1** is shown the immediate analysis of some biomass feedstocks.

Fuel analysis has been developed based on solid fuels, such as coal, which consists of chemical energy stored in two forms, fixed carbon and volatiles:

- the volatiles content, or volatile matter (VM) of a solid fuel, is that portion driven-off as a gas (including moisture) by heating (to 950 C for 7 min).

- the fixed carbon content (FC), is the mass remaining after the releases of volatiles, excluding the ash and moisture contents.

Elemental analysis of a fuel, presented as C, N, H, O and S together with the ash content, is termed the ultimate analysis of a fuel. **Table 2** gives the ultimate analyses for some biomass materials.

The significance of the VM and FC contents is that they provide a measure of the ease with which the biomass can be ignited and subsequently gasified, or oxidized, depending on how the biomass is to be utilized as an energy source.

The chemical breakdown of a biomass fuel, by either thermo-chemical or bio-chemical processes, produces a solid residue. When produced by combustion in air, this solid residue is called "ash" and forms a standard measurement parameter for solid and liquid fuels. The ash content of biomass affects both the handling and processing costs of the overall, biomass energy conversion cost. During biochemical conversion, the percentage of solid residue will be greater than the ash content formed during combustion of the same material.

Dependent on the magnitude of the ash content, the available energy of the fuel is reduced proportionately. In a thermo-chemical conversion process, the

Biomass	VM (%)	FC (%)	Ash (%)	HHV (MJ/Kg)
Sugar cane [41, 42]	85.49	12.39	2.12	18.73
Cassava [43, 44]	79.89	13.40	5.43	15.39
Corn stalk [45]	75.38	17.95	6.67	16.59

Table 1.
Immediate analysis of some biomass feedstocks (wt%).

Material	C	H	O	N	S
Sugar cane [41]	49.8	6.00	43.90	0.20	0.06
Cassava [42]	49.4	6.10	44.60	0.17	0.10
Corn stalk [43]	42.53	6.17	43.59	0.93	0.11

Table 2.
Ultimate analyses for typical biomass materials (wt%).

chemical composition of the ash can present significant operational problems. This is especially true for combustion processes, where the ash can react to form a "slag," a liquid phase formed at elevated temperatures, which can reduce plant throughput and result in increased operating costs.

4. Methods of generating electricity using biomass

4.1 Biomass in the Brazilian energy matrix

Global energy demand is still largely satisfied by non-renewable energy sources. According to the International Energy Agency [46], in the world, of the nearly 14 million toe (tons of oil equivalent) of the total primary energy supply in 2019, oil, coal, and natural gas together represented more than 80% of this demand, where each one corresponded to 32, 27, and 22% respectively, on the other hand, biomass-derived energy supplied only 10% of this amount [46].

The domestic energy supply in Brazil in 2019 reached 294 million toes, corresponding to a total of approximately 1.4 tons per inhabitant, an increase of 1.4% over the previous year [47].

Of this total, renewable sources corresponded to 46.1% of the total energy generation in the country, divided into biomass from sugarcane (18%), hydraulic (12.4%), firewood, and charcoal (8.7%) and other renewables (7%) [47]. Only biomass from sugarcane and firewood with charcoal were responsible for 26.7% of the domestic energy supply, however, it is known that among others renewables there are more biomass-derived energy sources such as leachate, biogas, biodiesel, and others, which still increase the share of biomass in the Brazilian energy matrix [47].

The current Brazilian electric scenario is even more centered on renewable energies, since of the domestic supply of electric energy in the country, approximately 651.3 TWh, during 2019, 83% was composed of renewable energy sources, with the generation of electricity through hydroelectric (64.9%), wind (8.6%), biomass (8.4) and solar (1%) [47].

4.2 Electricity production and cogeneration using biomass

The power production by biomass can be carried out through different technological routes, commonly the process consists of converting the feedstock into an intermediate product, which will be used for the operation of generating mechanical energy in a machine to drive an electric generator, which will produce electricity [48].

Traditionally, the industrial sectors that generate electricity by biomass also choose to have a cogeneration system, where two or more energy forms are produced from a single process for generating energy, such as heat and electricity [48].

Among the main technological routes, stands out the steam cycle with back pressure turbines, steam cycle with extraction condensing turbines, and the biomass integrated gasification combined cycle [48].

4.3 Steam cycle with back pressure turbines

In this process for generating power, the steam produced by the direct burning of biomass in the boilers is used in turbines coupled with generators, for the production of electrical energy or in turbines for the production of mechanical work and also the fraction that would be released into the atmosphere can be reused directly to meet the thermal needs of the process [48]. In general, the back-pressure steam turbines provide not only electricity but also steam to be used in the plant facilities [49].

The advantage of this process is that the back pressure turbines have few stages with simple structure and small exhaust parts, which results in low cost of the equipment [49], currently this route is the most used and is already well developed commercially, having in Brazil several producers for most equipment [48].

4.4 Steam cycle with extraction condensing turbines

This process is similar to the previous one, however, the steam is totally or partially condensed, after its use in the production process, therefore, its main difference is found in the presence of a condenser in the exhaust of the turbine and specific levels for heating the feedwater boiler supply [48].

In addition, extraction condensing turbines can independently change the production of electricity and process steam, through the control of valves [49], thus, this type of cycle has greater operational flexibility for power generation, concerning with the back pressure turbines [48, 49], also having higher global energy efficiency, allowing obtaining a larger volume of power produced [48].

However, the disadvantage of this type of process is the higher investments for its implementation in relation to the use of back pressure turbines [49] and simple condensing systems [48].

4.5 Biomass integrated gasification combined cycle

By biomass gasification, the fuel gas is obtained and can be used in thermoelectric plants operating on gas for power generation, and applied on a large scale, transforms biomass into an important feedstock for the large thermoelectric plants and through the use of combined cycles of gas and steam, increases the system efficiency [48].

However, it is still a technological route that is not yet commercially competitive, since its greatest difficulty for its application is the production of quality fuel gas, with reliability and safety, adapted to the parameters of biomass and operation [48].

4.6 Biomass thermoelectric plants in Brazil

The participation of thermoelectric plants operating on biomass plays an increasingly important role for the national panorama concerning the supply of electricity. The immense Brazilian land surface located mainly in tropical and rainy regions favors the production and energy use of biomass on a large scale [50].

Sugarcane bagasse is the most widely used biomass as fuel in Brazil for the production of electricity, corresponding to 82% of the electricity exported to the National Interconnected System (in Portuguese, Sistema Interligado Nacional or SIN), this is only possible, because its plants can be energetically self-sufficient [51]. As a result, the power generation costs are competitive with the conventional supply system, which makes the plants through cogeneration being energetic self-sufficient [50].

Sugarcane stands out among biomasses because its high productivity crops together with the gains from the transformation processes of sugar-alcohol biomass, make available an enormous amount of organic matter, mainly in the form of bagasse in the plants and distilleries, also, there is still an interesting complementary relationship between the electricity generated through sugarcane bagasse biomass and hydroelectric power plants, since the sugarcane harvest season coincides exactly in the dry months, thus, the generation of electricity through biomass acts to complement the electrical demand, **Figure 2** [50, 51].

The share of electricity generated through sugarcane in Brazil's energy matrix in 2019 was 3.8%, of which, out of 366 sugar-energy plants in operation in 2019, 220 exported to the network about in mean 2.6 GW [51].

Recently, there has been a significant increase in the export of power to the electrical system generated through other biomasses, especially black liquor, biogas, forest residues, rice husk and others, being important for energy security and reliability, given the seasonality of sugarcane biomass [51]. **Figure 3** shows the biomass thermoelectric plants in operation in Brazil and the potential installed by states in September 2003.

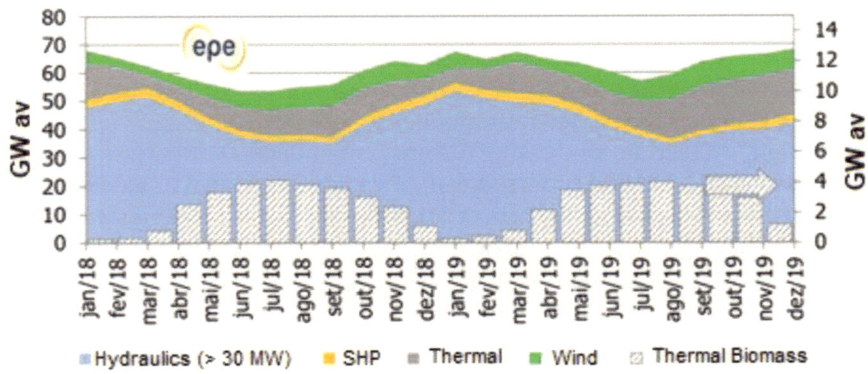

Figure 2.
Participation of sugarcane biomass in electricity generation from January 2018 to December 2019 [51].

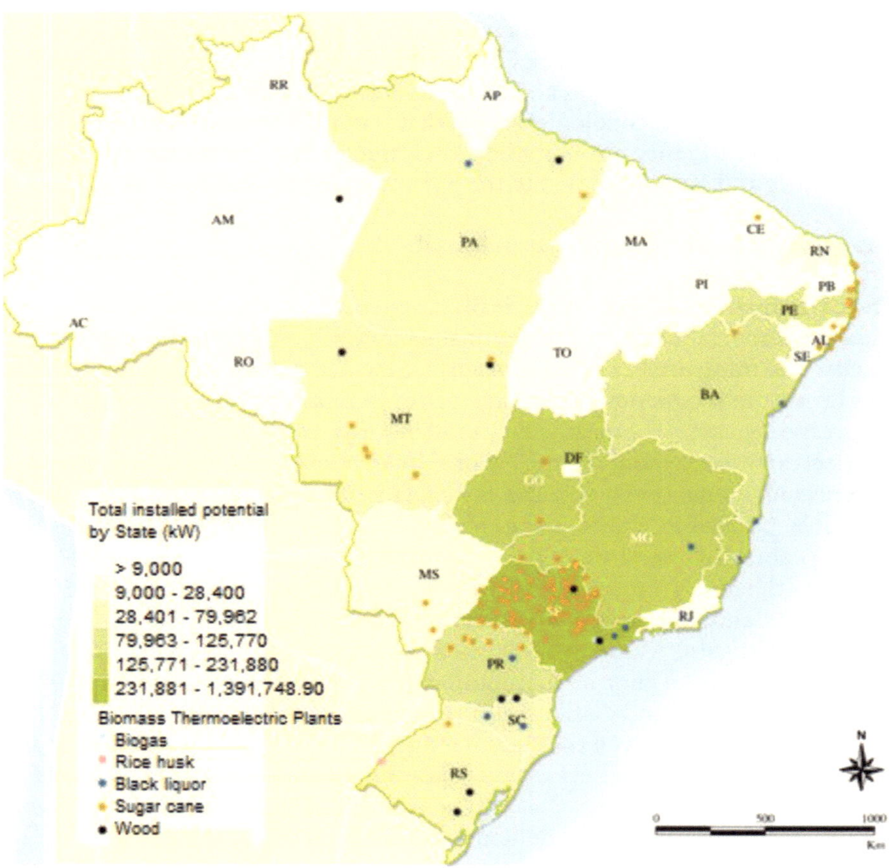

Figure 3.
Biomass thermoelectric plants in operation in Brazil by states in September 2003 [51].

5. Conclusions

The global energy market's dependence on fossil fuel energy sources such as coal, oil, and natural gas needs to give way to alternative and sustainable ways to meet this demand. One of the most promising alternatives is the use of biomass, the only natural and renewable resource based on carbon that is vast enough to be used as a substitute for fossil fuels

The biomass conversion process can generate solid, liquid, or gaseous fuels, the biomass energy being one of the renewable energies, the only one that can be converted into liquid fuel. In addition, biomass energy is renewable energy, an abundant and easily produced source. The use of biomass energy is an important ally in mitigating greenhouse gas emissions.

Brazil occupies a prominent position in the world production of cassava, alongside countries in subtropical regions. Cassava starch produces a significant amount of waste, which must be used rationally to minimize the environmental impact of agricultural activities. Brazil is also the largest producer of sugarcane. In addition, the cultivation of sugarcane has the potential to increase environmental benefits, increasing carbon sequestration and optimizing the agricultural production chain. Corn stands out among agricultural species with the potential to provide biomass for energy production, as it has a large acreage worldwide and is a biomass with high calorific value.

The inherent properties of the biomass source are that determine the choice of the conversion process and the possible processing difficulties. In this way, the choice of the biomass source is influenced by the way in which energy is needed, and it is the interaction between these two aspects that allows flexibility to be introduced in the use of biomass as an energy source. Biomass is converted into energy through two main energy conversion routes, namely, thermochemistry and biochemistry.

Due to its vast territory, Brazil has high potential or effective rates of waste from agro-industrial products, making the use of biomass a great possibility for increasing energy production in the Brazilian energy matrix. The results achieved in the year 2019, by 220 sugar-energy plants in operation, exported about 2.6 average GW to the network, showing that the participation of biomass thermoelectric plants assumes an increasingly important role for the national panorama about electricity supply. Sugarcane bagasse is the biomass most used as fuel in Brazil to produce electric energy, due to the high productivity of sugarcane in its crops to serve the sugar and alcohol sector. The energy generated by this biomass stands out, and due to the sugarcane harvest exactly coincides with the dry months, the generation of electricity from biomass acts as a complement to the electric demand produced by hydroelectric plants; there was also a significant increase in energy exports for the electrical system generated by other biomasses.

Author details

Cintia de Faria Ferreira Carraro[1,2*], André Celestino Martins[1], Ana Carolina da Silva Faria[1] and Carla Cristina Almeida Loures[1]

1 The Federal Center for Technological Education Celso Suckow da Fonseca (Cefet/RJ), Angra dos Reis, RJ, Brazil

2 Federal University Fluminense, Volta Redonda, RJ, Brazil

*Address all correspondence to: cintia.carraro@cefet-rj.br

References

[1] Kumar A, Samadder SR. A review on technological options of waste to energy for effective management of municipal solid waste. Waste Management. 2017;**69**:407-422

[2] REN21. Renewables 2018 e global status report (Paris: REN21 secretariat). Renewable Energy Policy Network for the 21st Century; 2018. p. 2018

[3] Bilandzija N, Voca N, Jelcic B, Jurisic V, Matin A, Grubor M, et al. Evaluation of Croatian agricultural solid biomass energy potential. Renewable and Sustainable Energy Reviews. 2018;**93**:225-230

[4] Rountree V. Nevada's experience with the renewable portfolio standard. Energy Policy. 2019;**129**:279-291

[5] Gutierrez AS, Eras JC, Hens L, Vandecasteele C. The energy potential of agriculture, agroindustrial, livestock, and slaughterhouse biomass wastes through direct combustion and anaerobic digestion. The case of Colombia. Journal of Cleaner Production. 2020;269

[6] Sikka M, Thornton TF. Sustainable biomass energy and indigenous cultural models of wellbeing in an Alaska forest ecosystem. Ecology and Society. 2013;**18**:38

[7] Bildirici M, Özaksoy F. An analysis of biomass consumption and economic growth in transition countries. Economic Research-Ekonomska Istraživanja. 2018;**31**:386-405. DOI: 10.1080/1331677x.2018.1427610

[8] Aydin M. The effect of biomass energy consumption on economic growth in BRICS countries: A country-specific panel data analysis. Renewable Energy. 2019;**138**:620-627

[9] Wang Z, Bui Q, Zhang B, Pham TL. Biomass energy production and its impacts on the ecological footprint: An investigation of the G7 countries. The Science of the Total Environment. 2020;**743**:140741

[10] Malico I, Pereira RN, Gonçalves AC, M 0 SA. Current status and future perspectives for energy production from solid biomass in the European industry. Renewable and Sustainable Energy Reviews. 2019;**112**:960-977

[11] Özdenkçi K, De Blasio C, Sarwar G, Melin K, Koskinen J, Alopaeus V. Techno-economic feasibility of supercritical water gasification of black liquor. Energy. 2019;189

[12] Carrillo-Nieves D, Rostro Alanís MJ, de la Cruz Quiroz R, Ruiz HA, Iqbal HMN, Parra-Saldívar R. Current status and future trends of bioethanol production from agro-industrial wastes in Mexico. Renewable and Sustainable Energy Reviews. 2019;**102**:63-74

[13] Ike M, Donovan JD, Topple C, Masli EK. The process of selecting and prioritising corporate sustainability issues: Insights for achieving the sustainable development goals. Journal of Cleaner Production. 2019;**236**:117661

[14] Ferreira LRA, Otto RB, Silva FP, De Souza SNM, De Souza SS, Ando Junior OH. Review of the energy potential of the residual biomass for the distributed generation in Brazil. Renewable and Sustainable Energy Reviews. 2018;**94**:440-455

[15] Famoso F, Prestipino M, Brusca S, Galvagno A. Designing sustainable bioenergy from residual biomass: Site allocation criteria and energy/exergy performance indicators. Applied Energy. 2020;**247**:115315

[16] Edreis EMA, Yao H. Kinetic thermal behaviour and evaluation of physical structure of sugar cane bagasse

char during non-isothermal steam gasification. Journal of Materials Research and Technology [Internet]. 2016;**5**(4): 317-326. DOI: 10.1016/j.jmrt.2016.03.006

[17] Gokcol C, Dursun B, Alboyaci B, Sunan E. Importance of biomass energy as alternative to other sources in Turkey. Energy Policy. 2009;**37**(2):424-431. DOI: 10.1016/j.enpol.2008.09.057

[18] Kitani O, Hall CW. Biomass Handbook (Gordon and Breach Science Publishers; 1989. p. 963

[19] Melin K, Hurme M. Lignocellulosic biorefinery economic evaluation. Cellulose Chemistry and Technology. 2011;**45**(7):443-454

[20] McKendry P. Energy production from biomass (part 1): Overview of biomass. Bioresource Technology. 2002;**83**(1):37-46. DOI: 10.1016/s0960-8524(01)00118-3

[21] Bodunrin MO, Burman NW, Croft J, Engelbrecht S, Goga T, et al. The availability of life-cycle assessment, water footprinting, and carbon footprinting studies in Brazil. The International Journal of Life Cycle Assessment. 2018;**23**(8):1701-1707. DOI: 10.1007/s11367-018-1484-2

[22] Caldeira-Pires A, Benoist A, Luz, da SM, Silverio VC, Silveira CM, Machado FS. Implications of removing straw from soil for bioenergy: An LCA of ethanol production using total sugarcane biomass. Journal of Cleaner Production. 2018;**181**:249-259. DOI: 10.1016/j.jclepro.2018.01.119

[23] Du C, Dias LC, Freire F. Robust multi-criteria weighting in comparative LCA and S-LCA: A case study of sugarcane production in Brazil. Journal of Cleaner Production. 2019;**2**:35. DOI: 10.1016/j.jclepro.2019.02.035

[24] Ren N-Q, Zhao L, Chen C, Guo W-Q, Cao G-L. A review on bioconversion of lignocellulosic biomass to h2: Key challenges and new insights. Bioresource Technology. 2016;**215**:92e9. DOI: 10.1016/j.biortech.2016.03.124

[25] de SouzaYes Fernandes D, dos Santos TPR, Fernandes AM, Leonel M. Harvest time optimization leads to the production of native cassava starches with diferent properties. International Journal of Biological Macromolecules. 2019;**132**:710-721. DOI: 10.1016/j.ijbiomac.2019.03.245

[26] Pandey A, Soccol CR, Nigam P, Soccol V, Vanderberghe LPS, Mohan R. Biotechnological potential of agro-industrial residues II: Cassava bagasse. Bioresourse Technol. 2000;**74**:81-87

[27] Sun L, Xu M, Sun RF. Behaviour of corn stalk in an indirectly heated pyrolysis reactor. In: Proceedings of Pyrolysis and Gasification of Biomass and Wastes. Strasbourg: CPL Press; 2002

[28] Zabaniotou A, Ioannidou O. Evaluation of utilization of corn stalks for energy and carbon material production by using rapid pyrolysis at high temperature. Fuel. 2008;**87**(6):834-843. DOI: 10.1016/j.fuel.2007.06.003

[29] FAO. Food and Agriculture Organization of the United Nations. 2016. Available from: http://ww.fao.org

[30] USDA - UNITED STATE DEPARTMENT OF AGRICULTURE. Near Record Safrinha Crop Increases Glogal Corn Supply. 2016. Available from: www.usda.gov

[31] Nogueira LA, Lora EES. Dendroenergy: Fundamentals and Applications. 2th ed. Rio de Janeiro: Interciência; 2003. p. 200

[32] Quirino WF et al. Calorific power of wood and lignocellulosic residues. Biomass and Energy. 2004;**1**(2):173-182

[33] Ioannidou O et al. Investigating the potential for energy, fuel, materials and chemicals production from corn residues (cobs and stalks) by non-catalytic and catalytic pyrolysis in two reactor configurations. Renewable and Sustainable Energy Reviews. 2009;**13**(4):750-762

[34] Pordesimo LO, Edens WC, Sokhansanj S. Distribution of aboveground biomass in corn Stover. Biomass and Bioenergy. 2004;**26**(4):337-343

[35] Varvel GE, Wilhelm WW. Cob biomass production in the western corn belt. BioEnergy Research. 2008;**1**(3-4):223-228

[36] Ambrosio R, Pauletti V, Barth G, Povh FP, da Silva DA, Blum H. Energy potential of residual maize biomass at different spacings and nitrogen doses. Ciência e Agrotecnologia. 2017;**41**(6):626-633. DOI: 10.1590/1413-70542017416009017

[37] Zeng K, He X, Yang H, Wang X, Chen H. The effect of combined pretreatments on the pyrolysis of corn stalk. Bioresource Technology. 2019;**02**:107. DOI: 10.1016/j.biortech.2019.02.107

[38] Bioenergy and Biofuels. International Energy Agency (IEA). Available from: http://www.iea.org/topics/renewables/bioenergy/

[39] Murugan S, Gu S. Research and development activities in pyrolysis – Contributions from Indian scientific community – A review. Renewable and Sustainable Energy Reviews. 2015;**46**:282-295

[40] Chintala V. Production, upgradation and utilization of solar assisted pyrolysis fuels from biomass – A technical review. Renewable and Sustainable Energy Reviews. 2018;**90**:120-130. DOI: 10.1016/j.rser.2018.03.066

[41] Vassilev SV, Baxter D, Andersen LK, Vassileva CG, Morgan TJ. An overview of the organic and inorganic phase composition of biomass. Fuel. 2012;**94**: 1-33. DOI: 10.1016/j. fuel.2011.09.030

[42] Channiwala SA, Parikh PP. A Unifiel correlation for estimating HHV of solid, liquis and gaseous fuels. Fuel. 2002;**81**:1051-1063

[43] De Faria Ferreira Carraro C, de Almeida Loures CC, da Silva LM, de Castro JA. Characterization of cassava biomass using differential scanning calorimetry and thermogravimetry for energy purposes. Journal of Thermal Analysis and Calorimetry. 2019;**8**:905. DOI: 10.1007/s10973-019-08905-2

[44] Cruz G, Rodrigues AP, Silva DF, Gomes WC. Physical–chemical characterization and thermal behavior of cassava harvest waste or application in thermochemical processes. Journal of Thermal Analysis and Calorimetry. 2020:1-12. DOI: 10.1007/s10973-020-09330-6

[45] Chen D, Cen K, Cao X, Li Y, Zhang Y, Ma H. Restudy on torrefaction of corn stalk from the point of view of deoxygenation and decarbonization. Journal of Analytical and Applied Pyrolysis. 2028;**09**:15. DOI: 10.1016/j.jaap.2018.09.015

[46] IEA - International Energy Agency. World Energy Balances: An Overview. Available from: https://www.iea.org/reports/world-energy-balances-overview

[47] EPE - Energy Research Company. National Energy Balance. Available from: https://www.epe.gov.br/pt/publicacoes-dados-abertos/publicacoes/balanco-energetico-nacional-ben

[48] ANEEL - National Electricity Agency. Atlas of Electrical Energy in Brazil. 3 ed. Brasília: ANEEL, 2008.

Available from: http://www2.aneel.gov.br/arquivos/PDF/atlas3ed.pdf

[49] OHJI A, HARAGUCHI M. Steam Turbine Cycles and Cycle Design Optimization: The Rankine Cycle, Thermal Power Cycles, and IGCC Power Plants. Woodhead Publishing, 2017. DOI: 10.1016/B978-0-08-100314-5.00002-6

[50] ANEEL - National Electricity Agency. Atlas of electrical energy in Brazil. 2nd ed. Brasília: ANEEL. 2005. Available from: http://www2.aneel.gov.br/arquivos/pdf/livro_atlas.pdf

[51] EPE - Energy Research Company. Biofuels Analysis. 2019. p. 64. Available from: https://www.epe.gov.br/pt/publicacoes-dados-abertos/publicacoes/analise-de-conjuntura-dos-biocombustiveis

Chapter 6

The Potential of Biomass in Africa and the Debate on Its Carbon Neutrality

Joan Nyika, Adeolu Adesoji Adediran, Adeniyi Olayanju, Olanrewaju Seun Adesina and Francis Odikpo Edoziuno

Abstract

To enhance the energy security and promote energy diversity, biomass sources of energy are viable resources worldwide. Bioenergy is an organic source of power derived from various feedstock including fuel wood, energy crops, solid wastes, and residues of plants. This book chapter explores the use of biomass in Africa and the technical and economic potential of these resources for energy supply in the continent. Findings of literature revealed that the potential of biomass is high in Africa due to availability of land, its preference due to limited electricity supply and the exorbitant nature of fossil fuels, the assorted variety of energy crops suitable for growth in the continent and the green nature associated with the resource. The chapter also established that bioenergy is renewable and not carbon neutral. As such, accurate computation of its resultant greenhouse gas emissions based on their sequestration and emission rates is strongly advised to optimize biomass for energy utility and sustainability compared to conventional energy sources.

Keywords: Africa, biomass, carbon emissions, environmental sustainability

1. Introduction

The global population is growing at a fast rate so that today's population is 200% more compared to the 1960s and is further projected to rise up to 9 billion by 2050 [1]. According to Jackson et al. [2], the global per capita energy use was rated to increase by 0.2% annually with consumption in developing countries such as India and China having an increased consumption rate of 3.4 and 1.6% per year, respectively. The European Union and USA recorded declined energy consumption rates of 0.3 and 0.2% per annum, respectively [2]. The increment trend has and is expected to increase the global energy demand particularly in urban areas of developing nations considering that these countries will account for 99% of the population growth and 50% of these individuals will be in major cities [3]. Energy increments are also attributed to industrial revolution and the need to realize the sustainable development goal number 17 on affordable and clean energy according to [4]. These anticipations though reasonable are against the current global efforts to mitigate climate change, which is a serious environmental crisis.

In response to these developments on accommodating accessibility of sufficient energy and mitigation of climate change effects, the global energy mix especially in urban areas is growing although many cities still rely heavily on conventional energy sources based on fossil fuels. The use of the energy sources has stirred a heated debate on energy sustainability since they are associated with environmental pollution and the apparent climate change state [1]. In China for instance, the exponential growth in use of natural gas resulted to a 2–7% increase in carbon dioxide emissions correspondent to extensive air pollution in China [2]. Evidence showing that cities are the greatest environmental polluters and climate change contributors from the 70% carbon dioxide emissions out of the total possible, most of which is anthropogenic-based confirms the need for alternative, reliable, easily accessible and low-carbon emitting energy sources [5]. Zaharia et al. [4] agreed with these sentiments claiming that prosperity, population and non-renewable energy consumption in developing economies of Asia and Africa are attributable to the rise in pollutant emissions.

Of these proposed alternatives in the energy mix is biomass, which is organic matter that is used as energy directly for heating and combustion or indirectly as biofuels [6]. Biofuel examples include wood shavings, sawdust, firewood, fruit stones (avocados, olives and nutshells) wastewater, manure, paper waste and pellets. Biomass especially from wood is a promising domestic energy source according to Bildirici and Ozaksoy [7] who reported that 81% of African population depend on it for economic, household and cooking activities. The wide availability of biomass obtained from agricultural and industrial processes' by-products justifies it high preference. Additionally, its direct and indirect uses to produce energy make its suitable in developing regions of Africa. However, it is worth noting that direct use of biomass is not always feasible and in some cases require additional treatment (biologically or physically) to prevent the effects of conventional fuels [1]. This book chapter focuses on the various sources of biomass in Africa and assesses their potential in addition to having a candid discussion on the carbon neutrality of biomass. Three categories of biomass including forestry biomass, energy crops and wastes or residues will be discussed. The prospects of the chapter will help in drawing a roadmap to providing reliable energy for socio-economic growth in Africa while at the same time, taking precautionary measures to conserve the environment.

2. Types of biomass

Biomass, which is sourced from organic matter from the biosphere (animal or plant origin) and through transformation of wastes, is a promising source of energy. This renewable energy source can be classified into three: (1) forestry biomass, (2) energy crops and (3) biomass from wastes and residues. These three forms of biomass will be discussed in the following sections.

2.1 Forestry biomass and residues

Forests as terrestrial ecosystems store and generate biomass, which justifies their applicability as energy sources since time immemorial [8, 9]. This biomass form differs based on topography, stand structure, site and management systems. Irrespective of the variations, forest is a primordial energy source due to its uniformity and availability globally as well its carbon neutrality [10, 11]. Forest biomass is removed as harvests or in silvicultural activities. Forest biomass is classified into two categories: (1) energy plantations and (2) timber systems where energy is produced as forest residues. Energy plantations are distinguished from agricultural crops from the ability to enhance their biodiversity, their variability globally, harvest flexibility, economic variability, low risk and their capacity to perform phytoremediation [12, 13]. Some countries such as China, Canada, USA and Europe have some of these plantations as documented by Goncalves et al. [14] and compared to developing countries. A number of factors such as the management practices, harvest cycle, rotation, density and the selection of species are considered in the growth of energy crops [14]. Forest residues include stumps, stems, limbs and tops of trees and their production depends on tree species, stem quality and stand structure [8].

The current share of forest biomass use is limited despite the known advantages of its use in energy production including the ability to convert it to transportation fuels, heat and electricity. The use of bioenergy and renewable wastes for energy supply accounted for 9.4% compared to all sources in 2015 [15]. Among these biomass supplies, 63.7% was from solid biofuels such as renewable municipal waste, biogas and liquid biofuels while other renewable biomass took the remaining share. Wood, wood fuel and wood residues produce heat and electricity and can be used indirectly by power plants combined with heat and power or directly by end users. Forest biomass contributed to 87% of biomass feedstock while 3 and 10% was from municipal waste and agricultural feedstock, respectively [16]. Examples of forest biomass sources include wood pellets, pine wood chips, pine bark, beech woos, willow wood, poplar and eucalyptus wood [17]. In sub-Saharan Africa, woody biomass is the main source of energy at domestic level and 81% of the population use it for economic, household and cooking activities [7]. This rate is by far higher compared to higher income developing countries of India and China. Although projections by the IEA as noted by Stecker et al. [18] claimed that wood biomass use for energy would reduce globally by 2035, it is noted that in Africa, this form of biomass will contribute to 51–57% of energy consumption. Wood biomass use in Africa varies with some countries such as Central African Republic, Burundi and Rwanda having a percentage use rate of 90% and above [18].

2.2 Energy crops

Energy crops are wild and cultivated crops, which produce biomass for various purposes. They exist as woody, herbaceous, perennial, or annual and generate raw materials for gaseous or liquid biofuels in addition to solid biomass. A number of factors including maintenance of land productivity, improved soil fertility, use of crop rotation systems, climate change adaptation and crop characteristics influence the successful production of energy crops [19]. Energy crops are used for three main purposes: 1) biodiesel, 2) bioethanol and 3) electric and thermal production [20]. Some of the crops used to produce biodiesel include *Cynara cardunculus,* cotton, *Glycine max, Helianthus annuus* and *Brassica napus*. Energy crops used in bioethanol production include *Beta vulgaris*, *Zea mays* and *Sorghum bicolor*, wheat among other cereals. *Miscanthus giganteus*, *Eucalyptus globulus* and *Arundo donax* are used in electric and thermal production. According to Lynd et al. [21], energy crops occur in four categories: (1) cellulosic such as trees, grass and a variety of wastes, (2) oil rich such as palm oil, soy, rapeseed and sunflower, (3) sugar rich including sugar beet and sugarcane and (4) starch rich crops such as sorghum, wheat and maize. A number of conversion technologies transform the crops to energy. These technologies include biological processes such as fermentation, lignocellulose hydrolysis and anaerobic digestion as well as non-biological processes such as transesterification, pyrolysis, gasification and combustion. African countries such as Kenya, Zimbabwe, South Africa, Tanzania, Ghana and Ethiopia have embraced the use of these biomass crops as energy sources in addition to the use of forest biomass, residues and other forms of wastes [21].

2.3 Biomass from wastes

Municipal solid waste commonly known as garbage comprises of leather and wood by-products, leaves, clippings from grass, food wastes, cardboard, paper and biogenic material from plants and animals. All these form biomass and can be transformed to energy for heating or steam for electricity generation. This has been done in developed countries such as the USA where in 2018, 14 billion kilowatt-hours of electricity from combusting 29.5 million tons of municipal solid waste was produced by 68 power plants [22]. More than 60% of the combustible waste consisted of biomass materials and accounted for the more than 50% of the generated power [22]. The remaining combustible weight was from non-biomass materials such as plastics. Landfill gas also made from biomass material is transformed to methane gas and used in energy production. In Africa, the use of municipal solid waste for energy production has high potential as Scarlat et al. [23] concluded in an evaluation of its potential especially in urban areas of the continent though it is done at small-scale levels.

3. An overview of biomass in Africa

Bioenergy from biomass is the primary source of energy for more than 2.7 billion people globally and serves a traditional role in Africa [24]. Organization for economic cooperation and development (OECD) [25] highlighted that more than 81% of the population accounting for 653 million Africans rely on biomass for their energy demands and the figure is expected to rise by 2030 to 720 million. The total energy demand in Africa is dominated by biomass that accounts for almost half (about 48%) of the total available sources (**Figure 1a**). A similar trend is evident in the sub-Saharan Africa as shown in **Figure 1b**. With the exclusion of South Africa, the rest of sub-Saharan Africa depends on biomass to a rate of more than

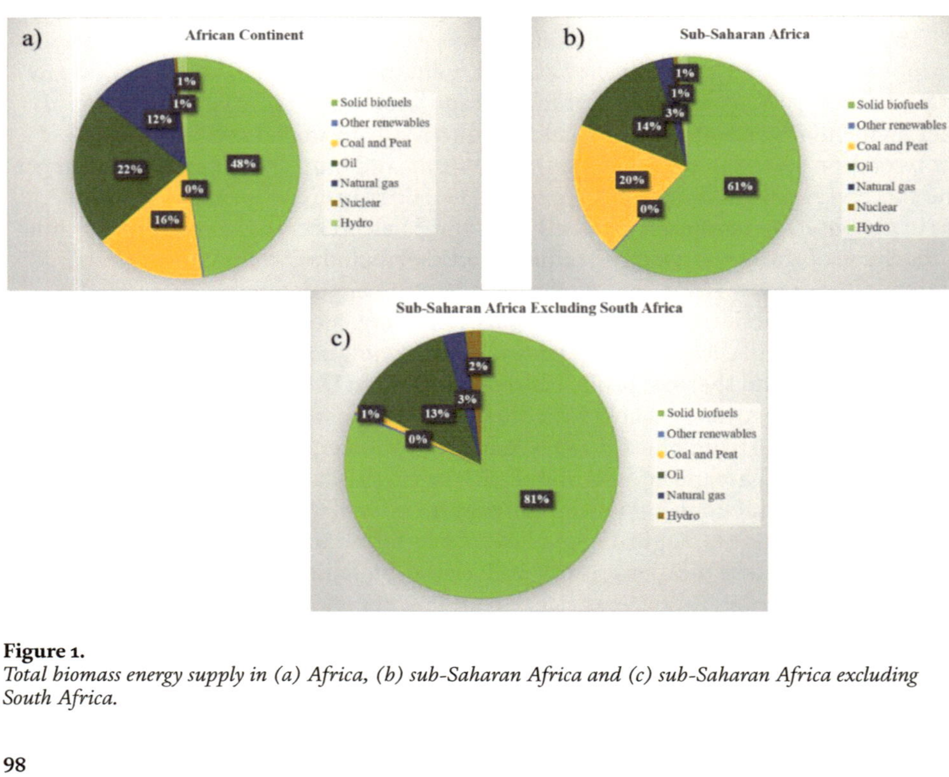

Figure 1.
Total biomass energy supply in (a) Africa, (b) sub-Saharan Africa and (c) sub-Saharan Africa excluding South Africa.

81%. (**Figure 1c**) Total biomass energy supply for the entire continent is at 28,177 petajoules (PJ) while in sub-Saharan Africa it is 21, 646 and 15,575 PJ including and excluding South Africa, respectively, according to the IEA data of 2009 [18, 26].

Apart from contributing to the primary energy demand in Africa, biomass also contributes significantly to the total final consumption. Although it is expected that this trend is on a reducing trend due to other competing uses of biomass such as animal feeds, organic sources and food, IEA [26] still projects that biomass sources will contribute to 51–57% of energy consumption by 2035 in the continent. In poorer countries of Africa especially those of sub-Saharan Africa excluding South Africa, the tendency to use biomass for energy is even higher according to Dasappa [27]. Usually the uses are traditional referring to the inappropriate use of animal dung, agricultural residues, animal dung, tree residues and fuel wood for space heating, lighting and cooking. This could be contrasted to modern biomass technical and effective use of energy characterized by high efficiency. Most of poor African population relies on traditional use of biomass for its energy uses despite the unsustainability of these trends, the rarity of quality biomass energy in these areas and the need for food security usually sourced from biomass sources [18]. The traditional uses of biomass via inefficient stoves is associated with indoor air pollution, soil degradation, forest degradation, ample time spent collecting firewood and ultimately, poverty [24]. These challenges necessitate a comprehensive analysis of biomass potential in Africa to find solutions towards having high quality, effective and efficient biomass. The following sections discuss the various biomass types with specific production levels in Africa and thereafter the potential of biomass in the continent.

Africa has more than 650 million hectares of forest cover, which accounts for 17% of the world's total area. The area covered is a fifth of the continent though the distribution of this resource is uneven with the Congo Basin and some areas of central and western Africa taking the largest share as shown in **Figure 2**. In the regions, production of wood products and round-wood is a key source of employment and African forests account for 0.85 ha per capita of population according to Dasappa [27]. Approximately 1% of the continent is characterized as forest plantation while the tropical rain forests account to 25% of such areas globally. Due to the lack of recent statistics, this study used the Food and Agriculture Organization [29] data to show the forest product statistics for some African countries as shown in **Table 1**. Summarizes the wooded and forested areas of Africa with statistics showing 645 Mha accounting for 21% of total area as having biomass cover. Regions of central, west, east and South Africa have larger forested and wooded regions compared to the north. This could be because the latter has a considerable share of fossil fuel resources compared to other African regions.

Round-wood is the major forest product at 237 million tons compared to charcoal, fuel wood and industrial products at 15, 52, and 207 million tons, respectively. The ratio of wood fuel to round-wood for some named African countries ranges from 0.9 to 1. In addition to wood, the processing of wood generates residues such as tops, lops, sawdust and cut-offs that are used as biomass. During forest and plant production, residues in the form of leaves, husks, cobs, shells and stalks are produced and serve as useful biomass too.

In the use of municipal solid waste biomass in Africa for energy, the section is largely unexploited according to Hafner et al. [30]. This trend is predominant in the continent despite the great potential of valorizing waste biomass to generate renewable and efficient energy in addition to dealing with the current waste disposal crises if conducted in large scale. The UN Environment Program [31] lauds Ethiopia for constructing a waste biomass-to-power plant, which is one of the first in large-scale capacity in the continent. Africa has also taken up the use of energy

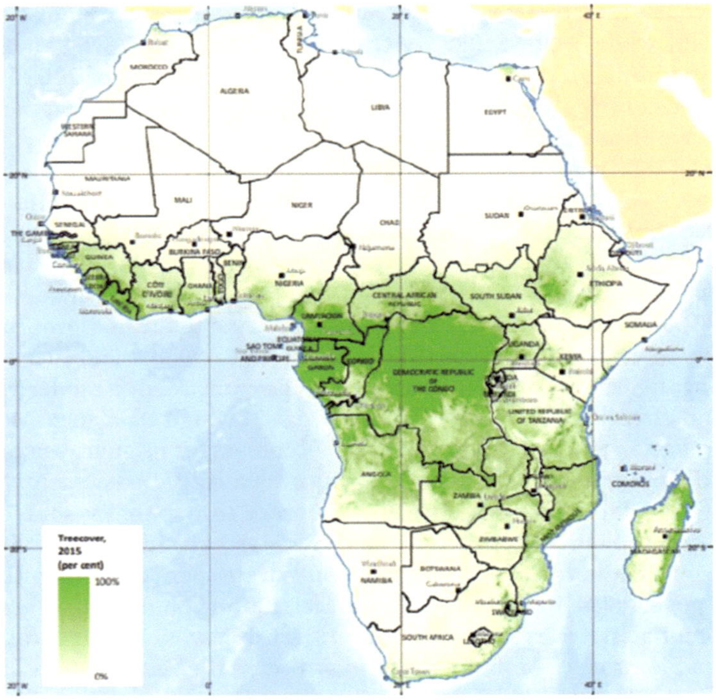

Figure 2.
Percentage of forest cover in Africa [28].

Region	Forested land area (1000 ha)	% Land area	Other wooded land (1000 ha)	Other land with tree cover (1000 ha)
Southern and Eastern Africa	226,534	27.8	167,023	10,345
Northern Africa	131,048	8.6	94,609	10,207
Central and Western Africa	277, 829	44.1	144,468	788
Total Area	645, 412	21.4	406,100	21,339

Table 1.
Forest and wooded areas in Africa according to the FAO 2005 statistics [27].

crops for biofuel production. The feedstock for such processes comes from: (1) first generation food crops such as cereals, sugarcane and vegetable oils, (2) from second generation crops such as wood, wastes and bagasse and (3) from third generation organisms such as algae. It is not easy to quantify the use of energy crops due in Africa due to their affiliated competition with food demands especially in famine prone areas of sub-Saharan Africa. Additional challenges including food-fuel competition exacerbated by corruption, weak governance, political instability and competition for land slow down efforts aimed at modernizing biomass for energy in most African countries [30]. IEA [32] expressed optimism that with the appropriate policies, African countries including Uganda, South Africa, Nigeria, Ghana and Mozambique could use biofuels to meet energy demands of their respective transport sectors. It is from this optimism that several examples of biomass use in Africa have been documented. These include bioethanol generation from sugarcane in Malawi, jatropha electrification in Mali, the use of sisal waste for biogas production in Tanzania and the production of ethanol from cassava in Benin [33–35]. In Zambia, Tanzania, South Africa, Sierra Leone, Liberia, Kenya, Ghana, Gambia, Cameroon, Burkina Faso and Botswana, policies on the use of bioenergy have been formalized and are in the implementation stages [36].

4. Potential of biomass in Africa

The potential of biomass in Africa has been examined in a number of studies especially in relation to available land [34, 35, 37]. These studies however focus on productive areas compared to arid and semi-arid regions. In Africa however, most of the area is largely arid of semi-arid characterized by mismanaged natural resources, low productivity and high vulnerability to climate change and soil erosion, which worsens the continent's poverty crises. The potential of biomass is therefore generalized using two aspects: (1) the availability of land and the viable production systems (technical potential) and (2) the expenditure and income resulting from biomass production (economic potential) that vary from humid to arid and semi-arid areas. Ultimately, with these considerations, the economic potential of bioenergy generation is affected. The next section focuses on Africa's biomass potential in relation to its technical and economic potential.

4.1 Technical potential

The technical potential of biomass is classified into two: (1) available land for bioenergy production and (2) viable biomass production systems. Available land defines the land left after current high biodiversity, agricultural and unsuitable areas are excluded. In this context, unsuitable areas include steep slopes, deserts and cities while high biodiversity areas include wetlands, forests, biodiversity hotspots and protected areas. In this context, Africa has a great technical potential of biomass as it has ample land for growth of bioenergy crops [27] and has serious electricity supply problems especially in rural areas steered up by poverty and

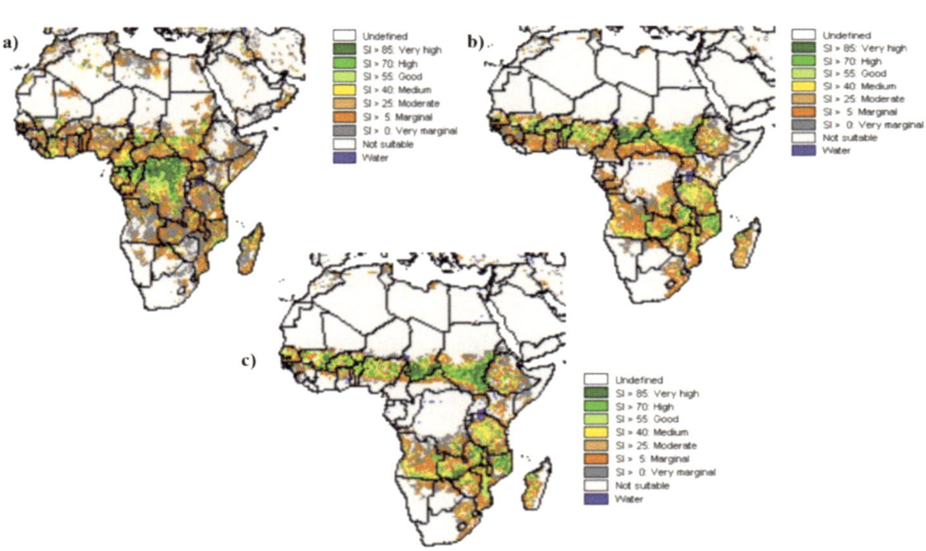

Figure 3.
The suitability of growing (a) sugarcane, (b) maize and (c) sorghum in Africa [40].

Bioenergy crop	Suitable conditions for optimal production	Yield for every hectare	Producing countries
Sugarcane	1600 meters (m) above sea level	4000 liters/hectare (l/ha) in Africa	Mauritius, Zimbabwe, Swaziland, Kenya, Sudan, South Africa
Corn	Can grow everywhere with enough watering	700 l/ha in Africa	Tanzania, Kenya, Ethiopia, Nigeria, South Africa
Sweet sorghum	2500 m attitude in dry temperate and tropical areas	3000–6000 l/ha	Burkina Faso, Sudan, Ethiopia, Nigeria
Cassava	Above 1000 m attitude in tropical climate	1750 l/ha in Africa	Angola, Ghana, Mozambique, DR Congo, Nigeria
Palm oil	Above 700 m attitude in humid tropic climate	3000 l/ha in Africa	Ghna, DR Congo, Cote d'Ivoire, Nigeria
Jatropha	Above 500 m attitude and as low as 300 mm rainfall in semi-arid and tropical climate	40–2200 l/ha oil	Tanzania, Mozambique, Mali, Ghana

Table 2.
Some of the bioenergy crops grown in Africa, their climatic conditions, estimated yield rates and producing countries [38–39].

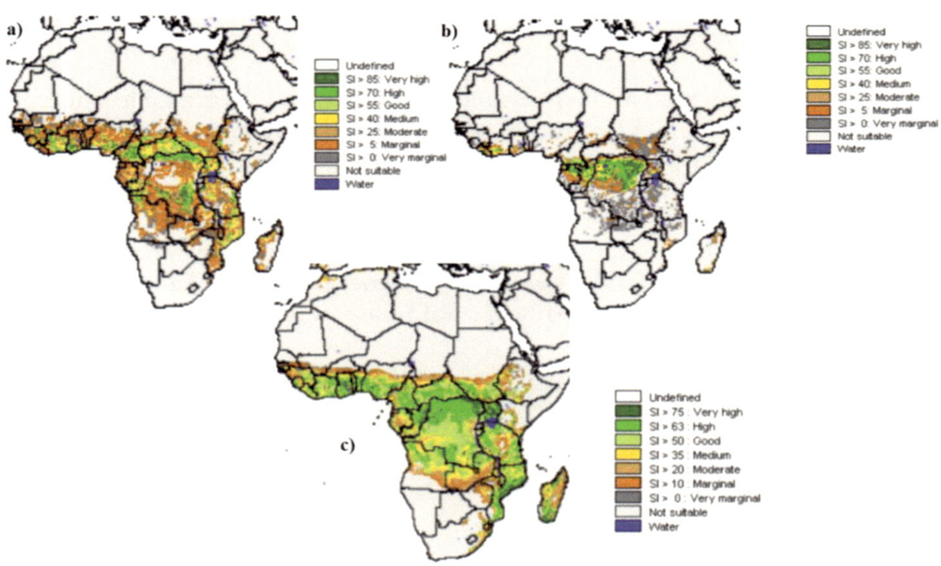

Figure 4.
The suitability of growing (a) cassava, (b) palm oil and (c) jatropha in Africa [40].

these factors could stimulate the use of biomass as an alternative energy source [30]. Kemausuor [38] supported the suggestion that Africa has high biomass potential by showing that its available land, harvested residues and bioenergy crops are higher compared to those of other parts of the world as shown in **Figure 3**. The figures on the available land by FAO also confirm the sufficiency of land for production of fuel wood and other bioenergy crops. However, the characteristics of African land such as its vulnerability to soil erosion, low productivity and misuse of natural resources coupled with traditional biomass uses are limiting factors to its optimal exploitation [24, 30]. Africa has many biofuel options from the many production systems of plants such as sugarcane, corn, sweet sorghum,

Country	Zambia	Tanzania	South Africa	Senegal	Mali	Kenya	Burkina Faso	Botswana
Transportation costs (US$$t^{-1}km^{-1}$)	0.07	0.07	0.05	0.08	0.08	0.07	0.08	0.05
Transport distance for cassava Arid areas (km)	-	-	636	40	92	35	164	121
Semi-arid areas (km)	36	35	69	57	37	26	40	39
Transport distance for fuelwood and jatropha Arid areas (km)	-	-	115	5	12	8	21	16
Semi-arid areas (km)	7	10	15	9	6	7	7	6
Land costs (US$ $ha^{-1}y^{-1}$)	20	20	93	22	22	20	22	93
Labour costs (US$ h^{-1})	0.3	0.3	4	0.4	0.4	0.3	0.4	1
Fertilizer costs for NPK (US$)	2102	2226	1521	2332	2332	1998	2332	2102
Yield rate of fuelwood Arid areas ($t\ ha^{-1}\ y^{-1}$)	-	-	1.1	5.7	2.7	8.9	6.3	0.7
Semi-arid areas ($t\ ha^{-1}\ y^{-1}$)	12.4	9.5	8.7	7.4	8.1	12.4	10	5.5
Yield rate of jatropha Arid areas ($t\ ha^{-1}\ y^{-1}$)	-	-	0.3	1.7	0.7	2	2.4	0.2
Semi-arid areas ($t\ ha^{-1}\ y^{-1}$)	2.7	2.5	2.3	2.7	2.6	2.4	3.1	2.5
Yield rate of cassava Arid areas ($t\ ha^{-1}\ y^{-1}$)	-	-	0.4	1.2	0.6	4.4	1.8	0.2
Semi-arid areas ($t\ ha^{-1}\ y^{-1}$)	4.9	8.9	4.8	2.8	3.4	7.5	3.8	2.3

Table 3.
The labour, land, transport and fertilizer costs of energy crops in some African countries in comparison to their yield rates [27].

cassava, palm oil and jatropha that are all energy crops [39]. The first three crops are collectively known as the ethanol crops while the last two are useful in biodiesel production. All the crops are economically and technically feasible in various parts of Africa based on their suitable conditions, yields from every hectare and some African producers summarized in **Table 2** [39].

Ethanol crops were initially developed for feed and crop production but their energy potential has suited the use of their biomass. Maize and sugarcane have greater potential since they are cultivated in many African countries at both small- and large-scale levels. Biodiesel crops include examples such as sunflower, castor oil, sesame, rapeseed, coconut, soya bean, jatropha and palm oil. However, for Africa palm oil and jatropha are focused on because of the high yield rates for every hectare and capacity to produce biofuel, respectively [39]. Areas where these energy crops are grown in Africa based on their suitability and according to the IIASA / FAI [40], statistics are shown in **Figures 3** and **4**.

4.2 Economic potential

Economic potential of biomass focuses on its production for profitable gains and with economic viability. To assess this biomass potential in Africa, costs of energy crop production such as inputs, labor, land and transportation costs from the farm until the last stage of energy conversions are considered. Other considerations according to Dasappa [27] include taxes, retail and wholesale margins, fertilizer and distribution costs. They help in comparing the economic viability of biomass energy with conventional energy prices. Some of these costs in eight named countries of Africa in comparison to the average yields of fuel wood, a biodiesel (jatropha) and ethanol (cassava) crop and according to literature are summarized in **Table 3** [27].

From the estimates of literature, the costs vary based on countries and there is need to adopt modern biomass uses that focus on efficiency and effectiveness even at the production levels [33, 34]. The costs in arid areas are higher compared to the semi-arid areas due to the challenges of land aforementioned in this chapter. The estimates are however, a simplification of the actual situation and more accurate and region specific estimates are needed as Dasappa [27] highlighted.

5. The carbon neutrality debate of biomass

Bioenergy or biomass energy has received a lot of attention globally as a viable alternative to conventional energy sources from fossil fuels because of its capacity to enhance energy security, result to economic growth and at the same time, cause minimal environmental impacts [41]. With this high attention drawn to biomass production and its subsequent conversion to bio power, researchers, government agencies, biomass feedstock generators and environmentalists are equally paying attention to its carbon neutrality issue. The carbon neutrality debate revolves around the ability of biomass production and conversion to energy processes resulting to zero increase in the greenhouse gas (GHG) levels in the atmosphere following a full life cycle basis. The debate influences future adoption to biomass sources and legislation on their use. During the contest, some bioenergy generators and biomass feedstock farmers support that associated energy resources are neutral since carbon released during biomass generation originates from feedstock that withdrew carbon from the atmosphere during growth. On the other hand, some environmentalists argue that bioenergy is not carbon neutral since the GHG emissions released in production of a unit of energy in a case such as combustion could even be higher than those of fossil fuels depending on the biomass type. Van Renssen [42] bases the debate on carbon neutrality of biomass energy sources to the inaccurate GHG emission assessment, which could result to long-term environmental issues.

To understand the debate around the carbon neutrality of biomass, this chapter does a summative focus on the carbon cycle. The cycle involves many pathways where carbon is exchanged between land, water and the atmosphere. Anthropogenic activities emit CO_2 and contribute to the carbon cycle. The contribution of CO_2 by humans is considerably small compared to other sources but once released to the environment; CO_2 is taken up by oceans, soils and vegetation at a slower rate compared to the emission rate [43]. Unless there are available CO_2 sinks in ocean and on land, the gas is likely to accumulate in the atmosphere causing modifications on the climatic conditions of the earth. Energy production is one of the human activities that releases significant amounts of CO_2. The net result of any energy production activity occurs in three ways: [43].

1. Carbon positivity, which defines activities that release CO_2 to the environment.

2. Carbon negativity, which defined activities that draw CO_2 more from the environment compared to the emission rates.

3. Carbon neutrality that defines activities leading to CO_2 absorption and release of equal measure.

To be carbon neutral, biomass has to meet the following four conditions according to Miner [44].

1. Compared to conventional energy sources, biomass sources should result to lower net increments of GHG emissions.

2. Emissions of biomass overall life cycle from the cultivation, harvesting and transportation processes should sum up to zero.

3. If biomass cultivation draws more atmospheric CO_2 compared to resultant emissions.

4. If by nature, biomass sources are carbon neutral then their products will be neutral too.

The suppositions by Miner [44] are contentious and escalate the carbon neutrality debate. For example, the assumption that biomass is carbon neutral naturally, fails to account for GHG emissions that occur during energy crop tendering processes such as fertilization. Additionally, the demand to remove CO_2 resulting from biomass growth equally means more planting of such crops. To assess the carbon neutrality of biomass compared to conventional fossil fuels, it is important to focus on their specific carbon cycles and identify differences as shown in **Figure 5**. Bioenergy has renewable sources of carbon in that plants can be re-grown and result to stable carbon concentrations compared to fossil fuel energy with finite sources of carbon that lead to additional CO_2 concentrations. Emissions from biofuels mainly occur from bio power technology type, feedstock production and transformation. This fact therefore suggests that the use of biomass as an alternative to conventional energy sources eliminates or reduces emissions from fossil fuels but also results to its own emissions and cannot possibly be carbon neutral as Bird et al. [45] suggested. The authors cited the example of combusting a metric ton of bone-dry wood that emits 1.8 tons of atmospheric CO_2. These differences coupled with the fact that feedstock growth consumes CO_2 could justify the ideologies of biomass as carbon neutral according to Bracmort [43].

A number of policies consider the burning of biomass as carbon neutral irrespective of their sources. Concurrently, the policies acknowledge the presence of carbon emissions using fossil fuels to process biomass but fail to narrow it down to CO_2 [45]. Through this error when computing emissions from bioenergy, they conclude that all biomass-based energy sources are carbon neutral. According to Haberl et al. [41] such policies are inaccurate. In another assumption, carbon neutrality is assumed since combustion of biomass releases the carbon that was initially drawn from the atmosphere as the plants were growing. This is a baseline error since the ideology fails to acknowledge that if energy crops were not harvested, they would

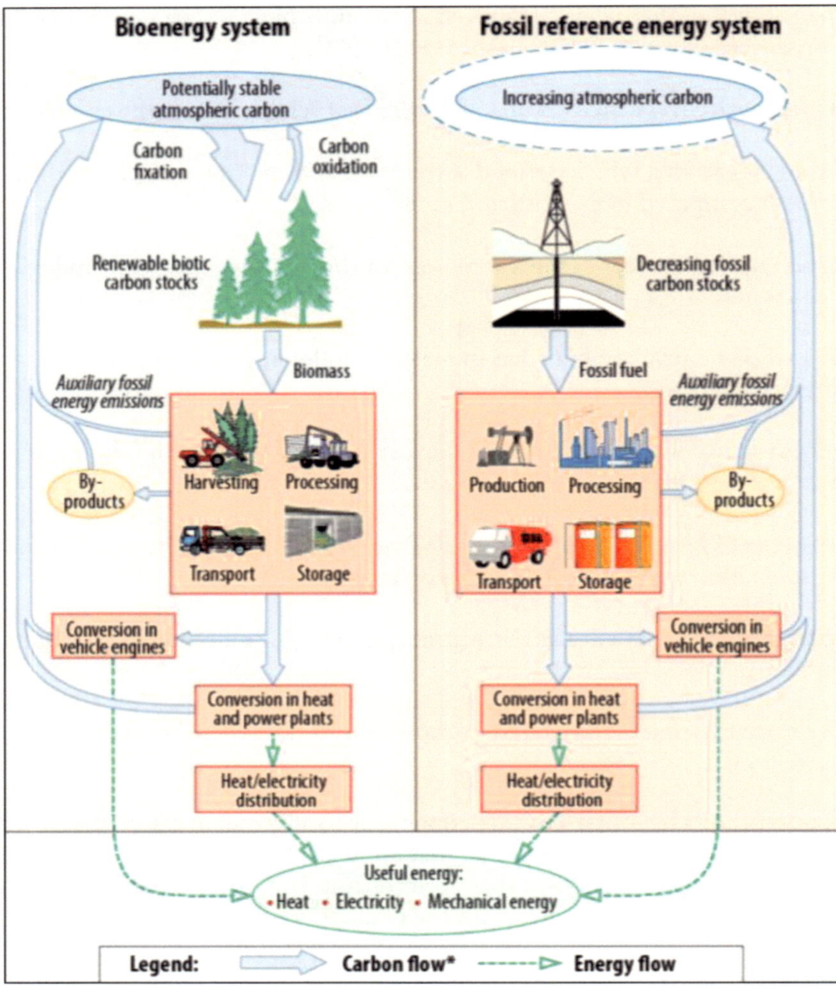

Figure 5.
Carbon cycles of bioenergy compared to fossil fuel sourced energy [43].

continue to absorb atmospheric CO_2. The resultant carbon reductions are included in the global estimates of CO_2 emissions in future and this in not precise since it results to double counting. Ritcher et al. [46] emphasized the computational error of carbon neutrality using the example of a hectare of cropped land that is left to reforest. In this case, the growing plants absorb atmospheric CO_2 to form biomass. Some of the biomass is eaten by microorganisms, fungi and animals and released to the atmosphere while the other is stored in soils and vegetation during growth processes. The overall effect would be reduced CO_2 emissions and a negative effect on global warming. On the other hand, if energy crops were cultivated to be combusted in power plants, fossil fuel based emissions would reduce but carbon emissions from the plants' chimneys would arise. Bird et al. [45] supported this line of thought claiming that for every unit of energy, CO_2 emitted from the power plants would even be higher that fossil fuels because (1) the efficiency of combusting biomass compared to fossil fuel is lower and (2) biomass has lower unit energy potential compared to natural gas or petroleum based power. Therefore growing energy crops draws CO_2 from the atmosphere but it foregoes the sequestration of this gas that would occur if the land was forested. The foregone CO_2 atmospheric withdrawals are not accounted in existent biomass GHG emission computation methods. The growth of forests in Ukrainian forests for instance after abandoning farmland resulted more carbon sinking at the rate of one ton per hectare of forested land annually [47]. The growth of energy crops causes more carbon to be sequestered in underground fossil fuels though the advantage has an opportunity cost of less carbon being stored in soils and plants. Biomass energy sources would reduce carbon emissions to be considered neutral if the former effect outweighs the latter.

The use of food crops such as maize, cassava, sorghum for energy crops is a perfect scenario to demystify the carbon neutrality debate. The process does not compensate the emissions from its use and does not directly lead to additional growth of plants [48, 49]. However, the energy crops can significantly reduce carbon emissions indirectly under the following circumstances:

1. The crops sequester carbon from the atmosphere for longer periods since humans and animals consume them and then return carbon during respiration. If the crops are not replaced, they result to net carbon reductions and their consumption emits less CO_2. However, the approach is not sustainable in reducing GHGs.

2. If more crops are concentrated per unit land, more carbon is absorbed. In the event more land is cultivated, carbon withdraws from the atmosphere are likely to increase.

In these two scenarios, carbon fluctuations due to land-use changes must be determined accurately. From the many considerations on biomass carbon neutrality made in this chapter, the main issue in the debate is the failure to consider the emissions that would result if bioenergy was produced from other alternatives apart from energy crops. This error results to incorrect GHG accounting [41]. Therefore, accurate GHG accounting should reflect the carbon stock losses during production of biomass, the energy consumed and consider the carbon withdrawals that would result if bioenergy was not used at all. In forested areas of countries at the northern hemisphere, biomass accumulation occurs [46, 50] resulting to more carbon sequestration. In events that the harvest of biomass does not surpass forest growth, carbon stocks are estimated to be constant and consequent GHG emission reductions can be realized [43, 51]. If forests are left to regrow following harvest, they realize the same carbon sequestration levels as the unharvested ones when carbon stock build up slows and stops at maturity. At that point, biomass use is considered carbon neutral. Such a realization could take many years and as such, atmospheric CO_2 is retained longer in the atmosphere before removal by plants, which is the cause of climate change [48, 49]. Increasing the harvest times for forests in the long term for sustainable fuel wood supply decreases the carbon stocks resulting to a carbon debt that is repaid after longer periods even if forest conservation occurs [51]. Holistic GHG emission accounting from biomass sources of energy should consider plant growth rate in the presence and absence of bioenergy generation and the changes in carbon storage in soils and plants as a result of the initiatives or otherwise.

6. Conclusions and recommendations

Biomass is a useful energy source in most African countries and is used for thermal applications in addition to cooking and producing electricity. As an alternative source of energy, it is essential as large part of the continent do not have direct access to electricity and other conventional energy sources. Additionally the use of fossil fuel based energy is associated with climate change among other environmental problems. Biomass is sourced from fuel wood, energy crops, municipal solid wastes and plant residues. This book chapter analyzed the technical and economic potential of biomass for energy in Africa based on literature. The findings showed that Africa has adequate land, climatic conditions, and a variety of suitable energy crops for biomass production. Evenly, the costs of biomass production though varied based on the country and climatic condition (humid, arid and semi-arid) are not as high. Biomass is therefore a potential driver to socio-economic growth of the continent through its capacity to enhance energy security. The chapter also explored on the carbon neutrality of biomass energy sources and laid the conditions for this realization. Additionally, the error in computing GHG emissions due to biomass production and use is discussed. Conclusively, biomass energy sources are renewable but not carbon neutral. This chapter therefore makes the following recommendations as efforts to realizing carbon neutrality through the use of biomass.

1. African countries and the rest of the world should formulate policies to encourage use of biomass for energy while reducing GHG emissions and not compromising ecosystems services of providing fiber and food.

2. Global expectations of bioenergy use potential and use should be modified to the earth's ability to produce more biomass without affecting natural ecosystems negatively.

3. Integrated biomass production that enhances food security should be encouraged through the preference to use biomass from residues, wastes and by-products unless needed in soil management for energy generation rather than fuel wood and food crops that have other competing needs.

4. Computation of GHGs resulting from biomass combustion should consider offsets from additional biomass cultivation, its reduced decomposition or otherwise in relation to CO_2 sequestration and release to the atmosphere.

Acknowledgements

The authors appreciate the support from Landmark University Center for Research, Innovation and Development (LUCRID) through the SDGs 9 Group-Industry, Innovation and Infrastructure.

Conflict of interest

The authors declare no conflict of interest.

Author details

Joan Nyika[1], Adeolu Adesoji Adediran[2*], Adeniyi Olayanju[2], Olanrewaju Seun Adesina[2] and Francis Odikpo Edoziuno[3]

1 Technical University of Kenya, Nairobi, Kenya

2 Innovation, Industry and Infrastructure (SDGs 9) Group, Landmark University, Omu-Aran, Kwara State, Nigeria

3 Department of Metallurgical Engineering, Delta State Polytechnic, Ogwashi-Uku (DSPG), Nigeria

*Address all correspondence to: adediran.adeolu@lmu.edu.ng; dladesoji@gmail.com

References

[1] Perea-Moreno M, Sameron-Manzano E, Perea-Moreno A. Biomass as renewable energy: Worldwide research trends. Sustainability. 2019;**11**:863

[2] Jackson R, Quere C, Andrew R, Canadell J, Korsbakken I, Liu Z, et al. Global energy growth is outpacing decarbonisation. Environmental Research Letters. 2018;**13**:120401

[3] International Renewable Energy Agency (IRENA). [Internet]. 2016. Available from: http://www.irena.org/publications/2016/Oct/Renewable-Energy-in-Cities [Accessed: 20 July 2020]

[4] Zaharia A, Diaconeasa M, Brad L, Ladaru G, Ioanas C. Factors influencing energy consumption in the context of sustainable development. Sustainability. 2019;**11**:4147

[5] Lu C, Li W. A comprehensive city-level GHGs inventory accounting quantitative estimation with an empirical case of Baoding. The Science of the Total Environment. 2018;**651**:601-613

[6] Petrie B. South Africa: A Case for Biomass? London: International Institute for Environment and Development (IIED); 2014. pp. 1-35

[7] Bildirici M, Ozaksoy F. Woody biomass energy consumption and economic growth in sub-Saharan Africa. Procedia Economics and Finance. 2016;**38**:287-293

[8] Roser D, Asikainen A, Stupak I, Pasanen K. Forest energy resources and potentials. In: Roser D, Asikainen A, Raulund-Rasmussen K, Stupak I, editors. Sustainable Use of Forest Biomass for Energy: A Synthesis with Focus on the Baltic and Nordic Region. Managing Forest Ecosystems. Dordrecht: Springer; 2008. pp. 9-28

[9] Kirby KJ, Buckley GP, Mills J. Biodiversity implications of coppice decline, transformations to high forest and coppice restoration in British woodland. Folia Geobotanica. 2017;**52**:5-13

[10] Perez S, Renedo CJ, Ortiz A, Mañana M, Delgado F, Tejedor C. Energetic density of different forest species of energy crops in Cantabria (Spain). Biomass and Bioenergy. 2011;**35**:4657-4664

[11] Dimitriou I, Rutz D. Sustainable Short Rotation Coppice a Handbook. Munich: WIP Renewable Energies; 2015. p. 104

[12] Guidi W, Pitre F, Labrecque M. Short-rotation coppice of willows for the production of biomass in eastern Canada. In: Matovic MD, editor. Biomass Now—Sustainable Growth and Use. Rijeka: InTech; 2013. pp. 421-448

[13] Verwijst T, Lundkvist A, Edelfeldt S, Albertsson J. Development of sustainable willow short rotation forestry in northern Europe. In: Matovic MD, editor. Biomass Now - Sustainable Growth and Use. Rijeka: InTech; 2013. pp. 479-502

[14] Goncalves A, Malico I, Sousa A. Solid biomass from forest trees to energy a review. In: Jacob-Lopes E, Zepka L, editors. Renewable Resources and Biorefineries. Vol. 2013. Rijeka: InTech; 2018. pp. 1-25

[15] IEA. Renewables Information: Overview. Paris: International Energy Agency; 2007. p. 11

[16] WBA. WBA Global Bioenergy Statistics 2017. Stockholm: World Biomass Association; 2017. p. 79

[17] ECN. Phyllis2, Database for Biomass and Waste. Available from:

https://www.ecn.nl/phyllis2 . Energy Research Centre of the Netherlands [Accessed: May 2, 2018]

[18] Stecher K, Brosowski A, Thran D. Biomass potential in Africa. International Renewable Energy Agency (IRENA). 2013:1-43

[19] Zegada-Lizarazu W, Monti A. Energy crops in rotation. A review. Biomass and Bioenergy. 2011;**35**:12-25

[20] Anatolioti V, Leontopoulos S, Skoufogianni G, Skenderidis P. A study on the potential use of energy crops as alternative cultivation in Greece. Issues of farmer's attitudes. In: International Conference on Food and Biosystems Engineering. Crete Island: FaBE; 2019. pp. 145-153

[21] Lynd L, Sow M, Chimphango A, Cortez L, Cruz C, Elmissiry M. Bioenergy and African Transformation. Biotechnology for Biofuels. 2015;**8**(18):1-18

[22] Energy Information Administration. Biomass Explained: Waste-to-Energy (Municipal Solid Waste) [Internet]. 2019. Available from: https://www.eia.gov/energyexplained/biomass/waste-to-energy.php [Accessed: 08 July 2020]

[23] Scarlat N, Motola V, Dallemand J, Monforti-Ferrario F, Mofor L. Evaluation of energy potential of municipal solid waste from African urban areas. Renewable and Sustainable Energy Reviews. 2015;**50**:1269-1286

[24] Wicke B, Smeets E, Watson H, Faaji A. The current Bioenergy production potential of semi-arid and arid regions in sub-Saharan Africa. Biomass and Bioenergy. 2015;**35**(7):2773-2786

[25] OECD IEA. World Energy Outlook 2010. Paris: Organization for Economic Co-operation and Development and International Energy Agency; 2010

[26] IEA. World Energy Outlook 2010. Paris: IEA; 2010. p. 736

[27] Dasappa S. Potential of biomass energy for electricity generation in sub-Saharan Africa. Energy for Sustainable Development. 2011;**15**(3):203-213

[28] African Development Bank. The New Deal on Energy for Africa. A Transformative Partnership to Light Up and Power Africa by 2025. Update of Implementation [Internet]. 2017. Avaialble from: https://www.afdb.org/fileadmin/uploads/afdb/Documents/Generic-Documents/Brochure_New_Deal_2_red.pdf. [Accessed: 21 July 2020]

[29] Food, Agriculture Organization (FAO). Rome, Italy: Food and Agriculture Organization; 2005 [Internet]. 2005. Available from: http://www.fao.org [Accessed: 1 July 2020]

[30] Hafner M, Tagliapietra S, de Strasser L. Prospects for Renewable Energy in Africa. In: Energy in Africa. Springer Briefs in Energy. Springer, Cham; 2018

[31] UN Environment Program. Ethiopia's Waste-to-Energy Plant Is a First in Africa. UN Environ [Internet]. 2017. Available from: http://www.unenvironment.org/news-and-stories/story/ethiopias-waste-energy-plant-first-africa [Accessed: 9 July 2020]

[32] International Energy Agency. Energy Access Outlook (World Energy Outlook Special Report); 2017

[33] Smeets E, Dornburg V, Faaij A. Traditional, Improved and Modern Bioenergy Systems for Semi-Arid and arid Africa—Experiences from the COMPETE Network. 2009. Available from: http://www.compete-bioafrica.net [Retrieved: 30 June 2020]

[34] Smeets E, Dornburg V, Faaij A. Report on Potential Projects for Financing Support—Experiences from

the COMPETE Network. 2009. Available from: http://www.competebioafrica.net [Retrieved: 30 June 2020]

[35] Watson HK. Potential impacts of EU policies on sustainable development in southern Africa. StudiaDiplomatica. 2009;**LXII**(4):85-102

[36] COMPETE Project. Competence Platform on Energy Crop and Agroforestry Systems for Arid and Semi-Arid Ecosystems—Africa. 2009. Available from: http://www.competebioafrica.net [Retrieved: 15 July 2020]

[37] Hoogwijk M, Faaij A, Eickhout B, De Vries B, Turkenburg W. Potential of biomass energy out to 2100, for four IPCC SRES land-use scenarios. Biomass and Bioenergy. 2005;**29**(4):225-257

[38] Kemausour F. Biomass Potential in Africa and African Experiences. AFRETEP 2nd Regional Workshop Ouagadougou, Burkina Faso. 2012

[39] Sielhorst S, Molenaar J, Offermans D. An Assessment of Risks and Benefits for African Wetlands. Wageningen, Netherlands: Wetlands International; 2008

[40] IIASA/FAI. Global Agro-Ecological Assessment for Agriculture in the 21st Century. 2002. Available from: http://www.iiasa.ac.at/Research/LUC/SAEZ/index.html

[41] Haberl H, Sprinz D, Bonazountas M, Cocco P, Desaubies Y, Henze M. Correcting a fundamental error in greenhouse gas accounting related to bioenergy. Energy Policy. 2012;**45**:18-23

[42] Van Renssen S. A biofuel conundrum. Nature Climate Change. 2011;**1**:389-390

[43] Bracmort K. Is Biopower Carbon Neutral? Congressional Research Service [Internet]. Available from: https://fas.org/sgp/crs/misc/R41603.pdf [Accessed: 13 July 2020]

[44] Miner R. Biomass Carbon Neutrality in the Context of Forest-Based Fuels and Products. Washington, DC: U.S. Department of Agriculture (USDA) Bioelectricty and GHG Workshop; 2010

[45] Bird DN, Pena N, Zanchi G. Zero, one, or in between: Evaluation of alternative national and entity-level accounting for bioenergy. Blackwell Publishing Ltd, GCB Bioenergy. 2011;**4**:576-587

[46] Richter D, de B, Houghton RA. Gross CO_2 fluxes from land-use change: Implications for reducing global emissions and increasing sinks. Carbon Management. 2011;**2**:41-47

[47] Kuemmerle T, Olofsson P, Chaskovskyy O, Baumann M, Ostapowicz K, Woodcock CE, et al. Post-soviet farmland abandonment, forest recovery, and carbon sequestration in western Ukraine. Global Change Biology. 2011;**17**:1335-1349

[48] Cherubini F, Peters GP, Berntsen T, Strømann AH, Hertwich E. CO_2 emissions from biomass combustion for bioenergy: Atmospheric decay and contribution to global warming. GCB Bioenergy. 2011;**3**:413-426

[49] Cherubini F, Strømman AH, Hertwich E. Effects of boreal forest management practices on the climate impact of CO_2 emissions from bioenergy. Ecological Modelling. 2011;**223**:59-66

[50] Pan Y, Birdsey RA, Fang J, Houghton R, Kauppi PE, Kurz WA, et al. A large and persistent carbon sink in the world's forests. Science. 2011;**333**:988-993

[51] Holtsmark B. Harvesting in boreal forests and the biofuel carbon debt. Climate Change. 2012;**112**:415-428

Chapter 7

A General Overview of Sweet Sorghum Genomics

Raymundo Rosas-Quijano,
Abraham Ontiveros-Cisneros, Noé Montes-García,
Arturo Díaz-Franco, Alfredo Vázquez-Ovando
and Didiana Gálvez-López

Abstract

Sorghum is one of the main cereal crops, its consumption is large, since it provides grain, fiber and biofuel. Likewise, its genome, with only 10 diploid chromosomes, makes it an attractive model for research and genetic improvement. Sorghum is the most studied C4 plant of its genus; several lines have been developed under three main characteristics: grain, forage and sugar biomass. Compared to other crops, sweet sorghum possesses high levels of highly fermentable sugars in the stem. Also, it has the ability of producing high production yields in marginal lands. These characteristics make it and attractive crop for the generation of biofuels. Molecular markers associated to several resistances and tolerances to biotic and abiotic factors have been described in literature. These allow the development of high-density linkage maps, which, along with the rising availability of sorghum genomes, will accelerate the identification of markers and the integration of the complete genome sequence. This will facilitate the selection of traits related to biofuels and the marker-assisted genetic improvement. Most of the information presented in this review is focused in *Sorghum bicolor* (L.) Moench. However, from the bioenergetics perspective, it is limited to sweet sorghum, which represents a promising opportunity for further studies.

Keywords: genetic differences, trait sequencing, biofuel, post-transcriptional regulation

1. Introduction

Sorghum (*Sorghum bicolor*) was originated in Africa; however, it is of high agricultural and economic importance in México. It is a multipurpose crop, since it serves as food, fiber and fuel. In addition, it can be grown in a wide range of agro-ecosystems, especially those of fragile conditions. It is also known as the "camel" of the crops, since to complete a productive cycle it requires less amount of water compared to other cereals of agricultural relevance. It has a remarkable ability to produce under low water availability regimes, as well as in other adverse abiotic conditions. Regarding the global production, it occupies the fifth place among the grains of economic importance [1]. Every year, sorghum production has been stabilized in 60 million tons in a producing area of 44 million hectares. México has the third place worldwide with 4.5 million tons produced in 2018 [2]. From this, 62% corresponds to grain production, while 38% to forage.

Sweet sorghum is a natural variant of the common grain sorghum, with high content of sugars in the stem. It is frequently regarded as a "smart crop", since it can produce food as well as fuel. Currently, there are no records of commercial production of sweet sorghum in México, since it is a crop of recent introduction in the country. The first appearance of this crop in México was in the INIFAP Campo Experimental Río Bravo in the year of 2006, with the varieties Dale, Topper 76–6, Theis and M-81E [3]. Since then, the INIFAP genetic improvement program for sorghum, has searched for varieties which can adapt to the arid and semi-arid regions of the northeast or the country. These conditions are met by the variety RB Cañero, which has been tested in different environments, where it has shown a remarkable development and a high potential to produce bioethanol sugars [4].

Despite sweet sorghum is still a new crop in Mexico, the productive potential and the area of exploitation are highly promising. In 2010, INIFAP estimated the area available in México for sweet sorghum production, being the northeast region suitable with more than 4.38 million hectares [3]. The latter, pictures México as a country with great expectative of self-sufficiency in the generation of its own biofuel. The use of biofuels, being a cleaner and more efficient way of producing energy, would help to reduce pollution and greenhouse gases, which destabilize climate [5]. Pursuing this goal, biotechnology is an effective tool to develop methods which can optimize the bioethanol production from sweet sorghum. This review presents some of the biotechnological advances, as well as the current state of the genetic and molecular studies in sorghum, related to new routes to achieve an efficient generation of biofuels from this crop. This knowledge makes clear the necessity for effort and economic investment on this field, to reach self-sufficiency in the generation of energy sources in the country.

2. Sorghum as a research model

Sorghum possesses a small genome, which makes it attractive as a model organism for studies focused on understanding the structure, function and evolution of the cereals' genome. Sorghum is a tropical crop with a typical C4 photosynthesis, which uses a complex specialized biochemical and morphological system for carbon assimilation when it is exposed to elevated temperatures. This is a unique feature among the species of the same family it belongs to.

Like in other crops, sorghum is compared to other Poaceae of agricultural relevance such as maize (*Zea mays*), rice (*Oryza sativa*), sugar cane (*Saccharum officinarum*) or wheat (*Triticum aestivum*). Sorghum has ten diploid chromosomes and has a low degree of genetic duplication, which if compared to maize, makes it an outstanding model to develop functional genomics studies. On the other hand, maize developed a duplication of its entire genome since its evolutionary divergence from sorghum, changing the chromosome number from $n = 5$ to $n = 10$, being one of the sub genomes very similar to sorghum than to any other cereal [6]. However, evidence shown in NCBI database indicate that homologous genes between these species vary mainly in gene size and protein coding [7]. In the case of rice, it possesses a bigger genome than sorghum and also more genetic duplication. However, some comparisons prove that between sorghum and rice homologous gene exist, but with differences that could be attributed to the particularities of each species. Some reports indicate that sorghum and maize split from a common ancestor 12 million years ago [6], whereas the rice lineage is about 42 million years old compared to these two crops [7]. Lastly, sugar cage, possibly the most important sugar producing crop worldwide, could have share ancestors with sorghum approximately 5 million years ago [8], since they conserve similar genes [9], and even it is viable to generate offspring from intergeneric crosses. Sugar cane contains at least two genomic duplications [9].

Sorghum represents an excellent model for research, since linkage mapping methodologies have been successfully implemented on it and possesses a wide mating system by self-pollination, which tends to preserve the association relationships for longer time periods compared to the self-pollination of cereals like maize, which facilitates the development of pure lines. Also, its genome sequence is available in several databases [10].

3. Differences between sweet sorghum and grain sorghum

Sweet sorghum plant produces sugars which can be directly fermented, together with its ability to produce high biomass volumes under adverse conditions, this crop is consider ideal for the generation of bioethanol of first and second generation. Also, its cultivation is suitable for marginal lands, avoiding competence for land with other food crops [11–13]. However, the genetics underlying these traits have been barely studied. The genetic differences between sweet and grain sorghum consist on a series of variations in the sequence and alterations of the genetic structure. The variations at sequence level are usually identified by single-nucleotide polymorphisms (SNPs), association sequences, genetic diversity and domestication [14–16].

S. bicolor has three subspecies: arundinaceum, bicolor y drummondii. All the feasible varieties belong to bicolor, which is divided into 5 races: bicolor, caudatum, durra, guinea y kafir. Sweet sorghum differs phenotypically from grain sorghum in several aspects: juicy stem rich in sugars, higher plant length, higher biomass production and less amount of grains in the spike [11]. These differences cannot be only explained by the genetic variation among these two varieties, which suggest DNA methylation and other epigenetic mechanisms are key factors to describe them [17].

	Trait	Grain sorghum	Sweet Sorghum
Phenotypic	Height (cm)	97.6	230–281
	Biomass (g plant^{-1})	67	605–1096
	Spike (dry weight, g) (g plant^{-1})	24.2	60–80
	Roots (dry weight, g) (g plant^{-1})	15.1	68–88
	Stem (dry weight, g) (g plant^{-1})	27	164
Genotypic	SNPs	85,041(14,782 genes)	
Line BTx623 (grain)	Indels	16,781 (7,977 genes)	
vs.	SVs	1,847 (2,071 genes)	
Line Keller (sweet)	Protein functional divergence	563(SNP), 287(Indels), 69 (SV)	

Grain sorghum data obtained from Assefa et al., *[20] and Wang* et al., *[21]. Sweet sorghum data obtained from Ekefre* et al., *[22]. Genotypic differences data obtained from Jiang* et al., *[17].*

Table 1.
Main differences between grain and sweet sorghum.

Sweet sorghum has been found on different races [18], which challenges its origin, selection and genetics. This also suggest high genetic variability between sweet and grain sorghum, which could be exploited for genetic improvement of sweet sorghum. Currently the BTx623 grain sorghum genome sequence is available [10], which provides a genomic base for comparative studies of the genome. Regardless this achievement, it is still difficult to access the information related to the hidden variability among genomes of the same species. Zheng et al., [19] studied the resequencing of the two sweet and one grain sorghum genomes, with the aim of identify polymorphism patterns of the sequences and structural variations, using BTx623 as a reference genome. This study allowed the identification of great differences in the number of SNPs, indels, copy number variations and structural variations (SV) among these genomes. The comparison of this genetic variation defined potential genomic regions and metabolic pathways associated to sweet sorghum and traits such as sugar production. **Table 1** presents phenotypic and genotypic differences between grain and sweet sorghum.

4. Sorghum genetic mapping

Building a linkage map is the fundamental step required for a detailed study of genetic improvement of crops by marker-assisted selection. Mapping of sorghum genome based on DNA markers started in the 90's, and nowadays there are several genetic maps available. It is important to mention that sorghum, particularly *S. bicolor*, possesses 10 chromosomes and has been classified as a diploid ($2n = 2x = 20$) [23]. However, it has been assumed that sorghum has a tetraploid origin, due to the large number of complementary gene *loci* and to some studies on meiotic mating among chromosomes as in *S. halepense*, which is $2n = 4x = 40$. Other studies with fluorescence *in situ* hybridization (FISH) have reached the same conclusion [24]). Nonetheless, [25] used the same FISH technique and other structural genomic resources, including genomic clones with large inserts in artificial bacterial chromosomes (BACs), and identified the 10 chromosomes simultaneously. Years after, Paterson *et al.,* [7] found identities and homologies among the linkage groups in metaphase state and this determined *S. bicolor* diploidy ($2n = 20$) as well as the genome length of 730 Megabases (Mb).

The first genetic maps built where based on DNA analogy tests based on corn genome Binelli *et al.*, [26]; Whitkus *et al.*, [27]; Melake-Berhan *et al.*, [28]; Pereira *et al.*, [27] After, maps were built from genomic DNA analysis Chittenden *et al.*, [29]; Ragab *et al.*, [30]; Xu *et al.*, [31]. Other published map was based on tests done in sugar cane and maize [32]. All these maps were built using restriction fragments length polymorphisms (RFLPs) and the majority of them used F_2 populations, while Dufour *et al.*, [32] used two populations of recombinant inbred lines (RILs). This last map was extended by Boivin *et al.,* [33] with the addition of a great number of RFLPs and AFLPs (Amplified Fragment Length Polymorphisms). On the other hand, [34] built a sorghum map using a RIL population and a variety of tests which include sorghum genomic DNA, corn and sugar cane DNA and cDNA, versus tests of other cereals, and 8 simple sequence repeat (SSRs) microsatellite *loci.* Subudhi and Nguyen [35] completed the alignment of the 10 linkage groups using RFLPs on a RIL population and completing the maps of de Chittenden *et al.* [29], Ragab *et al.* [30, 31] of corn and other cereals.

Kong et al. [36] mapped a RIL population with 31 SSR polymorphic *loci* obtained from 51 clones isolated from a *S. bicolor* genetic library, which was provided with four oligomers di- and trinucleotides radioactively labeled. Haussmann *et al.* [37] mapped molecular markers related to resistance of the hemiparasite

Striga hermonthica in two recombinant populations (RIP-1, −2) of F_{3-5} lines. RIP-1 and RIP-2 maps covered 1,498 cM and 1,599 cM respectively with 157 markers distributed in among the linkage groups.

Apart from these linkage maps, integrated maps have also been built. An integrated linkage map of SSRs and AFLPs from sorghum was reported by Kong *et al*. [36] using different sorghum lines. SSR *loci* were designed from clones isolated from two sorghum BAC libraries. The linkage map covered 1,406 cM and consisted in 147 SSR *loci* and 323 RFLP *loci*. Klein *et al*. [36] constructed an integrated physical and genetic map of sorghum genome (750 Mb) from PCR methods for the creation of BAC libraries and the localization of BAC clones in sorghum genetic maps. Also, Menz *et al*. [38] built a genetic map using AFLPs. The 1713 cM of the map covered 2,926 *loci* distributed among the 10 linkage groups, where 2,454 were AFLPs, 136 were SSRs previously mapped in sorghum and 203 were cDNAs and genomic clones coming from rice, barley, oat and maize. Another reported map was the one from [39], which consisted in 2,512 *loci* spaced in intervals of 0.4 cM on average, and it was based in 2,050 RFLPs, including 865 heterology tests from sugar cane, maize, rice, *Pennisetum setaceum* and *Arabidopsis thaliana*.

Recently, a high genetic density map was published by Ji *et al.* [40], where specific length amplified fragment markers (SLAFs) were utilized. This map was based on a F_2 population of 130 individuals originated from a cross between a grain sorghum variety, J204, and a sweet sorghum variety, Keter. Massive sequencing was used to cover the 52,928 SLAFs from the 43 million reads generated. From these markers, 12% appeared to be polymorphic and from 2,246 of these SLAFs a linkage map was built, covering the 10 chromosomes. The total length was 2,158 cM, which is 50% more compared to the previous maps available, which were constructed using RFLPs.

Another method used is the comparative genome mapping. This particular method is interesting for geneticists and evolutionary biologists to elucidate the mechanisms determining chromosome's evolution. Comparative genome mapping provides a powerful technique to study the way and the time where chromosomal evolution occurs [23]. This approach involves the use of molecular markers, such as RFLPs, to map the genomes of two species for a group of markers in common (*loci*). Even it is an expensive and intensive duty, this method can determine the reach and the nature of the chromosomes recombination in incompatible species crosses. The finding of small chromosomal regions which retain a similar gene order in sorghum and in two dicotyledon species (*Arabidopsis* y *Gossypium hirsutum*), suggest that comparative mapping can reach a major evolutive distance compared to what has been reached until now.

Among the *Andropogoneae* grass tribe, comparative mapping facilitates the understanding of sorghum genetics. At this point, several research groups have established a relationship between sorghum and maize genomes [27, 28, 32, 41, 42]. The high degree of conservation of the genes order between these two crops has limited the identification of chromosomal rearrangements between them. Apart from being compared with maize, sorghum has also been compared to rice, where certain apparent collinearity was also found.

Until 2015, more than 850 *loci* associated to traits relevant to biofuels production were identified in sorghum. These are traits regarding plant architecture (roots, leaves and stem), flowering time, and conversion rate of biomass into biofuels. These quantitative trait *loci* (QTLs) related to biofuels generation have been found in different mapped populations, which suggest the plasticity of these traits in different environments. This makes the genes located in these QTL regions could be potential targets to improve sweet sorghum yield for biomass and biofuels production [43].

Regardless of the multiple QTLs already reported, very few studies have been done with the aim of genetically improving these traits. In one of these, a quantitative gene (dw3), orthologous to branchytic 2 (br2) from corn, was cloned with the intention of reducing plant height. This gene is a P-glycoprotein which modules auxin transport in maize stems [44]. Another group of researchers cloned and sequenced, from the cultivar dulce Rio, homologous genes of the sucrose transporter proteins (SUTs), which were compared to the published sequence of BTX623 grain sorghum variety. It was possible to identify six SUTs in BTx623, along with nine differences in the amino acids sequence of SbSUT5 between the two varieties. Two of the five remaining SUTs exhibited unique variations in the amino acids sequences of SbSUT1 and SbSUT2, whereas the rest shared identical sequences. It was also proven that in a mutant of *Saccharomyces* (SEY6210), uncapable of growing with sucrose as the only available carbon source, sorghum SUTs are capable of transporting sucrose [45]. This showcases the relatively low knowledge of the genes underlying the traits associated to biofuels generation in sweet sorghum and bolsters the potential of sweet sorghum breeding to produce biofuel through the exploitation of its genetic resources.

5. Genome sequencing and sorghum functional genomics

Massive sequencing of the line BTx623 is nowadays completed and approximately 10.5 million of reads (8X coverage) have been deposited in the NCBI database. In the preliminary assembly, more than 97% of the genes codifying for proteins (Expressed Sequence Tag, EST) in sorghum were found in 250 large contigs. The majority was able to be joined, ordered and oriented using genetic and physical maps to reconstruct the full chromosomes. The preliminary alignment assembly for the sorghum sequence was based on methyl-filtrated sequences. Also, the assembly for sorghum, maize, sugar cane transcripts, as well as *Arabidopsis* and rice proteomes, confirmed the correct assembly of the bases and local structure. This allowed the approximate prediction of 30,000 to 50,000 *loci* which code for proteins. The conserve genetic synteny with rice is evident, as expected from the comparisons obtained from the maps [10].

The spatial structure of the genes in sorghum is represented by approximately 125,000 ESTs, which have been grouped in 22,000 unigenes, representing more than the 20 diverse libraries of different genotypes [46]. Around 50,000 methyl-filtrated reads, which provide an estimated coverage of 1X [47] have been assembled into contigs. Another representative strategy is the cloning and direct sequencing (Cot-Base cloning), which was used in sorghum in 2001 for the first time [48]. This method offers the potential to cover and increase this coverage more than could be achieved with ESTs and methyl-filtrated reads as demonstrated in maize.

The progress in transcriptomes' characterization has been parallel to the identification of differential genes expressing in response to biotic and abiotic factors, as well as to damage caused by insects, dehydration, high salt concentration, abscisic acid [49], methyl-jasmonate, salicylic acid and amino cyclopropane carboxylic acid [50].

6. Post-transcriptional regulation by miRNAs in sorghum

The micro-RNAs (miRNAs) are small RNA molecules of approximately 21 nucleotides, which play an important role in the post-transcriptional genetic regulation inhibiting the translation of the messenger RNAs (mRNAs) by blocking translation machinery or by excision of the mRNAs [51]. In plants, the majority of miRNAs promote the degradation of mRNA targets by perfect or almost perfect mating of the complimentary RNA strands [52]. miRNAs intervene in a variety of biological processes, such as development and identity of organs, metabolism and stress responses [53]. A substantial number of miRNAs has been identified in different plants, and recently the number of studies in sorghum has been increasing with respect to the identification of miRNAs and their target genes.

Recently, Katiyar et al. [54] showed the importance of studying miRNAs and other RNA molecules using RNA sequencing from the libraries created from genotypes of a variety tolerant to drought (M35–1) and one susceptible. These varieties were cultivated in controlled conditions as well as in drought stress. After sequencing the RNA profiles generated, it was possible to identify 96 miRNAs regulated by the stressed caused by drought conditions. This represents new perspectives for the genetic engineering regarding the potential of miRNAs to improve drought resistance as well as other types of abiotic stresses.

Following the same research line, in 2016, Hamza et al., used 8 deregulated miRNAs by abiotic stress in 11 elite varieties of sorghum under low water availability and drought [55]. This study showed that the miRNAs miR396, miR393, miR397-5p, miR166, miR167 and miR168 have a significative deregulation, being sbi-miR396 and sbi-miRNA398 the ones with higher overexpression for all the genotypes. This same research group has studied the effects of drought and salinity in the miRNAs profiles generated in *S. bicolor* [56]; these results confirm that the miRNAs expression patterns are related to the dose of stress the plants are subjected; however, every miRNA responded in a unique way in every of the six genotypes.

Other important trait to improve sweet sorghum is sugar accumulation, which has been already studied by Yu et al. [57], who propose mir-271 as a specific miRNA of the Rio sweet sorghum variety, related to cellulose synthesis and sugar accumulation. A full detailed list with most of the relevant miRNAs for the genetic improvement of sorghum in biofuels production was published by Dhaka et al. [58].

7. Transformation and reverse genetic in sorghum

Methods for sorghum transformation have been available since the beginning of the 90's, initially by protoplasts [59] and cell culture [60], and subsequently *in planta* [61, 62], using *Agrobacterium* and protocols based in microprojectiles which are now available and with substantially improved efficiencies [63–69]. Sorghum is a crop hard to transform, since it is a recalcitrant genus for tissue culture and the transformation protocols reported are scarce and not very reproducible. In the particular case of sweet sorghum, [70] proposed a transformation system based on optimizing tissue culture conditions using embrionary callus with a regeneration of 90% in 12 weeks. Also, hygromycin resistance selection conferred by the Ubi-*hpt* transgene was performed, followed by particle bombardment. This method proved to be highly reproducible with an efficiency of transformation of 0.09% in every embryo. In 2012, Liu and Godwin, published a method with a better transformation efficiency in *S. bicolor*, in which using pure line embryos (IEs) Tx430, reaching an efficiency of 20.7% in the three independent experiments [71]. The protocol, which involves the use of microprojectiles and transgenes regulated under the *ubi1* constitutive promoter, improves the conditions of the media culture for embryos, as well as the parameters for transformation with microprojectiles. In this experiment, the transgenes were inherited by the T1 generation.

After, Tien-Do *et al*. [72] developed a fast and efficient system for sorghum transformation using binary vectors and the AGL1 *Agrobacterium* strain instead of microprojectiles. With the public genotype P898012 and the *bar* gene as a selective marker, callus regeneration time was reduced to 7–12 weeks, producing 18 plants per callus. This experiment achieved a frequency of 14%, where 40–50% of the transformation events possessed a single copy of integrated T-DNA with a segregation of mendelian 3: 1 estimated by Southern blot. An example of the utility of genetic transformation of sweet sorghum is presented by Zhu *et al.* [73], where using *Agrobacterium* and induce early flowering, the gene Bt cry1Ah was introduced. BT or Cry proteins are produced naturally in aggregates or crystals by *Bacillus thuringiensis.* There proteins are specific for the digestive system of different insects. Protein Cry1Ah, confers resistance to *Ostrinia furnacalis.* In this study, the generated plants, after being selected for herbicide resistance to confirm transformation, showed a high resistance at T0. Apart from the resistance tests, transgene expression was confirmed by RT-PCR and the presence of BT proteins produced by the plant by Western blot and ELISA.

One of the main arguments against the use of transgenics is the use of selection markers such as herbicide tolerance and the fact that they stay inside the genome of the transformed plant. The main issue is the possibility of the cultivar's pollen to pollinate related weeds and therefore the resistance is inherited to undesired plants. Against this problematic, there are several efforts to generate marker-free transgenics. An example is presented by Lu *et al.* [74], where at the cost of reducing the selection pressure, they manage to obtain marker-free transgenics.

In other hand, sorghum offers the opportunity to complete what has been previously described in the reverse genetics of rice and maize, providing to the genetic and familiar studies, those genes which are hard to manipulate in these crops. This allows the directed functional analysis to specify the genes in sorghum related to traits such as hydric stress and the production of certain sugars by genetic association. To accelerate the specific direct identification of genes, mutant lines using ethyl methane sulfonate (EMS) have been created. For the genotype BTx623 there are 1,600 M3 annotations, individual pedigrees, which was characterized [75]. Currently, each of the inspected M3 lines is distinguishable from the original stock and some have multiple mutant phenotypes. The additional M2 mutants are available for the scientific community for the production of thousands of M3 additional lines.

Until few years ago, even with the genome sequencing technology for the elite line BTx623, the genetic sources and sorghum germplasm where limited, making hard the functional validation of the sequenced genes. In 2016, 4,600 M4 mutant pedigrees where created by EMS mutagenizing of BTx623 seed, which were later transformed in lines by single-seed descendant method [76]. The sequencing of 256 mutant lines revealed more than 1.8 million of induced canonical mutations, affecting 95% of the sorghum genome.

8. Perspectives

The studies here presented represent an introduction to the current state of sorghum genomics. Regardless these advances have contributed relevant achievements to what it is known about genetic diversity of this species, it is still necessary to develop further studies, which its aim is focused in sweet sorghum. However, the knowledge acquired in grain sorghum and other related species, constitute an important molecular base to continue developing research studies which allow to know sweet sorghum and its unreported genomic regions. In them could lie the key for the increased production of sugars, lignin and other traits of interest such as tolerance to new plague's appearances such as yellow aphids and/or diseases. It is also necessary to develop genetic maps which allow the localization of genetic codifying regions to certain traits of agronomic interest. Regarding molecular studies in Mexico, there are no reports of genetic maps or genomics performed in sweet sorghum. This represents an opportunity to develop research lines which allow to generate the country's own sweet sorghum genotypes carrying tolerances to adverse biotic and abiotic conditions predominant in the country. This would allow the growth on its production and of its sub products, focusing in alternative environment friendly energy sources.

Author details

Raymundo Rosas-Quijano[1], Abraham Ontiveros-Cisneros[2], Noé Montes-García[3], Arturo Díaz-Franco[3], Alfredo Vázquez-Ovando[1] and Didiana Gálvez-López[1]*

1 Bioscience Institute, Autonomous University of Chiapas, Tapachula, Chiapas, Mexico

2 Lund University, Lund, Skane, Sweden

3 INIFAP, Río Bravo Experimental Field, Río Bravo, Tamaulipas, Mexico

*Address all correspondence to: didiana.galvez@unach.mx

References

[1] FAOSTAT, Food Agricultural Organization. 2019. Consulted on may/2019 [fao.org/faostat/en/#data/QC/visualize].

[2] SIAP (Servicio de Información Agroalimentaria). 2019. Consulted on may/2019 [nube.siap.gob.mx/cierreagricola/].

[3] N. Montes-García, V. Pecina-Quintero, M. E. Cisneros-López, y M. A. García-García, "RB Cañero: Sorgo dulce (*Sorghum bicolor* (L.) Moench) para la producción de etanol.", *INIFAP SAGARPA Foll. Técnico*, núm. 49, p. 31, 2010.

[4] N. Montes-García, J. R. Salinas-García, A. González-Jiménez, R. Loredo-Pérez, y G. Díaz-Padilla, "Guía Técnica de Producción de Sorgo Dulce (*Sorghum bicolor* (L.) Moench) en Tamaulipas.", *INIFAP SAGARPA Foll. Técnico*, núm. 49, p. 35, 2010.

[5] A. Almodares y M. R. Hadi, "Production of bioethanol from sweet sorghum: A review", *African J. Agric. Res.*, vol. 4, núm. 9, pp. 772-780, 2009.

[6] Z. Swigonova, "Close Split of Sorghum and Maize Genome Progenitors", *Genome Res.*, vol. 14, núm. 10a, pp. 1916-1923, sep. 2004, doi: 10.1101/gr.2332504.

[7] A. H. Paterson, J. E. Bowers, y B. A. Chapman, "Ancient polyploidization predating divergence of the cereals, and its consequences for comparative genomics", *Proc. Natl. Acad. Sci.*, vol. 101, núm. 26, pp. 9903-9908, jun. 2004, doi: 10.1073/pnas.0307901101.

[8] B. W. S. Sobral, D. P. V. Braga, E. S. LaHood, y P. Keim, "Phylogenetic analysis of chloroplast restriction enzyme site mutations in the Saccharinae Griseb. subtribe of the Andropogoneae Dumort. tribe", *Theor. Appl. Genet.*, vol. 87, núm. 7, pp. 843-853, feb. 1994, doi: 10.1007/BF00221137.

[9] R. Ming *et al.*, "Detailed alignment of Saccharum and Sorghum chromosomes: Comparative organization of closely related diploid and polyploid genomes.", *Genetics*, vol. 150, núm. 4, pp. 1663-1682, 1998.

[10] A. H. Paterson *et al.*, "The *Sorghum bicolor* genome and the diversification of grasses", *Nature*, vol. 457, núm. 7229, pp. 551-556, ene. 2009, doi: 10.1038/nature07723.

[11] W. L. Rooney, J. Blumenthal, B. Bean, y J. E. Mullet, "Designing sorghum as a dedicated bioenergy feedstock", *Biofuels, Bioprod. Biorefining*, vol. 1, núm. 2, pp. 147-157, oct. 2007, doi: 10.1002/bbb.15.

[12] N. C. Carpita y M. C. McCann, "Maize and sorghum: genetic resources for bioenergy grasses", *Trends Plant Sci.*, vol. 13, núm. 8, pp. 415-420, ago. 2008, doi: 10.1016/j.tplants.2008.06.002.

[13] W. Vermerris, "Survey of Genomics Approaches to Improve Bioenergy Traits in Maize, Sorghum and SugarcaneFree Access", *J. Integr. Plant Biol.*, vol. 53, núm. 2, pp. 105-119, feb. 2011, doi: 10.1111/j.1744-7909.2010.01020.x.

[14] K. J. Schmid, "Large-Scale Identification and Analysis of Genome-Wide Single-Nucleotide Polymorphisms for Mapping in *Arabidopsis thaliana*", *Genome Res.*, vol. 13, núm. 6, pp. 1250-1257, may 2003, doi: 10.1101/gr.728603.

[15] F. A. Feltus, "An SNP Resource for Rice Genetics and Breeding Based on Subspecies Indica and Japonica Genome Alignments", *Genome Res.*, vol. 14, núm. 9, pp. 1812-1819, sep. 2004, doi: 10.1101/gr.2479404.

[16] M. A. Gore *et al.*, "A First-Generation Haplotype Map of Maize", *Science (80-.).*, vol. 326, núm. 5956, pp. 1115-1117, nov. 2009, doi: 10.1126/science.1177837.

[17] S.-Y. Jiang, Z. Ma, J. Vanitha, y S. Ramachandran, "Genetic variation and expression diversity between grain and sweet sorghum lines", *BMC Genomics*, vol. 14, núm. 1, p. 18, 2013, doi: 10.1186/1471-2164-14-18.

[18] K. B. Ritter, C. L. McIntyre, I. D. Godwin, D. R. Jordan, y S. C. Chapman, "An assessment of the genetic relationship between sweet and grain sorghums, within *Sorghum bicolor ssp. bicolor* (L.) Moench, using AFLP markers", *Euphytica*, vol. 157, núm. 1-2, pp. 161-176, ago. 2007, doi: 10.1007/s10681-007-9408-4.

[19] L.-Y. Zheng *et al.*, "Genome-wide patterns of genetic variation in sweet and grain sorghum (*Sorghum bicolor*)", *Genome Biol.*, vol. 12, núm. 11, p. R114, 2011, doi: 10.1186/gb-2011-12-11-r114.

[20] Y. Assefa y S. A. Staggenborg, "Phenotypic Changes in Grain Sorghum Over the Last Five Decades", *J. Agron. Crop Sci.*, vol. 197, núm. 4, pp. 249-257, ago. 2011, doi: 10.1111/j.1439-037X.2010.00462.x.

[21] M. Wang *et al.*, "Evaluation of Sweet Sorghum as a Feedstock by Multiple Harvests for Sustainable Bioenergy Production", *J. Sustain. Bioenergy Syst.*, vol. 02, núm. 04, pp. 122-137, 2012, doi: 10.4236/jsbs.2012.24019.

[22] D. E. Ekefre *et al.*, "Evaluation of three cultivars of sweet sorghum as feedstocks for ethanol production in the Southeast United States", *Heliyon*, vol. 3, núm. 12, p. e00490, dic. 2017, doi: 10.1016/j.heliyon.2017.e00490.

[23] S. Kaur y G. S. Dhillon, "Recent trends in biological extraction of chitin from marine shell wastes: a review", *Crit. Rev. Biotechnol.*, vol. 35, núm. 1, pp. 44-61, ene. 2015, doi: 10.3109/07388551.2013.798256.

[24] M. I. Gómez *et al.*, "FISH of a maize sh2 -selected sorghum BAC to chromosomes of *Sorghum bicolor*", *Genome*, vol. 40, núm. 4, pp. 475-478, ago. 1997, doi: 10.1139/g97-063.

[25] J. S. Kim *et al.*, "Integrated karyotyping of sorghum by in situ hybridization of landed BACs.", *Genome*, vol. 45, núm. 402-412, 2002.

[26] G. Binelli *et al.*, "Similarity of maize and sorghum genomes as revealed by maize RFLP probes", *Theor. Appl. Genet.*, vol. 84, núm. 1-2, pp. 10-16, jun. 1992, doi: 10.1007/BF00223975.

[27] M. G. Pereira, M. Lee, P. Bramel-Cox, W. Woodman, J. Doebley, y R. Whitkus, "Construction of an RFLP map in sorghum and comparative mapping in maize", *Genome*, vol. 37, núm. 2, pp. 236-243, abr. 1994, doi: 10.1139/g94-033.

[28] A. M. Berhan, S. H. Hulbert, L. G. Butler, y J. L. Bennetzen, "Structure and evolution of the genomes Of*sorghum bicolor* And*zea mays*", *Theor. Appl. Genet.*, vol. 86, núm. 5, pp. 598-604, jun. 1993, doi: 10.1007/BF00838715.

[29] L. M. Chittenden, K. F. Schertz, Y. R. Lin, R. A. Wing, y A. H. Paterson, "A detailed RFLP map of *Sorghum bicolor* x *S. propinquum*, suitable for high-density mapping, suggests ancestral duplication of Sorghum chromosomes or chromosomal segments", *Theor. Appl. Genet.*, vol. 87, núm. 8, pp. 925-933, mar. 1994, doi: 10.1007/BF00225786.

[30] R. A. Ragab, S. Dronavalli, M. A. S. Maroof, y Y. G. Yu, "Construction of a sorghum RFLP linkage map using sorghum and maize DNA probes", *Genome*, vol. 37, núm. 4, pp. 590-594, ago. 1994, doi: 10.1139/g94-084.

[31] G.-W. Xu, C. W. Magill, K. F. Schertz, y G. E. Hart, "A RFLP linkage map of *Sorghum bicolor* (L.) Moench", *Theor. Appl. Genet.*, vol. 89-89, núm. 2-3, pp. 139-145, oct. 1994, doi: 10.1007/BF00225133.

[32] P. Dufour *et al.*, "Construction of a composite sorghum genome map and comparison with sugarcane, a related complex polyploid", *Theor. Appl. Genet.*, vol. 94, núm. 3-4, pp. 409-418, mar. 1997, doi: 10.1007/s001220050430.

[33] K. Boivin, M. Deu, J.-F. Rami, G. Trouche, y P. Hamon, "Towards a saturated sorghum map using RFLP and AFLP markers", *Theor. Appl. Genet.*, vol. 98, núm. 2, pp. 320-328, feb. 1999, doi: 10.1007/s001220051076.

[34] Y. Z. Tao, R. G. Henzell, y C. L. McIntyre, "Construction of a genetic map in a sorghumRIL population using probes from different sources and its comparison with other sorghum maps.", *Aust. J. Agric. Sci.*, vol. 49, pp. 729-736, 1998.

[35] P. K. Subudhi y H. T. Nguyen, "Linkage group alignment of sorghum RFLP maps using a RIL mapping population", *Genome*, vol. 43, núm. 2, pp. 240-249, mar. 2000, doi: 10.1139/g99-112.

[36] L. Kong, J. Dong, y G. E. Hart, "Characteristics, linkage-map positions, and allelic differentiation of *Sorghum bicolor* (L.) Moench DNA simple-sequence repeats (SSRs)", *Theor. Appl. Genet.*, vol. 101, núm. 3, pp. 438-448, ago. 2000, doi: 10.1007/s001220051501.

[37] B. I. G. Haussmann *et al.*, "Genomic regions influencing resistance to the parasitic weed *Striga hermonthica* in two recombinant inbred populations of sorghum", *Theor. Appl. Genet.*, vol. 109, núm. 5, pp. 1005-1016, sep. 2004, doi: 10.1007/s00122-004-1706-9.

[38] M. A. Menz, R. R. Klein, J. E. Mullet, J. A. Obert, N. C. Unruh, y P. E. Klein, "A high-density genetic map of *Sorghum bicolor* (L.) Moench based on 2926 AFLP®, RFLP and SSR markers", *Plant Mol. Biol.*, vol. 48, núm. 5-6, pp. 483-499, 2002, doi: 10.1023/A:1014831302392.

[39] J. E. Bowers *et al.*, "Development of a BAC based physical map of sorghum.", *Plant Anim. Genome IX Conf, San Diego*, 2001.

[40] G. Ji *et al.*, "Construction of a high-density genetic map using specific-locus amplified fragments in sorghum", *BMC Genomics*, vol. 18, núm. 1, p. 51, dic. 2017, doi: 10.1186/s12864-016-3430-7.

[41] L. Grivet, A. D'Hont, P. Dufour, P. Hamon, D. Roques, y J. C. Glaszmann, "Comparative genome mapping of sugar cane with other species within the Andropogoneae tribe", *Heredity (Edinb).*, vol. 73, núm. 5, pp. 500-508, nov. 1994, doi: 10.1038/hdy.1994.148.

[42] A. H. Paterson *et al.*, "Convergent Domestication of Cereal Crops by Independent Mutations at Corresponding Genetic Loci", *Science (80-.).*, vol. 269, núm. 5231, pp. 1714-1718, sep. 1995, doi: 10.1126/science.269.5231.1714.

[43] S. E. Anami, L. Zhang, Y. Xia, Y. Zhang, Z. Liu, y H. Jing, "Sweet sorghum ideotypes: genetic improvement of the biofuel syndrome", *Food Energy Secur.*, vol. 4, núm. 3, pp. 159-177, oct. 2015, doi: 10.1002/fes3.63.

[44] D. S. Multani, "Loss of an MDR Transporter in Compact Stalks of Maize br2 and Sorghum dw3 Mutants", *Science (80-.).*, vol. 302, núm. 5642, pp. 81-84, oct. 2003, doi: 10.1126/science.1086072.

[45] R. J. Milne, C. S. Byrt, J. W. Patrick, y C. P. L. Grof, "Are sucrose transporter expression profiles linked with patterns

of biomass partitioning in Sorghum phenotypes?", *Front. Plant Sci.*, vol. 4, 2013, doi: 10.3389/fpls.2013.00223.

[46] L. H. Pratt *et al.*, "Sorghum Expressed Sequence Tags Identify Signature Genes for Drought, Pathogenesis, and Skotomorphogenesis from a Milestone Set of 16,801 Unique Transcripts", *Plant Physiol.*, vol. 139, núm. 2, pp. 869-884, oct. 2005, doi: 10.1104/pp.105.066134.

[47] J. A. Bedell *et al.*, "Sorghum Genome Sequencing by Methylation Filtration", *PLoS Biol.*, vol. 3, núm. 1, p. e13, ene. 2005, doi: 10.1371/journal.pbio.0030013.

[48] D. G. Peterson, "Integration of Cot Analysis, DNA Cloning, and High-Throughput Sequencing Facilitates Genome Characterization and Gene Discovery", *Genome Res.*, vol. 12, núm. 5, pp. 795-807, may 2002, doi: 10.1101/gr.226102.

[49] C. D. Buchanan *et al.*, "*Sorghum bicolor*'s Transcriptome Response to Dehydration, High Salinity and ABA", *Plant Mol. Biol.*, vol. 58, núm. 5, pp. 699-720, jul. 2005, doi: 10.1007/s11103-005-7876-2.

[50] R. A. Salzman *et al.*, "Transcriptional Profiling of Sorghum Induced by Methyl Jasmonate, Salicylic Acid, and Aminocyclopropane Carboxylic Acid Reveals Cooperative Regulation and Novel Gene Responses", *Plant Physiol.*, vol. 138, núm. 1, pp. 352-368, may 2005, doi: 10.1104/pp.104.058206.

[51] E. Huntzinger y E. Izaurralde, "Gene silencing by microRNAs: contributions of translational repression and mRNA decay", *Nat. Rev. Genet.*, vol. 12, núm. 2, pp. 99-110, feb. 2011, doi: 10.1038/nrg2936.

[52] S. Ossowski, R. Schwab, y D. Weigel, "Gene silencing in plants using artificial microRNAs and other small RNAs", *Plant J.*, vol. 53, núm. 4, pp. 674-690, feb. 2008, doi: 10.1111/j.1365-313X.2007.03328.x.

[53] M. J. Axtell, J. A. Snyder, y D. P. Bartel, "Common functions for diverse small RNAs of land plants", *Plant Cell*, vol. 19, núm. 6, pp. 1750-1769, 2007, doi: 10.1105/tpc.107.051706.

[54] A. Katiyar, S. Smita, S. K. Muthusamy, V. Chinnusamy, D. M. Pandey, y K. C. Bansal, "Identification of novel drought-responsive microRNAs and trans-acting siRNAs from *Sorghum bicolor* (L.) Moench by high-throughput sequencing analysis", *Front. Plant Sci.*, vol. 6, jul. 2015, doi: 10.3389/fpls.2015.00506.

[55] N. B. Hamza, N. Sharma, *A. Tripathi*, y N. Sanan-Mishra, "MicroRNA expression profiles in response to drought stress in *Sorghum bicolor*", *Gene Expr. Patterns*, vol. 20, núm. 2, pp. 88-98, mar. 2016, doi: 10.1016/j.gep.2016.01.001.

[56] R. S. El Sanousi, N. B. Hamza, A. A. Abdelmula, I. A. Mohammed, S. M. Gasim, y N. Sanan-Mishra, "Differential Expression of miRNAs in <i>*Sorghum bicolor*</i> under Drought and Salt Stress", *Am. J. Plant Sci.*, vol. 07, núm. 06, pp. 870-878, 2016, doi: 10.4236/ajps.2016.76082.

[57] H. Yu *et al.*, "Identification of differentially expressed microRNA in the stems and leaves during sugar accumulation in sweet sorghum", *Gene*, vol. 571, núm. 2, pp. 221-230, oct. 2015, doi: 10.1016/j.gene.2015.06.056.

[58] N. Dhaka y R. Sharma, "MicroRNAs as targets for engineering biofuel feedstock Sorghum", *Indian J. Plant Physiol.*, vol. 22, núm. 4, pp. 484-492, dic. 2017, doi: 10.1007/s40502-017-0332-x.

[59] M. Battraw y T. C. Hall, “Stable transformation of *Sorghum bicolor* protoplasts with chimeric neomycin phosphotransferase II and β-glucuronidase genes”, *Theor. Appl. Genet.*, vol. 82, núm. 2, pp. 161-168, ago. 1991, doi: 10.1007/BF00226207.

[60] T. Hagio, A. Blowers, y E. Earle, “Stable transformation of sorghum cell cultures after bombardment with DNA-coated microprojectiles”, *Plant Cell Rep.*, vol. 10, núm. 5, ago. 1991, doi: 10.1007/BF00232571.

[61] A. M. Casas *et al.*, “Transgenic sorghum plants via microprojectile bombardment”, *Proc. Natl. Acad. Sci.*, vol. 90, núm. 23, pp. 11212-11216, dic. 1993, doi: 10.1073/pnas.90.23.11212.

[62] A. M. Casas *et al.*, “Transgenic sorghum plants obtained after microprojectile bombardment of immature inflorescences”, *Vitr. Cell. Dev. Biol. - Plant*, vol. 33, núm. 2, pp. 92-100, abr. 1997, doi: 10.1007/s11627-997-0003-0.

[63] Z. Zhao *et al.*, “Agrobacterium -mediated sorghum transformation”, núm. 1991, pp. 789-798, 2000.

[64] P. B. Devi y M. B. Sticklen, “In vitro culture and genetic transformation of sorghum by microprojectile bombardment”, *Plant Biosyst. - An Int. J. Deal. with all Asp. Plant Biol.*, vol. 137, núm. 3, pp. 249-254, ene. 2003, doi: 10.1080/11263500312331351491.

[65] Y. Tadesse, L. Sági, R. Swennen, y M. Jacobs, “Optimisation of transformation conditions and production of transgenic sorghum (*Sorghum bicolor*) via microparticle bombardment”, *Plant Cell. Tissue Organ Cult.*, vol. 75, núm. 1, pp. 1-18, 2003, doi: 10.1023/A:1024664817800.

[66] C. H. S. Carvalho *et al.*, “Agrobacterium-mediated transformation of sorghum: factors that affect transformation efficiency”, *Genet. Mol. Biol.*, vol. 27, núm. 2, pp. 259-269, 2004, doi: 10.1590/S1415-47572004000200022.

[67] Z. Gao, J. Jayaraj, S. Muthukrishnan, L. Claflin, y G. H. Liang, “Efficient genetic transformation of Sorghum using a visual screening marker”, *Genome*, vol. 48, núm. 2, pp. 321-333, abr. 2005, doi: 10.1139/g04-095.

[68] Z. Gao, X. Xie, Y. Ling, S. Muthukrishnan, y G. H. Liang, “Agrobacterium tumefaciens-mediated sorghum transformation using a mannose selection system”, *Plant Biotechnol. J.*, vol. 3, núm. 6, pp. 591-599, nov. 2005, doi: 10.1111/j.1467-7652.2005.00150.x.

[69] A. Howe, S. Sato, I. Dweikat, M. Fromm, y *T. Clemente*, “Rapid and reproducible Agrobacterium-mediated transformation of sorghum”, *Plant Cell Rep.*, vol. 25, núm. 8, pp. 784-791, ago. 2006, doi: 10.1007/s00299-005-0081-6.

[70] A. Raghuwanshi y R. G. Birch, “Genetic transformation of sweet sorghum”, *Plant Cell Rep.*, vol. 29, núm. 9, pp. 997-1005, sep. 2010, doi: 10.1007/s00299-010-0885-x.

[71] G. Liu y I. D. Godwin, “Highly efficient sorghum transformation”, *Plant Cell Rep.*, vol. 31, núm. 6, pp. 999-1007, jun. 2012, doi: 10.1007/s00299-011-1218-4.

[72] P. T. Do, H. Lee, M. Mookkan, W. R. Folk, y Z. J. Zhang, “Rapid and efficient Agrobacterium-mediated transformation of sorghum (*Sorghum bicolor*) employing standard binary vectors and bar gene as a selectable marker”, *Plant Cell Rep.*, vol. 35, núm. 10, pp. 2065-2076, oct. 2016, doi: 10.1007/s00299-016-2019-6.

[73] L. Z.-H. ZHU Li LI Gui-Ying, HE Kang-Lai, YUE Tong-Qing, ZHANG Jie, HUANG Da-Fang, “Introduction of Bt

cry1Ah Gene into Sweet Sorghum (*Sorghum bicolor* L. Moench) by Agrobacterium tumefaciens-Mediated Transformation", *Scientia Agricultura Sinica*, vol. 44, núm. 10. pp. 1989-1996, [En línea]. Disponible en: https://www.chinaagrisci.com.

[74] L. Lu *et al.*, "Development of marker-free transgenic sorghum [*Sorghum bicolor* (L.) Moench] using standard binary vectors with bar as a selectable marker", *Plant Cell, Tissue Organ Cult.*, vol. 99, núm. 1, pp. 97-108, oct. 2009, doi: 10.1007/s11240-009-9580-4.

[75] Z. Xin *et al.*, "Applying genotyping (TILLING) and phenotyping analyses to elucidate gene function in a chemically induced sorghum mutant population", *BMC Plant Biol.*, vol. 8, núm. 1, p. 103, 2008, doi: 10.1186/1471-2229-8-103.

[76] Y. Jiao *et al.*, "A Sorghum Mutant Resource as an Efficient Platform for Gene Discovery in Grasses", *Plant Cell*, p. tpc.00373.2016, jun. 2016, doi: 10.1105/tpc.16.00373.

Chapter 8

Current Situation and Future Outlook of Forest Biomass Production and Its Utilization in Japan

Takuyuki Yoshioka

Abstract

The current situation of forests and forestry as well as woody biomass utilization in Japan was described, and the future outlook for the use of forest biomass in Japan was presented. Many planted forests are now becoming mature, so the operational efficiency in forestry should be improved not only by the development of the forest infrastructure but also by the full mechanization of the logging system. The Kyoto Protocol adopted in 1997 promoted the energy utilization of waste woody biomass such as mill residues and wood-based waste materials, and the launch of the Feed-in Tariff scheme for renewable energy (FIT) in 2012 promoted the energy utilization of once-unutilized thinnings. In order to further expand the production of forest biomass and its utilization for energy, logging residues, small-sized trees, and short rotation woody coppices (SRWC) are promising. Thus, low-cost harvesting technologies should be developed as soon as possible, with reference to machines and systems operating in foreign countries where the utilization of such forest biomass is making steady progress.

Keywords: Forest Utilization, forest and forestry, woody bioenergy utilization, forest biomass harvesting, Japan

1. Introduction

The discipline of "Forest Utilization" in Japan was introduced with reference to the German "Forstbenutzung" at the end of the nineteenth century. The Forstbenutzung covers not only the processes of felling, processing, yarding/skidding, and transporting trees but also wood anatomy, wood physics, wood processing, and wood craft; therefore, it can be said that the Forstbenutzung pursues the rational utilization of forests and trees. In the case of Japan, the research subjects of wood anatomy, wood physics, wood processing, and wood craft went independent as the Wood Science after World War II; thus, the Forest Utilization has progressed to cover the fields of civil engineering, machine engineering, operational efficiency, and ergonomics in forestry. From this point of view, the current Japanese Forest Utilization is similar to the Forest Engineering developed in North America and the Forest Operations developed in Europe.

IntechOpen

Research and development (R&D) in the Forest Utilization is very important to make the Japanese forestry economically viable. Now, the stumpage price, i.e., the price of standing trees per m^3, Q' (USD/m^3), is expressed as:

$$Q' = (P - Q) \times \frac{n}{100} \tag{1}$$

where P (USD/m^3) is the market price of roundwood, Q (USD/m^3) is the logging cost, and n (%) is the utilization percentage to standing tree volume. Q' contains the costs of reforestation and tending; thus, the forestry itself cannot be economically justified when Q' is cheaper than a certain level. Then, Q is expressed as a function of yarding/skidding distance, x (m):

$$Q = a + b \times x + c \tag{2}$$

where a (USD/m^3) is the cost of felling and processing trees, b ($USD/m^3/m$) is the yarding/skidding cost per unit length of yarding/skidding distance, and c (USD/m^3) is the overhead cost. In order to increase the income of forest owners, Q' of Eq. (1) must be increased, that is, Q of Eq. (2) must be reduced; therefore, a and b of Eq. (2) must be diminished by developing efficient forestry machines as well as by improving logging methods, and x of Eq. (2) must be shortened by developing forest road networks. Improving the utilization percentage, n of Eq. (1), is also effective. Utilization of residual forest biomass such as logging residues, that is, tree tops and branches that are generated during limbing and bucking, and small-sized trees that are felled at pre-commercial stages can raise the utilization percentage, leading to increase in the income of forest owners.

In this chapter, with the aim of showing the R&D of forest biomass production and its utilization within the framework of the Japanese Forest Utilization, the current situation of forests and forestry as well as woody biomass utilization in Japan is described, and the future outlook for the use of forest biomass in Japan is presented.

2. Current situation of forests and forestry in Japan

The current total forest area in Japan is about 25 million ha. As shown in **Table 1**, in densely populated countries such as the USA, Canada, and Germany,

Region	Country	Land area [A]	Forest area [B]	B/A	Population
		(x 1,000 ha)	(x 1,000 ha)	(%)	(x 1,000)
Asia	Japan	36,450	24,958	68.5	126,573
North America	USA	916,192	310,095	33.8	321,774
	Canada	909,351	347,069	38.2	35,940
Central Europe	Germany	34,861	11,419	32.8	80,689
	Austria	8,244	3,869	46.9	8,545
Northern Europe	Sweden	41,034	28,073	68.4	9,779
	Finland	30,390	22,218	73.1	5,503

Table 1.
Comparison of land area, forest area, and population.

slightly more than 30% of national land area is covered with forest [1]. Japan has a population size similar to the abovementioned countries; however, forest occupies almost two-thirds of the national land, which is the same level as in sparsely populated Sweden and Finland. This can be attributed to the steepness of so much of the land area in Japan.

Figure 1 shows the changes in the breakdown of the Japanese forest area [2]. The total forest area has been constant for more than a half century. Historically, naturally regenerated forests were converted to planted forests. During and after World War II, quite large numbers of trees were felled and harvested, so afforestation was actively promoted from the 1950s to the 1970s in order to compensate. Ten million hectares of man-made planted forests, which equals almost 30% of the total national land area, was established in the 1980s.

On the other hand, after World War II, the more the Japanese economy grew, the worse the profitability of forestry became. Young people left rural areas, and then the self-sufficiency rate of wood decreased continuously. Thus, the incentive for afforestation was gradually diminished. As shown in **Figure 2**, consequently, the current age distribution of planted forest is extremely imbalanced, so that 65% of planted forests that are older than 45 years reach the time for being harvested finally [2].

Figure 3 shows the changes in the growing stock [2]. The growing stock keeps increasing mainly because of the dominance of young planted forests. The current average stem volume per 1 ha of planted forest, 324 m^3/ha, is similar to that of Austrian forests, 325 m^3/ha. This means that an operational efficiency equivalent to that of Austria might be expected if the forest infrastructure was well developed and the logging system was fully mechanized, but in practice, this is very difficult to achieve.

The changes in the wood supply and demand are shown in **Figure 4** [2]. As mentioned previously, the self-sufficiency rate continuously decreased in inverse proportion to the economic growth during the latter half of the twentieth century. The trade in roundwood was completely liberalized in 1964, more than a half

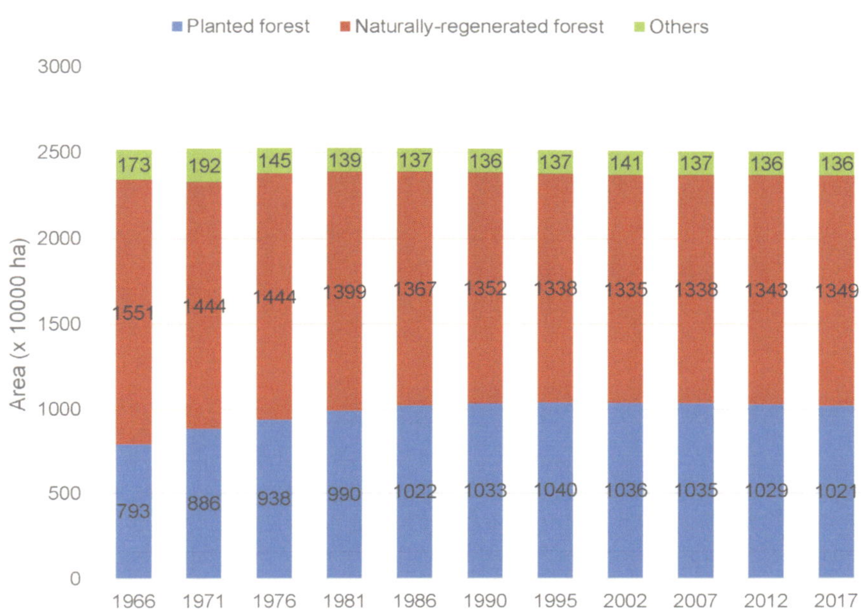

Figure 1.
Changes in the breakdown of the Japanese forest area.

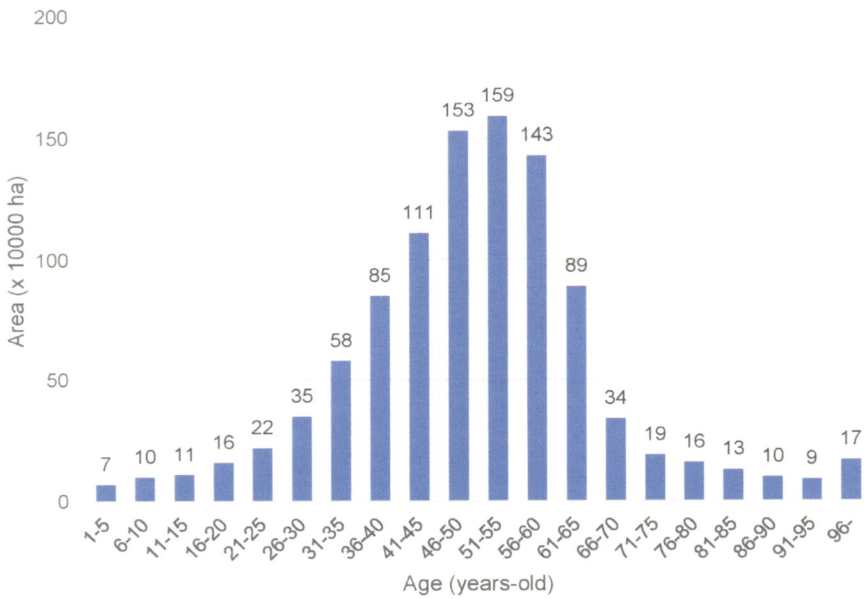

Figure 2.
Age distribution of planted forest (as of March 2017).

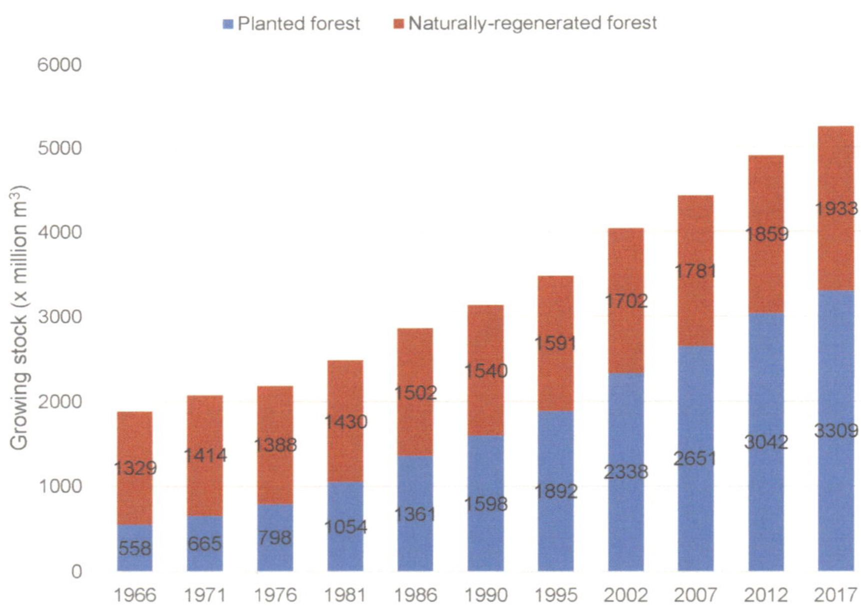

Figure 3.
Changes in the growing stock.

century ago. Since the middle of the 1990s, Japan has been in an economic slump, so the demand for wood is shrinking. On the other hand, planted forests are maturing, and thus, the supply of domestic wood is gradually increasing and the self-sufficiency rate itself is improving, the driving force for which will be explained later.

Figure 5 shows the number of workers that can be employed by 1 m^3 of standing Japanese cedar trees [3]. The number of workers is calculated by dividing the

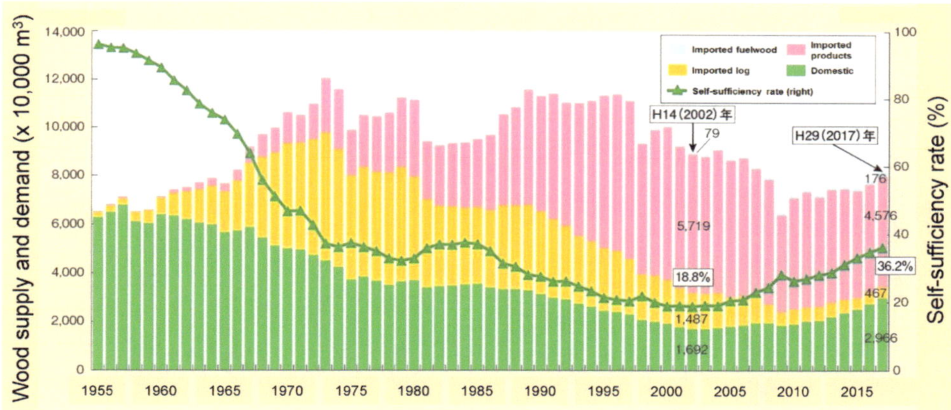

Figure 4.
Changes in the wood supply and demand.

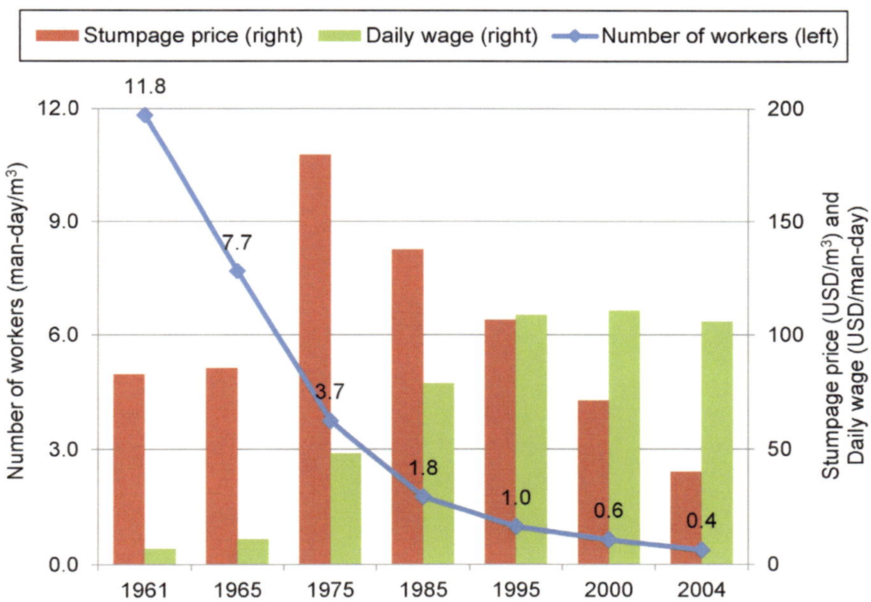

Figure 5.
Number of workers that can be employed by 1 m³ of standing Japanese cedar trees.

stumpage price of standing Japanese cedar trees by the daily wage of a forestry worker. It can be read from the graph that, even in Japan, labor-intensive work was possible during the 1960s.

The changes in the number of forestry workers are shown in **Figure 6** [2]. The number of forestry workers decreased by more than 100,000 in the past 35 years, and the percentage of aged workers (>65 years old) is relatively higher than that in other industries in Japan. On the other hand, forestry is the only industry in Japan in which the percentage of young workers (<35 years old) is increasing. The replacement of manual labor by mechanization in the limbing, bucking, and yarding/forwarding processes, but not in the felling process, seems to be contributing to this.

Figure 7 illustrates a typical mechanized logging system in Japan. The system is similar to that of Austrian mountainous areas. Nearly 10,000 advanced forestry

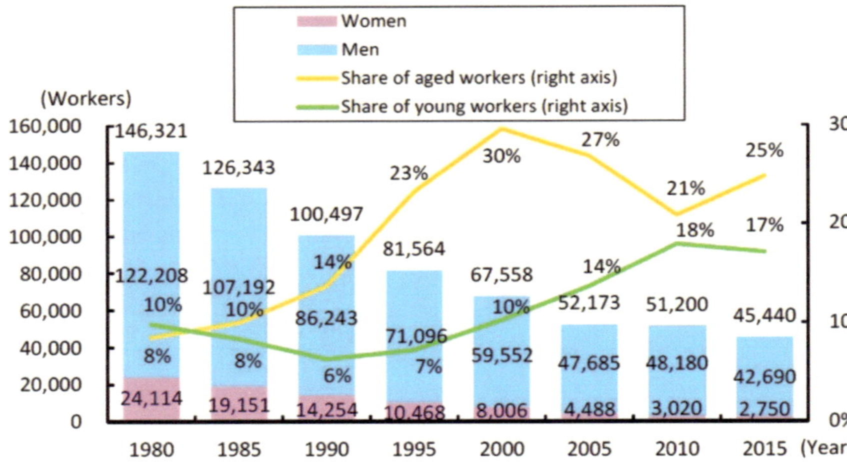

Figure 6.
Changes in the number of forestry workers.

Figure 7.
Typical mechanized logging system in Japan.

machines such as processors, tower yarders, and forwarders have been introduced, and 70% of the logs produced are processed by such forestry machines. However, there are not enough forest roads, so supplemental lower-grade operation roads have been constructed, and a forwarder uses them to haul logs to a forest road.

A comparison of productivity and logging costs is given in **Table 2** [4]. In Japan, the rate of operation of advanced forestry machines remains at a low level, which makes the productivity low and the logging cost high. It is said that, in Austria, intensive investments were made in the development of a forest road network in the 1960s, when the price of wood was relatively higher. Although the price of wood fell and the labor cost rose after that, the productivity was improved by mechanization. The forest road network density in Japan, 19.7 m/ha (13.1 m/ha for forest roads and 6.6 m/ha for operation roads), is less than a quarter of that in Austria, 89 m/ha (45 m/ha for forest roads and 44 m/ha for operation roads). The development of a forest road network is relatively delayed in Japan.

Country	Productivity (m^3/man-day)	Logging cost (USD/m^3)
Japan (final cutting)	4.00	57.7
Japan (thinning)	3.45	84.8
Sweden	30	11.8 (final cutting) 21.8 (thinning)
Austria	7–43	29.1–50.0

Table 2.
Comparison of productivity and logging costs.

3. Woody biomass utilization in Japan

After the Kyoto Protocol was adopted in 1997, renewable and carbon-neutral biomass attracted widespread attention for its potential as an ideal primary energy resource in a sustainable society. In 2001, the Japanese government officially defined biomass as one of the new energy resources in the "Law Concerning Special Measures for Promotion of the Use of New Energy" [5], and the government decided on the "Biomass Nippon Strategy" in 2002 [6]. As mentioned previously, forest resources are abundant in Japan, and thus, the energy utilization of woody biomass is expected to contribute to a revitalization of the forestry and forest products industries, which have long been depressed. The annual available amount of woody biomass resources is estimated to be 31.7 million dry-t/y [7], which has a calorific value of 634 PJ/y, corresponding to 2.8% of the national primary energy supply, 23.0 EJ/y, and woody biomass utilization was expected to promote the tending of planted forests, many of which were being neglected when the Biomass Nippon Strategy was adopted. This triggered the energy and material utilization of waste woody biomass such as mill residues (**Figure 8**), wood-based waste materials (**Figure 9**), and tree trimmings.

Figure 8.
Mill residue.

Figure 9.
Wood-based waste material.

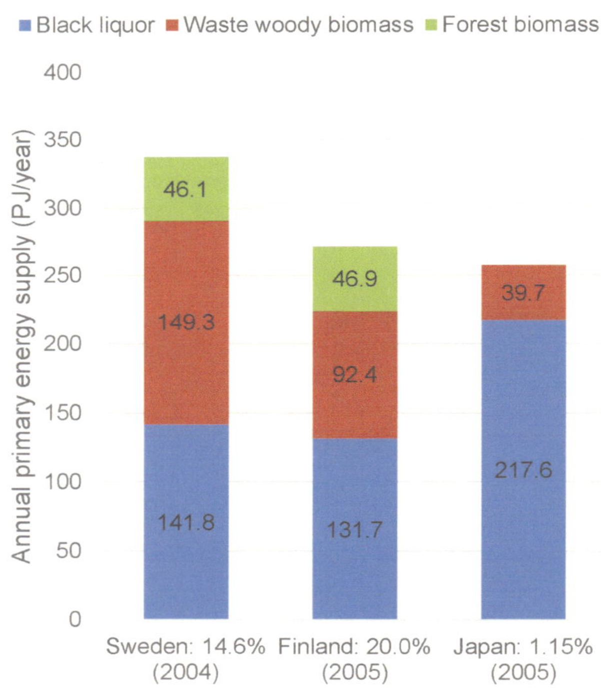

Figure 10.
Comparison of the woody bioenergy utilization around 2005 (Each percentage value means the share to the total domestic primary energy supply).

With respect to the actual situation of planted forests at that time, the thinning of largely established planted forests did not commonly take place because the trees were so small that there was little profitability on a business basis in thinning operations.

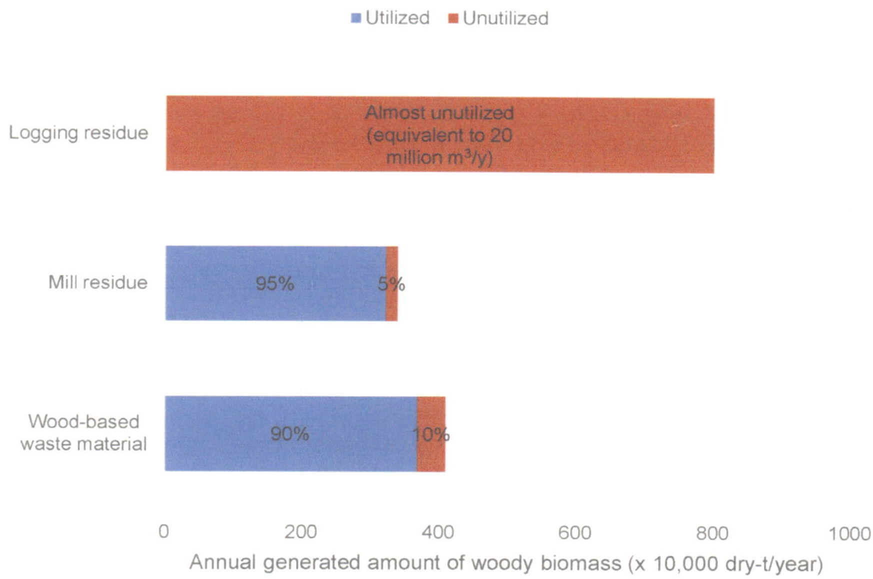

Figure 11.
Domestic woody biomass availability in 2010.

Figure 12.
Young planted forest contained large amounts of felled thinnings.

Figure 10 shows a comparison of the woody bioenergy utilization around 2005. From the point of view of the amount of woody bioenergy utilization itself, Japan was almost on a par with Sweden and Finland. This is because there has been a big pulp and paper industry in Japan, and the recycling of black liquor, that is, by-product from the kraft process when digesting pulpwood into paper pulp removing lignin, hemicelluloses, and other extractives from the wood to free the cellulose fibers, was promoted in the 1970s.

The domestic woody biomass availability in 2010 as estimated by the Japanese government is shown in **Figure 11** [8]. There was little available waste woody biomass such as mill residue or wood-based waste material, while logging residue went almost unutilized. It has been said since that time that logging residues should be utilized by developing dedicated harvesting machines as has been done in Sweden and Finland.

In these statistics, the term "logging residue" actually refers to unutilized thinning materials (**Figure 12**). In view of environmental conservation and global warming mitigation measures, huge amounts of subsidies were spent to fell trees for the purpose of thinning, but the felled trees were never harvested. Thus, the young planted forests contained large amounts of felled thinnings. So, the 20 million m^3/y of logging residue shown in **Figure 11** should have been classified as "forest biomass," most of which was composed of unutilized thinnings.

4. Launch of the FIT scheme 1 year after Fukushima

The Feed-in Tariff scheme for renewable energy (FIT) was launched in 2012, 1 year after the Fukushima nuclear disaster following the Great East Japan Earthquake, and the scheme increased the utilization of forest biomass. **Figure 13**

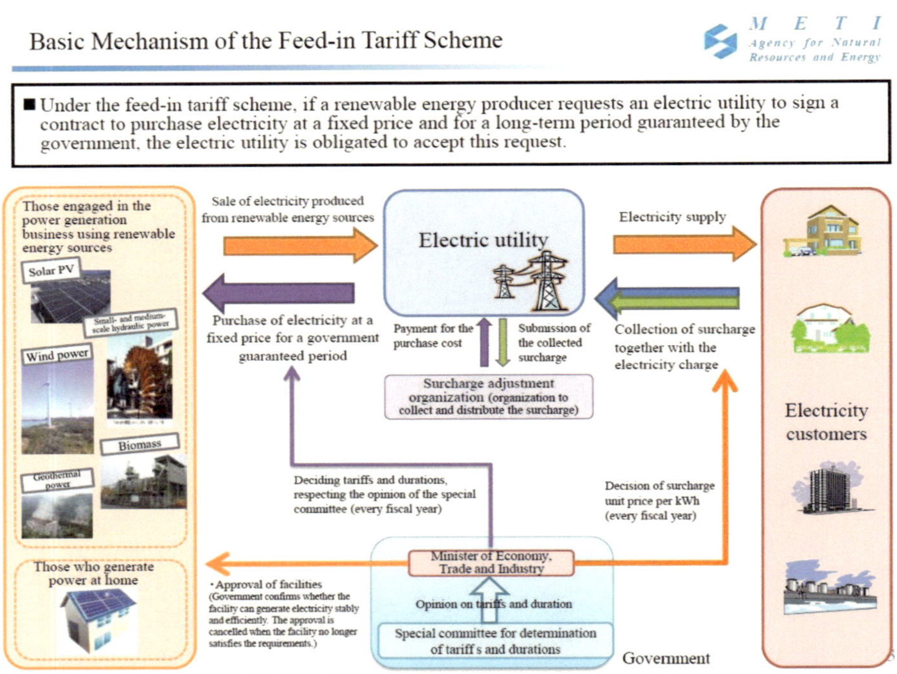

Energy source		Biomass				
Biomass type		Biogas	Wood fired power plant (Timber from forest thinning)	Wood fired power plant (Other woody materials)	Wastes (excluding woody wastes)	Wood fired power plant (Recycled wood)
Cost	Installation cost	3,920,000 yen/kW	410,000 yen/kW	410,000 yen/kW	310,000 yen/kW	350,000 yen/kW
	Operating and maintenance costs (per year)	184,000 yen/kW	27,000 yen/kW	27,000 yen/kW	22,000 yen/kW	27,000 yen/kW
Pre-tax IRR (Internal Rate of Return)		1%	8%	4%	4%	4%
Tariff (per kWh)	Tax inclusive	40.95 yen	33.60 yen	25.20 yen	17.85 yen	13.65 yen
	Tax exclusive	39 yen	32 yen	24 yen	17 yen	13 yen
Duration		20 years				

Figure 13.
Framework of the FIT.

shows the framework of the FIT [9]. In the case of biomass, the electric utilities have committed to buy the electricity derived from biomass at a higher price than the normal retail one for 20 years. However, this cost is passed down to the electricity consumers. Japanese public covered additional 18.8 billion US dollars in 2016 within the framework of the FIT. The price of electricity has already risen over 10% from the price before the implementation of the FIT.

With respect to the woody biomass resources for the FIT, "general wood" consists of mill residues and imported woods. From the point of view of power generation capacity, the construction project of a power generation plant which utilizes general wood as fuel accounts for the majority (**Figure 14**) [10]. There are plans to establish many large-scale plants with a power-generation capacity of more than 10 MW along seashores and to import a huge amount of woody biomass such as wood chips, wood pellets, and palm kernel shells (PKS).

Forest biomass utilization is also being done mainly in mountainous areas. The main source of forest biomass is the unutilized thinnings that were once abandoned in planted forests (**Figure 15**). Since the tariff on electricity derived from forest biomass is set to be hefty, the price of forest biomass as fuel is sometimes greater than that of forest biomass as pulpwood. In some areas, pulpwoods are also transported directly to power generation plants, and adjacent pulp mills located in such areas are obliged to import pulpwoods instead.

As a result, forest biomass utilization is rapidly increasing (**Figure 16**) [2]. The rapid increase in the utilized amount of forest biomass triggered by the FIT is the driving force behind the improvement of the self-sufficiency rate of wood.

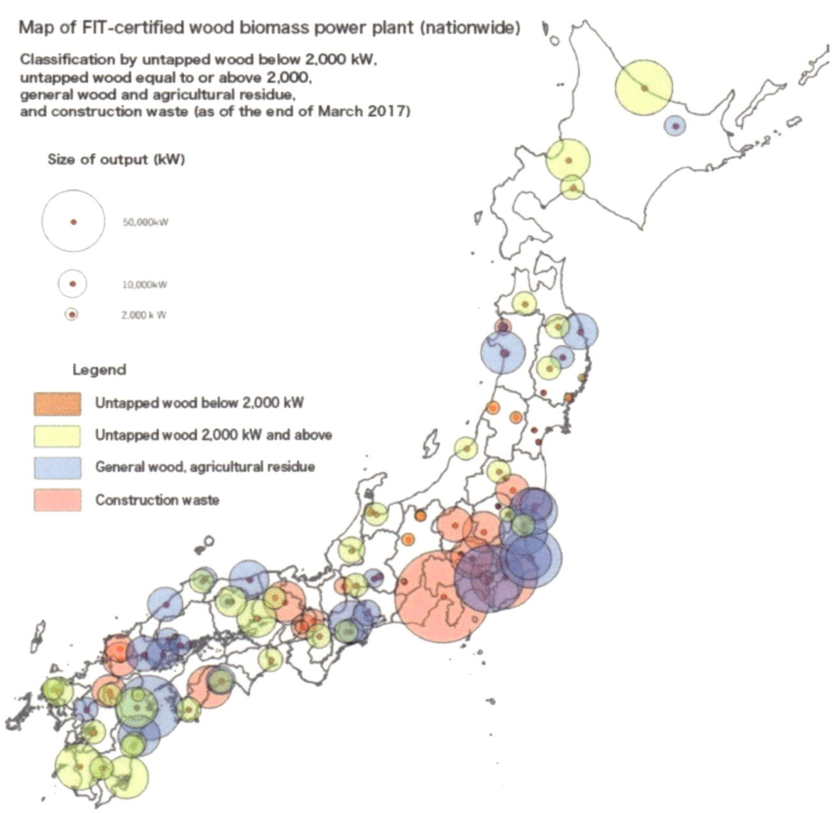

Figure 14.
Map of FIT-certified woody biomass power plants.

Figure 15.
Unutilized thinnings that were once abandoned in planted forests.

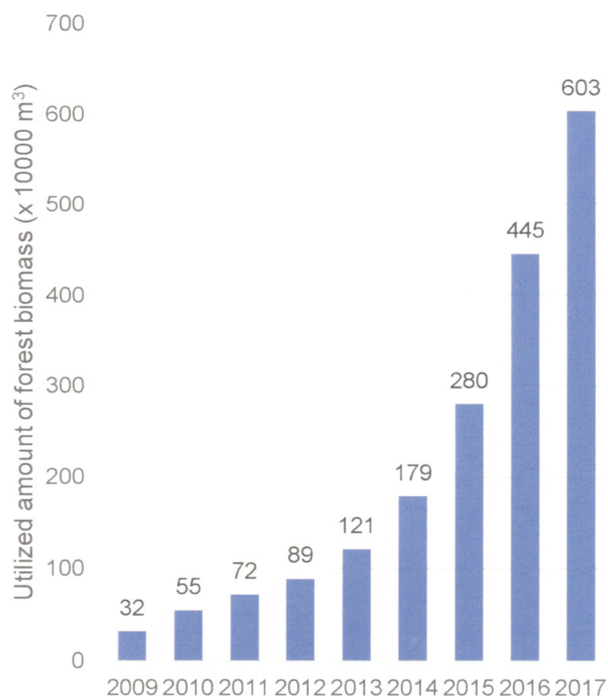

Figure 16.
Changes in the utilized amount of forest biomass.

The self-sufficiency rate of wood in 2017 as shown in **Figure 4** was 36.2%, but it goes down to 31.6% when the use of wood as fuelwood is excluded.

It is unclear whether unutilized thinnings can continue to be utilized as energy in the future. Planted forests are going to mature so that the value of the thinning material will increase and the amount available for energy use will decrease. On the other hand, "true" logging residues such as tree tops and limbs are not currently utilized much as energy sources. If this situation continues, the energy utilization of forest biomass might drop sharply after the completion of the FIT. Thus, a framework for the utilization of logging residues must be established as soon as possible.

5. Future perspectives on forest biomass in Japan

The use of the whole-tree logging system has increased through the spread of mechanization, such as the use of processors and harvesters. This situation makes logging residues easier to collect. So an efficient and low-cost harvesting, transporting, and chipping system for logging residues must be established. The author's

Figure 17.
Experimenting with a forwarder hauling of slashes.

Figure 18.
Comminution of logging residues with a tub grinder.

Figure 19.
Chipper-forwarder.

Figure 20.
Bundler.

research group experimented with the collection of logging residues by a forwarder (**Figure 17**) [11]. Comminution of logging residues was also investigated (**Figure 18**) [12], and the harvesting (collecting and comminuting) cost of logging residues was calculated as 76.0 USD/dry-t [13]. As compared to Sweden and Finland, where the energy utilization of forest biomass is making steady progress (see **Figure 10**), the calculated cost is relatively expensive, so that the development of dedicated machines such as the chipper-forwarder (**Figure 19**) [14] and bundler (**Figure 20**) [15] may be necessary [16].

The use of small-sized trees is also promising. The area covered by planted forests that have undergone final cutting and subsequent reforestation is now gradually increasing. Thus, a cleaning operation in young planted forests will be necessary 15–20 years from now, when the FIT will expire. So the development of

Figure 21.
Multi-tree feller-buncher.

Figure 22.
Harvesting small-sized trees with a truck-mounted multi-tree felling head.

efficient harvesting technology for small-sized trees will be necessary. An accumulative felling machine (**Figure 21**) may be effective [17, 18]. Harvesting small-sized trees with a truck-mounted multi-tree felling head was attempted (**Figure 22**), and the harvesting (felling, collecting, and comminuting) cost of small-sized trees was calculated as 99.4 USD/dry-t.

Figure 23.
Experiment of harvesting willow trees using a sugarcane harvester.

Short rotation woody coppices (SRWC) have a huge potential. Before and during World War II, an average of 50 million m^3/y of naturally regenerated forest was felled and harvested for energy use in the form of charcoal and firewood in Japan. The annual available amount of naturally regenerated broad-leaved trees used as SRWC is estimated to be 9 million dry-t/y [7]. The energy utilization of SRWC has already begun within the framework of the FIT. Moreover, the development of short rotation forestry in abandoned farmlands may be worth considering. Commercial willow plantations have been cultivated for bioenergy purposes in Sweden since the 1980s, and around 16,000 ha of short rotation willow plantations were established domestically from 1986 to 2000 [19]. In 2006, about 8,000 ha of the first commercial willow biomass crops in North America were started in upstate New York [20]. Growing and harvesting willow trees aimed at short rotation forestry was experimented with in northern Japan. A sugarcane harvester that was used in southern Japan was applied for harvesting willows during its agricultural off-season (**Figure 23**) [21]. The harvesting (growing, cutting, collecting, and comminuting) cost of SRWC was calculated as 136 USD/dry-t [22].

6. Conclusions

In this chapter, the current situation of forests and forestry as well as woody biomass utilization in Japan was described, and the future outlook for the use of forest biomass in Japan was presented. As a result, the following conclusions were drawn:

1. Many planted forests are now becoming mature, so the operational efficiency in forestry should be improved not only by the development of the forest infrastructure but also by the full mechanization of the logging system;

2. The Kyoto Protocol adopted in 1997 promoted the energy utilization of waste woody biomass such as mill residues and wood-based waste materials, and

then the launch of the FIT in 2012 promoted the energy utilization of once-unutilized thinnings;

3. In order to further expand the production of forest biomass and its utilization for energy, logging residues, small-sized trees, and SRWC are promising. Thus, low-cost harvesting technologies should be developed as soon as possible, with reference to machines and systems operating in foreign countries where the utilization of such forest biomass is making steady progress.

Acknowledgements

This chapter was financially supported in part by JSPS KAKENHI Grant Number JP20K06121.

Conflict of interest

The author declares no conflict of interest.

Author details

Takuyuki Yoshioka
Laboratory of Forest Utilization, Department of Forest Science, Graduate School of Agricultural and Life Sciences, The University of Tokyo, Tokyo, Japan

*Address all correspondence to: tyoshioka@fr.a.u-tokyo.ac.jp

References

[1] Forestry Agency. FY2018 Forest and Forestry Statistics Catalogue [Internet]. 2018. Available from: https://www.rinya.maff.go.jp/j/kikaku/toukei/youran_mokuzi2018.html [Accessed: 29 June 2020]

[2] Forestry Agency. FY2018 Annual Report on Forest and Forestry in Japan [Internet]. 2019. Available from: https://www.rinya.maff.go.jp/j/kikaku/hakusyo/30hakusyo/index.html [Accessed: 29 June 2020]

[3] Forestry Agency. FY2006 Annual Report on Forest and Forestry in Japan [Internet]. 2007. Available from: https://www.maff.go.jp/j/wpaper/w_rinya/h18/index.html [Accessed: 29 June 2020]

[4] Forestry Agency. FY2011 Annual Report on Forest and Forestry in Japan [Internet]. 2012. Available from: https://www.rinya.maff.go.jp/j/kikaku/hakusyo/23hakusyo/190411.html [Accessed: 29 June 2020]

[5] Yokoyama S. Present situation of Japanese biomass. Biomass and Bioenergy. 2005;**29**:303. DOI: 10.1016/j.biombioe.2005.06.001

[6] Anonymous. Appendix: Biomass-Nippon Strategy (Provisional Translation) Decided at the Cabinet Meeting, Government of Japan, 27 December 2002. Biomass and Bioenergy. 2005;**29**:375-398. DOI: 10.1016/j.biombioe.2005.06.002

[7] Yoshioka T, Hirata S, Matsumura Y, Sakanishi K. Woody biomass resources and conversion in Japan: The current situation and projections to 2010 and 2050. Biomass and Bioenergy. 2005;**29**:336-346. DOI: 10.1016/j.biombioe.2004.06.016

[8] Forestry Agency. FY2010 Annual Report on Forest and Forestry in Japan [Internet]. 2011. Available from: https://www.rinya.maff.go.jp/j/kikaku/hakusyo/22hakusho/190411.html [Accessed: 29 June 2020]

[9] Ministry of Economy, Trade and Industry. Feed-in Tariff Scheme in Japan [Internet]. 2012. Available from: https://www.meti.go.jp/english/policy/energy_environment/renewable/pdf/summary201207.pdf [Accessed: 29 June 2020]

[10] Japan Woody Bioenergy Association. Wood Biomass Energy Data Book 2018 [Internet]. 2018. Available from: https://www.jwba.or.jp/app/download/13159696992/Data+book+2018+%28English+version%29.pdf?t=1523342683 [Accessed: 29 June 2020]

[11] Yoshioka T, Iwaoka M, Sakai H, Kobayashi H. Feasibility of a harvesting system for logging residues as unutilized forest biomass. Journal of Forest Research. 2000;**5**:59-65. DOI: 10.1007/BF02762520

[12] Yoshioka T, Sakurai R, Aruga K, Nitami T, Sakai H, Kobayashi H. Comminution of logging residues with a tub grinder: Calculation of productivity and procurement cost of wood chips. Croatian Journal of Forest Engineering. 2006;**27**:103-114

[13] Yoshioka T, Kameyama S, Inoue K, Hartsough B. A cost breakdown structure analysis on the grinding of logging residues: A comparative study of Japanese and American operations. In: Proceedings of Grand Renewable Energy 2018: International Conference and Exhibition; 17-22 June 2018; Yokohama. Tokyo: Japan Council for Renewable Energy; 2018. 4 p. DOI: 10.24752/gre.1.0_216

[14] Spinelli R, de Francesco F, Eliasson L, Jessup E, Magagnotti N.

An agile chipper truck for space-constrained operations. Biomass and Bioenergy. 2015;**81**:137-143. DOI: 10.1016/j.biombioe.2015.06.017

[15] Andersson G. New technique for forest residue handling. In: Proceedings of FEG International Conference; 28-30 June 1999; Edinburgh. Bedford: Institution of Agricultural Engineers; 1999. p. 6

[16] Yoshioka T, Aruga K, Nitami T, Sakai H, Kobayashi H. A case study on the costs and the fuel consumption of harvesting, transporting, and chipping chains for logging residues in Japan. Biomass and Bioenergy. 2006;**30**:342-348. DOI: 10.1016/j.biombioe.2005.07.013

[17] Laitila J, Asikainen A, Nuutinen Y. Forwarding of whole trees after manual and mechanized felling bunching in pre-commercial thinnings. International Journal of Forest Engineering. 2007;**18**(2):29-39. DOI: 10.1080/14942119.2007.10702548

[18] Belbo H. Comparison of two working methods for small tree harvesting with a multi tree felling head mounted on farm tractor. Silva Fennica. 2010;**44**:453-464. DOI: 10.14214/sf.142

[19] Mola-Yudego B. Trends and productivity improvements from commercial willow plantations in Sweden during the period 1986-2000. Biomass and Bioenergy. 2011;**35**:446-453. DOI: 10.1016/j.biombioe.2010.09.004

[20] Volk TA, Luzadis VA. Willow biomass production for bioenergy, biofuels, and bioproducts in New York. In: Solomon BD, Luzadis VA, editors. Renewable Energy from Forest Resources in the United States. London and New York: Routledge; 2009. pp. 238-260

[21] Yoshioka T, Sugiura K, Inoue K. Application of a sugarcane harvester for harvesting of willow trees aimed at short rotation forestry: An experimental case study in Japan. Croatian Journal of Forest Engineering. 2012;**33**:5-14

[22] Yoshioka T, Inoue K, Hartsough B. Cost and greenhouse gas (GHG) emission analysis of a growing, harvesting, and utilizing system for willow trees aimed at short rotation forestry (SRF) in Japan. Journal of the Japan Institute of Energy. 2015;**94**:576-581. DOI: 10.3775/jie.94.576

Chapter 9

Location Analysis and Application of GIS in Site Suitability Study for Biogas Plant

Emmanuel C. Chukwuma, Chinenye F. Okey-Onyesolu, Daniel C. Anizoba and Joseph I. Ubah

Abstract

Proper livestock waste management and development of robust system for the treatment of the bio-waste has been emphasized and investigated by several searchers. Utilization of bio-waste for bio-energy production is advantageous for sustainable environment and socio-economic viewpoints. This study therefore is essential in providing critical strategy needed in situating bio-energy plants, consideration was made in the application of geospatial technology owing to it wide adoption and numerous advantages. Data for site analysis of biogas plant was obtained from GIS organizations and agency, the biomass generation and sites data was obtained from field survey. The biomass potential was based on paunch content generated in the various 43 abattoirs in the study area. The ArcGIS 10 software was used for all GIS operations and subsequent map production. The final suitability index map was obtained by overlaying the land use suitability map with the biomass spatial density layer. The suitable areas were divided into 4 classes: the Most Suitable, Highly Suitable, Moderate Suitable and Not Suitable. The study indicates that suitable sites are predominant in the East and central region of the study area, this study is essential in developing framework for siting biogas plant.

Keywords: biomass utilization, geospatial technology, location analysis, bio-energy plant, waste management

1. Introduction

Biogas technology is a renewable energy technique from which biogas is obtained from biomass by anaerobic digestion of substrates obtained from Industrial, agricultural and municipal wastes [1]. It is has been acclaimed as an appropriate technology and has received global massive attention recently, and has equally been recommended as a strategy to ease global energy and environmental problems [2]. The potential of biomass as energy source have been estimated by different experts and scientists, using various assumptions and scenarios. For instance, the European Biomass Association (AEBIOM) asserted that the European production of biomass based energy can be increased from the 72 million tones in 2004 to 220 Mtoe in 2020 [3]. As the global trend is advocating for a transition from fossil energy waste to Renewable Energy (RE) based on several socio-economic and environmental justification, the necessity to embark on a process that ensures that biogas plants are properly sited for energy production is inevitable [4–6].

Siting of biogas plants in strategic locations is a major means of combating some of the environmental challenges of bio-waste generation; that would also be convenient and economically advantageous [7]. One of the biggest barriers in utilizing bio-waste in several countries is the dispersion of livestock farms across a given geographical location. This often leads to generation of relatively small or inadequate bio-waste; also most farms lack the technical capability of operating a farm scale biogas plant. Therefore, based on technical feasibility and economic viability, centralized large scale biogas production has been advocated, however suitable location for the plant requires geospatial consideration and location modeling. Implementation of spatial information technologies such as remote sensing and GIS in addressing this issue have been receiving enormous attention recently, and has been described as appropriate methodology to be utilized in site selection and analysis for biogas plant [8, 9]. The application of GIS as an appropriate tool for site suitability analysis by several researchers is a strong indicator of its capability to resolve location issues [10–12]. This study therefore attempted to present logical framework that would serve as a guide in the process of identifying suitable sites for biogas plant using the power of geospatial technology.

2. Biogas plants for processing agricultural wastes

The biogas plants for processing agricultural wastes are considered as those plants which are utilized in processing feedstock that have agricultural origin. Common or notable feedstock types for this kind of plants are vegetable residues and vegetable, animal manure and slurries, dedicated energy crops, sewage sludge, various residues from food industries etc. [1]. The design and technology of biogas plants differ from one country to another, it depends on the climatic conditions and national frameworks, energy or biomass availability and affordability. Based on sizes, functions and locations, agricultural AD plants can be classified as [3]:

a. Family scale biogas plants (very small scale)

b. Farm scale biogas plants (small or medium to large scale)

c. Centralized/ joint co-digestion plants (medium to large scale)

3. The major driving factors in adoption of biogas technology

The benefits of biogas have increased the adoption rate amongst many countries. Biogas for instance can be utilized after treatment in numerous applications such as provision of electricity and heat generation, connection to the natural gas grid, or as biofuel in vehicles [13]. Several studies on biogas technology potentials and their adoption in various developing countries have shown that biogas technology has high potential in developing countries as an alternative energy source [15–17]. Further discussion on the benefits based on energy production, and environmental concern is discussed below:

3.1 Energy interest

After water, energy is a vital resource required for development globally, the demand is high especially in developing countries. To prevent further growth on the impact of climate change in most developing countries, it is argued that the energy market in such regions should be based on renewable sources [14]. Biogas as a renewable energy source, have received widespread adoption in Europe, they produce clean energy from organic waste and have framework for increase production [6]. The adoption of biogas technology based on energy interest in developing countries though faced with several barriers, critical information on viability of adoption of biogas technology is receiving attention currently by various researchers in these countries [7, 15–17].

3.2 Environmental interests

Biogas technology is suitable for recycling various types of biomass waste, however the operational conditions and parameters are of interest since this can inhibit the microbial operations necessary for the digestion and may restrict the end use of digested biomass as a bio-fertilizer [18]. It is of major environmental interest that nutrient losses from residue of anaerobic digestion process is minimized when applied to the soil. Additionally, anaerobic treatment in a biogas plant reduces odor nuisances during slurry application, this is a welcome issue by many farmers [19]. Furthermore, biogas technology tend to reduce the potency of greenhouse gases (GHG), by capturing and combusting methane during anaerobic digestion and utilization as cook gas or energy source, GHG has been reported to be 21 times as much atmospheric warming potential as CO_2 [20, 21]. Therefore, global attraction and adoption of biogas technology is hinged on the numerous environmental benefits associated with the technology.

4. Application of GIS in land/location suitability analysis

GIS is a computer system that enhances capturing, checking, storing, integrating, analyzing, and displaying data about the earth in a spatially referenced way. The application of GIS is cross-disciplinary and has been adopted and applied in several fields of science and engineering. Land suitability analysis involves the search for the best location of one or more facilities to support some desired function, it involves the process to ascertain whether the land resource is appropriate for some desired uses and to determine its suitability level. Land suitability refers to the inherent suitability of the land for some specific, persistent uses. Examples range from retail site location to the location of multiple ambulance dispatch points. Land suitability analysis or assessment is achieved by considering certain land features such as hydrology, geography, topography, geology etc. in an enabled environment using GIS technologies [22].

GIS technology has been applied by various researchers in biogas plant location suitability assessment. Few instances are stated here: To develop potential for collective biogas plants in France, GIS was used to geo-reference the bio-resources potential and also to locate the optimal sites at both national and regional scales for the country [23], the final suitability map from the study provided constraint map and the energy potential grid, synthesized in the form of a raster GIS file. The study provided successfully a suitability map using precise geo-location of farms obtained through the analysis of aerial photographs and Landsat imagery used in the identification of crop residues. Similarly, a study was done to determine the optimal locations, sizes and number of biogas plants in Southern Finland, this study analyzed the spatial distribution and amount of potential biomass feedstock for bio-methane production for the study area [24]. In addition to numerous studies on site suitability analysis for biogas plant, a GIS based spatial data mining approach was adopted to model the optimal location for distributed biomass power generation facilities in Tumkur district, India [25], these studies asserted that GIS is an appropriate and recognized spatial tool for location analysis [23–25].

5. Location and site suitability considerations

Suitable areas for biogas plants are evaluated to avoid close proximity to land features and uses that may be sensitive to the characteristics of utility-scale power production and waste streams. Sensitive land features to be avoided include surface water, wetlands, forests, public lands, highly sloped lands, and developed residential areas, with acceptable slopes of 14° or fewer [26]. Location and site suitability consideration for biogas plant includes:

Slope: Slope is usually derived from the Digital Elevation Model (DEM) of the study area. The lower the slope value, the flatter the terrain; similarly, the higher the slope value, the steeper the terrain, higher slopes is associated with higher cost of land preparation and grading, while an acceptable slope of 14% and below has been suggested [26]. The slope derived from the DEM is used to produce the slope layer.

Biomass availability within the region: Biomass availability is critical for the sustenance of biogas production plants, it is characterized with year to year variability and is subject to non-homogeneity [27]. Several studies have therefore embarked on assessment of biomass availability as preliminary study for biogas plant siting. The biomass resource potential is usually estimated using geospatial technology, the amount of agricultural biomass in the form of crop residues, wood and forestry products, animal waste production etc. are usually estimated.

Density of biomass production: clustering of biomass waste source are usually of economic advantage, it offers several benefits such as ability to maximize labor skills and professionals, reduction in transportation costs, easy access to common infrastructure for production and biomass resource. Areas with high clustering or density of biomass resource are usually considered potential areas for centralized biogas plants.

6. Application of GIS in siting biogas plant for abattoir biomass

An application of GIS in determination of suitable sites is applied here using a typical biomass data collected in Anambra state of Nigeria for demonstration purposes.

6.1 The study area and data collection

The study area for the application of GIS in biogas plant location analysis is Anambra State in South-east of Nigeria. The area is located between Latitudes $05^{0}42^{1}56^{11}$N and $06^{0}45^{1}34^{11}$N and Longitude $06^{0}37^{1}30^{11}$E and $07^{0}25^{1}30^{11}$E, it is surrounded by several states such as Delta State in the West, Imo State and Rivers State in the South, Enugu State in the East and Kogi State in the North. Data used in the study include primary and secondary data collected from various organizations, literatures and individuals. The flow chart of the research method is shown in **Figure 1**.

Figure 1 present the research procedure which usually involves data collection and analysis. Data for site analysis of biogas plant was obtained from GIS organizations and agency, the biomass generation and sites data was obtained from field survey. The figure also shows the major layers used for the study. The primary data was collected from field survey through visit to slaughter houses in the study area, the biomass potentials was determined using Global Positioning System (GPS) receiver (Handheld GARMIN 76S), the biomass potential was based on paunch content generated in the various 43 abattoirs in the state. The data on the biomass generation capacity of all the abattoirs is shown in the appendix. GPS was used to obtain the geographical co-ordinate of the biomass source for geo-coding in data analysis. ArcGIS 10 software was used for all GIS operations and subsequent map production. The GIS-based thematic maps used for the production of the suitability map include political boundary map layer, Land Use and Land Cover (LULC) map obtained from the Landsat imagery, slope layer and the biomass layer. The land use map for the study area, was generated from the Landsat-7 ETM+ image and then classified to extract the different land uses of the study area using maximum likelihood classification algorithm. The residential and reserved areas in the study area were termed constrained areas and were exempted in the suitability map, the constrain map was made considering several environmental and socio-economical factors. The data types, format, scale and sources is shown in **Table 1**.

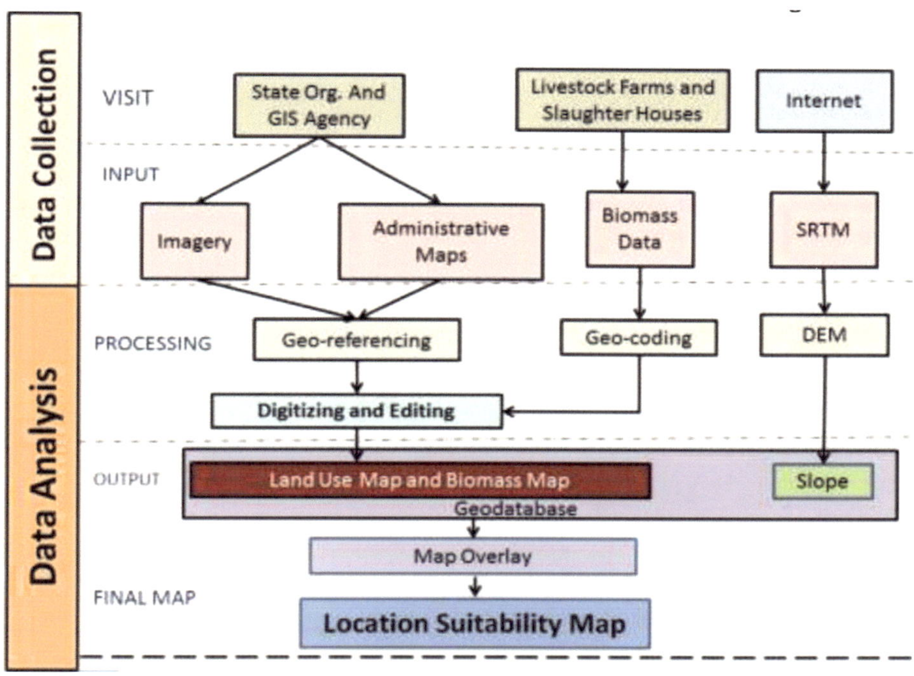

Figure 1.
Flowchart of geospatial modeling for biogas plant.

List of data	Format/Map scale	Source
Land use map	Landsat-7 ETM + imagery	National Remote Sensing Centre, Jos.
Administrative layer map	Arcinfo shapefile/digitalized from 1:50,000 scale map	Survey deparment, Ministry of Lands, Survey and Town Planning, Awka
Biomass source Location map	Arcinfo shapefile	Field trip to farms, Use of GPS
DEM	SRTM imagery at 30 m resolution of 2000	(http://www.landcover.org)

Table 1.
List of data sources and format.

The site suitability was assessed using Eq. 1 below:

$$S_i = \sum W_i X_i \quad (1)$$

Where Wi is the weighted score of the factor, Xi is the suitability rank of the factor, S is the suitability value for each factor and i is factor i.

7. Result and discussion

Data collection is critical in geospatial analysis, typical data used for this study include Land Use map classified from Remote sensing data source; geo-coded data of the biomass sources, this is usually in the form of point data, obtained using GPS device. The acquired data points and the value was transferred into Arcmap environment of ArcGIS and processed into vector map for the site suitability analysis. The result of the various data analysis and modeling of suitable sites for the biogas plant by excluding unwanted areas identified in the constrain map and overlaying the thematic maps is fully discussed below.

7.1 Land Use classification map

Based on prior knowledge of land use of some geographical co-ordinates points, six classes were categorized. They are agriculture areas, barren/open land, dense forest, sand, urban land, and water body. The classified land use map is shown in **Figure 2**.

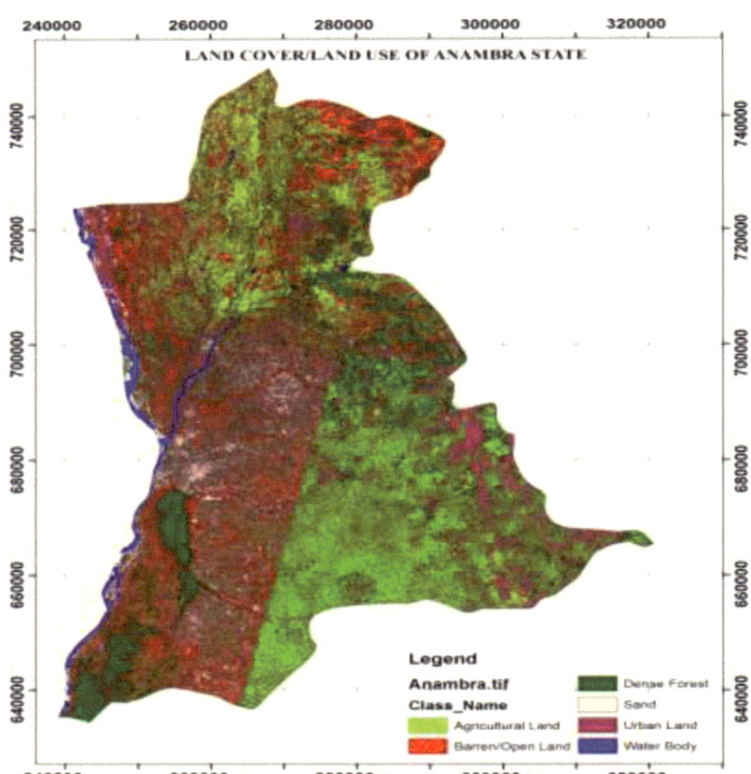

Figure 2.
Land cover and land use suitability map.

Assessment of classification accuracy was carried out using the scatter plot analysis in statistical toolbar in ArcGIS 10. All the training data were highlighted to compare the scatter plot of the six classes to each other. The classes were examined to detect any form of overlap (these are classes having different pixel value). This shown in **Figure 3**, the statistics for the training data was also used to assess the accuracy of the classification. The statistic are usually organized for each training area. The covariance statistics evaluates the correlation between the values of different bands and were adequate for the study.

The areas covered by each class of the LULC shows that urban land occupies 36.52% which represent 506896km^2 of landmass of the overall LULC while the least class is the sand class followed by water body, these feature classes occupies landmass of 13080km^2 and 14000km^2 respectively. The overall classification accuracy determined is 83%. The Table of LULC classification of Anambra State, area occupied in km^2 and percentage occupies by the various classes is shown in Table 4.1 (**Table 2**).

7.2 Biomass data of abattoir waste generating centers and map

One of the basics for site analysis of biogas plant is the biomass potential density; **Figure 4** shows the abattoir biomass data indicating areas where the bio-wastes are generated across the study area. **Figure 4** shows the towns and villages in the State that has abattoir centers. From the Figure, there are no abattoir centres in the

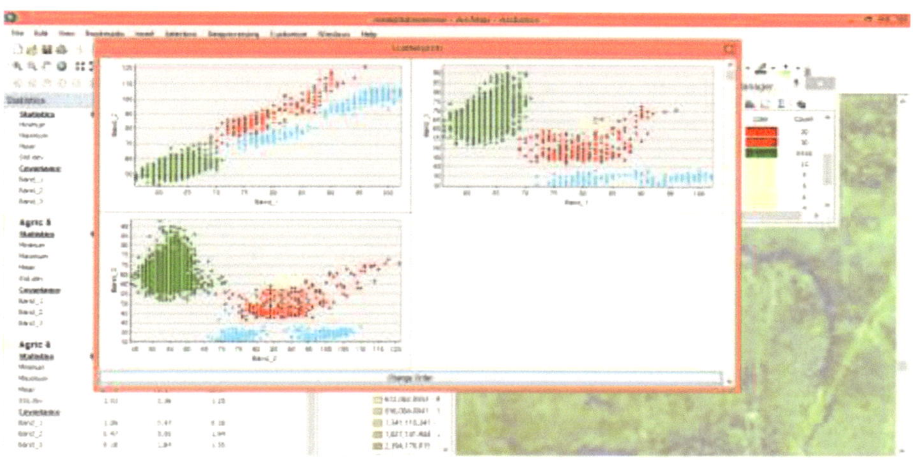

Figure 3.
Scatter plot of image classification in ArcGIS.

Class	Area(km)	Percentage (%)
WATER BODY	14000	1.00
SAND	13080	0.94
DENSE FOREST	257999	18.59
URBAN LAND	506896	36.52
AGRICULTURAL LAND	356430	25.68
BARREN/OPEN LAND	239400	17.25
TOTAL	1387805	100

Table 2.
Area occupation of various LULC classes.

Northern parts of the state. There are highest number of abattoir biomass within the central region of the state, and scarce generating centers towards the extreme of the Southern parts of the state.

Though the energy source from biomass in the study area could be utilized for biogas production and clean energy, this would be better to the current practices of burning wood in most homes and even in the abattoirs during meat processing operations. Biomass wastes generated during the slaughter of these animals includes blood, wastewater, ruminal content etc. The concentration of abattoirs in some of the areas in the study area is probably connected to the high population density of these areas. Since meat demand logically increases with increase in population. Areas with high concentration of biomass generation and clustering are best sites when considering proximity of waste sources as major criteria for bio-energy plant location. **Figure 4** was used to produce a vector map of biomass spatial density layer, used in the final suitability analysis.

7.3 Land use suitability map

The slope was derived from the digital elevation model of the study area. The slope of the study area was obtained through the slope function in spatial analyst tool in

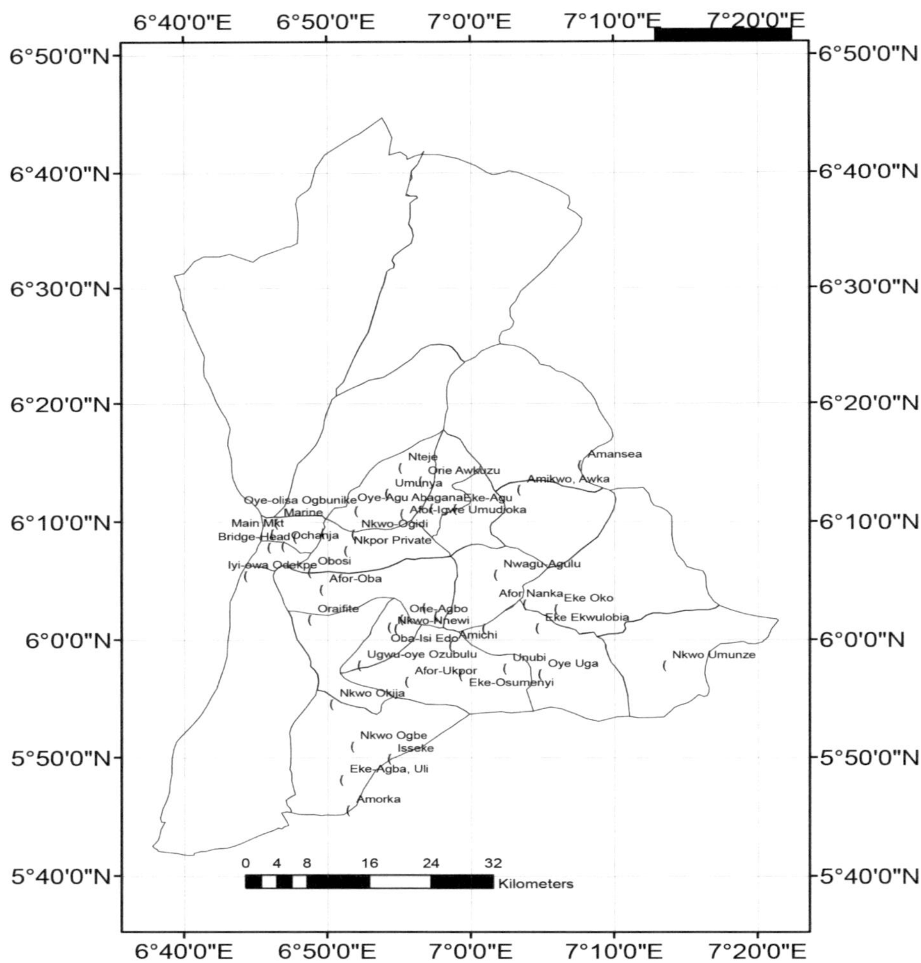

Figure 4.
Biomass potential density of the study area.

ArcGIS 10. The Slope command takes an input surface raster and calculates an output raster containing the slope at each cell, **Figure 5** shows the slope of the study area.

The output slope raster is in percent (percent rise) as shown in the figure.

The red color in the map represents areas in the study area that are less than 8%, thus they represent suitable locations to site biogas plant. Areas shown in the map within the range of 62–89% signifies high sloped areas, and should be avoided, this is of economic importance in siting and building of structures. Other land features such as hill-shade view for maximizing sunlight effect for increase in temperature for the biogas plant was created. Elevation layer was created from DEM to avoiding flood occurrence, all the layers was classified into 5 classes. The most suitable were sites with higher elevation, lower slope, and higher hill-view values. All the layers and land use layer obtained from Satellite imagery were overlaid using in ArcGIS using Weighted Overlay Tool. The result of such geospatial operation is the land use suitability map (not shown).

7.4 Final suitability map

The final suitability index map was obtained by overlaying the land use suitability map with the biomass spatial density layer. The output is shown in **Figure 6** below. The suitable areas were divided into 4 classes- the Most Suitable, Highly Suitable, Moderate Suitable and Not Suitable. **Figure 6** shows that the suitable sites are

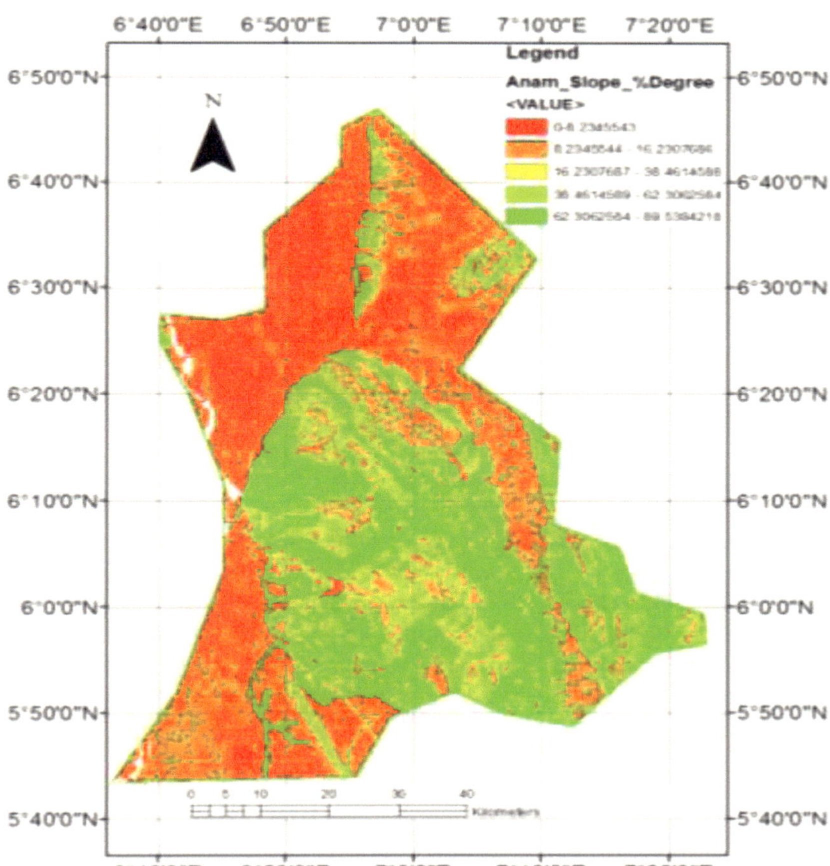

Figure 5.
Slope map of the study area.

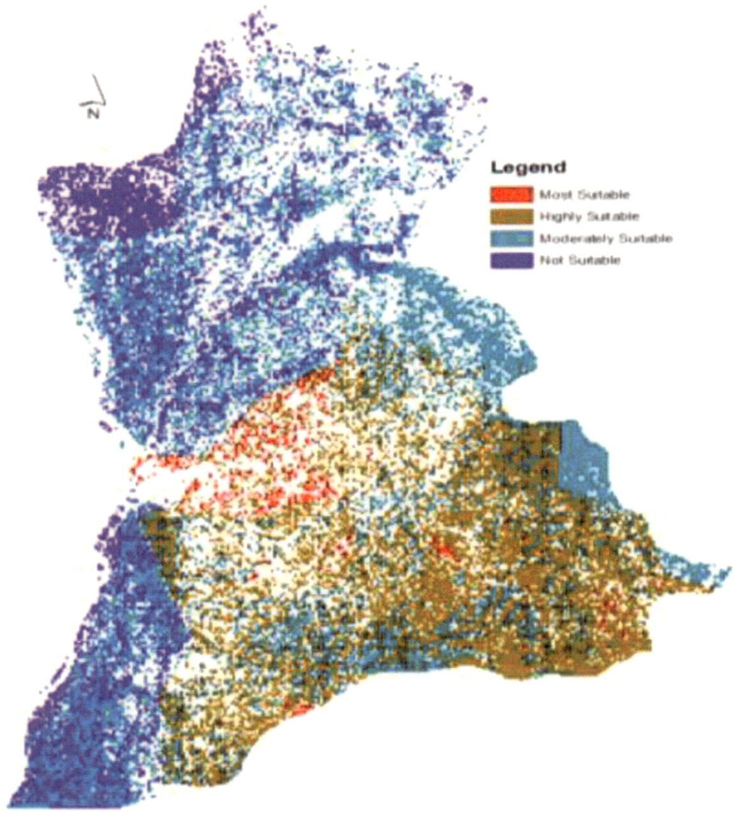

Figure 6.
Final suitability map. Source: [28].

predominant in the East and central region of the study area. Extraction of these areas based on the selected criteria indicates the power of GIS in extracting useful geospatial information for geographical data analysis.

8. Conclusion

This paper presents an application of GIS for site selection of bio-energy plant in Anambra State of Nigeria. Several environmental criteria and socio-economic factors were considered and used to obtain the land use suitability index map and biomass spatial density map. The final suitability index map was obtained by overlaying both maps. The suitable areas were divided into 4 classes- the Most Suitable, Highly Suitable, Moderate Suitable and Not Suitable and were predominant in the East and central region of the study area. This study therefore is a veritable framework tool in assessing and selecting suitable sites for siting of biogas treatment facilities.

Appendix

S/N	Locations	Aver. No. of cows Slaughtered daily	Amount Paunch (Kg/yr)
1	Nkwo Igboukwu	5	61137.5
2	Eke Ekwulobia	8.5	103933.75
3	Oye Uga	10	122275

S/N	Locations	Aver. No. of cows Slaughtered daily	Amount Paunch (Kg/yr)
4	Nwagu-Agulu	8	97820
5	Amikwo, Awka	18	220095
6	Amansea	23	281232.5
7	Afor-Igwe Umudioka	6.5	79478.75
8	Ugwu-oye Ozubulu	8	97820
9	Oraifite	3	36682.5
10	Nkwo-Ogidi	14.5	177298.75
11	Obosi	16	195640
12	Nkpor Private	5	61137.5
13	Nkpor	15	183412.5
14	Afor-Oba	7	85592.5
15	Afor-Nnobi	17	207867.5
16	Eke-Awka Etiti	35	427962.5
17	Eke-Agba, Uli	5	61137.5
18	Amorka	6	73365
19	Nkwo Ogbe	10	122275
20	Nkwo Okija	5	61137.5
21	Isseke	4	48910
22	Oye-Agu Abagana	6	73365
23	Eke-Agu	4	48910
24	Nkwo-Nnewi	10.5	128388.75
25	Orie-Agbo	3	36682.5
26	Oba-Isi Edo	14.5	177298.75
27	Amichi	4.5	55023.75
28	Afor-Ukpor	2	24455
29	Osumenyi Slaughter House	3	36682.5
30	Unubi Slaughter House	0.5	6113.75
31	Iyi-owa Odekpe	2	24455
32	Ochanja	70	855925
33	Bridge-Head	11.5	140616.25
34	Marine	26	317915
35	Ugwunabamkpa	1.5	18341.25
36	Main Mkt	20	244550
37	Afor Nanka	2	24455
38	Eke Oko	8	97820
39	Nkwo Umunze	4	48910
40	Nteje	13	158957.5
41	Oye-olisa Ogbunike	52.5	641943.75

S/N	Locations	Aver. No. of cows Slaughtered daily	Amount Paunch (Kg/yr)
42	Umunya	65	794787.5
43	Orie Awkuzu	12.5	152843.75
	Total	**565.5**	**6914651.25**

Source: [28].

Author details

Emmanuel C. Chukwuma[1,2]*, Chinenye F. Okey-Onyesolu[3], Daniel C. Anizoba[2] and Joseph I. Ubah[2]

1 Department of Agricultural and Biosystems Engineering, Landmark University, Omu-ara, Kwara State, Nigeria

2 Department of Agricultural and Bioresources Engineering, Nnamdi Azikiwe University, Awka, Nigeria

3 Department of Chemical Engineering, Nnamdi Azikiwe University, Awka, Nigeria

*Address all correspondence to: ec.chukwuma@unizik.edu.ng

References

[1] Achinas S, Jan G, Euverink W. A. Technological overview of biogas production from biowaste. Engineering. 2017;3:(3)299-307.

[2] Gómez D, Ramos-Suárez JL, Fernández B, Muñoz E, Tey L, Romero-Güiza M, Hansen F. Development of a modified plug-flow anaerobic digester for biogas production from animal manures. Energies. 2019;12: 2628.

[3] Al Seadi T, Rutz D, Prassl H, Köttner M, Finsterwalder T, Volk S, and Janssen R. Biogas Handbook. Esbjerg, Denmark: University of Southern Denmark. 2008

[4] Pérez-Orozco R, Patiño D, Porteiro J, Míguez JL. Novel test bench for the active reduction of biomass particulate matter emissions. Sustainability. 2020;12:422

[5] Hosseini KE, Azizi A, Lakeh BA, Hafez H, Elbeshbishy E. Comparison of liquid and dewatered digestate as inoculum for anaerobic digestion of organic solid wastes. Waste Manag. 2019;87:228-236.

[6] Blades L, Morgan K, Douglas R, Glover S, De Rosa M, Cromie T, Smyth B. Circular biogas-based economy in a rural agricultural setting. Energy Procedia. 2017;123:89-96

[7] Hren R, Petrovič A, Čuček L, Simonič M. Determination of various parameters during thermal and biological pretreatment of waste materials. Energies. 2020;13:2262.

[8] Chukwuma EC, Okey-Onyesolu FC, Ani KA, Nwanna EC. GIS bio-waste assessment and suitability analysis for biogas power plant: A case study of Anambra state of Nigeria. Elsevier Journal of Renewable Energy. 2021;163:1182-1194. https://doi.org/10.1016/j.renene.2020.09.046.

[9] Huang H, Li Q, Zhang Y. Urban residential land suitability analysis combining remote sensing and social sensing data: a case study in Beijing, China. Sustainability. 2019;11:2255. https://doi.org/10.3390/su11082255.

[10] Sadeghi M, and Karimi M. GIS-based solar and wind turbine site selection using multi-criteria analysis: Case study Tehran, Iran. Tehran's Joint ISPRS Conferences of GI Research, SMPR and EOEC. 2017;7-10

[11] Jahangeer A, Parrya JA, Ganaiea S A, Bhatb MS. GIS based land suitability analysis using AHP model for urban services planning in Srinagar and Jammu urban centers of J&K, India. Journal of Urban Management. 2019;7:46-56.

[12] Chukwuma EC. Facility location allocation modelling for bioenergy system in Anambra state of Nigeria: integration of GIS and location model. Journal of Renewable Energy. 2019;141:460-467.

[13] Rosén T, Ödlund L. System perspective on biogas use for transport and electricity production. Energies. 2019;12:4159.

[14] Plöchl M, and Heiermann M. Biogas farming in Central and Northern Europe: A strategy for developing countries. Agricultural Engineering International: the CIGR Ejournal. Invited Overview. 2006;8:8

[15] Sarker SA, Wang S, Adnan KM, Sattar NM. Economic feasibility and determinants of biogas technology adoption: Evidence from Bangladesh. Renewable and Sustainable Energy Reviews. 2020;123:109766

[16] Suberu MY, Bashir N, and Mustafa MW. Biogenic waste methane emissions and methane optimization

for bioelectricity in Nigeria. Renewable and Sustainable Energy Reviews. 2013;25:643-654

[17] Meryem SS, Ahmad SS, Aziz N. Evaluation of biomass potential for renewable energy in Pakistan using LEAP model. Int J Emerg Trends Eng Dev. 2013;1:243-9

[18] Abdullah Nsair A, Cinar SO, Ayah Alassali A, Qdais, HA, and Kerstin Kuchta K. Operational parameters of biogas plants: A review and evaluation study. Energies. 2020;13:3761. Doi:10.3390/en13153761.

[19] Hjort-Gregersen, K. Centralised biogas plants -integrated energy production, waste treatment and nutrient redistribution facilities. Danish Institute of Agricultural and Fisheries Economics. 1999

[20] Wang Q, Thompson E, Minghao L. Anaerobic digester systems (ADS) for multiple dairy farms: A GIS analysis for optimal site selection. Energy Policy. 2013;61:114-124

[21] EPA (United States Environmental Protection Agency). Waste transfer stations – A manual for decision making. 2010. Available from <http://www.epa.gov/garbage/pubs/r02002.pdf>.

[22] Manlun Y. Suitability analysis of urban green space system based on GIS. A thesis submitted to the international institute for geo-information science and earth observation enschede, The Netherlands. 2003

[23] Bioteau T, Boret F, Tretyakov O, Béline F, Balynska M, and Girault R. A GIS-based approach for optimizing the development of collective biogas plants. Global assessment for organic resources and waste management. Rennes France. 2012;88-95.

[24] Höhn JE, Rasi S, and Rintala J. A geographical information system (GIS) based methodology for determination of potential biomasses and sites for biogas plants in Southern Finland. Applied Energy. 2014;113:1-10

[25] Kaundinya DP, Balachandra P, Ravindranath NH, Ashok V. A GIS (geographical information system)-based spatial data mining approach for optimal location and capacity planning of distributed biomass power generation facilities: A case study of Tumkur district, India. 2013

[26] Ma JN, Scott S, degloria, and A. Lembo. Siting analysis of farm-based centralized anaerobic digester system for distributed generation using GIS, biomass and bioenergy. 2005;28(6):591-600

[27] Beccali, M., Pietro, C., Vincenzo, A., and Vincenzo, F. Assessment of bioenergy potential in Sicily: A GIS-based support methodology. Biomass and Bioenergy. 2009: 33:79-87.

[28] Chukwuma EC, Chukwuma GO, and Orakwe LC. GIS suitability analysis for anaerobic treatment facility for slaughter houses in Anambra State of Nigeria. Science Domain, Archives of Current Research International. 2016: 4(4): 1-10.

Chapter 10

Silvopastoral Systems for Energy Generation

Cid Marcos Gonçalves Andrade, Sérgio Inácio Gomes, Mauro Antônio da Silva Sá Ravagnani and Eugênia Leandro Almeida

Abstract

The silvipastoral systems are characterized by the association between tree crops, pastures and animals and can also constitute an efficient and sustainable means of supplying forest biomass for energy purposes such as electric, mechanical and thermal energy generation. It is an unconventional energy alternative and the evaluation of the energy potential offered by this productive system depends on several factors, such as management techniques, forest species, silvipastoral system characteristics and the design of the conversion and energy utilization process. In this context, it was developed a mathematical model to determine the energy efficiency of silvipastoril production system integrated with a cogeneration system for the production of thermal, mechanical and electrical energy. It can be concluded that these results are advantageous in relation to the conventional modalities of energy generation, taking into account the prices of electricity practiced in the market.

Keywords: cogeneration of energy, thermoeconomic analysis, exergoeconomic cost, exergoeconomic efficiency, silvipastoral system, modeling, simulation

1. Introduction

The procedures for the production and use of energy resources are the center of concern in the contemporary world, which requires the establishment of a more harmonious relationship between issues related to climate, energy, the environment and society [1].

According to the bibliographic review made by [2], several studies on future perspectives on the contribution of biomass to the global energy supply have reached very different conclusions. To exemplify, there are studies that indicate projections for the year 2050, below 100 EJ/year, while others, indicate them above 400 EJ/year. The major reason for it is that the parameters used are very uncertain, and subject to widely different opinions.

In any case, biomass of forest origin is a potential renewable resource, which can be planned and used as an energy alternative in view of the need to diversify the energy matrix [3].

According [4], energy is essential for individuals and populations to escape from poverty and move onto a path of greater well-being, security and prosperity. In view of this, a strategy is planned with a view to promoting pathways with the supply of energy to meet basic needs with the promotion of more modern and innovative ways of using biomass to generate income and reduce poverty, a strategy that is associated with other better objectives, such as the management, protection and improvement of productive ecosystems and landscapes, greater the use of sustainable and renewable bioenergy, which will mitigate climate change.

Consistent with the prospects for insertion of forest biomass for a more sustainable future, the silvopastoral system is an agroforestry modality with great potential [3].

According to [4], silvopastoral systems are agroforestry systems characterized by the association of tree crops, pastures and animals, constituting an efficient means of promoting the sustainable use of land.

According to [5], the commercial livestock activity is the main factor of deforestation in the world, with several negative environmental and social impacts.

Forest restoration, on the other hand, can increase soil productivity and fertility [6]. It can also improve the infiltration of water and its preferential drainage flow, since the trees in the pasture system reduce runoff in the face of greater rainfall intensities [7], making the silvopastoral system, a modality of high interest.

Many studies have demonstrated the environmental and economic benefits that can be obtained with the use of silvopastoral systems in agricultural activity.

Among others, we can pontuate [8], that evaluated the impacts of pasture afforestation systems on livestock activity in relation to meat quality.

Other studies have analyzed carbon stocks [9], soil quality [10] and the influence of grazing on the decomposition of tree stumps and roots [11].

In this way, silvopastoral systems have greater biodiversity and offer more environmental services when compared to conventional livestock systems. It can also offer environmental and economic benefits with the addition of a sustainable forest biomass production system and income generation for farmers [12].

Studies also shows that the configuration of tree planting has interactions with the environment, generating impacts on productivity, environmental characteristics and the soil, such as its hydrological properties [13].

In this context, this work aims to develop a mathematical model capable of making the thermoeconomic evaluation of a silvopastoral system for energy purposes.

The proposed simulation model is based on three principles, which are:

1. Mass balance;

2. Energy balance;

3. Thermoeconomic balance;

For the development of the work, cost factors and productive characteristics of silvopastoral systems, appropriate for the region of the Sandstone Caiuá, northwest of the State of Paraná, Brazil, will be considered.

It is noteworthy that the proposed system includes the silvopastoral system associated with a cogeneration process in which the biomass of forest origin is used as raw material. Cogeneration is defined as the production of two forms of energy simultaneously using a single fuel. The most common example is the use of a single thermal source for the production of thermal and electrical (or mechanical) energy.

The specificities and characteristics of the cogeneration system to be used is not part of the present work, it only assesses its efficiency.

1.1 Mass balance

The mass balance is based on the principle of conservation of mass, that is, the amount of mass that enters a process is equal to the amount of mass that comes out of it.

Therefore, as a starting point, it is necessary to quantify the average biomass produced annually by the silvopastoral system for energy purposes.

Thus, the proposed simulation starts from considering the configuration or spatial arrangement of the tree plantation for the silvopastoral system to be evaluated.

A silvopastoral system can consist of different arboreal spatial arrangements. It is a factor of great relevance when it is intended to carry out studies aimed at analysis of silvopastoral systems, since it is directly associated with the productivity of forest biomass, pasture and livestock.

The productivity of forest biomass is the result of many factors and variables, such as its edaphic characteristics, the water regime and soil nutrients, among many others.

However, the avaluation of the biomass productivity as a function of the spacing or arrangement of the trees without changing their density in the occupied area was the object of study by [14]. The authors concluded that the so-called "edge effect" has an influence on the growth of biomass.

The initial density of planting and the characteristics of the tree on growth, wood density and anatomical properties for the forest species were the object of study by [15].

According to the research by [16], it is important to assess the influence of the spatial arrangement of the silvopastoral system with regard to the quantity and quality of light and its effects on the production and chemical composition of the pasture.

This is a productive aspect relevant to livestock activity and compared the effect for a group of spatial arrangements. It was found that the 3.0 m x 2.0 m spacing offered the largest increase in dry matter production, but the denser spacing offered improvements in the composition of forages.

Thus, the spatial arrangement to be evaluated considering rows of trees with a width of 3 m x 2 m according [16], but any other arrangements can be simulated.

A certain area of pasture with the silvopastoral system can be characterized in terms of spatial arrangements, to be used in the mathematical modeling by defining five variables, which are:

1. area (in m^2);
2. distance between ranks, D_r (in m);
3. number of rows of trees in a rank, N_f (units);
4. distance between trees in a row, d_a (in m);
5. distance between rows, d_f (in m)

In our simulations, each species and each productive arrangement can be calculated using the matrix form of the equations. However, the present purpose is to present the fundamental mathematical relationships that are employed.

So, the density (or quantity) of trees in the silvopastoral system, depends on the spatial arrangement of afforestation in the planted area. It varies according to number of rows in a rank, the distances between rows, trees in a row or ranks, according to Eq. (1), expressed in (trees per hectare or trees per area).

$$number\ of\ trees\ in\ an\ area = \frac{N_f\ x\ area\ in\ (m2)}{d_a\ x\ \left(D_r + \left(N_f - 1\right)\ x\ d_f\right)} \tag{1}$$

The average annual volumetric increase in forest biomass, expressed in (m3/(ha. year)), is a statistical data on the average productivity of each forest species referred to a reference silvicultural system for the location or region.

So, the first variable to be considered as a reference in the analysis of the simulation will be the "total gross productivity of forest biomass" harvested in the monoculture system of silviculture with a forest species in a given area under the standard spatial arrangement (3 m x 2 m), at the end of a planting cycle.

So, based on the "total gross productivity of forest biomass", which consists of the final harvest of a known number of trees per hectare in conventional forestry, after a too known number of years, from planting to harvest, it is possible to obtain the average annual gross productivity per cultivated area (tonnes/(ha.year)).

Considering the basic density of dry forest biomass of that species, in (g/cm3) and the humidity factor on a wet basis in (%), we can calculate the average annual volumetric increase in forest biomass, I_{ma}, according to Eq. (2), expressed in (solid m3/(ha.year)).

$$I_{ma} = \frac{biomass\ moisture\ (wetb\%)}{basic\ density\ (g/cm3)} \ x\ \frac{gross\ productivity\ (tonnes)}{time\ planting\ to\ harvest\ (years)} \tag{2}$$

As seen by the studies by [14]. the "edge effect" has an influence on the growth of biomass as a function of the spacing or arrangement of the trees. So, it is possible to predict the establishment of a factor influencing the silvopastoral arrangement, (dimensionless) and the productivity of dry forest biomass, P_{mf}, obtained in a total cultivated área, in (ha), according to Eq. (3), expressed in (tonnes/year).

$$P_{mf} = silvopastoral\ factor\ x\ \frac{number\ of\ trees\ per\ hectare\ in\ the\ silvipastoral\ system}{number\ of\ trees\ per\ hectare\ in\ the\ conventional\ pasture}\ x\ I_{ma}\ x\ basic\ density\ x\ area \tag{3}$$

So, based on the mass balance, it is possible to estimate the average mass rate per forest species, referring to the average hourly flow of forest biomass, B_f, in (kg/h), which depends on the estimated average annual production of biomass, P_{mf} and the annual operating time, in (hours/year), according to Eq. (4), expressed in (kg/hour).

$$B_f = \frac{1000\ x\ annual\ production\ of\ biomass, P_{mf}\ (tonnes/year)}{annual\ operating\ time\ (hour)} \tag{4}$$

1.2 Energy balance

Like the mass balance, the energy balance is based on the principle of energy conservation, that is, energy cannot be created or destroyed, but transformed. Therefore, the energy entering the process must also be equal to the energy leaving it.

To understand the difference in terms of physical and quantitative meanings used in this simulation, it is necessary to explain the difference between energy and exergy.

According [17], to determine energy efficiency or performance in an open system it is necessary to make the mass balance and the energy balance that goes in and out control volume, constituting the thermal balance. However, the thermal balance does not provide the real values, since the full conversion of energy is considered without having to there, energetic destruction.

The exergetic method, however, allows to analyze the quality of the process in which the heat turns to work. Allows you to calculate energy losses, such as capacity carrying out work on the part of heat or steam and discovering its causes.

According to [18], exergy depends on the state of the fluid being considered and the state of the environment present.

According to the author, exergy of a system is defined as the maximum work capacity that can be performed by the compounds of the system in a reference environment. Therefore, exergy is defined as being the maximum possible useful work to be obtained by a flow of energy under conditions imposed by the surrounding environment.

It is also noteworthy that energy is conserved in any system or process. The energy cannot be destroyed, while exergy can be destroyed or lost according to [19].

Exergy incorporates concepts from first and second law of thermodynamics, but in real systems, exergy is never conserved. The analysis of a process or system through simultaneous use of the first and second laws of thermodynamics is, therefore, associated with the concept of exergy, or efficiency, or useful energy. According to [19] is therefore, the exergy and not the energy that can be valued as a merchandise.

So, based on the energy balance, it is possible to estimate the average exergy rate of entry of a process of energy transformation (cogeneration system) on a wet basis of the forest biomass.

The combustion of biomass itself implies the loss of its chemical exergy. Even if the efficiency in the boiler, for example, is high, say 90%, a good part of the exergy is lost in it.

According to [20, 21], the thermal exergy made available to the thermal cycle can be obtained according to the moisture content, the chemical composition contained in the forest biomass and the specific chemical exergy of each type.

A practical difficulty is to know the chemical composition of forest species and the relationship between the chemical exergy of biomass and the exergy effectively released in its complete combustion to be applied in the simulation. Thus, the exergy ratio to be released to the thermal cycle by each forest species used and its superior calorific value, according to the information provided by [22].

The superior and inferior calorific power of dry forest biomass of each forest species are expressed in (MJ/kg). They are tabulated and used in the simulation model in the form of a dimensionless exergy relationship for each species of forest biomass on a dry basis, R_{ex}.

An approximate value of R_{ex} is assigned, whenever the inferior calorific power of a given forest species is not available.

So, the rate of average exergy offered by each species of forest biomass on a dry basis at the beginning of the thermal cycle, E_{xibs}, will be given by the Eq. (5), expressed in (MJ/kg).

$$\text{dry } exergy\ IN = rate\ exergy\ x\ upper\ calorific\ value \tag{5}$$

The biomass moisture impacts its exergy and a standardization can be established based on [22], which outlined an experimental curve approximation.

For practical purposes, an equation can be used to estimate the average exergy rate offered by each species of forest biomass on a wet basis at the entrance of the thermal cycle, according to Eq. (6), expressed in (MJ/kg). Humidity factor is the dimensionless relationship between humidity and exergy of each forest species.

$$\text{wet } exergy\ IN = humidity\ factor\ x\ \text{dry } exergy\ IN \tag{6}$$

Therefore, the overall average exergy of the wet cycle thermal input, E_{xibm}, can be obtained based on the specific exergy on a wet basis, according to Eq. (7), expressed in (MJ/kg).

$$medium\ \text{wet}\ exergy\ IN = summation\ of\ all\ w\text{et}\ \text{exergy}\ IN \quad (7)$$

The effective input power of the cogeneration process, can be calculated by the Eq. (8), expressed in (MW).

$$input\ power = \frac{medium\ \text{wet}\ exergy\ IN\ x\ flow\ of\ biomass, B_f}{3600} \quad (8)$$

The difference between the output power and the input power can be understood as destroyed exergy.

As already mentioned, a considerable part of the loss of exergy in thermal power plants occurs in the boiler and not in condensers, which promote rejection of heat.

Therefore, an efficient energy use project must be planned for the use of the portion of thermal energy, whose utilization rate will depend on the characteristics of this energy use process.

According to [23], cogeneration may have better energy efficiency when compared to conventional energy conversion, since the thermal energy produced is underutilized or can be best used.

Thus, two concepts of exergetic efficiency can be defined, the efficiency of the cogeneration system, η_{COG}, and the efficiency of the electricity generation system, η_{EE}.

Once the energy efficiency of the cogeneration system is known, η_{COG}, the effective output power of the cogeneration process will be given by Eq. (9), expressed in [MW]. It represents the power available primarily in the form of thermal energy with additional capacity to perform mechanical work

$$output\ power = \eta_{COG}\ x\ input\ power \quad (9)$$

In the same way, we can calculate the active power of the electricity generation, in [MW], also as a function of the input power according to Eq. (10).

$$electricity\ active\ power = \eta_{EE}\ x\ input\ power \quad (10)$$

Therefore, the available power of thermal generation with possibilities of use can be obtained by the Eq. (11), expressed in [MW].

$$available\ thermal\ power = output\ power - electricity\ active\ power \quad (11)$$

The output power, in turn, can be understood as having three components, which would be:

1. Thermal losses;

2. Usable thermal exergy (or useful thermal exergy);

3. Usable kinetic exergy (in mechanical movement);

According [24], exergy may be associated with work or heat transfer. In the cogeneration system, kinetic exergy is associated with work transfer and thermal exergy is associated with heat transfer.

The kinetic exergy portion can be almost entirely converted into electrical energy, since its conversion efficiency is close to 100%, saved by reduced losses of exergy from a flow of mechanical energy (rotor of a generator in movement) that is converted into electrical energy, [23].

The power conversion rate at the input exergy to electricity is generally not much higher than 1/3 (it is, $\eta_{EEmax} \cong 33\%$).

However, the same does not happen from the point of view of temperature, that is, of thermal energy, on which it must be considered that the exergy of a thermal flow must be calculated according to the variation of the water temperature in relation to the environment, going from an initial temperature to a final temperature, in [Kelvin], according to what [23] called "temperature factor exergetic", according to Eq. (12), expressed in [%].

$$temperature\ factor\ exergetic = 1 - \frac{initial\ temperature}{final\ temperature} \quad (12)$$

Still according to [23], the temperature factor exergetic must be multiplied by the available thermal power produced from the outlet to obtain the useful thermal exergy flow (or useful thermal power), according to Eq. (13), expressed in [MW].

$$useful\ thermal\ power = temperature\ \text{factor exergetic}\ x\ available\ thermal\ power \quad (13)$$

Eq. (14) calculates the useful output power, expressed in [MW]

$$useful\ output\ power = electricity\ active\ power + useful\ thermal\ power \quad (14)$$

As the output and input powers refer to the annual average, the relationship between these corresponds to the overall exergetic efficiency of the cogeneration system, corresponding to the same relationship between the useful annual average exergy estimates, according to Eq. (15).

$$exergetic\ efficciency = \frac{useful\ output\ power}{input\ power} \quad (15)$$

The electrical power generated will be considered equal to the average power or guaranteed by the thermoelectric plant. Assured power is defined as the maximum power that a plant can supply during its worst cycle of raw material availability (fuel or primary energy).

The installed power or nominal power of the thermoelectric plant must be greater than the effective power of electricity generation. The capacity factor, is the relationship between the annual electricity supplied and the product of the installed power over time of annual operation.

This means that a plant that operates at full load full time, without operational intermittence, it will have the unit capacity factor.

In practice, the generation capacity factor is always less than the unit, the average of the generation factor being thermoelectricity capacity equal to 0.55 [25].

So, the nominal power of the electric generator of the thermoelectric can be dimensioned based on the Eq. (16), expressed in [MW].

$$nominal\ power\ of\ the\ generator = \frac{electricity\ active\ power}{load\ factor} \quad (16)$$

Thus, it is possible to estimate the average useful annual exergy generated in this process, which will be the integration of both types of average powers (thermal and electrical) developed over time (in hours) of the year, according to Eq. (17), expressed in [MWh/year].

$$useful\ annual\ exergy = useful\ output\ power\ x\ annual\ operating\ time \tag{17}$$

1.3 Termoeconomic balance

Thermoeconomy deals with the relationship between the thermal efficiency of the processes of conversion and energy use and the costs of investments and operation of these processes.

The central objective of thermoeconomics is to seek maximum thermal efficiency associated with the lowest economic cost, as long as they are adequately met with the requirements of operational reliability, thermodynamic restrictions, etc.

As pointed out by [18], the word thermoeconomics would be ambiguous, since it could refer to conventional energy analysis under the concept of the first Law of Thermodynamics, which does not consider the irreversibilities existing in all real energy conversion processes. In view of this, he proposed the use of the term exergoeconomics for analysis based on exergy under the concept of the second Law of Thermodynamics. The origin of the word comes from the Greek "ex" and "ergo", meaning "extraction of labor" and economics.

In the proposed simulation model, the thermoeconomic balance of the process has the objective to estimate the exergoeconomic efficiency of the cogeneration system from biomass forest produced by the silvopastoral system.

The associated costs are grouped into two categories: fixed costs and variable costs, when added together, make up the total costs of the process, according to Eq. (18), expressed in [$/year].

$$fixed\ costs = summation\ of\ all\ fixed\ costs\ to\ produce\ electric\ and\ thermal\ energy \tag{18}$$

The fixed costs for the production of electricity are dependent on the following variables that impact on fixed costs, which are:

1. average annual cost due to Operation and Maintenance (O & M) activities in function of the installed capacity for electric energy generation in cogeneration, C_{fgEE}, expressed in ($/(kW.year));

2. the nominal power of the generation (installed power), P_n, expressed in (MW);

3. average annual cost of investment in the cogeneration system for electricity generation, C_{iEE}, expressed in ($/year);

The fixed costs for the production of electric energy can be estimated by the Eq. (19).

$$C_{fEE} = C_{fgEE}\ x\ P_n + C_{iEE} \tag{19}$$

Similarly, the fixed costs for the production of thermal energy are dependent on the following variables that impact on fixed costs, which are:

1. average annual cost due to Operation and Maintenance (O & M) activities in function of the installed capacity for electric energy generation in cogeneration, C_{fgH}, expressed in ($/(kW.year));

2. the useful thermal power, expressed in (MW);

3. average annual cost of investment in the cogeneration system for useful thermal energyC_{iH}, expressed in ($/year);

The fixed costs for the production of useful thermal energy can be estimated by the Eq. (20).

$$C_{fH} = C_{fgH}\ x\ useful\ thermal\ power + C_{iH} \tag{20}$$

The average annual cost related to the portion of investment in generation of electrical energy, C_{iEE}, and thermal energy, C_{iH}, expressed in (1000$/year), as a function of average investment in installed capacity for the generation of electricity (and thermal energy) in a cogeneration system, in (1000$/year) and of the installed generation (nominal power of the generation or useful thermal energy), in (MW) and the depreciation time of the investment, T_{di}, in (years).

The average annual cost of investment in generation of electrical energy is estimated according Eq. (21) and (thermal energy), according to Eq. (22).

$$C_{iEE} = \frac{investment\ in\ installed\ capacity\ x\ P_n}{T_{di}} \tag{21}$$

$$C_{iH} = \frac{investment\ in\ thermal\ capacity\ x\ useful\ thermal\ power}{T_{di}} \tag{22}$$

Variable costs are composed of the sum of the costs of cogeneration of energy (electric and thermal), cost of transport and cost of forest biomass, according to Eq. (23).

$$variable\ costs = summation\ costs\ (cogeneration\ energy, transport\ and\ forest\ biomass) \tag{23}$$

The average annual variable costs for the generation of electrical and thermal energy cogeneration process, can be calculated, in [$/year], by Eqs. (24) and (25).

$$\begin{gathered} costs\ of\ cogeneration\ electrical\ energy = \\ electrical\ generation\ variable\ cost\ x\ annual\ electrical\ exergy \end{gathered} \tag{24}$$

$$\begin{gathered} costs\ of\ cogeneration\ thermal\ energy = \\ thermal\ generation\ variable\ cost\ x\ annual\ thermal\ exergy \end{gathered} \tag{25}$$

The cost of transporting forest biomass depends on the average transport distance, in [km], the average transport cost, in [$/km] and the volume of the total load to be transported, P_{vf}, in [m3/year].

To estimate the total volume of the load to be transported, it will be necessary to consider previously calculated data about the average annual productivity of forest biomass and convert it into volumetric terms (m3/year), according Eq. (26),

$$volumetric\ forest\ biomass, P_{vf} = \frac{productivity\ of\ forest\ biomass, P_{mf}}{basic\ density\ \left[\frac{g}{cm3}\right]} \tag{26}$$

So, the cost of transporting can be calculated, in [$/year], by Eq. (27).

$$transport\ cost = av.distance\ x\ av.transport\ cost\ x\ volumetric\ forest\ biomass \quad (27)$$

The cost of forest biomass is the sum of the total average annual cost of forest management by the silvopastoral system (or cost of biomass production in the silvopastoral system), with the average annual cost of remuneration for the use of pasture land, according Eq. (28), in [$/year].

$$cost\ of\ forest\ biomass = summation\ costs\ (management\ and\ remuneration\ land) \quad (28)$$

The cost of biomass production in the silvopastoral system (or management cost), in [$/year], depends on the planted area, in [hectare] and the cost of production of forest biomass (which is a statistic data of silvicultural activity), in [$/(hectare.year)]. The sum of the planted area is the total area multiplied by the relationship between the density of trees in the silvopastoral system and the density of trees in the monocultural reference system (with the same species).

So, the cost of biomass production can be calculated by Eq. (29).

$$\begin{gathered} management\ cost, C_{pfs} = \\ cost\ of\ production\ of\ forest\ biomass, C_{pf}\ x \\ summation\ of\ the\ planted\ area\ in\ the\ silvipastoral\ system \end{gathered} \quad (29)$$

The equation referring to the cost of remuneration for land use depends on the silvopastoral arrangement, the value of the land lease practiced in the region and the planted area, in [$/(hectare.year)], according to Eq. (30), expressed in [$/year].

$$\begin{gathered} remuneration\ land\ cost, C_{rem} = remuneration\ land\ per\ area\ x \\ summation\ of\ the\ planted\ area\ in\ the\ silvipastoral\ system \end{gathered} \quad (30)$$

Finally, the total cost of the integrated forest biomass production system with the energy cogeneration system, will be the sum of fixed and variable costs, according to the Eq. (31), expressed in [$/year].

$$total\ cost = summation\ of\ fixed\ costs\ and\ variable\ costs \quad (31)$$

According to [18], one of the objectives that can be obtained with exergoeconomic analysis is to calculate the costs associated with manufactured products.

In the present study, the product is the useful energy in thermal and electrical forms and its global exergoeconomic cost is determined by the Eq. (32), expressed in [kWh/$].

$$exergoeconomic\ cost = \frac{useful\ annual\ exergy}{total\ cost} \quad (32)$$

The exergoeconomic cost is an important indicator for the analysis of the economic viability of this unconventional modality of energy use.

2. Methodology and simulations

As mentioned, the present work proposes to evaluate the viability of an unconventional energy alternative by means of a mathematical simulation system.

It is the possibility of using silvopastoral systems for the production of forest biomass to be used as fuel (raw material) in an energy cogeneration system.

With this intention, a mathematical model was developed capable of evaluating the thermoeconomic viability of the system proposed.

The analysis procedure to be used in this work will be to evaluate the case studies that will be considered, in order to test and evaluate the effectiveness of the simulation model equations for which it was proposed.

2.1 Mass balance

The simulation model used in this work, evaluates the productive data from real cases. According [26], in terms of planted forests, the predominant genus in Brazil is the Eucalyptus, which had in 2015 a total area of 7.8 million hectares planted, where the main cultivated species indicated for the tropical and subtropical climates are, among others (Eucalyptus): camaldulensis, cloeziana, dunnii, grandis, saligna, tereticornis, urophylla, benthamii and the hybrid Urograndis (urophylla x grandis). *Eucalyptus grandis* is the most common, with almost 50% of the total area, followed by saligna and urophylla.

Therefore, for the purposes of the simulations that follow, these species (sp1, sp2, sp3) are considered with their typical characteristics and productive data, although it can apply to any other species.

In order to prospect perspectives on productive data in terms of forest biomass, with the silvopastoral system in the same local conditions, case studies are assumed in **Tables 1–4**. The data considered are compatible with those practiced in the evaluated region.

Several different situations regarding productive data can be assessed. In one of these case studies, the total gross productivity of the forest biomass (on a wet basis) was obtained, after 7 years from planting to harvest: 678 tons per "alqueire paulista", which is a measure of a very common area in Brazil and corresponds to 24200 m2.

It means that, the average productivity per hectare obtained after seven years was, 280.1 (tonnes/hectare).

The data referring to the spatial arrangements and productivity of forest biomass to be evaluated are real practices productive data of the region under evaluation.

Distance between trees	**Distance between ranks**	**Distance between rows**	**Number of rows**	**Tree density (trees / hectare)**
da (m)	**Dr (m)**	**df (m)**	**Nf (un)**	**Na (un)**
2	18	0	1	277,8
2	18	3	2	476,2
2,5	18	2,5	3	521,7
1,7	18	3	2	560,2
2	21	0	1	238,1
2,5	25	3	3	387,1
2,5	21	3	2	333,3
2	30	3	2	303,0
2,5	30	2,5	3	342,9

Table 1.
Data and results estimated by Eq. (1).

Forest species	Biomass moisture[2]	Average basic density of forest biomass[3]	Gross productivity of forest biomass	Time planting to harvest	Average annual volumetric increase in forest biomass
	(%)	(g/cm^3)	(tonnes/ hectare)	(years)	(m^3/(ha.year))
sp1	27,50	0,479	225	7	48,7
sp2	27,50	0,465	225	7	50,1
sp3	27,50	0,559	225	7	41,7
not specified[1]	30	0,50	517,6	20	36,2
Grevillea robusta[1]	25	0,6	125	7	22,3

[1]*Real cases.*
[2]*Wet basis in [%]. Refers to: one minus the average humidity measurement of forest biomass after post-harvest drying in the ambient condition (Humidity on a humid basis).*
[3]*Data source: [27].*

Table 2.
Data and results estimated by Eq. (2).

Silvopastoral Factor[1]	Number of trees per hectare in the silvipastoral system	Average annual volumetric increase in forest biomass	Average basic density of forest biomass (g/cm^3)	Cultivated area	Annual production of biomass
(dimension less)	Na (trees/ hectare)	(m^3/(hectare. year))	(g/cm^3)	(hectare)	(tonnes/ year)
1,2[2]	476,2	45,7	0,52	5,0	41
1,1[2]	476,2	45,7	0,52	5,0	37,4
1,3[2]	476,2	45,7	0,52	5,0	44
1,2[3]	333,3	45,7	0,52	5,0	28,5
1,2[3]	277,8	45,7	0,52	5,0	24

[1]*Silvopastoral factor: assigned value.*
[2]*Note 1: The objective of the three first line, is to assess the effect of the influence of the silvopastoral factor.*
[3]*Note 2: The objective of the fourth and fifth line, is to assess the effect of the influence of the density of trees per hectare and the spatial arrangement.*

Table 3.
Data and results estimated by Eq. (3).

Forest species[1]	Cultivated area (hectare)	Annual production of biomass (tonnes/year)	Average flow of forest biomass (kg/hour)[2]
sp1	150.0	1,198.9	136.9
sp2	120.0	959	109.5
sp3	80.0	639.6	73.0
summation:		2,797.5	319.4

[1]*The same genetic species as before (and the same basic density and average productivity); number of trees per hectare in the silvopastoral system: 476.2 (trees/hectare) with spatial arrangement: 18 m x 2 m x 3 m and silvopastoral factor: 1.2.*
[2]*Annual operating time considered: 8760 hour/year.*

Table 4.
Data and results estimated by Eqs. (1)–(4).

Table 1 shows and calculate the tree densities, according to Eq. (1), of some of the possible spatial arrangements that are commonly practiced in silvopastoral systems in the evaluated region.

Based on data on gross productivity in some areas, **Table 2** calculates the average annual volumetric increase in forest biomass.

Thus, according to Eq. (4), it is possible to estimate the average annual hourly flow of forest biomass, considering the annual time of operation of the energy conversion process, for each case study evaluated in **Tables 1–3**.

2.2 Energy balance

The energy balance applies to the energy conversion process. From the input data, Eqs. (5)–(8) calculate the input exergy on a dry and wet basis for each species of forest biomass and the average exergy.

Eqs. (4) and (5) are used to estimate the exergy provided by each forest species, as shown in **Table 5**, each with its exergy rate value, superior calorific value and humidity factor.

Although the forest species in **Table 5** are presented in a generic way, the magnitudes attributed are compatible with those of common species in the region.

In order to simulate the average exergy of entry into the cogeneration system, **Table 6** considers the same hypothetical data as the previous flow of forest biomass.

The mathematical resource for calculating species diversity is the use of matrix variables of the order 1 x n, where n is the number of forest species present in the process.

As long as the energy efficiency of the cogeneration process and the conversion to kinetic energy are known, Eqs. (9)–(11) make it possible to calculate the output powers available in the forms of electrical (or mechanical) and thermal energy, according to the data exemplified previously, shown in **Table 7**.

The exergetic temperature factor, according to [23], can be calculated as a function of the temperature variation. So, considering that the use of available

Forest species[1]	Exergy rate	Superior calorific power (MJ/kg)	Humidity factor	Dry exergy IN (MJ/kg)	Wet exergy IN (MJ/kg)
sp1	0.925	19.46	0.70	18.00	12.60
sp2	0.900	19.67	0.75	18.19	13.64
sp3	0.850	19.46	0.80	16.54	13.23

[1]*Assigned quantities are compatible with those of common species in the region.*

Table 5.
Data and results estimated by Eqs.(5) and (6).

Forest species	Flow of forest biomass (kg/hour)	Wet exergy IN (MJ/kg)	Specific wet exergy IN (MJ/kg)	Average exergy IN	Input power (MW)
sp1	136.87	12.26	1677.98		5.71
sp2	109.46	13.65	1494.16		5.67
sp3	73.02	13.23	966.03		3.55
summation:	319.35		4138.16	12.96	14.90

Table 6.
Data and results estimated by Eqs. (7), (8).

Input power	Cogeneration system efficiency	Output power	Kinetic energy conversion efficiency	Active power of the electricity generation	Available thermal power
(MW)	(%)	(MW)	(%)	(MW)	(MW)
14,90	65	9,68	30	4,47	5,21

Table 7.
Data and results estimated by Eqs. (9)–(11).

thermal energy aims to heat water from initial to final temperature, **Table 8** estimates the values for some case studies from the data previously exemplified, according to Eqs. (12)–(15).

To finalize the analysis of the energy balance, **Table 9** according to Eqs. (16) and (17) presents useful data for the estimation of the average annual flow of useful energy and for the dimensioning of the nominal generator power for the generation of electricity.

It was seen that the economic costs of the process are grouped into two categories: fixed costs and variable costs. **Table 10**, according to Eqs. (18)–(22) estimates the fixed cost of the process.

It can be seen that from **Tables 10–16**, in the columns on the left are the values considered for the variables and on the right the results obtained.

To assess the average annual variable cost for the production of electrical and thermal energy, it is necessary to define and consider a set of service cost conditions

Case Study	Start/end temperature	Temperature factor exergetic	Useful thermal power	Useful output power	Exergetic efficiency
	(K)	(%)	(MW)	(MW)	(%)
(1)	300/373	19,6%	1,02	5,49	36,8%
(2)	300/500	40,0%	2,09	6,55	44,0%
(3)	300/1000	70,0%	3,65	8,12	54,5%
(4)	300/1200	75,0%	3,91	8,38	56,3%

(1): Energy use with the variation of the ambient temperature up to 100°C.
(2): Energy use with the variation of the ambient temperature up to 227°C.
(3): Energy use with the variation of the ambient temperature up to 727°C.
(4): Energy use with the variation of the ambient temperature up to 927°C.

Table 8.
Data and results estimated by Eqs. (12)–(15).

Load factor[1]	Nominal power of the generator[2]	Useful annual exergy[3]		
		Thermal	Electric	Total
	(MW)	(GWh/year)		
0,55	8,13	34,25	73,40	107,65

[1]Load factor for thermoelectricity, according to [25].
[2]Nominal power of the generator considering active power of 4.47 MW electricity generation, according to Eq. (16).
[3]annual useful exergy considering full-time operating (8760 hours/year); thermal energy required for heating water from 300 K to 1200 K (3.91 MW) and active electric output power of 8,38 MW.

Table 9.
Data and results estimated by Eqs. (16) and (17).

Variables	Values considered	Units	Variables	Values obtained	Units
CfgEE	25,00	($/(kW.year))	CiEE[2]	447,15	(1000 $/kW)
CfgH	25,00	($/(kW.year))	CiH[3]	48,88	(1000 $/kW)
CigEE	1,10	(1000 $/kW)	CfEE[4]	650,40	(1000 $/kW)
CigH	0,25	(1000 $/kW)	CfH[5]	146,63	(1000 $/kW)
Pn	8,13	(MW)	Cf[6]	797,03	(1000 $/kW)
utp[1]	3,91	(MW)			
Tdi	20	(years)			

[1]*utp: useful thermal power.*
[2]*CiEE: average annual cost of investment in generation of electrical energy, according to Eq. (21).*
[3]*CiH: average annual cost of investment in generation of thermal energy, according to Eq. (22).*
[4]*CfEE: fixed costs for production of electric energy, according to Eq. (19).*
[5]*CfH: fixed costs for production of useful thermal energy, according to Eq. (20).*
[6]*Cf: fixed costs, according to Eq. (18).*

Table 10.
Data and results estimated by Eqs. (18)–(22).

associated with the production process in the silvopastoral system, which are described at the bottom of **Tables 11–16**.

Table 11 estimates the variable cost of forest biomass, according to Eqs. (28)–(30).

Variables	Values considered	Units	Variables	Values obtained	Units
Rem[1]	500,00	($/(ha.year))	Crem[3,5]	50.003,00	($/year)
Cpf[2]	250,00	($/(ha.year))	Cpfs[4,5]	25.001,50	($/year)
			Cbf[6]	75.004,50	($/year)

[1]*Rem: average annual cost for remuneration by land use for silvicultural activity.*
[2]*Cpf: average annual cost for the production of forest biomass by cultivation area in a monocultural silviculture system with a spatial arrangement of 3 m x 2 m.*
[3]*Crem: remuneration land cost, according to Eq. (30).*
[4]*Cpfs: cost of biomass production in the silvopastoral system, according Eq. (29).*
[5]*Note: Cpf and Cpfs apply to a silvopastoral system on 350 hectares with an arboreal spatial arrangement of 18 m x 3 m x 2 m and a monocultural reference arrangement of 3 m x 2 m.*
[6]*Cbf: cost of forest biomass, according Eq. (28).*

Table 11.
Data and results estimated by Eqs. (28)–(30).

Variables	Values considered	Units	Variables	Values obtained	Units
Dmed[1]	40	(km)	Pvf[5]	5.709,18	(m3/year)
Pmf[2]	2.797,5	(tonnes/year)	Ct[6]	57.091,84	($/year)
Ctm[3]	0,25	($/$m^3$. km)			
Dbm[4]	0,490	(g/cm^3)			

[1]*Dmed: average distance from the field to the thermoelectric plant.*
[2]*Pmf refers to the average annual productivity of forest biomass, estimated for a silvopastoral system in 350 hectares with an arboreal spatial arrangement of 18 m x 3 m x 2 m.*
[3]*Ctm: average cost for transporting biomass according to volume and distance.*
[4]*Dbm: average basic density of forest biomass.*
[5]*Pvf: global average annual volumetric productivity of biomass, according Eq. (26).*
[6]*Ct: average annual cost of transporting forest biomass, according Eq. (27).*

Table 12.
Data and results estimated by Eqs. (26) and (27).

Variables	Values considered	Units	Variables	Values obtained	Units
CvgEE[1]	0,90	($/MWh)	CvpEE[5]	66.060,00	($/year)
CvgH[2]	0,55	($/MWh)	CvpH[6]	18.837,50	($/year)
TE[3]	34,25	(GWh/year)			
EE[4]	73,40	(GWh/year)			

[1]*CvgEE: average annual variable cost depending on the generation of electric energy in cogeneration system.*
[2]*CvgH: average annual variable cost depending on the generation of thermal energy in cogeneration system.*
[3]*TE: Thermal useful annual exergy.*
[4]*EE: Annual electricity generated (useful annual electric exergy).*
[5]*CvpEE: average annual variable cost of electricity in the cogeneration system, according Eq. (24).*
[6]*CvpH: average annual variable cost of thermal energy in the cogeneration system, according Eq. (25).*

Table 13.
Data and results estimated by Eqs. (24) and (25).

Variables	Values considered	Units	Variables	Values obtained	Units
Cbf[1]	75.004,50	($/year)	Cv[5]	216.993,84	($/year)
Ct[2]	57.091,84	($/year)			
CvpEE[3]	66.060,00	($/year)			
CvpH[4]	18.837,50	($/year)			

[1]*Cbf: cost of forest biomass, according Eq. (28).*
[2]*Ct: average annual cost of transporting forest biomass, according Eq. (27).*
[3]*CvpEE: average annual variable cost for production of electricity, according Eq. (24).*
[4]*CvpH: average annual variable cost for production of thermal energy, according Eq. (25).*
[5]*Cv: average annual variable cost for production of electrical and thermal energy, according Eq. (23).*

Table 14.
Data and results estimated by Eq. (23).

Variables	Values considered	Units	Variables	Values obtained	Units
Fixed costs[1]	797,03	(1000 $/year)	total cost[3]	1.014,02	(1000 $/year)
Variable cost[2]	216,99	(1000 $/year)			

[1]*According to Eq. (18).*
[2]*According to Eq. (23).*
[3]*According to Eq. (31).*

Table 15.
Data and results estimated by Eq. (31).

Variables	Values considered	Units	Variables	Values obtained	Units
Useful annual exergy[1]	107,65	(GWh/year)	exergoeconomic cost[3]	106,16	(kWh/$)
Total cost[2]	1.014,02	(1000 $/year)			

[1]*According to Eq. (18).*
[2]*According to Eq. (31).*
[3]*According to Eq. (32).*

Table 16.
Data and results estimated by Eq. (32).

Table 12 estimates the variable cost of transporting forest biomass, according to Eqs. (26) and (27).

Table 13 estimates the average annual variable cost of electricity and thermal energy in the cogeneration system, according to Eqs. (24) and (25).

Table 14 estimates the average annual variable cost for production of electrical and thermal energy, according Eq. (23).

Table 15 estimates the total cost of the integrated forest biomass production system with the energy cogeneration system, which consists of the sum of fixed and variable costs, according to Eq. (31).

So, the exergoeconomic cost, is determined by Eq. (32), as shown in **Table 16**,

3. Conclusions

The parameter used in this chapter to assess the feasibility of the proposed agroenergetic alternative is the exergoeconomic cost.

The non-conventional alternative evaluated is the use of a silvopastoral system aimed at the production of forest biomass and its energy utilization in a thermal and electric energy cogeneration system.

The economic feasibility analysis is a cost/benefit analysis, which can be done based on tariff parameters practiced in the energy market.

For that, comparative measures with the values practiced in the energy sector can be used.

Just as an example, a comparison parameter is the value practiced from 2003 onwards for the average electricity supply tariff for the Brazilian electric system for all consumption classes and geographic regions of the country [28], which is (61.40 $/MWh), much higher than the value found in the present simulation (9.42 $/MWh).

Therefore, this study presents indications of good viability for this energy alternative as a possibility, which can be inserted among the renewable energy options in the energy matrix of the future.

Author details

Cid Marcos Gonçalves Andrade[1*], Sérgio Inácio Gomes[2],
Mauro Antônio da Silva Sá Ravagnani[1] and Eugênia Leandro Almeida[1]

1 UEM – State University of Maringá, Maringá, Brazil

2 IFPR – Federal Institute of Science and Technology of Paraná, Paranavaí, Brazil

*Address all correspondence to: cidmga@yahoo.com.br

References

[1] Göran B, Hoogwijk M, Hoogwijk R. The contribution of biomass in the future global energy supply: a review of 17 studies. Biomass and Bioenergy. 2003. 25: 1-28p. DOI: 10.1016/S0961-9534(02)00185-X

[2] Goerndt M, Aguilar F, Skog K. Resource potential for renewable energy generation from co-firing of woody biomass with coal in the Northern U.S. Biomass and Bioenergy. 2013. 59: 348-361p. DOI:10.1016/j.biombioe.2013.08.032

[3] Fey R, Malavasi U, Malavasi M. Silvopastoral system: A review regarding the family agriculture. Revista de Agricultura Neotropical. 2015. 2: 26-41p.

[4] Dobie P. Strategy on tree-based energy: clean and sustainable energy for improving the livelihoods of poor people. October 2015. Nairobi, Kenya. World Agroforestry Centre; 2015. p. 1-36

[5] Nepomuceno A, Silva I. Caracterização e avaliação de sistemas silvipastoris da região noroeste do estado do Paraná. Revista Floresta. 2009. 39:279-287. DOI: 10.5380/rf.v39i2.14556

[6] Chakravarty S, Ghosh C, Suresh A, Shukla G. Deforestation: Causes, Effects and Control Strategies. Global Perspectives on Sustainable Forest Management. 2012. 1: 1-27p.

[7] Stark H, Nothdurft A, Block J, Bauhus J. Forest restoration with Betula ssp. and Populus ssp. nurse crops increases productivity and soil fertility. Forest Ecology and Management. 2015. 339: 57-70p. DOI: 10.1016/j.foreco.2014.12.003

[8] Benegas L, Ilstedt U, Roupsard O, Jones J, Malmer A. Effects of trees infiltrability and preferential flow in two contrasting agroecosystems in Central America. Agriculture, Ecosystems and Environment. 2014. 183:185-196p. DOI: 10.1016/j.agee.2013.10.027

[9] Joele M, Lourenço L, Lourenço J, Araujo G, Budel J, Garcia A. Meat quality of buffaloes finished in traditional or silvipastoral system in the brazilian eastern Amazon. Journal of the Science of Food and Agriculture. 2017. 97:1740-1745p. DOI: 10.1002/jsfa.7922

[10] Mcgroddy M, Lerner A, Burbano D, Schneider L, Rudel T. Carbon stocks in silvipastoral systems: A study from four communities in southeastern Ecuador. Biotropica. 2015. 47(4):407-415p. DOI: 10.1111/btp.12225

[11] Martínez J, Cajas Y, León J, Osorio N. Robert, L. Silvipastoral systems enhance soil quality in grasslands of Colombia. Applied and Environmental Soil Science. 2014. 2014: 1-8p. DOI: 10.1155/2014/359736

[12] Mancilla-Leytón J, Sánchez-Lineros V, Martín A. Influence of grazing on the decomposition of Pinus pinea L. needles in a silvipastoral system in Doñana, Spain. Plant and soil. 2013. 373(1): 173-181p. DOI: 10.1007/s11104-013-1788-8

[13] Lunka P, Patil S. Impact of tree planting configuration and grazing restriction on canopy interception and soil hydrological properties: implications for flood mitigation in silvipastoral systems. Hydrological Process. 2016. 30 (6):945-958p. DOI: 10.1002/hyp.10630

[14] Ghezehei S, Everson C, Annandale J. Can productivity and post-pruning growth of Jatropha curcas in silvipastoral systems be regulated by manipulating tree 63 spacing/arrangement without changing tree

density. Biomass and Bioenergy. 2015. 74:233-243p. DOI: 10.1016/j.biombioe.2015.01.007

[15] Naji H, Bakar E, Soltani M, Ebadi S, Abdul-Hamid H, Javad S, Sahri M. Effect of initial planting density and tree features on growth, wood density, and anatomical properties from a Hevea brasiliensis trial plantation. Forest Products Journal. 2014. 64:41-47p. DOI: 10.13073/FPJ-D-13-00071

[16] Rodrigues C, Araújo S, Rocha N, Braz T, Delmar S, Villela J, Viana M. Light relations and performance of signal grass in silvopastoral system. Acta Scientiarum - Animal Sciences. 2014. 36 (2):129-136p. DOI: 10.4025/actascianimsci.v36i2.22398

[17] Costa F. Fuel gases as alternatives to electrothermy in direct heating and process heat in the Brazilian industrial sector. [Thesis]. São Paulo. University of São Paulo; 2014. DOI: 10.11606/T.86.2013.tde-13012014-164600

[18] Tsatsaronis G. Thermoeconomic Analysis and Optimization of Energy Systems. Progress in Energy and Combustion Science. 1993. 19:227-257p. DOI: 10.1016/0360-1285(93)90016-8

[19] Sciubba E, Wall G. A brief commented history of exergy from the beginning to 2004. International Journal of Thermodynamics. 2007.10 (1): 1-26p.

[20] Mendes, M. Metodologia para análise exergética-econômica de plantas a vapor para geração de eletricidade consumindo resíduo de biomassa amazônica. [Thesis]. Belém. University Federal of Pará; 2009

[21] Guohui S, Laihong S, Xiao J. Estimating specific chemical exergy of biomass from basic analisis data. Industrial & engineering chemistry research. 2011. 50:9758-9766p. DOI: 10.1021/ie200534n

[22] Lima E, Abdala E, Wenzel A. Influência da umidade da madeira no poder calorífico superior da madeira – Comunicado técnico 220. Embrapa – Empresa Brasileira de Pesquisa Agropecuária. 2008. 1:1-3p.

[23] Dincer I, Rosen M. Exergy: Energy, Environment and Sustainable Development. 2nd ed. Elsevier Science; 2012

[24] Kotas T. The Exergy Method of Thermal Plant Analysis. 1nd ed. Krieger Publishing Company; 1995

[25] Tolmasquim, M.T., Energia Termelétrica, Gás Natural, Biomassa, Carvão, Nuclear, EPE (coord), Rio de Janeiro, 2016; 417p., available in https://www.epe.gov.br/sites-pt/publicacoes-dados-abertos/publicacoes/PublicacoesArquivos/publicacao-173/Energia%20Termel%C3%A9trica%20-%20Online%2013maio2016.pdf; access in may_11_2021

[26] Gontijo, G.A., Silvicultura do Eucalipto: Principais espécies cultivadas no Brasil e suas Características, UFPR, Curitiba, Pr, 2018.

[27] Oliveira, J.T.S., Hellmeister, J.C., Filho, M.T., Variation of the moisture content and specific gravity in the wood of seven eucalypt species; Scielo, Revista Árvore, vol.29 no.1 Viçosa Jan./Feb. 2005

[28] Relatórios de Consumo e Receitas da Distribuição, Relatório da Tarifa Média por Classe de Consumo e por Região, disponível em https://www.aneel.gov.br/relatorios-de-consumo-e-receita, acess in may_06_2021.

Section 3

Alternative Biomass Sources

Chapter 11

Market Prospecting and Assessment of the Economic Potential of Glycerol from Biodiesel

Alisson Dias da Silva Ruy, Ana Luíza Freitas Ferreira, Antônio Ésio Bresciani, Rita Maria de Brito Alves and Luiz Antônio Magalhães Pontes

Abstract

Glycerol from biodiesel is a potential raw material for synthesis of several products with high added value. The world demand and the market value of these products are important information for defining the best investment for the implantation of a biorefinery. The information is available on websites of social associations, production companies and market consulting companies and can be mined, free of charge. The International Trade Center (ITC), with information on world trade and websites linked to the foreign trade agencies of every country, such as Comex Stat, in Brazil, are relevant search sources. In this context, this work presents procedures and search techniques for prospecting such information. Such a procedure is illustrated through a case study for which a search of market parameters for glycerol and its derivatives was carried out for use in the process design and economic evaluation of an industrial plant. It was found that crude glycerol had a market price close to US$ 170/ton, in 2019. Among its derivatives, acrylic acid, acrylonitrile and 1,3-propanediol have great potential for the development of new processes, within the scope of a biorefinery. Industrially, acrylic acid (US$ 1100/ton) and acrylonitrile (US$ 1500/ton) are produced from propene (US$ 880/ ton) and 1,3-propanediol (US $ 2000/ton) comes from glucose (US$ 460/t) or ethylene oxide (US$ 1200/t), which encourages the development of new sustainable processes.

Keywords: biomass, glycerol, market, economic analysis, industrial processes

1. Introduction

Vegetable biomass contains organic matter from the photosynthesis process and conversion of biological origin compounds into chemicals [1] has attired attention of researchers worldwide. The United States Department of Energy presented a series of chemical products, called platform molecules, which may be synthesized from biomass, generating more complex molecules of interest to society [2]. Biomass is also a raw material for the production of biofuels, representing a renewable alternative, easily accessible and more environmentally sustainable compared to fossil fuels. Among the biofuels, biodiesel has attracted attention since it is produced from simple and abundant raw materials, such as vegetable oils, animal fat and its residues [3].

Biodiesel is one of the most important energy commodities in the world. The USA (6.5 million m^3), Indonesia (6.2 million m^3) and Brazil (5.9 million m^3) were the largest producers in 2019. Germany is the largest producer of biodiesel in the European community (3.2 million m^3), followed by France and Spain. Other countries such as Argentina, Poland, Malaysia and Thailand are also important players in this market, producing more than 1 million m^3 [4]. The production of biodiesel has grown significantly worldwide, which results also in an increase in the amount of glycerol produced as a by-product [5]. Glycerol has great potential for application in the production of high added-value chemicals due to four main factors: its availability; its low commercial value; for being a renewable raw material; and for providing more economically viable alternatives for the biodiesel industry [6]. This may lead to a decrease in biodiesel prices and could improve the glycerol market. Therefore, there is an opportunity to produce value-added molecules, such as acrylic acid, acrylonitrile and 1,3-propanediol, among others [7, 8].

The development of industrial processes that contemplate the conversion of raw materials from biomass into chemical products requires an economic analysis to define the molecules that are of greatest interest to the market. Data can be obtained free of charge from portals such as the International Trade Center (ITC), from the official portals of trade agencies in each country, such as Comex Stat in Brazil and, partially, from websites of market consulting companies. The ITC was developed to provide access to export statistical data and specifications from several countries. As market consulting companies publish detailed reports of the worldwide market assessment of various products, which take into account price, demand and market potential. However, these reports are expensive and difficult to be acquired by researchers at universities and many research institutions.

In this context, this work presents methods and tools, easily accessible and free of charge, to search for information and parameters, analysis of the current market and the economic potential of products. In particular, data on glycerol and three of its main derivatives are analyzed: acrylonitrile, acrylic acid and 1,3-propanediol. Also, a simplified process design is presented for each of the selected derivatives, using the ASPEN PLUS tool.

2. Methods

In order to carry out an economic analysis of a given product of interest, it is important to obtain market data for the base country, where the industrial plant can be implanted, as well as for other countries, for assessment of international trade, threats and opportunities.

The market parameters of glycerol and each of its derivatives to be evaluated were obtained from the International Trade Center (ITC) portal and the Comex Stat portal. The future scenarios were obtained on the websites of consulting companies. This information must be carefully analyzed due to the differences in the assessment among different companies.

ITC is the joint agency of the World Trade Organization and the United Nations that aims to promote business between countries, mainly developing countries, simplify the economy [9], and provide several market analysis tools, covering 220 countries and territories and 5300 products found in the Harmonized System (HS).

The Trade Map is one of the most important and provides, through tables, graphs and maps, performance indicators for imports and exports, international demand, alternative and competitive markets, in addition to a directory of importing and exporting companies [10]. Comex Stat is a portal to access statistical data of foreign trade in Brazil. Every country has a free access consultation tool that quantifies national imports and exports, e.g. Data Web (United States - https://dataweb.usitc.gov/), Stat Can (Canada - https://www5.statcan.gc.ca/cimt-cicm/home-accueil?lang=eng) and ETCN (China - http://www.e-to-china.com).

The chemical products are found on the Trade Map and Comex Stat by using the Harmonized Commodity Description and Coding System, or simply Harmonized System (HS), and the Mercosur Common Nomenclature (NCM), in the specific case of Brazil. The Harmonized System (SH) is a multipurpose international nomenclature for the classification of products, published by the World Customs Organization (WCO). The HS allows participating countries to classify traded goods on a common global basis for customs purposes. It comprises more than 5000 commodity groups; each identified by a six-digit code, from SH2 to SH6, arranged in a legal and logical structure and is supported by well-defined rules to achieve uniform classification (WCO, 2020). The greater the number of digits, the more specific the HS classification becomes, so each country, or group of countries, has defined a common classification, which can have up to ten digits. These nomenclatures can represent a single substance or a group of them, having specific chemical characteristics in common. Usually the commodities and products of great commercialization are dealt with separately; however, the analysis of molecules grouped in the same SH is more difficult. The system is used by more than 200 countries and economies as a basis for their Customs tariffs and for the collection of international trade statistics. Over 98% of the merchandise in international trade is classified in terms of the HS. The HS contributes to the harmonization of customs and trade procedures, and the non-documentary trade data interchange in connection with such procedures, thus reducing the costs related to international trade.

MERCOSUR Common Nomenclature (NCM) is based on the Harmonized Commodity Description and Coding System. The NCM adopts WCO tariff subheadings (six digits) and adds two more digits to provide a greater level of detail, resulting in eight-digit codes. The MCN has been adopted by the countries that integrate the Argentina, Brazil and Uruguay Block to foster international trade growth, make the creation and comparison of statistics easier, in addition to elaborating freight tariffs and providing other relevant information to international trade.

The ITC Trade Map uses a classification of up to 6 digits, while Comex Stat uses not only the SH, but also the NCM, with 8 digits, specifying more the product under analysis.

The availability of the market parameters in these databases allows the user correlating and interpreting the information in order to economically analyze the product under evaluation. Import and export values, volumes, growth rates, market shares, etc. are obtained from the Trade Map/ITC (worldwide) and the Comex Stat (for Brazil), while the growth forecasts are obtained from the websites of market consulting companies. This procedure was applied to gather information on the market for glycerol, acrylic acid, acrylonitrile and 1,3-propanediol.

The correlation of parameters displayed in graphs and tables is necessary for better evaluation and interpretation of the results. The data can be evaluated by regions or countries with greater market share for a specific product, allowing the visualization of the marketing behavior of these substances during the period under study.

2.1 Search in the international trade Centre

The Trade Map from ITC is accessed by the link https://www.trademap.org/Index.aspx. **Figure 1** shows an example of the search page.

It is possible to perform the search in different ways depending on the purpose of the analysis. One may insert the product code (isolated or group) or the location (country or region), choosing whether the study will be about import or export. The search can be carried out considering one of the five aspects: "Trade Indicators", "Yearly", "Quarterly", "monthly time series" or "Companies". In "Trade Indicators", more specific market data are obtained. It is also possible to choose the way in which the values will be arranged, yearly, quarterly or monthly, and, finally, the main companies that commercialize the product. The monthly, quarterly and yearly trade flows are available from the most aggregated level to the tariff line level. An example is given in the **Figure 2**, considering the option "Yearly Time Series" for the glycerol.

As shown in **Figure 2**, the search can take into account all countries or a specific country. Other criteria can be redefined using the same filters shown in **Figure 1**, such as operations (import, export or exchange balance), time (trade indicators, year, quarter, or month), category (country, product or service). Also, it can be filtered by market parameters (volume, value and others) and unit (depending on the parameter defined by the researcher).

Data such as export values, among others, are generated for each country and period, being initially listed in decreasing order of the values from the last year. According to **Figure 2**, crude glycerol is included in SH6 1520.00, which covers also water and glycerol lyes. In the example, the output data were obtained for the world, specifying the values in dollars of the exporting countries annually. In "Map", the world map shows the expressiveness of each country in different color scales, depending on the commercialized values of the product. In "Graph", the output data can be generated in graphs. In "download", the table of values can be exported to a spreadsheet or text files.

2.2 Search in the Comex Stat

The tool Comex Stat is accessed at "General Exports and Imports" through the link: http://comexstat.mdic.gov.br/pt/home. Results are shown in **Figure 3**, according to the type of operation (exports or imports), period of time and filters of interest. The information may be filtered according to the product code (SH or NCM), country or economic region, means of transport used and how the results should be displayed. If no input is given related to region, the search is carried out all over the world, The option "Net Weight" in Values filter shows the US$ FOB (freight on board) value, which is used to indicate when liability and

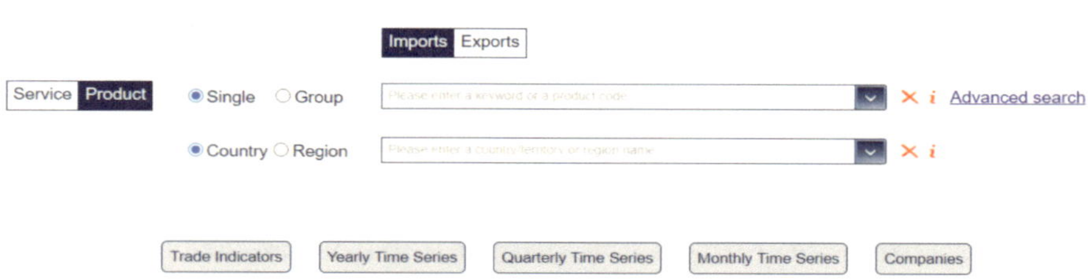

Figure 1.
Search of products in the trade map/ITC.

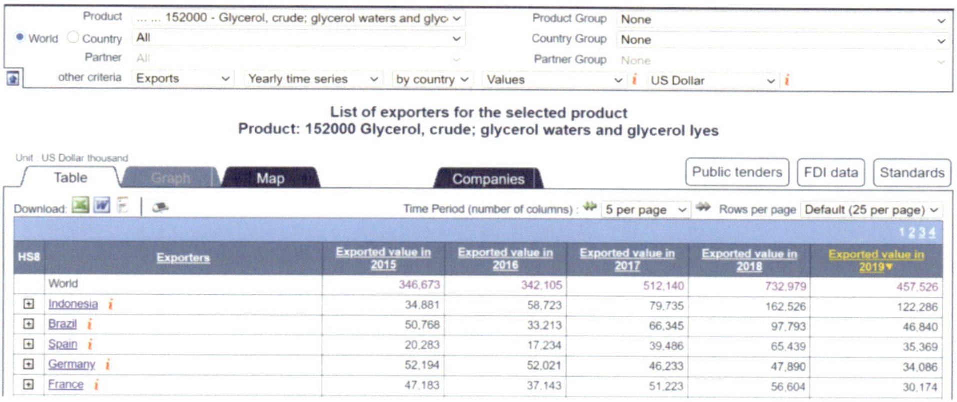

HS8	Exporters	Exported value in 2015	Exported value in 2016	Exported value in 2017	Exported value in 2018	Exported value in 2019
	World	346,673	342,105	512,140	732,979	457,526
	Indonesia	34,881	58,723	79,735	162,526	122,286
	Brazil	50,768	33,213	66,345	97,793	46,840
	Spain	20,283	17,234	39,486	65,439	35,369
	Germany	52,194	52,021	46,233	47,890	34,086
	France	47,183	37,143	51,223	56,604	30,174

Figure 2.
Example of results for glycerol and yearly time series option.

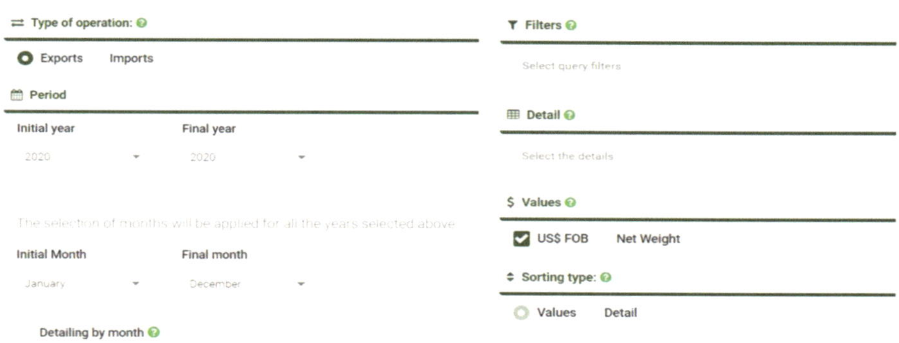

Figure 3.
Search by filters in the Comex Stat tool.

ownership of goods is transferred from a seller to a buyer, and also the amount commercialized in kilogram.

As an example of using COMEX STAT, information for pure glycerol (NCM: 29054500) is evaluated. The results, i.e. the data for the Brazilian annual exports are shown in **Figure 4**.

These data may be exported to a graph to evaluate the trends in imports or exports for the product over the years. Moreover, dividing the US$ FOB by Net Weight, the average annual value for the product is obtained (US$/weight), and it is possible to evaluate the trend of valorization or devaluation over the years. The data may also be exported to a spreadsheet.

2.3 Search in reports of market consulting companies

Websites of most consulting companies present market reports, from which it is possible to get an idea of the future trend analysis for the main producing companies and product participation by region. Typically, these websites can be found using the keywords *global market share* and the name of the desired product.

The market growth forecast is defined based on the Compound Annual Rate of Growth (CAGR), which is the average annualized rate of revenue growth between two given years, assuming that growth takes place at an exponentially compound rate. As the world market is sensitive to constant changes, the forecast of the CAGR may vary yearly and according to the evaluation techniques used by a given

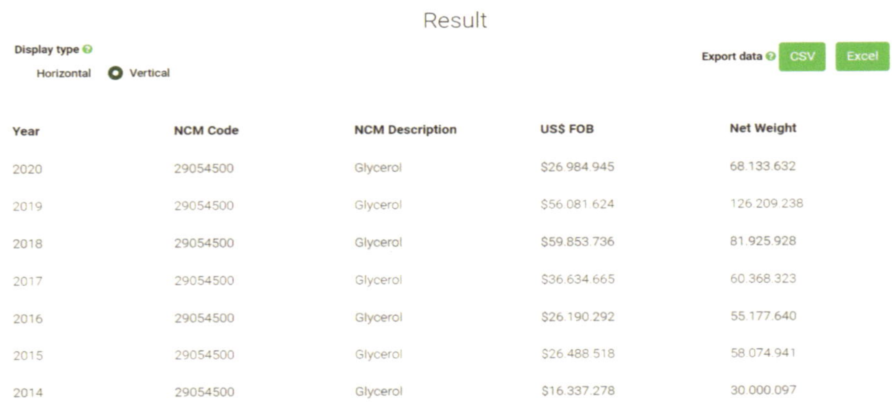

Result

Display type

Horizontal Vertical

Export data CSV Excel

Year	NCM Code	NCM Description	US$ FOB	Net Weight
2020	29054500	Glycerol	$26.984.945	68.133.632
2019	29054500	Glycerol	$56.081.624	126.209.238
2018	29054500	Glycerol	$59.853.736	81.925.928
2017	29054500	Glycerol	$36.634.665	60.368.323
2016	29054500	Glycerol	$26.190.292	55.177.640
2015	29054500	Glycerol	$26.488.518	58.074.941
2014	29054500	Glycerol	$16.337.278	30.000.097

Figure 4.
Brazilian annual exports for pure glycerol.

company. Particularly for 2020, the influence of COVID-19 on the growth of product markets has been reported. Due to decreasing in the world growth, the demand for fuels decreased momentarily in 2020 and may return more slowly in 2021.

3. Routes for glycerol conversion

The annual production of biodiesel worldwide has increased in the last two decades. As a result, a large amount of the surplus crude glycerol has been generated. Currently, 90% of glycerol is produced from biodiesel [8] through the reaction of transesterification, which produces biodiesel and glycerol at mass ratio of 10: 1, i.e., for every 10 kg of biodiesel, 1 kg of crude glycerol is produced [11].

The glycerol produced from transesterification contains several impurities, such as: water, soaps, fatty acids and their esters, methanol and catalyst, usually sodium or potassium hydroxide. The removal of these impurities is necessary for the application of glycerol in different kind of industries, such as pharmaceutical, food and cosmetics [8]. Glycerol must reach a specific purity depending on its application, and also on the specification of the contaminants that must be removed through more sophisticated separations [12]. For pharmaceutical industries, glycerol must reach USP (United States Pharmacopeia) grade, which contains 99.5% by weight, while for applications in the food industry, the FCC (Food Chemical Codex) grade is required, with 99.7% by mass [13]. In addition, for application in the production of high added-value molecules, the impurities must be removed as they can significantly inhibit bacterial growth and fermentation (biotechnological processes) or the activity of heterogeneous catalysts (chemical processes) [14].

Glycerol can be converted into chemicals by several different reactions, such as oxidation, dehydration, hydrogenation, acetylation, esterification, etherification, ammoxidation, among others, as shown in **Table 1** [7, 14, 15].

As shown in **Table 1**, several studies have been carried out to convert glycerol into value-added chemicals through catalytic routes. For each reaction, specific homogeneous and heterogeneous catalysts were studied as an active phase, which is responsible for promoting the conversion of glycerol. Although glycerol is a promising alternative to reduce the dependence or even replace conventional production of petrochemical products [14], the good selectivity to desired products at high conversions is still a technological barrier. This is due to glycerol hydroxyls to have similar reactivity, unknown reaction conditions and the lack of efficient catalysts [7].

Conversion of glycerol	Chemical derivatives	Catalysts and active phases
Oxidation	Glycolic acid, hydroxypyruvic acid, tartaric acid, oxalic acid, mesoxalic acid, propionic acid, acrylic acid, citric acid, succinic acid, fumaric acid	Pt/C, Pd/C and Au, Pt-Bi catalysts
Dehydration	Acrolein, acetaldehyde, hydroxypropanone, hydroxypropane, propanaldehyde, adducts, acetone, dihydroxyacetone and polyaromatic compounds,	SiO_2/Al_2O_3, $ZnSO_4$ H_2SO_4, H_3PO_4/ (TiO_2 and SiO_2) Al, V, Sb and Nb oxides Zeolites
Hydrogenation/ Reduction	1,2-Propanediol, 1,3-propanediol, propanol, ethylene glycol, lactic acid, acetol, dichloro-2-propanol	Pt, Pd, Rh and Ru Ni and Cu
Acetylation	Acetins (glycerol mono-, di- and triester)	ZrO_2/SiO_2/ME, ZrO_2/SiO_2/SG, $HClSO_3/ZrO_2$ S-ZrO_2 and H_2SO_4/ZrO_2
Esterification	Mono, di and tri glycerates, glycerol carbonate, polyesters and branched nylons	NaOH, KOH, $NaOCH_3$, $KOCH_3$ $H_3PW_{12}O_{40}/SiO_2$/C Niobic acid /ZrO_2
Etherification	Glycerol tert-butyl ether, methyl butyl ether, 1,3-di-tert-butyl glycerol, 1,2-di-tert-butyl glycerol and 1,2,3-tri-tert-butyl glycerol,	Ru/S, $CuCr_2O_4$ and Cu/Zn CH-SO_3H La and Mg catalysts
Amoxidation	Acrylonitrile	Mo, Bi, Sb, V, Sn, W, Zr, Ti, Ni, Al, P, G and Nb

Sources: Adapted from [2, 7, 14].

Table 1.
High added-value chemicals from glycerol.

The production of 1,3-propanediol, acrylic acid and acrylonitrile are promising products from glycerol. 1,3-propanediol has great economic potential and a recent market, which can be expanded using glycerol as raw material, while acrylic acid and acrylonitrile come from consolidated petrochemical processes, which can be replaced by sustainable glycerol processes. In this context, it is important to evaluate the market and economics of these derivatives, in order to define a proposal for a potential process of production of these molecules from glycerol.

4. Economic analysis

4.1 Glycerol

Glycerol is found in the ITC under the codes SH6 1520.00 and 2905.45. SH 1520.00 refers to crude glycerol, water and glycerol lye. These nomenclatures present glycerol with different concentrations depending on the producing company or country and, therefore, the value of the final sale price is different.

The excess of crude glycerol has promoted a continuous decrease in its market value, reaching US $ 170 / t in 2019. This is one of the main factors to enable the implementation of a process for the production of high added value product. As previously mentioned, the production of 1,3-propanediol, acrylic acid and acrylonitrile are promising products from glycerol. Currently, the industrial production of acrylonitrile and acrylic acid use propylene as raw material, which price was approximately US $ 880/t, in 2019 [10]. The price difference between crude glycerol and propylene opens up a great opportunity for the production of these derivatives from glycerol [16]. 1,3-propanediol is obtained mainly from glucose, by fermentative processes, and ethylene oxide, by catalytic routes [17], whose prices, on average, were US$ 460/ton and US$ 1200/ton, respectively, in 2019 [10]. Thus, there is a good opportunity for developing an economically competitive and sustainable industry, using glycerol as raw material in an integrated biorefinery.

Figure 5 shows the exported values and volumes of crude glycerol considering the five largest exporters: Indonesia, Brazil, Spain, Germany and France. The data were obtained using SH6 1520.00, which includes crude glycerol, water and glycerol lye.

The amount of crude glycerol (SH6 1520.00) exported worldwide reached 2.2 million tons in 2019, representing only US$ 457 million. In **Figure 5a**, it can be seen that Indonesia has the largest export revenue for glycerol, followed by Brazil. Besides these countries, The United States are also a large producer. Germany, Spain and France are also major producers in Europe. Spain has increased its exports considerably, surpassing France and Germany. Due to the great availability of glycerol, there was a considerable drop in prices, represented by the exported values, as shown in **Figure 5b**.

These countries have programs for addition of biodiesel in the conventional diesel due to their strategies for environmental improvement, mainly Indonesia and Brazil. Indonesia represents 26.7% of the world exported value and has the world's daring program for replacing fossil fuel, producing 30% biodiesel in 2020, and an increasing of 40% is expected in the coming years. Brazil will increase domestic consumption by 50% by increasing the percentage of biodiesel in the diesel from 10–15% between 2019 and 2023 [18].

Another SH6 present in the ITC is related to purified glycerol (SH6 290,545), which covers all glycerol that is subjected to different degrees of purification, from the simplest to more complex purification treatment, such as double distilled glycerol. In 2019, purified glycerol was traded at an average of $ 895/ton.

The global market for purified glycerol was $ 2.6 billion in 2019. Asia-Pacific represented the largest market, with approximately 35% of the total, followed by Europe, with 29%, and North America, with about 21%. The growth in these regions can be attributed to the growing adoption of healthy eating styles and the expansion of end-use industries. Glycerol sales are forecast to reach US$ 3.5 billion by 2027, expanding at a 4.0% CAGR. However, due to COVID-19, a severe impact on the glycerol supply chain is expected since the production of biodiesel has been reduced, which may result in the closure of several manufacturing companies.

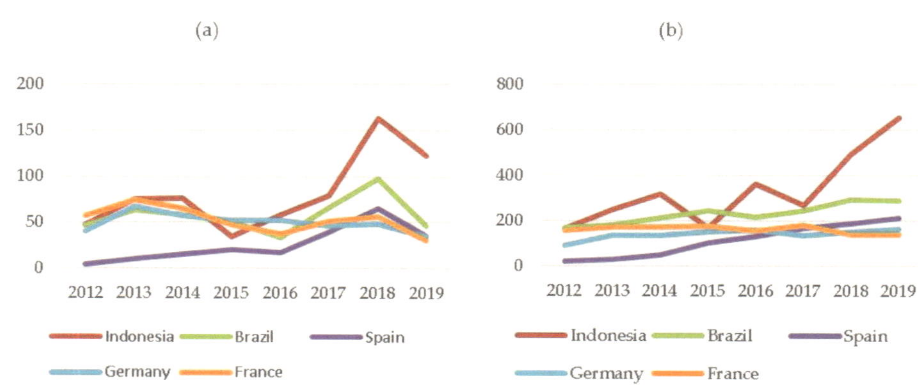

Figure 5.
Exported values and volumes of crude glycerol: (a) Exported values of crude glicerol (million US$/year); (b) Exported volumes of crude glycerol (thousand ton/year).

Europe will be the most affected region, especially in countries such as Italy and United Kingdom, adversely affecting the market growth in the region [19].

Purified glycerol has been largely commercialized worldwide, reaching US $ 1.2 billion in 2019. The main exporters of purified glycerol are shown in **Figure 6**.

Malaysia, since 2012, is the country with the highest exported value and volume (**Figure 6a**). Wilmar International, a large producer of biodiesel, has glycerin purification plants in Malaysia and Indonesia. However, Germany has been standing out in the export of purified glycerol, being the second largest exporter in values in 2019, exporting US $ 236 million, surpassing Indonesia, with US $ 232 million.

Indonesia surpassed Germany in 2018, but in the following year, it saw a drastic reduction in its numbers, exporting smaller values, in US$/year. In 2018, there was the highest peak in the export of purified glycerol, reaching a total 1.8 million tons, with a market value of US $ 1.6 billion. However, it is possible to observe a fall in the values and volumes exported by all countries in 2019, demonstrating the same behavior of crude glycerol (**Figure 6**).

Brazil has reduced considerably the imports of purified glycerol, with a 61% drop in the imported volume from 2014 to 2019. This is due to the implementation of new crude glycerol refining plants from 2014, which generated higher quantities of purified glycerol for internal consumption. Besides being a producer, the country has become one of the major exporters of purified glycerol, tripling the exported volume from 2014 to 2019, reaching US$ 56 million and 126 thousand tons. Comparing the average import and export prices, the country imports a higher cost glycerol, US$ 1275/ton, while exporting about US$ 440/ton. This difference can be justified by the different levels of quality and purity of the product, in addition to the continuous increase of its offer in the world [20].

4.2 1,3-Propanediol

1,3-propanediol (1,3-PDO) is produced from two main routes: biotechnology, developed by DuPont Tate & Lyle, through genetically modified microorganisms that convert glucose into 1,3-PDO; and catalytic, through the hydroformylation of ethylene oxide over cobalt catalysts at high hydrogen pressures. Another way to obtain this compound is from glycerol, however, still without an industrial plant in operation [21].

1,3-PDO is used for the production of polyesters and polyurethanes, being polytrimethylene terephthalate (PTT) the most used, representing 30% of the market. These polymers are widely used in the industry for the production of textile fibers and foam for mattresses, car seats and thermal insulation [22].

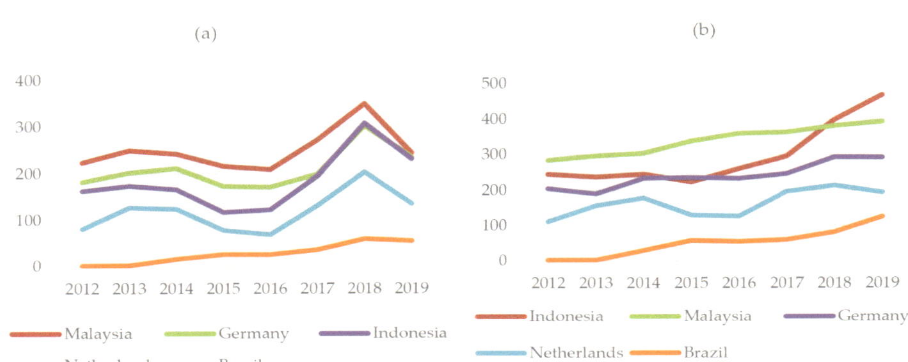

Figure 6.
Exported values and volumes of purified glycerol (SH6: 290545): (a) Exported values of purified glicerol (million US$/year); (b) Exported volumes of purified glycerol (thousand ton/year).

Comparing to polyethylene terephthalate (PET) and butylene terephthalate (PBT), PTT has a high strength, softness and elasticity as differential, and can be used in several segments of clothing, carpets, thermoplastics and monofilaments. 1,3-PDO also has applications in other sectors, such as production of solvents, adhesives, laminates, resins, cosmetics and personal hygiene and cleaning products [17].

North America is the most prominent region concerning the global market of 1,3-PDO, leading the market and having large growth perspective for the coming years. This is due to the high consumption of PTT and the high demand for polyurethane. The 1,3-PDO market in the year 2020 has been valued at US $ 401.7 million with an estimated growth of US$ 292.4 million from 2019 to 2024, presenting a CAGR of 11.4%, already considering the impact caused by the COVID-19. North America is responsible for 74% of this growth [23, 24].

1,3-PDO is not an isolated product in the ITC database, being grouped in SH6 2905.39, which contains the group of the heaviest diols, such as butanediols, hexanediols, but does not consider ethylene glycol and 1,2-propanediol, which are found in SH6 2905.31 and 2905.32, respectively. Thus, the analysis of the 1,3-PDO was performed considering also these compounds.

Since 2013, Germany leads exports of diols, exporting US$ 387 million, which represented 165 thousand tons in 2019, with Merck KGgA and BASF being the main producing companies in the country, mainly of heavier diols. The main world producer of 1,3-PDO is the United States, where DuPont Tate & Lyle is located, which represents the majority of exports of this product in the country. According to the **Figure 7**, the United States was the second most expressive country in the diol market, having exported US $ 355 million and 176 thousand tons in the same year.

China has been highlighted in the export of diols found in SH 2905.39 since 2013, more than doubling its exports. Until 2019, it handled 96 thousand tons, representing US$ 161 million. However, the country also stands out as the largest importer since 2013, having imported US$ 339 million and 220 thousand tons in the last year, as shown in **Figure 8**.

The great representativeness of China in the global scenario is explained by the high demand for diols and their derivatives, including 1,3-PDO, for which it has factories in the country, with emphasis on Zhangjiagang Glory Biomaterial Co., Ltd. and Haihang Industry Company Ltd.

Brazil has no expressiveness in the 1,3-PDO market, which is an isolated product, with NCM 2905.39.20.

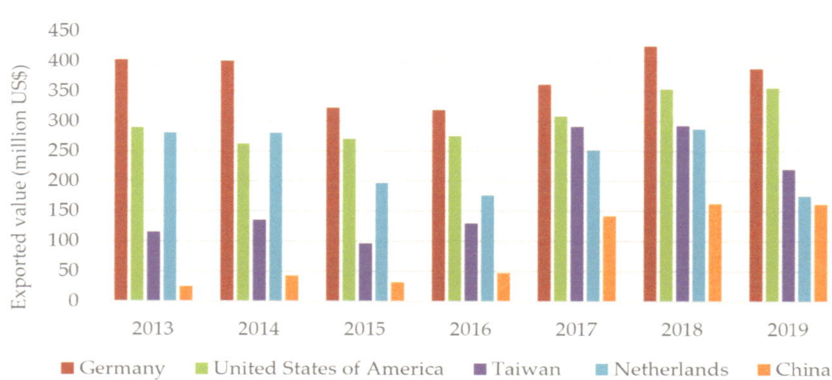

Figure 7.
Main exporters of SH6 2905.39 related products.

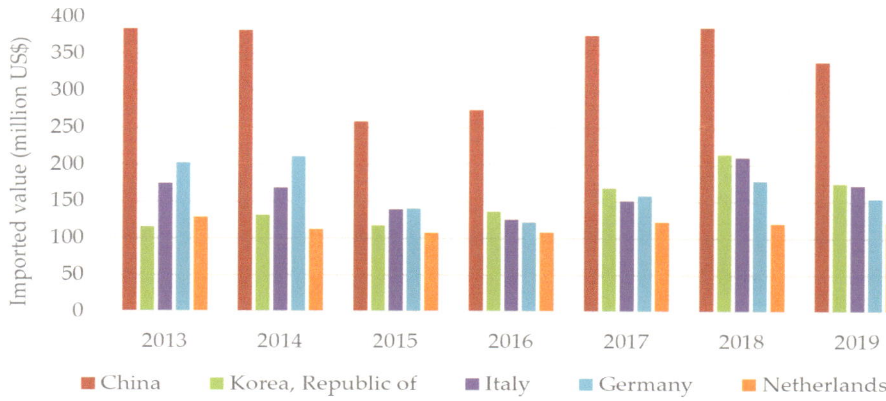

Figure 8.
Main importers of diol alcohols, except ethylene glycol and 1,2-propanediol (SH6 2905.39).

4.3 Acrylonitrile

Acrylonitrile is produced on a commercial scale through ammoxidation of propylene by catalytic routes, using bismuth-molybdenum oxides as catalysts [25]. Acrylonitrile is used in the manufacture of thermoplastics and textile fibers. Its main applications include ABS thermoplastic resin, acrylic fibers, carbon fibers, nitrile rubber, among others [26]. In 2019, acrylic fibers were responsible for most of the use of acrylonitrile, followed by ABS, a thermoplastic resin used in the construction, automotive and consumer goods industries.

Before COVID-19, growth forecasts for the acrylonitrile market were more promising, showing a 4.3% growth in CAGR for the coming years compared to the US$ 11.8 billion market in 2019 [27]. However, information released in [28] shows a lesser prospect of growth after the outbreak of the pandemic, forecasting a market of US$ 10.9 billion in 2020 and forecast to reach US$ 12.4 billion in 2024, with 2.2% CAGR. The growing demand for acrylic fibers and ABS and composites, such as carbon fiber, are the drivers of the growth of this market.

Figure 9 shows the largest exporters of acrylonitrile in the world. Acrylonitrile is isolated in SH6 2926.10. According to the data available at the ITC, 1.6 million tons of acrylonitrile was exported worldwide in 2019, with an average export value of US $ 1.486/ton, representing a trade of US $ 2.4 billion.

The United States is the largest exporter of acrylonitrile since 2013, with US$ 533 million in 2019. In 2018, the country exported more than 504 thousand tons of the product, representing 31% of global exports.

In 2019, INEOS Nitriles, one of the main producing companies, announced the closure of the plant located in Teesside, United Kingdom, with capacity of 300 thousand tons [29]. With the decline in trade in 2020 due to the pandemic, the withdrawal from the market of this expressive quantity may contribute to a smaller drop in the value in US $ /weight of acrylonitrile in 2020 and 2021.

China, since 2013, is the largest importer of acrylonitrile, with US$ 484 million in 2019 (**Figure 10**). According to the ITC, the average import price for acrylonitrile this year was US$ 1622/ton. Between 2014 and 2018, there was a drastic decrease in Chinese imports, from 152 thousand to 59 thousand tons.

In Comex Stat, acrylonitrile is an isolated product with NCM 2926.10.00 [20]. Brazil is an exporter of acrylonitrile, ranking eighth worldwide in 2019 [10]. In 2018, Brazil exported 83 thousand tons of acrylonitrile, representing about US$ 150 million. Brazil's representativeness in this market is due to the presence of Unigel,

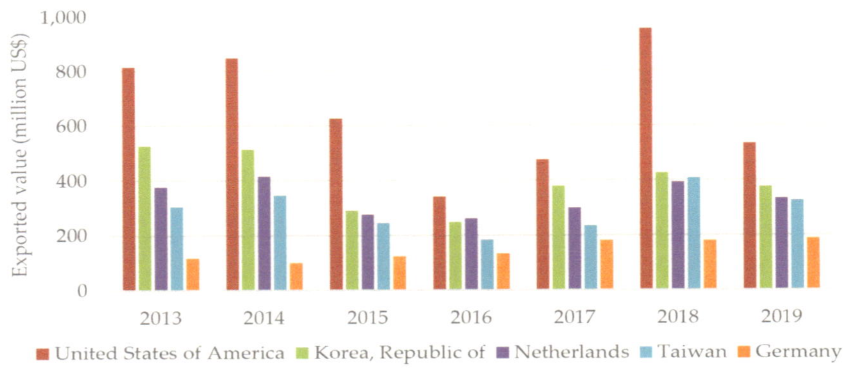

Figure 9.
The largest exporters of acrylonitrile.

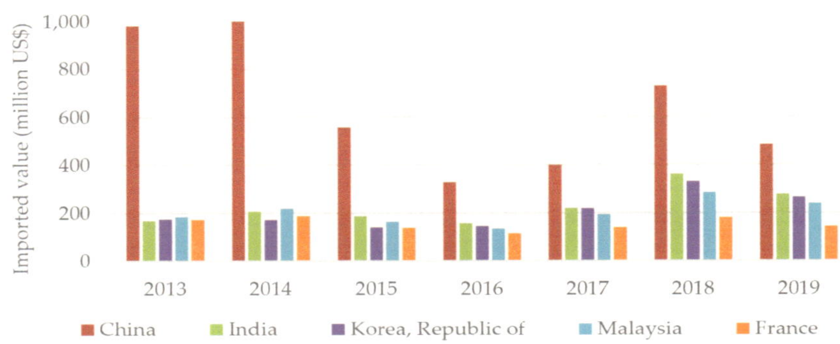

Figure 10.
Main importers of acrylonitrile.

which allocates a large part of its production for export. Unigel is the only producer in the southern hemisphere of the product with a nominal capacity of 100 thousand tons/year [30].

4.4 Acrylic acid

Acrylic acid is produced by the oxidation of propene in gas phase, over bismuth and molybdenum catalysts [16]. However, there is a potential to obtain acrylic acid from glycerol, being a more sustainable alternative to the use of petrochemical compounds [31].

Acrylic acid is applied in the production of polymers, latex, paints, leather finishing and paper coatings, in the textile sectors, in the production of surfactants, among others [32]. Superabsorbent polymers are the most represented segment in the market, being used in diaper production, water treatment and crude oil extraction.

The acrylic acid market may reach US$ 21 billion in 2027 [33]. In 2019, the volume handled in the acrylic acid market was 6.3 million tons and is expected to reach 8.3 million tons in 2025 [34]. The acrylic acid market mainly covers the Asia-Pacific region, with India and China accounting for 70% of product demand. The market has been driven by the growth of construction sectors, superabsorbent polymers and automobile industries, mainly in China, India and Brazil.

In the ITC portal, acrylic acid and its salts are included in SH6 2916.11. In 2019, world trade reached 774 thousand tons at an average price of US$ 1100/ton.

Saudi Arabia has become a major producer and has become the largest exporter of the product since 2015, surpassing China and Germany, selling 195 thousand tons in 2019, which represented US$ 217 million (**Figure 11**). This is due to the implementation of the new petrochemical complex in the country, which produces large amounts of acrylic acid [35]. Petroleum & Chemical Corporation and BASF SE, located in China and Germany, respectively, are responsible for a large part of the production of acrylic acid in the world and are highly representative of the market.

Taiwan was the largest importer in 2019, a position occupied in previous years by the United States and Belgium. The country imported 77 thousand tons, moving a value of US$ 71 million (**Figure 12**). China is also a major importer of acrylic acid, occupying the fourth position in the world ranking in 2019, with a large domestic demand for its derivatives. Some of these countries may not be producers, only selling acrylic acid.

Belgium has been the lead in imports of acrylic acid since 2013, losing its position to Taiwan, United States and China in 2019, since it has opened new plants producing acrylic acid and superabsorbent polymers [36].

In Brazil, acrylic acid is found in NCM 2916.11.10. Since 2014, there has been a significant decrease in external dependence on acrylic acid, with a reduction of 80% in the imported values, reaching 11 thousand tons and US$ 1293/ton in 2019.

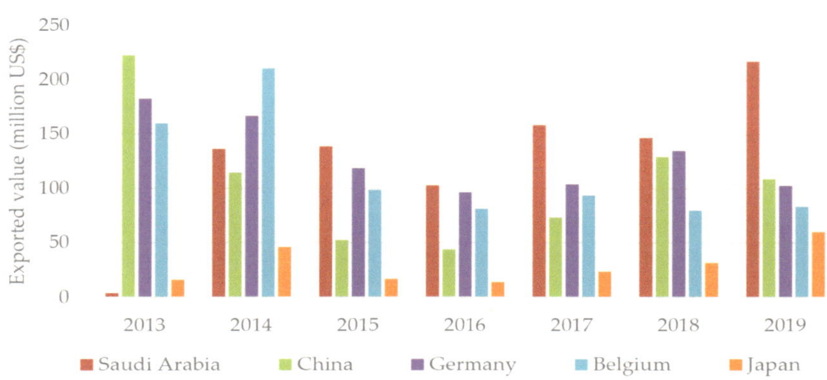

Figure 11.
Main exporters of acrylic acid and its salts.

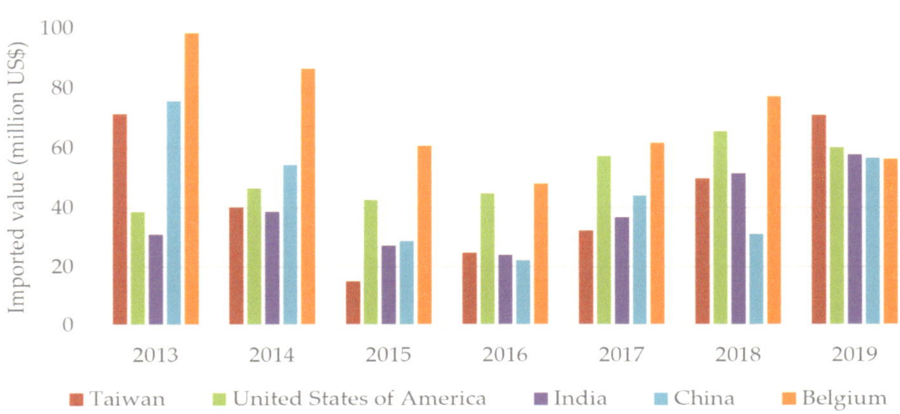

Figure 12.
Main importers of acrylic acid and its salts.

The decrease in the imported volume was due to the entry of Brazil into the acrylic acid market since 2014. This scenario was influenced by the investments made by BASF and Braskem, which, in 2014, carried out a large acrylic acid production project, adding domestic market and positively influencing Brazil's trade balance. In this sense, there was a considerable increase in the exported value and volume of acrylic acid in the country. In 2014, there were no records of exports of this product in the country, starting in 2015, when it reached the quantity of 7 thousand tons. In 2019, Brazil exported 18 thousands tons with an export value of US$ 26 million at an average price of US$ 1466/ton.

5. Conclusions

The growing worldwide use of biodiesel, an alternative renewable fuel, has generated large amounts of crude glycerol, a byproduct in its production, which presents large opportunities for application, after purification. Purified glycerol is a high-value and commercial chemical that may be used in a large number of applications, including its conversion into value-added products. Among them, 1,3-PDO, acrylonitrile and acrylic acid are promising products.

Commercial information about these products can be obtained from open websites of social associations, producing companies and market consulting companies free of charge. The careful analysis of these information shows there is high added-value when producing 1,3-PDO, acrylonitrile and acrylic acid from glycerol due to the difference between the prices of the products generated and the raw material.

This work shows that there is great potential for economic feasibility for the development of processes, in an integrated biorefinery, for the industrial production of chemicals from glycerol.

In addition, the effective use of glycerol as raw material for the production of chemicals, conventionally obtained from fossil raw material, may contribute to the reduction of using non-renewable natural resources and to the development of a greener chemistry.

Acknowledgements

The authors gratefully acknowledge the support of the RCGI – Research Centre for Gas Innovation, hosted by the University of São Paulo (USP) and sponsored by FAPESP – São Paulo Research Foundation (2014/50279-4) and Shell Brasil.

This study was financed in part by the Coordenação de Aperfeiçoamento de Pessoal de Nível Superior - Brasil (CAPES – Personnel Coordination of Improvement of Higher Level) - Finance Code 001.

The authors also gratefully acknowledge the financial support of the FAPESB – Bahia Research Foundation and CAPES National Exchange Program in partnership with Universidade Federal da Bahia and Universidade de São Paulo.

Conflict of interest

The authors declare that there are no conflicts of interest to disclose. The funding sources were not involved in the analysis and interpretation of data, neither in the writing of the manuscript, nor in the decision of submission for publication. This decision was based entirely on technical relevance and conjuncture pertinence of the study, based solely on the judgment of the authors.

Author details

Alisson Dias da Silva Ruy[1,2], Ana Luíza Freitas Ferreira[1], Antônio Ésio Bresciani[2], Rita Maria de Brito Alves[2]* and Luiz Antônio Magalhães Pontes[1]

1 Departamento de Engenharia Química, Universidade Federal da Bahia, Rua Prof. Aristides Novis, no. 02, 2nd floor, Federação, 40210-630, Salvador, Bahia, Brazil

2 Departamento de Engenharia Química, Universidade de São Paulo, Escola Politécnica, Av. Prof. Luciano Gualberto, no. 380, Tv. 3, 05508-010, São Paulo, SP, Brazil

*Address all correspondence to: rmbalves@usp.br

References

[1] Lietti L. Glycerol valorization towards a biorefinery [thesis]. Politecnico Di Milano; 2017.

[2] Werpy T, Petersen G. Top Value Added Chemicals from Biomass Volume I: Results of Screening for Potential Candidates from Sugars and Synthesis Gas. National Renewable Energy Lab., Golden, CO (US). 2004. DOI: 10.2172/15008859.

[3] Singh D, Sharma D, Soni SL, Sharma S, Sharma PK. A review on feedstocks, production processes, and yield for different generations of biodiesel. Fuel. 2020;262:116553. DOI: 10.1016/j.fuel.2019.116553.

[4] Tiseo I. Global biodiesel production by country 2019 [Internet]. 2020. Available from: https://www.statista.com/statistics/271472/biodiesel-production-in-selected-countries/ [Accessed: 2020/07/28]

[5] Gebremariam SN, Marchetti JM. Economics of biodiesel production: Review. Energy Convers Manag. 2018;168:74-84. DOI: 10.1016/j.enconman.2018.05.002.

[6] Quispe CAG, Coronado CJR, Carvalho JA. Glycerol: Production, consumption, prices, characterization and new trends in combustion. Renew Sustain Energy Rev. 2013;27:475-93. DOI: 10.1016/j.rser.2013.06.017.

[7] Bagheri S, Muhd N, Yehye WA. Catalytic conversion of biodiesel derived raw glycerol to value added products. Renew Sustain Energy Rev. 2015;41:113-27. DOI: 10.1016/j.rser.2014.08.031.

[8] Wang Y, Xiao Y, Xiao G. Sustainable value-added C3 chemicals from glycerol transformations: A mini review for heterogeneous catalytic processes. Chinese J Chem Eng. 2019;27:1536-42. DOI: 0.1016/j.cjche.2019.03.001.

[9] ITC. International Trade Centre (ITC) [Internet]. 2020. Available from: https://www.tfafacility.org/international-trade-centre

[10] ITC. International Trade Centre - Trade Map [Internet]. 2020. Available from: https://www.trademap.org/

[11] Vivek N, Sindhu R, Madhavan A, Anju AJ, Castro E, Faraco V, et al. Recent advances in the production of value added chemicals and lipids utilizing biodiesel industry generated crude glycerol as a substrate – Metabolic aspects, challenges and possibilities: An overview. Bioresour Technol. 2017;239:507-17. DOI: 10.1016/j.biortech.2017.05.056.

[12] Kaur J, Sarma AK, Jha MK, Gera P. Valorisation of crude glycerol to value-added products: Perspectives of process technology, economics and environmental issues. Biotechnol Reports. 2020;27:e00487. DOI: 10.1016/j.btre.2020.e00487.

[13] Ardi MS, Aroua MK, Hashim NA. Progress, prospect and challenges in glycerol purification process: A review. Renew Sustain Energy Rev. 2015;42:1164-73. DOI: 10.1016/j.rser.2014.10.091.

[14] Hejna A, Kosmela P, Formela K, Piszczyk Ł, Haponiuk JT. Potential applications of crude glycerol in polymer technology – Current state and perspectives. Renew Sustain Energy Rev. 2016;66:449-75. DOI: 10.1016/j.rser.2016.08.020.

[15] Veluturla S, Archna N, Rao DS, Hezil N, Indraja IS, Spoorthi S. Catalytic valorization of raw glycerol derived from biodiesel: a review.

Biofuels. 2016;9:305-14. DOI: 10.1080/17597269.2016.1266234.

[16] Braga ER, Mustafa G de S, Pontes D de A, Pontes LAM. Economic analysis and technicalities of acrylic acid production from. Chem Ind Chem Eng Q Available. 2020;26:59-69. DOI: 10.2298/CICEQ180111025B.

[17] Ruy AD da S, Alves RM de B, Hewer TLR, Pontes D de A, Teixeira LSG, Pontes LAM. Catalysts for glycerol hydrogenolysis to 1,3-propanediol: A review of chemical routes and market. Catal Today. 2020. DOI: 0.1016/j.cattod.2020.06.035.

[18] International Biofuels. Indonesian B40 biodiesel plan back on track as palm oil prices improve [Internet]. 2020. Available from: https://biofuels-news.com/news/indonesian-b40-biodiesel-plan-back-on-track-as-palm-oil-prices-improve/ [Accessed: 2020/08/08]

[19] Grand View Research. Glycerol Market Size, Share & Trends Analysis Report By Source (Biodiesel, Fatty Acids, Fatty Alcohols, Soap), By Type (Crude, Refined) By End Use (Food & Beverage, Pharmaceutical), By Region, And Segment Forecasts, 2020-2027 [Internet]. 2020. Available from: https://www.grandviewresearch.com/industry-analysis/glycerol-market [Accessed: 2020/07/26]

[20] Comex Stat. [Internet]. 2020. Available from: http://comexstat.mdic.gov.br/pt/home

[21] Samudrala SP. Glycerol Transformation to Value-Added 1,3-Propanediol Production: A Paradigm for a Sustainable Biorefinery Process. In: Glycerine Production and Transformation-An Innovative Platform for Sustainable Biorefinery and Energy. IntechOpen; 2019. p. 24. DOI: 10.5772/intechopen.83694.

[22] Shi G, Xu J, Song Z, Cao Z, Jin K, Xu S, et al. Selective hydrogenolysis of glycerol to 1,3-propanediol over Pt-WOx/SAPO-34 catalysts. Mol Catal. 2018;456:22-30. DOI: 10.1016/j.mcat.2018.06.018.

[23] Markets and Markets. 1,3-Propanediol (PDO) Market worth $870 million by 2024 [Internet]. 2019. Available from: https://www.marketsandmarkets.com/Market-Reports/1-3-propanediol-pdo-market-760.html [Accessed: 2020/08/10]

[24] Business Wire. COVID-19 Impacts: 1,3-Propanediol (PDO) Market Will Accelerate at a CAGR of Over 11% Through 2020-2024 | Growing Polyester Applications Across Various Industries to Boost Growth | Technavio [Internet]. 2020. Available from: https://www.businesswire.com/news/home/20200814005012/en/COVID-19-Impacts-13-Propanediol-PDO-Market-Accelerate-CAGR [Accessed: 2020/08/15]

[25] Liebig C, Paul S, Katryniok B, Guillon C, Couturier JL, Dubois JL, et al. Glycerol conversion to acrylonitrile by consecutive dehydration over WO3/TiO2 and ammoxidation over Sb-(Fe,V)-O. Appl Catal B Environ. 2013;132-133:170-82. DOI: 10.1016/j.apcatb.2012.11.035.

[26] Rezaie F, Pirouzfar V, Alihosseini A. Technical and economic analysis of acrylonitrile production from polypropylene. Therm Sci Eng Prog. 2020;16:100463. DOI: 10.1016/j.tsep.2019.100463.

[27] Grand View Research. Acrylonitrile Market Size, Share & Trends Analysis Report By Application (Acrylic Fibers, Adiponitrile, Styrene Acrylonitrile, Acrylonitrile Butadiene Styrene, Acrylamide, Carbon Fiber), By Region, And Segment Forecasts, 2020-2027

[Internet]. 2020. Available from: https://www.grandviewresearch.com/industry-analysis/acrylonitrile-acn-market [Accessed: 2020/07/21]

[28] Market Watch. Acrylonitrile Market 2020 is expected to see magnificent spike in CAGR with Global Industry Brief Analysis by Top Countries Data which includes Driving Factors by Manufacturers Growth and Forecast 2024 [Internet]. 2020. Available from: https://www.marketwatch.com/press-release/acrylonitrile-market-2020-is-expected-to-see-magnificent-spike-in-cagr-with-global-industry-brief-analysis-by-top-countries-data-which-includes-driving-factors-by-manufacturers-growth-and-forecast-2024-2020-08-10 [Accessed: 2020/08/13]

[29] Ineos Nitriles. INEOS announces launch of consultation on the potential closure of Seal Sands Acrylonitrile Plant [Internet]. 2019. Available from: https://www.ineos.com/news/shared-news/ineos-announces-launch-of-consultation-on-the-potential-closure-of-seal-sands-acrylonitrile-plant/ [Accessed: 2020/08/01]

[30] Unigel. Acrílicos [Internet]. 2020. Available from: https://www.unigel.com.br/index.php/acrilicos/ [Accessed: 2020/08/01]

[31] Li X, Zhang Y. Highly Efficient Process for the Conversion of Glycerol to Acrylic Acid via Gas Phase Catalytic Oxidation of an Allyl Alcohol Intermediate. ACS Catal. 2016;6:143-50. DOI: 10.1021/acscatal.5b01843.

[32] Buitelaar MM, Van Daatselaar E, Van Teijlingen DG, Stokvis HI, Wendt JD, De Sousa Ribeiro RJ, et al. Process Designs for Converting Propylene Glycol to Acrylic Acid via Lactic Acid and Allyl Alcohol. Ind Eng Chem Res. 2020;59:1183-92. DOI: 10.1021/acs.iecr.9b04334.

[33] Globe News Wire. Acrylic Acid Market To Reach USD 20.19 Billion By 2027 | Reports and Data [Internet]. 2020. Available from: https://www.globenewswire.com/news-release/2020/08/12/2076817/0/en/Acrylic-Acid-Market-To-Reach-USD-20-19-Billion-By-2027-Reports-and-Data.html [Accessed: 2020/08/12]

[34] Expert Market Research. Global Acrylic Acid Market to Grow at a CAGR of 4.7% Over the Forecast Period of 2020-2025 [Internet]. 2020. Available from: https://www.expertmarketresearch.com/pressrelease/acrylic-acid-market [Accessed: 2020/08/02]

[35] Volkova M. Tasnee cuts feedstock to affiliates SEPC, SPC by 41% after drone attacks [Internet]. 2019. Available from: http://www.mrcplast.com/news-news_open-358733.html [Accessed: 2020/08/02]

[36] Alperowicz *N. Nippon* Shokubai inaugurates acrylic acid, SAP units in Belgium [Internet]. Chemical Week. 2018. Available from: https://chemweek.com/CW/Document/98987/Nippon-Shokubai-inaugurates-acrylic-acid-SAP-units-in-Belgium [Accessed: 2020/08/09]

Chapter 12

Collagen: From Waste to Gold

Safiya Noorzai and Casparus J.R. Verbeek

Abstract

Industrial processing of bovine hides into leather results in many unusable hide off-cuttings, shavings and trimmings. This waste raw material is under-utilised and presents a waste valorisation opportunity to derive a high-value product such as collagen. Collagen is a highly sought-after protein which consists of three polypeptide chains, comprising 30% of the mammalian body's protein, being the main component of skin, connective tissue and cartilage. The demand for collagen is rising at approximately 20% annually and global collagen-based biomaterials market is predicted to reach US$5 billion by 2025. This chapter presents a waste valorisation opportunity to extract collagen from waste bovine hide off-cuttings. Further, it discusses collagen extraction method optimization and methods used to investigate physicochemical properties of collagen are reviewed.

Keywords: collagen extraction, bovine collagen, bovine hide, acid-enzyme, solubilisation, extraction, waste valorisation

1. Introduction

Increased environmental awareness has led researchers to find alternative solutions to replace petroleum-based materials in a sustainable manner. With an increase in world's population, one of the most concerning problems the planet is currently experiencing is the cumulative waste from various industries. The world's population produce an astounding 3.6 million tonnes of municipal solid waste each day. It is projected to rise to 6.1 million tonnes per day by the year 2025. It is adversely affecting health, contaminating our air, landscape, fresh water and ocean life. Waste valorisation is one method of managing waste in a sustainable manner and in return deriving a high-value product. The meat industry constitutes many by-products, which are under-exploited, from which a large number of valuable proteins, fats and chemicals can be derived from. Specifically related to the meat industry are tanneries and rendering plants, which process bovine and cattle hides for leather and fat production.

Hide off-cuttings, shavings and finished leather scrap are generated as waste in tanneries. These are currently disposed of in landfill sites and they have high landfilling costs per mass unit due to their low density and present low compaction ability. At best, the hide off-cuttings and shavings are converted to animal feed providing little or no economical or sustainable value, despite their content in valuable biopolymers. Bovine hides are rich in the valuable protein collagen, especially in the corium layer of the skin.

Considering the high cost of collagen and the vast number of applications and industries it can be of value, a more sustainable and a waste valorisation option would be to recover as much collagen as possible from hide off-cuttings.

IntechOpen

Collagen is a structural protein, which provides strength, stability, and flexibility and is a major constituent of the skin tissue. Hence, bovine hides contain an abundance of collagen. The collagen molecule is a triple-helix comprised of three distinct alpha chains of repeating units of $(GLY\text{-}X\text{-}Y)_N$ amino acids, where X is often proline and Y is often hydroxyproline [1, 2].

Collagen is a highly sought-after protein, finding use in regenerative medicine, in cosmetics, used as casings, in supplements, films, pharmaceuticals, as a precursor to biodegradable materials, for use in tissue engineering and more recently in 3D printing [3–8]. The demand for collagen is rising at approximately 20% annually and global collagen-based biomaterials market is predicted to reach US$5 billion by 2025. Specifically extracting bovine collagen has many advantages over other potential sources, such as having a higher denaturation temperature in comparison to collagen from marine sources, extracting fish and porcine collagen present limitations; applications of fish collagen are limited because of its lower hydroxyproline content [9] and porcine products are prohibited by Muslim and Jewish communities [10].

This chapter aims to represent a background on waste generation in tanneries, to use the tannery waste bovine hide off-cuttings for extraction of high value collagen. Further, collagen extraction methodologies are discussed in detail and finally methods used to investigate physicochemical properties of collagen are reviewed.

2. Tannery processing: waste production

In recent years waste valorisation has attracted a significant amount of attention with the sole aim of managing waste in the most sustainable way. Waste from various industries is one of the most concerning problems the planet is currently experiencing and will increase with the increase in population and needs to be addressed. The meat industry constitutes many by-products that are under-exploited, from which a large number of valuable proteins, fats and chemicals can be derived from.

Tanneries and rendering plants process bovine and cattle hide for leather and fat production. Casualty and cattle used for meat consumption result in a large quantity of waste and one of the most valuable by-products is the bovine and cattle skins or hides.

Industrial rendering separates animal by-products into value-added products such as animal protein meal and rendered animal fat and tanneries aim to process hides into leather, however, a substantial amount of waste is still produced from these processes that can be used to derive high-value products. Collagen is such a product that can be extracted from hide off-cuttings that is additional waste generated during leather preparation steps. Considering the high cost of collagen and its vast number of applications, extraction of such high-value product from bovine hide off-cutting is both sustainable and economical.

2.1 Impact of tannery waste on the environment

As much as this sector is considered to play a vital role because it recycles and reuses the by-products of the meat industry, the processes carried out in the different stages have a serious environmental impact. Environmental concerns that result from tanneries are due to resource consumption such as water, chemicals, energy and the generation of emissions such as volatile organic compounds, wastewater and solid waste. Moreover, hide off-cuttings, trimmings, hair and fleshings are removed from the hides during the tanning process. Only about 25% by weight of raw salted hides results in the finished leather [11]. Furthermore, other solid wastes are also produced from wastewater and sludge treatment.

Figure 1 shows the stages carried out in a tannery and post tanning in order to convert hides into leather. These steps result in the release of corrosive gasses into the atmosphere and in large quantities contaminated wastewater. Though leather is used for many applications, from furniture to bags and shoes and is economical in many industries, some bovine hides such as bull-hides are often too thick to process and requires additional processing steps for thinning of hides.

During the conversion of bovine hides into leather (**Figure 2**) a vast number of chemicals are released into the environment and waste products are generated at each stage. **Table 1** is showing chemicals used and wasted generated at each stage of leather production.

2.2 Use of bovine hides for collagen production

As bovine hides are being converted to leather, additional waste is generated during the preparation steps. Collagen-rich hide off-cuttings, trimmings and defected parts end up in landfill or at best as animal protein feed which is of low value considering the processing costs.

Bovine hide off-cuttings, trimmings and potentially bull-hides that are too thick to process for leather production and calf-hides that have defects can be used for collagen extraction. Collagen is the most abundant protein found in the mammalian body, making up approximately 30% of the total body protein. This structural protein which provides strength, stability and flexibility is a major constituent of skin tissue [14] and hence bovine hides are rich in collagen, especially in the corium section of the hide [15].

Hide off-cuttings can come from various bovine sources, such as bull, cow, ox, calf and even bovine face-piece hides. Additional to bovine hide off-cuttings, bull-hides that are too thick to process and require additional thinning processes can also be used for collagen extraction. This reduces extra processing costs and can directly be used for collagen extraction.

Bovine hide off-cuttings can be processed for collagen extraction. This collagen can be used by various industries for many applications from biodegradable films, pharmaceuticals to cosmetics. Several methods and techniques can be applied to extract collagen from bovine hide off-cuttings and the most efficient, economical and environmentally favourable methods can be worked with in order to reduce chemical and solid waste. Further, the market value of collagen is a lot more than leather, ranging from $37 per gram to as high as $1000 per gram for native lab-grade collagen [16].

3. Collagen

Collagen is the most abundant structural protein found in the vertebrate body. Collagen is a rigid, inextensible, fibrous protein that is the principal component of connective tissue in animals, including tendons, cartilage, bones, teeth, skin and blood vessels. As a structural protein it is mainly used to give strength to structures in the body, however, it has different functions depending on the location of the body [17]. One-third of the total protein content in the mammalian body is collagen and accounts for three-quarters of the dry weight of the skin.

3.1 Collagen structure

The triple-helix of collagen consists of three distinct alpha chains coiled around each other and this is termed as tropocollagen. The tropocollagen units are arranged

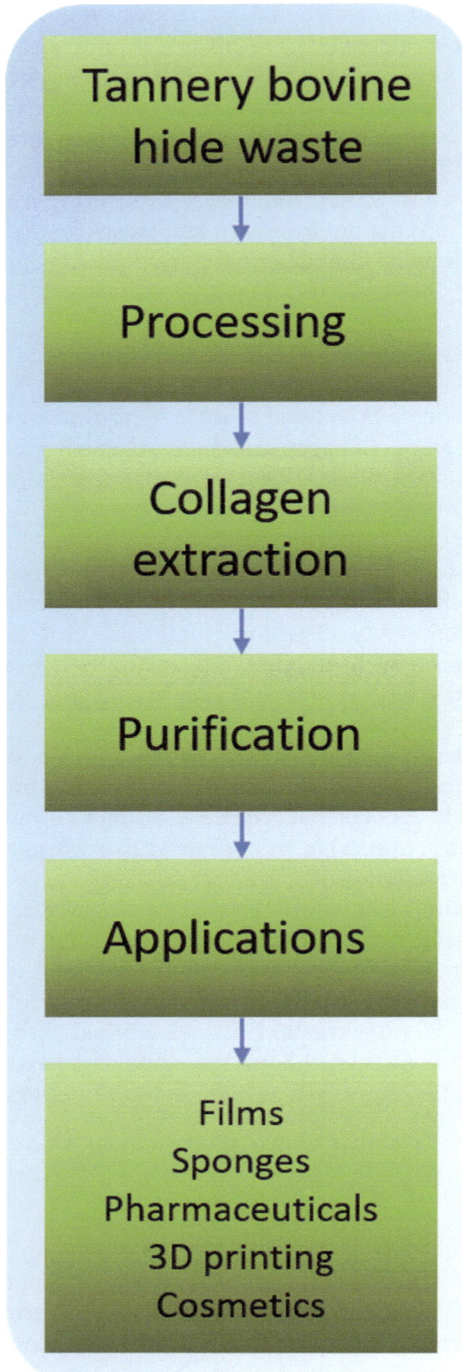

Figure 1.
Process flow of waste valorisation from tanneries to collagen extraction and possible collagen-based applications.

as fibres or sheets. A tropocollagen unit is about 285 kDa, 3000 Å in length and 15 Å in diameter. The triple helix is composed of repeating units of $(Gly\text{-}X\text{-}Y)_N$ amino acids, where X and Y are any amino acids, however, often X is proline and Y is hydroxyproline. The individual polypeptide chains of collagen each contain

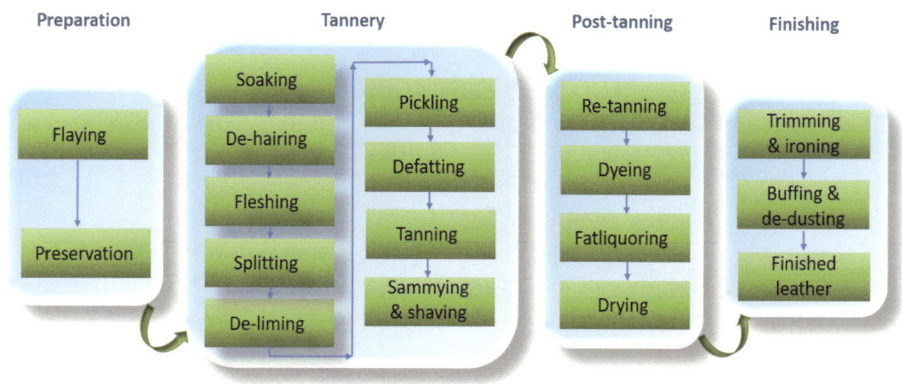

Figure 2.
Process flow of transformation of hides into leather [12].

Tanning step	Chemicals	Wastes generated
Preservation		
	Salt	Contaminated salt, raw hide trimmings
Soaking		
	Water, surfactants, and enzymes	Salted and contaminated wastewater
De-hairing		
	Water, sodium sulphide, and enzymes	Hair, alkaline water
Fleshing		
	Water, mechanical processes	Flashings, alkaline water
Splitting		
	Skin/hide	Limed hide
De-liming		
	Water, ammonium sulphate and weak acids	Acidic wastewater

Table 1.
Chemicals used at each stage of hide to leather conversion and wastes generated [13].

approximately 1000 amino acid residues. The accurate folding of these chains requires a glycine residue to be present in every third position of the polypeptide chain [1]. One-third of the amino acids in collagen in glycine and it always occupies the first position of the triplet. This is due to glycine being a small and an uncharged amino acid near the axis of the collagen triple helix. Glycine is a very crucial part of collagen molecule inherent characteristic as substitution of a single glycine for another amino acid disrupts the triple helix and results in skeletal deformities such as ontogenesis imperfect.

Imine acids make up approximately 25% of the residues in the collagen triple-helix. Imine acids – proline and hydroxyproline are typically found around the outside of the trip helix and the pentagon structure of these two amino acids includes the amine nitrogen and the α-carbon of the backbone chain. These limit the possible rotation in the amino acid (**Figure 1**) and hence forcing each collagen chain to form a left-handed helix. The high content of these imine acids makes the α-helix and β-sheet arrangements (generally found in proteins) unstable. Collagen triple-helix is held together by hydrogen bonding between chains. The NH group in

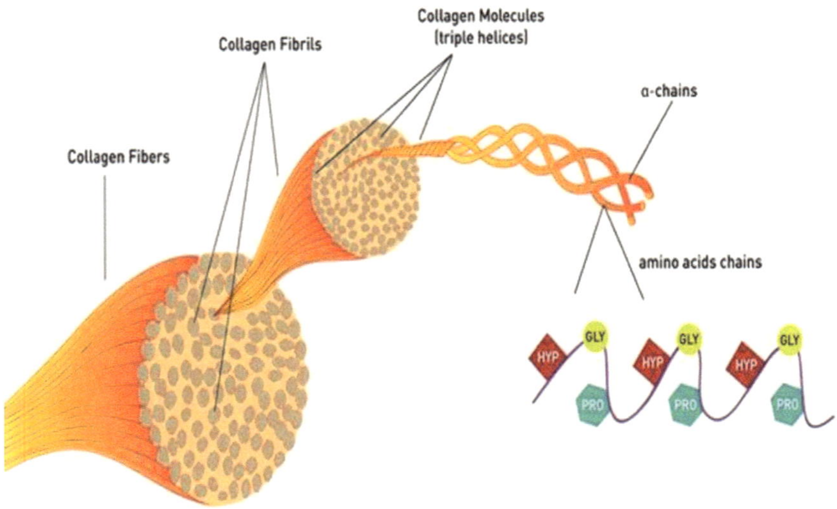

Figure 3.
Collagen structure being broken down to fibre, fibril, triple helix and an alpha chain respectively [20] (used with permission).

glycine in polypeptide chains forms H-bonds with adjacent peptide CO groups of the other chains.

After the formation of the collagen polypeptide chain, proline in the third position of the triplet in the amino acid sequence is hydroxylated by the enzyme propyl hydroxylase. The hydroxyl groups of the hydroxyproline and water molecules form hydrogen bonds that stabilise the triple-helix. Inhibition of hydroxylation causes diseases such as scurvy (caused by a lack of vitamin C in the diet) which is the inability of the triple-helix to form at body temperature (37°C) [18]. A decrease in imine acids (proline and hydroxyproline) content lowers the thermal stability of collagen as collagen loses its helical structure and shrinkage or denaturation occurs [18]. Avian and mammalian collagen have very similar amounts of hydroxyproline at 13.5% of the total amino acids. In comparison, aquatic animals have a lower level of hydroxyproline at approximately 10.3% [19].

The alpha-triple helix of collagen is shaped into a right-handed helix. The alpha chains each are shaped into a left-handed symmetry (the opposite direction), and then three of these alpha coiled strands get together to form a right-handed triple helix so when under strain, the chains twist into each other, giving strength and preventing unravelling. Each alpha helix is approximately 1.4 nanometres in diameter and 300 nanometres in length (approx. 1000 amino acids). The collagen molecule can be composed of either three identical alpha chains (homotrimers), or two or three different alpha chains (heterotrimers), however, the chain configuration depends on the collagen type being synthesised [2]. The hierarchical structure of collagen is zoomed-in starting from the alpha chains coiling together to form the triple helix is shown in **Figure 3**.

Cross-links that are covalent bonds occur between the ends tropocollagen before the formation of the collagen fibre. The triple helix and the cross-linking give rise to a collagen material that is very rigid, inextensible and stable. Since collagen on the primary level is composed of repeating units of Gly-X-Y amino acids, it is therefore rich in carboxylic acid groups, hydroxyl groups, amide and amine groups. The triple helix structure is stabilised by inter-chain hydrogen bonding and triple helix (tropocollagen) molecules parallel to each other are covalently cross-linked with each other through their aldehyde and amino groups, forming collagen fibrils. There are multiple types of hydrogen bonding patterns found in the triple-helix. These include, i) direct hydrogen bonding among the peptides (i.e. the NH group in glycine in each polypeptide chain forms H-bonds with adjacent peptide CO groups of the other chains), ii) water-mediated hydrogen bonding linking carbonyl groups, and iii) water-mediated hydrogen bonding, which links hydroxyproline OH groups and carbonyl groups. Collagen self-organisation forms bundles or a meshwork that determines the tensile strength and the elasticity and geometry of the tissue.

The various collagen types are distinguished by the ability of their helical and non-helical regions to associate into fibrils and to form sheets or to cross-link different collagen types. For example, a two-dimensional network of type IV collagen is unique to the basal lamina. Most collagen is fibrillar and is composed of type I molecules (**Figure 4**) [2].

3.2 Collagen synthesis

Tropocollagen is produced by fibroblasts found in connective tissue in mammals and birds. The collagens α-chains are translated on the rough endoplasmic reticulum (RER). Inside the ER hydroxylation of the specific proline and lysine residues occurs, however lack of vitamin C will hinder this step. Inside the Golgi apparatus glycosylation of pro-α-chain lysine residues and formation of procollagen occurs. Procollagen molecules are exocytosed into extracellular space. The rest of the synthesis steps occur outside the fibroblasts. Procollagen peptidases cleave terminal regions of procollagen, transforming procollagen into insoluble tropocollagen. Many staggered tropocollagen molecules are reinforced by covalent lysine-hydroxylysine cross-linkage (by Lysyl oxidase) to make collagen fibrils. Lysyl oxidase requires copper (Cu++) for its activity [22].

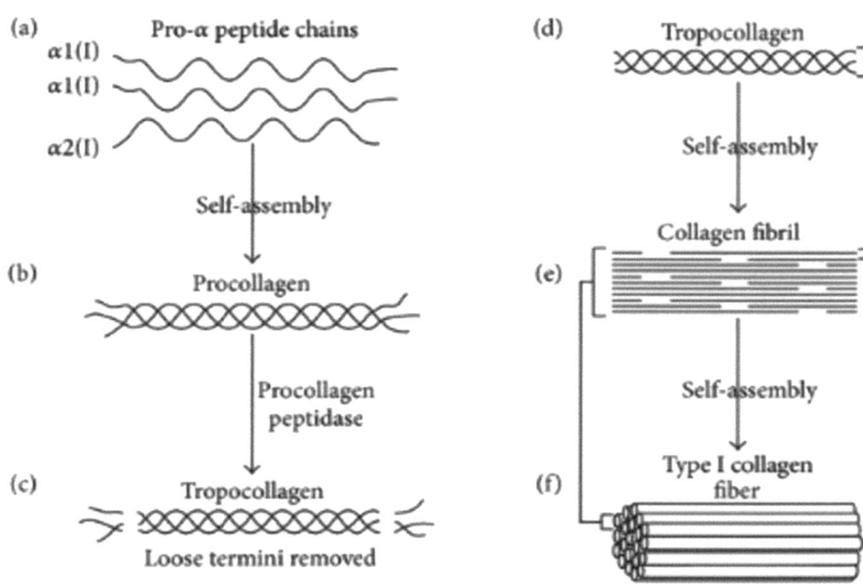

Figure 4.
structure of collagen, with b) procollagen (loose ends), triple-helix wound together and c) collagen subunit tropocollagen (loose terminal removed) for final self-assemble of the collagen fibril and fibre (d-f) [21] (used with permission).

3.3 Collagen fibres

The assembly of collagen fibrils into parallel bundles forms collagen fibres that have high strength and flexibility. When tropocollagen is assembled into collagen, it forms fibrous or sheet-like staggered structures. These fibrous structures have striations every 680 Å consisting of a dense-packed region where fibres overlap, and a loose-packed region is formed (**Figure 5**). In one single row, tropocollagen units are separated by 400 Å gaps, and these gaps are found in the loose-packed region. If the tropocollagen rows are aligned next to each other, each adjacent row is offset by 680 Å, forming a structure that repeats every five rows.

Hydrophobic and charged amino acid residues along the length of tropocollagen cause the staggered arrangement of tropocollagen. Tropocollagen units are aligned where the sum of the hydrophobic and charged region interaction between two units is strongest, hence the 680 Å staggering between units.

3.4 Collagen maturation

Inter-and intra-molecular covalent cross-links are formed between and within tropocollagen (collagen triple-helix) units giving strength to collagen fibres. Intra-molecular cross-links form between adjacent lysine groups and within individual triple-helix units and intermolecular cross-links occur between two triple-helix units comprising of two hydroxylysine groups and a lysine group.

The enzyme Lysyl oxidase converts the NH_3^+ group on the lysine and hydroxylysine sidechains to an aldehyde that then undergoes a condensation reaction forming an adol cross-link with other converted lysine sidechains. In each tropocollagen unit, four groups can contribute in the intermolecular cross-linking; lysines near the amino and carboxyl ends in the non-helical regions and hydroxylysines in the helical region. A hydroxyl-pyridinium cross-link is formed between one lysine and two hydroxylysine between residues near the amino-acid

Figure 5.
Collagen fibre showing the striations where tropocollagen is densely packed (light sections) [23] (used with permission).

end of one tropocollagen unit and the residues near the carboxyl-end of an adjacent tropocollagen unit. The enzyme Lysyl oxidase is small enough to fit between the 400-Å gaps between the triple-helix molecules to initiate the intermolecular cross-linking.

Collagen maturity or the amount of cross-linking increases drastically with age of the tissue and depends on the type and function of the tissue where collagen is found.

3.5 Collagen types

Collagen has a wide range of structural roles in mammalian and aquatic tissue. It is the major constituent of skin, bone, tendon, cartilage, blood vessels and teeth. Collagen is found in almost every organ of the body, starting from skin to the cornea of the eye. To serve functions in such diverse tissues, there are different types of collagen that differ in how they interact with each other and with other tissue.

There are more than 28 types of collagen identified. Collagen types I, II, III are the most abundant and most investigated for various applications. However, over 90% of the collagen found in the body is type I. The variations are due to the differences in the assembly of basic polypeptide chains, different lengths of the helix, and differences in the terminations of the helical domains [24].

Each collagen molecule is composed of three different polypeptide chains (α1, α2, and α3). Each chain is identified by its amino acid composition (**Table 2**). Collagen type I, for example, is identified for its constitution of α1 (I) and/or α2 (I) chains. The most commonly occurring variant of type I collagen consists of two α1 (I) and one α2 (I) chain. The alpha symbol is used to indicate a single chain component seen after collagen denaturation and the letter β, γ, and δ have been used to indicate covalently linked dimers, trimers or tetramers of the alpha chain.

The most common types of collagen are:

- Collagen type I: found in skin, tendon, organs and bone tissues.
- Collagen type II: main component of cartilage.
- Collagen type III: the main component of reticular fibres, alongside type I.
- Collagen type IV: Forms the bases of the cell basement membrane.
- Collagen type V: the main component of cell surfaces, hair and placenta.

Function	Description
Structural integrity	Collagens within the body serve largely for the maintenance and structural integrity of tissues and organs.
Entrapment and storage	The collagen within the body fulfils the role of entrapment, local storage, delivery of growth factors and cytokines and hence it plays an important role during organ development, wound healing and tissue repair.
Biodegradable	Collagen possesses the feature of being biodegradable and low immunogenicity.
Variety of applications	Collagen has been used in many industries, from the biomedical, cosmetic, pharmaceutical, leather, film industry to tissue engineering.

Table 2.
Collagen and its features [25].

3.6 Collagen sources

As collagen is one of the most abundant proteins on earth, it can be extracted from various sources. Collagen can be extracted from almost every living animal, including alligators and kangaroos. However, common sources of collagen for the food industry and tissue engineering applications include bovine skin and tendons, porcine skin and rat-tail. Collagen can also be extracted from marine life; it can be extracted from sponges to fish and jellyfish. All collagen sources are worth investigating as each source differs in the collagen type in terms of characteristics.

3.6.1 Bovine collagen

Collagen is extracted from many different sources; however, bovine collagen is seen to be the most used collagen type in a variety of different applications, such as the food industry, cosmetics, and medical applications. As the name implies, bovine collagen is a by-product of cows, mainly from the hides. It is a naturally occurring substance found in the skin, muscle, bones and tendons of cows. In the 1970s, the research on bovine collagen gained momentum, as researchers developed a system of extracting collagen and processing it in a liquid form [26].

The natural, unbleached skin and hair of cattle is the bovine hide (skin). Bovine hides are a by-product of the food industry from cattle. Bovine hides without complex processing can be manufactured into leather, which in turn can be used in the shoes and clothing industry. However further complex processing of the hides can be carried out to obtain the corium section of the hide for a variety of different medical and scientific applications [27]. One of the main applications of the corium is in the production of collagen.

Animal hide constitutes 60–65% water, 25–30% protein and 5–10% fats. The protein is mainly collagen [28]. Raw hides have four main parts; epidermis (6–10%), grain (less than 10%), corium (55–65%) and flesh and the thickness vary all over the animal (**Figure 6**) [29].

The epidermis and flesh layers are removed during tanning leaving the grain and corium layers. The grain is made up of collagen and elastin protein fibres. The corium is packed with collagen protein fibres. The thickness of corium also increases with age [30].

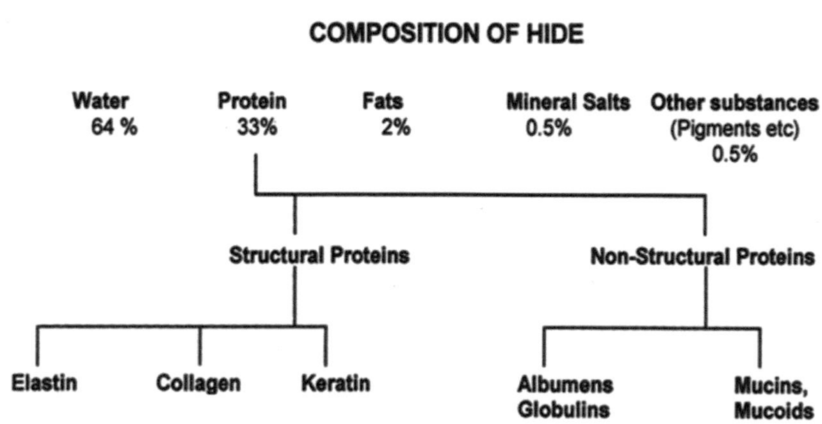

Figure 6.
The approximate composition of bovine hide [28] (used with permission).

3.6.1.1 Properties of bovine hides

Each section of the animal hide for its properties is discussed further [29] (**Figure 7**) [29]:

- *Epidermis:* There are two epidermis layers; one being the thin protective layer of cells during the life of the animal and the other being the flesh remain which is removed during tanning (leather production) by a process called liming.
- *Grain:* This layer is composed of elastin and collagen protein fibres. This layer is mainly used in the cosmetic industry for moisturisers and facial creams.
- *Corium:* The corium layer is made of collagen fibres, arranged in bundles and interwoven to give the structure strength, favourable elasticity and durability. Calf hides corium layer is thinner and smoother than the hides of mature animals; this is because the thickness of the corium increases with age.

3.6.2 Collagen from fish

Collagen from aquatic animals have been used as a safe substitute for bovine collagen, this is due to collagen from bovine sources have shown to be contaminated with some diseases. Fish solid wastes constitute 50–70% of the original raw material; however, this depends on the method of meat extraction [6].

Shark type I collagen forms fibrils under different conditions compared to bovine and porcine collagen [32]. For example, shark type I collagen gels and membranes have stronger rigidity and higher affinity to water vapour than those of porcine collagen, thus indicating the potential for utilising shark collagen as a new type I collagen material for various uses such as cell culture and medical technology [33].

3.6.3 Porcine collagen

Pigskin is a by-product of the pork production industry. Collagen extracted from pigskin or bone is not favourable to be a component of foods or pharmaceuticals

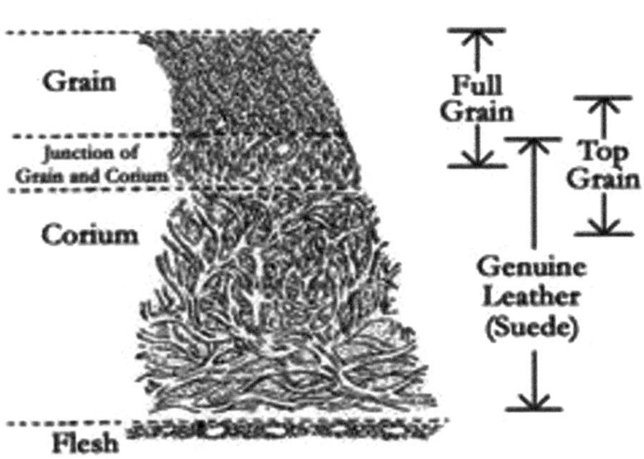

Figure 7.
Structure of bovine hide [31] (used with permission).

due to religious objections. Porcine collagen type I is extracted from pig hides, and in the medical field. Porcine collagen sheet material has proven to be useful as an implant for reconstructive surgery [34].

4. Collagen market and its applications

4.1 Collagen industries

There are many collagen-producing companies in around the world. However, not all of them produce 100% pure collagen but rather gelatine (hydrolysed collagen). These companies lack further innovation with the collagen, thus distributing the collagen in powder or liquid form to pharmaceutical and research industries. Therefore, extracting collagen from bovine hides and using this collagen to investigate high value applications would possibly generate huge economic potential for a product that is derived of waste materials.

Collagen plays an important role both in the mammalian and the non-mammalian body and in its extracted form. Due to collagen's high mechanical strength, it finds applications in several different industries, ranging from biomedical to the food industries.

4.1.1 Gelita

Gelita is the world's leading supplier of hydrolysed collagen proteins for the food, health and pharmaceutical industries. Gelita is based in numerous locations around the world with its headquarters in Germany [35]. However, the collagen Gelita produces is not 100% native collagen but hydrolysed collagen, in other terms it is gelatine.

4.1.2 Southern lights biomaterials

Based in Napier New Zealand, Southern Lights Biomaterials was founded in 2003. They provide high-quality processed and semi-processed biomaterials to medical device manufacturers across the globe. One of their flagships processed products is polymeric collagen, which is delivered to contracted customers [36].

The polymeric collagen produced by Southern Lights Biomaterials is type I collagen derived from bovine tendon and is naturally cross-linked [36]. They do not take advantage of using cattle hides or face-pieces. Their collagen is sold to independent contractors without further processing.

4.1.3 Revolution fibres

Revolution Fibres produce and market nano-fibre and nano-fibre products. Based in Auckland New Zealand, Revolution Fibres has developed its own technology for the industrial production of nano-fibre. This technology is called electrospinning [37]. Revolution Fibres manufacture biodegradable air filters from nano-particle sized fibres that are 'electro-spun' from collagen extracted from Hoki fish skins. They have launched a skincare range using collagen fibres to deliver plant extracts into the skin [38].

4.1.4 Waitaki biosciences

Waitaki Biosciences based in Christchurch New Zealand manufactures speciality nutritional supplement ingredients from natural, biological sources. Waitaki

Biosciences aims to target joint and bone health, immune and digestive support, along with skin and hair care. Marine collagen, natural collagen and chondroitin complex are some of their products [39]. The marine collagen produced by Waitaki Biosciences is in powder form, with a blend of ingredients selected from marine species. This marine collagen is designed for use as an oral supplement to support skin, nail and hair health [40].

Observing the collagen suppliers in New Zealand, there is a clear shortage in further innovation with the extracted collagen. Most of the above collagen suppliers distribute the collagen in a powder form or a liquid solution and export to external markets or distribute to local contractors. This collagen once supplied to contractors is usually blended in cosmetic products or encapsulated as pills in the pharmaceutical industry.

4.2 Collagen applications

Collagen has been widely used in a range of applications in cosmetic, biomedical, pharmaceutical, film industries, tissue engineering and recently in 3D/bio-printing.

4.2.1 Biomedical uses of collagen

i. Collagen sponges

   The collagen sponges act as a biological absorbance material. They have been useful in the treatment of severe burns and as a dressing for pressure sores, leg ulcers and donor sites. Collagen sponges can absorb large quantities of tissue exudate, smooth adherence to the wet wound bed with preservation of low moist climate as well as shielding against mechanical harm and bacterial infection [41].

   Collagen sponges have also been found to be effective as drug delivery systems. For example, the collagen sponges were found to be suitable for short term delivery of antibiotics, such as gentamicin [42].

ii. Collagen shields

   Originally, collagen shields were designed for bandage contact lenses. However, it's mostly used as a delivery device and has led to the development of drug delivery systems for ophthalmic applications [43]. For example, the collagen corneal shield is produced from porcine sclera tissue that closely resembles collagen molecules of the human eye. The collagen corneal shield promotes epithelial healing after corneal transplantation [44].

iii. Collagen mini pellets

   A mini pellet made from collagen is usually a rod with a diameter and length of 1 mm and 1 cm respectively. These are very useful as a drug delivery device. This is because the mini pellet (rod) is small enough to be injected into the subcutaneous space through a syringe needle and still spacious enough to contain large molecular weight protein drugs, such as interferon [42].

iv. Skin replacement

   Collagen has been widely used as vehicles for transportation of cultured skin cells or drug carrier for skin replacement and burn wounds [45]. Type

I collagen is suitable for skin replacement and burn wounds due to their mechanical strength and biocompatibility [7].

v. Bone substitutes

Collagen has been previously used as implantable carriers for bone inducing proteins [41]. Due to osteo-inductive activity; collagen itself has recently been used as bone substitutes [42]. Collagen combined with other polymers has been used for orthopaedic defects. Demineralised bone collagen in combination with hydroxyapatite was used as a bone graft material to treat acquired and congenital orthopaedic defects in rats [46].

vi. 3D printing and collagen

3D printing is the process of converting digital designs to three-dimensional solid objects. 3D printing works by initially designing a 3D image of the desired object, with computer-aided design (CAD) [47]. The object is divided into digital cross-sections by the program so that the printer can build the object layer-by-layer. Once the specified design is sent to the 3D printer, a specific material can be chosen. Depending on the printer type, this material can be rubber, plastics, paper, metals and more [48]. However, in the case of bio-printing; bio-ink (cells) and bio-paper (collagen, nutrients) are required [49].

4.2.2 Collagen and rheumatoid arthritis and osteoarthritis

Collagen has shown to have positive effects on rheumatoid arthritis and osteoarthritis [50]. Published studies [51] have reported that ingestion of type II collagen relieves joint discomfort associated with osteoarthritis and rheumatoid arthritis. The authors also conducted a randomised trial involving 60 patients with severe active rheumatoid arthritis; a decrease in the number of swollen joints and tender joints occurred in subjects fed with type II collagen [51].

4.2.3 Cosmetic applications of collagen

Collagen has great tensile strength and being rich in proline and hydroxyproline, it is the main component of fascia, cartilage, ligaments, tendons, bone and skin. Having these properties, it is responsible for skin strength and elasticity. Its degradation leads to wrinkles that accompany ageing. Collagen has become a valuable ingredient of many cosmetic formulations. Cosmetic uses include skin and hair products. Collagen type III is predominant in young skin; it is referred to as "restructuring" collagen as it appears during the wound healing process [7]. With ageing collagen type III decreases leading to wrinkles and lines, thus moisturising creams and cosmetic injects containing collagen have become in high demand [52].

Bovine collagen has been the most widely used source for cosmetic applications. Recently, collagen from other sources such as fish skin, pigskin, and range of cattle skin has been used in the cosmetics industry. However, collagens from various sources differ in their physiochemical properties. For example, they all have different thermal stabilities, and this can affect the formulation or the shelf life of the products [3].

4.2.4 Collagen films

Thin films or biodegradable films are flexible, transparent and often strong materials derived from natural polymers such as whey protein, collagen, starch, gelatine and many other natural renewable polymers [53, 54]. Due to rising environmental concerns, biodegradable films have attracted considerable attention especially from the food and drug packaging industries as they in constitution with other natural polymers can potentially replace plastic films which are derived from synthetic polymers [55].

Due to collagen being a biodegradable, biocompatible and a non-toxic polymer it has been used in the meat industry to form edible films and coatings through extrusion [56]. Collagen-based films in constitution with other biodegradable materials have been prepared in several studies to be used as packaging materials. Collagen's high tensile strength and the added advantage of biodegradability makes it an ideal agent for natural polymer films.

One of the main applications of collagen films in the biomedical industry is as a barrier membrane. These collagen films have been used for slow-release drug delivery and they have been used for the treatment of tissue infection, such as infected corneal tissue or liver cancer [42].

Edible films and coatings are a category of packaging materials. They differ from other bio-based packaging materials, and conventional packaging, by being formed from edible ingredients. These films and coatings may be used to reduce the amount of synthetic packaging used in a product or allow conversion from a multi-layer, multi-component packaging material to a single component material. The purpose of edible films and coatings may be to inhibit migration of moisture, oxygen, carbon dioxide and or to improve the mechanical integrity or handling characteristics of the food. Edible films may also be used to separate different components in multi-component foods, thereby improving the quality of the product. Edible films may also help to maintain food quality by preventing moisture and aroma uptake or loss after opening of the synthetic packaging.

The use of natural polymers such as collagen for film preparation has many advantages over synthetic and petroleum-based polymers. Biopolymer films for the purpose of packaging materials have the advantages of biodegradability, renewability, and environmental compatibility. Collagen also has good film-forming properties, high tensile strength, good thermal stability, and the fact that the collagen is derived from waste hide off-cuttings presents a sustainable solution. One drawback of collagen-based films is the inflexibility of films. However, this can be overcome by the addition of plasticizers to improve the flexibility and elongation (%) properties of the films. The use of plasticizers has been shown to provide improvement of films in terms of flexibility and elongation; however, this is generally at the expense of strength and stiffness. The effect of plasticiser concentration should, therefore, be investigated to identify best concentration results in the optimum mechanical, thermal and physical properties.

Biopolymer films made for the food industry as coatings or packaging needs to be transparent, have desirable tensile strength and elongation, it should be edible and possibly have a high resistance to transmission of liquids, gases and fats and oils. However, the above criteria will vary depending on the food industry application of the film.

Sionkowska et al. [57] prepared biopolymer films based on blends of collagen and silk fibroin. Films were prepared by method solution casting and characterised for their mechanical properties and structure. Film blends of collagen and silk fibroin showed better mechanical properties than for pure silk fibroin films. Sionkowska et al. [57] concluded that the better mechanical properties of the blend films were due to molecular interactions between collagen and silk fibroin. No plasticizing agent was added in the preparation of collagen and silk fibroin blend films. This would result in a very brittle and stiff film due to interactions between protein chains through hydrogen bonding, electrostatic forces and hydrophobic interaction [58]. Hence the per cent elongation values of the film blends were very low (0.30–5.10%) [57].

Not all collagen extraction methods result in a collagen product that will be suitable for film preparation. Hence, to develop a collagen film with desired properties, it is necessary to investigate the various processes to prepare acid/alkaline/enzyme/acid-enzyme collagen that could easily be used as a raw material for extruded or casting of collagen-based films. O'Sullivan [6] reported that hydrochloric acid solubilisation extraction method of collagen is not favourable for the fabrication of edible films. However, acetic acid solubilisation with further processing gave a suitable collagen product as a raw material for the fabrication of edible film fabrication.

5. Methods used to extract collagen from bovine hides

5.1 General extraction procedure requirements

Every bovine collagen extraction procedure is restricted to the following four variable conditions:

- De-hairing, cleaning and storage of the hide section off-cutting.
- Cutting the de-haired hide section into approximately 1 cm x 1 cm pieces.
- Extraction temperature: For bovine tissue, the extraction procedure can be carried out at room temperature, as collagen denaturation temperature for bovine is ~39°C. However, it is preferable to extract collagen at a temperature of ~4°C to prevent contamination.
- Solubilisation: acid solubilisation, acid and enzyme solubilisation, or modified methods combining acids and enzymes.

Prior to collagen extraction, the sample is chopped to increase the extraction surface area and to speed up the extraction process. However, the temperature of the sample needs to be monitored, as high temperatures will unravel the tropocollagen making it soluble in solution, resulting in gelatine (denatured collagen). This greatly reduces the value of the protein, thus if native collagen is desired, any heating or denaturation of collagen should be avoided at every step of the process. Bovine collagen extraction is mostly carried out at temperatures of approximately 4°C to prevent bacterial contamination [9].

Collagen from juvenile sources (e.g. new-born calves or chicken embryos) will readily swell and dissolve in a low concentration of acetic acid solution and can be recovered by precipitating out the collagen by adding 1 to 5 M NaCl. However different types of collagen from different tissues will precipitate at different NaCl concentrations [59].

The older the animal/tissue sample, the greater the amount of lysine-hydroxylysine covalent cross-links that form between tropocollagen units. These cross-links typically form between the unwound part of a tropocollagen strand and another part of another tropocollagen unit, improving structural strength and chemical resistance of collagen, making the sample largely insoluble in acetic acid. The amount of cross-linking depends on the type of tissue (i.e. tendons are highly cross-linked to give strength) and age of the tissue (i.e. mature sources, such as bull-hides have high cross-linking in comparison to younger sources such as calf-hides) [60].

In order to dissolve mature collagen, pepsin enzyme can be added to the acetic acid solution, which attacks and cleaves the unwounded part of tropocollagen, allowing the tropocollagen units to separate and dissolve [59].

The following sub-sections discuss the main extraction steps/parameters or variables in more detail.

5.2 Temperature control

To prevent collagen denaturation and contamination, majority of the researchers carry out the collagen extraction process at approximately 4°C. Contamination occurs due to thermal denaturation or microbial degradation (**Table 3**).

5.3 Fat removal and demineralisation

Once the collagen source is de-haired, sized and cleaned it is then processed for defatting. Majority of collagen extraction processes defat the tissue of interest with an organic solvent or detergent prior to extraction (**Tables 4** and **5**).

5.4 Non-collagenous protein removal

Contaminating proteins need to be removed after defatting and demineralization. Most collagen extraction methods utilise salt or alkali solutions to solubilise the contaminants. Collagen is a lot more chemically resistant than most other proteins therefore, it is much less likely to be degraded or solubilised by a weak salt (**Table 6**).

Collagen source	Temperature (°C)	Reference
Bovine	4 °C	[14, 61–65]
Fish	4–9°C	[61, 62, 66–72]

Table 3.
Processing temperatures used to extract collagen.

Source	Solvent	Reference
Bovine	Acetone	[19, 65]
Fish	0.5% detergent 10% butyl alcohol 15% Butyl Alcohol	[69, 73] [72, 74] [75]

Table 4.
Solvents used for de-fatting of collagenous tissue in literature.

Collagen source	Chemical	Reference
Bovine	0.5% HCl 0.5 M EDTA	[64] [76]
Fish	0.5 M EDTA	[74]

Table 5.
Chemicals used for demineralisation in literature.

Collagen source	Chemical	Reference
Bovine	0.5 M NaCl	[77]
	1 M NaCl	[78]
	$K_2HPO_4^-$	[79]
	0.1 M NaOH	[65]
Fish	0.1 M NaOH	[70–74]

Table 6.
Chemicals used for non-collagenous protein removal in literature.

6. Possible collagen extraction methods

There are various methods to extract collagen from different animal tissues. The methods used to extract collagen from bovine or any other tissue such as fish skin; pigskin, rat tail, tendons etc. vary slightly, differing in enzyme concentration, acid concentration, salt concentration or pre-treatment period [6]. These variations can be studied and the most optimal method for bovine hide extraction can be obtained. However, acid extraction which results in acid-soluble collagen (ASC), pepsin extraction that gives pepsin solubilised collagen (PSC) and salt extractions. Some of the main extraction procedures found in literature are discussed in detail below [80].

6.1 The salting-out method

This method is seen as the least favourable method of collagen extraction. Collagen proteins, like general proteins have the property of being salt soluble. Different types of collagen proteins can be separated using the relationship between different collagen sources and salt concentrations. Neutral salt solutions are usually used, such as NaCl, Tris–HCl, phosphate, or citrate. In the salting-out method, the concentration of salt is the key factor to control, if for example, the concentration of NaCl is less than 1 mol/L in the neutral solution, its suitable for dissolution of type I collagen, however, if the concentration is bigger than 1 mol/L, it will precipitate the type I collagen. Since mature sources of collagen are less soluble because most collagen protein molecules have cross-linked, the salting-out method is not an efficient method alone to extract collagen [80].

6.2 The alkali method

The main chemicals used in the alkali method of collagen extraction are sodium hydroxide and monomethylamine [81]. This extraction method is not favoured as the main extraction method due to similar reasons as the salting-out method.

Hattori et al. [81] prepared collagen from bovine hides by alkaline solubilisation with 3.0% NaOH and 1.9% monomethylamine. The study also extracted bovine hide collagen by acid and enzymatic methods for comparison. These methods were carried out on animals of different ages. The amount of collagen extracted through this method was estimated by comparing the hydroxyproline content in the whole hide with that in the extracted collagen.

6.3 The alkali-enzyme method

The alkali-enzyme method is not as effective as the acid-enzyme method. This is because alkali is such as NaOH does not have the ability to fully solubilise collagen and disrupt the cross-linking in a collagen molecule. This method is more preferred for gelatine production [80].

6.4 The acid-alkali oxidation method

A series of repetitive steps having acid then alkali soaking of samples for a long period can be used to extract collagen. However, this method requires a very long period and the reaction time is very slow. It does not work for mature tissues as it is near impossible for acid and alkali alone to disrupt the cross-linking developed in mature tissue, thus an enzyme is a must requirement. The collagen yield extracted decrease or increase for the same tissue type depending on the literature. These differences are due to denaturation of protein during the process of extraction, the difference in environmental temperature and the solubilisation method used to extract the collagen.

The yield of collagen by the different acid (HCl, citric acid, acetic acid) is dependent on the reaction time. The longer the period of solubilisation, the greater the yield of collagen being extracted. For example, Skierka [82], concluded that during a 24 hour of collagen extraction in acid, about 33% of collagen was solubilised, and after 72 hours, about twice as much collagen was solubilised.

The solubility of collagen in acids depends up the enzyme concentration. A low concentration of enzyme with an acid can completely solubilise collagen; however, it will also depend on the type of acid. For example, enzyme concentration on the solubility of collagen in citric acid and HCl gave a maximum yield of 75% for citric acid and 85% for HCl [82].

6.5 The acid method

Acids such as acetic acid, citric acid and hydrochloric acid (HCl) of low concentration can be added to collagen-containing samples. Acids at a pH of 2–3 and a concentration of approximately 0.5 mol/L can be used to solubilise collagen. In acid extraction of collagen, the acids swell collagen, disrupting the hydrophobic and electrostatic interactions between the tropocollagen units, and release the acid-soluble collagen (ASC). Yang et al. [80] concluded that citric acid has the best effect to extract collagen, second being acetic acid and last being hydrochloric acid. However, according to Skierka [82] and Higham [83], the most effective acid for collagen solubilisation was acetic acid and the least effective solvent was HCl. In order to achieve a sound conclusion, experiments need to be carried out to investigate the solubilisation efficiencies of each acid.

The acid molecules disrupt the collagen cross-linking in order to solubilise the collagen by allowing ligand substitution for each peptide side chain, causing disassociation of the cross-link. Thus, swelling the collagen and solubilising it out of the tissue and into solution [73].

The acid method is seen to be corrosive to the experimental equipment in terms of large-scale production. However, using a low concentration of acid in combination with an enzyme will avoid equipment corrosion and achieve a high yield product (**Table** 7).

6.6 The enzyme method

The enzyme method is seen to be as the ideal method of collagen extraction. The three commonly used enzymes for collagen extraction are pepsin, papain and tryptase [80]. The enzyme acts on the non-helical peptide chains of the collagen protein, having no effect on the helix peptide chains of the collagen protein. The enzyme has better reaction selectivity and it is less destructive to the collagen

Collagen source	Acid type and concentration	Reference
Bovine	0.5 M acetic acid	[14, 62, 81, 84–87]
	10% acetic acid with 0.2% chlorhydric acid	[88]
Fish	0.5 M acetic acid	[62, 66, 67, 69, 70]
	0.15 M HCl	[73, 82]
	Citric acid	[73, 82]

Table 7.
Acids used for collagen extraction.

Collagen source	Enzyme type and conc.	Reference
Bovine	1% Trypsin	[90]
	Pepsin	[62, 86–88]
Fish	1% (w/w) pepsin	[62, 66, 69]

Table 8.
Enzyme extraction methods used for collagen extraction.

protein, resulting in a protein whose triple helix structure is better preserved. Thus, the extracted collagen will have a better purity, and retain stable physical and chemical properties. The enzyme method also provides mild reaction conditions that avoid equipment corrosion and less energy consumption. However, reaction time may be long, depending on the type of enzyme used [89].

The enzyme solubilisation method works by disrupting the cross-linking that occurs in collagen. The chosen enzyme cleaves to the amino telopeptides from the tropocollagen molecule thus disrupting the cross-linking and allowing solubilisation of the collagen molecule. Enzyme solubilisation is mostly required when extracting collagen from mature tissue, this is due to the cross-links forming keto-imines which are increasingly difficult to disrupt as they contain strong intermolecular bonds. However, the enzyme method has the disadvantage of not only breaking the collagen molecule but also resulting in the scission of other proteins may occur too, hence, causing protein contamination as a result [83]. Enzymes have been used in collagen extraction, McClain et al. [78] used papain at 0.1% in buffers containing 0.02 M phosphate and 0.003 M EDTA as a solubilisation method for collagen.

The enzyme method is usually combined with the acid method to enhance the extraction process (**Table 8**).

6.7 The acid-enzyme method

The enzyme-acid solubilisation method is seen to be the most effective way to extract collagen. Both acids (citric acid, hydrochloric acid, acetic acid) and enzymes have the capability to disrupt the cross-links in a collagen molecule and make collagen soluble in solution. Addition of both an acid and an enzyme speed up the reaction time and results in a collagen protein well-kept in its triple helix structure [80]. Concentration and acid/enzyme type greatly depend on the collagen tissue and method optimization.

7. Collagen purification via dialysis and filtration

7.1 Collagen purification via dialysis

Dialysis is a preferred method of purification for collagen extraction, however, scaling up this technique for commercialisation has proven to be difficult. Dialysis tubes are utilised with different cut off molecular weights to separate pure collagen from other solvents, salts and enzymes and other impurities.

7.2 Collagen purification via filtration

Ultrafiltration can be applied to remove the non-collagenous material prior to lyophilisation of collagen. Ultrafiltration utilises positive pressure to force a liquid through a semi-permeable membrane to separate species in an aqueous solution by molecular size, shape or charge. Ultrafiltration has the advantages of having a high throughput, cost-effective and large-scale purification is possible without being limited to lab-scale purification by dialysis.

Ultrafiltration enables the removal of solvents and salts of lower molecular weight from a solution (permeate). Thus, this results in the enrichment of the retained molecule (pure collagen). Ultrafiltration membranes can retain molecules in the range of 10 kDa to 1 MDa, thus concentration and purification of collagen (300 kDa) can be successfully achieved through this process.

- Crossflow/tangential flow filtration: the incoming feed passes parallel across the surface of a semi-permeable membrane. A permeate and a retentate stream are generated, where the permeate is the portion of the fluid that passes through the membrane and the remainder of the feed stream, which does not pass through the cross-flow membrane, is known as the retentate stream.

- Dead-end filtration: The feed moves towards the filter membrane. The particles that can be filtered are settled on the filter surface, however, this type of filtration is not sustainable as the accumulated solids need to be removed periodically or the filter needs to be replaced.

8. Collagen preservation

In order to preserve extracted collagen, it is usually freeze-dried and stored at conditions not exceeding −4°C. However, some researchers use hydrogen peroxide (0.3–3%) to disinfect collagen after extraction especially from fish sources.

9. Collagen extraction research and its applications

The popularity of collagen extraction continues to increase due to many reasons. It is a high strength protein, bio-derived, has excellent biocompatibility, biodegradability, and has weak antigenicity. Another main reason that relates to waste valorisation and sustainability is the fact that collagen can be extracted from almost any mammalian skin, bones, cartilage, fish skin and even chicken feet. Most often, the meat industry results in these by-products that can end up in landfill. These advantageous characteristics have made collagen one of the most useful biomaterials.

Research to this day is being carried out to improve extraction methods in terms of efficiency and economics. In addition to improving extraction methodologies, research is being carried out on collagen to enhance its use in several industries (**Table 9**).

Period	Collagen extraction research
1960–1969	• Bakerman [91] extracted human skin collagen with age via the acid-solubilisation method. No defatting or demineralization steps were carried out, citric acid was used as the solubilisation agent. There was no mention of methods of collagen content analysis, only extracted yield was reported.

Period	Collagen extraction research
	• In 1968 Rigby [92] analysed the amino acid composition and thermal stability of ice-fish skin. The fish skin was swollen in 0.1 M HCl for extraction purposes. Td was found to be in the range 5.5–6°C. • Young et al. [93] extracted cod skin collagen with mild solvents in the pH range of 3.4–8.7 at 3–90°C. • Grant et al. [94] studied the carbohydrate content of bovine collagen. It was shown that crude preparations of collagen were contaminated with mannose, fucose and hexosamine. • Bronstein et al. [95] studied human collagen and the relation between intra and intermolecular cross-linking. • Miller et al. [96] extracted and characterised chick bone collagen with acetic acid. Specific methodology is not given, and the focus of this study was to understand collagen compositional changes with ageing via chromatography. During this period, research was mainly carried out to understand collagen as a protein. There was no research on method optimization or investigation of different extraction methodologies.
1970–1979	• Anderson et al. [97] extracted bovine nasal collagen with 4 M guanidinium chloride or 1.9 M $CaCl_2$ and examined the structure by studying their scanning electron microscopy images. • Pierson et al. [98] studied the effect of post-mortem ageing, time and temperature on pH, tenderness and soluble collagen fractions in bovine Longissimus muscle. Salt and acid-soluble collagen were not affected by temperature nor length of post-mortem ageing. • Maekawa et al. [99] extracted collagen from the skin of mice. Extraction was carried out with 0.5 M acetic acid at 40°C for different times. • Francis et al. [100] extracted collagen from biopsies of human skin. The study concluded to show that polymeric collagen of normal and diseased human skin from biopsies was feasible. • Uitto et al. [101] analysed the solubility of skin collagen in normal human subjects and in patients with generalised scleroderma. Extraction was carried out with 0.14 M NaCl and number of extractions were varied. • In 1976, Trelstad et al. [102] applied differential separation to separate native collagen types I, II, and III. The main precipitants were ammonium sulfate, sodium chloride and ethanol. • Riemschneider et al. [103] extracted collagen from cow placenta via pepsin solubilisation. The focus of collagen extraction in this period remained to be for medical purposes. The main sources of collagen were of human skin and rat skin.
1980–1989	• Merkel et al. [104] studied the content of type I and III collagen of healing wounds in fatal and adult rats. Collagen was extracted with 0.5 M acetic acid with pepsin. Collagen content ratio was estimated from densitometer scans of electrophoretically separated α-chains. • Graham et al. [105] extracted and quantified types of collagen both in control intestine and as well as in both in inflamed and structured intestine resected from patients with Crohn's disease. This study mainly focused on differences in collagen types between the controlled and both inflamed and structured intestines. • Murata et al. [106] studied the changes in collagen types in various layers of the human aorta and their changes with the atherosclerotic process. Collagen from human aortas was extracted by repeated pepsin digestion and the collagen types were identified by SDS-PAGE analysis. • Elstow et al. [107] extracted and characterised type V foetal calf skin collagen. Neutral salt solutions (pH 9.2) with phosphate-buffered saline (PBS) were used to extract collagen and SDS-PAGE analysis was used to characterise and identify type V collagen. • Laurent et al. [108] showed a simplified method for quantification of the relative amounts of type I and type III collagen in rabbit lung samples. This extracted involved repetitive homogenisation of the collagenous tissue in 2% sodium dodecyl sulfate and dried acetone powder. • Van Amerongen et al. [109] analysed the concentration and extractability of collagen in human dental pulp. Premolar and third molar dental pulps were studied for their

Period	Collagen extraction research
	collagen content and acetic acid or neutral salts were used to extract collagen. Using SDS-PAGE analysis, 42.6% of extracted collagen to be type III. • Kurita et al. [110] analysed the changes in collagen types during the healing of rabbit tooth extraction wounds. Collagen type was identified by use of SDS-PAGE analysis and hydroxyproline analysis was applied to observe collagen content. Collagen quantification and understanding types of collagen present in diseased tissue vs. normal human tissue was the focus of collagen research in this period.
1990–1999	• Montero et al. [111] extracted collagen from Plaice skin and analysed its functional properties. Acetic acid was the main solubilising agent in this study and homogenisation was carried out with 0.4 M NaCl. • Ambrose et al. [112] extracted and characterised collagen from bone and teeth for isotopic analysis. Carbon to nitrogen ratios of bone and teeth collagen was analysed the use of purification procedures that removed acid and alkaline-soluble contaminants were recommended. • Nomura et al. [113] extracted and analysed properties of type I collagen from fish scales. Collagen was extracted with 0.5 M acetic acid and it was concluded that a large portion (80% of collagen remained insoluble which was further denatured to gelatine to be used for food purposes. • Ciarlo et al. [114] extracted collagen from hake skin. Acetic acid was used for solubilisation and collagen was characterised for its viscosity and collagen type (SDS-PAGE). • Bishop et al. [115] extracted and characterised collagen types II and IX from bovine vitreous. Centrifugation and precipitation with 4.5 M NaCl were applied and the collagen types were identified. In this era, extraction of collagen from waste materials such as fish skin began, however, the methodologies were mainly focused on salt extractions, which is not very efficient, and the study of collagen in human tissue was still predominant.
2000–2009	• In 2007, Nalinanon et al. [59] extracted collagen from the skin of bigeye snapper using pepsin. Acid-extracted collagen resulted in lower collagen yields in comparison to pepsin-solubilised collagen. • Woo et al. [116] extracted collagen from yellowfin tuna skin. Methodology was optimised by varying NaOH concentration, treatment time and pepsin concentration. The objective of this study was to determine the optimum conditions for extracting collagen from yellowfin tuna skin and characterisation was carried out by SDS-PAGE, FTIR, and solubility analysis. • In 2003, Sadowska et al. [73] isolated collagen from the skin of Baltic cod. The aim of this investigation was also to determine optimum conditions for the extraction of collagen from cod skin. Acetic acid and citric acid were used as the main collagen solvents and within the two solvent extractions, time of treatment and digestion were varied. • Jongjareonrak et al. [117] extracted and characterised collagen from bigeye snapper. Acid and pepsin solubilised collagen were isolated and characterised for their properties. The Td of the acid-solubilised and pepsin solubilised collagen varied slightly with, pepsin solubilised collagen having a higher Td (30.87°C). • Zhang et al. [69] extracted and characterised collagen from the skin of grass carp via the method of pepsin-solubilisation. A collagen yield of 46.6% (dry-basis) was obtained and SDS-PAGE showed that the extracted collagen was type I and collagen Td was found to be 24.6°C. • Nalinanon et al. [118] used pepsin from the stomach of tuna fish to use it for collagen extraction of threadfin bream skin. Pepsin from different tuna species were obtained to use for collagen extraction and to determine the differences in collagen extraction efficiency. • Cao et al. [119] extracted and characterised type II collagen from chick sternal cartilage. Pepsin was the main solubilising agent in this study and NaCl was used for collagen precipitation. SDS-PAGE confirmed the presence of collagen type II and the amino acid composition of the type II collagen extracted was very close to the reference collagen type II obtained from Sigma Aldrich. In this period, a vast number of different fish species were analysed for their collagen extractability. The main reason for this was due to adding value to waste fish skin and due to a large acceptance of fish collagen by a diverse group of people (Jewish, Hindu and Muslim religions not accepting certain meat products).

Period	Collagen extraction research
	Method optimization and comparison of different acids and enzymes for extraction efficiency had started in this period, due to a high demand for collagen from various markets.
2010–2015	• In 2010, Uriarte-Montoya et al. [120] extracted and characterised collagen from Jumbo squid and analysed its potential to be formed into a composite film with chitosan. Acid-solubilised collagen was extracted, and film blends of chitosan-collagen were prepared by casting. The films were characterised for their thermal and mechanical properties. The purpose of these films was to be used as bio-friendly packaging materials. • Muralidharan et al. [71] extracted collagen from skin, bone and muscle of both trash fish and leather jacket fish. The collagen was characterised for their properties. Three methods of extraction were applied, and each method resulted in different collagen yields with the highest collagen yield being 71%. It was concluded that collagen from both trash fish and leather jacket fish could be used to extract collagen use it for potential pharmaceutical and biomedical applications. • In 2012, Liu et al. [121] extracted collagen from fins, scales, skin, bones, and swim bladders of bighead carp. The aim of this study was to characterise pepsin-solubilised collagen from the five sources for simultaneous comparison purposes. It was concluded that all five tissues could be used as a potential substitute for mammalian collagen. • Matmaroh et al. [122] extracted collagen from the scale of spotted golden goatfish via acid and pepsin solubilisation. SDS-PAGE showed both methods had revealed type I collagen and FTIR confirmed the presence of collagen triple helical structure. The main purpose of this study was to study collagen from the scale of spotted golden goatfish. • Kittiphattanabawon et al. [123] extracted and characterised collagen from the skin of brown-banded bamboo shark. Both acid soluble and pepsin soluble collagen were extracted and the collagen yield with pepsin solubilisation was slightly lower than with acid solubilisation. It was concluded that collagen from skin of brownbanded bamboo shark could serve as an alternative source of collagen. • Singh et al. [124] isolated collagen from skin of striped catfish. Once again, both methods of acid and pepsin solubilisation were applied to extract collagen. Collagen yield with pepsin-solubilisation was slightly higher (7.7%) in comparison to acid solubilised collagen (5.1%). In this period and currently, most research being carried out on collagen extraction is based on method improvement and investigating novel tissues for possibility of collagen extraction. Sources that were not investigated and sources that potentially be environmentally friendly are being analysed for collagen extraction in order to find cheaper and efficient means of extraction.

Table 9.
Timeline of advancements in collagen extraction (1960s – 2015).

10. Methods used to investigate the physiochemical properties of collagen and collagen-based films

There are at least 27 collagen types with 42 distinct polypeptide chains identified. Types I to XXVII collagen are fibril-forming collagens, containing triple-helix structures that can bundle into fibrils. Some collagen types are only present in certain tissues, for example, collagen types II, IX and XI are mostly found in cartilage tissues. Collagen types I to III are the ones mostly present in all collagen-containing tissues, type I being mainly present in skin tissue. Collagen characterisation is carried out to acquire information on structure, denaturation temperature, quantity, quality, thermal stability and fibril arrangement. Understanding the properties of each type of collagen will result in a better picture of what applications it can further be applied in.

The properties of the extracted collagen can be characterised by Sodium dodecyl sulphate polyacrylamide gel electrophoresis (SDS-PAGE), Fourier transform infrared spectroscopy (FTIR), thermal stability (thermogravimetric analysis (TGA), differential scanning calorimetry (DSC)), morphology analysis, such as scanning electron microscopy (SEM) and transmission electron microscopy (TEM); collagen moisture content, and hydroxyproline analysis.

The results from these analyses can be compared to standard collagen found in the market to compare yields and quality. The investigation of physiochemical properties of collagen through these characterisation methods is a way of optimising future collagen extraction methods.

10.1 Collagen molecular stability

10.1.1 Sodium dodecyl sulphate polyacrylamide gel electrophoresis (SDS-PAGE)

SDS-PAGE analysis can be used to differentiate between the different collagen types and their individual chains. SDS-PAGE patterns of the extracted collagen can be obtained through any electrophoresis device such as the Mini-Protean or a PhastGel system. The collagen sample is boiled in SDS, resulting in collagen to break down into its polypeptide chains so that the α and β components of the collagen molecule can be analysed.

Though SDS electrophoresis has been utilised for preparative separation of collagen [125], it has been mainly used to compare collagen from different tissue types and to identify collagen types and polypeptide chains. Wu et al. [126] extracted bovine collagen and applied SDS-PAGE to identify the different collagen types present.

10.1.2 Fourier transform infrared spectroscopy (FTIR)

In order to assess the collagen for abnormal formation and organisation and changes in its secondary structure, Fourier Transform Infrared Spectroscopy (FTIR) can be applied to reveal the collagen bio-distribution. FTIR has been used to study collagen denaturation [127], collagen cross-linking [128], and thermal self-assembly [129].

The vibrational bands characteristic of peptide groups and side chains provide information on protein structures. Spectral changes in amide A, amide I (1636–1661 cm^{-1}), amide II (1549–1558 cm^{-1}), and amide III (1200–1300 cm^{-1}) regions are indicative of changes in collagen secondary structure [127]. An increase in the intensity of amide III and broadening of amide I are related with increased intermolecular interactions via hydrogen bonding in collagen. Among these, the amide I band (peptide bond C=O stretch) is especially sensitive to secondary structures. A reduction in the intensity of amide A, I, II and III peaks and narrowing of amide I band are associated with collagen denaturation (Td) [127].

An FT-IR spectrophotometer can be used to obtain a spectrum for collagen. Approximately 2–4 mg of collagen in 100 mg potassium bromide (kBr) can be used to obtain spectra from 4000 to 1000 cm^{-1}.

10.2 Collagen content: hydroxyproline analysis

Hydroxyproline is an amino acid found in collagen, comprising about 13% of the collagen molecule, this amino acid is not found in any other proteins apart from elastin. Thus, determining the hydroxyproline content in a specified tissue enables the calculation of the total amount of collagen present. Experimentally, the amount of hydroxyproline content in a sample for mammals [130] is multiplied by 7.46 to give the amount of collagen in the sample.

Many studies on collagen extraction have applied hydroxyproline analysis to calculate collagen content [81, 87, 131, 132]. Researchers have developed methods to effectively measure hydroxyproline concentration of collagen using calorimetric assays [19, 133], high-performance liquid chromatography, and enzymatic methods [133]. Calorimetric methods usually require complete hydrolysis of collagen to its individual amino acids, oxidising hydroxyproline present to a pyrrole, and then reacting the pyrrole with a colour forming agent. This colour change is measured using a UV/Vis spectrophotometer and compared against calibration data to determine hydroxyproline concentration [19]. To obtain the amount of collagen in a sample for mammals, the amount of hydroxyproline in the sample (mg) is multiplied by a factor of 7.46 [67].

10.3 Collagen denaturation temperature (td) and thermal stability

10.3.1 Differential scanning calorimetry (DSC) and thermal denaturation temperature (T_d)

Any DSC calorimeter brand can be used, such as a Perkin Elmer DSC7. The thermal behaviour; stability of the native molecular structure and denaturation of collagen can be determined by carrying out differential scanning calorimetry (DSC). Denaturation temperature is obtained from the transition in the baseline in the 30–80°C region by taking the inflexion point reading. Total denaturation enthalpy (ΔH) can be estimated by measuring the area in the DSC thermogram.

Collagen denaturation temperature (Td) depends on collagen water content, collagen extraction method, collagen source, degree of collagen cross-linking and hydroxyproline content. Thermal stability of the collagen triple helix depends on hydrogen bonds (inter- and intra-hydrogen bonding) which further influences the folding and unfolding process when hydrogen bonds are broken and connected [134, 135]. Hence, the thermal stability of collagen depends on the cross-linking of collagen molecules (inter and intra).

Due to the polymeric nature of collagen, the thermal-induced denaturation of collagen is usually complicated. Heating collagen in wet or dry state reveals a series of thermal transitions. Thermal denaturation of collagen occurs due to hydrogen bonds breaking and hence the unfolding of the triple helices forming random polypeptide coils [136]. Cross-linking among the collagen molecules increase and mature with age and provides further stability. The age-related accumulation of cross-links increases the thermodynamic stability of collagen by increasing the activation energy required for collagen denaturation. However, the maturity of collagen cross-linking is limited to the functionality of the tissue. Post-mortem cross-linking of collagen can increase to the point where the tissue may become brittle [137].

Within the collagen fibril, there are complex interactions within and between the packed molecules. In addition to inter, intramolecular cross-links, and different forms of cross-linkages, there are several additional hydrophobic and ionic interactions that must be accounted for regarding collagen denaturation. The presence of non-collagenous components in the extracted collagen sample can cause variations in thermal denaturation [138].

Due to the domain structure of the triple helix, not all parts of the collagen molecule may denature at the same rate and it is almost impossible to define a definite equilibrium Td. Studies have also shown an increase in Td with an increase in hydroxyproline content [61, 139].

10.3.2 Thermogravimetric analysis (TGA)

Thermal stability of extracted collagen is investigated using a gravimetric analyser. Approximately 5–10 mg of the sample can be used. The mass loss is recorded while the sample is heated from room temperature up to 800°C at a rate of 10°C per minute. The first derivative of percentage mass change versus temperature can also be calculated to investigate temperature regions where mass loss was occurring.

Ramanathan et al. [140] used TGA to assess the thermal stability of fish skin collagen which was extracted via acid-solubilisation. They report using samples of approximately 5 mg and heating samples at 10°C/min in the temperature range of 0–800°C. The acid-solubilised collagen showed two weight loss steps on the TGA thermogram, relating the first stage to the loss of structural and bound water and stage two to thermal degradation of the polypeptide chain. The study concluded to show that the two peaks observed on the TGA differential curve were of collagen denaturation and collagen degradation respectively.

10.4 Collagen morphology

10.4.1 Scanning electron microscopy (SEM)

The protein morphology of the extracted collagen can be studied using SEM. The morphology of the extracted collagen can be compared to the standard bovine

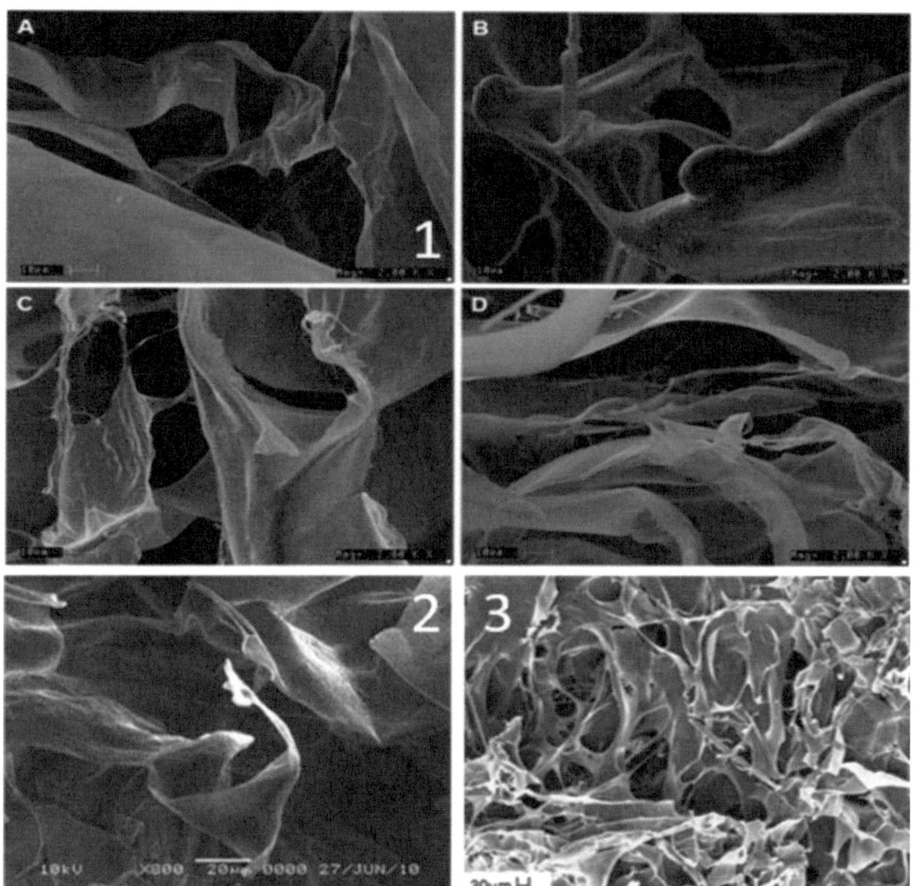

Figure 8.
SEM images of extracted collagen, with 1) acid-soluble collagen of Catla fish (a), pepsin-soluble collagen of Catla fish (B), acid-soluble collagen of Rohu fish (C) and pepsin-soluble collagen of Rohu fish (D) [139], (2) being from buffalo skin [14] and 3) being SEM image of porcine skin collagen [142] (used with permission).

collagen available in the market. The expected microstructure of collagen from SEM images would be to observe collagen sheets which would be a combination of collagen fibrils and fibres that are bundled together to form a fibril network and dense sheet-like structure.

Ramanathan et al. [140] used SEM to observe the surface morphologies of freeze-dried acid-solubilised fish skin collagen. The images showed a smooth surface texture, in two of the images, a layer-by-layer structure was observed (no definite fibres), and this was related to the intertwining of collagen fibres. Similarly, Rizk et al. [14], Tziveleka et al. [141], Rodrigues et al. [142], Pal et al. [139] all carried out SEM to assess the surface morphology of extracted collagen and all showed SEM images to have smooth or slightly wrinkled surfaces or sheet-like structures (**Figure 8**).

10.4.2 Transmission electron microscopy (TEM)

Transmission electron microscopy is usually carried out to observe collagen fibril structure and it is uniformity in a much deeper level. SEM only provides limited information on collagen morphology. **Figure 8** is showing an electron transmission image of mammalian lung tissue collagen at a magnification of 50 nm, while is showing a TEM image of collagen fibrils and fibres.

The preparative steps of collagen TEM are very specific and usually requires a technician to carry out each step carefully in order to observe the fibrillar structure of collagen. The Karnovsky fixative is mostly used as a preparative method prior to taking TEM images.

11. Conclusion

Collagen has risen its rank to be an integral material and an element of importance both in biomedical and non-biomedical sectors. In conclusion, research has shown that collagen can be effectively extracted from bovine and cattle hides. Using waste valorization concepts, collagen containing waste materials can be utilized to derive high-value products.

Author details

Safiya Noorzai[1]* and Casparus J.R. Verbeek[2]

1 School of Engineering, Department of Science and Engineering, University of Waikato, Hamilton, New Zealand

2 Plastics Centre of Excellence, University of Auckland

*Address all correspondence to: safiya.noorzai1@gmail.com

References

[1] Shoulders, M.D. and R.T. Raines, *Collagen structure and stability.* Annual review of biochemistry, 2009. **78**: p. 929–958.

[2] Gelse, K., E. Pöschl, and T. Aigner, *Collagens—structure, function, and biosynthesis.* Advanced drug delivery reviews, 2003. **55**(12): p. 1531–1546.

[3] Peng, Y., et al., *Evaluation for collagen products for cosmetic application.* International Journal of Cosmetic Science **26**(6): p. 313.

[4] Simelane, S. and Z. Ustunol, *Mechanical Properties of Heat-cured Whey Protein-based Edible Films Compared with Collagen Casings under Sausage Manufacturing Conditions.* Journal of food science, 2005. **70**(2): p. E131-E134.

[5] MBG, P.M.C., et al., *An oral supplementation based on hydrolyzed collagen and vitamins improves skin elasticity and dermis echogenicity: a clinical placebo-controlled study.* Clin Pharmacol Biopharm, 2015. **4**(142): p. 2.

[6] O'Sullivan, O., et al., *Extraction of Collagen from Fish Skins and Its Use in the Manufacture of Biopolymer Films.* Aquatic Food Product Technology 2006. **15**(3): p. 21–32.

[7] Meena, C., S. Mengi, and S. Deshpande. *Biomedical and industrial applications of collagen*. in *Proceedings of the Indian Academy of Sciences-Chemical Sciences*. 1999. Springer.

[8] Pereira, R.F. and P.J. Bártolo, *3D bioprinting of photocrosslinkable hydrogel constructs.* Journal of Applied Polymer Science, 2015. **132**(48).

[9] Aberoumand, A., *Comparative study between different methods of collagen extraction from fish and its properties.* World Applied Sciences Journal, 2012. **16**(3): p. 316–319.

[10] Schmidt, M., et al., *Collagen extraction process.* International Food Research Journal, 2016. **23**(3).

[11] The European IPPC Bureau, *Best available techiques (BAT) reference document for the tanning of hides and skins*. 2013: Seville.

[12] Wallace Corporation Limited. *Wallace Corporation Limited*. 2015 [cited 2015 5/05/2015]; Available from: http://www.wallace.co.nz/.

[13] Palacios, M.J.E.L., Maria Angeles Perez; Aran, paqui, *From leather waste to functional leather* 2016, European Commission through the LIFE+ Programme Elda, Spain.

[14] Rizk, M.A. and N.Y. Mostafa, *Extraction and Characterization of Collagen from Buffalo Skin for Biomedical Applications.* Oriental Journal of Chemistry, 2016. **32**(3): p. 1601–1609.

[15] Maffia, G., et al., *Collagen processing.* Journal of the American Leather Chemists Association, 2004. **99**(4): p. 164–169.

[16] SigmaAldrich. *Calf Skin collagen* 2018; Available from: http://www.sigmaaldrich.com/catalog/product/sigma/c3511?lang=en&region=NZ.

[17] Yang, L., *mechanical properties of collagen fibrils and elastic fibres explored by AFM*. 2008, University of Twente Enschede The Netherlands

[18] Stryer, L., *Biochemistry*. 1975, New York: W. H. Freeman and Company.

[19] Neuman, R.E. and M.A. Logan, *The determination of collagen and elastin in tissues.* Journal of Biological Chemistry, 1950. **186**(2): p. 549–556.

[20] Naked Nutrition. *Collagen synthesis*. 2020 10/02/20]; Available from: https://

blog.nkdnutrition.com/collagen-synthesis/.

[21] Kruger, T.E., A.H. Miller, and J. Wang, *Collagen scaffolds in bone sialoprotein-mediated bone regeneration.* The Scientific World Journal, 2013. **2013**.

[22] Lippiello, L., D. Hall, and H.J. Mankin, *Collagen synthesis in normal and osteoarthritic human cartilage.* Journal of Clinical Investigation, 1977. **59**(4): p. 593.

[23] Howard, L. *Fibers of Collagen Type I - TEM*. 2014 10/02/20]; Available from: https://commons.wikimedia.org/wiki/File:Fibers_of_Collagen_Type_I_-_TEM.jpg.

[24] Bornstein, P. and H. Sage, *Structurally distinct collagen types.* Annual review of biochemistry, 1980. **49**(1): p. 957–1003.

[25] Maynes, R., *Structure and function of collagen types*. 2012: Elsevier.

[26] Armstrong, A.M. *What is bovine collagen*. 2010 [cited 2015; Available from: www.ehow.com/facts_bovine-collagen_.html.

[27] Mokrejs, P., *Extraction of collagen and gelatine from meat industry by-products for food and non food uses.* SAGE Journal **27**(1): p. 31–37

[28] Assignment Point. *Histological Structure of Cowhide*. 2015 [cited 2015 4/06/2015]; Available from: http://www.assignmentpoint.com/business/economics/histological-structure-cowhide.html.

[29] All about leather. *What is leather*. 2015 [cited 2015 2/03/2015]; Available from: http://www.all-about-leather.co.uk/what-is-leather/what-is-leather.htm.

[30] Uesaka, S., *Morphological and histological studies on cattle hide and leather VI Effects of bull castration on the hide and leather in histological meaning.* Nihon Chikusan Gakkaiho, 1951. **22**(2–4): p. 66–69.

[31] Khan, C. *Why does patio furniture cost more than leather furniture*. 2019 10/02/20]; Available from: https://www.patioproductions.com/blog/features/why-does-patio-furniture-cost-more-than-leather-furniture/.

[32] Nomura, Y. and N. Kitazume, *Use of shark collagen for cell culture and zymography.* Bioscience, biotechnology, and biochemistry, 2002. **66**(12): p. 2673–2676.

[33] Yoshimura, K., et al., *Physical properties of shark gelatin compared with pig gelatin.* Journal of agricultural and food chemistry, 2000. **48**(6): p. 2023–2027.

[34] Shevchenko, R.V., et al., *Use of a novel porcine collagen paste as a dermal substitute in full-thickness wounds.* Wound repair and regeneration, 2008. **16**(2): p. 198–207.

[35] Gelita NZ. *Solutions and Products*. 2015 [cited 2015 22/05/2015]; Available from: http://www.gelita.com/solutions-and-products.

[36] SouthernLights Biomaterials. *Biomaterials for regenerative medicine*. 2015 [cited 2015; Available from: http://www.slbio.com/.

[37] RevolutionFibres. *Nanofibre products from concept completion*. 2015 21/05/2015]; Available from: http://www.revolutionfibres.com/.

[38] Callaghan Innovation. *Revolution Fibres putting NZ on the map*. 2012 [cited 2015.

[39] Waitaki Biosciences. *Waitaki Biosciences*. 2015 [cited 2015 25/05/2015]; Available from: http://www.waitakibio.

com/natural-ingredients/company-profile.

[40] Waitaki Biosciences. *Marine Collagen Powder*. 2015 [cited 2015 25/05/2015]; Available from: http://www.waitakibio.com/manufacturer/marine-collagen.

[41] Geiger, M., R. Li, and W. Friess, *Collagen sponges for bone regeneration with rhBMP-2.* Advanced drug delivery reviews, 2003. **55**(12): p. 1613–1629.

[42] Lee, C.H., A. Singla, and Y. Lee, *Biomedical applications of collagen.* International journal of pharmaceutics, 2001. **221**(1): p. 1–22.

[43] Poland, D.E. and H.E. Kaufman, *Clinical uses of collagen shields.* Journal of Cataract & Refractive Surgery, 1988. **14**(5): p. 489–491.

[44] Willoughby, C., M. Batterbury, and S. Kaye, *Collagen corneal shields.* Survey of ophthalmology, 2002. **47**(2): p. 174–182.

[45] Wisser, D. and J. Steffes, *Skin replacement with a collagen based dermal substitute, autologous keratinocytes and fibroblasts in burn trauma.* Burns, 2003. **29**(4): p. 375–380.

[46] TAKAOKA, K., et al., *Ectopic Bone Induction on and in Porous Hydroxyapatite Combined with Collagen and Bone Morphogenetic Protein.* Clinical Orthopaedics and Related Research, 1988. **234**: p. 250–254.

[47] Lee, W., et al., *Three-dimensional bioprinting of rat embryonic neural cells.* Neuroreport, 2009. **20**(8): p. 798–803.

[48] Petronzio, M. *How 3D printing actually works*. 2013 15/5/2015]; Available from: http://mashable.com/2013/03/28/3d-printing-explained/.

[49] Funk, J. *3D Bio-Printing; becoming economically feasible* 2013; Available from: http://www.slideshare.net/Funk98/3d-bioprinting-becoming-economically-feasible.

[50] Charni, N., F. Juillet, and P. Garnero, *Urinary type II collagen helical peptide (HELIX-II) as a new biochemical marker of cartilage degradation in patients with osteoarthritis and rheumatoid arthritis.* Arthritis & Rheumatism, 2005. **52**(4): p. 1081–1090.

[51] Trentham, D., et al., *Effects of oral administration of type II collagen on rheumatoid arthritis* Science 1993. **261**(5129): p. 1727–1730.

[52] Fishman, H.M., *Collagen in cosmetics.* Household & personal products industry, 2009. **46**(8).

[53] Fairley, P., et al., *Mechanical properties and water vapor permeability of edible films from whey protein isolate and sodium dodecyl sulfate.* Journal of Agricultural and Food Chemistry, 1996. **44**(2): p. 438–443.

[54] Paschoalick, T., et al., *Characterization of some functional properties of edible films based on muscle proteins of Nile Tilapia.* Food Hydrocolloids, 2003. **17**(4): p. 419–427.

[55] Cao, N., X. Yang, and Y. Fu, *Effects of various plasticizers on mechanical and water vapor barrier properties of gelatin films.* Food hydrocolloids, 2009. **23**(3): p. 729–735.

[56] Miller, A.T., *Collagen sausage casing.* 1983, Google Patents.

[57] Sionkowska, A., et al., *Polymer films based on silk fibroin and collagen-the physico-chemical properties.* Molecular Crystals and Liquid Crystals, 2016. **640**(1): p. 13–20.

[58] Sommer, I. and P.M. Kunz, *Collagen-based packaging films for nonfood applications and their*

comparison with commercial polyolefin (polypropylene block copolymer) films. Journal of Plastic Film & Sheeting, 2012. **28**(1): p. 30–62.

[59] Nalinanon, S., et al., *Use of pepsin for collagen extraction from the skin of bigeye snapper (Priacanthus tayenus).* Food Chemistry, 2007. **104**(2): p. 593–601.

[60] Ko, Y.-G., et al., *Preparation of novel collagen sponges using an ice particulate template.* Journal of Bioactive and Compatible Polymers, 2010. **25**(4): p. 360–373.

[61] Lin, Y.K. and D.C. Liu, *Comparison of physical–chemical properties of type I collagen from different species.* Food Chemistry, 2006. **99**(2): p. 244–251.

[62] Park, S.-H., et al., *Comparative analysis of collagens extracted from different animal sources for application of cartilage tissue engineering.* International Journal of Precision Engineering and Manufacturing, 2012. **13**(11): p. 2059–2066.

[63] Li, D., W. Yang, and G.y. Li, *Extraction of native collagen from limed bovine split wastes through improved pretreatment methods.* Journal of chemical technology and biotechnology, 2008. **83**(7): p. 1041–1048.

[64] Zhang, Z., G. Li, and B. Shi, *Physicochemical properties of collagen, gelatin and collagen hydrolysate derived from bovine limed split wastes.* Journal society of leather technologists and chemists, 2006. **90**(1): p. 23.

[65] Ran, X.-G. and L.-Y. Wang, *Use of ultrasonic and pepsin treatment in tandem for collagen extraction from meat industry by-products.* Journal of the Science of Food and Agriculture, 2014. **94**(3): p. 585–590.

[66] Chen, Y., R. Ye, and Y. Wang, *Acid-soluble and pepsin-soluble collagens from grass carp (Ctenopharyngodon idella) skin: a comparative study on physicochemical properties.* International Journal of Food Science & Technology, 2015. **50**(1): p. 186–193.

[67] Kittiphattanabawon, P., et al., *Characterisation of acid-soluble collagen from skin and bone of bigeye snapper (Priacanthus tayenus).* Food Chemistry, 2005. **89**(3): p. 363–372.

[68] Yan, M., et al., *Characterization of acid-soluble collagen from the skin of walleye pollock (Theragra chalcogramma).* Food Chemistry, 2008. **107**(4): p. 1581–1586.

[69] Zhang, Y., et al., *Isolation and partial characterization of pepsin-soluble collagen from the skin of grass carp (Ctenopharyngodon idella).* Food chemistry, 2007. **103**(3): p. 906–912.

[70] Tylingo, R., et al., *Isolation and characterization of acid soluble collagen from the skin of african catfish (Clarias gariepinus), salmon (Salmo salar) and baltic cod (Gadus morhua).* J Biotechnol Biomater, 2016. **6**(234): p. 2.

[71] Muralidharan, N., et al., *Skin, bone and muscle collagen extraction from the trash fish, leather jacket (Odonus niger) and their characterization.* Journal of food science and technology, 2013. **50** (6): p. 1106–1113.

[72] Kiew, P.L. and M.M. Don, *The influence of acetic acid concentration on the extractability of collagen from the skin of hybrid Clarias sp. and its physicochemical properties: A preliminary study.* Focus Mod. Food Ind, 2013. **2**: p. 123–128.

[73] Sadowska, M., I. Kołodziejska, and C. Niecikowska, *Isolation of collagen from the skins of Baltic cod (Gadus morhua).* Food Chemistry, 2003. **81**(2): p. 257–262.

[74] Nagai, T. and N. Suzuki, *Isolation of collagen from fish waste material—skin,*

bone and fins. Food Chemistry, 2000. **68** (3): p. 277–281.

[75] Zhang, M., W. Liu, and G. Li, *Isolation and characterisation of collagens from the skin of largefin longbarbel catfish (Mystus macropterus).* Food Chemistry, 2009. **115**(3): p. 826–831.

[76] Bowman, S.M., et al., *The tensile behavior of demineralized bovine cortical bone.* Journal of biomechanics, 1996. **29** (11): p. 1497–1501.

[77] Fukuta, S., et al., *Identification of types II, IX and X collagens at the insertion site of the bovine achilles tendon.* Matrix Biology, 1998. **17**(1): p. 65–73.

[78] McClain, P.E., *Chemistry of collagen crosslinking: relationship to aging and nutrition*, in *Protein Crosslinking*. 1977, Springer. p. 603–618.

[79] Gross, J., J.H. Highberger, and F.O. Schmitt, *Extraction of collagen from connective tissue by neutral salt solutions.* Proceedings of the National Academy of Sciences, 1955. **41**(1): p. 1–7.

[80] Yang, H. and Z. Shu, *The extraction of collagen protein from pigskin.* Journal of Chemical and Pharmaceutical Research 2014. **6**(2): p. 683–687.

[81] Hattori, S., et al., *Alkali-Treated Collagen Retained the Triple Helical Conformation and the Ligand Activity for the Cell Adhesion via α2β1 Integrin.* Journal of Biochemistry, 1999. **125**(4): p. 676–684.

[82] Skierka, E. and M. Sadowska, *The influence of different acids and pepsin on the extractability of collagen from the skin of Baltic cod (Gadus morhua).* Food Chemistry, 2007. **105**(3): p. 1302–1306.

[83] Higham, C., *Quantification of collagen, chondroitin and elastin in TBE1 process streams*, in *Science and Engineering*. 2010, University of Waikato Hamilton.

[84] Gómez, K.K., et al. *Extraction and characterization of collagen from different biological tissues*. in *MEDICAL PHYSICS: Twelfth Mexican Symposium on Medical Physics*. 2012. AIP Publishing.

[85] Mu, C., et al., *Temperature induced denaturation of collagen in acidic solution.* Biopolymers: Original Research on Biomolecules, 2007. **86**(4): p. 282–287.

[86] Komsa-Penkova, R., et al., *Thermal stability of calf skin collagen type I in salt solutions.* Biochimica et Biophysica Acta (BBA)-Protein Structure and Molecular Enzymology, 1996. **1297**(2): p. 171–181.

[87] Mocan, E., O. Tagadiuc, and V. Nacu, *Aspects of collagen isolation procedure.* Clin. Res. Studies, 2011. **2**: p. 3–5.

[88] Rodrigues, C., et al., *Characterization of a bovine collagen–hydroxyapatite composite scaffold for bone tissue engineering.* Biomaterials, 2003. **24** (27): p. 4987–4997.

[89] Munasinghe, K.A., J.G. Schwarz, and A.K. Nyame, *Chicken Collagen from Law Market Value By-Products as an Alternate Source.* Journal of Food Processing, 2014. **2014**.

[90] Kopp, J., M. Bonnet, and J. Renou, *Effect of collagen crosslinking on collagen-water interactions (a DSC investigation).* Matrix, 1990. **9**(6): p. 443–450.

[91] Bakerman, S., *Quantitative extraction of acid-soluble human skin collagen with age.* Nature, 1962. **196** (4852): p. 375.

[92] Rigby, B., *Amino-acid composition and thermal stability of the skin collagen of the Antarctic ice-fish.* Nature, 1968. **219** (5150): p. 166.

[93] Young, E.G. and J. Lorimer, *The acid-soluble collagen of cod skin.* Archives of biochemistry and biophysics, 1960. **88**(2): p. 373–381.

[94] Grant, M. and D. Jackson, *Carbohydrate content of bovine collagen preparations.* Biochemical Journal, 1968. **108**(4): p. 587–591.

[95] Bornstein, P. and K.A. Piez, *A biochemical study of human skin collagen and the relation between intra-and intermolecular cross-linking.* The Journal of clinical investigation, 1964. **43**(9): p. 1813–1823.

[96] Miller, E., et al., *Characterization of chick bone collagen and compositional changes associated with maturation.* Journal of Biological Chemistry, 1967. **242**(23): p. 5481–5489.

[97] Anderson, H.C. and S.W. Sajdera, *The fine structure of bovine nasal cartilage: Extraction as a technique to study proteoglycans and collagen in cartilage matrix.* The Journal of cell biology, 1971. **49**(3): p. 650–663.

[98] Pierson, C. and J. Fox, *Effect of postmortem aging time and temperature on pH, tenderness and soluble collagen fractions in bovine longissimus muscle.* Journal of Animal Science, 1976. **43**(6): p. 1206–1210.

[99] Maekawa, T., et al., *Changes in collagen with age—I: The extraction of acid soluble collagens from the skin of mice.* Experimental Gerontology, 1970. **5**(2): p. 177–186.

[100] Francis, M. and D. Macmillan, *The extraction of polymeric collagen from biopsies of human skin.* Biochimica et biophysica acta, 1971. **251**(2): p. 236–245.

[101] Uitto, J., K. Øhlenschläger, and I. Lorenzen, *Solubility of skin collagen in normal human subjects and in patients with generalised scleroderma.* Clinica chimica acta, 1971. **31**(1): p. 13–18.

[102] Trelstad, R.L., V.M. Catanese, and D.F. Rubin, *Collagen fractionation: separation of native types I, II and III by differential prectipitation.* Analytical biochemistry, 1976. **71**(1): p. 114–118.

[103] Riemschneider, R. and M. Abedin, *Pepsin-solubilized collagen from cow placenta.* Die Angewandte Makromolekulare Chemie: Applied Macromolecular Chemistry and Physics, 1979. **82**(1): p. 171–186.

[104] Merkel, J.R., et al., *Type I and type III collagen content of healing wounds in fetal and adult rats.* Proceedings of the Society for Experimental Biology and Medicine, 1988. **187**(4): p. 493–497.

[105] Graham, M.F., et al., *Collagen content and types in the intestinal strictures of Crohn's disease.* Gastroenterology, 1988. **94**(2): p. 257–265.

[106] Murata, K., T. Motayama, and C. Kotake, *Collagen types in various layers of the human aorta and their changes with the atherosclerotic process.* Atherosclerosis, 1986. **60**(3): p. 251–262.

[107] Elstow, S.F. and J.B. Weiss, *Extraction, isolation and characterization of neutral salt soluble type V collagen from fetal calf skin.* Collagen and related research, 1983. **3**(3): p. 181–193.

[108] Laurent, G.J., et al., *A simplified method for quantitation of the relative amounts of type I and type III collagen in small tissue samples.* Analytical biochemistry, 1981. **113**(2): p. 301–312.

[109] Van Amerongen, J., I.G. Lemmens, and G. Tonino, *The concentration, extractability and characterization of collagen in human dental pulp.* Archives of oral biology, 1983. **28**(4): p. 339–345.

[110] Kurita, K., et al., *Changes in collagen types during the healing of rabbit tooth extraction wounds.* Journal of dental research, 1985. **64**(1): p. 28–32.

[111] Montero, P., et al., *Plaice skin collagen extraction and functional properties.* Journal of Food Science, 1995. **60**(1): p. 1–3.

[112] Ambrose, S.H., *Preparation and characterization of bone and tooth collagen for isotopic analysis.* Journal of archaeological science, 1990. **17**(4): p. 431–451.

[113] Nomura, Y., et al., *Preparation and some properties of type I collagen from fish scales.* Bioscience, biotechnology, and biochemistry, 1996. **60**(12): p. 2092–2094.

[114] Ciarlo, A.S., M.E. Paredi, and A.N. Fraga, *Isolation of soluble collagen from hake skin (Merluccius hubbsi).* Journal of Aquatic Food Product Technology, 1997. **6**(1): p. 65–77.

[115] Bishop, P., et al., *Extraction and characterization of the tissue forms of collagen types II and IX from bovine vitreous.* Biochemical Journal, 1994. **299** (2): p. 497–505.

[116] Woo, J.-W., et al., *Extraction optimization and properties of collagen from yellowfin tuna (Thunnus albacares) dorsal skin.* Food Hydrocolloids, 2008. **22**(5): p. 879–887.

[117] Jongjareonrak, A., et al., *Isolation and characterization of collagen from bigeye snapper (Priacanthus macracanthus) skin.* Journal of the Science of Food and Agriculture, 2005. **85**(7): p. 1203–1210.

[118] Nalinanon, S., et al., *Tuna pepsin: characteristics and its use for collagen extraction from the skin of threadfin bream (Nemipterus spp.).* Journal of Food Science, 2008. **73**(5): p. C413-C419.

[119] Cao, H. and S.-Y. Xu, *Purification and characterization of type II collagen from chick sternal cartilage.* Food chemistry, 2008. **108**(2): p. 439–445.

[120] Uriarte-Montoya, M.H., et al., *Jumbo squid (Dosidicus gigas) mantle collagen: Extraction, characterization, and potential application in the preparation of chitosan–collagen biofilms.* Bioresource technology, 2010. **101**(11): p. 4212–4219.

[121] Liu, D., et al., *Extraction and characterisation of pepsin-solubilised collagen from fins, scales, skins, bones and swim bladders of bighead carp (Hypophthalmichthys nobilis).* Food Chemistry, 2012. **133**(4): p. 1441–1448.

[122] Matmaroh, K., et al., *Characteristics of acid soluble collagen and pepsin soluble collagen from scale of spotted golden goatfish (Parupeneus heptacanthus).* Food chemistry, 2011. **129**(3): p. 1179–1186.

[123] Kittiphattanabawon, P., et al., *Isolation and characterisation of collagen from the skin of brownbanded bamboo shark (Chiloscyllium punctatum).* Food Chemistry, 2010. **119**(4): p. 1519–1526.

[124] Singh, P., et al., *Isolation and characterisation of collagen extracted from the skin of striped catfish (Pangasianodon hypophthalmus).* Food Chemistry, 2011. **124**(1): p. 97–105.

[125] Deyl, Z., V. Rohlicek, and M. Adam, *Separation of collagens by capillary zone electrophoresis.* Journal of Chromatography A, 1989. **480**: p. 371–378.

[126] Wu, J., et al., *Extraction and isolation of type I, III and V collagens and their SDS-PAGE analyses.* Transactions of Tianjin University, 2011. **17**(2): p. 111.

[127] Muyonga, J., C. Cole, and K. Duodu, *Fourier transform infrared (FTIR) spectroscopic study of acid soluble collagen and gelatin from skins and bones of young and adult Nile perch (Lates niloticus).* Food Chemistry, 2004. **86**(3): p. 325–332.

[128] Paschalis, E., et al., *Spectroscopic characterization of collagen cross-links in bone.* Journal of Bone and Mineral Research, 2001. **16**(10): p. 1821–1828.

[129] Jakobsen, R., et al., *Intermolecular interactions in collagen self-assembly as revealed by Fourier transform infrared spectroscopy*. Science, 1983. **220**(4603): p. 1288–1290.

[130] Cui, X., et al., *Direct human cartilage repair using three-dimensional bioprinting technology*. Tissue Engineering Part A, 2012. **18**(11–12): p. 1304–1312.

[131] Kittiphattanabawon, P., et al., *Characteristics of pepsin-solubilised collagen from the skin of splendid squid (Loligo formosana)*. Journal of Chemistry, 2015. **2015**.

[132] Cheng, F.-Y., et al., *Effect of different acids on the extraction of pepsin-solubilised collagen containing melanin from silky fowl feet*. Food Chemistry, 2009. **113**(2): p. 563–567.

[133] Kesava Reddy, G. and C.S. Enwemeka, *A simplified method for the analysis of hydroxyproline in biological tissues*. Clinical biochemistry, 1996. **29**(3): p. 225–229.

[134] Bischof, J.C. and X. He, *Thermal stability of proteins*. Annals of the New York Academy of Sciences, 2006. **1066**(1): p. 12–33.

[135] Shanmugam, G. and P.L. Polavarapu, *Structural transition during thermal denaturation of collagen in the solution and film states*. Chirality: The Pharmacological, Biological, and Chemical Consequences of Molecular Asymmetry, 2009. **21**(1): p. 152–159.

[136] Miles, C.A., T.V. Burjanadze, and A.J. Bailey, *The kinetics of the thermal denaturation of collagen in unrestrained rat tail tendon determined by differential scanning calorimetry*. Journal of molecular biology, 1995. **245**(4): p. 437–446.

[137] Badea, E., T. Usacheva, and G. DellaGatta, *The use of differential scanning calorimetry to characterise collagen deterioration in parchment*. Rossiiskii Khimicheskii Zhurnal—Zhurnal Rossiiskogo Khimicheskogo Obshchestva im. DI Mendeleeva (Russian Chemistry Journal), 2015. **59**(1): p. 28–41.

[138] Burjanadze, T.V. and A. Veis, *A thermodynamic analysis of the contribution of hydroxyproline to the structural stability of the collagen triple helix*. Connective tissue research, 1997. **36**(4): p. 347–365.

[139] Pal, G.K., T. Nidheesh, and P. Suresh, *Comparative study on characteristics and in vitro fibril formation ability of acid and pepsin soluble collagen from the skin of catla (Catla catla) and rohu (Labeo rohita)*. Food Research International, 2015. **76**: p. 804–812.

[140] Ramanathan, G., et al., *Extraction and characterization of collagen from the skin of Arothron stellatus fish—A novel source of collagen for tissue engineering*. Journal of Biomaterials and Tissue Engineering, 2014. **4**(3): p. 203–209.

[141] Tziveleka, L.-A., et al., *Collagen from the marine sponges Axinella cannabina and Suberites carnosus: Isolation and morphological, biochemical, and biophysical characterization*. Marine drugs, 2017. **15**(6): p. 152.

[142] Rodrigues, F.T., V.C. Martins, and A.M. Plepis, *Porcine skin as a source of biodegradable matrices: alkaline treatment and glutaraldehyde crosslinking*. Polímeros, 2010. **20**(2): p. 92–97.

Chapter 13

Composting of Pig Effluent as a Proposal for the Treatment of Veterinary Drugs

Robson Evaldo Gehlen Bohrer,
Mastrângello Enivar Lanzanova, Ramiro Pereira Bisognin,
Marciel Redin, Eduardo Lorensi de Souza,
Daniela Mueller de Lara, Divanilde Guerra,
Danni Maisa da Silva, Maiara Figueiredo Ramires
and Diego Armando Amaro Da Silva

Abstract

Pig farming currently occupies a prominent place in the southern states of the Brazil, owning approximately 50% of the national squad, estimated at 42 million pig heads. However, the swine activity contributes significantly to the generation of environmental impacts on the environment. Recently, the greatest need for animal protein has exerted pressures on the current animal production system and one of the alternatives has been to the use of veterinary medicines, which have several uses ranging from therapeutic use, preventive in the treatment of various diseases and as growth promoters. Its indiscriminate and uncontrolled use is currently endangering the environmental balance of producing sites through effluent contamination. Many producers have been using contaminated slurry as a biofertilizer. In this sense, further studies on techniques and processes of treatment of organic effluents contaminated by veterinary drugs are necessary. Alternative low-cost and environmentally viable treatment systems are needs to minimize the entry into the environment of these contaminants. Therefore, the composting process that can defined as a process of aerobic microbial decomposition of organic matter and nutrient recycling can be an alternative for the treatment of effluents contaminated by veterinary drugs.

Keywords: antibiotics, contamination, impacts, human health, resistance, resistance

1. Introduction (Composting)

The composting technique emerged around the year 1920, when a researcher named Albert Howard developed the process in India [1]. This process happens naturally next to the environment, where the biological degradation of the compounds occurs [2], this technique can still be characterized as a process of treatment of different types of residues and origins, among them (urban, industrial, forestry and agricultural), where a diverse population of microorganisms (bacteria, fungi) act [3].

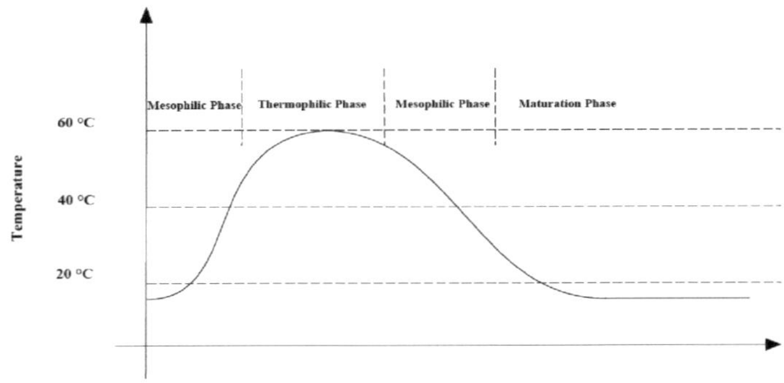

Figure 1.
Composting phases according to the temperature in the system. Source: Adapted from [3].

During the composting process, four fundamental stages for the biodegradation of the compounds occur: initial mesophilic phase, thermophilic, mesophilic cooling and maturation. In the initial phase, mesophilic, the decomposition process accelerated by the gradual increase of existing microorganisms, as well as the increase in temperature. In the thermophilic phase, the temperature of the composting system can reach more than 60°C, resulting from the action and thermophilic microorganisms, generating heat and water vapor. In the mesophilic phase of cooling, the most resistant compounds degraded by mesophilic microorganisms, which after this degradation eventually decrease their activity. Finally, the maturation phase, where the decomposition is low or zero and where the humid are released, generating by end, a stabilized compound [1]. The temperature profile during the composting phases are showing in **Figure 1**.

During the composting phases, CO_2 detachment and heat release occur in the composted material (mesophilic phase), reaching the temperature peaks, where the vapor (thermophilic phase) is generated, which also contributes to inertization in the system, all based on the metabolism of microorganisms [2].

Composting has several benefits for the management of waste produced by pig producers, as it can eliminate up to 90% of the effluent volume [4]. The process also has numerous advantages: minimizes the generation of greenhouse gases, reduces the proliferation of vectors, reduces odors, has environmental technical feasibility, elimination of costs for disposal and transport of waste, technical feasibility to expand the current pig production systems [2–5].

2. Physical-chemical composting parameters

For the composting process to succeed, some aspects such as heat transfer, airflow, steam and finally moisture balance must observed [6]. The main parameters to be evaluated during the composting process are: oxygen/aesthesia; pH; Moisture content; Temperature; C/N ratio, in addition to the presence of microorganisms [6].

Oxygen is directly interconnected with microbial activity, due to the composting process, being aerobic. The obtaining of oxygen from the composting system can be obtained by mechanical, physical or even forced aeration [7]. Its consumption and supply is tied, the humidity of the composting system, having optimum humidity in the range of 55%. The higher the microbial activity, the higher the demand for oxygen and the lower the humidity in the composting system [7]. Also for the author, systems with humidity below 40%, can inhibit the activity of bacteria, predominantly the higher activity of fungi.

The aeration has equal importance in the process, because with greater revolving and aeration of the system, the rate of decomposition of organic matter is increased. In his experiment of mechanized composting [5], he made the aeration during the injection of effluents in the trees and after 2 days, where he made only the revolving of organic matter.

Moisture in the composting system is of fundamental importance, being correlated with the rate of decomposition of organic matter by aerobic microorganisms. As already mentioned above, the ideal rate varies between 50% [3], and 55% [7]. Its control can was established by revolving, because at high humidity, anaerobiosis may occur in the system [3]. This, can also was controlled by the relationship between effluent injection and the amount of dry mass (DM). The author [5] used an initial rate of 1.47 liters of scum per kg (DM) in the first days up to a rate of 0.21 liters of scum per kg (DM) at the end of the experiment. Other aspects need observed, which are the phases of the composting process, mesophilic and thermophilic phase, both related to the elevation and decrease of temperature and water evaporation.

The state of organic matter decomposition can be measured from the hydrogenic potential (pH) found in the composting system, because for each phase of the process, the pH variation will occur. The initial pH values can be in the order of 4–5 [7], this due to the release of minerals acids and carbonic gas, where at this time the bacteria acidophilus and fungi act in the decomposition of cellulose [3]. Subsequently, the pH can reach values between 7 and 12, due to the formation of organic acids by alkalophiles bacteria [7], when there is stabilization of the compound [3].

With the ease of measuring throughout the process, the temperature profile plays a crucial role in composting. The temperature can range from 10–70°C, mesophilic phase and thermophilic respectively [3]. Its elevation and decrease linked with humidity and microbial activity, as demonstrated in the temperature profile in **Figure 1**.

Another important role of this parameter is the inertization and maturation of the compound over time. Pathogen microorganisms have their inactivation from three days at temperatures higher than 55°C [8], but in composting cells where revolving often occurs; the minimum is 15 days with temperatures above 55°C [4]. Aspects such as raw material, composting system configuration, presence of microorganisms, moisture, oxygen, directly affect temperature. Still the temperature can define the amount of revolving of the composting system. In his work [1], he observed the temperature variation in a process of composting of cattle slaughter residues, and in another study, [4], in a process of composting of pig effluents; both studies presented maximum temperatures did not exceed 55°C. In relation to **Figure 3(a)**, it is also noted that when revolving the composting system, the internal temperature increases significantly. Temperature profiles vary from research to research, [5] observed 72°C in the composting system at 22 days of experiment, [2], observed values not exceeding 52°C, already [9], obtained the maximum temperature of 61°C.

To prepare the composting system it is necessary to note the Carbon/Nitrogen ratio. Both compounds are important in the process, where carbon acts as an energy source and nitrogen as a respirometric source of microorganisms [3], also acting, in cell growth, formation of proteins, amino acids and nucleic acids [2]. The C/N ratio has been observed by several authors [1–4, 6, 9–11], who evaluated that this ratio should not exceed 30/1 in the initial phase, because in high relationships, the degradation of the compounds is delayed, already at the end, when the compound is matured, the ratio may reach 10/1. In this process, much of the carbon is released into the atmosphere in the form of CO_2 [6], and nitrogen can be released, when the ratio is low, in the form of ammonia, characterizing bad odor [2].

3. Use of veterinary drugs

The indiscriminate use of veterinary medicinal products in animal husbandry, especially in pig farming, has become the gateway of these pollutants to the environment [12]. Antibiotics used since prevention, therapeutic use, helping in the treatment of diseases such as infections, diarrhea, still being able to act as growth promoters [13–16]. Currently, there is a great concern in the academic community, with residues derived from veterinary drugs, due to their potential contaminant, but also, by the non-absorption completely, by the animal organism. Several authors, such as [15–19] point out that on average 60% of the veterinary medicinal product dosed to animals are excreted through urine and feces [20], and the group of drugs most commonly used today are antibiotics [15, 21], and among them, tetracyclines, sulfonamides, lincosamides, β-lactamines and macrolides [22].

After their non-absorption by the body, veterinary drugs can reach soil and water resources [13, 16, 22] as unaltered substances or metabolites. The use of pig effluents as fertilizer may be emphasizing the spread and increase of antibiotic residues in the environment [12, 20]. Several studies have proven environmental contamination around the world through antibiotics, such as tetracycline in Brazil, Canada and China, [20, 23, 24], by sulfonamines in Bolivia, Czech Republic and USA, [25–27], macrolides in South Korea, United Kingdom and Hong Kong [28–30], fluroquinolonas in Poland, Austria and Thailand [31–33], and kilane in China and Malaysia [25, 34].

In addition, to soil contamination by heavy metals [35], contamination of water resources [36], the application of biofertilizers, directly or indirectly in the soil [34], has caused change in the soil biotic community [37]. Another point that draws attention, and more significant is the resistance of microorganisms to antibiotics, several authors have found evidence of the resistance of microorganisms [38–42] thus enhancing the risk to human health [34, 38, 43].

The availability and use of pig effluent are justified by the size of the production chain. Brazil is among the four largest producers of pigs in the world, behind China, the United States and the European Union, with an estimated herd of 42 million heads, representing US$ 1 billion annually in meat sales [44, 45]. The Southern Region of the country, where the states of Paraná, Santa Catarina and Rio Grande do Sul are located, account for 49.3% of the national production [46]. With a well-established production chain, a large amount of waste generated. Point out that a pig in the finishing phase can produce up to 7.6 liters of manure/day [47], often causing failures and overload in effluent treatment systems, which mostly treated by biological ponds.

Many of these existing environmental problems and pressures are due to traditional organic effluent treatment systems, widespread in pig-producing units, which are not efficient in the treatment of these pollutants [48]. Compliance with environmental laws, as well as the feasibility of managing waste produced in farms, generated the need for alternatives for the treatment of effluents.

With this, composting emerged as a treatment proposal, which is a natural process of nutrient recycling, through aerobic microbial decomposition of organic matter [4], under favorable conditions of temperature, pH, oxygen, humidity, presence of chemical substances, raw material and C/N ratio [3, 47], resulting in a material with relative stability and quality [49]. The advantages range from minimizing the volume of effluents of about 90% [4], reducing the emission of greenhouse gases and the proliferation of vectors. Another important factor is the technical feasibility, to expand the current pig production systems [2–5], showing be a practical proposal, low cost [23, 25, 50], still classified as a clean and viable method [28], for the correct management of waste. Another advantage is the inactivation and immobilization of pathogens, nutrients and veterinary drugs [13, 48, 51, 52], thus becoming a potential proposal for the treatment of veterinary antibiotics [23, 25, 48, 53].

Studies indicate that composting has potential in the treatment and decay of drug concentrations [3, 13, 48, 49, 51, 54–56]. Antibiotic residues such as florfenicol, sulfadimetoxin, sulfametazin and tylosin reached 95–99% decline in 21 days of composting [51]. Tetracycline, sulfonamides and macrolides had 99%, 96% and 95% decaine, respectively, through composting [53] in 35 days of testing. [48], they did not detect the presence of a group of sulfonamides (Sulfametazin (SMZ) and Sulfamethoxazole (SMX)) in the final compound.

4. Composting as a proposal for treatment of swine effluents

Technologies that seek to reduce residues of veterinary medicinal products (RMV), mainly veterinary antibiotics (AVs) found in organic and industrial effluents disposed as fertilizer in the soil is a necessity to minimize the environmental impacts generated by these compounds [25]. Traditional organic effluent treatment systems, widespread in pig-producing units, are not efficient in the treatment of these pollutants [48].

Among the various technologies and treatment systems for different origins and compositions of organic effluents, including pigs, composting has been shown to be a practical proposal, low cost [23, 25, 50], classified as a clean and viable method [28] for the correct management of waste. This technique can be developed as an alternative for the treatment of effluents in small properties, located in regions with high concentration of pigs and with little agricultural area available for final disposal [4], as well as proposal for treatment of veterinary antibiotics [23, 25, 48, 53].

Composting can be defined as a process of aerobic microbial decomposition of organic matter, being a natural process of nutrient recycling, used since ancient civilizations [4], under favorable conditions of temperature, pH, oxygen, humidity [47], presence of chemicals, raw material and C/N ratio [3], resulting in a material with relative stability and quality [49]. Treatment by composting reduces the volume of effluents, inactivates and immobilizes pathogens, nutrients and veterinary drugs [13, 48, 51, 52], and finally produces a by-product (substrate), with economic and agronomic value [49, 50]. **Figure 2**, shows the cycle of inputs and outputs during the composting process.

This treatment proposal has been shown to be effective in the management of organic waste from production processes confined to pigs, poultry and cattle, and has the potential to treat emerging organic pollutants (POEs) [57]. The decay of the concentration of veterinary medicinal products through composting has been researched by several authors [13, 48, 49, 51, 54–56], for different types of effluents and organic residues.

The decline of 27% OF CTC was observed in swine effluents [57] and 92% in poultry manure in a composting system for 42 days. When analyzing the decline of

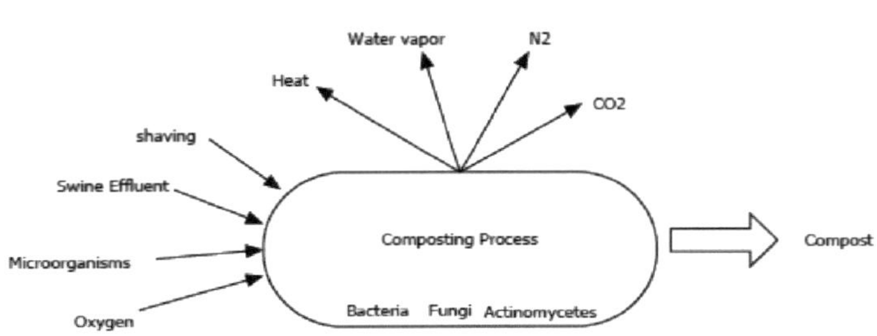

Figure 2.
Flow of entry and exits in the composting process. Source: Author (2017).

Figure 3.
Mechanized composting system of swine effluents (a) [58], shaving bed after effluent injection and revolving (b) municipality of Concordia, SC.

4 antibiotics (florfenicol, sulfadimetoxin, sulfametazin and tylosin) [51] during the composting process of domestic effluents, approximately 95–99% of antibiotics were degraded after 21 days of testing. Antibiotic decline [53] was evaluated (tetracycline 96%, 99% sulfonamides and macrolides 95%) during the composting process, [48], after 35 days of bench-scale composting, they did not detect the presence of antibiotics from the Sulfonamide Group (Sulfametazin (SMZ) and Sulfametoxazole (SMX)).

Figure 3 shows one of the processes of composting of existing pig effluents, the mechanized one, which consists of mixing the waste produced by pigs in the rearing systems, with shavings, sawdust or straw in beds/beds [50].

However, this process as a proposal for the treatment of veterinary medicines of different classes, has yet been developed in the country, justifying the proposal, having presented good results in research already developed, its application becomes important in the search for new alternatives to minimize the potential environmental risks caused by these contaminants, since in contact with environmental matrices can be accumulated in the soil, as well as being leached for water resources [56].

5. Use of composting in the treatment of veterinary drugs

One of the determining points for the development of research and its scientific relevance is the potential for contamination by veterinary antibiotics. Currently the pig production chain in the south of the country is estimated at 20.5 million heads. Considering only the state of Rio Grande do Sul (7 million heads), and assuming that the main group of antibiotics, tetracycline [15, 21], which is given in the order of 400 mg/animal/80 kg. The medicated with the main antibiotic group, and 70% of the dose is excreted by urine and feces [18]. If this residue is deposed in current treatment systems, which can reduce its concentration by 50%, this would result in 0.140 g/tetracycline/animal, representing 0.98 tons of antibiotics that would be dumped into the soil annually along the effluent in the form of biofertilizer [59].

The search for technical alternatives for the treatment of pig manure contaminated with residues of veterinary drugs was decisive for the accomplishment of the work, considering the size of the production chain in the country, due to the lack of research at the national level, but mainly in minimizing the potential damage, they can cause to the environment. The results observed in the research point a potential for chronic contamination and disturbances at the environmental level, but also at the social level, very expressive, but also point to the need for research aimed at the search for technological alternatives for the treatment of these residues, often left aside.

Based on the results obtained [58], it was observed that composting proved to be effective in the degradation of 19 veterinary drugs, divided into 8 groups. The decay/degradation rate ranged from 33.7–100% in 150 days. The antibiotics sulfatiazole, tetracycline and chlortetracycline showed 100% decay. The mean degradation of antibiotics was 97.2%, proving composting as a technique for the treatment of swine effluents contaminated by antibiotics, however, at the end of composting, some antibiotics presented residues in the order of milligrams per kg in the final compound. Therefore, further research on the behavior of these compounds during composting would elucidate whether these compounds are actually degraded or if they generate some kind of metabolites or other substance.

Regarding the community of microorganisms for Bacteria and Fungi [58], a great diversity in the level of phylum and genera observed in both kingdoms throughout the composting. Regarding phyla and genera of bacteria, 7 phylum and more than 70 genera of bacteria were observed over time (0, 15, 30, 45, 60, 75, 90, 120 2150 days). Fungal diversity at phylum and gender level was 2 phylum and 16 genera. This abundance and diversity may be related to the proposed identification methodology, new generation sequencing, which proved to be able to identify a wide range of the micro biota found during composting. In this context, a correlation between environmental variables and antibiotics with microorganisms was, also observed, proving through redundancy analysis that the main factors to have significance in the bacterial community were humidity, but not influencing the fungal community. Veterinary antibiotics (Tilcomisin and Ciprofloxacin) showed a positive correlation in the vast majority of bacteria genera, an effect not observed in fungi. In the fungal genera, the antibiotic Tilmicosin has a positive correlation with the genera *(Apiotrichum* and *Penicillium)* and Ciprofloxacin has a positive relationship with the genera (*Tricosporium, Parascedosporium, Petriella* and *Cryptococcus*).

6. Microbial communities of the composting process

The application of animal waste contaminated with residues of veterinary medicinal products has become the gateway to the expansion of several types of antibiotic resistance genes, caused by the indiscriminate use of antibiotics in the production of animal protein [60, 61]. In the composting process, there are different types of microorganisms, among them the predominance of bacteria, fungi and actinomycetes, divided into aerobic, thermotolerant and mesophilic [48], which are responsible for about 95% of microbial activity [3]. One of the most important parameters for the proliferation of these microorganisms is temperature, which should not exceed 65°C, for fungi and actinomycetes, and for bacteria, temperatures should be higher than 40°C [48].

Another important aspect is the presence of microorganisms, which are capable of contaminating the environment [9] including *E. coli and* other pathogens. Also according to the authors, during the experiment, carried out with composting of swine effluents, the average presence of 2 to 5 ($\log^{10}$ NMP g^{-1}) of total coliforms was found.

It found 39 species of fungi in the composting process [50], many of which were identified only at the beginning or at the end of the experiment (**Table 1**).

Fungi identified at the beginning of the process	Fungi identified at the end of the process
Absidia Hesseltinii	*Alternaria alternata**
Alternaria alternata	*Arthrobotrys sp.*
Candid's aspergillus	*Fumigatus aspergillus*
Flavus aspergillus	*Niger Aspergillus*
Fumigatus aspergillus	*Aspergillus sp.*
Niger Aspergillus	*Aureobasidium floccosum**
Aspergillus	*Botryosphaeria sp.*
Parasiticus aspergillus	*Cephalosporium sp.**
Versicolor aspergillus	*Cladosporium sp.*
Aureobasidium sp.	*Colletotrichum sp.*
Bipolaris Maydis	*Curvularia bezel*
Botryodiplodia theobromae	*Dreschlera sp.*
Acremonium	*Fusarium oxysporum**
Cephalosporium spp.	*Gibberella sp.*
Chaetomium globosum	*Helminthosporium spp**
Cladosporium cladosporioides	*Hysterium sp.*
Clavata curvularia	*Mycosphaerella sp.*
Drechslera carbonum	*Nigrospora sp.*
Oxysporum fusarium	*Septoria sp.*
Solan's Fusarium	*Stenocarpella Maydis*
Graminearum fusarium	*Tetraploa sp.*
Moniliform fusarium	*Viride Trichoderma*
Helminthosporium spp.	
Phaseoline macrophomine	
Monilinia sp.	
Mucor sp.	
Oryzae nigrospora	
P. funiculosum	
Penicillium citrinum	
Penicillium sp.	
Penicillium spp.	
Phoma herbarum	
Oryzae	
R. stolonifer	
Solani rhizoctonia	
Oligosporum rhizopus	
Syncephalastrum racemosum	
Harzianum Trichoderma	

*Source: [50] * genres that occurred throughout the process.*

Table 1.
Fungi identified during the process of composting of swine effluents with residues of treated seeds.

It is observed that of the total fungi, five species were found since the beginning of the process (*Alternaria alternata* Aureobasidium floccosum* Fusarium oxysporum* Helminthosporium spp**). In a system of composting of poultry, waste found 3 phylos (kingdoms) in greater quantity: Betaproteobacteria; Firmicutes and Bacteria [49].

Evaluating the resistance of microorganism genes to the antibiotic Oxytetracycline (OTC) [13], observed the predominance in 95.3% of the bacteria found, with the following phylos (kingdoms) Actinobacteria, Bacteroidetes, Chloroflexi, Firmicutes and Proteobacteria, and that of these kingdoms, 50 generos (*Clostridium sensu stricto, Aquamicro Aquabium; Paenibacillus; Azoarcus; Jonesia; Gracilibacillis; Devosia; Celivibrio; Marinobacter; Tepidimicrobium; Ornithinibacillus; Paracoccus; Pelagibacterium; Turicibacter; Streptomyces; Leucobacter; Vulgatibacter; Steroidobacter; Bordetella; Chelatococcus; Trupera; Nonomuraea; Thermovum; Brumimicrobium; Caldalkalibacillus; Ornithinimicrobium; Jeotgalicoccus; Ureibbacillus; Sphaerobacter; Saccharibacteria_genera_incertae_sedis; Pseudomonas; Corynebacterium; Dietzia; Clostridium XI; Sphingobacterium; Pusillimonas; Luteimonas; Flovobacterium; Actinomadura; Rhodopirellula; Verrucosispora; Nocardioids; Bacillus; Ammoniibacillus; Planifilum; Georgenia; Idiomamarina; Saccharomonospora and Thermobifida)* during aerobic composting of bovine effluent, as well as the increase in OTC resistance in some genders [13].

Also in relation to the resistance of microorganisms with antibiotics, [57] they state that CTC inhibited the growth of 12 soil bacteria at different concentrations. In addition to the increased intake of antibiotics in the environment, this can pose risks to human health, such as increased allergy to antibiotics and increased resistance to antibiotics, as many foods develop in places with inadequate effluent disposal and transfer a contaminated load.

Microorganisms such as *E. coli have* shown antibiotic resistance in several studies [58, 62–64]. *E. coli resistance* was tested from wastewater and wastewater treatment system in 24 antibiotics [64], distributed in 6 classes (Penicillins, Cephalosporins, Chynomas, AminoGlycosides, Sulfonamides and Tetracycline). The results showed that the groups of antibiotics with the highest resistance were, Penicillin; Cephalosporin; Kilonomonas; Sulfonamides and Tetracycline.

In another study [14] they found 14 tetracycline-resistant genes and three antibiotic resistance genes Sulfonamines, which modified ribosomal protection proteins, enzymatic inactivation proteins. These results can also be confronted by the high persistence and accumulation capacity that antibiotics have when they are in environmental matrices, especially in soils. Evaluated the persistence of 5 antibiotics in the soil (Tetracycline, Sulfametazin, Norfloxacin, Erythromycin and Chloramphenicol), where the highest rate of antibiotic adsorption in the soil was: Tetracycline > Norfloxacin > Erythromycin> Chloramphenicol> Sulfametazin, thus increasing the risk to the environment [61].

7. Conclusions

In the end, we can admit that the composting process presented itself as an alternative to the current treatment systems, since it combines, at the same time, the treatment of swine effluent, but it has the capacity to degrade antibiotic residues found in swine effluents, minimizing their effects. Impacts on the environmental matrices (soil and water), and still at the end, generate a product (compost) with agricultural potential superior to the use of effluents directly in the soil.

Author details

Robson Evaldo Gehlen Bohrer[1*], Mastrângello Enivar Lanzanova[1], Ramiro Pereira Bisognin[1], Marciel Redin[1], Eduardo Lorensi de Souza[1], Daniela Mueller de Lara[1], Divanilde Guerra[1], Danni Maisa da Silva[1], Maiara Figueiredo Ramires[2,3] and Diego Armando Amaro Da Silva[4]

1 Universidade Estadual do Rio Grande do Sul – Uergs/Brazil

2 Dr. em Ciência do Solo – Universidade Federal de Santa Maria – UFSM/Brazil

3 Graduanda em Agronomia – Universidade Estadual do Rio Grande do Sul – Uergs/Brazil

4 Engenheiro Agrônomo – Universidade Federal de Santa Maria (UFSM), Especialista em Segurança Alimentar e Agroecologia (UERGS - Unidade Três Passos), Mestre em Agronomia - Agricultura e Ambiente (UFSM - Campus Frederico Westphalen)-Brazil

*Address all correspondence to: robson-bohrer@uergs.edu.br

References

[1] BARWINSKI, M. J. B. Legal framework and technology for composting slaughterhouse waste. 2015. 146 p. Thesis (Master in Agroecossitemas) - Federal University of Santa Catarina, Florianópolis, SC, 2015.

[2] JAHNKE, D.S. Treatment of organic waste from small cattle slaughterhouses through composting. 2012. 109 p. Dissertation (Master in Animal Science) - Federal University of Pelotas - UFPEL, Pelotas, RS, 2012.

[3] SILIA, H. D. N. Studies on ETE slat composting with a view to application on agricultural soil. 2015. 64 p. Course Completion Work (Environmental and Urban Engineering Course) - Federal University of ABC, UFABC, Santo André, SP, 2015.

[4] CAVALETTI, L. B. Evaluation of the mechanized composting system for pig manure. 2014. 84 p. Course Completion Work (Environmental Engineering Course) - University of Vale do Taquari - UNIVATES, Lajeado, RS, 2014.

[5] DONEDA, A. Acidification of liquid pig manure affects ammonia and greenhouse gas emissions in the automated composting process. 2014. 99 f. Thesis (Doctorate in Soil Science) - Federal University of Santa Maria - UFSM, Santa Maria, RS, 2014.

[6] TEIXEIRA, C. Sludge hygiene from sewage treatment plant by thermophilic composting for agricultural use. 2012. 139 p. Thesis (Master's degree in Agroecosystems) - Federal University of Santa Catarina, Florianópolis, SC, 2012.

[7] ANGNES, G. Gas emissions in the process of compostegm of pig manure. 2012. 133 p. Thesis (Master in Agroecosystems) - Federal University of Santa Catarina, Florianópolis, SC, 2012.

[8] USDA/USEPA. Unified National Strategy For Animal Feeding Operations, Washington, March 9, 1999. Available from: <http://www.epa.gov/npdes/pubs/finafost.pdf>. Accessed: 10 Jan. 2017.

[9] SÁ, M. F. *et al.* Dynamics of the coliform population during automated composting of swine liquid manure. Brazilian Archive of Veterinary Medicine and Animal Science, v. 66, n. 4, p. 1197-1206, 2014.

[10] SARDÁ, L. G. Composting as an alternative for the treatment of pig manure and the reduction of the emission of polluting gases. 2009. 84 p. Thesis (Master in Agroecosystems) - Federal University of Santa Catarina, Florianópolis, SC, 2009.

[11] ZANOTELLI, L. *et al.* Mesophilic and thermophilic fungi important in mechanized composting of pig manure. Research Book ,v. 23, n. 2, p. 43-53, 2014.

[12] NGUYEN, T. K. X. et al. Tiamulin removal by wood-rot fungi isolated from swine farms and role of ligninolytic enzymes. International Biodeterioration & Biodegradation, v. 116, p. 147-154, 2017.

[13] QIAN, X. et al. Variable effects of oxytetracycline on antibiotic resistance gene abundance and the bacterial community during aerobic composting of cow manure. Journal of Hazardous Materials, v. 315, p. 61-69, 2016b.

[14] BEN, W. et al. Dissemination of antibiotic resistance genes and their potential removal by on-farm treatment processes in nine swine feedlots in Shandong Province, China. Chemosphere, v. 167, p. 262-268, 2017.

[15] TASHO, R. P.; CHO, J. Y. Veterinary antibiotics in animal waste, its

distribution in soil and uptake by plants: A review. Science of the Total Environment, v. 563-564, n. 3, p. 366-376, 2016.

[16] GUO, C. et al. Development of a modified QuEChERS method for the determination of veterinary antibiotics in swine manure by liquid chromatography tandem mass spectrometry. Journal of Chromatography B: Analytical Technologies in the Biomedical and Life Sciences, v. 1027, p. 110-118, 2016.

[17] GELBAND, H. et al. The State of the world's antibiotics. Center for Disease Dynamics, Economics & Policy. 2015. p.79, CDDEP: Washington, D.C.

[18] DAGHRIR, R.; DROGUI, P. Tetracycline antibiotics in the environment: A review. Environmental Chemistry Letters, v. 11, n. 3, p. 209-227, 2013.

[19] DU, L.; LIU, W. Occurrence, fate, and ecotoxicity of antibiotics in agro-ecosystems. The review. Agronomy for Sustainable Development, v. 32, n. 2, p. 309-327, 2012.

[20] PULICHARLA, R. et al. Degradation of chlortetracycline in wastewater sludge by ultrasonication, Fenton oxidation, and iron-sonication. Ultrasonics Sonochemistry, v. 34, p. 332-342, 2017.

[21] KIM, J. et al. Multiresidue analytical method for the determination of antimicrobials, preservatives, benzotriazole UV stabilizers, flame retardants and plasticizers in fish using ultra high performance liquid chromatography coupled with tandem mass spectrometry, Journal of Chromatography A, vol. 1218, no. 22, pp. 3511-3520, 2011.

[22] SOLLIEC, M. et al. Fractionation and analysis of veterinary antibiotics and their related degradation products in agricultural soils and drainage waters following swine manure amendment. Science of the Total Environment, v. 543, p. 524-535, 2016.

[23] WEI, R. et al. Occurrence of 13 veterinary drugs in animal manure-amended soils in Eastern China. Chemosphere, v. 144, p. 2377-2383, 2016.

[24] PINHEIRO, A. et al. Veterinary antibiotics and hormones in water from application of pig slurry to soil. Agricultural Water Management, v. 129, n. 375, p. 1-8, 2013.

[25] HO, Y. B. et al. Degradation of veterinary antibiotics and hormone during broiler manure composting. Bioresource Technology, v. 131, p. 476-484, 2013.

[26] ARCHUNDIA, D. et al. Antibiotic pollution in the Katari subcatchment of the Titicaca Lake: major transformation products and occurrence of resistance genes. Science of The Total Environment,v. 576, p. 671-682, 2017.

[27] KOBA, O. et al. Antibiotics degradation in soil: A case of clindamycin, trimethoprim, sulfamethoxazole and their transformation products. Environmental Pollution, p. 1-13, 2016.

[28] WU, J. *et al.* Effect of precursors combined with bacteria communities on the formation of humic substances during different materials composting. Bioresource Technology, v. 226, p. 191-199, 2017.

[29] KIM, Y.; LEE, B.B.; CHOI, K. Effect of runoff discharge on the environmental levels of 13 veterinary antibiotics: A case study of Han River and Kyungahn Stream, South Korea. Marine Pollution Bulletin,p. 1-13, 2015.

[30] PETRIE, B. et al. Multi-residue analysis of 90 emerging contaminants

in liquid and solid environmental matrices by ultra-high-performance liquid chromatography tandem mass spectrometry. Journal of Chromatography A, v. 1431, p. 64-78, 2015.

[31] GBYLIK-SIKORSKA, M. et al. Chemosphere Liquid chromatography – tandem mass spectrometry multiclass method for the determination of antibiotics residues in water samples from water supply systems in food-producing animal farms. Chemosphere, v. 119, p. 8-15, 2015.

[32] MARTÍNEZ-CARBALLO, E. et al. Environmental monitoring study of selected veterinary antibiotics in animal manure and soils in Austria. Environmental Pollution, v. 148, n. 2, p. 570-579, 2007.

[33] RICO, A. et al. Use, fate and ecological risks of antibiotics applied in tilapia cage farming in Thailand. Environmental Pollution, v. 191, p. 8-16, 2014.

[34] LI, C. et al. Occurrence of antibiotics in soils and manures from greenhouse vegetable production bases of Beijing, China and an associated risk assessment. Science of the Total Environment, v. 521-522, p. 101-107, 2015.

[35] SILVA, D. M. Biota of soil in pasture under application. 2015. 64 f. Thesis (PhD in Soil Science) - Federal University of Santa Maria - UFSM, Santa Maria, RS, 2015.

[36] SEGURA, P. A. et al. Global occurrence of anti-infectives in contaminated surface waters: Impact of income inequality between countries. Environment International, v. 80, p. 89-97, 2015.

[37] BILA, D. M.; DEZOTTI, M. Endocrine disruptors in the environment: effects and consequences. New chemistry, v. 30, n. 3, p. 651-666, 2007

[38] BONDARCZUK, K.; MARKOWICZ, A.; PIOTROWSKA-SEGET, Z. The urgent need for risk assessment on the antibiotic resistance spread via sewage sludge land application. Environment International, v. 87, p. 49-55, Feb. 2016.

[39] CENTNER, T. J. Recent government regulations in the United States seek to ensure the effectiveness of antibiotics by limiting their agricultural use. Environment International, v. 94, p. 1-7, 2016.

[40] LUCAS, D. et al. Fungal treatment for the removal of antibiotics and antibiotic resistance genes in veterinary hospital wastewater. Chemosphere, v. 152, p. 301-308, 2016.

[41] RHOUMA, M.; BEAUDRY, F.; LETELLIER, A. Resistance to colistin: what is the fate for this antibiotic in pig production? International Journal of Antimicrobial Agents, v. 48, n. 2, p. 119-126, 2016.

[42] WANG, J. et al. Dissemination of veterinary antibiotics and corresponding resistance genes from a concentrated swine feedlot along the waste treatment paths. Environment International, v. 92-93, p. 317-323, 2016a.

[43] HE, L. Y. et al. Discharge of swine wastes risks water quality and food safety: Antibiotics and antibiotic resistance genes from swine sources to the receiving environments. Environment International, v. 92-93, p. 210-219, 2016.

[44] ABCS. Brazilian Association of Pig Breeders. ABCS and agro-industry representatives meet to discuss Integrated Production Bill. Available in http://www.abcs.org.br/informativo-abcs. January 2014.

[45] MAP. Ministry of Agriculture, Livestock and Supply. Brazil, 2012.

Available in: http://www.agricultura.gov.br/animal/especies/suinos. Accessed 18 October, 2015.

[46] BRAZILIAN INSTITUTE OF GEOGRAPHY AND STATISTICS - IBGE. Production of Municipal Livestock, v. 43, p. 1-49. 2015. Available in: http://biblioteca.ibge.gov.br/index.php/biblioteca-catalogo?view=detalhes&id=784. Accessed January 14, 2016.

[47] HACHMANN, T. L. et al. Poultry and swine waste : potential. Green Journal of Agroecology and Sustainable Development,v. 8, n. 5 p. 59-65, 2013.

[48] LIU, B. et al. Bioresource Technology Effects of composting process on the dissipation of extractable sulfonamides in swine manure. Bioresource Technology, v. 175, p. 284-290, 2015.

[49] OMAR, F. N. et al. Investigation of physico-chemical properties and microbial community during poultry manure co-composting process. JES, p. 1-14, 2014.

[50] GHISLENI, R. Composting of pig manure with substrate of seeds treated with pesticides and sawdust. 2013. 71 p. Dissertation (Master in Environmental Technology) - University of Santa Cruz do Sul - UNISC, Santa Cruz do Sul, RS, 2013.

[51] MITCHELL, S. M. et al. Antibiotic Degradation During Thermophilic Composting. Water Air Soil Pollut, v. 225, n. 13, p. 1-12, 2015.

[52] ZHANG, Z. *et al.* Evaluation of aerobic co-composting of penicillin fermentation fungi residue with pig manure on penicillin degradation, microbial population dynamics and composting maturity. Bioresource Technology, p. 3-30, 2015b.

[53] KIM, K. et al. Decline in extractable antibiotics in manure-based composts during composting. Waste Management, v. 32, p. 110-116, 2012.

[54] MORTIER, N.; VELGHE, F.; VERSTICHEL, S. Organic Recycling of Agricultural Digestion. cap. 4. p. 69-124, Elsevier Inc., http://dx.doi.org/10.1016/B978-0-12-803622-8.00004-5. 2016.

[55] WANG, R. et al. Effect of red mud addition on tetracycline and copper resistance genes and microbial community during the full scale swine manure composting. Bioresource Technology, v. 216, p. 1049-1057, 2016b.

[56] ZHANG, H. et al. Residues and risks of veterinary antibiotics in vegetable soils following application of different manures. Chemosphere, v. 152, p. 229-237, 2016a.

[57] BAO, Y. *et al.* Depletion of chlortetracycline during composting of aged and spiked manures. Waste Management, v. 29, n. 4, p. 1416-1423, Apr. 2009.

[58] BOHRER, R. E. G. Swine effluent composting as a proposal for the treatment of veterinary drugs. Thesis, Course of the Postgraduate Program in Civil Engineering, Concentration Area Water Resources and Environmental Sanitation, at the Federal University of Santa Maria (UFSM, RS), p. 112, 2018.

[59] WU, M. H. *et al.* Occurrence, fate and interrelation of selected antibiotics in sewage treatment plants and their receiving surface water. Ecotoxicology and Environmental Safety, v. 132, p. 132-139, 2016.

[60] CHENG, W. et al. Behavior of antibiotics and antibiotic resistance genes in eco-agricultural system: A case study. Journal of Hazardous Materials,v. 304, p. 18-25, 2016.

[61] PAN, M.; CHU, L. M. Adsorption and degradation of five selected antibiotics in agricultural soil. Science of the Total Environment, v. 545-546, p. 48-56, 2016.

[62] BOUKI, C.; VENIERI, D.; DIAMADOPOULOS, E. Detection and fate of antibiotic resistant bacteria in wastewater treatment plants: A review. Ecotoxicology and Environmental Safety, v. 91, p. 1-9, 2013.

[63] KÜMMERER, K. Resistance in the environment. Journal of Antimicrobial Chemotherapy, v. 54, n. 2, p. 311-320, 2004.

[64] REINTHALER, F. *et al.* Antibiotic resistance of *E. coli* in sewage and sludge. Water Research, v. 37, n. 8, p. 1685-1690, Apr. 2003.

Chapter 14

A Comparative Study of MSW to Emery in Oman

Tariq Umar

Abstract

The adverse impact of the energy production from fossil fuels is now well recognized globally; therefore, the move toward renewable and sustainable energy has become an integral part to achieve the United Nations Sustainable Development Goals (SDGs). This chapter presents a comparative study considering a waste-to-energy plant to produce electricity in Oman. A research strategy that includes both qualitative and quantitate research methods were adopted to evaluate the MSW generation and emissions, electricity consumption and emissions, public participation in waste segregation, and to estimate the reduction in emission by considering a 5000 tons/day waste-to-energy plant in Oman. The results show that the current emission from fossil fuels to meet the electricity requirement of 70,633.37 Million kWh/year is 161.781 Million tonnes (CO_2/year). Similarly, the emissions from MSW which currently stood at 2.159 million tons/year are 3,424,247 tons CO_2/year. A 5000 ton per day waste-to-energy plant will not only produce 29.30 million kWh daily but will also enable an annual reduction of 24,527 million kg CO_2. Such an initiative will help Oman to improve its sustainability performance in energy, climate change, waste reduction, and economic growth and will pave the road to achieve the relevant SDGs by 2030.

Keywords: energy, sustainability, waste management and disposal

1. Introduction

Energy is an integral part of today's modern life, but the way most the energy is produced around the world creates several environmental and sustainability issues. Environmental sustainability is the core issue that needs to be addressed for development to focus on human well-being and yet stay within the limitations of the planet's capacity. Environmentally sound waste management is one of the key elements for sustainable development. The idea of sustainability developed in the early 1980s as reported in the International Geosphere-Biosphere Program can be defined as "meeting fundamental human needs while preserving the earth natural environment" [1]. Since the earth's population is increasing, it is putting pressure on the earth's resources. According to the World Economic Forum, it is estimated that food production will need to double by 2050 to feed 10 billion people on the earth [2]. Today, sustainability has three essential pillars, including environmental protection, social development, and economic growth; sustainable development can be been defined as a development that meets the needs of the present without compromising the ability of future generations to meet their own needs [3]. The need for sustainable development is truly recognized by all countries, and thus in 2015, the United Nations (UN) was able to introduce 17 Sustainable Development Goals (SDGs) to be achieved by 2030 [4]. The UN under Goal 12 (Responsible Consumption and Production) of its SDGs aims to substantially reduce waste generation through prevention, reduction, recycling, and reuse. Data from 214 cities or municipalities in 103 countries show that about three quarters of MSW generated is collected (**Figure 1**). In sub-Saharan Africa, less than half of all MSW generated is collected, with adverse effects on the health of residents. Moreover, even when waste is collected, it is often not treated and disposed of in a sustainable and environmentally sound manner. Managing such wastes continues to be a major challenge facing urban areas in several regions. Appropriate waste management is important for conserving local and global environments. Improvement of waste management in developing countries is directly related to preventing environmental pollution and expanding public health services. Appropriate waste management contributes to reducing not only the emission of water/atmospheric pollutants and odors but also the emission of greenhouse gases. In this regard, some studies reported that Green House Gases (GHG) emissions from the waste sector contribute to 3–4% of total global GHG emissions [7]. The rapidly increasing amount of municipal waste in cities around the globe is connected with economic development, as an increase in the city population creates many major challenges associated with economic development [8].

Gulf Cooperation Council (GCC) member countries (Saudi Arabia, Oman, United Arab Emirates, Kuwait, and Qatar) are considered as major consumers of the natural resources, which results in a huge amount of emissions [9, 10]. Similarly, the annual solid waste generation in the GCC region has exceeded 150 million tons. GCC countries feature among the world's top 10 per capita waste generators (**Figure 2**). Similarly, the annual solid waste generation in the GCC region has exceeded 150 million tons. Comparatively, this is lower than the waste generated in the UK, as GCC has a lower population (=54 million) than the UK (66 million); however, at the same time; the UK recycles more than 45% of its waste [12]. The recycling of waste in the GCC is almost zero. Lack of legal and institutional frameworks has been a major stumbling block in the progress of the waste management sector [11]. The per capita production of municipal waste in top GCC cities, such as Riyadh, Doha, Abu Dhabi, and Dubai, is more than 1.5 kg per day which is among the highest worldwide [13]. Some recent studies which considered situations of waste in the whole gulf region indicate that the recycling sector is underdeveloped and hardly 10–15% of the waste is recycled [14]. This chapter considers the MSW in Oman to produce electricity. Currently, the MSW

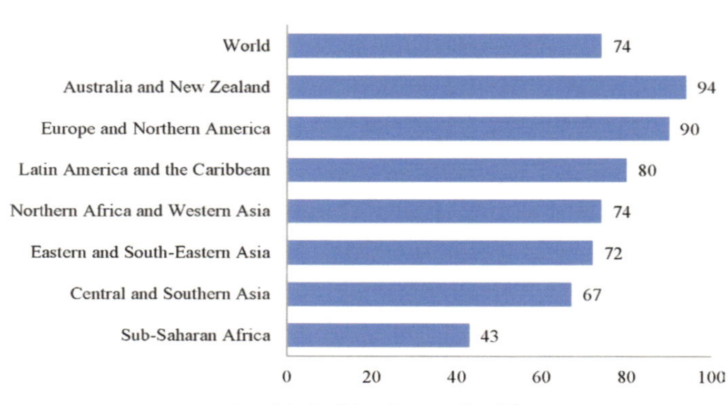

Figure 1.
Proportion of the municipal solid waste generated that is collected, 2001–2015 (data from 214 cities/municipalities in 103 countries) [5, 6].

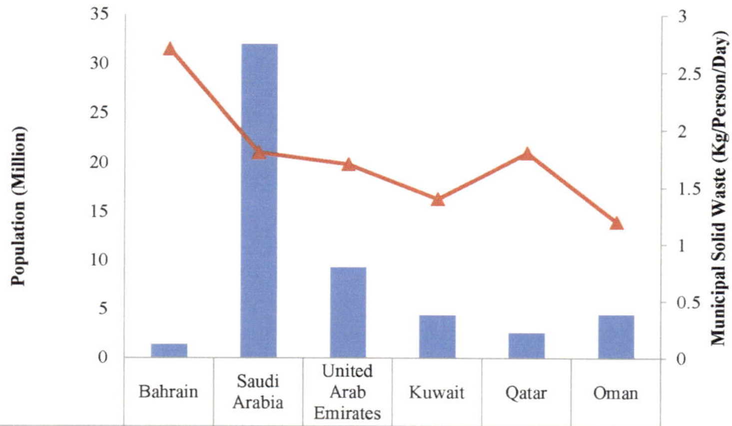

	Bahrain	Saudi Arabia	United Arab Emirates	Kuwait	Qatar	Oman
Population (Million)	1.4	32	9.3	4.4	2.6	4.5
Municipal Solid Waste (Kg/Person/Day)	2.7	1.8	1.7	1.4	1.8	1.2

Figure 2.
GCC population versus municipal waste generation [6, 11].

in Oman is deposited in more than 300 landfills/dumpsites managed by municipalities. Overall, most of the solid waste is sent to authorized and unauthorized dumpsites for disposal, which is creating environmental and health issues. There are several dumpsites which are located in the midst of residential areas or close to catchment areas of private and public drinking water bodies. Similarly, these landfill stations occupied a large area of land which can be utilized for some other purposes. For instance, as per the Ministry of Housing regulations, each Omani is eligible to get a plot of 500 sqm for the house after reaching a certain age [15]. Of course, this is a non-sustainable approach, but this is the current policy of the government, and thus the reduction in landfills will help Oman to fulfill such commitment in a more appropriate manner. Similarly, a survey conducted by the Be'ah, a company established under the Royal Decree 46/2009, shows that solid waste in Oman is characterized by a very high percentage of recyclables, primarily paper and cardboard (15%), plastics (20.9%), metals (1.8%), and glass (4%) [16]. Some of the newspaper reports show that currently 100% of the MSW in Oman goes to landfills [17].

To reduce GHG emissions from MSW and to reduce the burden of landfills on the earth, proper disposal and recycling of MSW are important. One of the modern methods that are recently adopted by many countries is to use MSW in a plant to produce energy. This type of plant is commonly known as waste-to-energy plants. The aim of this research is therefore to explore the opportunities to use MSW for electricity production in Oman. Such opportunities, however, cannot be understood well without knowing the composition of the MSW, public participation, and cooperation in activities related to recycling. This research, therefore, incorporates both quantitative and qualitative approaches, commonly known as mixed method, were adopted to accomplish the aims and objectives set for the research. The next section provides a literature review, covering Oman energy situation, electricity consumption, and the types of waste-to-energy plants.

2. Literature review

Although energy has become an integral requirement of today's modern life and it is considered as a fundamental element for social and economic growth, however, the United Nations report indicated that 13% of the earth's populations still have no access to modern electricity. Similarly, more than 3 billion people are still using wood, coal, charcoal, or animal waste for cooking their food and heating purposes. Energy is considered as the dominant contributor to climate change, which is estimated to be around 60% of total global greenhouse gas emissions. Similarly, it is estimated that in 2012, the indoor air pollution from using combustible fuels for household energy caused 4.3 million deaths around the world [18]. Overall, most of the current energy production is based on conventional resources such as oil, gas, and coal which, on the one hand, are non-sustainable but also, on the other hand, these resources produce greenhouse gases. These gases are considered a threat to the earth due to its contribution to global warming and climate change. The main gas which highly contributes to global warming and climate change is CO_2. The emission of the CO_2 to the earth's atmosphere has been significantly increased since 1950 which has reached a level of 400 parts per million (ppm). The CO_2 emission during the past 800,000 years until 1950 was below the level of 300 ppm [19].

Majority of the greenhouse gases are regarded as manmade gases in which the major role has been played by the recent industrialization. Although, the issue of global warming and climate change is well regarded as a threat to human life on the earth and there have been several efforts to control the emissions which cause global warming and climate change, the data from different sources reflect that these emissions are still increasing. For instance, the global CO_2 emission from fossil fuels in 2010 was 33.1 gigatons which have increased to 37.1 gigatons in 2018, representing a total increase of 12.08%. This emission quite alarming and if not tackled properly, and if it is increased at the same level, it would reach 41.58 gigatons by 2028 [20].

Goal 7 of the UN SDGs is "Affordable and Sustainable Energy" under which the member countries agreed to ensure access to affordable, reliable, sustainable, and modern energy. This goal is further supported by five global targets as mentioned below:

Target 1: By 2030, ensure universal access to affordable, reliable, and modern energy services.

Target 2: By 2030, increase substantially the share of renewable energy in the global energy mix.

Target 3: By 2030, double the global rate of improvement in energy efficiency.

Target 4: By 2030, enhance international cooperation to facilitate access to clean energy research and technology, including renewable energy, energy efficiency, and advanced and cleaner fossil-fuel technology, and promote investment in energy infrastructure and clean energy technology.

Target 5: By 2030, expand infrastructure and upgrade technology for supplying modern and sustainable energy services for all in developing countries, in particular, least developed countries, Small Island, developing States, and land-locked developing countries, in accordance with their respective programs of support.

To effectively understand the energy requirement of GCC countries, the electricity consumption and CO_2 emission in these countries are considered in the first instance. As the gulf region is rich in oil and gas reserves, therefore this region is considered as the main producer and supplier of the energy. The oil and gas revenue constitutes a major portion of the gross domestic product (GDP) in most of the GCC countries and remained support for the government and industrial sectors [21]. At the same time, the region is also a main consumer of energy as compared with other countries around the world. Similarly, due to the high rate of electricity consumption in these countries, the CO_2 emission is also high as presented in **Figure 3**. High electricity consumption and CO_2 emission could be justified due to the climatic condition of the region where the temperature in summer reaches 50°C as reported by Umar and Egbu [24], however, such

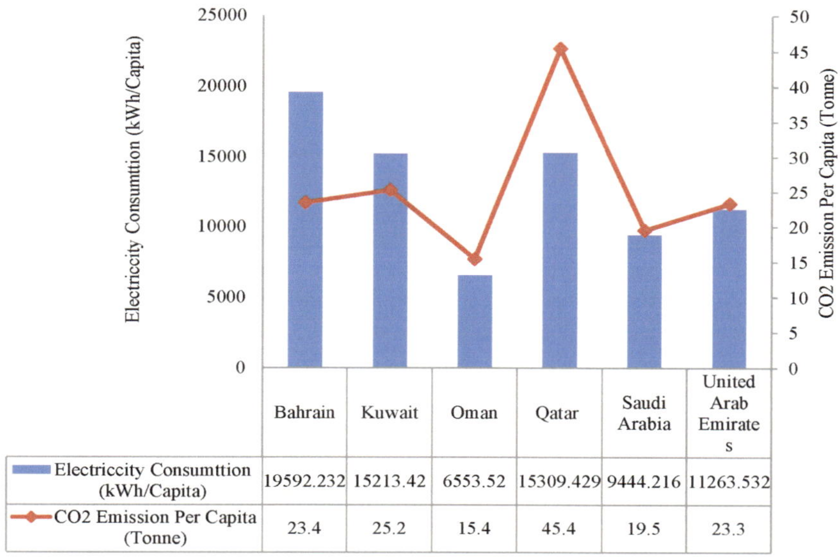

Figure 3.
Electricity consumption and CO_2 emission per capita in GCC countries [22, 23].

consumption and emissions are more than the double when compared with other countries. For instance, the average electricity consumption in GCC countries [=12,896.058 kilowatt-hours (kWh)] as of 2014, is more than three times greater than the electricity consumption in China (=3927 kWh) [22]. Similarly, the GCC electricity consumption per capita is more than double the consumption in the United Kingdom (=5130 kWh). It is very difficult to justify so high consumption of electricity in the GCC region based on the argument that it has a hot climatic condition. if this argument is considered to be true, then a high consumption is to be expected from the United Kingdom as well, as it has a cold climatic condition, however, the consumption in the united kingdom is far low than the average consumption of the GCC. Similarly, the average CO_2 emission is GCC countries (=25.36 tons) is more than three times greater than the CO_2 emission per capita in China (=7.5 tons) and almost four times greater than the CO_2 emission per capita in the United Kingdom (=6.5 tons) [23]. The main reason for this high CO_2 emission in GCC countries is that most of the electricity in these countries is produced by oil and gas. For instance, in Saudi Arabia, a total of 330.5 billion kWh of electricity was generated in 2016. 40.30% of this electricity was produced from oil, 59.6% was produced by gas, and only 1% was produced from renewable resources [25]. Similarly, according to the British Petroleum report, 68% of the electricity in Oman is produced by gas while the remaining 32% is produced by oil. Overall, the share of GCC countries in the renewable section is almost negligible. At the same, in other parts of the world, research is in progress to explore how to meet the full energy requirement of cities through renewable sources [26, 27]. Thus, it is obviously clear that a huge emission could be expected from these countries when all of its energy requirements will be met from fossil fuels.

Different estimates show that 0.0016 barrels of oil is required to produce 1 kWh of electricity and one barrel of oil produces 0.43 tons of CO_2 [28, 29]. To clarify the situation more effectively, the total electricity consumption per capita in the GCC region is 77376.349 kWh and the total CO_2 emission in this region is 152.2 ton (=152,200 kg). In other words, the CO_2 production per capita per kWh in the GCC region stands at 1.97 kg (1.97 kg/kWh). This further reveals that the CO_2 production for China (=1.90 kg/kWh) and United Kingdom (1.26 kg/kWh) is lower than the GCC region. As shown in **Figure 4**, the high consumption of natural resources has derailed the progress of both the relevant goals of UN SDGs, Goal 12 (Responsible Consumption and Production) and Goal 13 (Climate Action). On the other hand, Denmark which is ranked first in terms of progress toward UN SDGs has achieved a score of 90.20 (out of 100) in Goal 13 [30].

The move of the GCC region toward renewable energy may change these figures and could enable the region to reduce its emission per kWh. Oman has particularly low oil and gas reserves, therefore, it is important for the country to take advantages from other resources that are available and can be used for electricity production so that that the pressure on its oil and gas reserves can be reduced and as such can stay for long [31]. Despite the fact that Oman has comparatively low oil and gas reserves, its progress toward renewable recourses is low [32]. Similarly, both the Omani visions 2020 and 2040 stress to reduce the dependency on the oil and gas revenue and on the action plan mentions that improve MSW collection service. These visions also emphasize that improved MSW collection and disposal system are mandatory to reduce the impact on the natural environment [33, 34].

One of the resources that have been used by many countries to produce electricity is waste. For instance, Japan is using 72% of its waste to produce energy, while the United States is using approximately 13% of the waste for the same purpose. The top leading countries around the world which are using waste for energy production along with the percentage of the waste they are using for such purpose as shown in **Figure 5**. There is an opportunity for Oman to take advantage of the waste-to-energy technique and produce a share of its energy requirement from the MSW as other countries are doing. This will not only help to reduce the burden of the Oman oil and gas reserves but will also help to minimize MSW impact on the natural environment.

There are three types of combustion technologies that can be used to produce energy from MSW [36]. The mass-burn facilities are the most common type of waste-to-energy plants installed in the United States [37]. In some types of plants, it is necessary to segregate the MSW before it is moved to the combustion chamber while in other it is not necessary. The segregation of waste before entering the burn unit allows extracting the recyclable materials from the waste [38]. In most cases, the mass-burn facilities are made to burn the waste in one burning chamber allowing excess air. In the combustion process, excess air must be allowed to promote turbulence and mixing so that air can reach all waste. This process is important due to the inconsistent composition of MSW. In a common mass-burn plant, the MSW is

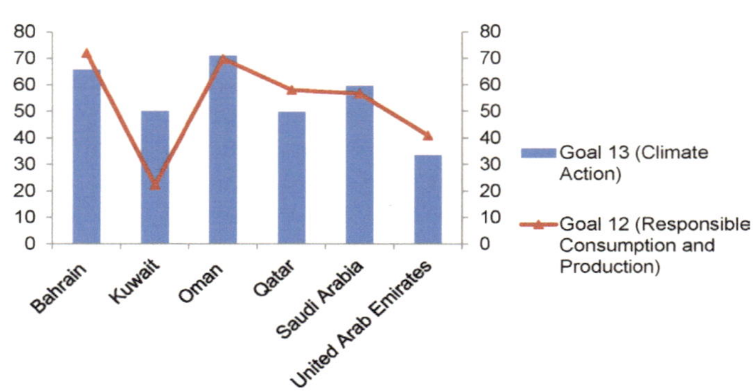

Figure 4.
Goals 12 and 13 score of different GCC countries.

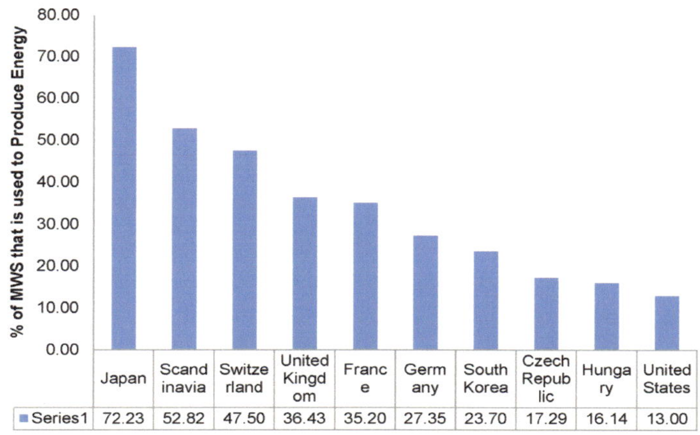

Figure 5.
Percentage of waste that is used for energy production in different countries [35].

burned in a sloping and moving grate. The vibration allows the blending of MSW and helps it to mix with air. Similarly, the modular systems are made to combust the unprocessed and mixed waste. Such plants are quite smaller when compared with the mass-burn unit and therefore can be moved from one site to another site easily [39]. The third type of waste-to-energy plants are known as "Refuse Derived Fuel Systems." Such plants apply mechanical systems to segregate the MSW and allow only combustible materials and mixtures to be used in the furnace or in a conventional boiler system [40, 41].

The research methodology used in this research is explained in the next section.

3. Research approach

Both the qualitative and quantitative research strategies were employed to obtain the aims and objectives set for this research. Since the quantity and the composition of the MSW are important to know energy content and emissions, therefore, the samples of MSW were collected from different households. To know the composition of the MSW in Oman, a total of 238 samples collected from 25 residential houses (175 samples), four restaurants (28 samples), three shopping markets (21 samples), and two hotels (14 samples) as shown in **Table 1**. The houses were selected in a way so that the reliability could be achieved. Thus, the houses with the family members of 2, 3, 4, 5, and 6 were considered for data collection. Samples of waste were collected from each house on a 24-hour basis. The same approach was adopted to collect the samples from restaurants (four numbers), shopping markets (three numbers), and hotels (two numbers). The samples were collected on each day of the week. The samples were deposed in the municipality collection point after recording the required data. Every morning the samples were collected, segregated, and measured. The data collection was completed in 2 months, from June 27, 2019 to August 26, 2019. The segregation method adopted was according to the criteria followed in the 3R projects implemented by Japan International Cooperation Agency (JICA) in Hanoi [42–44].

Additionally, the residents' willingness and participation in recycling activities were captured through an interview conducted at the time of sample collection. A total of 34 interviews, 25 from residential houses, four from restaurants, three from shopping markets, and two from hotels were conducted. The sample used to know

Type of entity	Number of samples per day (24 hours)	Number of days samples were collected	Number of entitles used for sample collection/interview	Total samples
Residential houses	1	7	25 (one sample from each houses with family members of 2, 3, 4, 5, and 6)	175
Restaurants	1	7	4	28
Shopping markets	1	7	3	21
Hotels	1	7	2	14
Total:			34	238

Table 1.
MSW samples collection approach.

the residents' willingness and participation was considered appropriate as there is evidence in the existing literature that similar studies were conducted with much smaller samples. For instance, Mason [45], in his research entitled "Sample Size and Saturation in Ph.D. Studies Using Qualitative Interviews," reported the result of 560 studies and noted that the most common size of the sample in these studies was 20. Similarly, Umar [46], while developing an integrated approach to promote sustainability in university campuses, used a sample of 20 respondents. Although the questionnaire used for this purpose was prepared in English, due to the diversity of the respondents, the interview process was deemed fit to the respondents. Overall, 45% of the residents in Oman are expatriates and the majority of them belong to some Asian countries [47]. The participants in the data collection were therefore interviewed in the local language so that there could be no communication barriers.

The environmental performance of the current approach of MSW management was measured through the emissions produced by landfills. The Intergovernmental Panel on Climate Change (IPCC) 2006 model is used to calculate GHG emissions from landfills [48]. IPCC model is an international model used by The United Nations Framework Convention on Climate Change (UNFCCC) member countries to report the national GHG inventory. Eq. (1) is used to calculate the CH_4 emission from landfills. Similarly, for the environmental performance of waste-to-energy approach, different parameters such as emissions per tons, reduction in landfills, and energy production and social indicators such as employment were considered. For instance, Eq. (2) was used to determine the emissions from crude oil when used in electricity production. In the comparison of emissions from both scenarios, landfilling and waste-to-energy approach, the emissions produced from the transportation of waste were ignored. Such emissions include the GHG emission from the transportation of waste from the collection point to the recycling facility for which trucks are commonly used [49].

$$CH_4 = \left[\left(MSW_{(\text{Land Fill})} \times MCF \times DOC \times DOC_f \times F \times (16/12) - R\right) \times (1 - OX)\right] \quad (1)$$

where CH_4 = methane emission in Gg/year (1 Gg = 10^9 g; 1 Gg = 1000 ton), $MSW_{(\text{Land Fill})}$ = total amount of MSW in the landfill in wet weight basis (Gg/year), MCF = CH_4 correction factor—value used in the calculation = 0.6, DOC = the fraction of degradable organic carbon in MSW (Gg C/Gg MSW)—value used in the calculation = 0.2455, DOC_f = the fraction of DOC that can decompose (fraction)—value used in the calculation = 0.77, F = the fraction of CH_4 in generated landfill gas—value used in the calculation = 0.5, R = the recovered CH4 (Gg/year), 16/12 is the molecular weight ratio CH_4/C—value used in the calculation = 0, and OX = the oxidation factor—value used in the calculation = 0.

$$E = A\,x\,B\,x\,C\,x\,D \tag{2}$$

where E = CO_2 emission per barrel (tons), A = the average heat content of the crude oil (=5.80 mmbtu per barrel; mmbtu = one million British Thermal Units), B = the average carbon coefficient of crude oil (=20.31 kg carbon per mmbtu), C = the fraction oxidized (=100% [48]), and D = the ratio of the molecular weight of carbon dioxide to the carbon (=44 kg CO_2/12 kg C) [9, 50].

Similarly, the electricity content and emissions from the MSW was established considering the existing literature. Different keywords such as "electricity content in MSW," "energy production from MSW," "waste-to-energy plant," and "emissions from energy to waste" were used in the main search engines. The Preferred Reporting Items for Systematic Reviews and Meta-Analyses (PRISMA) guidelines were adhered to during the review process [51]. Different results associated with the electricity content and emissions from waste-to-energy plants were obtained from the systematic review. For instance, in the United States, there are 68 waste-to-energy power plants that produce about 14 billion kilowatt-hours (kWh) of electricity using 26.76 tons of combustible MSW [52]. In other words, the energy or electricity content per tons of MSW in the United States is approximately 586 kWh. The waste-to-energy plant installed in Qatar which has a capacity of 2300 tons per day produces 50 Mega Volt Amp (MVA) which is equal to 40,000 kWh per day [53]. Similarly, the statistics issued by Waste Management World indicates that electricity production from MSW can be up to 875 kWh per ton of MSW processed in a waste-to-energy plant [54]. The calculation presented in this chapter, however, considers a value of 586 kWh per MSW used in a waste-to-energy plant, keeping in mind that this value is from the United States which uses the same types of waste to plant as proposed in this research.

The total waste which is burned for energy recovery in the United States currently stood at 12.70% of the total waste. Apart from producing energy, the burning of waste can reduce the volume of waste by 90%. Similarly, a recent waste-to-energy plant constructed in Ethiopia uses the same approach of burning waste to produce electricity. The collected waste is kept for 5 days to allow the moisture to seep out and then burning the waste at 1000°C to turn it to heat energy to run the steam turbine [55]. The latest and modern waste-to-energy plant can reduce the emissions (CO_2) ranging from 100 to 350 kg CO_2 equivalent per ton of the MSW used [56]. An average emission reduction value of 225 kg CO_2 equivalent to per ton of MSW can be used to measure such reduction in emission; however, the Environmental Protection Agency (EPA) of the United States reveals that the burning of 1 ton MSW in a waste-to-energy plant results in 1 ton of less CO_2 when compared to the common practice of landfilling [57]. The potential reduction in emissions between landfilling and waste-to-energy plants investigated by Wang et al. [58] in China noted that such reduction can be more than 1000 kg per ton. Similarly, the study conducted by Obermoser et al. [59] to establish a reliable CO_2 value from waste-to-energy plant noted that CO_2 emission can be in the range of 30–67 kg CO_2 per Giga Joule which can be translated into 0.175 kg CO_2 per kWh.

The next section describes the results and analysis made from the data collected in Section 3.

4. Results and analysis

Considering the different aspects of the results and analysis, this section is divided into different subsections. The first subsection describes the results and analysis of the MSW composition and the public participation in the MSW segregation.

4.1 MSW composition and public willingness

As noted in Section 3, a total of 175 samples were collected from the residential houses. These samples were used to determine the MSW generation per capita. The results show that the mean weight of the MSW samples was 1.3 ± 0.28 kg/per capita. Currently, as of May 2020, the total Oman population stands at 4,613,726, which can be translated into a total of 5998 tons MSW generation per day in Oman [60]. Similarly, the whole samples collected from the residential houses, restaurants, shopping markets, and hotels were used to determine the MSW composition in Oman. This composition is reported in **Figure 6**. Overall, the composition of the MSW represents a good percentage of waste that can be combustible and suitable to be used in a waste-to-energy plant to produce electricity.

During the MSW sample collection process, the inhabitants were asked about their cooperation and participation in the waste segregation activities. The willingness of the residents was considered important as this may be helpful to transfer only the MSW that is suitable to be used in the waste-to-energy plant. In other words, when the residents will segregate the waste at their own, the process at the waste-to-energy plant will become more straightforward as it will receive only the waste which could be used in the plant. A total of 34, consisting of 67.64% male and 32.35% female face-to-face interviews were conducted during this stage. All the participants appraised the idea of using MSW for electricity generation. A large number of the interviewees (70.58%) agreed that they are willing to participate in the segregation of the MSW at their doorsteps. The remaining participants did not answer as no but they were somehow not sure how they can do such segregation.

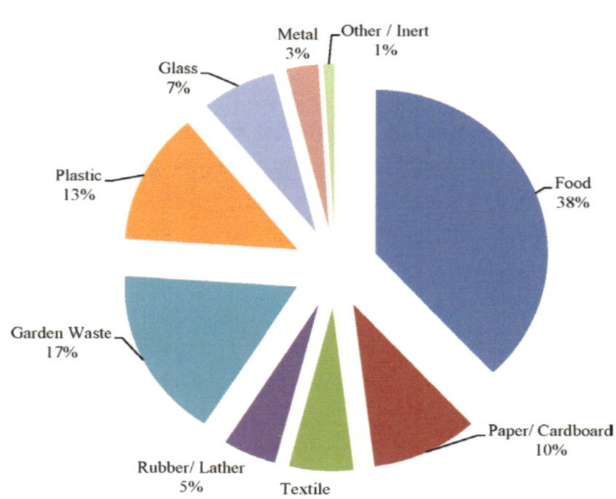

Figure 6.
Composition of MSW in Oman.

Overall, the majority of the respondents (79.94%) noted that they required some sort of training and tools to do such tasks in their homes.

The next section outlines the emission from MSW considering the landfilling scenario.

4.2 Emissions from MSW

To calculate the emissions from MSW using Eq. (1), DOC for different waste materials was obtained from the IPCC and the Atmospheric Brown Clouds Emission Inventory Manual (ABC EIM). Based on these two documents, the DOC paper = 0.25, DOC food = 0.4, DOC textile = 0.25, and the DOC rubber = 0.39 were considered. The average DOC value was calculated based on the above values and waste fraction as shown in **Figure 6**. The final DOC value used in the calculation, therefore, stands at 0.2455. The total MSW waste considered in this calculation was equal to 2,159,219 tons per year or 2159 Gg per year. This MSW produces a total of 163,060 Gg/year CH_4 which is equal to 3,424,247 tons/year CO_2 equivalent. This can be translated into the emissions produced 1 ton, which can be equal to 1.58 ton per year CO_2 equivalent per ton of MSW.

The next section describes the electricity consumption, production, and emissions produced by such consumption and production.

4.3 Electricity production and emissions

In terms of electricity production, Oman is using both oil and gas to meet its electricity requirement. As noted in **Figure 3**, the current consumption of electricity per capita in Oman is 15,309.4 kWh per year that can be translated into total energy consumption by multiplying this figure with the total population of Oman (15,309.4 × 4,613,726), which gives a total consumption of 70,633.37 million kWh per year. To calculate the emissions from electricity production through oil and gas, 70% of the electricity production is considered to be from oil, and 30% is considered to be from natural gas. These percentages were taken from the study conducted by the Authority of Electricity Regulation in Oman [61]. The EIA guidelines were used to establish the emissions both from oil and gas when used for electricity production. As per these guidelines, to produce 1 million British thermal units (btu) energy which is equal to 29.31 kWh from oil, a total of 161.30 pounds (=73.16 kg) of CO_2 is produced. Similarly, if the same amount of energy is produced from gas, then the total emissions will be equal to 117.0 pounds (53.07 kg). The emissions from electricity production in Oman considering both oil and gas are therefore calculated as follows:

Emissions from oil: Electricity produced from oil = total electricity consumption × percentage produced from oil = 70,633.37 million kWh × 70% = 49,443.35 million kWh per year.

Emissions for 29.31 kWh from oil = 73.16 kg CO_2.

Emission for 49,443.35 million kWh = $\dfrac{49{,}443.35\,x\,73.16}{29.31} = 1.234 \times 10^{11}$ kg = **123.414 million tons CO_2 per year.**

Emission from gas: Electricity produced from oil = total electricity consumption × percentage produced from oil = 70,633.37 million kWh × 30% = 21,190.01 million kWh per year.

Emissions for 29.31 kWh from gas = 53.07 kg CO_2.

Emission for 21,190.01 million kWh = $\dfrac{21{,}190.01\,x\,53.07}{29.31} = 3.836 \times 10^{10}$ kg = **38.367 million tons CO_2 per year.**

Total Emission from Electricity Production in Oman = **123.414** + **38.367** = **161.781 million tons CO_2 per year.**

Overall, the emission from MSW and electricity production in Oman is therefore equal to 165.205 million tons CO_2 per year.

The next section presents different parameters of energy production from MSW in Oman.

4.4 Energy production from MSW

The composition of MSW presented in **Figure 6** shows that more than 50% of the MSW can be classified as combustible materials suitable for use in waste-to-energy plants. This indicates that a total of 1,079,610 tons per year (~3000 tons per day) of MSW can be used in waste-to-energy plants. The United States Energy Information Administration statistics indicate that 85% of the MSW can be burned in a waste-to-energy plant to produce electricity. Similarly, in some cases, the segregations of the waste are also not required as some of the latest plants known as the mass-burn waste-to-energy plant can process all the waste together. As shown in **Figure 7**, such plants have the capacity to segregates the waste such as metals, ash. The pant can also segregate the food waste and other organic waste that can be used in landfilling or for composting. Similarly, the separated ash can also be used as aggregates in construction works. These plants can also be classified based on their daily capacity. After reviewing the total waste-to-energy plants manufactured and installed by Deltaway Energy, these plants are found to have a capacity from 68 tons per day to 4900 tons per day [62]. To ensure a realistic estimate for the waste-to-energy plant, the MSW produced in different governorates were considered. As noted in **Table 2**, Muscat governorate, which is also the capital of Oman, is on the top of the waste production having a waste production capacity of 1905 tons per day. If Muscat, North AlBatinah, Al-Dakhiliya, South AlBatinah, South AlSharqiya, North AlShariqiya, and Al-Dhahirah governorates which are somehow close to each other (as shown in **Figure 8**) are considered, then the total production of waste will be equal to 5000 tons per day. Of course, these are 2017 data, and the population

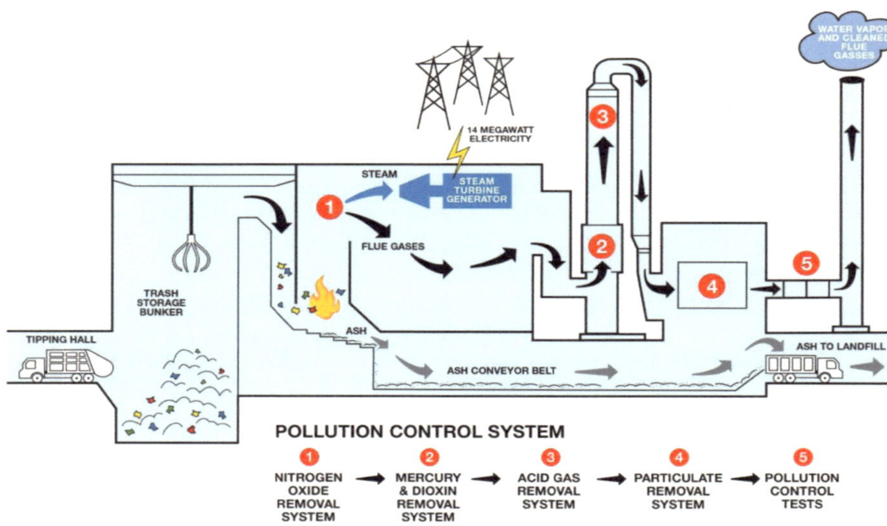

Figure 7.
A typical waste-to-energy plant.

Governorates	Waste production ton per year (2017)	Waste production ton per day (2017)
Muscat	685,654	1905
North AlBatinah	336,791	936
Al-Dakhiliya	330,185	917
Dhofar	330,884	919
South AlBatinah	150,038	417
South AlSharqiya	98,372	273
North AlShariqiya	98,372	273
Al-Dhahirah	106,055	295
Al-Buraiymi	154,374	429
Al-Wusta	9799	27

Table 2.
Waste generation in different governorates of Oman [63].

of these governorates will increase, the MSW production in these governorates will also increase. For instance, based on World Bank statistics, Oman's population increased at a rate of 3.4% per year [64]. Based on this indicator, the 2020 waste production in different governorates can be calculated using Eq. (3).

$$MSW_{(future)} = MSW_{(Current)}\ \mathrm{x}\left(1+\mathrm{i}\right)^{\mathrm{n}} \tag{3}$$

where $MSW_{(future)}$ = the MSW generation in future (for instance in 2020) for Muscat governorate, $MSW_{(Current)}$ = the MSW generation in future (for instance in 2017) for Muscat governorate, i = annual growth rate (decimal)—in this case, it can be 0.034, and n = number of years projected into future—in this case, it will be 3 years.

Based on the above parameters and using the equation, the Muscat governorate MSW production in 2020 will be 757,996 tons per year, which was 685,654 tons per year in 2017.

As noted in Section 3, 1 ton of MSW in the United States has the potential to produce 586 kWh, thus 5000 tons daily capacity waste-to-energy plant will be able to produce 29.30 million kWh daily. Similarly, this will help to reduce the volume of the waste that directly goes to landfills and produce emissions. Considering different parameters, the reduction of emissions through waste-to-energy plant is calculated below:

a. CO_2 emissions from 29.30 million kWh per day when produced by oil considering emissions for 1 kWh from oil = 2.5 kg CO_2 = 29.30 × 2.5 = **73.25 million kg CO_2 = 26,370 million kg CO_2 per year.**

b. Emission from 29.30 million kWh per day when produced by waste-to-energy plant considering 0.175 kg CO_2 per kWh [59] = 29.30 × 0.175 = **5.12 million kg CO_2 per day = 1845.90 million kg CO_2 per year.**

c. Reduction of waste emissions from landfilling, considering a reduction of 1 ton per ton of MSW.

Figure 8.
Oman map showing different governorates.

Emissions per tons of MSW when disposed in landfilling = 1.58 ton per year CO_2 equivalent per ton of MSW.

Reduction in emissions per tons = 1.58 – 1.0 = 0.58 ton CO_2 per year.

Emissions from 5000 tons per year = 5000 × 0.58 = **2900 tons CO_2 per year = 2.9 million kg CO_2 per year.**

Total reduction in emissions = A – B + C = 26,370 – 1845.90 + 2.90 = 24,527 million kg CO_2 per year.

Apart from the environmental performance of waste-to-energy plants, initial cost, operational cost, and return on investment are the key factors that the government or the investors considered in their decision. The initial cost of a waste-to-energy plant can be established considering its daily waste capacity. The Waste-to-Energy Research and Technology Council (WTERTC) has established the initial cost of a waste-to-energy plant at a rate of US$200,000 per daily ton of capacity. As the plant considered in the research has a capacity of 5000 tons per day, thus the initial cost of the plant can be estimated at US$1000 million [65]. Similarly, a 1000 ton daily capacity of the plant would require a total of 60 personnel, thus at this rate, a 5000 ton daily capacity plant would be able to generate employment for 300 workers. The operating costs of waste-to-energy plants in China are calculated at approximately US$30 per tons [66]. Depending on the life span of the plant, the operation cost can be up to 85% of the plants' total costs. If the plant has an estimated life of 40 years, then the cost of supply and construction can be 14% and the management and feasibility cost can be up to 1% of the total costs of the plant [67]. The research conducted by Carneiro and Gomes [68] established a levelized cost of electricity production from waste-to-energy plant at US$64–89 per MWh. Similarly, the profit margin of a waste-to-energy plant can be up to 25% while the return on investment can be up to 18%. The payback period of such a plant is normally 13 years with an internal rate of return up to 11% [66]. The research conducted by Kaplan et al. [69] in the United States estimated the average cost of electricity production from MSW at a rate of four cents per kWh with average revenue of US$25 per tons of MSW used in a waste-to-energy plant.

The next section provides a discussion and conclusion of the research.

5. Discussion and conclusion

It is been now well recognized that the earth resources need to be utilized in a sustainable manner as there is no other plant to live in. The UN SDGs and the Paris Agreements are some of the main indicators which reflect the commitment of world leaders toward sustainability. The main sustainability indicators such as energy and wastes are recognized so importantly that they have been placed among the 17 goals that the UN aims to achieve by 2030. The access to clean and modern energy does not mean that the people on the earth should be able to cook their food with gas or electric oven rather than burning the wood. This is one of the aspects, but the scope of clean energy is quite vast. It is not only to ensure access to clean and affordable energy but also to ensure the sustainability of such energy. For instance, making energy from fossil fuel is not sustainable because of two reasons. First, fossil fuels are not guaranteed to be available forever, and second, the emissions produced by such resources have other negative impacts that cause climate change and global warming. Even though, there is still doubt among the society that an increase of 1°C in the earth temperature is not a big issue. But in reality, such an increase creates a big difference by melting the glacier in the north and south poles. Such melting of glaciers not only expands the sea and but also disturbs the natural distribution of the dry and water portions on the earth's surface. The change in the natural distribution of wet and dry portions on the earth can cause the load variations on the earth plates which can increase the earthquakes. Apart from this phenomenon, the glacier on the earth's surface helps to reduce the temperature of the lower plates of the earth and thus reduce the chances of volcanic actions. It is therefore important to move toward renewable and sustainable resources for energy. Globally, some countries have reflected good progress in adopting renewable and sustainable energy resources, but in other countries including the GCC region, such progress is quite low.

Similarly, the UN under its Goal 12, which is related to the consumption of earth resources and production, aims to minimize the effect of such consumption and production. The waste produced during consumption and production has somehow similar effects as fossil fuels have. Such waste, if not properly disposed and recycled, will produce emissions and will utilize a large area on the earth's surface that can be used for some other purposes. Currently, the waste produced per capita in different countries is not only non-sustainable, but in most countries, there is no proper arrangement of recycling of such waste. In this regard, the GCC countries not only produce the highest amount of waste per capita (~1.77 kg per capita per day), but in most of these countries, landfilling is the common practice to dispose such waste. In all GCC countries, there is only one waste-to-energy plant in Qatar which has a capacity of 2300 tons per day, while the productions of MSW in all GCC countries currently stand at 93,430 tons per day.

This chapter, therefore, attempted to present a comparative study by considering the electricity production from waste-to-energy plants, considering the current electricity production and MSW generation. Both qualitative and quantitative research methods were utilized to achieve the aims and objectives of this research. The samples of the MSW collected from different entities including residential houses, shopping markets, hotels, and restaurants indicate a good volume (~50%) of combustible waste that can be used in a waste-to-energy plant. More than 70% of the interviewees confirmed that they are willing to segregate their MSW. The results further show that the current MSW generation in Oman stands at 1.3 ± 0.28 kg/per capita. This value and the current population of Oman are used to determine the total daily MSW generation in Oman. Since landfilling is used to dispose the total waste in Oman, the IPCC and ABC EIM guidelines were used to estimate the emissions from the total MSW. These calculations indicate the emission of 3,424,247 tons CO_2/year or 1.58 tons CO_2 per year per ton of MSW. Currently, Oman is producing 70% of electricity from oil and 30% from natural gas. The emissions from current electricity consumption (~70,633.37 million kWh per year) is estimated at 161.781 million tons CO_2 per year. If the emissions from the MSW are also added to this emission, then the total emission from electricity consumption and MSW generation in Oman will be equal to 165.205 million tons CO_2 per year. Considering the current energy consumption, MSW generation, and emissions from these variables, a waste-to-energy plant that has a capacity of 5000 tons per day is proposed to use the waste from Muscat, North AlBatinah, Al-Dakhiliya, South AlBatinah, South AlSharqiya, North AlShariqiya, and Al-Dhahirah governorates. Apart from producing 29.30 million kWh daily, this plant will be able to significantly reduce the emissions from the MSW and electricity production in Oman. The reduction in emission from this waste-to-energy plant is estimated at 24,527 million kg CO_2 per year. Similarly, this plant will be able to provide jobs for at least 300 personnel. The literature review suggests that the initial costs of such a plant with a capacity of 5000 tons per day can be equal to the US$1000 million. Similarly, the operating costs can be up to the US$30 per tons of waste used in the plant. Currently, the progress of Oman toward a number of UN SDGs is not satisfactory. Such an initiative of waste-to-energy plants will help Oman to improve its performance in a number of areas, including energy, climate change, waste management, and economic growth.

Author details

Tariq Umar
Department of Civil Engineering, Surveying and Construction Management, School of Engineering and the Environment, Kingston University, London, United Kingdom

*Address all correspondence to: tariqumar1984@gmail.com

References

[1] IGBP (International Geosphere-Biosphere Programme). Global Analysis, Integration and Modelling. Durham, NH: IGBP, International Council of Scientific Unions, (ICSU); 1999, 2000. Available from: http://gaim.unh.edu/. Schellnhuber HJ. Nature 402, C19. National Research Council, Committee on Global Change Research, Global Environmental Change: Research Pathways for the Next Decade. Washington, DC: National Academy Press; 1999. p. 531. Available from: http://www.nap.edu/catalog/5992.html [Accessed: 23 April 2020]

[2] WEF (World Economic Forum). This Is How to Sustainably Feed 10 Billion People by 2050. Cologny, Switzerland: World Economic Forum; 2018. Available from: https://www.weforum.org/agenda/2018/12/how-to-sustainably-feed-10-billion-people-by-2050-in-21-charts/ [Accessed: 23 April 2020]

[3] Sachs JD. The Age of Sustainable Development. New York, United States: Columbia University Press; 2015

[4] UN-SDGs (United Nations Sustainable Development Goals). About the Sustainable Development Goals. New York, United States: United Nations; 2015. Available from: https://www.un.org/sustainabledevelopment/sustainable-development-goals/ [Accessed: 23 April 2020]

[5] TSDGR (The Sustainable Development Goals Report). The Sustainable Development Goals Report. New York, United States: United Nations; 2018. Available from: https://unstats.un.org/sdgs/files/report/2018/TheSustainableDevelopmentGoalsReport2018-EN.pdf [Accessed: 25 April 2020]

[6] Umar T, Egbu C, Ofori G, Honnurvali MS, Saidani M, Shibani A, et al. UAE's commitment towards UN sustainable development goals. In: Proceedings of the Institution of Civil Engineers-Engineering Sustainability. Thomas Telford Ltd; 2020. pp. 1-18. DOI: 10.1680/jensu.19.00036

[7] IPCC (Intergovernmental Panel on Climate Change). Summary for policymakers. In: Metz B, Davidson OR, Bosch PR, Dave R, Meyer LA, editors. Climate Change 2007: Mitigation. Contribution of Working Group III to the Fourth Assessment Report of the Intergovernmental Panel on Climate Change. Cambridge, United Kingdom and New York, NY, USA: Cambridge University Press; 2007. Available from: https://www.ipcc.ch/report/ar4/wg3/ [Accessed: 04 April 2020]

[8] Chen X, Geng Y, Fujita T. An overview of municipal solid waste management in China. Waste Management. 2010;**30**(4):716-724. DOI: 10.1016/j.wasman.2009.10.011

[9] Umar T, Egbu C. Global commitment towards sustainable energy. In: Proceedings of the Institution of Civil Engineers-Engineering Sustainability. Vol. 172. Thomas Telford Ltd; 2018. pp. 315-323. DOI: 10.1680/jensu.17.00059

[10] Umar T, Wamuziri S. Briefing: Conventional, wind and solar energy resources in Oman. Proceedings of the Institution of Civil Engineers: Energy. 2016;**169**(4):143-147. DOI: 10.1680/jener.16.00011

[11] Zafar S. Waste management outlook for the Middle East. In: Palgrave Handbook of Sustainability. Cham, Basingstoke, United Kingdom: Palgrave Macmillan; 2018. pp. 159-181

[12] DEFRA (Department for Environment Food and Rural Affairs). UK Statistics on Waste. London, UK:

Department for Environment Food and Rural Affairs; 2019. Available from: https://assets.publishing.service.gov.uk/government/uploads/system/uploads/attachment_data/file/784263/UK_Statistics_on_Waste_statistical_notice_March_2019_rev_FINAL.pdf [Accessed: 27 March 2020]

[13] Zyoud SH, Al-Jabi SW, Sweileh WM, Al-Khalil S, Zyoud SH, Sawalha AF, et al. The Arab world's contribution to solid waste literature: A Bibliometric analysis. Journal of Occupational Medicine and Toxicology. 2015;**10**(35): 1-9. DOI: 10.1186/s12995-015-0078-1

[14] Zafar S. Solid Waste Management in Saudi Arabia. Doha, Qatar: EcoMENA; 2015. Available from: http://www.ecomena.org/solid-waste-management-in-saudiarabia/ [Accessed: 16 April 2020]

[15] MOH (Ministry of Housing). Frequently Asked Questions: Common Questions about Land Sector. Muscat, Oman: Ministry of Housing; 2019. Available from: https://eservices.housing.gov.om/eng/Pages/FAQs.aspx [Accessed: 05 May 2019]

[16] Be'ah (Oman Environmental Service Holding Company S.A.O.C). Waste Characterization and Quantification Survey. Muscat, Oman: Be'ah; 2013

[17] Muscat Daily. Oman Completes Privatization of Municipal Waste Management. Muscat, Oman: Muscat Daily; 2017. Available from: https://muscatdaily.com/Archive/Oman/Oman-completes-privatisation-of-municipal-waste-management-54ut [Accessed: 04 April 2020]

[18] WHO (World Health Organization). Household Air Pollution and Health. Geneva, Switzerland: World Health Organization; 2018. Available from: https://www.who.int/news-room/fact-sheets/detail/household-air-pollution-and-health [Accessed: 07 August 2019]

[19] Lindsey R. Climate Change: Atmospheric Carbon Dioxide. Maryland, United States: National Oceanic and Atmospheric Administration; 2018. Available from: https://www.climate.gov/news-features/understanding-climate/climate-change-atmospheric-carbon-dioxide [Accessed: 18 April 2020]

[20] WRI (World Resource Institute). New Global CO_2 Emissions Numbers Are In. Washington, D.C., United States: World Resource Institute; 2018. Available from: https://www.wri.org/blog/2018/12/new-global-co2-emissions-numbers-are-they-re-not-good [Accessed: 18 April 2020]

[21] Umar T. Briefing: Cost of accidents in the construction industry of Oman. In: Proceedings of the Institution of Civil Engineers-Municipal Engineer. Vol. 170. Thomas Telford Ltd; 2016. pp. 68-73. DOI: 10.1680/jmuen.16.00032

[22] WB (World Bank). Electric Power Consumption (kWh Per Capita). Washington, D.C., United States: The World Bank; 2014a. Available from: https://data.worldbank.org/indicator/EG.USE.ELEC.KH.PC?end=2014&name_desc=false&page=2&start=2014&view=map&year=2014 [Accessed: 18 April 2020]

[23] WB (World Bank). CO_2 Emissions (Metric Tons Per Capita). Washington, D.C., United States: The World Bank; 2014b. Available from: https://data.worldbank.org/indicator/en.atm.co2e.pc?year_high_desc=true [Accessed: 18 April 2020]

[24] Umar T, Egbu C. Heat stress, a hidden cause of accidents in construction. In: Proceedings of the Institution of Civil Engineers–Municipal Engineer. Vol. 173. Thomas

Telford Ltd; 2020. pp. 49-60. DOI: 10.1680/jmuen.18.00004

[25] King Abdullah Center for Atomic and Renewable Energy (K.A. CARE). Energy Sustainability for Future Generations. Riyadh, Saudi Arabia: King Abdullah Center for Atomic and Renewable Energy; 2016. Available from: https://www.kacare.gov.sa/en/FutureEnergy/Pages/vision.aspx [Accessed: 18 April 2020]

[26] Calvillo CF, Sánchez-Miralles A, Villar J. Energy management and planning in smart cities. Renewable and Sustainable Energy Reviews. 2016;**55**:273-287. DOI: 10.1016/j.rser.2015.10.133

[27] Strielkowski W, Veinbender T, Tvaronavičienė M, Lace N. Economic efficiency and energy security of smart cities. Economic Research-Ekonomska Istraživanja. 2020;**33**(1):788-803. DOI: 10.1080/1331677X.2020.1734854

[28] EIA (Energy Information Administration). How Much Coal, Natural Gas, or Petroleum Is Used to Generate a Kilowatt-Hour of Electricity. Washington, USA: US, Energy Information Administration; 2017. Available from: https://www.eia.gov/tools/faqs/faq.php?id=667&t=6 [Accessed: 05 October 2019]

[29] EPA (Environmental Protection Agency). Greenhouse Gases Equivalencies Calculator - Calculations and References. Washington, USA: Environmental Protection Agency; 2018. Available from: https://www.epa.gov/energy/greenhouse-gases-equivalencies-calculator-calculations-and-references [Accessed: 04 October 2019]

[30] Sachs J, Schmidt-Traub G, Kroll C, Lafortune G, Fuller G. Sustainable Development Report 2019. New York: Bertelsmann Stiftung and Sustainable Development Solutions Network (SDSN); 2019. Available from: https://s3.amazonaws.com/sustainabledevelopment.report/2019/2019_sustainable_development_report.pdf [Accessed: 21 April 2020]

[31] EIA (Energy Information Administration). International Statistics. Washington, D.C., United States: Energy Information Administration; 2018. Available from: https://www.eia.gov/beta/international/ [Accessed: 18 April 2020]

[32] Oxford Business Group. Discovery of Oil Reserves and New Concessions in Oman Leverage Growth on Global Indices. London, United Kingdom: Oxford Business Group; 2020. Available from: https://oxfordbusinessgroup.com/overview/pumped-reserve-discoveries-new-concessions-and-efficient-technology-leverage-steady-growth-global [Accessed: 04 April 2020]

[33] SCP (Supreme Council for Planning). Oman Vision, 2020. Muscat, Sultanate of Oman: Supreme Council for Planning; 1995. Available from: https://www.scp.gov.om/en/Page.aspx?I=14 [Accessed: 02 October 2019]

[34] SCP (Supreme Council for Planning). Oman Vision, 2040. Muscat, Sultanate of Oman: Supreme Council for Planning; 2019. Available from: https://www.scp.gov.om/en/Projects.aspx?I=1&Search=2040 [Accessed: 02 May 2020]

[35] EIA (Energy Information Administration). Waste-to-Energy Around the World. Washington, D.C., United States: Energy Information Administration; 2020. Available from: https://www.eia.gov/energyexplained/biomass/waste-to-energy-in-depth.php [Accessed: 14 April 2020]

[36] EPA (Environmental Protection Agency). Wastes - Non-Hazardous Waste - Municipal Solid Waste: Basic

Information. Washington, D.C., United States: Environmental Protection Agency; 2016. Available from: https://archive.epa.gov/epawaste/nonhaz/municipal/web/html/basic.html [Accessed: 31 May 2020]

[37] Psomopoulos CS, Bourka A, Themelis NJ. Waste-to-energy: A review of the status and benefits in USA. Waste Management. 2009;**29**(5):1718-1724. DOI: 10.1016/j.wasman.2008.11.020

[38] Phongphiphat A, Ryu C, Finney KN, Sharifi VN, Swithenbank J. Ash deposit characterisation in a large-scale municipal waste-to-energy incineration plant. Journal of Hazardous Materials. 2011;**186**(1):218-226. DOI: 10.1016/j.jhazmat.2010.10.095

[39] Kramer R, Pelter L, Kmiotek K, Branch R, Colta A, Popa B, et al. Modular waste/renewable energy system for production of electricity, heat, and potable water in remote locations. In: 2011 IEEE Global Humanitarian Technology Conference. IEEE; 2011. pp. 30-35. DOI: 10.1109/GHTC.2011.13

[40] Hernandez-Atonal FD, Ryu C, Sharifi VN, Swithenbank J. Combustion of refuse-derived fuel in a fluidised bed. Chemical Engineering Science. 2007;**62**(1-2):627-635. DOI: 10.1016/j.ces.2006.09.025

[41] Násner AML, Lora EES, Palacio JCE, Rocha MH, Restrepo JC, Venturini OJ, et al. Refuse derived fuel (RDF) production and gasification in a pilot plant integrated with an Otto cycle ICE through Aspen plus™ modelling: Thermodynamic and economic viability. Waste Management. 2017;**69**:187-201. DOI: 10.1016/j.wasman.2017.08.006

[42] JICA (Japan International Cooperation Agency). Laos Pilot Program for Narrowing the Development Gap toward ASEAN Integration, Progress Report 1: Supplement 1 (Vientiane). Vientiane, Lao PDR: Japan International Cooperation Agency (JICA); 2012. Available from: http://open_jicareport.jica.go.jp/pdf/12245338.pdf [Accessed: 04 April 2020]

[43] Otoma S, Hoang H, Hong H, Miyazaki I, Diaz R. A survey on municipal solid waste and residents' awareness in Da Nang city, Vietnam. Journal of Material Cycles and Waste Management. 2013;**15**(2):187-194. DOI: 10.1007/s10163-012-0109-2

[44] Taniguchi Y, Yoshida M. Public involvement and mobilization for promoting 3R initiative in Hanoi City: Lessons from 3R initiative project in Hanoi City 2006-2009. In: Proceedings of the 8th Expert Meeting on Solid Waste Management in Asia and Pacific Islands (SWAPI). Vol. 21. 2011. p. 23

[45] Mason M. Sample size and saturation in PhD studies using qualitative interviews. Forum Qualitative Sozialforschung/Forum: Qualitative Social Research. 2010;**11**(3):1-19. DOI: 10.17169/fqs-11.3.1428

[46] Umar T. An integrated sustainability management approach for universities. In: Proceedings of the Institution of Civil Engineers-Engineering Sustainability. Thomas Telford Ltd; 2020. pp. 1-12. DOI: 10.1680/jensu.19.00014

[47] Umar T. Causes of delay in construction projects in Oman. Middle East Journal of Management. 2018;**5**(2):121-136. DOI: 10.1504/MEJM.2018.091132

[48] IPCC (Intergovernmental Panel on Climate Change). 2006 IPCC Guidelines for National Greenhouse Gas Inventories. Japan: IPCC; 2006. Available from: https://www.ipcc-nggip.iges.or.jp/public/2006gl/ [Accessed: 04 April 2020]

[49] Braschel N, Posch A. A review of system boundaries of GHG emission inventories in waste management. Journal of Cleaner Production. 2013a;**44**:30-38. DOI: 10.1016/j.jclepro.2012.10.050

[50] EPA (Environmental Protection Agency). Greenhouse Gases Equivalencies Calculator - Calculations and References. Washington, USA: Environmental Protection Agency; 2017. Available from: https://www.epa.gov/energy/greenhouse-gases-equivalencies-calculator-calculations-and-references [Accessed: 04 April 2020]

[51] Moher D, Liberati A, Tetzlaff J, Altman DG, Prisma Group. Preferred reporting items for systematic reviews and meta-analyses: The PRISMA statement. PLoS Medicine. 2009;**6**(7):e1000097. DOI: 10.1371/journal.pmed.1000097

[52] EIA (Energy Information Administration). Biomass Explained Waste-to-Energy (Municipal Solid Waste). Washington, USA: US, Energy Information Administration; 2019. Available from: https://www.eia.gov/energyexplained/biomass/waste-to-energy.php [Accessed: 28 May 2020]

[53] Seghers K. Waste-to-Energy Plants. Doha, Qatar: Keppel Seghers Engineering; 2020. Available from: http://www.keppelseghers.com/en/content.aspx?sid=3028#A4-DSWMC [Accessed: 30 May 2020]

[54] WMW (Waste Management World). Waste to Energy: The Carbon Perspective. Vienna, Austria: Waste Management World; 2015. Available from: https://waste-management-world.com/a/waste-to-energy-the-carbon-perspective [Accessed: 31 May 2020]

[55] Abebe MA. Challenges of waste to energy facility in Reppi (koshe), Addis Ababa City. Journal of Pharmacy and Medical Sciences (IRJPMS). 2018;**1**(4):9-16

[56] Rogoff MJ, Screve F. Waste-to-Energy: Technologies and Project Implementation. 3rd ed. Netherland: Elsevier; 2019

[57] EPA (Environmental Protection Agency). Energy Recovery from the Combustion of Municipal Solid Waste (MSW). Washington, D.C., United States: Environmental Protection Agency; 2019. Available from: https://www.epa.gov/smm/energy-recovery-combustion-municipal-solid-waste-msw [Accessed: 31 May 2020]

[58] Wang Y, Yan Y, Chen G, Zuo J, Yan B, Yin P. Effectiveness of waste-to-energy approaches in China: From the perspective of greenhouse gas emission reduction. Journal of Cleaner Production. 2017;**163**:99-105. DOI: 10.1016/j.jclepro.2015.09.060

[59] Obermoser M, Fellner J, Rechberger H. Determination of reliable CO2 emission factors for waste-to-energy plants. Waste Management and Research. 2009;**27**(9):907-913. DOI: 10.1177%2F0734242X09349763

[60] NCSI (National Center of Statistics and Information). Population Clock. Muscat, Oman: National Center of Statistics and Information; 2020. Available from: https://www.ncsi.gov.om/Pages/NCSI.aspx [Accessed: 28 May 2020]

[61] AER (Authority of Electricity Regulation). Study on Renewable Energy Resources, Oman. Muscat, Oman: Authority of Electricity Regulation; 2008. Available from: https://www.aer.om/pdf/studyreport.pdf [Accessed: 01 February 2020]

[62] Deltaway Energy. Projects WTE and Biomass Power Plant Experience. San Francisco, California, United States:

Deltaway Energy; 2020. Available from: https://deltawayenergy.com/projects-list/ [Accessed: 30 May 2020]

[63] Be'ah (Oman Holding Company for Environment Services SAOC). MSW Generation in Different Governorates in Oman. Muscat, Oman: Be'ah (Oman Holding Company for Environment Services SAOC); 2017. Available from: https://www.beah.om/Knowledge-Center/Data [Accessed: 30 May 2020]

[64] WDI (World Development Indicators). World Development Indicators (WDI) Is the World Bank's Premier Compilation of Cross-Country Comparable Data on Development. Washington, D.C., United States: World Bank; 2020. Available from: http://datatopics.worldbank.org/world-development-indicators/ [Accessed: 21 May 2020]

[65] WTERTC (Waste to Energy Research and Technology Council). The ABC of Sustainable Waste Management (SWM). New York, United States: Waste to Energy Research and Technology Council; 2020. Available from: http://www.seas.columbia.edu/earth/wtert/faq.html [Accessed: 31 May 2020]

[66] Xin-gang Z, Gui-wu J, Ang L, Yun L. Technology, cost, a performance of waste-to-energy incineration industry in China. Renewable and Sustainable Energy Reviews. 2016;**55**:115-130. DOI: 10.1016/j.rser.2015.10.137

[67] UNEP (United Nations Environment Programme). Waste to Energy Considerations for Informed Decision-Making. Nairobi, Kenya: United Nations Environment Programme; 2019. Available from: https://www.developmentaid.org/api/frontend/cms/uploadedImages/2019/08/WTEfull-compressed.pdf [Accessed: 31 May 2020]

[68] Carneiro MLN, Gomes MSP. Energy, exergy, environmental and economic analysis of hybrid waste-to-energy plants. Energy Conversion and Management. 2019;**179**:397-417. DOI: 10.1016/j.enconman.2018.10.007

[69] Kaplan PO, Decarolis J, Thorneloe S. Is it better to burn or bury waste for clean electricity generation? Environmental Science & Technology. 2009;**43**(6):1711-1717. DOI: 10.1021/es802395e

Section 4

Biochemical Products

Chapter 15

Production Pathways of Acetic Acid and Its Versatile Applications in the Food Industry

Gunjan Deshmukh and Haresh Manyar

Abstract

Acetic acid is a commodity chemical with the global demand of approximately 15 million tons per year with several applications in the chemical and food industry. The production of acetic acid can be widely categorized into chemical and fermentative routes, with the chemical route being the predominant one in the current industrial practice. In this chapter, we have reviewed the most recent developments in acetic acid production and applications over past two decades, including process intensification and catalysis by keeping the main emphasis on process sustainability. Acetic acid is used in several industrial sectors such as chemical, pharmaceutical, textile, polymer and paints, food and beverages. Furthermore, acetic acid has several applications in food industry and is traditionally known as vinegar. In addition, it is an acidulant, which is used to give a characteristic flavor profile to food. It can be used for microbial decontamination of meat and as a mild descaling agent in the food industry. More recently, acetic acid is reported to be used as an antimicrobial edible food coating agent. The diversified food culture has a significant demand in the development of such kind of innovation and acetic acid can be an efficient solution.

Keywords: acetic acid production, acetification, acidulant

1. Introduction

The bridge between chemistry and the day-to-day human life is always growing wider and stronger, and acetic acid is one of the perfect examples. Acetic acid is a clear liquid with a pungent odour, sharp taste, melting point of 16.73°C and boils at 117.9°C. Acetic acid, traditionally known as 'vinegar' is widely used as a food preservative, first discovered (c. 5000 BC) when unattended grape juice turned into wine. A famous physician Hippocrates II (c. 420 BC) used acetic acid to clean the wounds [1]. With direct and indirect applications of acetic acid, it has diversified into several chemical sectors such as food, pharma, chemical, textile, polymer, medicinal, cosmetics etc. Since then, acetic acid is proven to be a multi-application chemical building block resulting in ever-increasing demand. The production of acetic acid is expected to reach 18 million ton with an average growth of 5% per year [2, 3].

The overall routes for production and the applications of acetic acid are shown in **Figure 1**. Currently, the manufacturing demand is fulfilled via two main production routes, which are chemical and fermentative. Among the chemical manufacturing processes, the key processes are Cavita process (carbonylation of methanol), oxidation of aldehyde and oxidation of ethylene. The major players are BP chemicals and BASF, which follow carbonylation route. The major consumption of acetic acid mainly comes from the preparation of vinyl acetate monomer (VAM), acetic anhydride and C1-C4 acetates and it is used as a solvent in synthesis of terephthalic acid (PET). VAM is one of the main ingredients used in polymer industry with application as emulsifier, resins, as intermediate in surface coating agent, acrylic fiber and polymer wires. It is also used in textile industry to generate synthetic fibers as a result of condensation reaction. The other condensation reaction of acetic acid produces acetic anhydride used as typical acetylation agent, which is subsequently utilized to produce cellulose acetate, used in synthetic textiles and for silver-based photographic films. Most derived esters of acetic acid are ethyl acetate, n-butyl acetate, isobutyl acetate and propyl acetate, which are frequently used as solvents for inks, paints and coatings. Glacial acetic acid is an excellent polar protic solvent that is frequently used as a solvent for recrystallization to purify organic compounds. Several researchers are working on developing a sustainable process with the simple design to produce acetic acid that meets current demand.

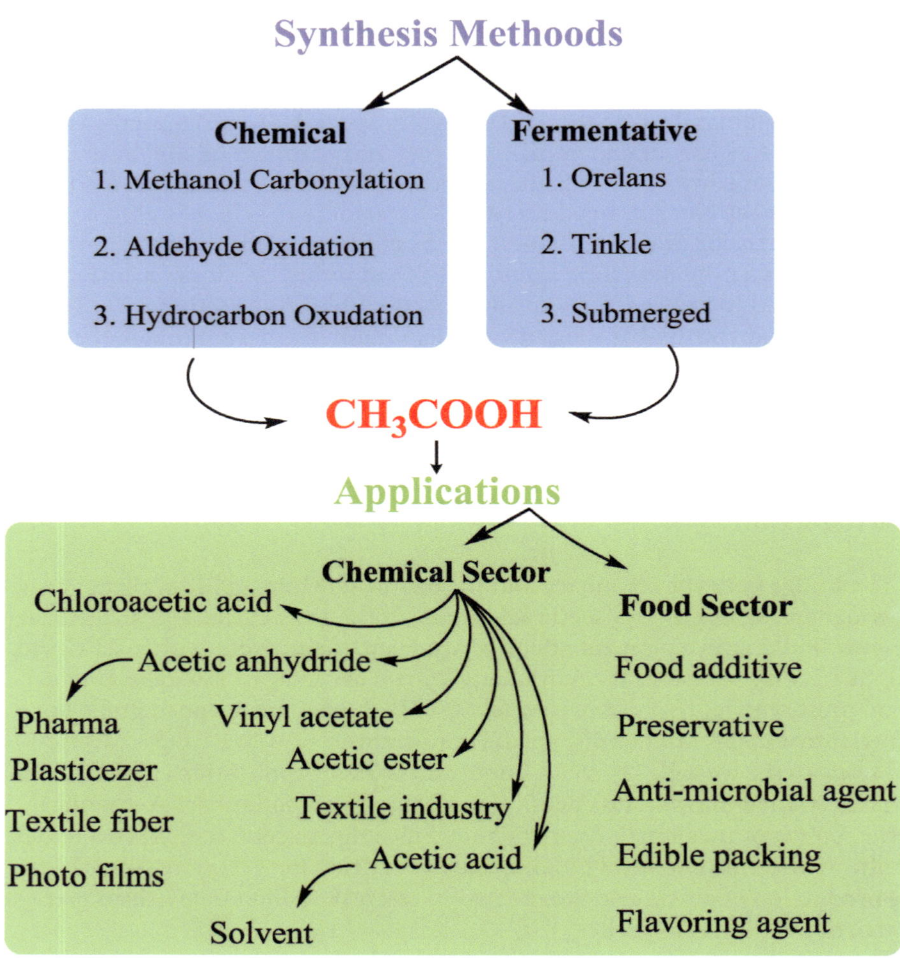

Figure 1.
Commercial routes for synthesis of acetic acid and applications.

Several homogeneous as well as heterogeneous catalytic systems are reported for the production of acetic acid with carbonylation process [4].

Acetic acid produced via fermentation route is mainly utilized in the food industry in the form of vinegar. Use of vinegar is more diversified these days, with more innovative ways to adjust and suit the current lifestyle and food culture. The different concentrations of acetic acid are used to sharpen the taste of food with a longer shelf life period and as a food preservative. Some new applications have also come such as edible and non-edible antimicrobial coating [5, 6].

This chapter reviews the current commercial processes for the synthesis of acetic acid to meet an ever-increasing global demand. The chapter also gives insight into the pros and cons associated with the process available and then how should we design a sustainable strategy to develop a simple commercial process. Further, the state of art to produce vinegar is discussed with exploitation as a multiapplication tool in the modern food industry.

2. Production of acetic acid

Acetic acid is mainly produced via chemical route that involves homogeneous as well as heterogeneous catalytic methods. The carbonylation of methanol via Monsanto process is the most adopted route, which further evolved as Cavita process with a choice of catalysts and process intensification. In the recent decade, the fermentative approach has also gained attention; however the commercial approach is not established yet. The current trends in sustainable manufacturing demand an urgent paradigm shift to develop and pursue more sustainable routes to reduce environmental burden. An approach is also made with the development of membrane-based technology, which offers a very simple design with eco-friendly production [7].

2.1 Conventional process

2.1.1 Methanol carbonylation process

Carbonylation process is a most employed commercial route for synthesis of acetic acid, also known as Monsanto process (**Figure 2**). Methanol and carbon monoxide are reacted in liquid phase in the presence of rhodium (Rh)-based catalyst at 150–200°C temperature and 30–50 bar pressure to produce acetic acid with 95% selectivity and 5% side products such as formic acid and formaldehyde [8]. Hydrogen iodide is used as an alkali promoter in this process. The reaction proceeds in liquid phase with methyl acetate as solvent using homogeneous catalyst. Controlled amount of water is required for the reaction, which is generated *in situ* by reaction of methanol with hydrogen iodide. The rate of reaction in the Monsanto process depends on the concentration of water. CO_2, H_2 and methanol are obtained

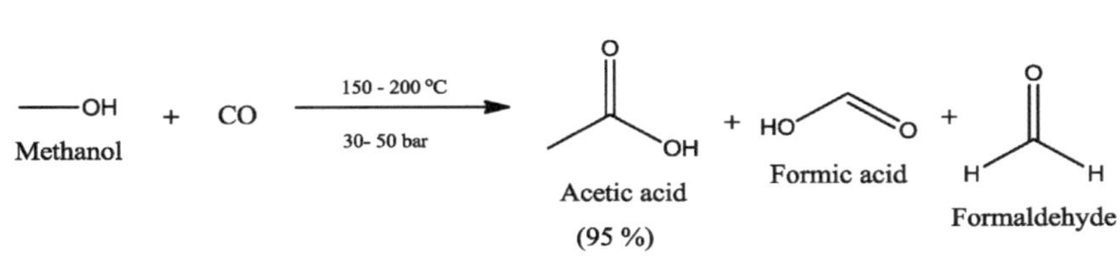

Figure 2.
Production of acetic acid by carbonylation method.

as by-products in the reaction. The generated methanol in the reaction is recycled. The process has evolved with time and different strategies have been adopted to separate pure acetic acid from a mixture of water and by-products. This process was modified by BP chemicals replacing rhodium-based catalyst with iridium (Ir) catalyst known as Cavita process [4]. The choice of Ir as a coordination metal is relatively more economic process than rhodium. The use of an iridium catalyst improves the overall rate of reaction.

The safety and the environmental hazards arising from the current methods are a serious concern. Acetic acid is highly corrosive, and the production processes need to be more sustainable and environmentally benign by reducing the amount of energy required in production and subsequent separation technologies as well as using heterogeneous catalysts. The Japanese firm Chiyodo developed a heterogeneous Rh catalysed process, wherein Rh metal was immobilized on the vinylpyridine resin. The use of heterogeneous catalyst prevails the loss of catalyst in the liquid phase and facilitates easy separation from the reaction mixture. The amount of water used in the reaction is very low and thus the separation of water from acetic acid is more energy-efficient compared to the other processes mentioned.

2.1.2 Acetaldehyde oxidation process

Acetaldehyde oxidation was the predominant process followed for the synthesis of acetic acid, wherein acetaldehyde is first prepared by oxidation of ethylene using palladium and copper chloride and it was further oxidized to form acetic acid (**Figure 3**). The same process is reported using cobalt and chromium-based catalyst at 55 bar pressure and 150°C temperature. The one-step process for conversion of ethylene to acetic acid is also practised using lead and lead-platinum based catalyst at high pressure compared to the acetaldehyde oxidation process with a low yield of acetic acid [9].

2.1.3 Hydrocarbon oxidation process

Hydrocarbons derived from petroleum stock such as butane and naphtha are utilized to generate acetic acid using cobalt acetate and chromium acetate catalyst (**Figure 4**). The reaction proceeds at a comparatively higher temperature range (150–230°C) and pressure (50–60 bar). The process involves petroleum feedstock, which contains hydrocarbon mixture, which leads to the formation of other by-products such as acetone, formic acid, propionic acid along with acetic acid. Thus,

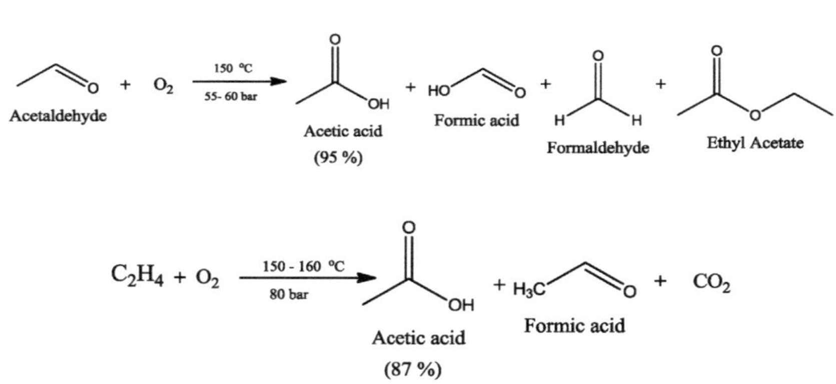

Figure 3.
Production of acetic acid by acetaldehyde oxidation.

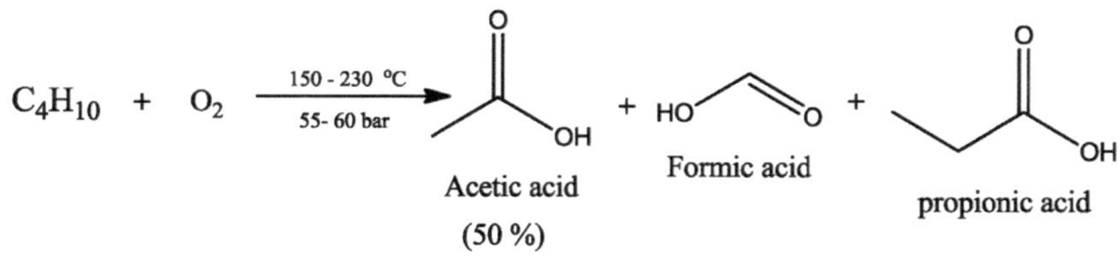

Figure 4.
Production of acetic acid by hydrocarbon oxidation.

this process fails to give pure acetic acid. This process is more suitable for manufacturing a mixture of volatile fatty acids.

2.2 Fermentation route

Fermentative route is mostly adapted for the generation of food-grade acetic acid that is vinegar. This process mainly involves the use of renewable carbon resources such as apple, grape, pears, honey, cane, coconut, date, syrup cereals, hydrolysed starch, beer and wine [10]. The fermentation process is mainly divided into two steps: the treatment with yeast followed by acetic acid bacteria (AAB). Commercial production of vinegar is done via oxidative fermentation using AAB. *Acetobacter* and *Gluconacetobacter* are most used species among ten classified genera. *Acetobacter pasteurianus* is traditionally used for commercial production of vinegar with concentration not exceeding 6% (v/v), whereas, *Gluconacetobacter europaeus* is utilized to produce high-concentration vinegar (10% v/v). The price of the vinegar varies with the kind of source used and the region where it is generated.

2.2.1 Orleans method

This method is well established, traditional and preferred for low-volume production of acetic acid. Derived from the French word Orléans, wooden barrels are used to ferment the feed in this process. This method is followed to prepare exotic brands of vinegar in different regions of the world with specific raw material available in the specific season. The traditional balsamic vinegar is produced in different parts of the world such as sherry from Spain, oxos from Greece, and Modena in Italy.

2.2.2 Trickling process

This process was developed to overcome the slow rate of acetification in Orleans process [11]. The process intensification was done to improve the acetic acid bacteria and substrate interaction. The alcoholic substrate was sprayed over the fermentation in continuous loop to achieve the desired concentration of acetic acid. The heat of the reaction was controlled by passing the air through the system. The process has the drawback of accumulating gelatinous material on the surface the membrane, which reduces the rate of reaction over the period.

2.2.3 The continuous submerged process

This modern fermentation method is followed to produce vinegar in masses. This is the most widely method and has a high yield along with a fast rate of oxidation as compared to the previous method. This method is 30 times faster than the Orleans method with higher efficiency for production of acetic acid. This process requires comparatively small space with higher yields. The Fringe fermenter is used for this process to increase the rate of the acetification. The yield of acetic acid is 98%. The pure substrates are required to achieve the high quality of acetic acid. This fermentation process is much economical, of simple design with easy process control.

The Fermentation process for acetic acid is economically feasible with comparatively simple operations. The application of this process is very limited to the present global demand. Whereas, the conventional process involves several steps such as fractional distillation, condensation and crystallization, which add to the high machinery cost. The operating conditions are harsh considering the process temperature and pressure along with the corrosive nature of acetic acid [12]. The purification of acetic acid from water is a multi-step process consuming a high amount of energy, which makes overall process complex and critical. In addition to this, the process requires huge manpower with stringent safety protocols and norms.

3. Need for development of novel sustainable technologies

Looking at the ever-increasing threats of global warming and ever-increasing global demand of acetic leads to an urgent need to develop a novel technological approach and sustainable feedstock for the generation of acetic acid. Even though many processes and technological developments are reported recently, they fail to sustain the production cost to profit margins. The separation of acetic acid remains the key issue to overcome the economical and energy consumption barriers. The different operations such as distillation, evaporation, absorption, filtration crystallization and alkali neutralization are time and energy consuming. Even though these processes involve multiple steps, the ever-growing demand forces to follow this path. On the other hand, fermentation process is reliable but cannot match the scale of current demand. Thus, the development of a novel route for generation or process intensification in separation can drastically reduce the overall production cost of acetic acid. Utilization of CO and CO_2 as feedstock generated from natural gas can offer long-term sustainability of acetic acid production. This technology offers high purity of acetic acid with eco-friendly production. Furthermore, membrane-based separation processes can provide efficient way to produce acetic acid. The pathways are discussed briefly.

3.1 CO and CO_2 as valuable feedstock

Utilization of CO_2 and syngas can offer sustainable alternatives to produce acetic acid. BP has announced the breakthrough process, wherein, acetic acid will be manufactured from syngas as a feedstock derived from natural gas. This will give an alternative to SaaBre process that produces acetic acid in three integrated steps. The production of acetic acid from syngas will avoid the purification of CO and purchase of methanol. Though the technology is not fully developed, it provides better alternatives in terms of sustainability. Similarly, acetic acid can be synthesized via reacting CO_2 and H_2 to give methanol followed by subsequent carbonylation step. This route gives liberty to utilize CO_2 as value-added feedstock.

3.2 Membrane-based technologies

The membrane technology can offer the separation of liquid, vapour and gas selectively with controlled mass transfer rates. These processes are easy to operate and simple to design. The technology can offer development on energy intensification. Several types of processes are reported based on the pore size of the membrane for separation of different components. These are namely microfiltration, ultrafiltration and nanofiltration membrane. The operating pressure (varying from 1 to 20 bar) of the system varies according the pore size of the membrane are used. Reverse osmosis is another membrane technology with non-porous membrane. This process operates at pressure more than 20 bar. The membrane technologies collectively can be applied in downstream processing for separation of acetic acid in chemical process as well as fermentation processes. The combination of fermenter with acetic acid permeable membrane can help in separation of acetic acid to avoid the self-inhibition.

4. Application of acetic acid in food industry

Direct applications of acetic acid are reported from ancient times. It was used as a medicine and food preservative. Over the period, applications of acetic have diversified as per the demands of modern life. Using different concentrations, it is utilized in food additives, food preservation, antimicrobial agent, acidulant, flavour and taste enhancer, edible packaging material, artificial food ripening agent, etc. Some of the applications such as acidulant and as acetification agents are described in detail here.

4.1 Acetification

Acetification is simply the bacterial oxidation of ethanol to produce acetic acid and water (**Figure 5**). The process is also termed as oxidative fermentation. The rate of the reaction in acetification mainly depends on the type of microorganism used to catalyse and the concentration of available oxygen in the media [13].

There are different types of microorganisms that occur naturally in food and are responsible for the different natural processes such as acetification, alcoholism, proteolysis and enzymatic reactions, which alter the natural condition of the food. This bioprocess technology is studied and systematically utilized to improve the quality of food s in terms of texture, taste, mouthfeel, colour and prolonged shelf life. The overall concept has grown into generating different types of food and beverages produced in a cheap and sustainable way.

Acetification of different food categories using acetic acid bacteria (AAB) has led to the production of several food products [14]. AAB are naturally found on fruits, flowers, and plants, which naturally react and convert carbohydrate sugars into organic acids in the presence of oxygen. The same concept is biotechnologically utilized to prepare a diverse variety of food and beverages.

4.2 Flavouring agent

Different parts of the world have utilized the acetification process to generate a variety of foods and beverages. The famous Lambic beer is produced from malted barley, aged dry hops and unmalted wheat. The different AAB and yeast are responsible for the generation of this beer, which is matured for over a period of 3 years. The typical acidic flavour of the beer is achieved with the help AAB

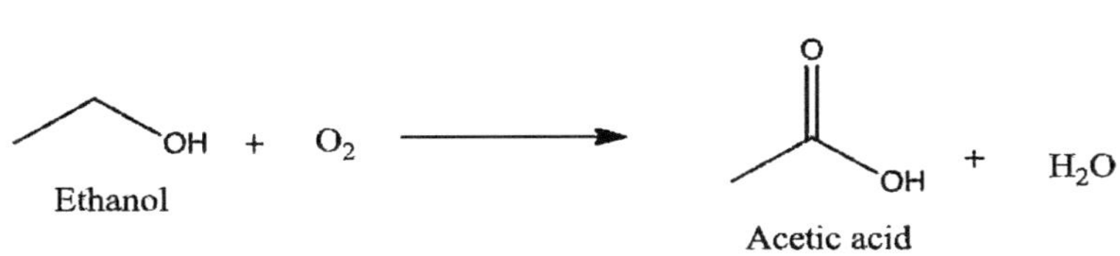

Figure 5.
Production of acetic acid by fermentative oxidation of alcohol.

together with lactic acid. The sparkling water is another famous example, which gives typical acidic and fruity flavour via fermentation of water and natural sucrose. Water kefir is one of the examples of such type. Kombucha is another type of beverage produced by oxidative fermentation. It is prepared from Kombucha (tea fungus), water and sugar. Similarly, Cocoa is fermented from cocoa beans with the help of AAB and yeast, which is used as raw material for chocolate production.

4.3 Acidulant

Acidulants are essential ingredients or additives that are generally used to improve the taste of food and make it sharper. There are naturally occurring acidulants such as acetic acid, citric acid, malic acid, fumaric acid, lactic acid, tartaric acid, succinic acid, phosphoric acid etc. having different taste profiles. Many fruits such as orange, lemon, apples, tomatoes and yogurt contain natural acids with the most common example being citric acid. Citric acid comes with lemon flavour, acetic acid with strong familiar vinegar flavour, tartaric acid gives sharp taste and lactic acid comes with a smooth taste [10]. Apart from taste enhancement, acidulants also act as a food preservative. The choice of the acidulant is usually made based on its characteristic flavour and the physical state and solubility. Some food formulations require solid acidulants. In general, inorganic acids such as sulphuric acid, phosphoric acid, monosodium orthophosphate and diphosphates are used as dry acidulants in controlled concentrations. The composition of the acidulants is based on their selection and different concentrations calculated by total titratable acid. Acetic acid is mainly used in the form of vinegar with the pungent smell. As it appears in the liquid state, it is used as a preservative in pickles. It is also used in the manufacture of cheese to improve the shelf life period, good mouthfeel and taste [15].

4.4 Edible packing

Acidulants are also used as food coating, which may be edible or non-edible to prevent food from contamination with the surrounding environment, to protect it from bacterial infection and to improve the shelf life of the food. These films are easily biodegradable. The water-soluble non-edible coating is used for the packing of food [16]. The edible coating is used for breath freshening agent, in drug delivery and as flavour. Acetic acid is used in edible films to enhance sour flavour. Various compositions of acetic acid are used to develop antimicrobial food coating to stop the outgrowth of bacterial and fungal cells. It is also used in meat coating and preservation of meat products. The chitosan-based edible food coatings along with aqueous acetic acid are used to enhance anti-listerial activity.

4.5 Antibacterial agent

Acetic acid is commonly used in medicine since ancient times. The low concentrations (3%) of acetic acid can be used as a local antiseptic against various microorganisms. Acetic acid is always considered as an alternative. It can be utilized as *in vitro* antimicrobial agent combined with other antiseptics. Acetic acid covers the wide range of spectrum with Gram-positive as well as Gram-negative bacteria.

5. Conclusion

Acetic acid has remained one of the key chemical molecules associated with human life. It is one of the main building blocks for developing several chemical entities. Cavita process shares a major part of the production to meet the global demand. The process utilizes methanol as a raw material, which is obtained from biogas. Though the process utilizes bio-derived feedstock, it utilizes high energy and manpower with multiple separation steps. The innovative and simple technologies for separation of acetic acid can improve the overall process. The other well-known process, that is, fermentative route, is slow and commercially unsuitable to meet the global demand. The fermentation process is globally followed to generate the food-grade acetic acid commonly known as vinegar. The demand for acetic acid will always keep growing, which necessitates the development of an eco-friendly process. Utilization of CO_2 and syngas may offer excellent alternatives as a sustainable feedstock to develop innovative technologies to develop commercial processes. This offers development of 100% bio-derived feedstock process. Further, the modern food industry has come up with different innovative applications of the acetic acid in food preservations and improved quality of food.

Acknowledgements

The authors gratefully acknowledge the financial support from the UK Catalysis hub and EPSRC for the project, 'Catalysis at the Water-Energy Nexus: Energy and Fuels from wastewater', R1043CCE, for postdoctoral funding for GD.

Conflict of interest

The authors declare no conflict of interest.

Author details

Gunjan Deshmukh and Haresh Manyar*
School of Chemistry and Chemical Engineering, Queen's University Belfast, Belfast, UK

*Address all correspondence to: h.manyar@qub.ac.uk

References

[1] Johnston CS, Gass CA. Vinegar: Medicinal Uses and Antiglycemic Effect. Medscape General Medicine. 2006;**8**(2):61-71. Available from: https://www.ncbi.nlm.nih.gov/pubmed/16926800

[2] Medrano-García JD, Ruiz-Femenia R, Caballero JA. Revisiting Classic Acetic Acid Synthesis: Optimal Hydrogen Consumption and Carbon Dioxide Utilization. Elsevier Masson SAS; 2019. DOI: 10.1016/B978-0-12-818634-3.50025-4. Available from: https://www.sciencedirect.com/science/article/pii/B9780128186343500254

[3] Nayak J, Pal P. A Green Process for Acetic Acid Production, ICCEES' 2015, Pattaya (Thailand). 2015. DOI: 10.13140/RG.2.1.4879.0883

[4] Kalck P, Le Berre C, Serp P. Recent advances in the methanol carbonylation reaction into acetic acid. Coordination Chemistry Reviews. 2020;**402**:213078. DOI: 10.1016/j.ccr.2019.213078

[5] Möller H, Grelier S, Pardon P, Coma V. Antimicrobial and physicochemical properties of chitosan—HPMC-based films. Journal of Agricultural and Food Chemistry. 2004;**52**:6585-6591. DOI: 10.1021/jf0306690

[6] Sánchez-Ortega I, García-Almendárez BE, Santos-López EM, Amaro-Reyes A, Barboza-Corona JE, Regalado C. Antimicrobial edible films and coatings for meat and meat products preservation. Scientific World Journal. 2014;**2014**:1-18. DOI: 10.1155/2014/248935

[7] Pal P, Sikder J, Roy S, Giorno L. Process intensification in lactic acid production: A review of membrane-based processes. Chemical Engineering and Processing Process Intensification. 2009;**48**:1549-1559. DOI: 10.1016/j.cep.2009.09.003

[8] Maitlis PM, Haynes A, Sunley GJ, Howard MJ. Methanol carbonylation revisited: Thirty years on. Journal of the Chemical Society, Dalton Transactions. 1996:2187-2196. DOI: 10.1039/dt9960002187

[9] Sano KI, Uchida H, Wakabayashi S. A new process for acetic acid production by direct oxidation of ethylene. Catalysis Surveys from Japan. 1999;**3**:55-60. DOI: 10.1023/A:1019003230537

[10] Solieri L, Giudici P. Vinegars of the world. Vinegars of the World. 2009:1-297. DOI: 10.1007/978-88-470-0866-3

[11] Lu HM, Zhang LP, Zhang BF, Jiang XL, Dai LX. Study on the control of gelatinous membrane in the process of vinegar trickling fermentation. Advances in Materials Research. 2013;**739**:226-232. DOI: 10.4028/www.scientific.net/AMR.739.226

[12] Yoneda N, Kusano S, Yasui M, Pujado P, Wilcher S. Recent advances in processes and catalysts for the production of acetic acid. Applied Catalysis A: General. 2001;**221**:253-265. DOI: 10.1016/S0926-860X(01)00800-6

[13] Bourgeois JF, Barja F. The history of vinegar and of its acetification systems. Archives Des Sciences. 2009;**62**:147-160

[14] De Roos J, De Vuyst L. Acetic acid bacteria in fermented foods and beverages. Current Opinion in Biotechnology. 2018;**49**:115-119. DOI: 10.1016/j.copbio.2017.08.007

[15] Continuous cheese production process, US20140242248A1, n.d. Available from: https://patents.google.com/patent/US20140242248?oq=Patent+US2006%2F0062885+A1. US Patent No WO2013016154A1

[16] Acidulent film and method of making same, US 2005/0075432 A1, n.d. Available from: https://patents.google.com/patent/US20050075432A1/en?oq=US+2005%2F0075432+A1. US Patent No US20050089548A1

Chapter 16

Xylanase and Its Industrial Applications

Abdul Basit, Wei Jiang and Kashif Rahim

Abstract

Lignocellulosic biomass is a renewable raw material. Industrial interest with new technology has grown to take advantage of this raw material. Different microbial enzymes are treated with biomass to produce the desired products under ideal industrial conditions. Xylanases are the key enzymes that degrade the xylosidic linkages in the xylan backbone of the biomass, and commercial enzymes are categorized into different glycoside hydrolase families. Thermophilic microorganisms are an excellent source of thermostable enzymes that can tolerate the extreme conditions of industrial processing. Thermostability of xylanases from thermophilic microorganisms has given the importance for a specific activity at elevated temperatures and distinction due to biochemical properties, structure, and mode of action. Optimized xylanases can be produced through genetic engineering: a novel xylanase is isolated from an extreme environment and then genetically modified to improve suitability for industrial contexts. Recombinant protein techniques have made it possible to engineer and express thermostable xylanases in bacteria, yeasts, and filamentous fungi. We will discuss the biotechnological potential of xylanases from thermophilic microorganism and the ways they are being optimized and expressed for industrial applications.

Keywords: xylanases, thermophilic xylanases, genetic engineering, applications

1. Introduction

Increasing energy costs and environmental concerns have pushed the global demand for sustainable renewable fuels. The impacts of plant biomass gain particular interest due to their availability as the most abundant raw material worldwide and certainly play an important role because of its significant role as a renewable source of energy [1]. Plant biomass refers to lignocellulosic biomass and its growing demand for the more effective utilization appears to evolve as one of the most important area with great industrial interest. Lignocellulose is composed of the structural polymers of cellulose, hemicellulose, pectin, and lignin. Cellulose is the most abundant polysaccharide in nature followed by hemicelluloses. Xylan is a major structural component of hemicullose and makes up 20–40% of total plant biomass [1].

Various hydrolyzing enzymes are required to degrade lignocellulosic biomass. Carbohydrate-active enzymes, the second largest group of industrial enzymes are sourced from animals, plants, and microorganisms. The hydrolytic enzymes make up nearly 75% of the total extent of industrial enzymes worldwide [2]. Efficient

IntechOpen

degradation of plant biomass remains a major challenge that requires the availability of enzymes for the hydrolysis of lignocellulosic biomass. Biomass can be efficiently degraded by combining multiple enzymes that hydrolyze complex polysaccharides into fermentable sugars under ideal industrial conditions. These enzymes must be active during a wide range of conditions, such as high temperature and pH [3]. Xylanases are a crucial group of depolymerizing enzymes used for the hydrolysis of the xylan that is a major component of hemicellulose. There is a critical need for thermophilic xylanases operating more efficiently at higher temperatures than current commercially available ones. Xylanase producing thermophilic microorganisms are of prime importance and well-suited for industrial applications. Therefore, there is growing interest in exploring thermophilic microorganisms from extreme environment for biotechnological applications toward biomass degradation. Because of their biotechnological characteristics, xylanases are most often produced from microorganisms for commercial applications. Recently, there has been much industrial interest in xylanases for biofuel production, chemical and pharmaceutical industries, wood pulp bioleaching, papermaking, the manufacture of food and beverages, and animal nutrition. Let us look briefly at the genesis and some promising industrial uses of xylanases.

2. Structure of xylan and xylanases

Lignocellulosic biomass is a potential raw material and its growing demand for the more effective utilization appears to evolve as one of the most important technologies with great industrial interest. Hemicellulose is a complex of polymeric carbohydrates including xylan, xyloglucan (heteropolymer of D-xylose and D-glucose), glucomannan (heteropolymer of D-glucose and D-mannose), galacto-glucomannan (heteropolymer of D-galactose, D-glucose and D-mannose) and arabinogalactan (hetero-polymer of D-galactose and arabinose) [4]. Xylan is a major structural polysaccharide found in the cell walls of plants, in which they may constitute more than 30% of the dry weight [5].

The structure of xylan holds a backbone of β-1,4-linked xylopyranosyl residues with an equatorial configuration and a diversity of substituted groups such as arabinose, acetyl, glucuronic acids, ferulic acid, and p-coumaric acid [6]. However, the structure of xylan is variable, ranging from linear 1,4-β-linked polyxylose sugars other than D-xylose, but the main chain of xylan is analogous to that of cellulose, composed of D-xylose instead of D-glucose. Moreover, xylan interacts with lignin and polysaccharides through covalent and non-covalent linkages, respectively. Different types of covalent linkages are responsible to interconnect xylan and lignin in the secondary cell wall [7]. These covalent interactions include glycosidic linkages between xylopyranosyl and p-coumaric acid and ester linkages between arabinofuranosyl residues and p-coumaric acid or ferulic acid. Feruloyl residues in arabinoxylans are known to be bridging units between xylan and lignin. The distribution pattern of the substituents on xylan affects a series of functional properties, including their solubility, interactions with other polymeric cell wall substances, degradability by enzymes, and solution behavior [8]. In addition, a relationship between the chemical structure of xylan results in a certain degree of complexity of xylan-containing materials, that may have several different xylan polymers of related structures but differ by more or less important features [8].

Xylanases are a crucial group of depolymerizing enzymes used for the hydrolysis of the xylan that is a major component of hemicellulose. Enzymatic depolymerization of hemicellulose to monomer sugars needs the synergistic action of multiple enzymes. These enzymes include endo-xylanase (endo-1,4-β-xylanase), β-xylosidase (xylan-1,4-β-xylosidase), α-glucuronidase (α-glucosiduronase), α-arabinofuranosidase (α-L-arabinofuranosidase) and acetylxylan esterase. Among them endo-xylanases and β-xylosidases are the two key enzymes responsible for the hydrolysis of xylan. Endo-xylanases randomly cleaves the xylan backbone into xylo-oligosaccharides, while β-xylosidases further hydrolyze the xylo-oligosaccharides from non-reducing end into xylose monomers. The removal of side groups is catalyzed by α-D-glucuronidases, α-L-arabinofuranosidases, ferulic acid esterases, acetylxylan esterases and p-coumaric acid esterases [9, 10].

3. Classification of xylanases

Initially xylanases were classified into two groups, those with low molecular weight (less than 30 kDa) and basic isoelectric points (pI), secondly those with high molecular weight (greater than 30 kDa) and acidic pI. However, this classification system was unable to classify most of the recently discovered xylanases [11]. Afterward, another classification system was introduced that were based on the comparisons of primary structure of the catalytic domains and these enzymes were grouped into families based on related sequences. This classification system now considered the standard means for the classification of enzymes including xylanases. In addition, this classification system gave an extra edge that classifies the glycosidases in general [11]. The most extensive group of enzymes is "Glycoside hydrolases" that refers to catalyze the glycosidic bond cleavage between carbohydrates or between carbohydrate and non-carbohydrate moiety. In glycoside hydrolases (GH) families, some family protein folds are more conserved than their amino acid sequences, and these families are further grouped into clans. Presently, 14 different clans have been proposed (GH-A to GH-N), with most clans encompassing two or more than two families [11].

According to the information provided in the Carbohydrate-Active Enzymes Database (CAZy), xylanases have been classified into 13 families, however only the GH10 (formerly F) and GH11 families (formerly G) with exclusive activities for endo-β-xylanase in them. The difference between these two families based on sequence, different catalytic properties, substrate specificity, three-dimensional structure and mechanism of action [11, 12]. Besides the GH family 10 and 11, xylanases activity are also found in families of GH5, GH7, GH8, GH16, GH26, GH43, GH52 and GH62 [11, 13]. For the reason that some bifunctional enzymes are containing two catalytic domains, for example xylanases having domain of family GH10 or GH11 and it contains a domain of glycosidase as well. Among the other families, GH8 xylanases act solely on xylan whereas GH5, GH7, and GH43 xylanases also show activities as endo-glucanases, licheninases or arabino-furanosidases. Therefore, the enzymes with xylanase activity are solely not only confined to families GH10 and GH11 but also expanded to include other families like GH5, GH7, GH8, GH16, GH43, GH52, GH62 [11].

Xylanases belongs to GH families 10 and 11, which hydrolyze glycosidic bonds by acid base-assisted catalysis through a double displacement mechanism leading to retention of anomeric configuration at the cleavage site [14]. The xylanases from GH family 10 belongs to clan GH-A and the crystal structures display an $(\alpha/\beta)_8$ barrel fold or "salad bowl" shape with extended loops creating a catalytic cleft that contains at least four to seven xylose-binding subsites [15]. The catalytic site contains two glutamate residues, one acting as a nucleophile and the other as an acid/base catalyst. Catalytic amino acids and enzymatic mechanism are conserved, presenting a domain for catalysis of 250–450 amino acids. From the biochemical point of view, most of them have high molecular weight though there are reports of low molecular weight enzymes [16]. The values of their pI are generally alkaline (8.0–9.5), however, some also have acid values and all of them sustain the same three-dimensional structure. Most of the substrate binding subsites are highly conserved in xylanases, but the affinity differences between these subsites significantly affect their mode of action, as well as substrate and product preferences [17]. As heat stability has great concern in commercial usages of xylanases. For this purpose, a number of studies analyzed the crystal structures of thermostable xylanases. Intra- and intermolecular interactions in structural topography such as disulfide bond and hydrogen bond, compact the overall fold and stabilized N and C terminal end, fusion with CBM (carbohydrate-binding motif) and lower B-factor have been proposed to bestow the enzyme for increased heat stability [18].

The xylanases from the GH11 family belongs to clan GHC. It displays exclusive substrate specificity toward xylose containing substrates and a preference for insoluble polymeric substrates. The structure of GH11 is highly homologous and contains a single major α-helix and two extended pleated β-sheets which form a jelly-roll fold [19]. The structural features include a compact globular structure and a thumb-like structure as an 11-residue long loop that connects β-strands β8 and β7, and a long cleft that spans the entire molecule and contains the active site [20]. The catalytic machinery is composed of two glutamate residues, acting as a nucleophile and an acid/base catalyst, located in the middle of the long cleft [19]. Moreover, catalytic amino acids and enzymatic mechanism of GH11 are conserved and presenting domains for catalysis of 180–200 amino acids that fold into β-sheet conformation curved on itself.

GH family	Fold	Clan	Mechanism	Nucleophile/ proton donor	Xylanase Activity	PDB No.*
GH5	(β/α)8	CH-B	Retaining	Glu/Glu	Endo-β-1,4-xylanase (EC 3.2.1.8), Arabinoxylan-specific endo-β-1,4-xylanase (EC 3.2.1.-)	2Y8K 5G56 4U3A
GH7	β-jelly roll	CH-B	Retaining	Glu/Glu	Endo-β-1,4-glucanase (EC 3.2.1.4), Endo-β-1,3–1,4-glucanase (EC 3.2.1.73)	1EG1 3OVW
GH8	(α/α)8	CH-M	Inverting	Asp/Glu	Endo-1,4-β-xylanase (EC 3.2.1.8)	1H13 1XW2
GH10	(β/α)8	CH-A	Retaining	Glu/Glu	Endo-1,4-β-xylanase (EC 3.2.1.8), Endo-1,3-β-xylanase (EC 3.2.1.32)	4QCE 1NQ6 1 W32
GH11	β-jelly roll	CH-A	Retaining	Glu/Glu	Endo-1,4-β-xylanase (EC 3.2.1.8), Endo-1,3-β-xylanase (EC 3.2.1.32)	3WP3 1YNA 1XNK
GH30	(β /α) 8	GH-A	Retaining	Glu/Glu	Endo-β-1,4-xylanase (EC 3.2.1.8), Endo-β-1,4-xylanase (EC 3.2.1.136), β-xylosidase (EC 3.2.1.37)	4FMV 4FMV
GH43	5-fold β-propeller	GH-F	Inverting	Asp/Glu	Xylanase (EC 3.2.1.8), β-xylosidase (EC 3.2.1.37)	5GLN 2EXJ

PDB, Protein data base number.

Table 1.
Characteristics of different glycoside hydrolase family containing enzymes with a demonstrated xylanase activity.

GH 5 is the largest glycoside hydrolase family with varying activities including endo-1,4-β-xylanase. It hydrolyzes the β-1,4 xylan chain at a specific site directed by the position of an α-1,2-linked glucuronate moiety. The structural analysis XynA (of the family 5 xylanase) showed that, the catalytic domain displayed a common $(\beta/\alpha)_8$ barrel fold [21]; whereas, the β-barrels aligned well with those of another family 5 enzyme. The α-helices and loops were different, showing variances in the positioning, length and orientation. The xylanases belongs to family GH8 are classified in clan CH-M also contains endo-1,4-β-xylanase along with other glycoside hydrolase enzymes. It has also the aptitude to hydrolyze the β-1,4 xylan chain and exhibits the $(\alpha/\alpha)_6$ barrel structure formed by six inner and six outer α helices [22]. Similarly, the GH26 are the member of the clan CH-A and exhibits the (β/α)8 structure. This family contains different glycoside hydrolase enzymes including β-1,3-xylanase, capable of hydrolyzing β-1,3-xylan. Activity, mechanisms and the structure of other member of glycoside hydrolase enzymes are listed in **Table 1**.

4. Sources/genesis of xylanases

Demystifying definitional issue for common understanding, the xylanases are enzymes commonly found in microorganisms, marine algae, protozoans, snails, crustaceans, insects, seeds, plants, and other natural sources [23]. Recently, there has been much industrial interest in xylanases for wood pulp bioleaching, papermaking, the manufacture of food and beverages, animal nutrition, and bioethanol production. Because of their biotechnological characteristics, xylanases are most often produced from microorganisms for commercial applications.

Nature is replete with myriad microorganisms producing enzymatic complexes that degrade cellulose and hemicellulose releasing sugars, used for attainment of products with high economical value [24]. Microbial xylanases are of prime importance in industrial application. Most of commercial enzymes are accrued from mesophilic microorganisms. The thermostable enzymes from thermophilic micro-organisms can better meet the need of high temperatures in the industrial processes for preparing end products. Of course there is a growing interest for multiple studies in exploring the importance of enzymes producing thermophilic microorganisms in relationship with biotechnological application. The microorganisms being extremophilic in nature can survive and thrive in extreme environments on account of which thermo-stability is provided to industrial processes. Biological sources including bacteria, fungi and yeasts have been reported as xylanase producing organisms in a natural process.

4.1 Bacterial source

Xylanase producing thermophilic bacteria are found in variegated environments and the recent one was isolated in Tunisian hot springs. Various thermophilic *Bacillus* strains isolated and the identification of *Bacillus* strains was based on the phenotypic characteristics of *Bacillus* genus and phylogenetic analysis of the 16S rDNA sequence. Activity tests of these *Bacillus* strains confirmed the xylanase producing strains [25]. A thermophilic anaerobic bacteria *Caldicoprobacter algeriensis* TH7C1(T), isolated from the hydrothermal hot spring has been reported as extracellular thermostable xylanase (XYN35) producing organism [26]. The isolation of thermophilic gram positive strain Rxl, a member of genus *Thermoanaerobacterium* with xylan degrading ability was reported from hot springs in Baoshan of Yunnan Province, China. The successful cultivation of the bacterium was made through utilization of xylan, starch and wide range of monosaccharide and polysaccharides [27]. Others like xylanase genes, was cloned from bacterial strain *Planococcus* sp., SL4, was isolated from the sediment of Soda Lake Dabusu of high alkalinity nature [28].

4.2 Fungal source

In comparison with the bacteria, the filamentous fungi have been in use as most potent industrial enzyme producers for the last five decades. Filamentous fungi are exuberant producers of xylanolytic enzymes in medium being used for the purpose. The genomes of lignocellulolytic fungi like for example *Trichoderma reesei*, *Aspergillus niger*, and *Myceliophthora thermophila* are producing diversity of enzymes that breakdown the complex cell wall components [29–31]. *M. thermophila* known as a powerful cellulolytic organism was used in new advanced technologies for industrial enzyme production, like the biomass-derived fuels. Due to its peculiarities *M. thermophila* distinguishes from rest of xylanase producers such *A. niger* and *T. reesei*. It is best source of gene encoding extracellular thermophilic xylanases. It has presence of a relatively high number of (glucurono) arabinoxylan degrading enzymes. It has lignocellulolytic enzymes that synthesize a complete set of enzymes necessary for the breakdown of cellulose. On the sequence analysis and the genome of the *M. thermophilia* has revealed a large repertoire of genes responsible for the production of thermostable lignocellulolytic enzymes such as carbohydrate-active enzymes, proteases, oxidoreductases, lipases and xylanase [30, 32, 33]. *M. thermophila* has 9110 genes, organized in 7 chromosomes, sequenced and annotated and also consists of 250 genes encode carbohydrate-active enzymes of which 180 are potential glycoside hydrolases. Thirteen out of 180 genes has ability to encode xylanases [34, 35].

Other thermophilic fungus like genus *Humicola* a nonpathogenic and nontoxic fungus also produces a wide range of hemicellulases and cellulases. Thermophilic *Humicola insolens* Y1 is an excellent producer of xylanolytic enzymes, including the thermophilic xylanases from family GH10 and GH11 [36]. More to the list is the *Thermoascus aurantiacus* another potential fungus with the ability to produce thermostable cellulases and xylanase, reported from Aravali forest area of University of New Delhi [37]. The fungus was able to produce antioxidant compounds as byproduct of its inoculum preparation process, which could be used for exploiting crop residues for biofuel production.

4.3 Xylanases from archaeal domain

In the field of biotechnology, the thermophilic micro-organisms from archeal domain have been reported/isolated with ability to express enzymes that can tolerate high temperatures (80–115°C), extreme pH, and high salt concentration [38]. These thermophilic enzymes with attribute of hydrolyzing lignocellulosic biomass were characterized, cloned and expressed in various hosts. *Sulfolobus solfataricus* is thermoacidophile that can live in acidic volcanic hot springs and grows optimally up to 87°C and pH 2–4. It produces enzymes with carbohydrate depolymerizing activities, such as endoglucanases and xylanases, as well as β-glucosidases/xylosidases involved in the degradation of plant-derived complex polysaccharides [39]. The genome of *S. solfataricus* has been sequenced, and three open reading frames (sso1354, sso1949, and sso2534) coding for putative extracellular endo-glucanases have been identified. These enzymes belong to a GH12 of glycoside hydrolases family and member of clan C [40]. Sources of microbial xylanases with demonstrated xylanase activity are listed in **Table 2**.

Sources	Gene	Substrate	Xylanase activity	References
Humicola insolens	Xyn11B	Beechwood Xylan	382.0 U/mg	[36]
Streptomyces sp.	XynA	Beechwood Xylan	250.69 U/mg	[41]
Streptomyces sp.	—	Birchwood Xylan	5098.28 U/mg	[42]
Schizophyllum commune	XynA	Beechwood Xylan	5768 U/mg	[43]
Streptomyces sp.	XynBS27	Oat spelt Xylan	3272.0 U/mg	[44]
Aspergillus niger	XAn11	Birchwood Xylan	909.4 U/mg	[45]
Aspergillus. niger	XAn11	Birchwood Xylan	415.1 U/mg	[45]
Planococcus sp.	XynSL4	Birchwood Xylan	244.7 U/mg	[28]
Acrophialophora nainiana	Xyn6	Oat spelt xylan	172 mg/L	[46]
Trichoderma reesei	Xyn2	Birchwood Xylan	1600 U/mg	[47]
Myceliophthora thermophila	MYCTH_56237	RBB-Xylan*	1533.7 U/mg	[48]
Myceliophthora thermophila	MYCTH_49824	RBB-Xylan*	1412.5 U/mg	[48]
Thermothelomyces thermophila	MYCTH_39555	Birchwood Xylan	105.42 U/mg	[49]

RBB-Xylan, Remazol brilliant blue-Xylan.

Table 2.
Sources of microbial xylanases with demonstrated activity.

5. Expression systems for xylanases

To acquire a pure form of a particular enzyme from a given source is challenging. Also it is inconvenient to have cultivation of bacteria or fungi for large scale protein production that often leads to many interfering enzymes. It might need multiple purification steps to get the intended enzymes purified from a pool of proteins which in turn will increase the cost. Therefore, recombinant DNA technology is recommendable for application with success prospects for desired object [50]. Recombinant DNA technology allows large scale expression of enzymes in both homologous and heterologous protein expression. The genes of enzymes with industrial importance were reportedly cloned and expressed in expression hosts in order to enhance specific enzymes production plus improvement in substrate utilization, and other commercially useful properties. Likewise, genes encoding thermophilic xylanases from different sources have been cloned with the objectives of overproduction of the xylanases and changing its properties to suit commercial applications [9].

5.1 Bacterial expression system

Escherichia coli are the most promising host for cloning and expression of heterologous recombinant proteins. Success of this platform as a recombinant expression host mainly due to the ease of is attributed toward some factors such as wide choice of cloning vectors, rapid growth, inexpensive media and simple techniques required for transformation, secretion of heterologous proteins into the culture medium and avoid the difficulties associated with purification of the recombinant protein [9]. *E. coli* expression systems been used for recombinant proteins production both intracellularly and extracellularly. In spite of the many advantages of using *E. coli* as expression host, there are certain limitations such as upon gene over expression, recombinant protein aggregates to form inclusion bodies in the cytoplasm. In order to reduce the inclusion bodies, several strategies could be used such as regulation of the protein synthesis rate, co-expression of chaperone genes and empowering the secretion of proteins into the periplasm. However, the control rate of protein synthesis can be achieved by altering the promoter to regulate the level of expression, fusing the target gene to another gene, and adjusting the growth conditions, such as pH and temperature of the medium [9].

Although previously it has been reported that the expression of the xylanase genes usually cannot be functionally expressed in *E. coli* due to some factors including the repetitive appearance of rare codons and the requirement for specific post translational modifications such as disulfide bond formation and glycosylation [51] and also it require N-glycosylation whereas *E. coli* can only perform simple O-glycosylation [52]. However, recently recombinant xylanase of family GH10 (XYN) from *Thermoanaerobacterium thermosaccharolyticum* (DSM 571) was successfully overexpressed in *E. coli* (strain BL21) [53]. Similarly, another xylanase gene (xyn10B) encoding the endo-xylanase from *Thermotoga thermarum*, was successfully cloned and expressed in *E. coli* (strain BL21) and exhibits the thermostability at high temperature [54]. These finding indicated that *E. coli* might be an effective and suitable host for the expression of xylanases. Furthermore, xylanase (gene xynA) from thermophilic fungus *Thermomyces lanuginosus* exhibits the activity endo-xylanase of GF11 and the expression of optimized sequence of xynA in *E. coli* was found to be a high level. However, the recombinant XynA was mainly found in inclusion bodies, and only a small proportion was soluble and active [55]. For this purpose, a strategy was exposed to overcome inclusion-body formation, an expression plasmid named pHsh exhibit a synthetic heat-shock (Hsh) promoter, in which gene expression is regulated by an alternative sigma factor (σ32). pHsh derivative was constructed by fusing a signal peptide to xynA2 gene, eases to export the recombinant protein to periplasm and xylanase was successfully produced in a soluble form [56].

5.2 Fungal expression system

Filamentous fungus is the promising organism for protein expression and its production by fermentation has a long history in industrial area. Even developed other expression systems for recombinant protein expression, fungal expression system also considered an appropriate candidate for the expression [9]. Natural capability of fungal expression system to secrete large amounts of proteins into the medium gave an advantage to this expression system. Furthermore, it has feasibility for functional expression of other xylanases from remote sources by using of native xylanase expressing machinery [7, 9]. Most of the xylanase genes have been expressed in fungi under homologous expression system and frequently used fungus as expression hosts are *T. reesei*, *A. niger* and *Aspergillus oryzae* [57]. *T. reesei* system relies on the integration of the transforming DNA into the fungal genome, which results in excellent stability of transformants. The vectors construct provided a variety of N and C-terminal modifications that facilitate gene product processing and purification. In *T. reesei*, the most frequent choice of a promoter used for recombinant gene expression is the *cbh1* (cellobiohydrolase 1) gene encoding the cellulose. For high level expression of recombinant proteins in *T. reesei* is to use a variety of strong promoters that simultaneously transcribe the target gene, instead of one promoter with multiple copies of genes, which might be leads to the depletion of specific transcription factors for the promoter. Recently *T. reesei* strain expressed the recombinant bacterial xylanase XynB under the promoters of *egl2* (endoglucanase 2), *xyn2* and *cbh2*. Promoter of the *T. reesei xyn2* gene encodes the endo-1,4-β-xylanase II (Xyn2). Gene expression cassettes with the *xyn2* and *cbh2* promoters were introduced simultaneously into a *T. reesei* strain (EC-21), which produced XynB under the egl2 promoter [58]. Expression of the *xynB* gene under all three different promoters resulted in improvements of the enzyme [58]. Two novel genes of family GH11 xylanases *xyn5* and *xyn6*, isolated from the thermophilic filamentous fungus *Acrophialophora nainiana,* were successfully expressed in an industrially-exploited fungal host *T.* reesei [46]. Moreover, the beneficial aspect of this fungus use for recombinant gene expression is the secretion of proteins into the growth medium and consequent gene products achieved comparatively straight-forward. However, degradation of recombinant gene products also occurs due to the secreted acidic proteases into the cultivation medium [58].

5.3 Yeast expression system

Yeasts considered as excellent and attractive host for the expression of heterologous proteins and offer many advantages over the other established expression systems especially in protein maturation [59]. The methylotrophic yeast *Pichia pastoris* is an established protein expression host for the production of industrial enzymes. It can be grown to very high cell densities, produces high titer of recombinant proteins and ability to secrete proteins into fermentation media thus it is a very useful expression host, especially when scaling up to industrial process [60]. *P. pastoris* can be expressed intercellularly and provides extra benefits over the other expression systems such as ability to perform eukaryotic post-translational modifications, glycosylation, proper folding of the proteins [61]. Moreover, the most significant feature the of this expression host are due to the availability of strong and regulatory promoter of alcohol oxidases AOX1, involved in the methanol utilization pathway which provided exceptionally high levels of heterologous recombinant protein [62]. Because of all such features, the expression of xylanase genes in *P. pastoris* preferred mostly and provides high yield of recombinant xylanases expression under methanol induction. The enzyme activity of xylanase was reported 3676 U mL^{-1} for the gene product of xylB from *A. niger*, when expressed under AOX1 in *P. pastoris*. In fact, this is one of the highest expressions of recombinant xylanase expressed from *P. pastoris* reported [63]. Similarly, Cheng et al. [61] and Chantasingh et al. [64] also attained high xylanase activity (342.2 U mL^{-1} and 238.5 mg mL^{-1}, respectively) under this promoter (67-fold and 4-fold) higher recombinant xylanase activity, compared to the native fungal xylanases. The gene coding xylanase (*xynS14*) from a thermophilic xylan degrading actinomycetes *Actinomadura* S14, were expressed in both *E. coli* and *P. pastoris* [65]. The specific activity of purified recombinant xylanase from *P. pastoris* transformants was approximately 2.4-fold higher than that of purified recombinant xylanase from *E. coli* transformants, suggesting that *P. pastoris* is a better host for expression of recombinant XynS14. Although both recombinant XynS14 showed approximately the same basic properties, such as substrate specificity, optimal pH and temperature, stability for pH and temperature, and effects of EDTA and metal ions, whereas XynS14 (*P. pastoris*) showed higher specific activity and kinetic values (V_{max} and K_{cat}) than XynS14 (*E. coli*) [65]. These finding suggested the glycosyl chains present in XynS14 (*P. pastoris*) stabilized the enzyme and the enzymes were folded properly in *P. pastoris*. Cloning and expression of another xylanase gene belongs to family GH11 from *T. fusca* NTU22, also reported higher yield and thermostability than the original strain [61].

6. Xylanases: genetic engineering and optimization

In most of the cases xylanases need to undergo some genetic modifications in order to enhance expression level, enzymes activity and that might have some influence on substrate specificity and stability to high temperature and pH. The gene encoding cellulolytic and xylanolytic enzymes are usually regulated by a repressor/inducer system in fungi. Xylanolytic transcriptional regulators have been reported in thermophilic fungi. The strong promoter MtPpdc (pyruvate decarboxylase) recently used for the overexpression of xylanases from *M. thermophila* ATCC42464 (MtXyr1). The extracellular xylanase activity of the recombinant was reported higher as compared to the wild type indicating the MtXyr1 is a positive regulatory factor for xylanase gene expression and its feasibility of improving xylanase production by overexpressing Mtxyr1 in *M. thermophila* represented an effective approach to increase total xylanase productivity [66]. Similarly, a new N-glycosylated site was created in the coding sequence by amino acid replacements for the expression of endoxylanase in *T. reesei* M2C38. For this purpose, amino acid Asn at position 131 associated with Thr/Ser at position 133 was inserted that resemble a conserved feature for family GH11 xylanases. The new created N-glycosylation site Asn-Xaa-Thr/Ser displayed 40% enhanced protein expression in comparison with wild type [67]. The downregulation of Cre1 plays an important role in enhancing enzyme production and also the silencing of CRE1 improves cellulase and xylanase expression. In *T. reesei* and other fungi, the key regulator of CCR (Carbon catabolite repression) is the Cys_2His_2-type transcription factor CRE1. The role of CRE1 in *M. thermophilia* was verified through RNAi and suggested the feasibility of improving cellulase production by modifying the expression of regulators in thermophilic fungi [68].

In order to fulfill the demands of industrial requirements, gene mining and protein engineering are applied to develop thermostable xylanases. Although some of thermophilic xylanases were produced from thermophiles but their lower expression levels and specific activities making them unable to be applied efficiently. The higher specific activity with enhanced thermostability of xylanases is therefore needed through genetic engineering. Recently the thermostability of mesophilic xylanase (AuXyn10A from *Aspergillus usamii* E001), was improved through elucidation of some local structures affecting the thermostability of mesophilic xylanases in corresponding to thermophilic *Thermoascus aurantiacus*. The temperature optimum of the mutant was 10°C higher than that of AuXyn10A [69]. The thermostability and alkalophilicity of another endo-xylanase from *T. reesei* was improved by replacement of amino acids at different positions and the replacement of NH_2 terminal amino acid sequence of *Thermomonospora fusca* along with the addition of some extra amino acids selected from N-terminus of *Clostridium acetobutylicum* xynB. All these strategies increased the thermophilicity and alkalophicility of the enzyme from 55 to 75°C and pH 7.5 to 9.0 respectively [70]. Similarly, the thermostability of mutant xylanase (Xyn10A_ASPNG) was improved by 17.4°C [71].

7. Factors affecting xylanase activity

7.1 Substrates

The enzymatic degradation reaction of insoluble polysaccharides had great importance. However, glycoside hydrolase enzymes often inaccessible to the active site of the appropriate enzymes, such polysaccharides relatively inefficiently. In order to overcome these problems, many of the glycoside hydrolases that utilize insoluble substrates are modular, comprising catalytic modules appended to one or more non-catalytic carbohydrate-binding modules (CBMs) [72]. Two xylanase genes belongs to family GH11 (xynC81 and xynC83) from a thermophilic strain

Achaetomium sp. Xz-8, with substantial xylanase activity. Substrate specificity and the hydrolysis analysis of purified recombinant revealed that XynC81 and XynC83 were moderate on beechwood xylan (67 and 69%, respectively) and birchwood xylan (45 and 52%, respectively), and weak on barley β-glucan (18 and 14%, respectively). Only XynC81 had detectable activity against insoluble wheat arabinoxylan (22%). Although enzymatic properties of XynC81 and XynC83 were similar but XynC81 with CBM 1 had activity against insoluble wheat arabinoxylan (22%), whereas XynC83 had not. This result further implied the importance of CBM in enzyme activity toward insoluble substrate. As all of the GH11 xylanases characterized so far, about 25% of them carry at least one CBM [73]. Unlike the serine/threonine/asparagine-rich linker sequence found in other fungal xylanases, the XynC81 linker sequence is extremely glycine-rich. Therefore, it could be predicted that CBMs take part in the action of cellulolytic enzymes toward insoluble substrates [74].

Moreover, some xylanases bears CBMs specific for cellulose, which probably assist indirectly localization of xylanase to the xylan substrate, since it is in close association with cellulose. The number of characterized fungal xylanases harboring CBM1 module is relatively reduced. It includes xylanases from *Penicillium funiculosum* XynB [75, 76], *Neocallimastix patriciarum* XynS20 [77], *Lentinula edodes* Xyn11A [78] and *Phanerochaete chrysosporium* XynB/XynB-1 [79]. Recently, a Xyl-11 from *Podospora anserina* harboring a C-terminal CBM1 efficiently supplemented the industrial cocktail produced by *T. reesei* by improving significantly the release of reducing sugars upon hydrolysis of wheat straw [80].

7.2 Metals and chemicals

The metal ion and chemical reagents had been proved to be one of the critical factors which affected the enzyme activity of xylanases. The effect of metal ions and chemical reagents on the xylanase activities has been determined on various metal ions (Na^+, K^+, Ca^{2+}, Li^+, Co^{2+}, Cr^{3+}, Ni^{2+}, Cu^{2+}, Mg^{2+}, Fe^{3+}, Zn^{2+}, Pb^{2+}, and Ag^+) and chemical reagents (SDS, β-mercaptoethanol, ethanol, Triton X-100, and EDTA) at the standard condition. Ca^{2+}, Pb^{2+}, K^+, Mn^{2+}, EDTA, β-ME, Cu^{2+} and Triton X-100 were reported to enhance enzyme activity by 6.4–29.9% [81], while Fe^{3+}, Cd^{2+}, Hg^{2+}, and Ba^{2+} completely suppressed the xylanase activity. Besides this, the enzyme had certain ability to resist the Fe^{2+}, Mg^{2+}, Ag^{2+}, SDS, ethanol and SDS. Xylanase activity was not inhibited by chelating reagents such as EDTA and EGTA. Moreover, it is predicted that Ca^{2+} and Mg^{2+} ions enhance the enzyme activity by stabilizing the enzyme–substrate complex. In contrast, EDTA is a chelating agent and it removes ions from the enzymes, thus inhibits the enzyme activity [7]. More detailed studies are needed in order to understand the mechanistic effect of metal ions on enzyme activity. Similarly, a xylanase activity isolated from *Planococcus* sp. *SL4* was enhanced by Ca^{2+} and β-mercaptoethanol. K^+, Cr^{3+}, Li^+, and Na^+ showed little or no effect on xylanase activity. Ag^+, Cu^{2+}, Hg^{2+}, Pb^{2+}, Zn^{2+}, Fe^{3+}, and Cr^{3+} resulted in an almost complete loss of activity [28].

7.3 Temperature and pH

As thermophilic enzymes are preferred over the mesophilic enzymes complements because of high temperatures, which had a great influence on many factors such as decreases contamination risk and viscosity of substrate [82]. In a study carried out with the effect of xylanase activity from *A. niger* DFR-5 on different temperature (between 20 and 60°C). The xylanase activity was increased with increase in temperature with maximum activity of at 40°C. On further increase in

Species	Strain	pH	Temperature	References
Caldicoprobacter algeriensis	TH7C1(T)	11.0	70°C	[26]
Microcella alkaliphila	JAM-AC0309	8.0	65°C	[86]
Planococcus sp.	SL4	7.0	70°C	[28]
Geobacillus sp.	WBI	7.0	65°C	[87]
Caldicellulosiruptor sp.	F32	6.6	75°C	[88]
Actinomadura sp.	Cpt20	10.0	80 °C	[89]
Thermofilum pendens	Tpbgl	3.5	95°C	[90]
Sulfolobus solfataricus	MT4	7.0	90°C	[91]
Thermococcus zilligii	AN1	6.0	75°C	[92]
Penicillium occitanis	Pol6	3.0	65°C	[93]
Saccharopolyspora pathumthaniensis	S582	6.5	70°C	[94]
Paecilomyces thermophila	J18	7.0	75°C	[95]
Bispora sp.	MEY-1	4.5	85°C	[96]
Malbranchea cinnamomea	S168	6.5	80 °C	[97]
Penicillium oxalicum	B3–11(2)	5.0	50 °C	[98]

Table 3.
List of xylanase producing microbial species/strains with demonstrated pH and temperature.

temperature, the enzyme activity declined gradually and at 60°C, enzyme exhibited 35.5% of the maximum activity [83]. Similarly, the temperature effect of purified xylanase (xynZF-2) from *A. niger*, was evaluated [84]. The optimum temperature recombinant enzyme was 40°C and the enzyme activity was observed relatively stable on the temperature below 40°C, however enzyme activity decreased rapidly with rise of temperature (above than 40°C). The activity became completely lost on incubation at 50°C for 15 min [84].

Evaluation of the pH stability is a vital part of any enzyme characterization before it can be exploited commercially. Similarly effect of pH had great influence on the xylanase activity. The effect of pH on extracellular xylanase from *A. niger* DFR-5 was evaluated [83]. That suggested the formation of an improper ionic form of the xylanase and/or substrate between pH 4.0 and 5.0 and between pH 5.0 and 6.5. Because the activity of pre-incubated xylanase at pH >6.5 or pH <4.0, was full not recovered at pH 5.0 again. The decline activity at pH above than 6.5 and pH below than 4.0 resulted from irreversible enzyme inactivation [83]. Many researchers have reported the pH stability of xylanase but without interpretation. Xylanase isolated from *Planococcus* sp. SL4 was highly active and stable over the neutral and alkaline pH range from 6 to 11, with maximum activity at pH 7 and more than 60% activity at pH 11 [28]. Xylanase purified from *Arthrobacter* sp. MTCC 5214 was found stable in a narrow pH range of 7.0–8.0 [85]. List of xylanase producing microbial species/strains with demonstrated pH and temperature are given in **Table 3**.

8. Application of xylanases

Because of their biotechnological characteristics, xylanases are most often produced from microorganisms for commercial applications. The environmental hostility and lethality to most of living organisms due to extreme hot environments is obvious but such factors are counter-productive in the presence of specialized microorganisms. These microorganisms from harsh environments can tolerate high or low temperatures, extremes pH and high concentrations of salts. Based on the unique stability of their enzymes at high temperature, extremes of pH they are expected to be a powerful tool in industrial biotransformation processes that run at harsh conditions. Enzymes derived from extremophiles have great importance to the local enzymes because they can perform industrial processes even under harsh conditions, under which conventional proteins are completely denatured. The extremophile research, the high demands of the biotech industries for tailor-made novel biocatalysts, and the rapid development of new techniques such as genomics, proteomics, metabolomics, directed evolution and gene shuffling will stimulate the development of new industrial processes on the basis of biocatalysts from extremophiles.

Enzymatic hydrolysis of xylan has become attractive due to its biotechnological applications in the food, animal feed, waste treatment, ethanol production, textile, and pulp and paper industries [11]. Xylanases gained its importance for the saccharification process in degradation of lignocellulosic biomass, because xylanases assist the hydrolysis of xylan and ultimately leads to the hydrolysis of hemicellulose and cellulose, to obtain industrially important products [99]. In addition, biofuel production from lignocellulosic biomass has great concern in industrial sectors worldwide. However, biofuel production is still a bottleneck, because the initial conversion of biomass into sugars requires multiple enzymes (including xylanases) with high activity and excellent enzyme properties. Therefore, recent development in biotechnological applications xylanases toward the biomass conversion is on way [100]. Moreover, highly stable enzymes, active under high temperature and wide range of pH, displayed several advantages over the enzymes from mesophilic or neutrophilic origin. For example, highly active xylanases under extreme conditions reduce the contamination risk due to fast reaction rate. In second generation ethanol industries, mild pretreatment of biomass requires to reduce the total costs and sustain the high contents of hemicellulose. However, high dosage of other enzymes is still needed for the complete conversion of hemicellulose. Therefore, the biotechnological application of xylanases especially thermophilic xylanases and its interaction with other enzymes such as cellulases, xylosidases, arabinofuranosidase seems to be an effective in saccharification process.

The use of xylanases, as accessory enzymes, has shown to improve the hydrolysis of xylan and cellulose, and it has contributed to the reduction of enzyme dosage, but it seems to be a substrate dependent reaction [101]. Recently, less severe pretreatments of lignocellulosic biomass are gaining popularity, thus, hemicellulase characterization and studies regarding synergism of cellulases and xylanases, can contribute to reducing pretreatment severities and enhancing glucose and xylose release [102]. Xylanases are very important in bioconversion of xylan into value added products, such as xylitol. Xylitol is used in soft drinks, candies, ice cream, chewing gum and various pharmaceutical products. Xylitol have key role in sweeten food products and used as a natural sweetener in toothpaste. Xylanases have great importance in the bio-bleaching of wood pulp and in the bioprocessing of textiles.

Xylanases are very useful in manufacturing of animal feed. Xylanases reduce the viscosity and enhance the absorption rate by degrading the starch polysaccharides in rich fiber and barley based feeds. Pre-treatment of agricultural silage and grain feed employed with xylanases increases the nutritional value and improves the feed digestion in ruminants. Similarly, feed supplementation combined with xylanases for broiler diets, not only increases growth performance like weight gain but also helpful in the yield production of poultry products.

Conclusively, the xylanases have potential applications in a wide range of industrial processes, covering all sectors of industrial enzymes markets. Arguably, these are attractive for research studies and deeper investigation with aims to explore dividends and utility of these useful products in industrial context in relation to industrial enzymes producing organisms.

9. Summary

Microbial xylanases are gaining importance in industrial applications. The thermophilic microorganisms are of interest in the field of biotechnology because of their ability to express enzymes that tolerate high temperature and pH. The thermophilic bacteria, and an archaeal species, there are also filamentous fungi that grow in diverse environments and produce xylanase. New recombinant DNA technology can be used to express xylanases in both homologous and heterologous host organisms. Compared to *E. coli, P. pastoris* considered the most suitable and excellent host for the heterologous proteins expression including xylanases. Xylanases generally must be genetically modified in order to increase the expression level with improved xylanase activity, but these alterations can change substrate specificity and stability at high temperatures and pH. Many xylanases have xylan-binding domains that can affect the thermostability of the enzymes and their ability to bind insoluble substrates. In conclusion, the lignocellulosic biomass and enzyme-producing microorganisms are compelling subjects for research studies aiming to explore the growing importance of xylanases in various industrial products.

Acknowledgements

This study was supported by National Natural Science Foundation of China (31570067).

Conflict of interest

The authors declare no conflict of interest.

Author details

Abdul Basit[1]*, Wei Jiang[1]* and Kashif Rahim[2]

1 Beijing Advanced Innovation Center for Food Nutrition and Human Health, State Key Laboratory of Agro-Biotechnology, College of Biological Sciences, China Agricultural University, Beijing, China

2 Institute of Biochemistry and Molecular Biology, College of Life Sciences, Beijing Normal University, Beijing, China

*Address all correspondence to: abdul_9090@yahoo.com
and jiangwei01@cau.edu.cn

References

[1] Basit A, Liu J, Rahim K, Jiang W, Lou H. Thermophilic xylanases: From bench to bottle. Critical Reviews in Biotechnology. 2018a;**38**(7):989-1002

[2] Uday US, Choudhury P, Bandyopadhyay TK, Bhunia B. Classification, mode of action and production strategy of xylanase and its application for biofuel production from water hyacinth. International Journal of Biological Macromolecules. 2016;**82**: 1041-1054

[3] McClendon SD, Mao Z, Shin HD, Wagschal K, Chen RR. Designer xylanosomes: Protein nanostructures for enhanced xylan hydrolysis. Applied Biochemistry and Biotechnology. 2012; **167**:395-411

[4] Khandeparker R, Numan MT. Bifunctional xylanases and their potential use in biotechnology. Journal of Industrial Microbiology & Biotechnology. 2008;**35**:635-644

[5] Van Gool MP, van Muiswinkel GCJ, Hinz SWA, Schols HA, Sinitsyn AP, Gruppen H. Two novel GH11 endo-xylanases from *Myceliophthora thermophila* C1 act differently toward soluble and insoluble xylans. Enzyme and Microbial Technology. 2013;**53**: 25-32

[6] Scheller HV, Ulvskov P. Hemicelluloses. Annual Review of Plant Biology. 2010;**61**:263-289

[7] Juturu V, Wu JC. Microbial xylanases: Engineering, production and industrial applications. Biotechnology Advances. 2012;**30**:1219-1227

[8] Ebringerová A, Heinze T. Xylan and xylan derivatives biopolymers with valuable properties, naturally occurring xylans structures, isolation procedures and properties. Macromolecular Rapid Communications. 2000;**21**:542-556

[9] Ahmed S, Riaz S, Jamil A. Molecular cloning of fungal xylanases: An overview. Applied Microbiology and Biotechnology. 2009;**84**:19-35

[10] Knob A, Carmona EC. Purification and characterization of two extracellular xylanases from *Penicillium sclerotiorum*: A novel acidophilic xylanase. Applied Biochemistry and Biotechnology. 2010; **162**:429-443

[11] Collins T, Gerday C, Feller G. Xylanases, xylanase families and extremophilic xylanases. FEMS Microbiology Reviews. 2005;**29**:3-23

[12] Motta FL, Andrade CCP, Santana MHA. A review of xylanase production by the fermentation of xylan: Classification, characterization and applications. In: Sustainable Degradation of Lignocellulosic Biomass Techniques, Applications and Commercialization. North America: InTech; 2012. DOI: 10.13140/RG.2.1.2781.6724

[13] Lombard V, Golaconda Ramulu H, Drula E, Coutinho PM, Henrissat B. The carbohydrate-active enzymes database (CAZy) in 2013. Nucleic Acids Research. 2014;**42**:490-495

[14] Pell G, Szabo L, Charnock SJ, Xie H, Gloster TM, Davies GJ, et al. Structural and biochemical analysis of *Cellvibrio japonicus* xylanase 10C: How variation in substrate-binding cleft influences the catalytic profile of family GH-10 xylanases. The Journal of Biological Chemistry. 2004;**279**:11777-11788

[15] Pollet A, Delcour JA, Courtin CM. Structural determinants of the substrate specificities of xylanases from different glycoside hydrolase families. Critical Reviews in Biotechnology. 2010;**30**: 176-191

[16] Alvarez-Cervantes J, Diaz-Godinez G, Mercado-Flores Y, Gupta VK,

Anducho-Reyes MA. Phylogenetic analysis of beta-xylanase SRXL1 of *Sporisorium reilianum* and its relationship with families (GH10 and GH11) of Ascomycetes and Basidiomycetes. Scientific Reports. 2016;**6**:24010

[17] Bar M, Golan G, Nechama M, Zolotnitsky G, Shoham Y, Shoham G. A new crystal form of XT6 enables a significant improvement of its diffraction quality and resolution. Acta Crystallographica Section D: Structural Biology. 2004;**60**:545-549

[18] Zheng Y, Li Y, Liu W, Chen CC, Ko TP, He M, et al. Structural insight into potential cold adaptation mechanism through a psychrophilic glycoside hydrolase family 10 endo-beta-1,4-xylanase. Journal of Structural Biology. 2016;**193**:206-211

[19] Paes G, Cortes J, Simeon T, O'Donohue MJ, Tran V. Thumb-loops up for catalysis: a structure/function investigation of a functional loop movement in a GH11 xylanase. Computational and Structural Biotechnology Journal. 2012;**1**(2): e201207001

[20] Sermsathanaswadi J, Pianwanit S, Pason P, Waeonukul R, Tachaapaikoon C, Ratanakhanokchai K, et al. The C-terminal region of xylanase domain in Xyn11A from *Paenibacillus curdlanolyticus* B-6 plays an important role in structural stability. Applied Microbiology and Biotechnology. 2014;**98**:8223-8233

[21] Parkkinen T, Hakulinen N, Tenkanen M, Siika-aho M, Rouvinen J. Crystallization and preliminary X-ray analysis of a novel *Trichoderma reesei* xylanase IV belonging to glycoside hydrolase family 5. Acta Crystallographica Section D: Structural Biology. 2004;**60**:542-544

[22] Van Petegem F, Collins T, Meuwis MA, Gerday C, Feller G, Van Beeumen J. The structure of a cold-adapted family 8 xylanase at 1.3 Å resolution. Structural adaptations to cold and investigation of the active site. The Journal of Biological Chemistry. 2003;**278**:7531-7539

[23] Shahana Ehsan SS, Bushra K, Syeda MH, Munawar I. A rapid solid supported synthesis of 4-amino-5-(pyridin-4-YL)-4H-1,2,4-triazol-3-thiol and its derivatives. Chemistry International. 2016;**2**:262-266

[24] Intriago P. Marine microorganisms: Perspectives for getting involved in cellulosic ethanol. AMB Express. 2012;**2**: 1-12

[25] Thebti W, Riahi Y, Gharsalli R, Belhadj O. Screening and characterization of thermo-active enzymes of biotechnological interest produced by thermophilic *Bacillus* isolated from hot springs in Tunisia. Acta Biochimica Polonica. 2016;**63**:581-587

[26] Amel BD, Nawel B, Khelifa B, Mohammed G, Manon J, Salima KG, et al. Characterization of a purified thermostable xylanase from *Caldicoprobacter algeriensis* sp. noval strain TH7C1(T). Carbohydrate Research. 2016;**4**(19):60-68

[27] Shang SM, Qian L, Zhang X, Li KZ, Chagan I. *Themoanaerobacterium calidifontis* sp. a novel anaerobic, thermophilic, ethanol-producing bacterium from hot springs in China. Archives of Microbiology. 2013;**195**: 439-445

[28] Huang X, Lin J, Ye X, Wang G. Molecular characterization of a thermophilic and salt- and alkaline-tolerant xylanase from *Planococcus* sp. SL4, a strain isolated from the sediment of a Soda lake. Journal of Microbiology and Biotechnology. 2015;**25**:662-671

[29] Andersen B, Poulsen R, Hansen GH. Cellulolytic and xylanolytic activities of

common indoor fungi. International Biodeterioration & Biodegradation. 2016;**107**:111-116

[30] Martinez D, Berka RM, Henrissat B, Saloheimo M, Arvas M, Baker SE, et al. Genome sequencing and analysis of the biomass-degrading fungus *Trichoderma reesei* (syn. *Hypocrea jecorina*). Nature Biotechnology. 2008;**26**:553-560

[31] Pel HJ, de Winde JH, Archer DB, Dyer PS, Hofmann G, Schaap PJ, et al. Genome sequencing and analysis of the versatile cell factory *Aspergillus niger* CBS 513.88. Nature Biotechnology. 2007;**25**:221-231

[32] Vanden Wymelenberg A, Gaskell J, Mozuch M, Kersten P, Sabat G, Martinez D, et al. Transcriptome and secretome analyses of *Phanerochaete chrysosporium* reveal complex patterns of gene expression. Applied and Environmental Microbiology. 2009;**75**: 4058-4068

[33] Vanden Wymelenberg A, Gaskell J, Mozuch M, Sabat G, Ralph J, Skyba O, et al. Comparative transcriptome and secretome analysis of wood decay fungi *Postia placenta* and *Phanerochaete chrysosporium*. Applied and Environmental Microbiology. 2010;**76**: 3599-3610

[34] Berka RM, Grigoriev IV, Otillar R, Salamov A, Grimwood J, Reid I, et al. Comparative genomic analysis of the thermophilic biomass-degrading fungi *Myceliophthora thermophila* and *Thielavia terrestris*. Nature Biotechnology. 2011;**29**:922-U222

[35] Hinz SWA, Pouvreau L, Joosten R, Bartels J, Jonathan MC, Wery J, et al. Hemicellulase production in *Chrysosporium lucknowense* C1. Journal of Cereal Science. 2009;**50**:318-323

[36] Shi P, Du Y, Yang H, Huang H, Zhang X, Wang Y, et al. Molecular characterization of a new alkaline-tolerant xylanase from *Humicola insolens* Y1. BioMed Research International. 2015;**2015**:149504

[37] Jain KK, Bhanja Dey T, Kumar S, Kuhad RC. Production of thermostable hydrolases (cellulases and xylanase) from *Thermoascus aurantiacus* RCKK: A potential fungus. Bioprocess and Biosystems Engineering. 2015;**38**:787-796

[38] Carolina MMCA, Nei PJ, Garo A. Extremely thermophilic microorganism and their polymeric hydrolytic enzymes. Reviews Microbiology. 1999;**30**:287-298

[39] Huang YH, Huang CT, Hseu RS. Effects of dockerin domains on *Neocallimastix frontalis* xylanases. FEMS Microbiology Letters. 2005;**243**:455-460

[40] Girfoglio M, Rossi M, Cannio R. Cellulose degradation by *Sulfolobus solfataricus* requires a cell-anchored endo-beta-1-4-glucanase. Journal of Bacteriology. 2012;**194**:5091-5100

[41] Xu Y, Wu J, Zheng K, Wu D. A xylanase from *Streptomyces* sp. FA1: Heterologous expression, characterization, and its application in Chinese steamed bread. Journal of Industrial Microbiology & Biotechnology. 2016;**43**:663-670

[42] Fu XY, Zhao W, Xiong AS, Tian YS, Peng RH. High expression of recombinant *Streptomyces* sp. S38 xylanase in *Pichia pastoris* by codon optimization and analysis of its biochemical properties. Molecular Biology Reports. 2011;**38**:4991-4997

[43] Song Y, Lee YG, Choi IS, Lee KH, Cho EJ, Bae HJ. Heterologous expression of endo-1,4-beta-xylanase A from *Schizophyllum commune* in *Pichia pastoris* and functional characterization of the recombinant enzyme. Enzyme and Microbial Technology. 2013;**52**:170-176

[44] Li N, Shi P, Yang P, Wang Y, Luo H, Bai Y, et al. Cloning, expression, and

characterization of a new *Streptomyces* sp. S27 xylanase for which xylobiose is the main hydrolysis product. Applied Biochemistry and Biotechnology. 2009; **159**:521-531

[45] Elgharbi F, Hmida-Sayari A, Zaafouri Y, Bejar S. Expression of an *Aspergillus niger* xylanase in yeast: Application in bread making and in vitro digestion. International Journal of Biological Macromolecules. 2015b;**79**: 103-109

[46] Salles BC, Te'o VS, Gibbs MD, Bergquist PL, Filho EX, Ximenes EA, et al. Identification of two novel xylanase-encoding genes (xyn5 and xyn6) from *Acrophialophora nainiana* and heterologous expression of xyn6 in *Trichoderma reesei*. Biotechnology Letters. 2007;**29**:1195-1201

[47] Jun H, Bing Y, Keying Z, Xuemei D, Daiwen C. Sequencing and expression of the xylanase gene 2 from *Trichoderma reesei* Rut C-30 and characterization of the recombinant enzyme and its activity on xylan. Journal of Molecular Microbiology and Biotechnology. 2009; **17**(3):101-109

[48] Basit A, Liu J, Miao T, Zheng F, Rahim K, Lou H, et al. Characterization of two endo-β-1,4-xylanases from *Myceliophthora thermophila* and their saccharification efficiencies, synergistic with commercial cellulase. Frontiers in Microbiology. 2018;**9**:233

[49] Basit A, Miao T, Liu J, Wen J, Song L, Zheng F, et al. Highly efficient degradation of xylan into xylose by a single enzyme. ACS Sustainable Chemistry & Engineering. 2019;**7**: 11360-11368

[50] Korona B, Korona D, Bielecki S. Efficient expression and secretion of two co-produced xylanases from *Aspergillus niger* in *Pichia pastoris* directed by their native signal peptides and the *Saccharomyces cerevisiae* α-mating factor. Enzyme and Microbial Technology. 2006;**39**:683-689

[51] Juturu V, Teh TM, Wu JC. Expression of *Aeromonas punctata* ME-1 exo-xylanase X in *E. coli* for efficient hydrolysis of xylan to xylose. Applied Biochemistry and Biotechnology. 2014; **174**:2653-2662

[52] Messner P. Prokaryotic protein glycosylation is rapidly expanding from "curiosity" to "ubiquity". Chembiochem. 2009;**10**:2151-2154

[53] Li X, Shi H, Ding H, Zhang Y, Wang F. Production, purification, and characterization of a cellulase-free thermostable endo-xylanase from *Thermoanaerobacterium thermosaccharolyticum* DSM 571. Applied Biochemistry and Biotechnology. 2014; **174**:2392-2402

[54] Shi H, Zhang Y, Zhong H, Huang Y, Li X, Wang F. Cloning, over-expression and characterization of a thermo-tolerant xylanase from *Thermotoga thermarum*. Biotechnology Letters. 2014;**36**:587-593

[55] Yin E, Le Y, Pei J, Shao W, Yang Q. High-level expression of the xylanase from *Thermomyces lanuginosus* in *Escherichia coli*. World Journal of Microbiology & Biotechnology. 2008;**24**:275-280

[56] Le Y, Peng J, Wu H, Sun J, Shao W. An approach to the production of soluble protein from a fungal gene encoding an aggregation-prone xylanase in *Escherichia coli*. PLoS One. 2011;**6**:e18489

[57] Nevalainen KM, Te'o VS, Bergquist PL. Heterologous protein expression in filamentous fungi. Trends in Biotechnology. 2005;**23**:468-474

[58] Miyauchi S, Te'o VSJ, Nevalainen KMH, Bergquist PL. Simultaneous expression of the bacterial *Dictyoglomus thermophilum* xynB gene under three different *Trichoderma reesei*

promoters. New Biotechnology. 2014; **31**:98-103

[59] Siekstele R, Veteikyte A, Tvaska B, Matijosyte I. Yeast *Kluyveromyces lactis* as host for expression of the bacterial lipase: Cloning and adaptation of the new lipase gene from *Serratia* sp. Journal of Industrial Microbiology & Biotechnology. 2015;**42**:1309-1317

[60] Spohner SC, Muller H, Quitmann H, Czermak P. Expression of enzymes for the usage in food and feed industry with *Pichia pastoris*. Journal of Biotechnology. 2015;**202**:118-134

[61] Cheng YF, Yang CH, Liu WH. Cloning and expression of *Thermobifida* xylanase gene in the methylotrophic yeast *Pichia pastoris*. Enzyme and Microbial Technology. 2005;**37**:541-546

[62] Tsai CT, Huang CT. Overexpression of the *Neocallimastix frontalis* xylanase gene in the methylotrophic yeasts *Pichia pastoris* and *Pichia methanolica*. Enzyme and Microbial Technology. 2008;**42**: 459-465

[63] Ruanglek V, Sriprang R, Ratanaphan N, Tirawongsaroj P, Chantasigh D, Tanapongpipat S, et al. Cloning, expression, characterization, and high cell-density production of recombinant endo-1,4-β-xylanase from *Aspergillus niger* in *Pichia pastoris*. Enzyme and Microbial Technology. 2007;**41**:19-25

[64] Chantasingh D, Pootanakit K, Champreda V, Kanokratana P, Eurwilaichitr L. Cloning, expression, and characterization of a xylanase 10 from *Aspergillus terreus* (BCC129) in *Pichia pastoris*. Protein Expression and Purification. 2006;**46**:143-149

[65] Sriyapai T, Somyoonsap P, Matsui K, Kawai F, Chansiri K. Cloning of a thermostable xylanase from *Actinomadura* sp. S14 and its expression in *Escherichia coli* and *Pichia pastoris*. Journal of Bioscience and Bioengineering. 2011;**111**:528-536

[66] Wang J, Wu YN, Gong YF, Yu SW, Liu G. Enhancing xylanase production in the thermophilic fungus *Myceliophthora thermophila* by homologous overexpression of Mtxyr1. Journal of Industrial Microbiology and Biotechnology. 2015;**42**:1233-1241

[67] White T, Giroux Genevieve R, Wallace Katie EA. Modified Xylanases Exhibiting Improved Expression. 2005 (IOGEN BIO-PRODUCTS CORPORATION)

[68] Yang F, Gong Y, Liu G, Zhao S, Wang J. Enhancing cellulase production in thermophilic fungus *myceliophthora thermophila* ATCC42464 by RNA interference of cre1 gene expression. Journal of Microbiology and Biotechnology. 2015;**25**(7):1101-1107

[69] Wang J, Tan Z, Wu M, Li J, Wu J. Improving the thermostability of a mesophilic family 10 xylanase, AuXyn10A, from *Aspergillus usamii* by in silico design. Journal of Industrial Microbiology & Biotechnology. 2014; **41**:1217-1225

[70] Sung T, Sung WL: Thermostable xylanases. 2006; Patent, US 7060482.

[71] Song L, Tsang A, Sylvestre M. Engineering a thermostable fungal GH10 xylanase, importance of N-terminal amino acids. Biotechnology and Bioengineering. 2015;**112**:1081-1091

[72] Boraston AB, Bolam DN, Gilbert HJ, Davies GJ. Carbohydrate-binding modules: Fine-tuning polysaccharide recognition. The Biochemical Journal. 2004;**382**:769-781

[73] Paes G, Berrin JG, Beaugrand J. GH11 xylanases: Structure/function/ properties relationships and applications. Biotechnology Advances. 2012a;**30**:564-592

[74] Nimlos M, Matthews J, Crowley M, Walker R, Chukkapalli G, Brady J, et al.

Molecular modeling suggests induced fit of family I carbohydrate-binding modules with a broken-chain cellulose surface. Protein Engineering, Design & Selection. 2007;**20**:179-187

[75] Brutus A, Villard C, Durand A, Tahir T, Furniss C, Puigserver A, et al. The inhibition specificity of recombinant *Penicillium funiculosum* xylanase B towards wheat proteinaceous inhibitors. Biochimica et Biophysica Acta. 1701;**2004**:121-128

[76] Guais O, Borderies G, Pichereaux C, Maestracci M, Neugnot V, Rossignol M, et al. Proteomics analysis of "Rovabio™ excel", a secreted protein cocktail from the filamentous fungus *Penicillium funiculosum* grown under industrial process fermentation. Journal of Industrial Microbiology & Biotechnology. 2008;**35**:1659-1668

[77] Liu JR, Duan CH, Zhao X, Tzen JT, Cheng KJ, Pai CK. Cloning of a rumen fungal xylanase gene and purification of the recombinant enzyme via artificial oil bodies. Applied Microbiology and Biotechnology. 2008;**79**:225-233

[78] Lee CC, Wong DW, Robertson GH. Cloning and characterization of the xyn11A gene from *Lentinula edodes*. The Protein Journal. 2005;**24**:21-26

[79] Decelle B, Tsang A, Storms RK. Cloning, functional expression and characterization of three *Phanerochaete chrysosporium* endo-1,4-beta-xylanases. Current Genetics. 2004;**46**:166-175

[80] Couturier M, Haon M, Coutinho PM, Henrissat B, Lesage-Meessen L, Berrin JG. *Podospora anserina* hemicellulases potentiate the *Trichoderma reesei* secretome for saccharification of lignocellulosic biomass. Applied and Environmental Microbiology. 2011;7:237-246

[81] Zhao J, Guo C, Tian C, Ma Y. Heterologous expression and characterization of a GH3 beta-Glucosidase from thermophilic fungi *Myceliophthora thermophila* in *Pichia pastoris*. Applied Biochemistry and Biotechnology. 2015;**177**:511-527

[82] Badieyan S, Bevan DR, Zhang C. Study and design of stability in GH5 cellulases. Biotechnology and Bioengineering. 2012;**109**:31-44

[83] Pal A, Khanum F. Efficacy of xylanase purified from *Aspergillus niger* DFR-5 alone and in combination with pectinase and cellulase to improve yield and clarity of pineapple juice. Journal of Food Science and Technology. 2011;**48**: 560-568

[84] Fu G, Wang Y, Wang D, Zhou C. Cloning, expression, and characterization of an GHF 11 xylanase from *Aspergillus niger* XZ-3S. Indian Journal of Microbiology. 2012;**52**:682-688

[85] Khandeparkar R, Bhosle NB. Purification and characterization of thermoalkalophilic xylanase isolated from the *Enterobacter* sp. MTCC 5112. Research in Microbiology. 2006;**157**: 315-325

[86] Kuramochi K, Uchimura K, Kurata A, Kobayashi T, Hirose Y, Miura T, et al. High molecular weight, alkaline, and thermostable beta-1,4-xylanase of a subseafloor *Microcella alkaliphila*. Extremophiles. 2016;**20**: 471-478

[87] Mitra S, Mukhopadhyay BC, Mandal AR, Arukha AP, Chakrabarty K, Das GK, et al. Cloning, overexpression, and characterization of a novel alkali-thermostable xylanase from *Geobacillus* sp. WBI. The Journal of Basic Microbiology. 2015;**55**:527-537

[88] Ying Y, Meng D, Chen X, Li F. An extremely thermophilic anaerobic bacterium *Caldicellulosiruptor* sp. F32 exhibits distinctive properties in growth and xylanases during xylan hydrolysis.

Enzyme and Microbial Technology. 2013;**53**:194-199

[89] Taibi Z, Saoudi B, Boudelaa M, Trigui H, Belghith H, Gargouri A, et al. Purification and biochemical characterization of a highly thermostable xylanase from Actinomadura sp. strain Cpt20 isolated from poultry compost. Applied Biochemistry and Biotechnology. 2012;**166**:663-679

[90] Li D, Li X, Dang W, Tran PL, Park SH, Oh BC, et al. Characterization and application of an acidophilic and thermostable beta-glucosidase from *Thermofilum pendens*. Journal of Bioscience and Bioengineering. 2013; **115**:490-496

[91] Maurelli L, Giovane A, Esposito A, Moracci M, Fiume I, Rossi M, et al. Evidence that the xylanase activity from *Sulfolobus solfataricus* O-alpha is encoded by the endoglucanase precursor gene (sso1354) and characterization of the associated cellulase activity. Extremophiles. 2008;**12**:689-700

[92] Uhl AM, Daniel RM. The first description of an archaeal hemicellulase: The xylanase from *Thermococcus zilligii* strain AN1. Extremophiles. 1999;**3**: 263-267

[93] Driss D, Driss Z, Chaari F, Chaabouni SE. Immobilization of His-tagged recombinant xylanase from *Penicillium occitanis* on nickel-chelate Eupergit C for increasing digestibility of poultry feed. Bioengineered. 2014;**5**: 274-279

[94] Sinma K, Khucharoenphaisan K, Kitpreechavanich V, Tokuyama S. Purification and characterization of a thermostable xylanase from *Saccharopolyspora pathumthaniensis* S582 isolated from the gut of a termite. Bioscience, Biotechnology, and Biochemistry. 2011;**75**:1957-1963

[95] Fan G, Katrolia P, Jia H, Yang S, Yan Q, Jiang Z. High-level expression of a xylanase gene from the thermophilic fungus *Paecilomyces thermophila* in *Pichia pastoris*. Biotechnology Letters. 2012;**34**(11):2043-2048

[96] Luo H, Yang J, Li J, Shi P, Huang H, Bai Y, et al. Molecular cloning and characterization of the novel acidic xylanase XYLD from *Bispora* sp. MEY-1 that is homologous to family 30 glycosyl hydrolases. Applied Microbiology and Biotechnology. 2010;**86**:1829-1839

[97] Fan G, Yang S, Yan Q, Guo Y, Li Y, Jiang Z. Characterization of a highly thermostable glycoside hydrolase family 10 xylanase from *Malbranchea cinnamomea*. International Journal of Biological Macromolecules. 2014;**70**: 482-489

[98] Wang J, Mai G, Liu G, Yu S. Molecular cloning and heterologous expression of an acid-stable endoxylanase gene from *Penicillium oxalicum* in *Trichoderma reesei*. Journal of Microbiology and Biotechnology. 2013;**23**:251-259

[99] Beg QK, Kapoor M, Mahajan L, Hoondal GS. Microbial xylanases and their industrial applications: A review. Applied Microbiology and Biotechnology. 2001b;**56**:326-338

[100] Lynd LR, Laser MS, Bransby D, Dale BE, Davison B, Hamilton R, et al. How biotech can transform biofuels. Nature Biotechnology. 2008;**26**:169-172

[101] Hu J, Arantes V, Pribowo A, Saddler JN. The synergistic action of accessory enzymes enhances the hydrolytic potential of a “cellulase mixture” but is highly substrate specific. Biotechnology for Biofuels. 2013;**6**:112

[102] Harris PV, Xu F, Kreel NE, Kang C, Fukuyama S. New enzyme insights drive advances in commercial ethanol production. Current Opinion in Chemical Biology. 2014;**19**:162-170

Chapter 17

Chemical Modification of Xylan

Djamila Kerrouche, Nabila Rouba and Vincent Chaleix

Abstract

Our study is part of the general context of valuing by-products from the wood industry, which consists of the chemical modification of xylan by synthesis of branched copolymers such as xylan-g-PLLA. The used xylan is extracted from chestnut and 4-dimethylaminopyridine (DMAP) is the catalyst. In fact, the synthesis of xylan-graft-poly (L-lactide) copolymer starting from natural and renewable resource products xylan and L-lactide is performed under different conditions. The results of the grafting reaction are unfavorable due to longer time because of depolymerization reactions. Another result is the solubility and insolubility of the copolymers synthesized in water. This first result indicates that there is a change in the properties of xylan. Moreover, the solubility of the xylan-*g*-PLLA copolymers is different from one study to another. Grafting of PLLA onto xylan was confirmed by Fourier transform infrared (FT-IR) and ^{1}HNMR analyzes. The dynamic mechanical analysis showed that the xylan-g-PLLA plastic materials have interesting thermomechanical properties.

Keywords: extraction, graft copolymer, poly (L-lactide), valorization, xylans

1. Introduction

Xylan is a natural biodegradable polymer, a major component of hemicellulose which is one of the most abundant polysaccharides in nature after cellulose and starch. Hemicellulose can be extracted by alkaline solution from plants, including agricultural and forestry products [1, 2]. Xylan is a major component of the hemicellulose of straw, grasses and agricultural residues. It has a skeleton consisting of β (1 → 4) xylopyranose units, and generally has individual arabinose units attached to some of the C3 positions of the xylan skeleton as the main substituent, and small amounts of 4-O-methylglucuronic acid residues essentially linked to C2 position [3]. The solubility of xylan is favored by the presence of charged substituents such as uronic acids, which increase the hydrophily of the polymer and the intermolecular electrostatic repulsions. Its original physico-chemical properties bring it closer to hydrocolloids. It is used in particular in the food, cosmetic and pharmaceutical sectors, as a thickener, emulsifier or gelling agent. In addition, its binding properties have been exploited as additives in the preparation of paper pulps. They provide better flexibility to fibers and improve the mechanical resistance of paper. In recent years, increasing attention has been paid to poly (lactic acid) due to increasing environmental concerns and the decrease in fossil resources [4–7]. In addition, this polymer comes from renewable plant resources and has excellent properties such as biodegradability, mechanical strength, transparency and biocompatibility. It has considerable market potential in the fields of packaging, agriculture and biomedical and it is considered as a substitute for polymers of

IntechOpen

non-degradable basic products [8, 9]. In order to improve the specific, thermal and mechanical properties of xylan, modification is one of the ways that best responds to this limitation of properties and that can enhance it. For example, acetylation of xylan increases its hydrophobicity and thermal stability [10, 11]. Indeed, the grafting of poly (L-lactide) is one of the modifications which gives a product derived entirely from nature. The concept of grafting PLLA on polysaccharides is already studied. Recently, a study has shown that PLLA-g-hemicellulose is a good accounting for a mixture between wood hydrolysates and PLA [12]. The blocks or grafted copolymers such as cellulose-graft-PLA, xylans-grafts-PLA, PLA-grafted starch copolymers, having hydrophobic and hydrophilic segments, have been reported to form different types of microstructures and have been applied as biomaterials [13]. Several researchers have studied cellulose graft copolymers such as poly (lactide) grafted cellulose or poly (ε-caprolactone) grafted cellulose, as biodegradable plastics, and have indicated that chains of poly (lactide) or poly (ε-caprolactone) acted as internal plasticizers for the polysaccharide and it was found that the properties of the xylan-g-PLA copolymers strongly depend on the length of the PLA chains grafted on the xylan [14–17]. In fact, the grafting reaction depends on the method used according to the case of the modification of the xylan or of the PLLA [18] and also of several parameters such as: temperature, reaction time, type and quantity of catalysts, etc. Trimethylamine is the catalyst used for grafting PLLA on chitosan with a yield of less than 50%. Other catalysts are used for this type of reaction such as DMAP which can open the lactide cycle [12], carbine [19]. In a study [20], the catalyst used is triazobicyclodecene (TBD) at low temperatures and for short periods of time. This part is concerned with the structural analysis of xylans extracted from chestnut sawdust as well as their transformation into plastic films. After extraction and purification, the structure of these polysaccharides was characterized by IR and NMR. In a second part, the xylans were modified by the grafting of the PLLAs. Finally, a study of the physical and thermomechanical properties of the synthesized copolymers is carried out [21].

2. Industrial upgrading of xylans

Cellulose and starch are the main plant-based polysaccharides for industrial use. Hemicelluloses in general, xylans in particular, although they currently represent a modest volume of exploitation, are nevertheless receiving increasing attention [22]. Their original physico-chemical properties bring them closer to hydrocolloids. The latter denote polysaccharides of natural origin or their derivatives, which dissolve or disperse in water to form viscous solutions or suspensions. Hydrocolloids, which have a great affinity for water, affect the texture of the medium to which they are added and modify the consumer's perception of a product, hence their use in particular in the food, cosmetic and pharmaceutical sectors, as a thickener, emulsifier or gelling agent [23]. These polymers have in solution a low viscosity under stress and a high viscosity at rest at high concentration. These characteristics are those sought after for commercial naturally occurring thickeners such as xanthan gum and locust bean gum. In addition, their binding properties have been exploited as additives in the preparation of paper pulps. They provide better flexibility to the fibers and improve the mechanical strength of the paper [24]. Finally, their nutritional properties are not negligible. Xylans are in fact used as dietary fibers. They are not degraded by human digestive enzymes and thus accelerate intestinal transit. In addition, their ingestion would significantly decrease the accumulation of lipids in the liver and the blood cholesterol level [25]. Xylans can therefore be used in the native state, on the basis of their physicochemical properties, but the main ways of upgrading these polysaccharides are based either on their hydrolysis to form precursors used in the chemical industry, or on their functionalization which allows to consider new applications.

2.1 A new way of chemical upgrading of plastic xylans of plant origin

The petrochemical plants of the world consume 270 million tonnes of oil and gas every year in the production of plastics [26]. Fossil fuels provide the energy and raw material needed to transform crude oil into materials such as polystyrene, polyethylene or polypropylene. The progressive scarcity of these organic materials will inevitably be accompanied by a significant increase in their cost. Biotechnologies and organic chemistry can nevertheless provide solutions and literature offers numerous references on this subject [27, 28]. Biotechnologies currently seem to favor three approaches to replace plastics with products derived from plants: the direct production of plastics by micro-organisms, by cultivated plants or the transformation of sugars. It was in 1977 that the American companies Cargill and Dow Chemical joined their efforts to produce, after fermentation sugars of plant origin into lactic acid and polymerization of the latter, a plastic called polylactic acid [26] or PLA. A few years later, Imperial Chemical Industries marketed another plastic obtained after fermentation of sugars of plant origin [29], Biopol, which is a copolymer of the family of poly-3-hydroxyalkanoates (PHA). This bioplastic is however significantly more expensive than its synthetic counterparts derived from fossil fuels. Its only advantage is its biodegradability. Faced with high production costs, scientists have directed their research towards the direct synthesis of plastics by plants. The objective here is to modify the genetic heritage of cultivated plants in order to make them synthesize plastic directly. However, these works face a series of problems linked to: the physiology of the plant: the chloroplasts of the leaves which are the seat of photosynthesis seem to be a privileged place for the production of plastics by the plant. Too much synthesis at this level lowers the yields of photosynthesis and therefore the amount of produced plastic; to the methods of extraction and purification of plastics from the plant: These methods require the use of huge amounts of solvent; to public opinion: the dissemination of genetically modified organisms (GMOs) in our environment is currently causing sometimes violent controversies. The recent statement of the precautionary principle and the tightening of the regulations linked to this type of manipulation undoubtedly constitute a serious obstacle to the development of such technologies since they impose strictly controlled cultivation conditions incompatible with large-scale production. Organic synthesis provides different responses. One strategy is to chemically modify plant polymers. Remember that the first synthetic polymers were obtained by chemical modification of cellulose, such as nitrocellulose or cellulose acetate, used among others as thermoplastics. The esterification of the hydroxyl groups of the cellulosic fibers by aliphatic chains profoundly modifies their properties, and in particular the thermoplasticity and the hydrophobic character, but also the biodegradability, the solubility, the inflammability, etc. The properties of the material obtained then depend on the length of the grafted chain, as well as on the degree of substitution of the esterified polymer or DS;ie, the number of esterified hydroxyl functions per anhydroglucose unit in the case of cellulose. Cellulosic esters with a short carbon chain (less than six carbon atoms) currently represent an important industrial market, used and marketed in fields as varied as textile fibers, films, film substrates and membranes, coatings and varnishes, thermoplastic materials or composite materials. Cellulose esters, such as cellulose acetate (CA) or mixed esters such as cellulose acetate propionate (CAP) and acetate butyrate (CAB), all made from highly purified microcrystalline cellulose, compete with plastics derived from the petrochemical industry such as polyethylene, polypropylene, polyethylene terephthalate, polycarbonates, nylons, etc., for which the basic products, ethylene, propylene, xylene or even ethylene glycol are still very economically attractive. In addition, obtaining cellulose esters requires the prior solubilization of the cellulose, which generates the implementation of a heavy and costly methodology, unlike the synthetic polymers derived from polycondensations. Their properties remain original, however, and they continue to satisfy certain markets. The thermomechanical qualities of cellulose esters remain their main handicap. Indeed, due to the fragility of the polysaccharide chains at high temperatures, the glass transition temperature (Tg) and the material decomposition temperature are often very close. The glass transition temperature corresponds to the transition from a glassy state, in which the polymer is hard and brittle, to a rubbery state for which the polymer is soft and flexible. From this temperature, the material is more flexible and easier to work. This is the reason why a plasticizer is often added in order to lower the glass transition temperature and therefore widen the field of application of thermoplastics. Several plasticizers are commonly used, including on an industrial scale. This is the case for triethylcitrate for cellulose acetate and dioctyladipate for CAP. However, the use of plasticizer can sometimes be inconvenient for certain applications. These small molecules tend to evaporate over time, which can change the performance of the material in the short or medium term. Another solution for lowering the Tg of a polymer is to graft long chain substituents, so as to increase the free volume and therefore reduce the interactions between the polymer chains. The presence of bulky and flexible side substituents such as fatty chains, by removing the polysaccharide chains, lowers their Tg and therefore influences the field of use of these thermoplastics. Studies have shown on a series of cellulose esters ranging from acetate to palmitate (C16), that Tg decreases significantly with the increase in the length of the grafted fatty chains [30]. Thus, without adding plasticizer, it is possible to obtain hydrophobic plastics that are more flexible, more deformable, less brittle, which suggests new areas of application, such as packaging or plasticulture and, in particular, agriculture mulch films [31, 32]. All of these works, initiated from a model substrate, cellulose, makes it possible to envisage new ways of upgrading for polysaccharides, and in particular for wood xylans. The objective of this study is the synthesis of new plastic materials by grafting poly (lactic acid) on xylan (4-O-methylglucuronoxylanes) extracted from chestnut sawdust.

2.2 Xylan extraction from wood

The main problem facing the experimenter is that of the chemical richness of the plant cell wall, which is reflected in the great structural diversity of the macromolecules, essentially polysaccharides, which are represented there. The protocols to be used must therefore be sufficiently selective to allow the extraction of a category of macromolecules; they must also be concerned with protecting the integrity of molecular structures by limiting the degradation of the latter. A detailed study of literature in the field of hemicellulose extraction reveals the existence of a large number of protocols. This observation can only be explained by the wide variety of hemicellulosic structures identified within the plant cell wall. The development of a specific extraction protocol for xylans with a view to their commercial exploitation is the subject of numerous studies. The question to be answered by the experimenter is what type of xylans to extract, from which plant material, for what properties, for what applications and using which protocols. The presence of a polyphenolic frame formed by lignin, as well as the existence of chemical bonds between the different hemicelluloses which constitute the wall, and between the hemicelluloses and lignin limits the extraction of xylans. It is

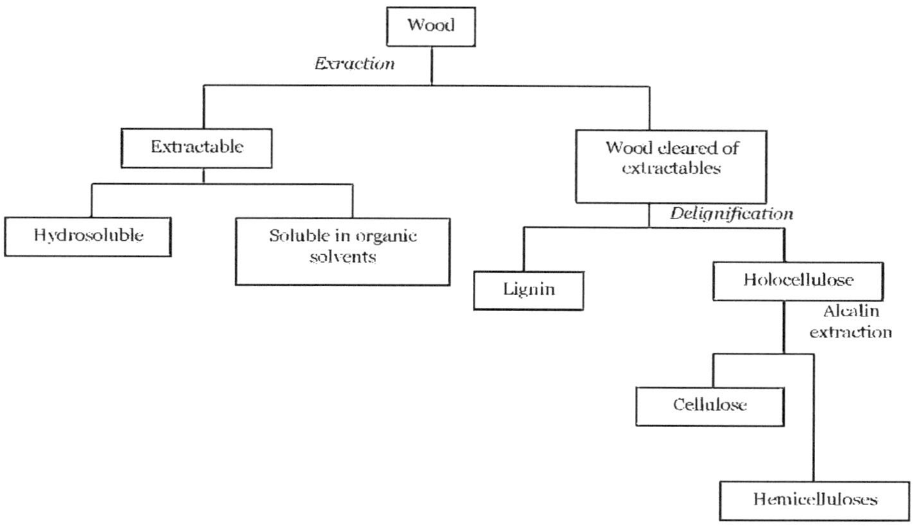

Figure 1.
Separation of wood constituents.

then impossible to extract a type of polysaccharide without breaking these bonds and therefore without modifying the polymer with respect to its state in situ. It is therefore necessary to apply sufficiently strong extraction conditions to allow the rupture of these bonds, without degrading the extracted molecules. In the case of extraction from wood, the xylans are conventionally collected, after delignification in the presence of sodium chlorite, by an alkaline extraction. Beforehand, it is necessary to eliminate the extractables, so that they do not interfere in the analyses later. A schematic representation of the separation of the constituents of wood is illustrated in **Figure 1**.

3. Preparation of the copolymers

The DMAc/LiCl solution is prepared for 1 hour at a temperature of 80° C with mechanical stirring. In a flask equipped with a condenser, we mixed a quantity of the solution of DMAc/LiCl, L-lactide (2, 4 or 8 equiv./ OH) and DMAP catalyst (0.1, 0.5 or 1 equi/OH) then heated it to 80° C with magnetic stirring at different reaction times (8, 16, 24 h). At the end of the reaction, a quantity of ethanol is added in order to precipitate the xylan-g-PLLA. The product obtained is purified with the solvent dimethyl sulfoxide (DMSO) and precipitated in ethanol. The obtained copolymer is dried at room temperature and is characterized by different methods.

3.1 Xylan extraction

Chestnut wood glucuronoxylans were extracted by this procedure (**Figure 2**).

They were obtained from chlorite-delignified sawdust by aqueous KOH extraction. The MGX has been extracted with a yield of 19%. The obtained fraction is characteristic of a classical MGX with more than 99% of xylose and 4-*O*-methylglucuronic acid (typical markers of glucuronoxylans). 4-*O*-methylglucuronic acid contents were 16.6% (molar ratio), which is in agreement with the data obtained by the colorimetric assays (17.4, mass ratio) and the

Xyl/4-*O*-MeGlcA ratio was found: 6.1. However, concerning the size of the extracted molecules, a DP of about 200 was found which is a typical result for this kind of products. This MGX is represented in **Figure 3**.

3.2 RMN

The result of ^1H *NMR spectra of xylan* (D_2O) *is presented in* **Table 1** and **Figure 4**.

3.3 Infrared analysis

The IR spectra of xylans (**Figure 5**), and more particularly of 4-Omethylglucuronoxylans, have been studied in detail by Marchessault and Liang (1962) (see [33]) and more recently by Kacurakovaet*coll*. (2000) (see [34]). The absorption bands appearing in the figure and listed in the table are characteristic of xylans of type (1 → 4), with a maximum absorption band at 1044 cm^{-1}, corresponding to the vibrations of elongation of the ring bonds and C-OH bonds. The weak, but very clear band at 896 cm^{-1} is characteristic of glycosidic type bonds [35].

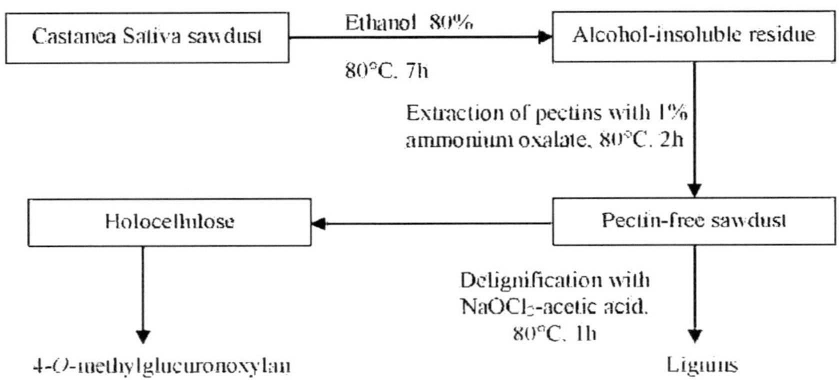

Figure 2.
Scheme of fractionalizationof 4-O-methylglucuronoxylans (MGX) from chestnut wood.

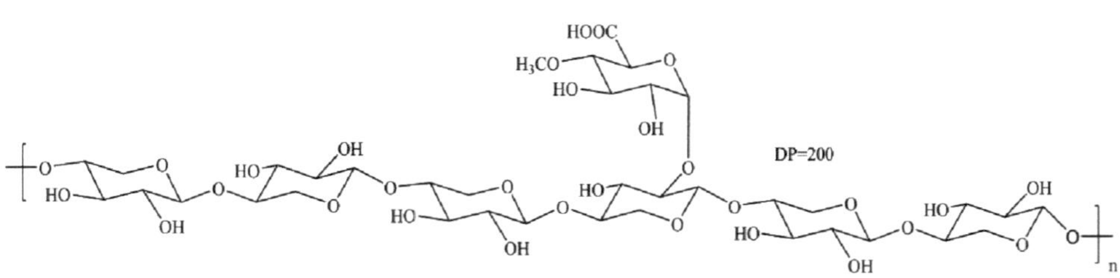

Figure 3.
Structure of 4-O-methylglucuronoxylan extracted from Castanea sativa.

Glycoside residues		Chemical shifts in ppm (J en Hz)				
		1	2	3	4	5
(1 → 4)- D-Xylp	^1H	4,48 (7,5)	3,29 (8,2)	3,55 (9,0)	3,79	4,10 (4,5; 11,5); 3,38 (11,0)

Table 1.
Chemical shifts of glycosidic residues of xylan (D_2O).

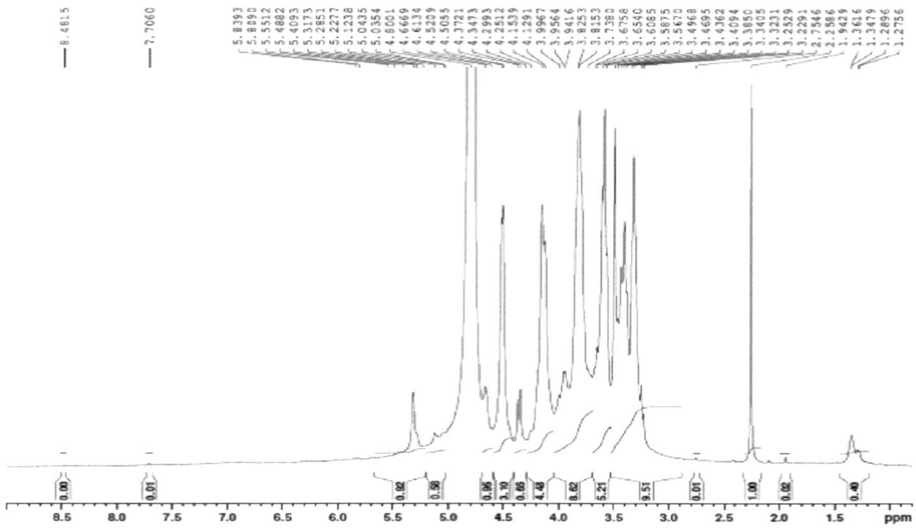

Figure 4.
^1H NMR spectra of xylan (D_2O).

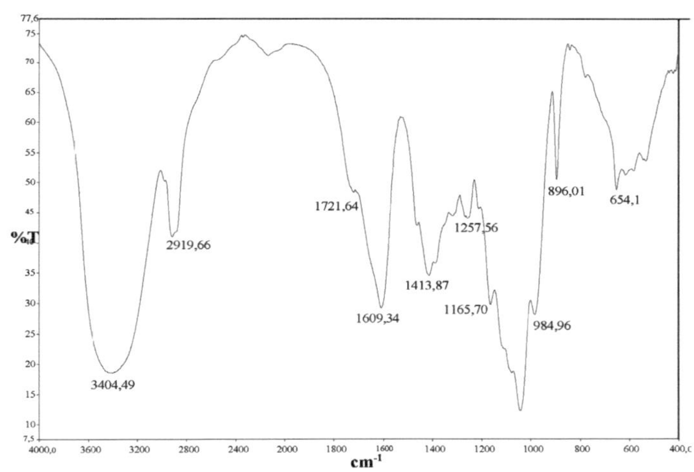

Figure 5.
FT-IR spectra of xylan.

4. The results

The synthesis of the biodegradable xylan-g-PLLA copolymers is carried out at atmospheric pressure by opening the L-lactide ring and the DMAc/LiClxylan solution (0.2 g/l) in the presence of DMAP as catalyst. The degree of substitution and the degree of polymerization express the efficiency of the grafting reaction. They are calculated from ^1H NMR spectra. The medium DS is defined as the medium number of grafted PLLA chains per repeat unit which is calculated by comparing the integration of the 3 protons of the terminal CH_3 with the equatorial proton H_5 of xylose. The medium DP, meanwhile, is calculated by comparing the integration of internal CH_3s with terminal CH_3s. The yield is defined as the ratio between the weight of plastic film and the initial weight of polysaccharide.

It should be noted that the DS and the DP show that the latter both increase with the amount of L-lactide and the DMAP and decrease beyond the 16 h time.

In addition, it has been proven that DMAP is used as a hyper nucleophilia catalyst which is 104 times more active than pyridine [10], this is more favorable when the reaction time is longer, favoring depolymerization reactions. The maximum conditions for having a larger DS and DP are: 16 h, 8 eq/OH, 1 eq/OH.

4.1 Characterization of copolymers

One of the properties verified after the grafting reaction is the solubility of the copolymers and some have been found to be soluble in water and others not; which confirms the modification of the properties of xylan. A copolymer is dissolved in water in a petri dish after evaporation of the solvent at 40° C in an oven, a casting plastic film was obtained (**Figure 6**).

The IR analysis of xylan, PLLA and xylan-g-PLLA is presented in the **Figure** 7. All the xylan absorption bands of type (1 → 4) are presented in the **Table 2**. An absorption band of the groups of hydroxyl is observed at 3544 cm^{-1}, another absorption band at 897 cm^{-1} of glycosidic C-H (β). [36] Other bands at 1096 cm^{-1} and 1044 cm^{-1} of the C-O, C-C and C-OH groups of the ring. A strong absorption band at 1784 cm^{-1} is observed on the chemically modified xylan characteristic of ester functions (C = O) [36, 37]. Other bands that characterize CH and CH_3 are seen at 2984 and 2934 cm^{-1}. All of these absorption bands confirm the modification of the xylan structure.

The ^{1}H NMR spectrum (**Figure 8**) shows intense signals corresponding to the protons of the xylose units of the unsubstituted main chain, as well as less intense signals attributed to the uronic acid units and to the xylose units which carry them in position 2. The intensity of these signals depends on the degree of substitution by uronic acid. The doublets attributed to the anomeric protons of the xylose units, substituted (4.5 ppm) or not (4.6 ppm), are associated with a coupling constant of about 7 Hz, characteristic of an osidic bond. The coupling constant of the doublet corresponding to the anomeric protons of uronic acid, at 5.3 ppm, is in the order of 2 Hz, which characterizes an osidic bond. It is also possible to observe on the different ^{1}H spectra of xylans, a fine singlet around 3.4 ppm, which is to be associated with the existence of methyl groups carried, given the integration, by the uronic acid. This last remark confirms the presence of 4-O-methylglucuronic acid. This is fixed in position 2 of the xylose, which

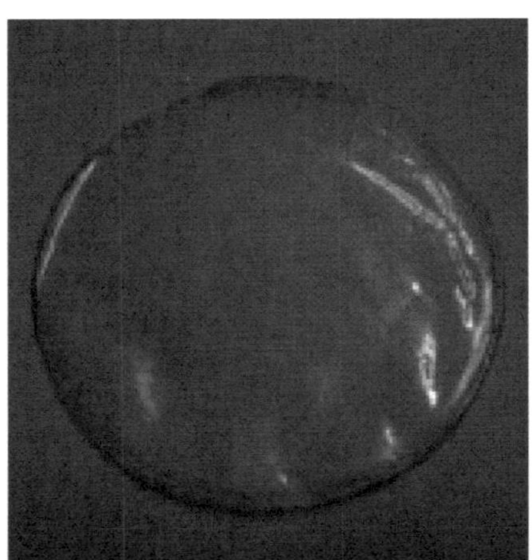

Figure 6.
Photograph of a xylan-g-PLLA film. Note: Xylan-g-PLLA, xylan-graft-poly(L-lactide).

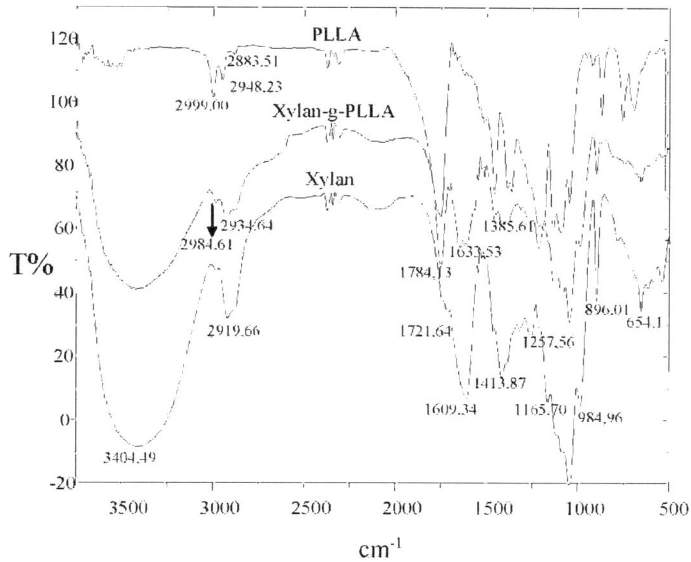

Figure 7.
Fourier transform infrared spectra of xylan, PLLA, xylan-graft- poly (L-lactide) copolymer.

Absorptions (cm^{-1})	Relative Intensity	Type of vibration
3404	w	ν(O-H)
2919	m	ν(C-H)
1721	m	ν(C=O) uronic acid
1609	e	ν(C=O)acid salt, anti-symmetric vibration
1465	f	δ(CH_2) symmetric
1413	f	ν(C=O) acid salt, symmetric vibration
1257	f	δ(CH_2)
1165–1149	m	ν(C-O-C), ν(C-C)
1131–1121	e	ν(C-O-C), ν(C-C)
1096	s	ν(C-O), cycle, ν(C-C)
1044	s	ν(C-O), ν(C-C)
984	m	ν(C-O), δ(OH), cycle vibrations
896	m	δ(CH) glycosidic (β), cycle vibrations

w: wide, m: medium, f: weak, s: strong, e: shoulder; ν: valence vibrations; δ: deformation vibrations.

Table 2.
Absorption bands characteristic of glucuronoxylans.

results in a greater deshielding of the H2 compared to the unsubstituted xylose units. We can distinguish on the ^{1}H NMR spectrum of the modified xylan, signals between 1 and 2.5 ppm of the protons of the aliphatic chains shows a signal at 5.20 ppm which corresponds to the C<u>H</u> group, another signal at 1.47 ppm of the proton of the internal methyl group $C\underline{H}_3$ and at 1.27 ppm for the proton of $C\underline{H}_3$ at the end of the chains of PLLA (Guang-Xin, 2006). These intense signals corresponding to the protons of the xylose units of the main chain and to the protons of the PLLA groups confirm the chemical modification of xylan.

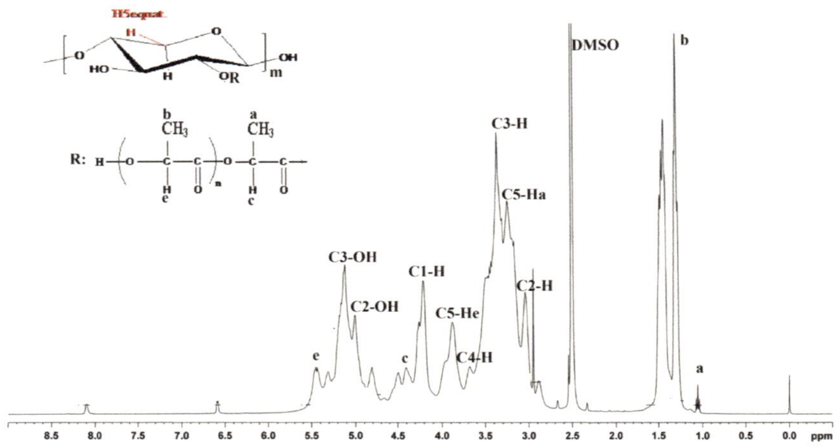

Figure 8.
^{1}H RMN spectra of xylan-graft-poly (L-lactide) copolymer (DMSO).

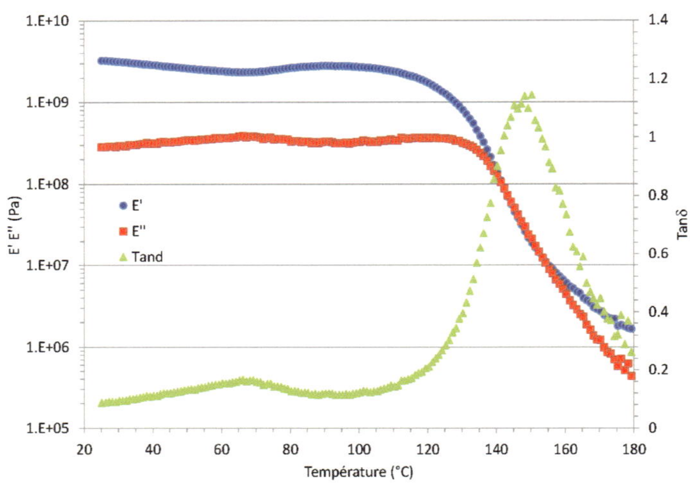

Figure 9.
Evaluation of the tangent of the loss angle and the loss modulus E" of xylan-g-PLLA as a function of temperature.

For medium DP of fixed PLLA, it did not vary a lot (about 2 or 3). It confirms that the chains are very short but long enough to enhance the filmogenic character of materials [21].

4.2 Mechanical properties

The mechanical properties of PLA have been studied by various researchers [38, 39]. PLA has mechanical properties similar to those of polystyrene [40]. Dynamic mechanical analysis (**Figure 9**) showed that the glass transition temperature of the xylan-g-PLLA film is 147°C. For the new PLLA-g-xylan material, the results of mechanical tests (**Figure 10**) show that the film has a nominal strain $\varepsilon_{rupture} = 8 \pm 2.5\%$, its Young's modulus E = 1.2 ± 0.1 GPa (1200Mpa) and a nominal stress of about 60 MPa. These mechanical properties are close to those of polypropylene [41].

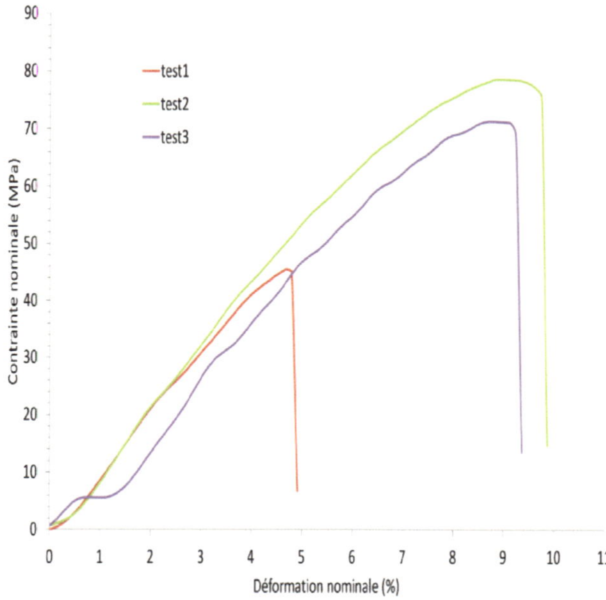

Figure 10.
Elongation at breakaccording to thenominalstrain (5%).

5. Conclusion

The modification of xylan by grafting the PLLA on the xylan extracted from the sawdust of chestnut was carried out in this work. The synthesis of branched xylan-g-PLA co-polymers is carried out from L-lactide and xylan using DMA as solvent for the xylan and 4-dimethylaminopyridine (DMAP) as catalyst at a temperature of 80° C. These copolymers are characterized and infrared analysis has shown the appearance of the characteristic band of the esters at 1784 cm^{-1} and of other bands of the group CH_3 on the xylan-g-PLLA spectrum. Following the ^1H NMR analysis, the appearance of the PLLA aliphatic protons was observed on the spectrum of the grafted xylan. The xylan copolymers are insoluble in water. The higher the DS, the more difficult the solubility is in water. Dynamic mechanical analysis has shown that Tg of the xylan-g-PLLA film is 147°C.

Author details

Djamila Kerrouche[1*], Nabila Rouba[1] and Vincent Chaleix[2]

1 Laboratory of Organic Materials, Department of Engineering Processing, Faculty of Technology, Abderrahmane Mira University, Bejaia, Algeria

2 Laboratory of Chemistry of Natural Substances, University of Limoges, Limoges, France

*Address all correspondence to: kerrouchedjamila@yahoo.fr

References

[1] Saake B, Erasmy N, Kruse T, Schmekal E, Puls J. Isolation and characterization of arabinoxylan from oat spelts. Hemicelluloses: Science and Technology. 2004;864:52-65

[2] Yukiko E.R, Tadahisa I. Synthesis of xylan-graft-poly(l-lactide) copolymers via click chemistry and their thermal properties. Carbohydrate Polymers. 2012;87:1933-1940

[3] Puls J, Schroder N, Stein A, Janzon R, Saake B. Xylans from oat spelts and birch kraft pulp. Macromolecular Symposia. 2006;232:85-92

[4] Rezgui F, Swistek M, Hiver J.M, G'Sell C, Sadoun T. Deformation and damage upon stretching of degradable polymers (PLA and PCL).Polymer. 2005; 46: 7370-7385

[5] Bogaert J.C and Coszach P. Poly(lactic acids): a potential solution to plastic waste dilemma.Macromolecular Symposia. 2000;153: 287-303

[6] Garlotta D. A literature review of poly (lactic acid). Journal of Polymers and the Environment. 2001; 9:63-84

[7] Yu L, Dean K, Li L. Polymer Blends and Composites from Renewable Resources. Progress in Polymer Science. 2006;31:576-802

[8] Sadoun T, Rouba N, Boutagrabet N, KerroucheDj, Zadi S, Mimi N. Thermo-oxidation and biodegradation study of low-density polyethylene/starch films by IR spectroscopy. Iranian Journal of Chemistry and Chemical Engineering. 2015; 34:4

[9] Auras R, Harte B, Selke S.An Overview of Polylactides as Packaging Materials.Macromolecular Bioscience. 2004; 4:835-864

[10] Fang J.M, Sun R.C, Tomkinson J, Fowler P. Acetylation of wheat straw hemicellulose b in a new non-aqueous swelling system. Carbohydrate Polymers. 1999;41(4): 379-387

[11] Fundador N.G.V, Enomoto-Rogers Y, Takemura A, Iwata I. Acetylation and characterization of xylan from hard wood kraft pulp. Carbohydrate Polymers. 2012;87(1):170-176

[12] Saadatmand S, Edlund U, Albertsson A.C. Compatibilizers of a purposely designed greft copolymer for hydrolysate/PLLA blends. Polymer. 2011;52(21):4648-4655

[13] Ouchi T, Miyazaki H, Arimura H, Tasaka F, Hamada A, Ohya Y. Synthesis of biodegradable amphiphilic AB-type diblock copolymers of lactide and depsipeptide with pendant reactive groups. Journal of Polymer Science, Part A: Polymer Chemistry. 2002;40:1218

[14] Teramoto Y, Nishio Y, Cellulose diacetate-graft-poly (lactic acid) s: synthesis of wide-ranging compositions and their thermal and mechanical properties. Polymer.2003;44(9):2701-2709

[15] Teramoto Y, Yoshioka M, Shiraishi N, Nishio Y. Plasticization of cellulose diacetate by graft copolymerization of epsilon-caprolactone and lactic acid. Journal of Applied Polymer Science.2002;84(14):2621-2628

[16] Vlcek P, Janata M, Latalova P, Dybal J, Spirkova M, Toman L. Bottlebrush-shaped copolymers with cellulose diacetate backbone by a combination of ring opening polymerization and Atrp. Journal of Polymer Science, Part A: Polymer Chemistry.2008; 46(2):564-573

[17] Yan C.H, Zhang J.M, Lv Y.X, Yu J, Wu J, Zhang J. Thermoplastic cellulose-graft-poly (l-lactide) copolymers

homogeneously synthesized in an ionic liquid with 4-dimethylaminopyridine catalyst. Biomacromolecules.2009;10 (8):2013-2018

[18] Yukiko E.R, Tadahisa I. Synthesis of xylan-graft-poly(l-lactide) copolymers via click chemistry and their thermal properties. Carbohydrate Polymers. 2012;87: 1933-1940

[19] Nyce G.W, Lamboy J.A, Connor E.F, Waymouth R.M, Hedrick J.L. Expanding the catalytic activity of nucleophilic n-heterocyclic carbenes for transesterification reaction. Organic Letters.2002;4(21):3587-3590

[20] Persson J, Dahlman O, Albertsson A-C. Birch xylan grefted with PLA branches of predictable length. BioRessources.2012;7 (3):3640-3655

[21] Djamila Kerrouche, Tahar Sadoun, Nabila Rouba, G. Sotclet, Vincent Sol, Vincent Gloaguen & Vincent Chaleix 'Synthesis and characterization of xylan-graft-poly(Llactide), International Journal of Polymer Analysis and Characterization. 2018; 23:3, 193-206.

[22] Kačuráková M, Wellner N, EbringerovaA,HromidkovaZ,WilsonR.H, and Belton P. S. Characterization of xylan-type polysaccharides and associated cell wall components by FT-IR and FT-Raman spectroscopies. Food Hydrocolloids. 1999;13(1):35-41

[23] Whistler K.B, Shah R.N. Modified Cellulosics. Academic Press, New York.1978; 341-356

[24] El-Ashmawy A.E, Mobarak F, El-Kaliobi S. Hemicelluloses as additives in papermaking. III. Effect of hemicelluloses isolated from plant raw materials compared to the effect of those isolated from pulps on paper properties. Cellulose Chemistry and Technology. 1976;10(5): 637-640

[25] Chanliaud E. Extraction, caractérisation et propriétés fonctionnelles des hétéroxylanes de son de maïs [thesis]. Universités Paris VII, Paris IX, et E.N.S.I.A; 1995:117

[26] Gerngross T.U and Slater S. How green are green plastics?, Scientific American. 2000;37, 5

[27] Paillet M, Peguy A. New biodegradable films from exploded wood solutions.Journal of Applied Polymer Science. 1990;40(3-4):427-433

[28] Gandini A, Belgacem M.N. Recent contributions to the preparation of polymers derived from renewable resources. Journal of Polymers and the Environment. 2002;10(3):105-114

[29] Kemmish D.In: Ching C, Kaplan D, Thomas E. (Eds). Biodegradable polymers and packaging. Technomic Publishing AG, Basel: Switzerland; 1993. 225

[30] Edgar K.J, Buchanan C.M, Debenham J.S, Rundquist P.A, Seiler B.D, Shelton M.C.,Tindall D. Advances in cellulose ester performance and application. Progress in Polymer Science. 2001; 26(9):1605-1688

[31] Joly N, Granet R, Krausz P, Crosslinking of cellulose by olefin metathesis.Journal of Carbohydrate Chemistry. 2003; 22(1):47-55

[32] Fredon E. Obtention de films plastiques hydrophobes à partir de polysaccharides contenus dans des co-produits agricoles, [thesis]. Université Paul Sabatier de Toulouse; 2001: 179

[33] Marchessault R.H, Liang C.Y. The infraredspectra of crystalline polysaccharides, VIII Xylans. Journal of Polymer Science. 1962; 59:357-378

[34] Kacurakova M, Capek P, Sasinkova V, Wellner N, Ebringerova A.

FT-IR study of plant cell wall model compounds: pectic polysaccharides and hemicelluloses. Carbohydrate Polymers. 2000;43(2):195-203

[35] Gupta S, Madan R.N, Bansal M.C. Chemical composition of Pinuscaribaea hemicelluloses. Tappi Journal. 1987; 70(8):113-114

[36] Wang N, X. S. Wu, H. Lujan-Upton, E. Donahue, and A. Siddiqui. Synthesis, characterization, biodegradation, and drug delivery application of biodegradable lactic/glycolic acid oligomers: I. Synthesis and characterization. Journal of Biomaterials Science, Polymer Edition. 1997;8(12):905-917

[37] Gorden S. H., X. Cao, A. Mohamed, and J. L. Willett. Infrared spectroscopy method reveals hydrogen bonding and intermolecular interaction between components in polymer blends. Journal of Applied Polymer Science. 2005; 97:813-821

[38] Donald G. A literature review of poly(lactic acid). Journal of Polymers and the Environment. 2001;9:63-84

[39] Gupta A. P, Kumar V. New emerging trends in synthetic biodegradable polymers-polylactide.European Polymer Journal. 2007;43:4053-4074

[40] Södegard A. Lactic acid based polymers for packaging materials for the food industry. The Food Biopack Conference, Copenhagen, Denmark, 2000;27-29

[41] Schindler A, Harper D. Polylactide. II. Viscosity-molecular weight relationships and unperturbed chain dimensions. Journal of Polymer Science Part A: Polymer Chemistry. 1979;17:2593-2599

Chapter 18

The Application of Solid State Fermentation for Obtaining Substances Useful in Healthcare

Łukasz Wajda and Magdalena Januszek

Abstract

In the current review we summarised the research involving solid state fermentation (SSF) for the production of compounds that could be used in healthcare (terpenoids, polyphenols, fibrinolytic enzymes, mycophenolic acid and others). We described several groups of obtained agents which hold various activity: antimicrobial, anti-inflammatory, immunosuppressive, anticoagulant and others (e.g. anticancer or anti-diabetic). It seems that especially terpenoids and polyphenols could be useful in that field, however, other substances such as enzymes and fatty acids play important role as well. We described main groups of microorganisms that are applied in SSF of those compounds, particularly *Bacillus* genus and fungi, and where possible provided information regarding genes involved in those processes. We also compared various approaches toward optimisation of SSF.

Keywords: solid state fermentation, healthcare, agricultural waste, *Bacillus*, fungi

1. Introduction

Solid state fermentation (SSF) is a process during which microorganisms (in the presence of small amounts of water) transform agro-industrial waste into valuable compounds [1]. Based on the literature research, wheat bran was commonly used for those processes (**Table 1**). It is composed of about 53% of dietary fibre (xylans, lignin, cellulose, and galactan, fructans) and contains variety of phenolic acids e.g. ferulic acid, vanillic acid, coumaric acid, caffeic acid, and chlorogenic acid [71]. Researchers also applied other materials rich in polysaccharides (e.g. rice, whole grain wheat, millet, barley) or simple sugars (e.g. fruit pomace) (**Table 1**). Other substrates which were utilised for SSF are not only sources of carbohydrates, but also protein e.g. soybeans, lentil flour, silkworm larvae, fish meal, cuttle fish waste and king oyster mushroom (**Table 1**). The selection of waste products used in SSF should ensure the proper balance of nutrients to allow microbial growth and production of terpenoids, polyphenols, enzymes, biosurfactants, short chain fatty acids or others. Therefore, industrial waste with a high content of carbohydrates, protein, pectin or lipids is a suitable substrate.

There were various review papers regarding SSF but in the current chapter we focused only on selected substances which could be used in healthcare and demonstrate antimicrobial, anti-inflammatory properties or/and are immunosuppressants, anticoagulants and anticancer agents, e.g. enzymes, surfactants, terpenoids,

Name of the substance	Microorganism	Agricultural waste	Reference
	Antimicrobial properties		
Nonactin, monactin, dynactin, trinactin	*Streptomyces cavourensis* TN638	Immobilised bacterial spores (XAD-16) on potato dextrose agar	[2]
Surfactin homologues	*Bacillus natto* NT-6	Potato dextrose medium	[3]
Biosurfactant	*Bacillus subtilis* SPB1	Millet	[4]
Surfactin	*Bacillus pumilus* UFPEDA 448	Okara and sugarcane bagasse	[5]
Biosurfactant	*Tremetes versicolor* TV-6	Two-phase olive mill waste, wheat bran and olive stone	[6]
Biosurfactant	*Aspergillus niger*	Wheat bran and corncob	[7]
Not specified	*Pediococcus acidilactici* KTU-05-7	Milk thistle seeds	[8]
Not specified	*Bacillus licheniformis*	Wheat bran, soybean meal, yeast, fish meal	[9]
Sambacide	*Fusarium sambucinum* B10.2	Potato	[10]
γ-Decalactone	*Yarrowia lipolytica* W29 (ATCC 20460)	Luffa sponge, cellulose sponge, corncob, castor seed	[11]
Phenolic acids	*Pleurotus sapidus*	Sunflower seed hulls, golden rice straw and husks	[12]
Curcumin	*Trichoderma* strains	Turmeric	[13]
Coumarins and oxylipins	*Aspergillus oryzae* KCCM 12698	Malt extract agar	[14]
Phenolic compounds	*Lentinus edodes*	Cranberry pomace	[15]
Phenolic compounds	*Trichoderma* strains	Commercial turmeric	
Phenolic compounds	*Trichoderma* strains	Ginger powder	[16]
Phenolic compounds	*Trichoderma reesei*	Garden cress seeds	[17]
Phenolic compounds	*Aspergillus oryzae* NCH 42	Chinese cucumber, Chinese sage, houpu magnolia, liquorice root	[18]
Phenolic compounds	*Bacillus clausii*	Spent coffee grounds (Arabica)	[19]
	Anti-inflammatory agents		
Phenolic compounds	*Trametes versicolor* TV-6	Grape pomace	[20]
Not specified	*Taiwanofungus camphoratus* obtained by SSF		[21]
Rutin	*Rhizopus oligosporus* NRRL 2710	Buckwheat groats	[22]
Betulinic acid	*Inonotus obliquus*	The spent substrate of king oyster mushroom, grain including corn, rice grain, white birch and mulberry powder	[23]

Name of the substance	Microorganism	Agricultural waste	Reference
Phenolic compounds	*Xylaria nigripes*	Wheat bran	[24]
limonene-1,2-diol, α-terpineol, (−)-carvone, α-tocopherol, dihydrocarveol and valencene	*Diaporthe sp. (Phomopsis sp.)*	Orange peel and bagasse	[25]
Phenolic compounds, lignans	*Pediococcus acidilactici* LUHS29	Grounded barley by-products	[26]
α-Pinene	*Saccharomyces cerevisiae* AXAZ-1 and *Kluyveromyces marxianus* IMB3	Mixed solid and liquid food industry wastes	[27]
δ-Octalactone γ-Undecalactone γ-Dodecalactone δ-Dodecalactone	*Trichoderma viride* EMCC-107	Dried and grounded sugarcane bagasse	[28]
cis-Linaloloxide, Phenanthrene	*Trichoderma viride Pers. ex Fr.; Aspergillus niger van Tieghem*	Pu-erh tea	[29]
Various terpenes	*Antrodia camphorata*	Millet	[30]
Neochlorogenic acid), chlorogenic acid, rutin, 6″acetyl-glucoside	*Aspergillus niger* ATCC-6275 *Rhizopus oligosporus* ATCC-22959	Stones and pomace from fully ripened apricot	[31]
Quercetin and phenolic acids: gallic, vanillic, p-hydroxybenzoic, ferulic	*Lactobacillus plantarum* CECT 748 ATCC 14917	Cowpeas	[32]
Phenolic compounds	*B. subtilis* BCRC 14715	Black soybeans	[33]
Daidzin, daidzein, genistin and genistein	*Eurotium cristatum* YL-1	Soybeans seeds	[34]
Gallic acid	*Rhizopus oryzae* (RO IIT RB-13, NRRL 21498) *Aspergillus foetidus* (GMRB013 MTCC 3557)	Powdered fruits of Myrobalan and Teri pod	[35]
4-hydroxybenzoic acid, 4-hydroxy-3-methoxybenzoic acid and unidentified compounds	*Rhizopus oryzae* RCK2012	Whole grain wheat	[36]
3,4-di-hydroxybenzoic acid, ferulic acid, vanillic acid, quercetin	*Aspergillus oryzae* LBA01, *A. niger* LBA02	Lentil flour	[37]
	Immunosupresants		
Mycophenolic acid	*Penicillium brevicompactum* DSM 2215	Rice bran-potato peel mixture	[38]
Mycophenolic acid	*Penicillium brevicompactum* ATCC 16024 (AFI 668)	Pearl barley	[39]
Mycophenolic acid	*Penicillium brevicompactum* ATCC 16024	Wheat bran	[40]
Mycophenolic acid	*Penicillium brevicompactum* MTCC 8010	Rice bran	[41]
Mycophenolic acid	*Penicillium brevicompactum* (various strains)	Various agricultural waste	[42]
Mycophenolic acid	*Penicillium roqueforti* (AG101 and LG109)	Sugarcane bagasse	[43]

Name of the substance	Microorganism	Agricultural waste	Reference
Cyclosporin A	*Tolypocladium inflatum* MTCC 557	Hydrolysed wheat bran flour and coconut oil cake	[44]
Cyclosporin A	*Tolypocladium inflatum* MTCC 557	Wheat bran flour and coconut oil cake	[45]
Cyclosporin A	*Tolypocladium inflatum* ATCC 34921	Wheat bran	[46]
Cyclosporin A	*Tolypocladium inflatum* DRCC 106 (mutated srain)	Wheat bran	[47]
Cyclosporin A	*Tolypocladium* sp.	Wheat bran	[48]
Tacrolimus	*Streptomyces hygroscopicus*	Various agricultural waste	[49]
	Anticoagulant agents		
Halotolerant chitinase	*Citrobacter freundii* str. nov. *haritD11*	Wheat bran with fish scale	[50]
Halotolerant Chitinase	*Citrobacter freundii* str. nov. *haritD11*	Wheat bran with shrimp shellfish	[51]
Fibrynolytic enzyme	*Bacillus amyloliquefaciens* LSSE-62	Chickpeas	[52]
Fibrinolytic enzyme	*Bacillus* sp. IND6	Wheat bran	[53]
Fibrinolytic enzyme	*Bacillus sp.* IND12	Cow dung	[54]
Nattokinase	*Bacillus subtilis* natto	Soybean	[55]
Fibrinolytic enzyme	*Bacillus subtilis* XZI125	Soybean meal	[56]
Fibrinolytic enzyme	*Bacillus subtilis* WR350	Corn steep	[57]
Fibrinolytic enzyme	*Bacillus halodurans* IND18	Wheat bran	[58]
Fibrinolytic enzyme	*Bacillus cereus* GD55	Apple pomace	[59]
Fibrinolytic enzyme	*Bacillus cereus* IND5	Cuttle fish waste and cow dung	[60]
Fibrinolytic enzyme	*Paenibacillus* sp. IND8	Wheat bran	[61]
Fibrinolytic enzyme	*Pseudoalteromonas* sp. IND11	Sun-dried cow dung	[62]
Fibrinolytic enzyme	*Xanthomonas oryzae* IND3	Cow dung	[63]
Fibrinolytic enzyme	*Bacillus firmus* NA-1	Soybean grits	[64]
Fibrinolytic enzyme	*Mucor subtillissimus* UCP 1262	Wheat bran	[65]
Fibrinolytic enzyme	*Fusarium oxysporum*	Rice chaff	[66]
Fibrinolytic enzyme	*Fusarium oxysporum*	Rice chaff	[67]
	Anticancer agents		
Short chain fatty acids	*Aspergillus kawachii* KCCM 32819	Silkworm larvae powder	[68]
Putative phytoestrogen	*Aspergillus fumigatus* F-993 or *A. awamori* FB-133	Defatted soybean	[69]
Andrastin A and C	*Penicillium expansum* KACC 40815	Malt extract agar	[70]

Table 1.
Examples of substances produced by solid state fermentation that could be used in healthcare.

polyphenols and short chain fatty acids. We also described main groups of microorganisms that were involved in cited studies and compared various approaches for optimising SSF.

2. Main properties of substances obtained from SSF

2.1 Antimicrobial properties

In the majority of cited studies (**Table 1**) authors did not verify which particular compound contributed to antimicrobial properties. In most of cases they concluded that polyphenols contributed to that phenomenon [13, 15–19] because in comparison to control groups, extracts obtained after SSF demonstrated stronger antimicrobial effects containing more phenolic compounds (PC) at the same time. In the paper written by Mohamed et al. [13, 72] authors did not carry out detailed qualitative and quantitative analysis of fungal metabolites – they assumed that only curcumin would be the substance demonstrating antibacterial properties.

Some studies involved detailed analysis of polyphenol profiles and authors assigned antibacterial and antifungal properties to phenolic acids which concentration was increased by *Pleurotus sapidus* [12]. Others indicated that antimicrobial activity was achieved due to the occurrence of coumarins and oxylipins detected in post-fermentation extracts when *Aspergillus oryzae* KCCM 12698 was used for SSF [14]. Kaaniche et al. [2] additionally analysed structures of obtained bioactive compounds and they proved that four most potent antimicrobials produced by *Streptomyces cavourensis* TN638 were macrotetrolides. Similar approach was applied to identify antimicrobial compounds produced by *Fusarium sambucinum* B10.2 and proved it was sambacide [10]. When surfactants produced by various *Bacillus* strains were tested for antimicrobial properties, researchers additionally tested their properties like emulsification activities [4] or tensioactive activity [5]. Except for latest reports regarding surfactin, we did not include antibiotics in our chapter because currently there are various resistant strains so some alternatives are required.

The majority of identified antimicrobial compounds demonstrated activity equal to [2, 13, 17] or greater [10, 13, 15, 16] than well-known antibiotics. In some cases authors did not provide results for control samples so it was not possible to assess how those substances were effective, however, inhibition zones in diffusion disk method were very prominent [4, 9, 14, 15]. In other studies MIC (Minimum Inhibitory Concentration) of extracted substances were not higher than for antibiotics, however, since those substances were obtained from agricultural waste which is a cost effective substrate, they still could be considered as potential antimicrobials [2, 12, 18, 19]. Only metabolites produced by *Pediococcus acidilactici* KT-05-7 demonstrated very weak antimicrobial properties [8].

2.2 Anti-inflammatory agents

Anti-inflammatory properties of terpenoids were already described in various reviews [73, 74] but they were not investigated in cited papers so we did not discuss obtained results. It is worth mentioning that each extract obtained after SSF contained at least one compound that could demonstrate such activity: lactones which were produced by *Trichoderma viride* EMCC-107 [28]; limonene-1,2-diol, α-terpineol, (−)-carvone, α-tocopherol produced by *Diaporthe* sp. KY113119 [25]; 1-terpineol, L-linalool produced by *Antrodia camphorata* [30]; betulinic acid – *Inonotus obliquus* and [23]; α-pinene produced by *Saccharomyces cerevisiae* AXAZ-1 and *Kluyveromyces marxianus* IMB3 [27].

In the case of polyphenols we took the same approach – we only summarised the research that investigated anti-inflammatory properties of extracts obtained after SSF. When grape pomace was treated with *Trametes versicolor* TV-6 the concentration of phenolic acids, flavan-3-ols and rutin increased while the concentration of anthocyanins decreased. Those changes resulted in enhanced anti-inflammatory activity of obtained extracts which was measured by the inhibition of 5-lipoxygenase and hyaluronidase [20]. Also polyphenols produced by *Xylaria nigripes* demonstrated enhanced anti-inflammatory properties which were verified by the inhibition of cyclooxygenase-2 [24]. Additionally, in both cases, obtained extracts demonstrated neuroprotective properties. Studies carried out by Yin et al. [75] demonstrated that *A. niger* was able to release ferulic acid bound to various polysaccharides in the wheat bran. Obtained substances exhibited stronger anti-inflammatory activity than that of free ferulic acid which probably took place due to the presence of accompanying compounds in obtained extracts. Moreover, those released compounds could significantly inhibit intracellular malondialdehyde formation and the LPS-induced inflammation. It is difficult to assess which extracts demonstrated greater activity because authors of all abovementioned studies used different way to express results – the provided IC_{50} values or % of inhibition.

In the majority of cited papers anti-inflammatory properties of extracts containing polyphenols were not verified despite the fact that there were some compounds among them which demonstrate such activity [76, 77]. Research mainly focused on antioxidant properties [20, 22, 23, 31]. This applies to various studies involving organisms providing increase of particular phenolic compounds: *Rhizopus oligosporus* NRRL 2710 – rutin [22], *Prunus armeniaca* L – cinnamic acids and selected flavonols [31]; *Lactobacillus plantarum* CECT 748 – hydroxybenzoic acids and flavonols [78]; *Eurotium cristatum* YL-1 – isoflavones (daidzin, daidzein, genistin and genistein) [34]; *Rhizopus oryzae* and *Aspergillus foetidus* – tannins [35]; *Rhizopus oryzae* RCK2012 – phenolic acids [36]; *Aspergillus oryzae* LBA01 – 3,4-di- hydroxybenzoic acid, ferulic acid, vanillic acid and quercetin [37].

2.3 Immunosuppressants

Immunosuppressants could be obtained by SSF as well. It seems that one of the most common substance which was detected in studies involving SSF is mycophenolic acid (MA, **Table 1**). It works as a blocker in producing precursors for the synthesis of RNA and DNA, so as the result it blocks proliferative response of T and B lymphocytes [79]. In the research that involved SSF, authors did not verify properties of the obtained MA.

Another substance belonging to that group, classified as calcineurin inhibitor [80], is cyclosporin A (CA). This substance is used not only in transplant patients but also in treatment of glomerular disease. CA prevents calcineurin-dependent transcription in activated T cells. Based on studies that were aiming to produce CA we concluded that authors did not verify properties or safety of obtained substances but they focused on various methods for its extraction and purification. They mostly used butyl acetate [44, 48] or ethyl acetate [47]. Tacrolimus could be also produced by SSF by *Streptococcus hygroscopicus*. The whole process of bacterial cultivation and the extraction of that compounds was covered by the patent [49].

2.4 Anticoagulants

Another group of substances that could be produced by SSF is anticoagulants. Blood coagulation is a physiological process, which consists of a series of coagulation factors and proteolytic activation steps, which lead to the production of thrombin – the main coagulation enzyme. The majority of research was focused on fibrinolytic enzymes (FE, **Table 1**) – subtilisin, however, in one case it was nattokinase [55]. In the majority of cited papers authors verified anticoagulant properties of obtained enzymes and they used different methods – some of them applied spectrophotometric method which measured the increase of turbidity at 275 nm caused by added enzyme [53, 54, 58, 60, 62, 65] and others measured zones of clearance in solid media containing fibrinogen [64, 66, 67].

Another compound which could be considered as a putative anticoagulant is halotolerant chitinase. Its properties were confirmed [50, 51] by testing how this enzyme could dissolve fibrin in time. Moreover, Meruvu et al. [51] showed that it held antifungal activity.

2.5 Anticancer agents

Terpenoids are well-known for their cytotoxic activity and there were various studies investigating such activity against cancer cells. Anticancer mechanism of terpene or essential oils that contain them was described in various review papers [81–83]. Several compounds belonging to that class could be obtained by SSF (**Table 1**): limonene from orange waste [25]; linalool, geraniol and β-caryophyllene from millet [28, 30]; andrastin A and C on malt extract agar [70]. Properties of extracts that contained those and other terpenoids were not investigated in cited papers so we decided not to discuss that aspect in the current chapter.

Another group of bioactive compounds which was shown to hold anticancer activity is polyphenols [84, 85]. The following substances occurred in cited reviews and studies that we summarised in **Table 1**: cinnamic acids [22], daidzein and genistein [86], quercetin [35, 87], and tannins [34]. Since authors of cited research papers (**Table 1**) did not test obtained extracts against those properties, we did not discuss that phenomenon.

In the study of Cho et al. (2019) authors demonstrated that fatty acids detected in extracts obtained from silkworm larvae powder fermented by *Aspergillus kawachii* demonstrated such activity against human hepatocellular carcinoma [68]. It was shown that fermentation increased concentration of those compounds, especially oleic and linoleic acids. This phenomenon took place due to the enhancement of cell apoptosis and suppression of protein responsible for preventing the apoptosis. The value of that research is particularly significant because so far, polyunsaturated fatty acids (PUFAs) attracted most attention in the context of colorectal cancer [87].

3. Groups of microorganisms demonstrating greatest potential for the production of health-promoting properties in SSF

3.1 *Bacillus* genus

It was demonstrated that the representatives of the *Bacillus* genus were able to produce fibrinolytic enzymes by SSF [52–58, 60]. That ability is mostly assigned to the expression of *fibE* gene which encodes enzyme called subtilisin [88] which solubilises blood clots. Gene expression was not investigated in cited papers so the ability to produce those enzymes by tested strains could be the result of other genes expression.

Bacillus genus is successfully used in solid-state fermentation to improve antimicrobial activity of fermented food. Rochín-Medina et al. [19] who tried to determine optimal bioprocessing conditions for SSF of spent coffee grounds by *Bacillus clausii* achieved increase of flavonoid and total phenolic contents by 13 and 36%, respectively. SSF also enhanced antibacterial activity of obtained extracts. That phenomenon could be explained by the fact that *Bacillus sp*. strains could metabolise fibre which releases phenolic compounds as a result of lignocellulolytic activity, and demonstrate strong correlation between the increase of phenolic compounds and the synthesis of cellulases and pectinases [19]. Those enzymes can break down plant cell wall components which leads to the hydrolysis of ester bonds that bind phenolic compounds [19].

The representatives of the *Bacillus* genus are able to produce surfactants as well (**Table 1**). There are various enzymes involved in the production of surfactin which form multienzyme peptide synthetase complex. One of those proteins is Srf (which consists of three units A, B and C). There are also two other genes involved in that process – *sfp* and *com*X. However, the role and interactions of protein and genes was described in more details in other review papers [89], therefore, we did not discuss that phenomenon in details.

3.2 Other bacteria

There are also other bacteria that were tested against the production of substances holding the potential for application in healthcare. For instance, there are several species that are able to produce fibrinolytic enzymes which could serve as potential anticoagulants: *Paenibacillus* sp. IND8 [61], *Pseudoalteromonas* sp. IND11 [62], *Xanthomonas oryzae* IND3 [63] or *Citrobacter freundii nov. haritD11* [50, 51]. In the case of *Paenibacillus* its ability for producing such enzymes could be assigned to the expression of *PPFE-I* gene [90]. It seems that the synthesis of protein encoded by that gene could be stimulated by Zn^{2+}, Mg^{2+} and Fe^{+} so the concentration of these ions could be considered in future studies focusing on optimisation of enzyme production. In the case of *Citrobacter freundii*, molecular mechanism seems much simpler because so far, only *chiX* gene was assigned to its ability for chitinase production [91]. It is still unclear which genes are involved in the production of anticoagulant agent by *Pseudoalteromonas* or *Xanthomonas oryzae* so it is an aspect that could be investigated in the future, since the quantity of the enzyme was very prominent – up to 1,388 U/ml.

It was also demonstrated that *Streptococcus hygroscopicus* is able to produce immunosuppressant in SSF based on agricultural waste with added supplements [49] and it seems that *aroA*, *fkbN*, and *luxR* genes are mostly responsible for that phenomenon [92]. On the other hand, molecular mechanisms standing behind the ability of *Yarrowia lipolytica* W29 (ATCC 20460) to produce γ-decalactone [11] which analogues demonstrate antiviral and antifungal properties [93] is simpler because it involves only *POX2* overexpression [94]. Similarly, in the case of *Lactobacillus plantarum* CECT 748 [95] which increased the concentration of particular phenolic compounds, probably only *est_1092* gene was involved – it encodes phenolic esterase [96].

None of the cited paper evaluated molecular mechanisms involved in processes carried out by tested strains.

3.3 Fungi

The majority of research involving SSF is carried out with various fungi (**Table 1**). Many of analysed examples focused on the increase of phenolic compounds. One of the genera which were involved in that process was *Trichoderma* [13, 16, 17]. Those fungi are known to produce cellulolytic, ligninolytic and xylanolytic enzymes [97] and it has been demonstrated in other studies that cellulase could significantly increase concentrations of various polyphenols e.g. caffeic acid, vanillin, p-coumaric acid, and ferulic acid [98]. In fact, similar strategies might apply to *Aspergillus* spp. [22, 37, 99] because representatives of that genus could produce enzymes demonstrating such activities [98] as well. *Rhizopus oryzae* [36, 100] synthesises cellulase, xylanase and pectinase [101]. Other fungi applied in other cited studies could produce the following enzymes that participate in polyphenol increase: *Lentinus edodes* [15] – xylanase and cellulase [102], *Pleurotus sapidus* – ligninolytic enzymes [12], *Trametes versicolor* TV-6 – β-glucosidase [20].

Genetic background of the synthesis of abovementioned enzymes by *Trichoderma* spp. and *Aspergillus* spp. was revised by Amore et al. [103] while in the case of *R. oryzae* and *P. chrysopsorium* it was described by Battaglia et al. [104] so we did not provide details of those processes, especially that many genes are involved.

There are also some fungi which could be natural sources of polyphenols: *Taiwanofungus camphoratus* [21], *Inonotus obliquus* [23], and *Xylaria nigripes* [24]. In all cases SSF increased their anti-inflammatory properties by increasing concentration of particular phenolic compounds which could also originate from substrates that were used for their cultivation and those were: spent substrate of *Pleurotus eryngii*, sunflower seed hulls, corn and rice grain, white birch and mulberry in the case of *I. obliqus*; wheat bran in the case of *X. nigripes*.

Another significant group of bioactive compounds which concentration was increased by SSF was terpenes (**Table 1**) and those processes were carried out by various fungi: *Fusarium sambucinum* B10 [10], *Penicillium expansum* KACC 40815 [70], *Diaporthe* sp. KY113119 [25], *Antrodia camphorata* [30], *Saccharomyces cerevisiae* AXAZ-1 and *Kluveromyces marxianus* IMB3 [27], *Trichoderma viride* EMCC-107 [28], and *Aspergillus niger* van Tieghem [29]. *A. camphorate* is the natural source of terpenes and their concentration was increased by selecting millet as the main substrate for SSF. *K. marxianus* and *S. cerevisiae* could probably increase the concentration of tested compounds by releasing terpenes from their glycosidic forms by β-glucosidases. In the case of *Fusarium* spp., *Aspergillus* spp. and *Penicillium* spp. it was demonstrated that those genera could produce several sesquiterpene synthases [105]. As in the case of polyphenols, molecular background of terpene transformations are very complex so we decided not to describe it in the current chapter, but refer to the review of Quin et al. [105] instead. It must be highlighted that except for the study regarding *P. expansum* KACC 40815 demonstrating that terpenoid cyclase was mostly involved in described processes [70], none of the authors investigated enzymatic activity during terpene transformation nor determined gene expression.

Based on cited papers it might be stated that *Penicillium brevicopactum* is most common in the research regarding production of mycophenolic acid in SSF [38–43]. Those fungi produce polyketide synthase encoded by *mpaC* gene along with other enzymes such us: protein transacylase, β- ketoacylsynthase, acyltransferase, acyl carrier protein, and methyltransferase (MT) domains [106]. There is another taxon which is able to produce immunosuppressants, namely cyclosporin A, and it is *Tolypocladium inflatum* [44–48] which has got nonribosomal peptide synthetase that encodes for cyclosporin synthetase (*simA* gene) [107].

Mucor subtilissimus UCP 1262 [65] and *Fusarium oxysporum* [66, 67] were shown to produce fibrinolytic enzymes. It has been already demonstrated that *FP* gene is responsible for encoding fibrinolytic protease in *Fusarium* sp. [108] but in the case of *Mucor* sp. it remains unknown. Molecular background of *A. kawachii* KCCM 32819 production of short chain fatty acids could be the same as for *A. nidulans* and other filamentous ascomycetes – *farB* gene is mostly responsible for that ability, however, *farA* participates as well [109]. Genetic background of biosurfactant production in moulds is still unknown.

3.4 Modified strains

It has been demonstrated that in some cases the yield of microbial metabolites significantly increased when the strain was subjected to gamma ray – it increased the quantity of obtained mycophenolic acid produced by two strains of *Penicillium requeforti* [43] in comparison of other cited studies which involved unmodified *Penicillium brevicompactum* [39, 41]. On the other hand, UV radiation was used for the modification of *Tolypocladium inflatum* which was used for the production of Cyclosporin A [47], however, in the case of that substance there were other studies which resulted in much higher yields [44, 45].

4. Optimisation of SSF conditions

One of the approaches that were applied for the optimisation of the concentration of bioactive substances by SSF is to provide pre-treatment to the main substrate. Heat-treatment was one of the main methods e.g. autoclaving, moistening in boiling water, cooking with deionised water, drying, freezing, freeze-drying, vacuum-drying and roasting. It inactivates native microorganisms and enzymes.

Once the main substrate is prepared, there are other experimental conditions that need to be optimised. In cited papers the major factor contributing to obtained results was moisture which was at least 50% [69, 110]. The next crucial factor was medium composition. In few cases authors used solid media commonly used for the cultivation of microorganisms [2, 70]. However, in the majority of cases authors added some nutrients to agricultural waste to provide better results of releasing bioactive compounds. In the case of mycophenolic acid produced *P. brevicompactum* MTCC 8010 from rice bran those supplements were: peptone, KH_2PO_4, glycine and methionine [41]. On the other hand, in the case of *P. brevicompactum* ATCC 16024, the addition of mannitol or $(NH_4)_2HPO_4$ to pearl barley did not enhance the MA synthesis [39]. Surprisingly, the quantity of MA was higher in the latter case – 5.47 g/kg of the substrate in comparison to 4.5 g/kg under optimised conditions. It seems that pearl barley has got chemical composition which is more preferable for obtaining MA.

On the other hand, Plackett-Burman design was applied for the optimisation of cyclosporin A production by *Tolypocladium inflatum* MTCC 557 which resulted in 8,166 mg/kg [45] which is 1.26-fold higher than in studies carried out by Survase et al. [44, 45] who applied the same fungal strain or even 45.62-fold higher than reported by Nisha and Ramasamy [46] who used *T. inflatum* ATTC 34921 strain. In the last study authors firstly used Plackett-Burman design for the selection of nutrients and later on, they used half-factorial central composite rotatable design (CCRD) of response surface methodology (RSM) to select optimum concentrations of those substances which resulted in more than 10-fold increase of tested compound [47].

When the production of fibrinolytic enzyme was optimised with RSM, its activity increased 3 times in the case of *Pseudoalteromonas* sp. IND11 [62]; 4 times in the case of *Xanthomonas oryzae* sp. IND3 [63]; while when central composite design (CCD) was used, 4.5-fold increase was noted when *Paenibacillus* sp. IND8 was used [61]. In the case of *Xanthomonas oryzae* sp. IND3, CCD was additionally used for estimating optimal values of the following variables: sucrose, yeast extract, and pH of the medium [63]. Among all mentioned microorganisms *Paenibacillus* sp. IND8 produced greatest quantities of the enzyme – 4,418 U/ml.

Various *Bacillus* strains were used for the production of fibrinolytic enzyme (**Table 1**) and those proved to be more efficient enzyme producer that abovementioned bacteria. When authors used two-level factorial design in the case of *Bacillus* sp. IND12 examining the impact of moisture, sucrose, and $MgSO_4$ levels added to the cow dung, maximum enzyme activity reached 4,143 U/g [54]. Further, orthogonal design (corn steep powder, sucrose and $MgSO_4 \cdot 7H_2O$) provided even greater enzyme activity (5,865 U/ml) in 100 l fermenter when applying *Bacillus subtilis* WR350 [57]. However, the greatest activity was achieved when *Bacillus halodurans* IND18 was used. It produced 6,851 U/g when two-level full factorial design was applied and the optimum conditions were as follows: 1% peptone, 80% moisture and pH 8.32, using wheat bran as the main substrate [58].

Ghribi et al. [4] took a different approach for the optimisation of surfactin production – firstly, authors applied Plackett-Burman design to assess which of the five variables were the most important and then they optimised the process with CCD involving three selected variables. They found parameters (temperature – 37°C, inoculum age – 14 h, and moisture – 88%) that were the most favourable for the production of surfactin and increased its yield by 2-fold (up to 2 g/l). Sun et al. [3, 111] carried out step-by-step optimisation and found out that the addition of attapulgite by 1.96-fold (4.3782 g/kg). This would suggest that *B. natto* NT-6 was the most suitable for that application among all tested strains.

Optimization of halotolerant chitinase was carried out using RSM –Box Behnken method which involved *Citrobacter freundii* and that process slightly improved enzyme production in comparison to initial optimisation experiments from 112.43 U/g dry substance to 124.73 U/g dry substance [51]. That optimisation was mainly focused on the ratio of main substrates (wheat bran and shrimp shellfish), temperature and moisture content. Similarly, minor changes were observed when wheat bran and powdered fish scales were used for the statistical optimisation of chitinase production [50].

5. Conclusions

Solid-state fermentation could provide various substances useful in healthcare: antimicrobials, immunosuppressants, anticoagulants, substances holding anti-inflammatory properties and anticancer agents. It seems that polyphenols and terpenes are especially versatile in their applications. The majority of studies involved various fungi mainly due to their enzymatic activity which supports the release of bioactive compounds from agricultural waste. Molecular mechanisms of those processes are usually very complex; however, they remain unknown for some fungi. Further studies are necessary to assess which genes could be expressed during those processes because those could be used for modification of microorganisms to increase their yield. It is also important to bear in mind that SSF requires the presence of various supplements and fermentation could be optimised by statistical tools, especially Response Surface Methodology and Central Composite Design. In some cases step-by-step optimisation could be sufficient.

Conflict of interest

The authors declare no conflict of interest.

Author details

Łukasz Wajda[1*] and Magdalena Januszek[2]

1 CDC Poland sp. z o.o., Zagnańska 153 street, 25-516 Kielce, Poland

2 Department of Fermentation Technology and Microbiology, Faculty of Food Technology and Human Nutrition, University of Agriculture in Krakow, Balicka 122 street, 30-149 Kraków, Poland

*Address all correspondence to: lukasz.wajda8@gmail.com

References

[1] Verduzco-oliva R, Gutierrez-uribe JA. Beyond Enzyme Production: Solid State Fermentation (SSF) as an Alternative Approach to Produce Antioxidant Polysaccharides. Sustainability. 2020;12: 1-14. doi:10.3390/su2020495

[2] Kaaniche F, Hamed A, Elleuch L, Chakchouk-mtibaa A, Mellouli L. Purification and characterization of seven bioactive compounds from the newly isolated Streptomyces cavourensis TN638 strain via solid-state fermentation. Microb Pathog. 2020; 104106. doi:10.1016/j.micpath.2020.104106

[3] Sun D, Liao J, Sun L, Wang Y, Liu Y, Deng Q, et al. Effect of media and fermentation conditions on surfactin and iturin homologues produced by Bacillus natto NT - 6: LC – MS analysis. AMB Express. 2019;9: 1-9. doi:10.1186/s13568-019-0845-y

[4] Ghribi D, Abdelkefi-mesrati L, Mnif I, Kammoun R, Ayadi I, Saadaoui I, et al. Investigation of Antimicrobial Activity and Statistical Optimization of Bacillus subtilis SPB1 Biosurfactant Production in Solid-State Fermentation. J Biomed Biotechnol. 2012;2012: 1-12. doi:10.1155/2012/373682

[5] Slivinski CT, Mallmann E, De Araújo JM, Mitchell DA, Krieger N. Production of surfactin by *Bacillus pumilus* UFPEDA 448 in solid-state fermentation using a medium based on okara with sugarcane bagasse as a bulking agent. Process Biochem. 2012;47: 1848-1855. doi:10.1016/j.procbio.2012.06.014

[6] Lourenço LA, Alberton Magina MD, Tavares LBB, Guelli Ulson de Souza SMA, García Román M, Altmajer Vaz D. Biosurfactant production by Trametes versicolor grown on two-phase olive mill waste in solid-state fermentation. Environ Technol (United Kingdom). 2018;39: 3066-3076. doi:10.1080/09593330.2017.1374471

[7] Kreling NE, Simon V, Fagundes VD, Thomé A, Colla LM. Simultaneous Production of Lipases and Biosurfactants in Solid-State Fermentation and Use in Bioremediation. J Environ Eng. 2020;146: 04020105. doi:10.1061/(ASCE)EE.1943-7870.0001785

[8] Juodeikiene G, Cizeikiene D, Ceskeviciute V, Vidmantiene D, Basinskiene L, Akuneca I, et al. Solid-State Fermentation of *Silybum marianum* L . Seeds Used as Additive to Increase the Nutritional Value of Wheat Bread. Food Technol Biotechnol. 2013;51: 528-538.

[9] Lin E-R, Cheng Y-H, Hsiao FS-H, Proskura, Witold S, Dybus A, Yu Y-H. Optimization of solid-state fermentation conditions of Bacillus licheniformis and its effects on *Clostridium perfringens* -induced necrotic enteritis in broilers. Rev Basiliera Zootec. 2019;48: 1-14. doi:10.1590/rbz4820170298

[10] Dong J, Cai L, Li X, Duan R, Shu Y, Chen F, et al. Production of a new tetracyclic triterpene sulfate metabolite sambacide by solid-state cultivated Fusarium sambucinum B10 . 2 using potato as substrate. Bioresour Technol. 2016. doi:10.1016/j.biortech.2016.07.014

[11] Try S, De-Coninck J, Voilley A, Chunhieng T, Waché Y. Solid state fermentation for the production of γ-decalactones by Yarrowia lipolytica. Process Biochem. 2018;64: 9-15. doi:10.1016/j.procbio.2017.10.004

[12] Pinela J, Omarini AB, Stojković D, Barros L, Postemsky PD, Calhelha RC,

et al. Biotransformation of rice and sunflower side-streams by dikaryotic and monokaryotic strains of Pleurotus sapidus: Impact on phenolic profiles and bioactive properties. Food Res Int. 2020;132. doi:10.1016/j.foodres.2020.109094

[13] Mohamed SA, Saleh RM, Kabli SA, Al-garni SM. Influence of solid state fermentation by Trichoderma spp. on solubility, phenolic content, antioxidant, and antimicrobial activities of commercial turmeric. Biosci Biotechnol Biochem. 2016;8451: 1-9. doi: 10.1080/09168451.2015.1136879

[14] Son SY, Lee S, Singh D, Lee N, Lee D, Lee CH. Comprehensive Secondary Metabolite Profiling Toward Delineating the Solid and Submerged-State Fermentation of Aspergillus oryzae KCCM 12698. Front Microbiol. 2018;9: 1-12. doi:10.3389/fmicb.2018.01076

[15] Vattem DA, Lin YT, Labbe RG, Shetty K. Phenolic antioxidant mobilization in cranberry pomace by solid-state bioprocessing using food grade fungus Lentinus edodes and effect on antimicrobial activity against select food borne pathogens. Innov Food Sci Emerg Technol. 2004;5: 81-91. doi:10.1016/j.ifset.2003.09.002

[16] Saleh RM, Kabli SA, Al-Garni SM, Al-Ghamdi MA, Abdel-Aty AM, Mohamed SA. Solid-state fermentation by Trichoderma viride for enhancing phenolic content, antioxidant and antimicrobial activities in ginger.pdf. Lett Appl Microbiol. 2018;67: 161-167. doi:https://doi.org/10.1111/lam.13003

[17] Abdel-Aty AM, Bassuiny RI, Barakat AZ, Mohamed SA. Upgrading the phenolic content, antioxidant and antimicrobial activities of garden cress seeds using solid-state fermentation by Trichoderma reesei. J Appl Microbiol. 2019;127: 1454-1467. doi:10.1111/jam.14394

[18] Wen Y, Yan L, Chen C. Effects of fermentation treatment on antioxidant and antimicrobial activities of four common Chinese herbal medicinal residues by Aspergillus oryzae. J Food Drug Anal. 2013;21: 219-226. doi:10.1016/j.jfda.2013.05.013

[19] Rochín-Medina JJ, Ramírez K, Rangel-Peraza JG, Bustos-Terrones YA. Increase of content and bioactivity of total phenolic compounds from spent coffee grounds through solid state fermentation by *Bacillus clausii*. J Food Sci Technol. 2018;55: 915-923. doi:10.1007/s13197-017-2998-5

[20] Bucić-Kojić A, Fernandes F, Silva T, Planinić M, Tišma M, Šelo G, et al. Enhancement of the anti-inflammatory properties of grape pomace treated by: Trametes versicolor. Food Funct. 2020;11: 680-688. doi:10.1039/c9fo02296a

[21] Liu DZ, Liang HJ, Chen CH, Su CH, Lee TH, Huang CT, et al. Comparative anti-inflammatory characterization of wild fruiting body, liquid-state fermentation, and solid-state culture of Taiwanofungus camphoratus in microglia and the mechanism of its action. J Ethnopharmacol. 2007;113: 45-53. doi:10.1016/j.jep.2007.03.037

[22] Wronkowska M, Honke J, Piskula Mariusz K. Effect of solid-state fermentation with Rhizopus oligosporus on bioactive compounds and antioxidant capacity of raw and roasted buckwheat groats. Ital J Food Sci. 2015;27: 424-431. doi:10.14674/1120-1770/ijfs.v373

[23] Chen HJ, Chen YS, Liu SL, Liou BK, Chen CS. The Increase of Bioactive Ingredients by Solid State Fermentation of Inonotus obliquus with Spent Substrate. Waste and Biomass Valorization. 2019. doi:10.1007/s12649-019-00911-7

[24] Divate RD, Wang CC, Chou ST, Chang CT, Wang PM, Chung YC.

Using wheat bran and soybean meal as solid state fermentation substances for the production of Xylaria nigripes with bioactivities. J Taiwan Inst Chem Eng. 2017;70: 127-133. doi:10.1016/j.jtice.2016.11.003

[25] Jucoski Bier MC, Medeiros ABP, De Kimpe N, Soccol CR. Evaluation of antioxidant activity of the fermented product from the biotransformation of R-(+)-limonene in solid-state fermentation of orange waste by Diaporthe sp. Biotechnol Res Innov. 2019;3: 168-176. doi:10.1016/j.biori.2019.01.002

[26] Bédard F, Hammami R, Zirah S, Rebuffat S, Fliss I, Biron E. Synthesis, antimicrobial activity and conformational analysis of the class IIa bacteriocin pediocin PA-1 and analogs thereof /631/45/611 /631/326/22 /639/638/549/978 /639/638/309/2419 /639/638/403/977 /140/131 /82/6 /82/16 article. Sci Rep. 2018;8: 1-13. doi:10.1038/s41598-018-27225-3

[27] Aggelopoulos T, Katsieris K, Bekatorou A, Pandey A, Banat IM, Koutinas AA. Solid state fermentation of food waste mixtures for single cell protein, aroma volatiles and fat production. Food Chem. 2014;145: 710-716. doi:10.1016/j.foodchem.2013.07.105

[28] Fadel HHM, Mahmoud MG, Asker MMS, Lotfy SN. Characterization and evaluation of coconut aroma produced by Trichoderma viride EMCC-107 in solid state fermentation on sugarcane bagasse. Electron J Biotechnol. 2015;18: 5-9. doi:10.1016/j.ejbt.2014.10.006

[29] Chun-hua Z, Zhi-guo S, Wen-xia Y, Ya-li LI, Ting-ting SUN. Effects on Aroma Composition of Pu-erh Tea under Solid State Fermentation with Different Beneficial Microorganisms. J Tea Sci. 2010;30: 251-258.

[30] Xia Y, Zhang B, Li W, Xu G. Changes in volatile compound composition of Antrodia camphorata during solid state fermentation. J Sci Food Agric. 2011;91: 2463-2470. doi:10.1002/jsfa.4488

[31] Dulf FV, Vodnar DC, Dulf EH, Pintea A. Phenolic compounds, flavonoids, lipids and antioxidant potential of apricot (*Prunus armeniaca* L.) pomace fermented by two filamentous fungal strains in solid state system. Chem Cent J. 2017;11: 1-10. doi:10.1186/s13065-017-0323-z

[32] Leite P, Silva C, Salgado JM, Belo I. Simultaneous production of lignocellulolytic enzymes and extraction of antioxidant compounds by solid-state fermentation of agro-industrial wastes. Ind Crops Prod. 2019;137: 315-322. doi:10.1016/j.indcrop.2019.04.044

[33] Juan MY, Chou CC. Enhancement of antioxidant activity, total phenolic and flavonoid content of black soybeans by solid state fermentation with Bacillus subtilis BCRC 14715. Food Microbiol. 2010;27: 586-591. doi:10.1016/j.fm.2009.11.002

[34] Chen Y, Wang Y, Chen J, Tang H, Wang C, Li Z, et al. Bioprocessing of soybeans (Glycine maxL.) by solid-state fermentation withEurotium cristatumYL-1 improves total phenolic content, isoflavone aglycones, and antioxidant activity. RSC Adv. 2020;10: 16928-16941. doi:10.1039/c9ra10344a

[35] Mukherjee G, Banerjee R. Biosynthesis of tannase and gallic acid from tannin rich substrates by Rhizopus oryzae and Aspergillus foetidus. J Basic Microbiol. 2004;44: 42-48. doi:10.1002/jobm.200310317

[36] Dey TB, Kuhad RC. Enhanced production and extraction of phenolic compounds from wheat by solid-state fermentation with Rhizopus oryzae RCK2012. Biotechnol Reports. 2014;4: 120-127. doi:10.1016/j.btre.2014.09.006

[37] Magro AEA, Silva LC, Rasera GB, de Castro RJS. Solid-state fermentation as an efficient strategy for the biotransformation of lentils: enhancing their antioxidant and antidiabetic potentials. Bioresour Bioprocess. 2019;6. doi:10.1186/s40643-019-0273-5

[38] Alpak I, Askin Uzel R, Sargin S, Yesil-Celiktas O. Supercritical CO2 extraction of an immunosuppressant produced by solid-state fermentation. J CO2 Util. 2018;27: 398-404. doi:10.1016/j.jcou.2018.08.014

[39] Alani F, Grove JA, Anderson WA, Moo-Young M. Mycophenolic acid production in solid-state fermentation using a packed-bed bioreactor. Biochem Eng J. 2009;44: 106-110. doi:10.1016/j.bej.2008.11.005

[40] Sadhukhan AK, Ramana Murthy M V., Ajaya Kumar R, Mohan EVS, Vandana G, Bhar C, et al. Optimization of mycophenolic acid production in solid state fermentation using response surface methodology. J Ind Microbiol Biotechnol. 1999;22: 33-38. doi:10.1038/sj.jim.2900597

[41] Patel G, Patil MD, Soni S, Chisti Y, Banerjee UC. Production of Mycophenolic Acid by Penicillium brevicompactum Using Solid State Fermentation. Appl Biochem Biotechnol. 2017;182: 97-109. doi:10.1007/s12010-016-2313-3

[42] Tiwari S, Santhyanathan D, Thomas E, Melarkode R, Ramavana G, Suryanarayan S. Fed batch solid state fermentation for the production of mycophenolic acid. India: World Intellectual Property Organization International, International Bureau; WO2003106690A1, 2003.

[43] El-Sayed ESR, Ahmed AS, Ismaiel AA. Agro-industrial byproducts for production of the immunosuppressant mycophenolic acid by Penicillium roqueforti under solid-state fermentation: Enhanced production by ultraviolet and gamma irradiation. Biocatal Agric Biotechnol. 2019;18: 101015. doi:10.1016/j.bcab.2019.01.053

[44] Survase SA, Shaligram NS, Pansuriya RC, Annapure US, Singhal RS. A novel medium for the enhanced production of Cyclosporin A by Tolypocladium inflatum MTCC 557 using solid state fermentation. J Microbiol Biotechnol. 2009;19: 462-467. doi:10.4014/jmb.0805.324

[45] Survase AS, Annapure US, Singhal RS. Statistical optimization for improved production of cyclosporin A in solid-state fermentation. J Microbiol Biotechnol. 2009;19: 1385-1392. doi:10.4014/jmb.0901.0035

[46] Nisha AK, Ramasamy K. Cyclosporin production in various solid substrate media by Tolypocladium inflatum (ATCC 34921). Biotechnology. 2008. pp. 357-359. doi:10.3923/biotech.2008.357.359

[47] Ramana Murthy M V., Mohan EVS, SadhukhanAK.Cyclosporin-Aproduction by Tolypocladium inflatum using solid state fermentation. Process Biochem. 1999;34: 269-280. doi:10.1016/S0032-9592(98)00095-8

[48] Sekar C, Rajasekar VW, Balaraman K. Production of cyclosporin A by solid state fermentation. Bioprocess Eng. 1997;17: 257-259. doi:10.1007/s004490050383

[49] Ul NA, Khandige S, Patil NS, Khedkar AP, Melarkode R, Gururaja R, et al. Solid state fermentation and fed batch for the production of an immunosuppressant. India: World Intellectual Property Organization International, International Bureau; WO2004069845A1, 2004.

[50] Meruvu H, Meruvu HBT. Anticoagulant activity of partially

purified chitinase produced by Citrobacter freundii str. nov. haritD11 by fermentation of wheat bran coupled with fish scales. SN Appl Sci. 2019;1: 1-9. doi:10.1007/s42452-018-0097-5

[51] Meruvu H, Meruvu HBT. Biosynthesis and Characterization of an Anticoagulant Chitinase from Fermented Wheat Bran & Shrimp Shells' Substratum. Trends Pept Protein Sci. 2018;3: 1-9. doi:https://doi.org/10.22037/tpps.v3i0.23165

[52] Wei X, Luo M, Xu L, Zhang Y, Lin X, Kong P, et al. Production of fibrinolytic enzyme from *Bacillus amyloliquefaciens* by fermentation of chickpeas, with the evaluation of the anticoagulant and antioxidant properties of chickpeas. J Agric Food Chem. 2011;59: 3957-3963. doi:10.1021/jf1049535

[53] Almalki MA, Dhas DSD, Vijayaraghavan P. Optimization of Fibrinolytic Enzyme Production from Bacillus sp. IND6 in Solid State Fermentation. J Sci Res Reports. 2017;16: 1-8. doi:10.9734/JSRR/2017/36748

[54] Vijayaraghavan P, Rajendran P, Gnana S, Vincent P, Arun A, Al-dhabi NA, et al. Novel Sequential Screening and Enhanced Production of Fibrinolytic Enzyme by Bacillus sp. IND12 Using Response Surface Methodology in Solid-State Fermentation. Biomed Res Int. 2017;2017: 1-13. doi:doi.org/10.1155/2017/3909657 Research

[55] Xin X, Ambati RR, Cai Z, Lei B. Development of universal purification protocols for fibrinolytic enzyme-producing bacilli. CyTA - J Food. 2019;17: 112-120. doi:10.1080/19476337.2018.1561521

[56] Bi H, Zhao H, Lu F, Zhang C, Bie X, Lu Z. Improvement of the nutritional quality and fibrinolytic enzyme activity of soybean meal by fermentation of Bacillus subtilis. J Food Process Preserv. 2014;39: 1-8. doi:10.1111/jfpp.12340

[57] Wu R, Chen G, Pan S, Ze J, Liang Z. Cost-effective fibrinolytic enzyme production by Bacillus subtilis WR350 using medium supplemented with corn steep powder and sucrose. Sci Rep. 2019;9: 1-10. doi:10.1038/s41598-019-43371-8

[58] Vijayaraghavan P, Gnana S, Vincent P, Valan M, Al-dhabi NA. Bioconversion of agro-industrial wastes for the production of fi brinolytic enzyme from Bacillus halodurans IND18: Purification and biochemical characterization. EJBT. 2016;20: 1-8. doi:10.1016/j.ejbt.2016.01.002

[59] Venkatanagaraju E, Divakar G. Production and Purification of Alkaline Fibrinolytic Enzyme from *Bacillus cereus* GD55 under Solid State. Curr Trends Biotechnol Pharm. 2015;9: 348-356.

[60] Devadhasan G, Arumugaperumal B. Bio-prospecting of cuttle fish waste and cow dung for the production of fibrinolytic enzyme from *Bacillus cereus* IND5 in solid state fermentation. 3 Biotech. 2016;6: 1-13. doi:10.1007/s13205-016-0553-0

[61] Vijayaraghavan P, Gnana S, Vincent *P. medium* Optimization for the Production of Fibrinolytic Enzyme by Paenibacillus sp. IND8 Using Response Surface Methodology. Sci World J. 2014;2014: 1-9.

[62] Vijayaraghavan P, Gnana S, Vincent P. Statistical optimization of fibrinolytic enzyme production by Pseudoalteromonas sp. IND11 using cow dung substrate by response surface methodology. Springerplus. 2014;3: 1-10.

[63] Vijayaraghavan P, Valan M, Rajan RA, Al-dhabi NA. Enhanced production of fibrinolytic enzyme by a new Xanthomonas oryzae IND3 using

low-cost culture medium by response surface methodology. Saudi J Biol Sci. 2019;26: 217-224. doi:10.1016/j.sjbs.2018.08.029

[64] Seo J-H, Lee S-P. Production of Fibrinolytic Enzyme from Soybean Grits Fermented by Bacillus firmus NA-1. J Med Food. 2004;7: 442-449.

[65] Pajeú T, Emmanuelle A, Souza C, Marcos R, Brandão P, Campos-takaki GM De, et al. Purification of a fibrinolytic protease from Mucor subtilissimus UCP 1262 by aqueous two-phase systems (PEG/sulfate). J Chromatogr B. 2016;1025: 16-24. doi:10.1016/j.jchromb.2016.04.046

[66] Tao S, Beihui L, Peng L, Deming L, Zuohu L. New solid-state fermentation process for repeated batch production of fibrinolytic enzyme by Fusarium oxysporum. Process Biochem. 1998;33: 419-422. doi:10.1016/S0032-9592(97)00096-4

[67] Tao S, Peng L, Beihui L, Deming L, Zuohu L. Solid state fermentation of rice chaff for fibrinolytic enzyme production by Fusarium oxysporum. Biotechnol Lett. 1997;19: 465-467.

[68] Cho HD, Min HJ, Won YS, Ahn HY, Cho YS, Seo K Il. Solid state fermentation process with Aspergillus kawachii enhances the cancer-suppressive potential of silkworm larva in hepatocellular carcinoma cells. BMC Complement Altern Med. 2019;19: 241. doi:10.1186/s12906-019-2649-7

[69] Ghanem KZ, Mahran MZ, Ramadan MM, Ghanem HZ, Fadel M, Mahmoud MH. A comparative study on flavour components and therapeutic properties of unfermented and fermented defatted soybean meal extract. Sci Rep. 2020;10: 1-7. doi:10.1038/s41598-020-62907-x

[70] Kim HY, Heo DY, Park HM, Singh D, Lee CH. Metabolomic and transcriptomic comparison of solid-state and submerged fermentation of Penicillium expansum KACC 40815. PLoS One. 2016;11: 1-14. doi:10.1371/journal.pone.0149012

[71] Onipe OO, Jideani AIO, Beswa D. Composition and functionality of wheat bran and its application in some cereal food products. Int J Food Sci Technol. 2015;50: 2509-2518. doi:10.1111/ijfs.12935

[72] Mohamed SA, Saleh RM, Kabli SA, Al-garni SM. Influence of solid state fermentation by Trichoderma spp . on solubility, phenolic content, antioxidant, and antimicrobial activities of commercial turmeric. Biosci Biotechnol Biochem. 2016;8451: 1-9. doi:10.1080/09168451.2015.1136879

[73] José Serrano Vega R, Campos Xolalpa N, Josabad Alonso Castro A, Pérez González C, Pérez Ramos J, Pérez Gutiérrez S. Terpenes from Natural Products with Potential Anti-Inflammatory Activity. Terpenes and Terpenoids. 2018. pp. 59-85. doi:10.5772/intechopen.73215

[74] Kim T, Song B, Cho KS, Lee IS. Therapeutic potential of volatile terpenes and terpenoids from forests for inflammatory diseases. Int J Mol Sci. 2020;21. doi:10.3390/ijms21062187

[75] Yin ZN, Wu WJ, Sun CZ, Liu HF, Chen WB, Zhan QP, et al. Antioxidant and Anti-inflammatory Capacity of Ferulic Acid Released from Wheat Bran by Solid-state Fermentation of Aspergillus niger. Biomed Environ Sci. 2019;32: 11-21. doi:10.3967/bes2019.002

[76] Joseph S V., Edirisinghe I, Burton-Freeman BM. Fruit Polyphenols: A Review of Anti-inflammatory Effects in Humans. Crit Rev Food Sci Nutr. 2016;56: 419-444. doi:10.1080/10408398.2013.767221

[77] Yahfoufi N, Alsadi N, Jambi M, Matar C. The immunomodulatory and

anti-inflammatory role of polyphenols. Nutrients. 2018;10: 1-23. doi:10.3390/nu10111618

[78] Dueñas M, Fernández D, Hernández T, Estrella I, Muñoz R. Bioactive phenolic compounds of cowpeas (*Vigna sinensis* L). Modifications by fermentation with natural microflora and with Lactobacillus plantarum ATCC 14917. J Sci Food Agric. 2005;85: 297-304. doi:10.1002/jsfa.1924

[79] Kitchin JES, Pomeranz MK, Pak G, Washenik K, Shupack JL. Rediscovering mycophenolic acid: A review of its mechanism, side effects, and potential uses. J Am Acad Dermatol. 1997;37: 445-449. doi:10.1016/S0190-9622(97)70147-6

[80] Wiseman AC. Immunosuppressive medications. Clin J Am Soc Nephrol. 2016;11: 332-343. doi:10.2215/CJN.08570814

[81] Huang M, Lu J-J, Huang M-Q, Bao J-L, Chen X-P, Wang Y-T. Terpenoids: natural products for cancer therapy. Expert Opin Investig Drugs. 2012;21: 1801-1818. doi:10.1517/13543784.2012.727395

[82] Cox-georgian D, Ramadoss N, Dona C, Basu C. Therapeutic and Medicinal Uses of Terpenes. In: Joshee N, Dhekney S, Parajuli P, editors. Medicinal Plants. Springer Nature Switzerland; pp. 333-359. doi:https://doi.org/10.1007/978-3-030-31269-5_15

[83] Blowman K, Magalhães M, Lemos MFL. Anticancer Properties of Essential Oils and Other Natural Products. Evidence-Based Complement Altern Med. 2018;2018: 1-12. doi:10.1155/2018/3149362 Review

[84] Fresco P, Borges F, Diniz C, Marques MPM. New insights on the anticancer properties of dietary polyphenols. Med Res Rev. 2006;26: 747-766. doi:10.1002/med.20060

[85] Mileo AM, Miccadei S. Polyphenols as Modulator of Oxidative Stress in Cancer Disease: New Therapeutic Strategies. Oxid Med Cell Longev. 2016;2016. doi:10.1155/2016/6475624

[86] Cao L, Guo X, Liu G, Song Y, Ho CT, Hou R, et al. A comparative analysis for the volatile compounds of various Chinese dark teas using combinatory metabolomics and fungal solid-state fermentation. J Food Drug Anal. 2018;26: 112-123. doi:10.1016/j.jfda.2016.11.020

[87] Ohara T, Mori T. Antiproliferative Effects of Short-chain Fatty Acids on Human Colorectal Cancer Cells via Gene Expression Inhibition. Anticancer Res. 2019;39: 4659-4666. doi:10.21873/anticanres.13647

[88] Ghasemi Y, Dabbagh F, Ghasemian A. Cloning of a fibrinolytic enzyme (subtilisin) gene from bacillus subtilis in *Escherichia coli*. Mol Biotechnol. 2012;52: 1-7. doi:10.1007/s12033-011-9467-6

[89] Krieger N, Neto DC, Mitchell DA. Production of microbial biosurfactants by solid-state cultivation. Adv Exp Med Biol. 2010;672: 203-210. doi:10.1007/978-1-4419-5979-9_15

[90] Lv F, Zhang C, Guo F, Lu Y, Bie X, Qian H, et al. Expression, purification, and characterization of a recombined fibrinolytic enzyme from endophytic *Paenibacillus polymyxa* EJS-3 in *Escherichia coli*. Food Sci Biotechnol. 2015;24: 125-131. doi:10.1007/s10068-015-0018-y

[91] Xu J, Yang Y, Liu Y, Ran C, Li J, He S, et al. Characterization of an exo-chitinase from a Citrobacter strain isolated from the intestine content of large yellow croakers. Acta Microbiol Sin. 2016;56: 1089-1104. Available: https://pubmed.ncbi.nlm.nih.gov/29733170/

[92] Wang C, Yang B, Wang J. Generation of Streptomyces hygroscopicus cell factories with enhanced ascomycin production by combined elicitation and pathway - engineering strategies. Biotechnol Bioeng. 2019; 1-14. doi:10.1002/bit.27158

[93] Kishimoto N, Sugihara S, Mochida K, Fujita T. In Vitro Antifungal Analogs and Antiviral Utilized as Activities of γ- and δ-Lactone Analogs Utilized as Food Flavoring. Biocontrol Sci. 2005;10: 31-36. doi:10.4265/bio.10.31

[94] Guo Y, Song H, Wang Z, Ding Y. Expression of POX2 gene and disruption of POX3 genes in the industrial Yarrowia lipolytica on the γ-decalactone production. Microbiol Res. 2012;167: 246-252. doi:10.1016/j.micres.2011.10.003

[95] Park MJ, General T, Lee SP. Physicochemical properties of roasted soybean flour bioconverted by solid-state fermentation using Bacillus subtilis and Lactobacillus plantarum. Prev Nutr Food Sci. 2012;17: 36-45. doi:10.3746/pnf.2012.17.1.036

[96] Esteban-Torres M, Landete JM, Reverón I, Santamaría L, de las Rivas B, Muñoz R. A Lactobacillus plantarum Esterase Active on a Broad Range of. Appl Environ Microbiol. 2015;81: 3235-3242. doi:10.1128/AEM.00323-15

[97] Arnau J, Yaver D, Hjort CM. Strategies and Challenges for the Development of Industrial Enzymes Using Fungal Cell Factories. 1st ed. In: Nevalainen H, editor. Grand Challenges in Fungal Biotechnology. 1st ed. Springer International Publishing; 2020. pp. 179-210. doi:10.1007/978-3-030-29541-7_7

[98] Chen D, Shi J, Hu X. Enhancement of polyphenol content and antioxidant capacity of oat (*Avena nuda* L .) bran by cellulase treatment. Appl Biol Chem. 2016;59: 397-403. doi:10.1007/s13765-016-0171-x

[99] Salgado JM, Abrunhosa L, Venâncio A, Domínguez JM, Belo I. Integrated use of residues from olive mill and winery for lipase production by solid state fermentation with Aspergillus sp. Appl Biochem Biotechnol. 2014;172: 1832-1845. doi:10.1007/s12010-013-0613-4

[100] Das P, Mukherjee S, Sen R. Genetic regulations of the biosynthesis of microbial surfactants: An overview. Biotechnol Genet Eng Rev. 2008;25: 165-186. doi:10.5661/bger-25-165

[101] Ghosh B, Ray RR. Current Commercial Perspective of Rhizopus oryzae: A Review. J Appl Sci. 2011;11: 2470-2486. doi:10.3923/jas.2011.2470.2486

[102] Mata G, Salmones D, Pérez-merlo R. Hydrolytic enzyme activities in shiitake mushroom (Lentinula edodes) strains cultivated on coffee pulp. Rev Argent Microbiol. 2016;48: 191-195. doi:10.1016/j.ram.2016.05.008

[103] Amore A, Giacobbe S, Faraco V. Regulation of Cellulase and Hemicellulase Gene Expression in Fungi. Curr Genomics. 2013;14: 230-249. doi:10.2174/1389202911314040002

[104] Battaglia E, Benoit I, Brink J Van Den, Wiebenga A, Coutinho PM, Henrissat B, et al. Carbohydrate-active enzymes from the zygomycete fungus Rhizopus oryzae: a highly specialized approach to carbohydrate degradation depicted at genome level. BMC Genomics. 2011;12: 38. doi:10.1186/1471-2164-12-38

[105] Quin MB, Flynn CM, Schmidt-Dannert C. Traversing the fungal terpenome. Nat Prod Rep. 2015;31: 1449-1473. doi:10.1039/c4np00075g.Traversing

[106] Regueira TB, Kildegaard KR, Hansen BG, Mortensen UH, Hertweck C, Nielsen J. Molecular basis for mycophenolic acid biosynthesis in Penicillium brevicompactum. Appl Environ Microbiol. 2011;77: 3035-3043. doi:10.1128/AEM.03015-10

[107] Bushley KE, Raja R, Jaiswal P, Cumbie JS, Nonogaki M, Boyd AE, et al. The Genome of Tolypocladium inflatum: Evolution , Organization, and Expression of the Cyclosporin Biosynthetic Gene Cluster. PLOS Genet. 2013;9: 20. doi:10.1371/journal.pgen.1003496

[108] Lu C, Chen S. Fibrinolytic Enzymes from Medicinal Mushrooms. In: Faraggi E, editor. Protein Structure. In Tech; 2012. pp. 338-362. doi:10.5772/38221

[109] Hynes MJ, Murray SL, Duncan A, Khew GS, Davis MA. Regulatory genes controlling fatty acid catabolism and peroxisomal functions in the filamentous fungus Aspergillus nidulans. Eukaryot Cell. 2006;5: 794-805. doi:10.1128/EC.5.5.794-805.2006

[110] Nascimento TP, Sales AE, Porto CS, Marcos R, Brandão P, Maria G, et al. Production and Characterization of New Fibrinolytic Protease from Mucor subtillissimus UCP 1262 in Solid-State Fermentation. Adv Enzym Res. 2015;3: 81-91.

[111] Sun D, Liao J, Sun L, Wang Y, Liu Y, Deng Q, et al. Effect of media and fermentation conditions on surfactin and iturin homologues produced by Bacillus natto NT-6: LC – MS analysis. AMB Express. 2019;9: 1-9. doi:10.1186/s13568-019-0845-y

Section 5

Microalgae Biomass

Chapter 19

Microalgae: The Multifaceted Biomass of the 21st Century

Donald Tyoker Kukwa and Maggie Chetty

Abstract

Microalgae are unicellular, eukaryotic organisms which possess unique qualities of replication, producing biomass as a precursor for biofuels, nutraceuticals, biofertilizer, and fine chemicals including hydrocarbons. Microalgae access nitrates and phosphates in wastewater from municipalities, industries, and agricultural processes to grow. Wastewater is, therefore, culture media for microalgae, and provides the needed nutrients, micronutrients, inorganic and organic pollutants to produce microalgae biomass. Suitable strains of microalgae cultivated under mesophilic conditions in wastewater with optimized hydrodynamics, hydraulic retention time (HRT), luminous intensity, and other co-factors produce biomass of high specific growth rate, high productivity, and with high density. The hydrodynamics are determined using a range of bioreactors from raceway ponds, photobioreactors to hybrid reactors. Carbon dioxide is used in the photosynthetic process, which offers different growth stimuli in the daytime and the night-time as the microalgae cultivation technique is navigated between autotrophy, heterotrophy, and mixotrophy resulting in microalgal lipids of different compositions.

Keywords: biomass production, autotrophy, heterotrophy, mixotrophy, wastewater treatment, pollutant sequestration, microalgal lipid production, biofuels, nutraceuticals, biofertilizer, photobioreactors, hybrid reactors

1. Introduction

Algae represent a highly diverse consortium of polyphyletic, thallophytic, photosynthetic, and cryptogamic organisms. *They* are morphologically simple, chlorophyll-containing, non-flowering, and typically aquatic plants of a large family with members including seaweeds and a range of microscopic and unicellular to very large multicellular organisms [1]. Algae are either prokaryotes or eukaryotes and lack vascular tissue, leaves, true stems, and roots. The prokaryotic algae are the blue-green algae, which are also referred to as Cyanobacteria or Cyanoprokaryota and belong to the kingdom Eubacteria. The eukaryotic algae belong to the kingdom Protoctista. Cyanobacteria also derive their energy through photosynthesis but do not have a nucleus and membrane-bound organelles, like chloroplasts (see **Figures 1** and **2**) and their prokaryotic nature describes the single-stranded deoxyribonucleic acid (DNA) in their formation, which confers the bacterial identity. On the other hand, eukaryotic algae have double-stranded DNA in their makeup and are equipped with a nucleus and chloroplast. The term "algae" is therefore exclusively reserved for the eukaryotic organisms; and this chapter considers and treats the prokaryotic cyanobacteria as bacteria [1, 2].

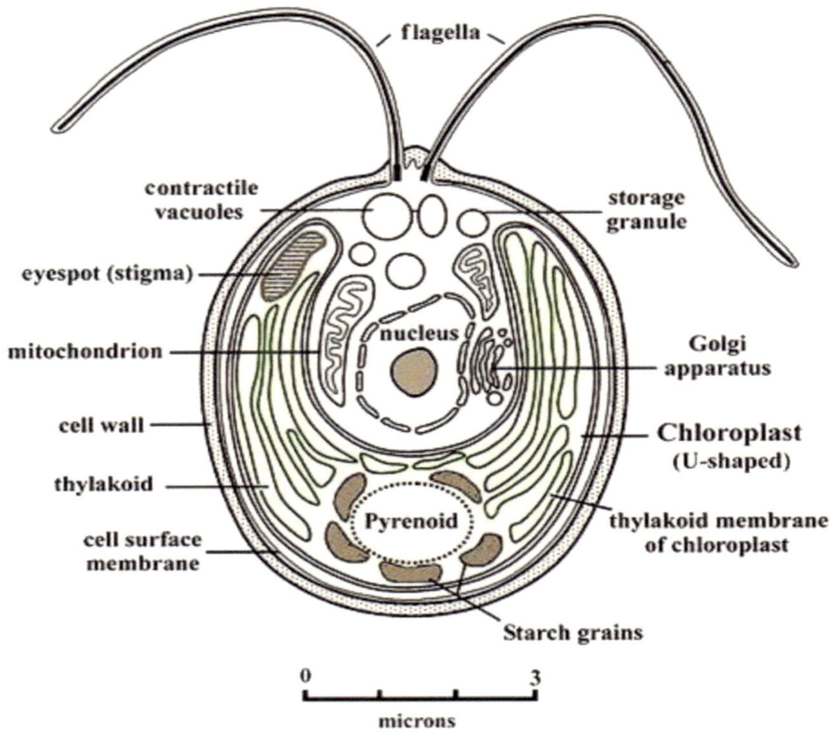

Figure 1.
The microalga Chlamydomonas reinhardtii's *cell structure [3].*

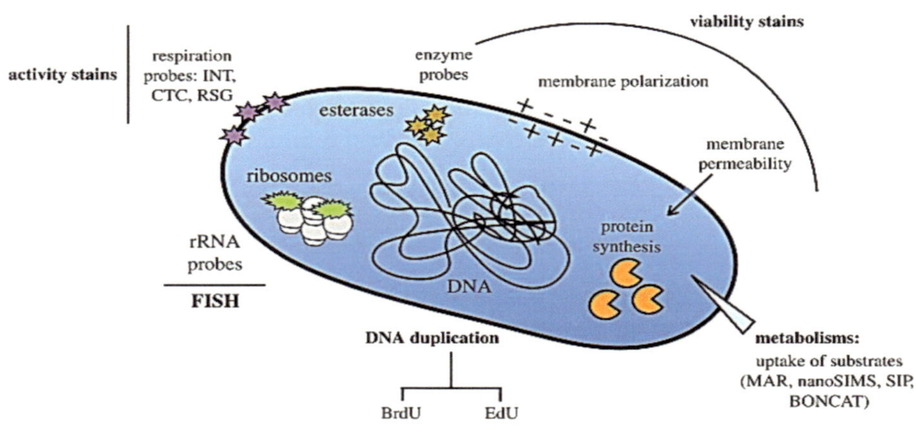

Figure 2.
Schematic of a prokaryotic cell with an indication of some of the methods used to probe cellular activity or growth [4].

Algae have six types of life cycles viz. haplontic, diplontic, isomorphic, heteromorphic, haplobiontic, and diplobiontic cycles; the exposition of these algal life cycles is discussed elsewhere [5]. The microscopic algae are the microphytes or microalgae and are typically found in freshwater and marine ecosystems at the benthic depths and in the water column. They are reported to be the chief converters of water and carbon dioxide to biomass and oxygen (see Eq. (1)) as they receive radiation from sunlight, and are therefore referred to as primary producers. Microalgae exist either individually, or in chains or groups; and depending on the species, their sizes are typically 3–30 μm, while the cyanobacteria are as small as 0.2–2 μm [2].

$$H_2O + CO_2 \xrightarrow{hv} CH_2O + O_2 \quad (1)$$

Aside from producing oxygen and availing themselves as food for a large number of aquatic animals, algae are a good resource base for fine chemicals, crude oil, food supplement for humans, and some pharmaceutical products and finished goods [5].

2. Photosynthetic pigments

Pigments are chemical compounds that reflect and transmit only certain wavelengths of visible light. This makes them appear as the colors perceived. More important than their reflection of light is the ability of pigments to absorb light of certain wavelengths. A photosynthetic pigment (accessory pigment; chloroplast pigment; antenna pigment) is a pigment that is present in chloroplasts of algae and other photosynthetic organisms and captures the light energy necessary for photosynthesis. The reaction of each pigment is associated with only a narrow range of the spectrum, and it is necessary to produce several kinds of pigments with different colors to capture more of the sun's energy. Five important pigments found in algae are (i) chlorophyll (ii) xanthophyll (iii) fucoxanthin (iv) phycocyanin and (v) phycoerythrin [6].

2.1 Chlorophyll

Algae and plants have chloroplasts in which the light-capturing chlorophyll is located, while in cyanobacteria the main light-capturing complex protein molecular assemblies are the phycobilisomes, which are located on the surface of thylakoid membranes [7]. Both chlorophyll and phycobilisomes absorb light most strongly between the high-frequency, high-energy wavelengths of 450 and 495 nm, which happen to be the blue region of the electromagnetic spectrum. Also, the photosynthetic pigments absorb the low-frequency, low-energy wavelengths between 620 and 750 nm, which is the red region of the electromagnetic spectrum. The chlorophyll pigment comes in different forms, and the structure of each type of Chlorophyll pigment is anchored on a chlorin ring with a magnesium ion at the centre. The side chain of each chlorophyll pigment type is different and they are so identified (see **Figure 3** and **Tables 1** and **2**) [7, 8].

Chlorophyll a with the molecular formula $C_{55}H_{72}O_5N_4Mg$ is the most common type of Chlorophyll. It is a green pigment with a chlorin ring having magnesium at the centre (see **Figure 3**). Chlorin is a tetrapyrrole pigment, which is partially hydrogenated porphyrin. The ring-shaped molecule is stable with electrons freely migrating around it to establish resonance structures [9]. It also has side chains and a hydrocarbon trail and contains only $-CH_3$ groups as side chains. The long hydrophobic tail anchors the molecule to other hydrophobic proteins on the surface of the thylakoid membrane. The chemical structural layout of chlorophyll shows a porphyrin ring attached to a protein backbone (see **Figure 3**). By substituting functional groups at positions C2, C3, C7, C8, and the C17-C18 bond, one can identify the structure of the desired chlorophyll (see **Tables 1** and **2**). Chlorophyll captures and absorbs blue, violet, and red light from the spectrum to transmit or reflect green, which is the color that the green algae exhibit [9, 10]. Oxygenic photosynthesis uses chlorophyll a to furnish electrons in the electron-transport chain. Photosystems I and II harbor many pigments that help to capture light energy.

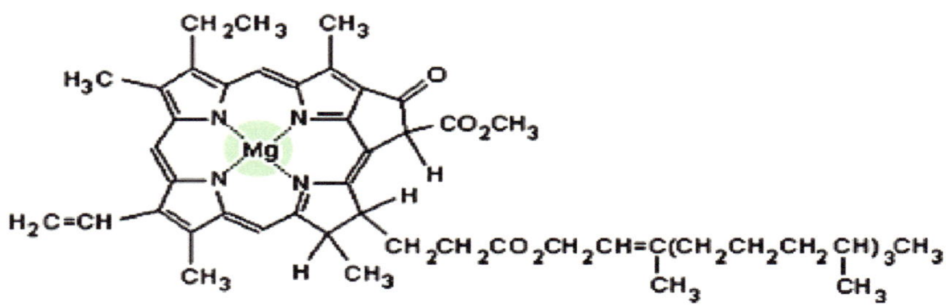

Figure 3.
Chlorophyll - a porphyrin ring structure attached to a protein backbone. The porphyrin is built up of pyrrole molecules – 5 membered aromatic rings which are made of four carbons and one nitrogen atom. This ring system acts as a polydentate ligand and has a magnesium cation at its Centre [8].

	Chlorophyll			
	a	b	c1	c2
Molecular Formula	$C_{55}H_{72}O_5N_4Mg$	$C_{55}H_{70}O_6N_4Mg$	$C_{35}H_{30}O_5N_4Mg$	$C_{35}H_{28}O_5N_4Mg$
C2 group	$-CH_3$	$-CH_3$	$-CH_3$	$-CH_3$
C3 group	$-CH{=}CH_2$	$-CH{=}CH_2$	$-CH{=}CH_2$	$-CH{=}CH_3$
C7 group	$-CH_3$	-CHO	$-CH_3$	$-CH_3$
C8 group	$-CH_2CH_3$	$-CH_2CH_3$	$-CH_2CH_3$	$-CH_2CH_3$
C17 group	$-CH_2CH_2COO$-Phytyl	$-CH_2CH_2COO$-Phytyl	-CH=CHCOOH	-CH=CHCOOH
C17-C18 bond	Single (chlorin)	Single (chlorin)	Double (porphyrin)	Double (porphyrin)
Occurrence	Universal	Plants	Algae	Algae

Table 1.
Chemical structure of chlorophyll.

	Chlorophyll	
	d	f
Molecular formula	$C_{54}H_{70}O_6N_4Mg$	$C_{55}H_{70}O_6N_4Mg$
C2 group	$-CH_3$	-CHO
C3 group	-CHO	$-CH{=}CH_2$
C7 group	$-CH_3$	$-CH_3$
C8 group	$-CH_2CH_3$	$-CH_2CH_3$
C17 group	$-CH_2CH_2COO$ – Phytyl	$-CH_2CH_2COO$ – Phytyl
C17-C18 bond	Single (Chlorin)	Single (chlorin)
Occurrence	Cyanobacteria	Cyanobacteria

Table 2.
Chlorophyll structural formulae.

A unique pair of pigment molecules are located at the reaction site of each photosystem. For photosystem I the unique pair is referred to as P700, while for photosystem II it is identified as P680. These reaction sites receive resonance energy released from chlorophyll a to sustain the redox reactions [10].

Chlorophyll b is found only in the green algae and in plants, and it absorbs most effectively at 470 nm (blue) but also at 430 nm and 640 nm. Molecular formula - $C_{55}H_{70}O_6N_4Mg$. It is an accessory photosynthetic pigment. The molecular structure consists of a chlorin ring with Mg centre. It also has side chains and a phytol tail. Pyrrole ring II contains an aldehyde group (− CHO). Chlorophyll b absorbs energy that chlorophyll a does not absorb. It has a light-harvesting antenna in Photosystem I [11].

2.2 Xanthophyll

Xanthophyll is one of the two major groups of the carotenoids group. Generally, it is a C_{40} terpenoid compound formed by condensation of isoprene units. Xanthophyll, with the formula $C_{40}H_{56}O_2$, contains oxygen atoms in the form of hydroxyl groups or as epoxides. Xanthophyll acts as an accessory light-harvesting pigment. They have a critical structural and functional role in the photosynthesis of algae and plants. They also serve to absorb and dissipate excess light energy or work as antioxidants. Xanthophyll may be involved in inhibiting lipid peroxidation [12].

2.3 Fucoxanthin

Fucoxanthin, with the formula $C_{42}H_{58}O_6$, is a xanthophyll carotenoid, being an accessory pigment that drives limited photosynthetic reactions in brown algae (phaeophytes) and other stramenopiles. It renders the brown or olive-green color to these seaweeds. Fucoxanthin captures the red light of the spectrum for photosynthetic activities. Some edible brown algae produce this pigment in abundance, and typical candidates in this category include *Sargassum incisifolium*, *Sargassum fulvellum, Undaria pinnatifida, Laminalia japonica,* and others. The alga *Sargassum incisifolium* has been used as a source of Fucoxanthin as a nutraceutical for its antiobesity effects and as much as 0.45 mg/g has been reported [12, 13]. Another rich source of Fucoxanthin is the South African brown alga *Zonaria subarticulata* and extracts as high as 0.50 mg/g have been reported, leading to preparations such as FucoThin™ [13]. The concentration of Fucoxanthin in any algal species may depend on geographical location, seasonal variations, life-cycle, and other factors.

2.4 Phycocyanin (PC)

Phycocyanin is a protein-pigment complex found in cyanobacteria as an accessory pigment to phycobilisomes. As a phycobiliprotein, phycocyanin is identified by the color it bears as blue phycocyanin. Depending on the cyanobacterial species, this can be phycocyanin, showing maximum absorbance at 620 nm and identified as C-PC, and allophycocyanin with maximum absorbance at 650 nm and identified as A-PC. From the red microalgae, phycocyanin is identified as R-PC [13]. The molecular structure of phycocyanin changes with the pH of the medium, exhibiting the $(\alpha\beta)_3$ trimeric structure at pH 7. However, at the pH range of 5–6, the much more available phycocyanin, C-PC, assumes the hexameric structural conformation $(\alpha\beta)_6$. Phycocyanin boosts the human and animal immune systems and protects against certain diseases. It exhibits hepatoprotection, cytoprotection, and neuroprotection. Persons undergoing chemotherapy and radiation for cancer are placed on Phycocyanin from spirulina as a dietary supplement to ease negative symptoms during treatment as well as rejuvenate post-treatment. Phycocyanin is used in the food industry as a food additive [12, 14].

2.5 Phycoerythrin (PE)

Phycoerythrin is an accessory pigment to the main chlorophyll pigment complex found in red algae and cryptophytes; it is part of a covalently bonded phycobilin chromophore in the family of phycobilins, typical of which is phycoerythrobilin, the phycoerythrin acceptor chromophore. Phycoerythrin is made up of (αβ) monomers aggregates. Except for phycoerythrin 545 (PE545), these monomer aggregates are assembled into $(\alpha\beta)_3$ trimers or $(\alpha\beta)_6$ hexamers with 3 or 32 symmetry and enclosing central channel [13, 14]. In red algae, they are attached to the stroma of thylakoid membranes of chloroplasts, whereas in cryptophytes, phycobilisomes are reduced and housed inside the lumen of thylakoids. Phycoerythrin captures light energy from the electromagnetic radiation and directs it to the reaction site through the phycobiliproteins, phycocyanin, and through A-PC. Each trimer and hexamer in the phycobilisome (PBS) has a minimum of one linker protein at the central channel. The α and β chains in B-phycoerythrin (B-PE) and R-phycoerythrin (R-PE) from the red algae also have γ sub-units conferring both link and light-capturing capabilities due to the presence of chromophores [14] (**Figure 4**).

The chloroplast of algal cell contains the water-soluble phycobilin pigments and while the same phycobilin pigments are found in the phycocyanin and phycoerythrin of Cyanobacteria and the red algae, the Rhodophyta. The algal chlorophyll has a structural difference from Bacteriochlorophylls (Bchl) of cyanobacteria, the latter having one of the porphyrin rings saturated, and absorbing longer wavelengths of light as opposed to chlorophylls. *Rhodopseudomonas viridis* has its bacteriochlorophyll b absorb 960 nm wavelength of light [15].

2.6 The Chromophore

The colors of pigments are the reflections of the electromagnetic spectrum from the pigments. A portion of the pigment molecule causes the formation of the color perceived, and this moiety is referred to as **chromophore**. A chromophore has two energy levels referred to as orbitals, and the difference in their energies lie within the visible spectrum of electromagnetic radiation. Thus, a photon of incident light can excite an electron from its ground-state orbital to the excited state. Chromophores are generally either **conjugated pi-systems** or transition **metal complexes** [16]. For **conjugated pi-systems,** electrons are excited between **pi**-orbitals distributed over alternating single and double bonds. Where conjugated systems are less than eight conjugated double bonds, absorbance occurs only in the ultraviolet region and is visible to the human eye. But blue or green compounds essentially do not rely on conjugated pi-bonds alone. Typical chromophores in this category are the azo compounds, lycopene, β-carotene, anthocyanin, and retinenes. **Metal complex chromophores** have transition metals whose d-orbitals are incomplete but are shared with the ligands. Chromophores in this category are the chlorophylls, hemoglobin, and hemocyanin [17].

In general, chromophores comprise four pyrrole rings; identified as (i) open-chain pyrroles with no transition metal involved – typically, carotenoids, phycobilins, and phytochromes, (ii) pyrroles arranged as a porphyrin ring with a central transition metal atom – typically, chlorophylls and bacteriochlorophylls ($C_{55}H_{74}MgN_4O_6$). Chlorophyll absorbs all other visible components of light except green, which is the color the human eye sees of plants in their leaves. Various chlorophylls and accessory pigments (as discussed in sections 2.1–2.5) have characteristic *absorption spectra*; and the *action spectrum* that drives photosynthetic

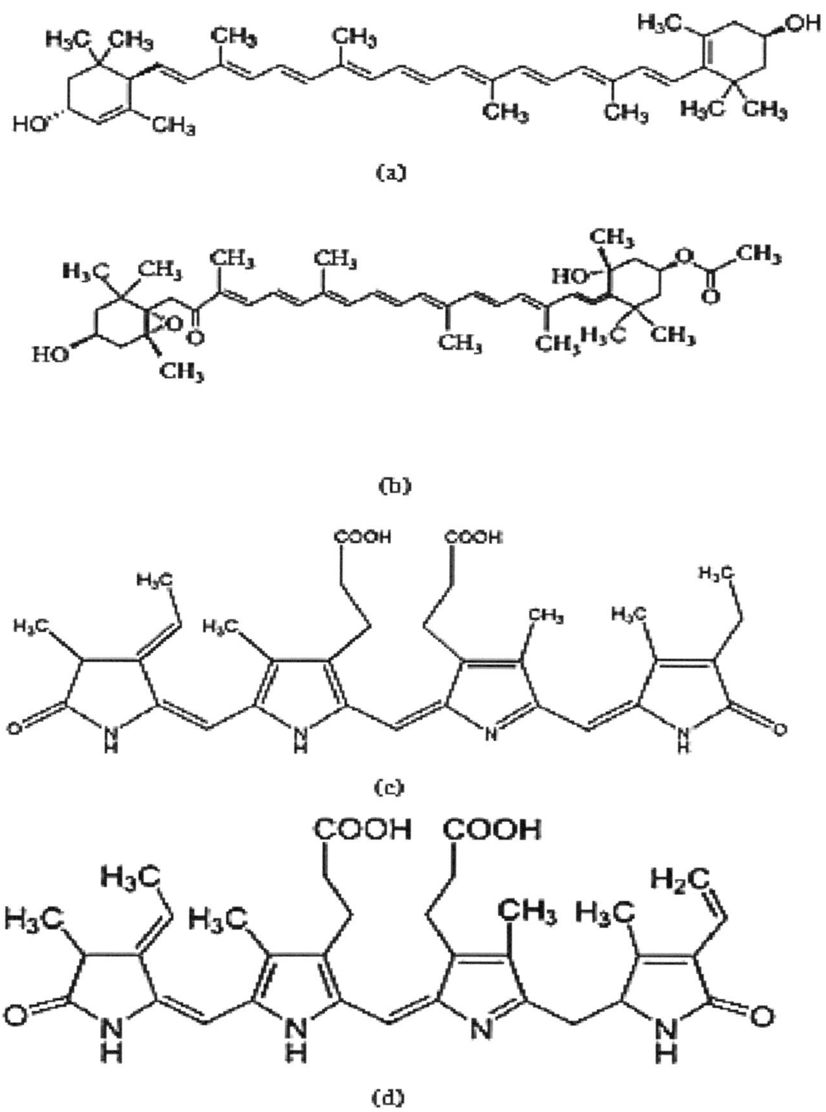

Figure 4.
The structure of the pigments: (a) xanthophyll (b) Fucoxanthin (c) Phycocyanin and (d) Phycoerythrin [12–14].

reactions relates proportionately to the different wavelengths of light (see **Figure 5**). On absorbing light energy, one or more of the following effects happen in a pigment: (i) light energy is transformed to heat energy, or (ii) there is fluorescence, signifying that light energy is re-emitted at longer wavelengths, or (iii) the quantum of **energy is passed** from an excited **molecule of chlorophyll to another molecule** in a process called **exciton** transfer, or (iv) the reaction centre (RC) chlorophyll absorbs the energy and gives up an excited electron to an electron acceptor and (v) the RC chlorophyll is unstable and wants to replace its missing electron, which creates concentration gradients, leading to the production of ATP and NADPH, which are fed into the Calvin cycle (see **Figure 6**) to produce carbohydrates [18].

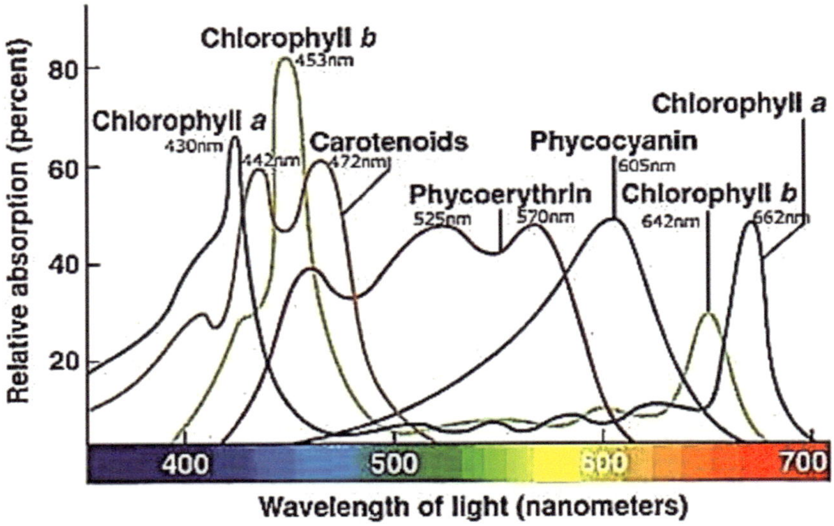

Figure 5.
Relative absorbance of photosynthetic pigments as a function of the wavelength of light [19].

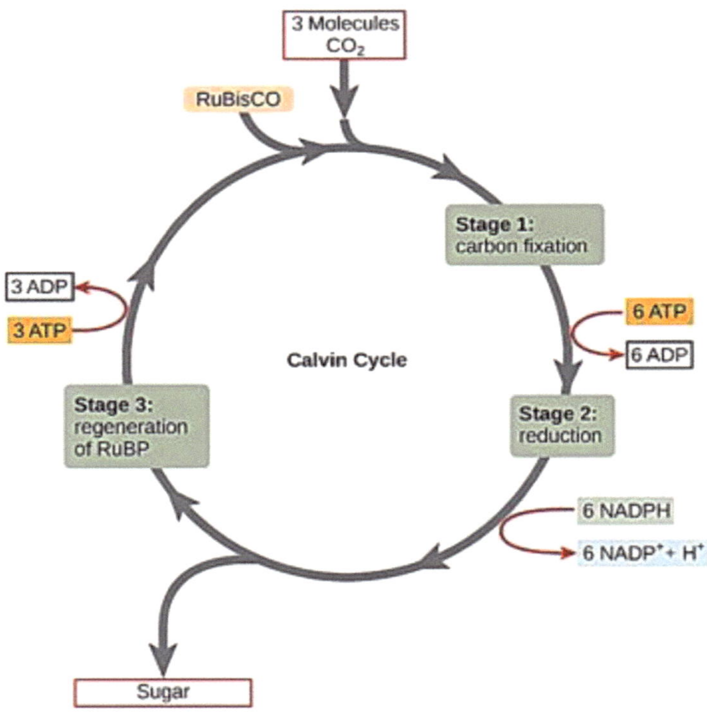

Figure 6.
The Calvin cycle [20].

2.7 The Calvin cycle

The dark reactions of photosynthesis occur in the stroma of the chloroplast and are referred to as the Calvin cycle. Although the Calvin cycle does not utilize light and can happen during the daytime or at night, they employ products of the

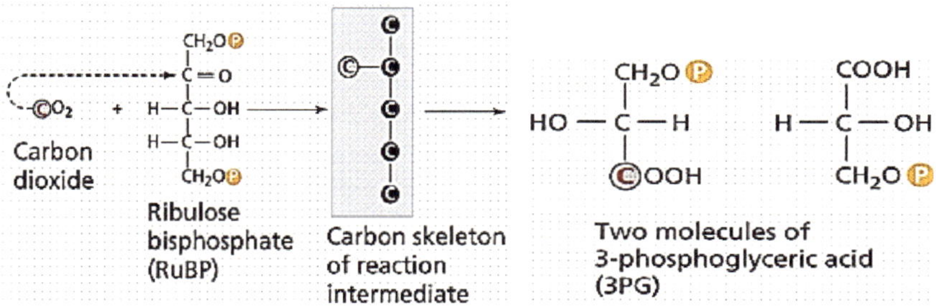

Figure 7.
The first reaction in the Calvin cycle: Carbon fixation.

light-dependent reactions to propagate. Products of the light-dependent reaction are ATP and reduced NADP; the energized electrons from the light-dependent reactions provide the energy to produce carbohydrates from carbon dioxide molecules.

Stage 1: Carbon fixation - Carbon-Fixing Reactions are also known as the Dark Reactions during which CO_2 gas diffuses through and dissolves in the water around the walls of mesophyll cells, diffuses through the cytoplasm and chloroplast membrane into the stomata. In the stroma, CO_2 undergoes a ribulose bisphosphate carboxylase (Rubisco) enzymatic catalyzed reaction with ribulose bisphosphate (RuBP) [19] (**Figure 7**).

The Calvin Cycle first produces phosphoglyceric acid (PGA), which is phosphorylated, using the energy carriers ATP and NADPH generated by the photosystems I and II, to produce 12 molecules of phosphoglyceraldehyde (PGAL). Two molecules of PGAL are ejected from the cycle in the form of a glucose molecule. The other ten molecules of PGAL are converted to 6 RuBP molecules, using the inherent energy in ATP and the cycle continues [19, 20].

Stage 2: Reduction and sugar production – The cell utilizes the high energy molecules ATP and NADPH and reduces 3-PGA to form triose phosphate, G3P.2G3P, which leaves the cycle to produce glucose, starch, cellulose, lipids, amino acids, and nucleotides [20].

Stage 3 Regeneration – The remaining then G3P (3-GPA) in the cycle are regenerated to RuBP, which is a 6-carbon molecule with 2 phosphates, and it requires energy to generate. This process utilizes the high energy ATP made during the light-dependent reactions. The RuBP molecule formed then interacts with more CO_2 from the atmosphere and generates more PGA to keep the cycle going [20].

The summary of the reactions in the Calvin cycle (see Eq. (2))

$$\begin{aligned} &3\,\mathrm{CO_2} + 6\,\mathrm{NADPH} + 6\,\mathrm{H^+} + 9\,\mathrm{ATP} \rightarrow 3 - \mathrm{PGAL} \\ &+6\,\mathrm{NADP^+} + 9\,\mathrm{ADP} + 3\,\mathrm{H_2O} + 8\,\mathrm{P_i}\ (\mathrm{P_i} = \text{inorganic phosphate}) \end{aligned} \tag{2}$$

2.8 The inherent energy of a photon

Light has properties of both waves and particles, from the quantum mechanics point of view [20]. The particulate behavior of light presents light as a stream of particles of energy, known as photons, which interact with electrons to cause the energy contained in the light to disappear and then reappear as the kinetic energy of the ejected electrons plus a work function.

$$E_{photon} = hv = KE_{electron} + \varphi \tag{3}$$

where E_{photon} is the energy of a photon, h is Planck's constant (6.626×10^{-34} J·s), and v is the frequency of the light wave, $KE_{electron}$ is the kinetic energy of electron and φ is the work function, which defines the minimum amount of energy that is necessary to induce photoemission of electrons from the surface of a metal, and the value of φ depends on the metal. We are dealing with biotic materials in the context of this chapter, so we may assume that $\varphi = 0$.

By definition, $v = \frac{c}{\lambda}$, where c is the velocity of light (3×10^8 m/s in a vacuum), and λ is the wavelength of light. It is important to note that the energy content of light of shorter wavelength is higher than that of longer wavelengths; and for one mole of photons, the energy is the total of the energies of all the particles in one mole, which is given in Eq. (4).

$$E = NE_{electron} \tag{4}$$

where N is the Avogadro's number (6.02×10^{23} molecules or photons/mol). Thus

$$E = NE_{electron} = hv = Nh\frac{c}{\lambda} \tag{5}$$

Thus for sunlight with a wavelength of 650 nm (650×10^{-9} m), the energy is computed in Eq. (6).

$$\begin{aligned} E &= \left(6.02\text{x}10^{23}\,\text{photons}/mol\right)\left(6.626\text{x}10^{-34}\,\text{J.s}\right)\left(\frac{3\text{x}10^{8}\,\text{m}/\text{s}}{650\text{x}10^{-9}\,\text{m}}\right) \\ &= 184100.86\ \text{J/mol} \\ &\cong 184.1\ \text{kJ/mol} \end{aligned} \tag{6}$$

If all this were to be used for synthesizing ATP from ADP and **Pi** it would be enough to synthesize several moles [20].

Chlorophylls b, c, d, and e are accessory pigments with xanthophylls, and carotenoids in algae and protistans, Pigments that are not accessory to chlorophyll absorb light energy at wavelengths that do not stimulate chlorophyll. Light energy absorbed by accessory pigments is channeled to the reaction site and is converted into chemical energy. The ability to absorb some energy from the longer, more penetrating wavelengths probably conferred an advantage to the benthic photosynthetic algae. Depending upon turbidity of the water, the shorter, high energy wavelengths penetrate very little in the euphotic zone (below 5 meters) in seawater [7, 8]. Chlorophyll molecules being the main producers of pigments are bound to proteins of the photosynthetic membranes and capture the sunlight in oxygenic plants, and convert light energy into chemical energy. This is facilitated by pigment-protein complexes known as Photosystem I (PSI) and Photosystem II (PSII) reaction sites [9]. In PS II water is *split* and the electrons are used to replenish excited electrons that are lost from the photosystem. The loss of electrons during the oxidation of water results in the formation of O_2 gas. In PS I the electron acceptor is first in an electron transport system in the thylakoid membrane. Electrons pass through the chain via a series of redox reactions until it hit the final electron acceptor. The final electron acceptor is $NADP^+$ which is reduced to NADPH. ATP is produced throughout the whole process via chemical osmosis, meaning using an H^+ gradient during electron transport (photophosphorylation). It has been shown that [6] the protein is composed of seven transmembrane helices with a retinal chromophore covalently bonded in the central region via a protonated Schiff base to a lysine residue (see **Figure 8**).

The most common chlorophylls are chlorophyll a, chlorophyll b, and chlorophyll c1, and chlorophyll c2. Each pigment registers a maximum signal at a particular wavelength of maximum absorption ($\lambda_{\max}$), this coupled with the selective scattering of light, microalgae are seen in their defined color. Consequently, in the blue region of the spectrum, $\lambda_{\max} = 440\,nm$ and in the red region of the spectrum, $\lambda_{\max} = 675\,nm$ (see **Figure 5**) [22]. As the microalgae cells are irradiated with light from a source, the photosynthetic process is initiated and propagated. The photosynthetic rate increases as the intensity of the irradiance are increased; and the level of irradiance is reached where the rate of photosynthesis attains a maximum and begins to retard (see **Figure 9**). At this stage, the cells are said to be in photoinhibition. Photoinhibition is thus a phenomenon that describes the inability of the photosynthetic organism to support photosynthesis in the presence of excess illumination from a light source [19].

The saturation irradiance ($I_k = \frac{P_m^B}{\alpha^B}$) and photo-inhibition (β) parameters indicate the half-saturation constant when the photosynthetic rate is half the maximum value of the photosynthetic rate of the system $\left(P = \frac{P_m^B}{2}\right)$. α^B is the measure of photosynthetic efficiency of solar energy conversion into chemical energy, and it takes into account that the light absorbed by the algal cell is

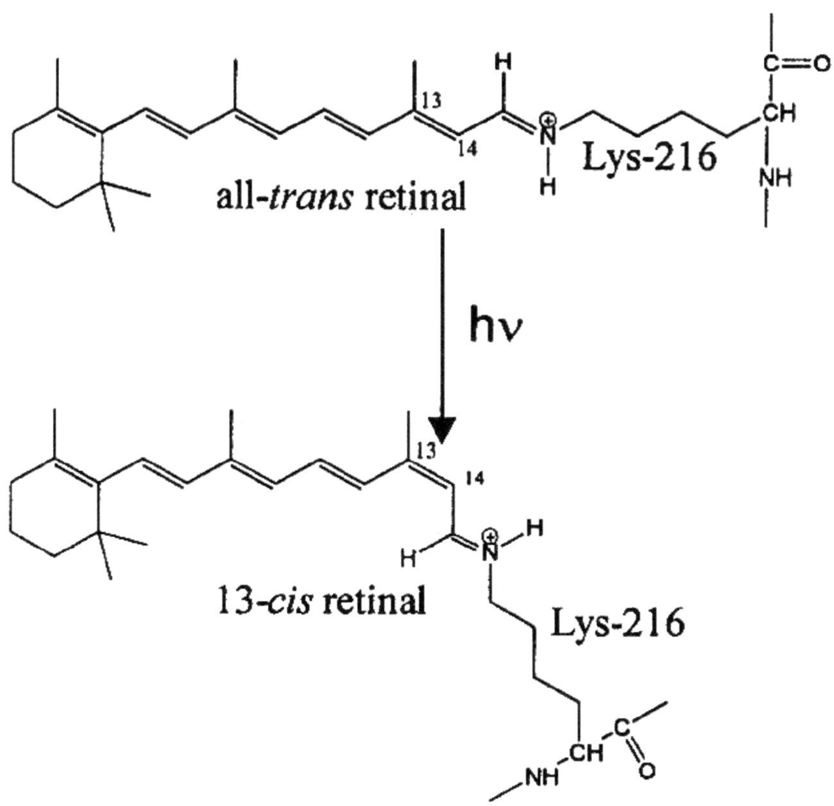

Figure 8.
Photo-isomerization of all-trans to 13-cis retinal in bR [21].

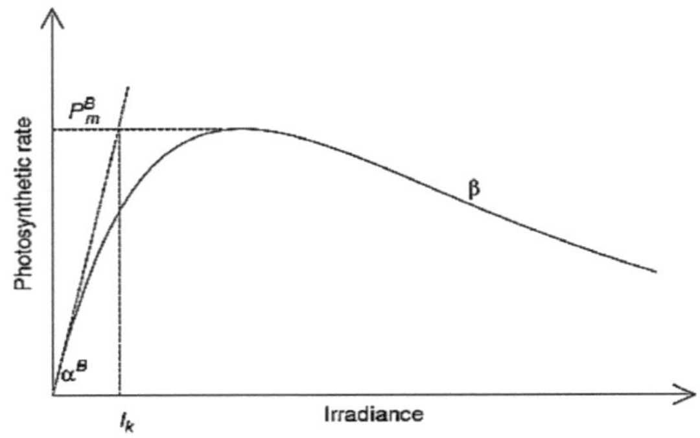

Figure 9.
Photosynthesis – Irradiance curve.

proportional to the functional absorption cross-section of the effective area that PS II presents to an incoming photon. P_m^B is the assimilation number which is the maximum photosynthetic rate [22].

3. The microalgae biomass

Microalgae is a promising renewable resource for biofuels, and optimization and control of the biomass growth production have gained economic and commercial interests. Algae do not compete with traditional food crops for space and resources [5]. Microalgae are highly diverse and differences within and between both species and populations lead to significant differences in biogeography and the environment. The macromolecular composition of the microalgae is of interest for understanding nutrient competition within microalgal communities, food web interactions, and developing algal systems for the development of biofuels, nutraceuticals, and mariculture [3]. Production of microalgae-derived metabolites requires processes for culturing the algae, recovery of the biomass, and further downstream processing to purify the metabolite. The cost of producing microalgal bioactive agents has to be weighed as the downstream recovery of the microalgal products can be substantially more expensive than the culturing of the microalgae [5]. Depending on their origin, algae are referred to as terrestrial algae, snow algae, seaweeds, and phytoplankton. Ubiquitous in marine, freshwater, and terrestrial habitats and possessing broad biochemical diversity, which is the basis for many biotechnological and industrial applications [3].

3.1 Algaculture (culture of microalgae in hatcheries)

Hatcheries are used to produce a range of microalgae biomass, which are used in a variety of ways for commercial purposes. Studies have adduced the success of a microalgae hatchery system to the following factors: (i) the dimensions of the container/bioreactor where microalgae are cultured, (ii) exposure to illumination, and (iii) concentration of microalgal cells within the reactor [23, 24]. Photosynthesis is one of the basic biochemical transformations of photosynthetic micro-organisms that convert solar energy into chemical energy. Many microalgae are autotrophs, which use photosynthesis to produce food. Some heterotrophic microalgae can grow in the dark by utilizing organic carbon. Some microalgae grow by combining both autotrophy and heterotrophy into a hybrid cultivation mode called mixotrophy [4, 6]. Diatoms and dinoflagellates are the two types of microalgae. Diatoms can be spheres, triangles, elliptical or stars. Many dinoflagellates have two flagella for their movement through the water. Both diatoms and dinoflagellates contain oils in their cells, helping them to swim. Both diatoms and dinoflagellates can grow very quickly and cause algal blooms [3].

3.1.1 Biomass by open pond cultivation

There are two main advantages of culturing microalgae using the open pond system. Firstly, an open pond system is easier to build and operate. Secondly, open ponds are cheaper than closed bioreactors because closed bioreactors require parts that are expensive to acquire. However, where the temperature is the growth or lipid accumulation limiting factor, using open pond systems may decrease the productivity of certain commercially important strains such as *Arthrospira sp.* Waste heat and CO_2 from industrial sources can be used to compensate for this [24]. Some organizations use the open raceway pond approach, employing foam fractionation to concentrate microalgal cells before they are lysed by the cavitation bubble collapse. Some commercial outdoor raceway ponds are located near power plants where 4–15% CO_2 from the flue gas is fed to the raceway ponds. 1.8 units of CO_2 are required to produce one unit of algal biomass, and the practical operation of open ponds has shown that dissolved CO_2 in water is not enough; therefore, the bubbling of air into water improves CO_2 dissolution [25]. Maintaining algae monocultures in open ponds poses serious challenges due to contamination with local algae species and invasion of algae predators. Some strategic operation models adopting higher salinity, pH, or temperature operating conditions have been proposed to provide a selective microenvironment to cultivate some commercial strains. In this regard, therefore, successes have been recorded in open ponds *Spiriluna* monoculture commercial cultivation at high pH values ranging from 9.0 to 11.0. In another operation, β-carotene is produced from *Dulaliella* sp. in open ponds wi high salinity values [25, 26] (**Figure 10**).

3.1.2 Biomass by vertical reactor systems

Many photobioreactors have been suggested for commercial production of algal biomass. However, only a few of them are suitable for practical application because of poor gas mass transfer. The vertical tubular photobioreactor provides a greater

Figure 10.
Algae raceway pond: The microalgae culture broth is constantly kept in motion with a powered paddle wheel [23].

surface area for the interaction of light and the algal cells, increasing the time of gas mass transfer in the culture broth, and the efficient uptake of nutrients. Most times, commercial cultivation of microalgae in vertical reactor systems and reactors of other configurations is not economically viable in batch mode, due to the time taken to load, unload, and clean the reactor systems. The vertical tubular reactor can be made of alveolar panels, polyethylene sleeves, or glass tubes and supported on steel frames (see **Figure 11**). The low productivity characterizing this reactor system is overcome when the surface area to volume ratio is increased. The O_2 gas mass transfer is aided by bubbling air through the culture broth [23].

3.1.3 Biomass by horizontal photobioreactors

This is an outdoor microalgal cultivation system, which has tubes laid on the ground to form a network of loops (see **Figure 12 (b)**). A pump is used to mix the microalgal suspended culture, which raises the culture vertically periodically into a photobioreactor. Pulsed mixing at intervals produces better results than continuous mixing. *Arthrospira sp.* used as a dietary supplement was reported to have higher productivity because of a better-suited temperature range and an extended cultivation period during warm weather periods [23]. The horizontal tubular photobioreactor was assembled with three loops of 80 m each (see **Figure 12**) connected via a manifold to a bubble column used for oxygen removal, temperature control, nutrient, and antifoam addition. The horizontal photobioreactor was operated at a

Figure 11.
Vertical tubular photobioreactors for culturing microalgae [23].

Figure 12.
Horizontal tubular photobioreactor of different orientations [27]: (A) standing tubular bioreactor. The cultivation area occupied is less compared to the floor-type. (B) floor tubular reactor. This occupies a large cultivation area.

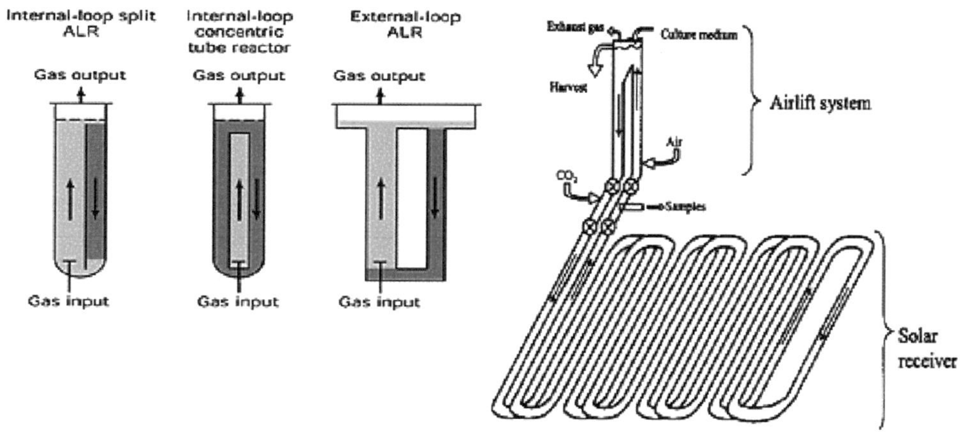

Figure 13.
Different types of airlift photobioreactor [26].

liquid velocity of 0.45 m s^{-1}. To prevent high concentrations of dissolved oxygen, a superficial gas velocity (vgs) of 0.04 m s^{-1} was used in the bubble column [23]. The productivity of *Nannochloropsis sp*. suffered a setback in 250% air-saturated dissolved oxygen solution; and as the dissolved oxygen concentration was raised above 300%, the growth of *Nannochloropsis sp.* was stalled, and that of *Neochloris oleoabundans* was inhibited. However, increasing the airflow rate in the reactor removed the growth inhibitory effect of dissolved oxygen [24].

3.1.4 Biomass by the air-lift method

This is an outdoor microalgal cultivation technique for the production of biomass and metabolites under a highly controlled environment. By this technique, the air is moved within the system to circulate the medium in which microalgae is growing. The culture is grown in transparent tubes that lie horizontally on the ground and are connected by a network of pipes (see **Figure 13**). Air is passed through the tube such that air escapes from the end that rests inside the reactor that contains the culture and creates an effect like stirring [28]. Other configurations of the airlift reactor are an improvement over this design. The external-loop ALR is a promising configuration for breakthrough scale-up *Scenedesmus sp.* biomass production [26].

4. Metabolic platforms for microalgal growth

Different microalgae strains acclimate in different environments, evolving their metabolic pathways to stimulate and propagate growth. However, the extent of growth depends on the composition of the culture media which can be enhanced by either inorganic or organic carbon metabolism or both. Other co-factors such as nutrient availability, pH, chemical oxygen demand (COD), and temperature also influence growth, and the accumulation of metabolites in microalgae (see **Table 3**) [29].

4.1 Autotrophic metabolism

The photosynthetic CO_2-fixation in microalgae suffices to possess a greater ability to fix CO_2. Photo trophy refers to an autotrophic mode of metabolism in which

Metabolic mode	Energy source	Carbon source	Light availability	Metabolism availability
Photo-autotrophic	Light	Inorganic	Obligatory	Fixed
Heterotrophic	Organic	Organic	Not required	Switch between sources
Photoheterotrophic	Light	Organic	Obligatory	Switch between sources
Mixotrophic	Light & organic	Inorganic & organic	Not obligatory	Simultaneous utilization

Table 3.
Microalgal metabolic requirements.

organisms can harness light energy with the help of photosynthetic pigments and convert it to chemical bond energy in the form of ATP (photophosphorylation).

Autotrophy is the ability of PMOs to use inorganic carbon in the form of CO_2 as the sole source of carbon to synthesize organic compounds necessary to build cell components. This is also referred to as carbon-autotrophy to distinguish the ability of some organisms to use molecular nitrogen as the sole source of nitrogen. Such organisms are referred to as nitrogen autotrophs. However, autotrophy as used in this chapter is carbon autotrophy. This is a property that is present primarily, in plants, algae, and phototrophic bacteria including cyanobacteria [30].

Aside from these organisms, all of which are photosynthetic, several groups of non-photosynthetic bacteria can grow using CO_2 as the sole source of carbon by their ability to oxidize inorganic compounds. Such organisms are chemoautotrophic or chemolithotrophic [31].

CO_2 is the end-product of aerobic respiration, a process that releases the energy of respiratory substrates. Carbon dioxide is, therefore, poor in energy content. In autotrophic metabolism, this energy-poor compound is used to build organic molecules which are much richer in energy content. Therefore, It is noted that the conversion of CO_2 to organic compounds requires the input of energy from an external source. The ultimate source in the case of photosynthesis is radiant energy and in the case of chemolithotrophy is the oxidation energy of inorganic chemical compounds. In either case, the immediate source of energy for driving the endergonic reaction involved in the conversion of CO_2 to organic compounds is ATP [32].

In photosynthesis, ATP is generated with the help of photosynthetic pigments through a process known as photophosphorylation. In chemoautotrophy, the energy of oxidation of inorganic compounds is channelized into the respiratory chain for ATP synthesis by oxidative phosphorylation.

Thus, autotrophic metabolism consists of two sets of reactions viz. (1) the ATP and the reducing force are generated and, (2) they are used for the reduction of CO_2 to organic compounds.

The reactions in (1) are different in phototrophic and non-phototrophic autotrophs. But the reactions in (2) are common between the two groups. In the majority of autotrophs, the reactions involved in the reduction of CO_2 proceed via a cyclic pathway, known as the reductive pentose phosphate pathway or, more commonly, as the Calvin-Benson cycle, or simply the Calvin cycle, although other pathways are also known to operate in some organisms, both in the phototrophic green plants and bacteria. The reduction of CO_2 to yield organic compounds is commonly known as CO_2-fixation [32, 33].

4.2 Heterotrophic metabolism

The supply of sufficient light for massive growth is the main goal and a limiting factor for microalgal cultivation. To ignore the requirement for illumination and present the possibility of high cell concentration, points at heterotrophic cultivation as a promising, efficient, and sustainable strategy for certain microalgae to produce metabolites of value by using carbon substances as the sole carbon and energy source. The optimized preliminary cell culturing of microalgae species is an important stage in culturing microalgae biomass at the commercial scale. The growth environment during the culturing process can be [32] either autotrophic (inorganic carbon) or heterotrophic (organic carbon) depending upon the nature of cells and their growth tendencies. Heterotrophic and mixotrophic microalgae are more capable of growing much faster with higher cellular oil accumulation as compared to autotrophic microalgae species. However, heterotrophic microalgae require organic carbon sources like glycerol, glucose, or acetate as a sole source of carbon for growth, which is responsible for about 80% of the costs of culture media [33]. The metabolism of respiration is applied to produce energy. The respiration rates, intimately geared to the growth and division, are determined by the oxidization of organic substrates of the given microalgae [32]. Glucose provides the organic carbon needed and it is preferred because of its high energy density compared to other sources. The oxidative assimilation of glucose employs either the Embden–Meyerhof–Parnas (EMP) pathway or the pentose phosphate (PP) pathway depending on the cycle position. During the dark cycle, PMOs assimilate and metabolize glucose via the PP pathway. However, during the daytime cycle, glycolysis in the cytosol is via the EMP pathway [34]. The growth rate, lipid content, and the ATP of microalgae under the heterotrophic metabolic strategy are higher compared to those under the photoautotrophic metabolic strategy but depend mainly on the PMO's species and strain used. The PMO's growth is steady and rapid in a nutrient-rich culture media using a high level of system control, to achieve biomass production of 50–100 g L^{-1} in heterotrophy which is higher than that achieved in photoautotrophy [35].

Heterotrophic metabolism eliminates the two main problems associated with autotrophic metabolism viz. (i) it allows the use of practically any vessel as a bioreactor, and (ii) low energy and high yield, as major outcomes, giving a significant reduction in costs for the process. Cost-effectiveness and relative simplicity of operations and daily maintenance are the main attractions of the heterotrophic growth approach. A significant benefit is that it is possible to obtain, heterotrophically, high densities of microalgae cells that provides an economically feasible method for large scale, mass production cultivation [34].

Heterotrophy has its drawbacks viz. (1) The microalgae species and strains that can grow by the heterotrophic strategy are limited; (2) Increasing energy expenses and costs by adding organic carbon substrate; (3) Contamination and competition with local microorganisms; (4) Inhibition of growth by excess organic substrate; and (5) Inability to produce light-induced metabolites [35]. Nonetheless, heterotrophic cultures are gaining increasing application for producing a wide variety of microalgal metabolites from bench experiments to commercial scale.

4.3 Mixotrophic metabolism

Mixotrophic cultivation of microalgae strategies provides both carbon dioxide and organic carbon simultaneously and both chemoheterotrophic and photoautotrophic metabolisms operate concurrently. Microalgae biomass produced by this approach has high density and contains high-value lipids, proteins, carbohydrates, and pigments; and the product range is very versatile [7–10]. These products range from high-value nutraceuticals, food supplements, and cosmetics to the lower value commodities biofuels, food, fertilizer, and application in wastewater treatment [10–12].

4.4 Microalgal metabolites

Microalgal biomass contains considerable amounts of bioactive molecules such as carotenoids (astaxanthins, β-carotenes, and xanthophylls), omega-3 fatty acids, polysaccharides, and proteins, which can be used in several applications as colorants, pharmaceuticals, food, food additives, and feed and as bioplastics.

4.4.1 Carotenoids

Microalgae produce carotenoids and all known xanthophylls found in terrestrial plants (e.g., zeaxanthin, lutein, antheraxanthin). Astaxanthin is a carotenoid pigment that occurs in microalgae, trout, yeast, and shrimp, among other sea creatures. It is found in abundance in Pacific salmon and the fish appears pinkish due to the presence of astaxanthin. Astaxanthin is an antioxidant; it is said to have many health benefits. Carotenoids as accessory pigments, capture light energy during photosynthesis and promote photoprotection. Stains of *Nannochloropsis sp., Rhodotorula glutinis,* and *Neochloris oleoabundans* have high contents of carotenoids. The red ketocarotenoid, astaxanthin (3, 30-dihydroxy-b,β-carotene 4,40-dione) is an antioxidant and the green microalga *Haematococcus pluvialis* is said to be a good natural source of astaxanthin [36–40].

4.4.2 Lutein

Lutein, a xanthophyll, is one of the many known naturally occurring carotenoids. Lutein is synthesized only by plants and is found in large quantities in green leafy vegetables like kale, spinach, yellow carrots, and in dietary supplements. The lutein-rich microalgae *Scenedesmus almeriensis* and *Desmodesmus sp.* could be considered as promising sources of lutein for their tolerance to harsh environmental growth conditions. It is a food colorant with the potential for preventing cancer. It is used for maintaining eye health and to reduce the risk of retinal macular degeneration. The performance of three *Chlorella* species on the production of biomass, lipid, and lutein showed high productivities, presenting the microalgae as a promising resource for these products [41].

4.4.3 Poly-unsaturated fatty acids

Microalgae are the dominant sources of polyunsaturated fatty acids in the marine food chain. *Schizochytrium sp.* is a type of marine microalgae with the natural capacity to produce oil extremely rich in docosahexaenoic acid (DHA) omega-3 fatty acids [42]. DHA-rich extracts from *Schizochytrium sp*. are presented as a feed supplement to swine for their muscle tissue development and as a raw material for the production of aquafeed. *N. oculata* and *P. tricornotum* have a favorable omega-3: omega-6 ratio that is adequate to enrich food [43]. The growth conditions are deliberately manipulated to achieve the desired fatty acid composition of the biomass. At low nutrient concentrations, the microalgal lipids accumulated are rich in triglycerides and are more suitable for biodiesel production; and high nutrients supply to the growth medium leads to the accumulation of long-chain unsaturated fatty acids [44]. DHA applications include healthcare, pharmaceutical, and food & beverage sectors. Within this segment, the pharmaceutical application holds a larger share of it [45].

The acetyl-CoA condensation to fatty acyls is one of the methods by which bio-hydrocarbons are produced in-situ biotic organisms. The second biohydrocarbon production pathway is the isopentenyl pyrophosphate (IPP) condensation to higher isoprenoids, which is responsible for the diverse isoprene derivatives, many of which are suitable for fuels or fuel additives due to their desirable cetane and pour point and other fuel properties [5]. The low-to-zero-oxygen content of isoprenoids results in energy densities similar to the alkanes in current diesel fuels and diversity of ring structures affords lower cloud points [46, 47]. Additionally, it has been found that slight modifications to enzymes involved in the final steps of higher isoprenoid synthesis can result in subtle product variants with distinct thermochemical and thermophysical properties [47]. The precursors for the majority of these compounds are metabolic intermediates in photosynthetic microorganisms (PMOs). Genetic engineering of microalgae and cyanobacteria would be required to enhance the productivity of PMOs [5].

4.4.4 Microalgal triglycerides

Triglycerides are lipids or waxes, formed by biochemically combining glycerol and fatty acids in the ratio of 1: 3 respectively. This combination may be a simple type or a mixed type. Triglycerides in which the glycerol backbone is attached to three molecules of the same fatty acid are referred to as simple triglycerides. Typical in this category is tripalmitin, $C_3H_5(OCOC_{15}H_{31})_3$. Only a few of the glycerides occurring in nature are of the simple type; most are mixed triglycerides (see **Figure 14**) [48]. Based on saturation and unsaturation of the attached fatty acids, triglycerides can be classified as saturated, monounsaturated, and polyunsaturated. In saturated triglycerides, all the fatty acids are saturated. Saturated fats abound in

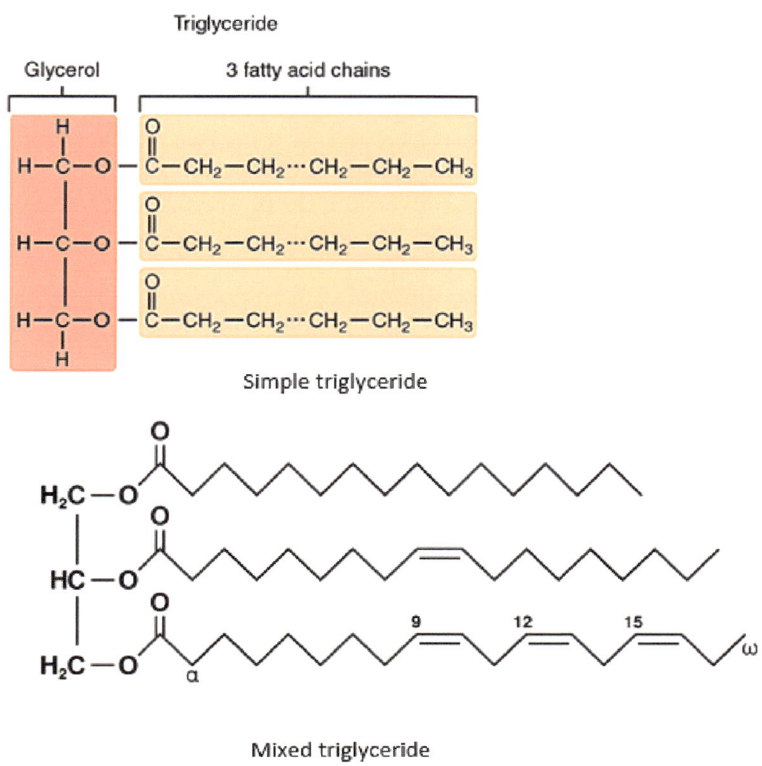

Figure 14.
The structure of triglyceride showing the simple and mixed types.

many animal products such as butter, cheese, cream, and fatty meats, ice cream, and whole milk. In monounsaturated triglycerides most of the fatty acids are monounsaturated. Vegetable oils such as canola oil, olive oil, peanut oil, and sesame oil have high levels of monounsaturated fats and polyunsaturated triglycerides. Omega-3 and omega-6 fatty acids are polyunsaturated.

Microalgae are a promising renewable resource for green production of triacylglycerols (TAGs), which can be used as a biofuel feedstock. Nitrogen starvation is the most effective strategy to induce TAG biosynthesis in microalgae [48]. One of the best microalgae for lipid production is *Botryococcus braunii* Kutzing, above 70% of lipid in its cell content. Whereas other microalgae like *Scenedesmus sp., Chlorella sp*., and *Nanochloropsis sp*. also produce lipid up to 40% [49–51].

4.4.5 Microalgal phospholipids

Phospholipids are made up of four components viz. fatty acids, a platform to which the fatty acids are attached, phosphate, and an alcohol attached to the phosphate. Phospholipids may be built on either glycerol or sphingosine framework. Phospholipids built on glycerol framework are called phosphoglycerides (or glycerophospholipids). A phosphoglyceride consists of a glycerol molecule, two fatty acids, a phosphate, and choline, which is an alcohol. Phosphoglycerides are the most abundant phospholipid molecules found in cell membranes. The phospholipids built on sphingosine framework are referred to as sphingolipids or glycolipids, depending on the number of glucose or galactose molecules they contain; and lipoproteins, which are complexes of cholesterol, triglycerides, and proteins that transport lipids in the aqueous environment of the bloodstream. These are complex lipids. The algae contain three major phospholipids, phosphatidylglycerol (PG), phosphatidylethanolamine (PE), and phosphatidylcholine (PC). Phospholipids are synthesized by both prokaryotic and eukaryotic organisms. They are the major component of most eukaryotic cell membranes, which play a fundamental role in compartmentalizing the biochemistry of life [52]. The hydroxyl groups at positions C-1 and C-2 in phosphoglycerides are esterified to the carboxyl groups of the two fatty acid chains. The hydroxyl group at position C-3 hydroxyl group of the glycerol backbone is esterified to phosphoric acid. At this extent of conversion, the product is phosphatidic acid, which is the simplest phosphoglyceride. Phosphatidic acid now serves as the backbone on which most phosphoglycerides are derived having moieties such as serine, ethanolamine, choline, glycerol, and the inositol. Consequently, we have phosphatidylserine, phosphatidylethanolamine, phosphatidylcholine, phosphatidylglycerol, and phosphatidylinositol respectively (see **Figure 15**) [52].

4.4.6 Vitamins and fine chemicals from microalgal biomass

Metabolites from both microalgae and cyanobacteria have attended to both human and animal health and food needs and these microorganisms have become attractive resources for bioactive natural products that have wide applications in pharmaceutical, food, and chemical industries. Algae-derived bioactive substrates are employed for drug screening, given their tremendous structural diversity and biological availability. Microalgae biomass has a wide range of physiological and biochemical characteristics and contains 50–70% protein compared to 50% in meat, and 15–17% in wheat, with 30% lipids, more than 40% glycerol, 8–14% carotene, and a reasonably high levels of vitamins B1, B2, B3, B6, B12, E, K, D, and others [54–56].

Microalgae that have been cultivated on commercial scales and are available include *Chlorella, Dunaliella, Nannochloris, Nitzschia, Crypthecodinium, Schizochytrium, Tetraselmis, Skeletonema,* etc. and the cyanobacterium, *Spirulina,* and

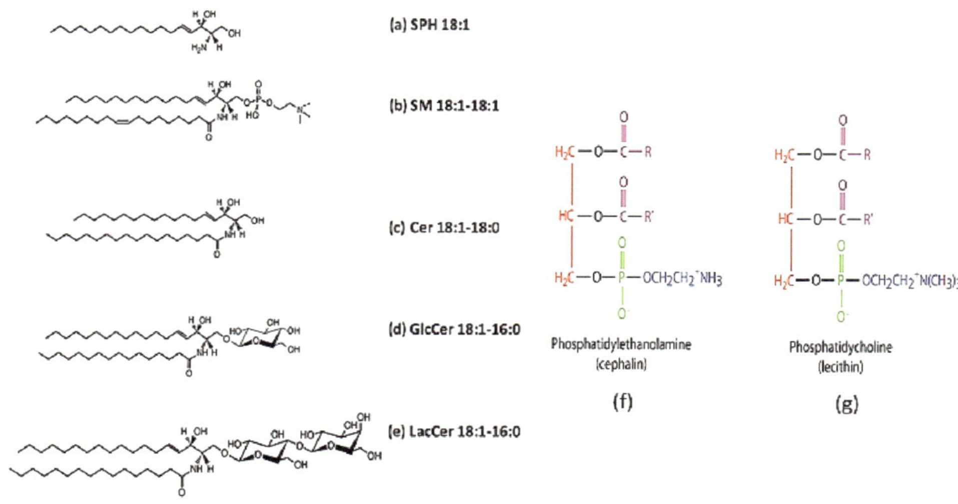

Figure 15.
Sphingolipids and phospholipids: The classification of sphingolipids is based on the group attached to the sphingosine (LCB) backbone (a). Sphingomyelin (b) and ceramides (c-e) differ in fatty acid length, unsaturation, and in the type of attached head group and hydroxylation. Phospholipids with glycerol framework: (f) phosphatidylethanolamine, (g) phosphatidylcholine [53].

a host of others. Most of the commercially produced microalgal biomass is presented to the market as a food supplement, and they are presented as tablets and capsules. Breakfast cereals, noodles, beverages, wines, and cosmetics now contain microalgae and their extracts. More than 75% of pharmaceutical product development is carried out by the microalgal food supplement production outfits. In the recent several years, microalgal and cyanobacterial research has explored diverse cultivation protocols aimed at improving growth rates, biomass yields, and accumulating metabolites for high nutritional value, and high-value chemicals (pigments and vitamins) [55]. Many more bioactive metabolites have been reported in microalgae. Dried microalgae biomass could be used as high-protein feeds for animals such as shrimp and fish, and microalgal biomass is a significant resource for cytotoxic agents with applications in cancer chemotherapy. The blooms of *Phaeocystis sp*., a marine microalga have antibiotic substances listed therein. *Phaeocystis pouchetii* produces acrylic acid, which makes up to 7.0% of its dry weight. The antibiotic metabolites so produced migrate in the food chain through the digestive system of some Antarctic marine animal species. Also, the alga *Dunaliella sp.* produces β-carotene and certain vitamins, which have boosted the Mariculture activities. Some cyanobacteria and microalgae such as *Ochromonas* sp. and *Prymnesium parvum* produce toxins, which may have the potential for pharmaceutical applications. These marine cyanobacteria produce bioactive metabolites such as acetogenins, bromophenols, fatty acids, terpenes, sterols, alkaloids, etc. with antibiotics, and antifungal activities. Diverse strains of cyanobacteria produce intracellular and extracellular metabolites with bioactive functions such as antitumor, anti-inflammatory, antialgal, antibacterial, antifungal, and antiviral activity [55, 56].

4.5 Microalgal biomass production limiting factors

Abiotic, Biotic, and process-related factors influence the growth of algae. Some of the abiotic factors are illumination and luminous intensity, daytime to night-time ratio, the temperature of the culture medium, nutrient availability, O_2, and CO_2 mass transfer, pH value, the hydraulic retention time (HRT), salinity, and presence of growth-inhibiting chemical agents [30]. Some of the biotic factors are the presence of pathogens (bacteria, fungi, viruses) and the presence of more than one algae strains. Each algae strain has a different capacity to assimilate nutrients, and in mixed cultures, there is competition for the available nutrients in the media, which may afferent the growth of some strains [36]. Process related factors that may influence algal growth are hydrodynamics of the culture broth, which is influenced by the choice of the bioreactor, the initial algal cell concentration in the reactor, and the related frequency of harvesting algal biomass [57, 58].

5. Conclusions

There is a major difference between microalgae and cyanobacteria in terms of their cell structure and this work has presented unmistakable evidence that microalgae have a nucleus and chloroplast, and their makeup includes their full identity in a two-stranded DNA. On the other hand, cyanobacteria are identified by one-stranded DNA and do not have a nucleus and neither a chloroplast. However, Microalgae and cyanobacteria do photosynthesize to produce their food.

It is seen from research as discussed in this chapter that value products aimed to meet pharmaceutical and food needs are obtainable by continuous availability of nutrients to the microalgae in the culture media. It is also seen that to accumulate lipid in the order of triglycerides for biodiesel production, microalgae must experience nutrients deficiency in the culture media at the stationary stage of growth.

The hydrodynamics of the microalgal culture broth depends on the choice of bioreactor for a particular cultivation activity and contributes to the algal growth factor.

The versatility of the microalgal biomass is expressed in the diversity of metabolites produced by manipulation of the growth factors in favor of the desired product. Also, the choice of the strain will drive towards the targeted product.

Acknowledgements

The authors appreciate the Institute for Water and Wastewater Technology (IWWT), Durban University of Technology, Durban for providing the platform for scientific investigations.

Conflict of interest

The authors have declared that they have no known competing financial interests or personal relationships that could have appeared to influence the work reported in this chapter.

DOI: http://dx.doi.org/10.5772/intechopen.94090

Author details

Donald Tyoker Kukwa* and Maggie Chetty
Department of Chemical Engineering, Durban University of Technology, Durban, South Africa

*Address all correspondence to: tyokerdoo@gmail.com

References

[1] Begum, H., F. M. Yusoff, S. Banerjee, H. Khatoon, and M. Shariff. "Availability and Utilization of Pigments from Microalgae." *Crit Rev Food Sci Nutr* 56, no. 13 (Oct 2 2016): 2209-22. https://doi.org/10.1080/10408398.2013.764841. https://www.ncbi.nlm.nih.gov/pubmed/25674822.

[2] Kostamo, Kirsi. "The Life Cycle and Genetic Structure of the Red Alga Furcellaria Lumbricalis on a Salinity Gradient." University of Helsinki, 2008. https://core.ac.uk/download/pdf/14917499.pdf (33).

[3] Bodénès, Pierre. "Study of the Application of Pulsed Electric Fields (Pef) on Microalgae for the Extraction of Neutral Lipids." Université Paris-Saclay,, 2017. https://tel.archives-ouvertes.fr/tel-01540436 (tel-01540436).

[4] M. Sebastian and J. M. Gasol, "Visualization is crucial for understanding microbial processes in the ocean," *Philos Trans R Soc Lond B Biol Sci,* vol. 374, no. 1786, p. 20190083, Nov 25 2019, doi: 10.1098/rstb.2019.0083.

[5] S. Bleakley and M. Hayes, "Algal Proteins: Extraction, Application, and Challenges Concerning Production," *Foods,* vol. 6, no. 5, Apr 26 2017, doi: 10.3390/foods6050033.

[6] P. Kumar and U. Mina, A. Kumar and J. Joseph, Eds. *Life Sciences Fundamentals and Practice Part I*, Fourth ed. New Delhi, India: Pathfinder Publication, 2014, p. 132.

[7] M. Mitra and S. Mishra, "A Biorefinery from Nannochloropsis spp. Utilizing Wastewater Resources," in *Application of Microalgae in Wastewater Treatment*, S. K. Gupta and F. Bux Eds. Switzerland Springer 2019, ch. Chapter 6, pp. 123-145.

[8] Z. K. Abdulsada, "Evaluation of Microalgae for Secondary and Tertiary Wastewater Treatment," Master in Applied Science in Environmental Engineering Master's Thesis, Department of Civil and Environmental Engineering, Carleton University, Ottawa, Ontario, 2014.

[9] Smith, Richard T., Krys Bangert, Stephen J. Wilkinson, and D. James Gilmour. "Synergistic Carbon Metabolism in a Fast Growing Mixotrophic Freshwater Microalgal Species Micractinium Inermum." *Biomass and Bioenergy* 82 (2015): 73-86. https://doi.org/10.1016/j.biombioe.2015.04.023.

[10] Abreu, A. P., B. Fernandes, A. A. Vicente, J. Teixeira, and G. Dragone. "Mixotrophic Cultivation of *Chlorella vulgaris* Using Industrial Dairy Waste as Organic Carbon Source." *Bioresour Technol* 118 (Aug 2012): 61-6. https://doi.org/10.1016/j.biortech.2012.05.055. https://www.ncbi.nlm.nih.gov/pubmed/22705507

[11] Chen, W. C., Y. C. Hsu, J. S. Chang, S. H. Ho, L. F. Wang, and Y. H. Wei. "Enhancing Production of Lutein by a Mixotrophic Cultivation System Using Microalga *Scenedesmus obliquus* Cwl-1." *Bioresour Technol* 291 (Nov 2019): 121891. https://doi.org/10.1016/j.biortech.2019.121891. https://www.ncbi.nlm.nih.gov/pubmed/31387049.

[12] O. Perez-Garcia and Y. Bashan, "Microalgal Heterotrophic and Mixotrophic Culturing for Bio-refining: From Metabolic Routes to Techno-economics," in *Algal Biorefi neries*, A. P. e. al Ed. Switzerland Springer, 2015, pp. 61-131.

[13] Ji, Min-Kyu, Hyun-Shik Yun, Young-Tae Park, Akhil N. Kabra, In-Hwan Oh, and Jaeyoung Choi. "Mixotrophic Cultivation of a Microalga *Scenedesmus*

obliquus in Municipal Wastewater Supplemented with Food Wastewater and Flue Gas CO_2 for Biomass Production." *Journal of Environmental Management* 159 (2015): 115-20. https://doi.org/10.1016/j.jenvman.2015.05.037.

[14] Ficner, Ralf, and Robert Huber. "Refined Crystal Structure of Phycoerythrin from *Porphyridium cruentum* at 0.23-Nm Resolution and Localization of the Y Subunit." *European Journal of Biochemistry* 218 (1993): 103-06.

[15] Canniffe, D. P., and C. N. Hunter. "Engineered Biosynthesis of Bacteriochlorophyll B in Rhodobacter Sphaeroides." *Biochim Biophys Acta* 1837, no. 10 (Oct 2014): 1611-6. https://doi.org/10.1016/j.bbabio.2014.07.011. https://www.ncbi.nlm.nih.gov/pubmed/25058304.

[16] Jeyaram, S., and T. Geethakrishnan. "Spectral and Third-Order Nonlinear Optical Characteristics of Natural Pigment Extracted from *Coriandrum sativum*." *Optical Materials* 107 (2020). https://doi.org/10.1016/j.optmat.2020.110148.

[17] J. Pruvost, B. Le Gouic, O. Lepine, J. Legrand, and F. Le Borgne, "Microalgae culture in building-integrated photobioreactors: Biomass production modelling and energetic analysis," *Chemical Engineering Journal,* vol. 284, pp. 850-861, 2016, doi: 10.1016/j.cej.2015.08.118.

[18] M. Wzorek, M. Płaczek, A. Patyna, S. Witczak, G. Królczyk, and A. Król, "Technical evaluation of photobioreactors for microalgae cultivation," *E3S Web of Conferences,* vol. 19, 2017, doi: 10.1051/e3sconf/20171902032.

[19] Akhtar, P., P. J. Nowakowski, W. Wang, T. N. Do, S. Zhao, G. Siligardi, G. Garab, *et al.* "Spectral Tuning of Light-Harvesting Complex Ii in the Siphonous Alga *Bryopsis corticulans* and Its Effect on Energy Transfer Dynamics." *Biochim Biophys Acta Bioenerg* 1861, no. 7 (Jul 1 2020): 148191. https://doi.org/10.1016/j.bbabio.2020.148191. https://www.ncbi.nlm.nih.gov/pubmed/32201306.

[20] J. Masojídek, G. Torzillo, and M. Koblížek, "Photosynthesis in Microalgae," in *Handbook of Microalgal Culture*, 2013, pp. 21-36.

[21] Jiang, L., and L. Wen. "Photonic Sensitive Switchable Materials." Chap. 4 In *Switchable and Responsive Surfaces and Materials for Biomedical Applications*, edited by Zheng Zhang, 324: Elsevier Ltd, 2015.

[22] Lehmuskero, Anni, Matilde Skogen Chauton, and Tobias Boström. "Light and Photosynthetic Microalgae: A Review of Cellular- and Molecular-Scale Optical Processes." *Progress in Oceanography* 168 (2018): 43-56. https://doi.org/10.1016/j.pocean.2018.09.002.

[23] Tredici, M.R., and R. Materassi. "From Open Ponds to Vertical Alveolar Panels: The Italian Experience in the Development of Reactors for the Mass Cultivation of Phototrophic Microorganisms." *Journal of Applied Phycology* 4 (1992): 221-31.

[24] D. Chaumont, "Biotechnology of algal biomass production: a review of systems for outdoor mass culture," *Journal of Applied Phycology,* vol. 5, pp. 593-604, 1993.

[25] Christenson, L., and R. Sims. "Production and Harvesting of Microalgae for Wastewater Treatment, Biofuels, and Bioproducts." *Biotechnol Adv* 29, no. 6 (Nov-Dec 2011): 686-702. https://doi.org/10.1016/j.biotechadv.2011.05.015. https://www.ncbi.nlm.nih.gov/pubmed/21664266

[26] Stephens, James, Kirk W. Dickinson, and Kelly Ogilvie. Apparatus for Containing, Cultivating. And

Harvesting Photosynthetic Marine Microorganisms within Water. United States 2009.

[27] de Vree, J. H., R. Bosma, M. Janssen, M. J. Barbosa, and R. H. Wijffels. "Comparison of Four Outdoor Pilot-Scale Photobioreactors." *Biotechnol Biofuels* 8 (2015): 215. https://doi.org/10.1186/s13068-015-0400-2. https://www.ncbi.nlm.nih.gov/pubmed/26689675.

[28] Fernandez, F. G. Acien, J. M. Fernandez Sevilla, J. A. Sanchez Perez, E. Molina Grima, and Y. Chisti. "Airlift-Driven External-Loop Tubular Photobioreactors for Outdoor Production of Microalgae: Assessment of Design and Performance." [In English]. *Chemical Engineering Science* 56, no. 8 (Apr 2001): 2721-32. https://doi.org/Doi 10.1016/S0009-2509(00)00521-2. <Go to ISI>://WOS:000168907800012.

[29] V. H. Work, "Metabolic and physiological engineering of photosynthetic microorganisms for the synthesis of bioenergy feedstocks: Development, characterization, and optimization," Doctor of Philosophy Civil and Environmental Engineering, Colorado School of Mines, 2014.

[30] Y. Wang, D. J. Stessman, and M. H. Spalding, "The CO2 concentrating mechanism and photosynthetic carbon assimilation in limiting CO2 : how Chlamydomonas works against the gradient " *Plant Journal: For Cell and Molecula Biology,* vol. 82, no. 3, pp. 429-448, 2015, doi: 10.1111/tpj.12829

[31] J. A. Raven, "Inorganic carbon acquisition by eukaryotic algae: four current questions " *Photsynthesis Research,* vol. 106, no. 1-2, pp. 123-34, 2010, doi: 10.1007/s11120-010-9563-7.

[32] Wang Y, Spalding MH. LCIB in the Chlamydomonas CO2-concentrating mechanism. Photosynthesis Research. 2014 Sep; 121(2-3):185-192. DOI: 10.1007/s11120-013-9956-5.

[33] H. Yang, Q. He, J. Rong, L. Xia, and C. Hu, "Rapid neutral lipid accumulation of the alkali-resistant oleaginous Monoraphidium dybowskii LB50 by NaCl induction," *Bioresour Technol,* vol. 172, pp. 131-137, Nov 2014, doi: 10.1016/j.biortech.2014.08.066.

[34] O. Perez-Garcia, F. M. Escalante, L. E. de-Bashan, and Y. Bashan, "Heterotrophic cultures of microalgae: metabolism and potential products," *Water Res,* vol. 45, no. 1, pp. 11-36, Jan 2011, doi: 10.1016/j.watres.2010.08.037.

[35] A. R. Memon, N. Soomro, and K. L. Khatri, "Growth Kinetics of Algae Cells Under Heterotrophic Conditions," *Sindh Univ. Res. Jour. (Sci. Ser.)* vol. 47, no. 4, p. 6, 2015.

[36] BP, Nobre, Villalobos F, Barragán BE, Oliveira AC, Batista AP, Marques PA, Mendes RL*, et al.* "A Biorefinery from Nannochloropsis Sp. Microalga--Extraction of Oils and Pigments. Production of Biohydrogen from the Leftover Biomass.". *Bioresource Technology* 135 (2013): 128-36. 23265815.

[37] Chen, L., L. Zhang, and T. Liu. "Concurrent Production of Carotenoids and Lipid by a Filamentous Microalga Trentepohlia Arborum." *Bioresour Technol* 214 (Aug 2016): 567-73. https://doi.org/10.1016/j.biortech.2016.05.017. https://www.ncbi.nlm.nih.gov/pubmed/27179952

[38] C. R. P. S. Paes, G. R. Faria, N. A. B. Tinoco, D. J. F. A. Castro, E. Barbarino, and S. O. Lourenco, "Growth, nutrient uptake and chemical composition of Chlorella sp. and Nannochloropsis oculata under nitrogen starvation," *Latin American Journal of Aquatic Research,* vol. 44, no. 2, pp. 275-292, 2016, doi: 10.3856/vol44-issue2-fulltext-9.

[39] Urreta, I., Z Ikaran, I Janices, E Ibañez, M Castro-Puyana, S Castañón, and S Suárez-Alvarez. "Revalorization of Neochloris oleoabundans biomass as source of biodiesel by concurrent production of lipids and carotenoids" *Algal research* 5, 1 (2014): 16-22. doi: 10.1016/j.algal.2014.05.001

[40] T.-Q. Shi, L.-R. Wang, Z.-X. Zhang, X.-M. Sun, and H. Huang, "Stresses as First-Line Tools for Enhancing Lipid and Carotenoid Production in Microalgae," *Frontiers in Bioengineering and Biotechnology,* vol. 8, 2020, doi: 10.3389/fbioe.2020.00610.

[41] B. Araya, L. Gouveia, B. Nobre, A. Reis, R. Chamy, and P. Poirrier, "Evaluation of the simultaneous production of lutein and lipids using a vertical alveolar panel bioreactor for three Chlorella species," (in English), *Algal Res,* vol. 6, pp. 218-222, Oct 2014, doi: 10.1016/j.algal.2014.06.003.

[42] P. K. Sarker, A. R. Kapuscinski, A. J. Lanois, E. D. Livesey, K. P. Bernhard, and M. L. Coley, "Towards Sustainable Aquafeeds: Complete Substitution of Fish Oil with Marine Microalga Schizochytrium sp. Improves Growth and Fatty Acid Deposition in Juvenile Nile Tilapia (*Oreochromis niloticus*)," *PLoS One,* vol. 11, no. 6, p. e0156684, 2016, doi: 10.1371/journal.pone.0156684.

[43] Â. P. Matos *et al.*, "Chemical Characterization of Six Microalgae with Potential Utility for Food Application," *Journal of the American Oil Chemists' Society,* vol. 93, no. 7, pp. 963-972, 2016, doi: 10.1007/s11746-016-2849-y.

[44] T. Fernandes, I. A. Fernandes, C. A. P. Andrade, and N. Cordeiro, "Changes in fatty acid biosynthesis in marine microalgae as a response to medium nutrient availability," *Algal Research,* vol. 18, pp. 314-320, 2016, doi: 10.1016/j.algal.2016.07.005.

[45] L. Moreno-Garcia, K. Adjallé, S. Barnabé, and G. S. V. Raghavan, "(Moreno-Garcia, 2017 #89)," *Renewable and Sustainable Energy Reviews,* vol. 76, pp. 493-506, 2017, doi: 10.1016/j.rser.2017.03.024.

[46] E. M. Grima, E.-H. Belarbi, F. G. A. n. Ferna'ndez, A. R. Medina, and Y. Chisti, "Recovery of microalgal biomass and metabolites: process options and economics," *Biotechnology Advances,* vol. 20, pp. 491-515, 2003, doi: 10.1016/S0734-9750(02)00050-2.

[47] G. V. Subhash, M. Rajvanshi, B. N. Kumar, S. Govindachary, V. Prasad, and S. Dasgupta, "Carbon streaming in microalgae: extraction and analysis methods for high value compounds," *Bioresource Technology,* vol. 244, no. Pt 2, pp. 1304-1316, Nov 2017, doi: 10.1016/j.biortech.2017.07.024.

[48] Peng, Baoxiang. "Transformation of Triglycerides and Fatty Acids into Biofuels with Sulfur-Free Catalysts." PhD, Technischen Universität München, 2012.

[49] L. Brennan and P. Owende, "Biofuels from microalgae—A review of technologies for production, processing, and extractions of biofuels and co-products," *Renewable and Sustainable Energy Reviews,* vol. 14, no. 2, pp. 557-577, 2010, doi: 10.1016/j.rser.2009.10.009.

[50] P. Feng, Z. Deng, L. Fan, and Z. Hu, "Lipid accumulation and growth characteristics of Chlorella zofingiensis under different nitrate and phosphate concentrations," *J Biosci Bioeng,* vol. 114, no. 4, pp. 405-10, Oct 2012, doi: 10.1016/j.jbiosc.2012.05.007.

[51] F. Lin, Y. Chen, R. Levine, K. Lee, Y. Yuan, and X. N. Lin, "Improving fatty acid availability for bio-hydrocarbon production in *Escherichia coli* by metabolic engineering," *PLoS One,* vol.

8, no. 10, p. e78595, 2013, doi: 10.1371/journal.pone.0078595.

[52] Baky, Hanaa H. Abd El, Farouk K. El Baz, Gamal S. El Baroty, Mohsen M. S. Asker, and Eman A. Ibrahim. "Phospholipids of Some Marine Microalgae: Identification, Antivirus, Anticancer and Antimicrobial Bioactivities." *Der Pharma Chemica* 6, no. 6 (2014): 9-18.

[53] Dembitsky, Valery M., and Olga A. Rozentsvet. "Phospholipid Composition of Some Marine Red Algae." *Phytochemistry* 29, no. 10 (1990): 3149-52.

[54] Ahmmed, Mirja Kaizer, Fatema Ahmmed, Hong Tian, Alan Carne, and Alaa El-Din Bekhit. "Marine Omega-3 (N-3) Phospholipids: A Comprehensive Review of Their Properties, Sources, Bioavailability, and Relation to Brain Health." *Comprehensive Reviews in Food Science and Food Safety* 19, no. 1 (2019): 64-123. https://doi.org/10.1111/1541-4337.12510

[55] Priyadarshani, I., and B. Rath. "Bioactive Compounds from Microalgae and Cyanobacteria: Utility and Applications." *Priyadarshani and Rath, IJPSR* 3, no. 11 (2012): 4123-30.

[56] Nicoletti, M. "Microalgae Nutraceuticals." *Foods* 5, no. 3 (Aug 22 2016). https://doi.org/10.3390/foods5030054. https://www.ncbi.nlm.nih.gov/pubmed/28231149.

[57] Wzorek, M., Małgorzata Płaczek, Agnieszka Patyna, Stanisław Witczak, G. Królczyk, and A. Król. "Technical Evaluation of Photobioreactors for Microalgae Cultivation." *E3S Web of Conferences* 19 (2017). https://doi.org/10.1051/e3sconf/20171902032.

[58] H. A. Najafabadi, M. Malekzadeh, F. Jalilian, M. Vossoughi, and G. Pazuki, "Effect of various carbon sources on biomass and lipid production of *Chlorella vulgaris* during nutrient sufficient and nitrogen starvation conditions," *Bioresour Technol,* vol. 180, pp. 311-7, Mar

Chapter 20

Microalgae Cultivation in Photobioreactors Aiming at Biodiesel Production

Mateus S. Amaral, Carla C.A. Loures, Fabiano L. Naves, Gisella L. Samanamud, Messias B. Silva and Arnaldo M.R. Prata

Abstract

The search for a renewable source as an alternative to fossil fuels has driven the research on new sources of biomass for biofuels. An alternative source of biomass that has come to prominence is microalgae, photosynthetic micro-organisms capable of capturing atmospheric CO2 and accumulating high levels of lipids in their biomass, making them attractive as a raw material for biodiesel synthesis. Thus, various studies have been conducted in developing different types of photobioreactors for the cultivation of microalgae. Photobioreactors can be divided into two groups: open and closed. Open photobioreactors are more susceptible to contamination and bad weather, reducing biomass productivity. Closed photobioreactors allow greater control against contamination and bad weather and lead to higher rates of biomass production; they are widely used in research to improve new species and processes. Therefore, many configurations of closed photobioreactors have been developed over the years to increase productivity of microalgae biomass.

Keywords: raw material for biodiesel production, biomass, microalgae, photobioreactors, biodiesel synthesis

1. Introduction

The continuous rise in global temperature caused by the emission of greenhouse effect gases, mainly carbon dioxide (CO_2) mostly due to anthropogenic activities such as the burning of fossil fuels, has led the search for alternative and renewable fuels, such as biofuels. Biofuels present a balance between the CO_2 emitted during burning and the subsequent absorption during the formation of biomass, a sustainable cycle.

Biofuels such as bioethanol and biodiesel are alternative substitutes to gasoline and diesel, respectively and have several conventional renewable sources of raw materials. The raw materials for the production of biodiesel are based on oils and fats, which may be of vegetable origin (soybean, peanut, cotton and sunflower oils), animal origin (fish oil, beef tallow, lard and fat from chicken) or residual frying oils and fats, all sources committed to the food chain.

However, another raw material, not competitive with the food chain, has received special attention for presenting many advantages when compared to conventional sources. These are microalgae which are photosynthetic microorganisms capable of accumulating high levels of oil in their cytoplasm which, in turn, can be extracted from biomass and converted into biodiesel.

Microalgae grow at a high speed (less than a week), requiring small territorial areas to develop, presenting high photosynthetic efficiency and, consequently, good absorption of atmospheric CO_2. They may use domestic and industrial effluents to develop (not requiring clean water), thus standing out as a promising raw material for the production of biodiesel.

Microalgae cultivations generate high concentration of biomass and, therefore, can be used in the production of biodiesel. The cultivation is generally conducted in illuminated bioreactors, called photobioreactors, which can have different shapes and configurations. Some configurations include columns of bubbles, tubulars, flat plates and open tanks in the shape of "raceways", the most well-known and widely spread configuration worldwide. However, the design and development of photobioreactors with new configurations and different lighting modes still play a crucial role in optimizing cell growth and increasing biomass formation.

In this context, this chapter will present the reasons for the use of microalgae as sources of biodiesel, the main photobioreactors used for the production of microalgae biomass and a brief description of the reaction synthesis of biodiesel.

2. Biodiesel sources

The average temperature of the planet has been increasing year by year and causing serious climatic imbalances, such as changes in the rain pattern and melting of the glaciers, resulting in losses in local agriculture and rising sea levels. Much of this rise in global temperature is due to the anthropogenic emission of carbon dioxide (CO_2) into the atmosphere, mainly from the burning of fossil fuels [1].

According to IPEA [2], the anthropogenic contribution to the emission of CO_2 in the atmosphere is mainly related to energy purposes, in which transportation is responsible for 67% of the emissions as a result of carbon oxidation during the combustion of the most varied fuels.

For energy purposes, the contribution of fossil fuels is higher than 80% of the anthropogenic emission of CO_2 into the Earth's atmosphere, corresponding to approximately 35 giga tons of CO_2 emitted in 2015 [2, 3].

Although technically there is a diversified energy matrix, including hydroelectric, thermoelectric, wind, solar and nuclear energy sources, the planet is highly dependent on thermoelectric plants, which burn coal (mainly) and natural gas, fossil fuels that are of finite origin and unsustainable, basically transferring terrestrial carbon to the atmosphere in its majority in the form of CO_2 [3].

In addition to the massive presence of fossil fuels in the generation of electric energy, they also dominate the transportation sector, overloaded with vehicles powered by combustion, with exponents busses and trucks fueled by diesel, a fossil fuel that highly contributes to emissions of CO_2 [4].

Biodiesel is the biofuel that can replace diesel and can be produced from vegetable oils (soy, palm, sunflower, cotton, peanuts and others), oils from microorganisms (cyanobacteria and microalgae), animal fat and the frying oil reuse [5, 6]. The current biofuel projections are based on sources that are also food commodities and resources suitable for conventional agriculture. Thus, the biodiesel production involves the use of lands already used to food production [7].

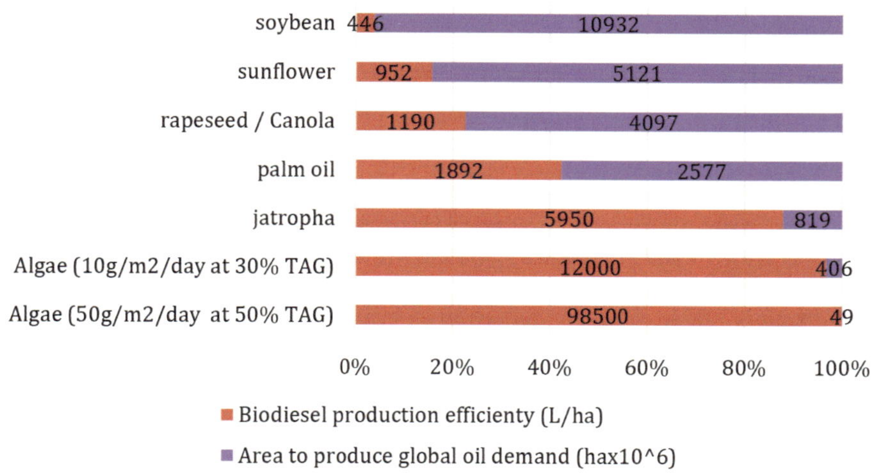

Figure 1.
Different sources of biodiesel and their respective production efficiencies and territorial demand.

In addition, the use of microalgae biomass for the production of biodiesel has been gaining prominence, as this type of microorganism has high growth rates even on smaller areas of cultivation, microalgae biomass can generate high amounts of biomass and oils extracted from its dry biomass compared to vegetable oilseeds as seen in **Figure 1**. The cultivation of microalgae promotes the biofixation of CO_2, allows the use of wastewater to treat industrial and domestic effluents and does not compete with agricultural food production [8, 9].

3. Microalgae

Microalgae are single-celled or multicellular organisms with microscopic dimensions, capable of photosynthesis due to the presence of chlorophyll [10]. Microalgae mainly inhabit aquatic environments such as: lakes, rivers and oceans, although they can live in solid humid environments such as wet soil and rocks. They have about 50% carbon in their composition, based on their dry biomass, as a result of photosynthesis [11, 12].

In the sea, microalgae constitute phytoplankton and are responsible for approximately 90% of photosynthetic activity in the oceans which can lead to an annual fixation of 45–50 Gton of carbon in the oceans [13]. It is estimated that 1.8 kg of absorbed CO_2, during photosynthesis, generates 1 kg of microalgal biomass [14, 15]. The photosynthesis reaction that represents this conversion of CO_2 into biomass can be represented in a simplified way by Eq. (1).

$$6\,CO_2 + 6\,H_2O \rightarrow C_6H_{12}O_6 + 6\,O_2 \quad (1)$$

The biological composition of microalgae varies widely between species and is strongly influenced by environmental factors such as temperature, lighting, photoperiod, pH of the culture medium, mineral nutrients, CO_2 supplement, etc. [15]. It is known that the concentration of nitrate in the microalgae culture medium significantly influences the production of lipids by the cell, especially when available in low concentrations inducing its accumulation inside the cell [16].

Microalgae synthesize lipids from the carbon source, whether inorganic such as CO_2 or organic (glucose, acetate, etc.). The components and levels of lipids in microalgal cells vary from species to species, being basically divided into neutral lipids (triglycerides and cholesterol) and polar lipids, such as phospholipids. Neutral lipids, such as triglycerides, are considered as the main material to produce biodiesel [17].

The oil content commonly found in microalgae is in the range of 15 to 50%, but these levels can be increased according to the manipulation of the crops aiming at the accumulation of lipids. In addition, microalgae of the Chlorella genus require less cultivation time and because of its high cell productivity and potential lipid extraction, microalgae can be widely used as a raw material in the production of biodiesel. The conversion to biodiesel occurs by the transesterification reaction using alcohol in the presence of acid or base as catalysts [12, 18].

Another group of substances abundant in the composition of most microalgae species are proteins. The high amount of proteins present in the cell composition in the most varied species of microalgae gives these microorganisms their recognition as an unconventional source of proteins. Proteins extracted from microalgal biomass for human nutrition are currently commercialized in various forms, such as tablets, capsules and liquids, which can be incorporated into pasta, cakes, sweets and drinks [19].

Along with proteins, the considerable composition of numerous essential vitamins in microalgae stands out: vitamin A, B1, B2, B6, B12, C, E, biotin, folic acid, among others, in levels that vary between species and according to cultivation techniques, as well as nucleic acids, which comprise from 1 to 6% of the cellular

Species	Proteins	Carbohydrates	Lipids	Nucleic acids
Scenedesmus obliquus	50–56	10–17	12–14	3–6
Scenedesmus quadricauda	47	—	1.9	—
Scenedesmus dimorphus	8–18	21–52	16–40	—
Chlamydomonas rheinhardii	48	17	21	—
Chlorella vulgaris	51–58	12–17	14–22	4–5
Chlorella pyrenoidosa	57	26	2	—
Chlorella minutissima	35.5	—	31	—
Spirogyra sp.	6–20	33–64	11–21	—
Dunaliella bioculata	49	4	8	—
Dunaliella salina	57	32	6	—
Euglena gracilis	39–61	14–18	14–20	—
Prymnesium parvum	28–45	25–33	22–38	1–2
Tetraselmis maculata	52	15	3	—
Porphyridium cruentum	28–39	40–57	9–14	—
Spirulina platensis	46–63	8–14	4–9	2–5
Spirulina máxima	60–71	13–16	6–7	3–4.5
Synechoccus sp.	63	15	11	5
Anabaena cylindrica	43–56	25–30	4–7	—

Fonte: [10].

Table 1.
Composition of microalgae.

composition of microalgae and play a vital role in cell growth and repair. Due to the presence of phosphorus and nitrogen in large amounts in nucleic acids, microalgal biomass can also be used as a fertilizer [10].

Carbohydrates, on the other hand, comprise the final product of photosynthesis in microalgae and can vary from 4 to 64% of their cell composition (**Table 1**) [10]. These are synthesized by fixing CO_2 in the form of glucose, disaccharides and starch as the primary source of energy for the cell [20].

In terms of microalgal biomass, carbohydrates can be extracted and converted into ethanol by the anaerobic fermentation process using yeasts (*Saccharomyces cerevisiae*, for example), or undergo degradation of organic matter in anaerobic conditions leading to the formation of methane and carbon dioxide [10].

All microalgae have one or more types of chlorophyll, among the four existing types, chlorophyll a, b, c and d, with chlorophyll being the most important pigment for photosynthesis playing a major role in the arrangement of photosystems to capture light energy. The other chlorophylls play secondary roles, contributing to increasing the total light absorbed by the microalgae. Commercially, chlorophylls are used mainly in the dye industry [20].

4. Photobioreactors

Photobioreactors are equipment designed to allow the use of light energy, whether natural or artificial, by the cells of the micro-organism present in it, with the purpose of providing the desired bioproduct. In the case of microalgae, it is desired that the photobioreactor helps promoting cell growth, and for this purpose several types of photobioreactors have been designed, such as open tanks, bubble columns, flat and tubular plates (conical or helical), among others [21].

According to Huang et al. [22], photobioreactors can be confined and the protective covered structures if built with transparent materials allow natural li lighting if the artificial lighting is not an option. They may also be available outdoors, subjected to weather conditions which randomly interfere in the development of the crop.

To minimize the interference associated with external factors, such as variations in the climate, impaired light by cloudy days or long rainy periods and possible contamination by insects, the photobioreactors that allow this direct contact of the culture medium with the external environment (called "open" photobioreactors), can be replaced by "closed" photobioreactors, in which the culture medium has no contact with the external environment, and if provided by artificial lighting they do not suffer from changes in radiance, ensuring greater control and predictability [23].

Among the open photobioreactors, we highlight the open ponds or ponds, known by the English term "Open ponds", which are cultivation containers made of plastic materials (PVC, for example), fiberglass or cement, and should contain smooth internal surfaces in order to reduce likely damages from friction and guarantee ease cleaning. In the construction of the tanks, the depth can vary from 10 to 50 cm approximately, so that it allows the diffusion of carbon dioxide from the atmosphere and the penetration of sunlight as well as allowing aeration by bubbling air in the culture medium [12, 20].

The circular ponds, or circular tanks, reach an average of 1000 m^2, 30cm deep, provided by large rotational arms that can extend up to 45 m in diameter, reaching yields that can vary from 1.5 to 16.5 g m^{-2} d^{-1} of dry biomass, while the "raceway pond" type gets its name from the format that resembles a racing circuit. It is the most used system at both pilot and commercial scale due to its ease of operation [24]. Productivities can vary from 0.19 to 23.5 g m^{-2} d^{-1} of dry biomass [10]. **Figure 2** shows open ponds photobioreactors.

Figure 2.
Open pounds photobioreactors. Raceway pond (A); circular pond (B). Source: authors.

According to Huang [22], in the class of "closed" photobioreactors, the bubble column type is widely used by both researchers on a laboratory scale and commercial purposes due to its versatility of operation, handling and construction, as they are basically cylinders with a high height/diameter ratio and made of transparent materials such as glass and mainly polymers (acrylics and polyvinyl chlorides for example).

According to Carvalho [21], the bubble column photobioreactor has a unique configuration with an air dispersion system in its internal structure, which allows controlling the size and release of air bubbles promoting a random pneumatic agitation of the medium of cultivation.

In order to promote an oriented movement of the liquid inside the cylinder, photobioreactors of "air-lift" models can be used, which are equipped with devices called "draft tube". These devices that allow the distinction of two regions: one that it presents an upward flow of medium where air bubbles are released (Riser), and another of downward liquid flow (Downcomer) [25].

Flat plate photobioreactors are characterized by cobblestones of glass, plastic or other material that allows the passage of solar radiation. The air, enriched with carbon dioxide necessary for the growth of biomass is injected by the base, and promotes turbulence so that all cells are affected by solar radiation [26].

The arrangement of flat plate photobioreactors can vary vertically, horizontally or at an angle. They have the advantage of having a large surface area, which allows better use of the received radiation favoring the photosynthetic activity of microalgae [10]. **Figure 3** shows bubbles column, air-lift and flat plate photobioreactors.

Tubular photobioreactors are the most popular; they have a high ratio of exposed surface to volume, high efficiency of CO_2 usage and sunlight. They consist of sets of transparent tubes, usually glass or plastic, whose diameters are less than 0.1 m, since light is required in the photosynthesis process. Such photobioreactors can be designed in several formats (**Figure 3**): serpentine, inclined, spiral, coils and in parallel [10, 21, 24, 27].

The photobioreactors of the agitated tank type, on the other hand, have a structure composed of cylindrical tubes of glass or material that allows the passage of natural or artificial light and occupy small areas. The reactor may have mechanical or pneumatic agitation systems (air distributors), or even both. They are known as mixing bioreactors, responsible for more than 90% of applications in the fermentation industry [10, 21]. **Figure 4** shows tubular and agitated tank type photobioreactors.

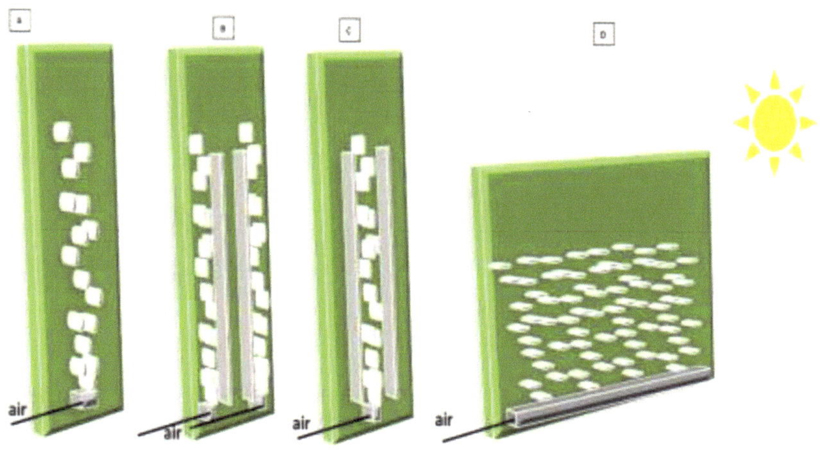

Figure 3.
Bubbles column (A), air-lift (B and C) and flat plate (D) photobioreactors. Source: authors.

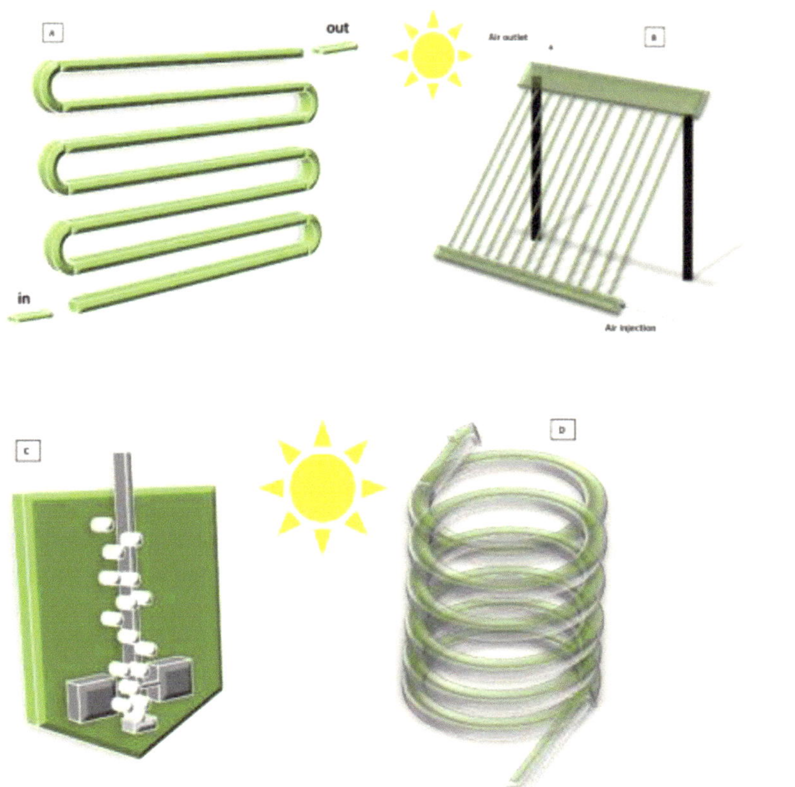

Figure 4.
Tubular photobioreactors: serpentine (A), inclined (B) and coil (D). Agitated tank photobioreactor (C). Source: authors.

In the municipality of Almeria, Spain, the microalgae *Scenedesmus almeriensis* is grown commercially in a 3000 L vertical tubular photobioreactor, owned by the CAJAMAR Foundation. The tubes are made of transparent polymeric material 90 mm in diameter and 400 m long, connected to a 3.5 m high bubble column [28].

Figure 5.
Tubular arrangements of photobioreactors: (A) serpentine tubular reactor; (B) serpentine tubular reactor with mirror at its base to increase the efficiency of light capture by reflecting the incident light. Source: authors.

Many researchers have been designing and developing new configurations or strategies for the arrangement of photobioreactors to stimulate cell growth and the accumulation of metabolites in microalgae. As an example, the photobioreactor developed by Liao et al. [29] (**Figure 5A**), who horizontally disposed a serpentine tubular reactor, with regions of the tube that do not allow the passage of light, creating light and dark regions in photobioreactor to induce the accumulation of lipids during cell growth. A similar strategy was devised and developed by Ilus and Abu-Goshi [30] (**Figure 5B**), with a similar photobioreactor, plus a mirror at its base to increase the efficiency of light capture by reflecting the incident light.

An unconventional photobioreactor was developed by Huang et al. [31], named as a rotating float consisting of bottles in hexagonal arrangement and used as paddles of a raceway pond for the cultivation of microalgae *Dunaliella tertiolecta*. Another interesting photobioreactor has been proposed by Pruvost et al. [32], in which a flat plate photobioreactor is integrated into a building, with the purpose of producing microalgae biomass. This photobioreactor takes advantage of the exhaust gases from the chimney of a commercial building, thereby increasing the production of biomass during cultivation. In addition, it provides thermal comfort to the building in which it is integrated, as it reduces the internal temperature of the place due to the absorption of sunlight by microalgae while growing in the photobioreactor, which is located next to the building. **Figure 6** shows flat plate photobioreactor integrated into a building and rotating float photobioreactor.

It seems that microalgal cultures can be conducted under the most diverse configurations of photobioreactors from open to closed tanks, bubble columns, airlift, tubular

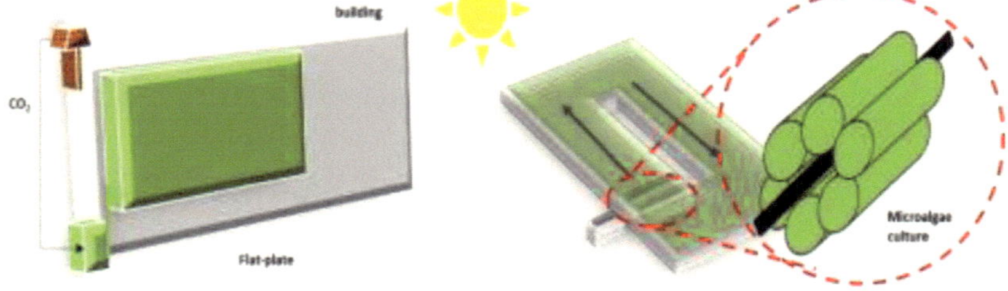

Figure 6.
Flat plate photobioreactor integrated into a building (A) and rotating float photobioreactor (B). Source: authors.

Configuration	Summary	Reference
Bubble column	The marine microalgae *Chlorella minutissima* was grown in a 20 L photobioreactor for the study of several controlled factors, such as temperature, CO2 and nutrients	Loures et al. [33]
Raceway	The cultivation of *Nannochloropsis* microalgae under nutritional stress conditions in an open tank in order to accumulate lipids to produce biodiesel.	Perrier et al. [34]
Tubular	The microalgae *Chlorella vulgaris* was cultivated in two stages as a strategy to accumulate lipids for the biodiesel production	Chen et al. [35]
Placas planas	The cultivation of the microalgae *Chlorella pyrenoidosa* was evaluated under different lighting gradients throughout the photobioreactor.	Huang et al. [36]

Table 2.
Use of traditional photobioreactors in research on microalgae cultivations.

(vertical, horizontal, inclined or spiral), combinations of more than one photobioreactor or even allied to architectural projects like the plate-type photobioreactor built next to an industrial building that rejects CO_2 in its chimney. **Table 2** summarizes some works carried out with several microalgae in the most common photobioreactors.

5. Harvesting biomass

After the microalgae cultivation period in the photobioreactors, the microalgal biomass must be harvested. This step is considered by many to be the key stage for the industrial production of biofuels from microalgae oils because they grow in very diluted cultures. The energy consumption is high due to the processing of large quantities of liquid which are needed to guarantee a feasible harvesting [37].

The main methods of harvesting microalgae are: flocculation, electroflocculation, biofloculation, gravimetric sedimentation, flotation, filtration and centrifugation [38].

The flocculation process involves the addition of chemicals capable of inducing the aggregation of microalgae cells either by neutralization, inversion of the electrical charges on the cell walls, or by the formation of bonds between the microalgae. It allows biomass recovery with lower economic costs when compared to the centrifugation process [37].

The main flocculating agents used for harvesting microalgae are salts such as $FeCl_3$, $Al2(SO_4)_3$ and $Fe_2(SO_4)_3$, polymers (chitosan) and bases such as NaOH or KOH, due to the change in pH [38].

The electrofloculation process, not widely disseminated, is based on the characteristic of microalgae to be electronegatively charged. Thus, by adapting a culture medium with the insertion of electrodes and allowing the passage of electrical current, it is possible to separate the microalgae cells of culture medium [38].

According to Barros et al. [38], in the biofloculation process, the flocculating agent is a microorganism, such as bacteria of the genus *Flavobacterium*, *Terrimonas* and *Sphingobacterium*, or fungi (*Rhizopus oryzae*, *Penicillium expansum*, *Aspergilus* and *Mucor circinelloides*), forming aggregates with the microalgal cells and favoring the microalgal cells gravimetric settling.

Biofloculation can eliminate the need for chemical flocculating agents, although for some cases it is necessary to add $CaCl_2$ to guarantee the efficiency of the process [38].

The flotation process can be defined as the "reverse" of sedimentation, where air bubbles promote the ascension of the cells of the culture by separating it from the liquid medium. However, cultures of marine microalgae are unlikely to undergo flotation, due to the high salinity of the medium [38].

The filtration method consists of passing the culture medium through a small pore membrane that retains the microalgal cells and allows only the passage of the liquid medium [37].

According to Lourenço [20], in the filtration of small volumes, a filtration apparatus such as a Kitasato flask or similar is applicable, however, on a large-scale cultivation, the separation of huge volumes of microalgae and biomass processing is only feasible if the species have large or filamentous cells, which are easily retained in the filter; a short process time.

Regarding the harvest time, gravimetric sedimentation is a slow and simple process which involves the separation of microalgae cells from the culture medium by the action of gravitational force. The downside of this process is the low efficiency [20].

On the other hand, centrifugation, cell sedimentation equipment that acts by the action of centrifugal force, is a fast process of biomass recovery, however, it requires high energy consumption. By this technique, biomass is concentrated without the addition of chemicals, preserving its original characteristics [20].

6. Extraction of lipids

Microalgae oils can be extracted similarly to other oilseed biomasses, which usually use physical extraction by adding a chemical solvent to improve the extraction process; solvents such as hexane and methanol are widely used [24].

The extraction of oil with solvent is a process of transferring soluble constituents (oil) from an inert material (biomass) to a solvent, in a purely physical process, without any chemical reaction. Solvent extraction is currently the most economical method and the use of hexane and chloroform has made it the fastest and most efficient method for lipid extraction from the most diverse biomasses. However, the presence of these chemical solvents can affect the lipid composition of the extracted oil [39].

Mechanical presses such as screw and piston, among others, can rupture the cells and release the oil, but these operation lead to low efficiency. On the other hand, when combined with the use of organic solvents, oil extraction efficiency can be reached in the order of 95% [10].

High concentrations of salts can cause a sudden change in the osmotic pressure, which can lead to the rupture of cells and the release of oil from inside; a method known as "osmotic shock" [39].

A method that has received notable attention is supercritical extraction, which uses fluids above the critical point of temperature and pressure, providing properties such as low viscosity and high diffusivity, allowing to achieve greater efficiency in the lipid extraction step. In this context, the use of CO_2 as a supercritical fluid is highly attractive due to its high selectivity, non-toxicity and short processing period. The only drawback is the high cost associated with installing adequate equipment to operate under safe conditions at high temperature and pressure [10].

In the search for environmentally friendly processes, enzymatic extraction is highly recommended, since it uses enzymes such as pectinase and cellulase, to degrade the cell wall, providing highly efficient extractions without affecting the microalgal lipid composition, unlike the method that uses organic chemical solvents [39]. On the other hand, these methods are still used only on a laboratory scale, due to the high costs involved with the acquisition of enzymes [24].

7. Synthesis of biodiesel

Biodiesel is the natural and renewable substitute fuel for petroleum diesel. It is produced from vegetable oils such as soy, palm, sunflower and jatropha and/or animal fat, such as beef tallow and chicken fat. In other words, any sources that have lipids can, in theory, be useful in the production of biodiesel [6].

First generation raw materials such as soy, palm and sunflower are not considered socially sustainable because they compete with the demand for food. In contrast, second generation raw materials such as beef tallow, residual frying oils and chicken fat, stand out as alternatives to produce biodiesel. Physic nut, which grows in semi-arid regions, where other sources of oilseeds are not able to grow and, therefore, do not compete with the food chain can also be used to produce biodiesel [6].

Microalgae, as well as cyanobacteria, belong to the group of third generation raw materials. These are a promising alternative, as they are able to accumulate high levels of lipids combined with high growth rates and biomass productivity compared to conventional oilseeds, and they do not compete with the food chain [40].

Biodiesel can be defined as a mixture of alkyl esters of fatty acids obtained by transesterification of triacylglycerides from vegetable oils or animal fat, or by the esterification reaction of free fatty acids resulting in alkyl esters of fatty acids and water. Both reactions occur in the presence of alcohols and catalyst [41].

Alcohols are considered to be transesterification agents, and may be methyl alcohol (methanol), ethyl alcohol (ethanol), propyl, butyl or amyl. Methanol is the most widely used alcohol due to its low cost and its physical–chemical properties (polarity and lower carbon chain), while ethanol has stood out for its potential for low toxicity and easy availability [37].

Transesterification is a multi-stage reaction, including three reversible stages in series, so that triglycerides are converted to diglycerides, then diglycerides are converted to monoglycerides and finally monoglycerides are converted to esters (biodiesel) and glycerol (co-product). Stoichiometrically the transesterification reaction requires 3 moles of alcohol for every mole of triacylglyceride, with an excess of alcohol being used to shift the balance towards the formation of the products, as it is a reversible reaction. The esterification reaction requires 1 mole of alcohol to 1 mole of fatty acid and requires excess alcohol to favor the direction of product formation [42].

The biodiesel synthesis reactions can be catalyzed homogeneously or heterogeneously. Homogeneous catalysts can be acidic (sulfuric, sulfonic and hydrochloric acid) or alkaline (sodium or potassium hydroxides), while heterogeneous catalysts can be enzymes or metallic compounds. In the catalyst determination, the acidity index is the main characteristic of the oil to be observed since high amounts of fatty acids do not allow alkaline catalysis (soap formation in the reaction medium). In this case, the use of acid catalysts is suggested [18].

On enzymatic catalysis, enzymes such as lipases, present in several organisms including animals, plants, fungi and bacteria have the biological function of accelerating the hydrolysis of fats and vegetable oils, releasing fatty acids, monoglycerides, diglycerides and glycerol [43].

Enzymes have some advantages when compared to the chemical catalytic process, such as high selectivity, reaction temperature in the range of 30–40°C and pH between 4 and 9. On the other hand, it requires long reaction periods to achieve high conversions and, therefore, costly [43].

Leung, Wu and Leung [44] explained that in general there will be advantages and disadvantages associated with the choice of the type of catalyst. The main

Microalgae	Catalyst	Condition	Reference
Chlorella minutissima	H_2SO_4	molar ratio (alcohol:lipid) 9:1, 90°C e 8 h	Loures et al. [45]
Chlorella protothecoides	H_2SO_4	Molar ratio (methanol:lipid) 56:1, 30 °C and 4 h	Miao, Wu [46]
Chlorella sp.	H_2SO_4	Molar ratio (alcohol:lipid) 30:1, 60°C and 4 h	Amaral et al. [47]

Source: authors.

Table 3.
Catalysts used in biodiesel reactions for Chlorella microalgae.

advantages of alkaline catalysts are high catalytic activity, low cost and moderate operating conditions. However, it requires a low acidity index to prevent soap formation. Acid catalysts have the advantage of preventing soap formation. However, these catalysts corrode the equipment and require long reaction times. Heterogeneous catalysts (enzymes and metallic compounds) are highly selective and allow recycling, but at high costs.

The oil extracted from marine microalgae has a high acidity index, indicating acid catalysis as the most suitable for biodiesel production, as can be seen in **Table 3** which highlights the research done in the process of biodiesel synthesis via acid catalysis [45].

The synthesis of biodiesel from microalgal oils is carried out in a reactor. The mixture of alcohol and catalyst reacts with the triglyceride and/or fatty acids present in the microalgal oil. After the reaction, the mixture is transferred to a separation tank to guarantee the formation of the upper layer consisting of methyl ester, excess of alcohol and catalyst (acid or base) and the lower layer, predominantly glycerol [24].

8. Reaction of biodiesel in situ

The biodiesel production process usually involves the extraction of oils or lipids from biomass, which will later be processed to produce biodiesel; thus called a two-stage process. Currently, industries still require new technologies in the oil extraction and purification of the biodiesel, which are estimated to impact 70 to 80% on production costs [48].

Direct or "in situ" transesterification, an expression of the Latin that means "on the spot", is the most prominent to perform lipid extraction and the transesterification reaction concurrently, in a single step of the process. Currently, the process has been called a Reactive Extraction Process. Therefore, the direct processing of biomasses seems to be an economically viable alternative as it results in a more economical use of resources in the production process. This direct use of biomass in the reactor is called Direct Transesterification or "In Situ". The "In situ" Transesterification process emerges as a viable alternative to this problem as it makes direct use of the raw material without the need to prioritize oil extraction (**Figure** 7) [49].

The TEIS of microalgae biomass is shown as a potential alternative in reducing the costs of microalgal biodiesel production by eliminating the pre-treatment of biomass, enabling lipid extraction and purification, which has been the main factor in preventing the advancement of this industry; it minimizes the high consumption of solvents and uptime [49]. The elimination of the lipid extraction step not only reduces production steps but also results in a lower initial investment cost, equipment installation and maintenance, and energy [50].

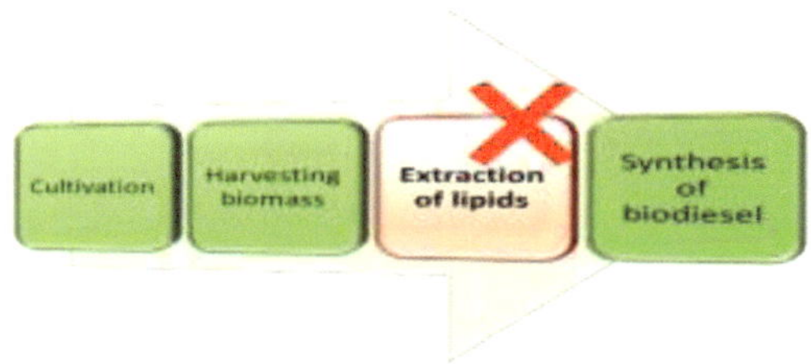

Figure 7.
The "In situ" Transesterification process use of the raw material without the need to prioritize oil extraction.

As reported by Skorupskaite et al. [50], in situ technology can be applied to almost any raw material of plant origin or waste. However, attention should be focused on the biomass characteristics such as humidity, particle size, oil composition, acid content and reaction conditions (characteristics of reagents and catalyst, reaction time, temperature, etc.).

The reaction system for the TEIS process is suitable if the agitation of the reaction medium is effective enough to keep the biomass suspended [42]. This implies that there is a homogeneous dispersion of the liquid solvent and the solid biomass in the TEIS process [50]. that is, a perfectly agitated reactor. However, several authors have proposed different reaction systems to produce biodiesel in situ from microalgae.

9. Conclusion

The high rate of cell growth, the high content of lipids in the biomass and the non-competition in the food chain place microalgae as sources of raw material suitable in the production of biodiesel. To maximize biomass productivity, photobioreactors are key equipment in the success of a microalgal biodiesel production chain, as the photobioreactors allow an accelerated growth of microalgae, making more lipids available to be converted in biodiesel. In this sense, different configurations of photobioreactors can be used to produce microalgae biomass, with raceways and tubular photobioreactors being the most used configurations.

Acknowledgements

We thank Coordenação de Aperfeiçoamento de Pessoal de Ensino Superior (CAPES) for financial support to conduct this work.

Author details

Mateus S. Amaral[1*], Carla C.A. Loures[2], Fabiano L. Naves[3], Gisella L. Samanamud[4], Messias B. Silva[5,6] and Arnaldo M.R. Prata[5]

1 Laboratory of Technological and Environmental Chemistry, Department of Chemistry, Institute of Exact and Biological Sciences (ICEB), Federal University of Ouro Preto, Campus Universitário Morro do Cruzeiro, Ouro Preto, MG, Brazil

2 Centro Federal de Educação Tecnológica Celso Suckow Fonseca – CEFET-RJ, Angra dos Reis, RJ, Brasil

3 Federal University of São João del Rei – UFSJ – Campus Ouro Branco, Ouro Branco, MG, Brazil

4 Department of Chemical and Materials Engineering, University of Kentucky, Paducah Extended Campus, Paducah, KY, USA

5 Engineering School of Lorena – EEL-USP, Universidade de São Paulo - USP, Lorena, SP, Brazil

6 Universidade Estadual Paulista - UNESP, Campus Guaratinguetá, Guaratinguetá, SP, Brasil

*Address all correspondence to: mateus.amaral@ufop.edu.br

References

[1] Yeong TK, Jiao K, Zeng X, Lin L, Pan S, Danquah MK. Microalgae for biobutanol production – Technology evaluation and value proposition. Algal Research. 2018;**31**:367-376. Available from: http://www.sciencedirect.com/science/article/pii/S2211926418300080

[2] Chen HM. IEA, CO2 emissions from fuel combustion Highlights 2017, A Report by the International Energy Agency, 2017. Vol. 53, Library Technology Reports; 2017

[3] Heck V, Gerten D, Lucht W, Popp A. Biomass-based negative emissions difficult to reconcile with planetary boundaries. Nature Climate Change. 2018;**8**(2):151-155. Available from: https://doi.org/10.1038/s41558-017-0064-y

[4] Zhu L., Huo, S., Qin, L. A microalgae-based biodiesel refinery: Sustainability concerns and challenges. International Journal of Green Energy. 2015;**12**(6):595-602. Available from: https://doi.org/10.1080/15435075.2013.867406

[5] Rós P, Silva C, Silva-Stenico M, Fiore M, & Castro H. Microbial oil derived from filamentous Cyanobacterium Trichormus sp. as feedstock to yield fatty acid ethyl esters by enzymatic synthesis. Journal of Advances in Biology & Biotechnology. 2017;**12**(4):1-14. Available from: https://doi.org/10.9734/jabb/2017/33045

[6] Ahmad AL, Yasin NHM, Derek CJC, Lim JK. Microalgae as a sustainable energy source for biodiesel production: A review. Renewable and Sustainable Energy Reviews. 2011;**15**:584-593

[7] Brownbridge G, Azadi P, Smallbone A, Bhave A, Taylor B, Kraft M. The future viability of algae-derived biodiesel under economic and technical uncertainties. Bioresour Technol. 2014;**151**:166-173

[8] Tan XB, Lam MK, Uemura Y, Lim JW, Wong CY, Lee KT. Cultivation of microalgae for biodiesel production: A review on upstream and downstream processing. Chinese Journal of Chemical Engineering. 2018;**26**(1):17-30

[9] Shuba ES, Kifle D. Microalgae to biofuels: 'Promising' alternative and renewable energy, review. Renewable and Sustainable Energy Reviews. 2018;**81**(April 2016):743-755

[10] Hattab M, Ghaly A. Production of biodiesel from marine and freshwater microalgae: A review. Advances in Research. 2015;**3**(2):107-155. Available from: http://www.sciencedomain.org/abstract.php?iid=695&id=31&aid=6322

[11] Mata TM, Martins AA, Caetano Nidia S, Mata TM, Martins AA, Caetano Nidia S. Renewable & sustainable energy reviews. Renewable and Sustainable Energy Reviews. 1997;**14**:217-232. Available from: https://EconPapers.repec.org/RePEc:eee:rensus:v:14:y:2010:i:1:p:217-232#.XxCxKFy3yaA.mendeley

[12] Chisti Y. Biodiesel from microalgae. Biotechnology Advances. 2007;**25**(3):294-306. Available from: http://www.sciencedirect.com/science/article/pii/S0734975007000262

[13] Spolaore P, Joannis-Cassan C, Duran E, Isambert A. Commercial applications of microalgae. American Journal of Bioscience and Bioengineering. 2006;**101**(2):87-96. Available from: http://www.sciencedirect.com/science/article/pii/S1389172306705497

[14] Yadav G, Sen R. Microalgal green refinery concept for biosequestration of carbon-dioxide vis-à-vis wastewater remediation and bioenergy production:

Recent technological advances in climate research. Journal of CO2 Utilization. 2017;**17**:188-206. DOI: 10.1016/j.jcou.2016.12.006

[15] Sánchez Mirón A, Cerón Garcıa MC, Garcıa Camacho F, Molina Grima E, Chisti Y. Growth and biochemical characterization of microalgal biomass produced in bubble column and airlift photobioreactors: Studies in fed-batch culture. Enzyme and Microbial Technology. 2002;**31**(7):1015-1023. Available from: http://www.sciencedirect.com/science/article/pii/S0141022902002296

[16] Ördög V, Stirk WA, Bálint P, Aremu AO, Okem A, Lovász C, et al. Effect of temperature and nitrogen concentration on lipid productivity and fatty acid composition in three chlorella strains. Algal Research. 2016;**16**:141-149. Available from: http://www.sciencedirect.com/science/article/pii/S221192641630073X

[17] Huang G, Chen F, Wei D, Zhang X, Chen G. Biodiesel production by microalgal biotechnology. Applied Energy. 2010;**87**(1):38-46. Available from: http://www.sciencedirect.com/science/article/pii/S0306261909002530

[18] Demirbas A. Use of algae as biofuel sources. Energy Conversion and Management. 2010;**51**(12):2738-2749. Available from: http://www.sciencedirect.com/science/article/pii/S0196890410002207

[19] Sathasivam R, Radhakrishnan R, Hashem A, Abd Allah EF. Microalgae metabolites: A rich source for food and medicine. Saudi Journal of Biological Sciences. 2019;**26**(4):709-722

[20] Lourenço SO. Cultivo de microalgas marinhas: princípios e aplicações. RiMa, editor. São Carlos; 2006. P. 588

[21] Carvalho JCM, Matsudo MC, Bezerra RP, Ferreira-Camargo LS, Sato S. Microalgae Bioreactors. In: Bajpai Rakesh K, Ales P, Zappi ME, editors. Algal Biorefineries. Dordrecht: Springer Netherlands; 2014. pp. 83-126. Available from: http://link.springer.com/10.1007/978-94-007-7494-0_4

[22] Huang Q, Jiang F, Wang L, Yang C. Design of Photobioreactors for mass cultivation of photosynthetic organisms. Engineering. 2017;**3**(3):318-329

[23] Singh RN, Sharma S. Development of suitable photobioreactor for algae production - a review. Renewable and Sustainable Energy Reviews. 2012;**16**(4):2347-2353

[24] Suali E, Sarbatly R. Conversion of microalgae to biofuel. Renewable and Sustainable Energy Reviews. 2012;**16**(6):4316-4342. Available from: http://www.sciencedirect.com/science/article/pii/S1364032112002304

[25] Mendes CE, Badino AC. Hydrodynamics of Newtonian and non-Newtonian liquids in internal-loop airlift reactors. Biochemical Engineering Journal. 2016;**109**:137-152. Available from: http://www.sciencedirect.com/science/article/pii/S1369703X16300079

[26] Arun J, Varshini P, Prithvinath PK, Priyadarshini V, Gopinath KP. Enrichment of bio-oil after hydrothermal liquefaction (HTL) of microalgae C. vulgaris grown in wastewater: Bio-char and post HTL wastewater utilization studies. Bioresource Technology. 2018;**261**(April):182-187. DOI: 10.1016/j.biortech.2018.04.029

[27] Wang B, Lan CQ, Horsman M. Closed photobioreactors for production of microalgal biomasses. Biotechnology Advances. 2012;**30**(4):904-912

[28] Solimeno A, Gabriel F, García J. Mechanistic model for design, analysis, operation and control of microalgae

cultures: Calibration and application to tubular photobioreactors. Algal Research. 2017;**21**:236-246

[29] Liao Q, Li L, Chen R, Zhu X. A novel photobioreactor generating the light/dark cycle to improve microalgae cultivation. Bioresource Technology. 2014;**161**:186-191

[30] Iluz D, Abu-Ghosh S. A novel photobioreactor creating fluctuating light from solar energy for a higher light-to-biomass conversion efficiency. Energy Conversion and Management. 2016;**126**:767-773

[31] Huang JJ, Bunjamin G, Teo ES, Ng DB, Lee YK. An enclosed rotating floating photobioreactor (RFP) powered by flowing water for mass cultivation of photosynthetic microalgae. Biotechnology for Biofuels. 2016;**9**(1):1-18

[32] Pruvost J, le Gouic B, Lepine O, Legrand J, le Borgne F. Microalgae culture in building-integrated photobioreactors: Biomass production modelling and energetic analysis. Chemical Engineering Journal. 2016;**284**:850-861

[33] Pereira FM, Loures CCA, Amaral MS, Gomes FM, Pedro GA, Machado MAG, et al. Evaluation of fatty acids production by Chlorella minutissima in batch bubble-column photobioreactor. Fuel. 2018;**230**(December 2017):155-162

[34] Perrier B, Crampon C, Guézet O, Simon C, Maire F, Lépine O, et al. Production of a methyl ester from the microalgae Nannochloropsis grown in raceways on the French west coast. Fuel. 2015;**153**:640-649. Available from: http://www.sciencedirect.com/science/article/pii/S0016236115002926

[35] Chen CY, Yeh KL, Chang HY, Chang JS. Strategies to improve oil/lipid production of microalgae in outdoor cultivation using vertical tubular-T ype photobioreactors. In: Energy Procedia. Elsevier Ltd; 2014. pp. 2755-2758

[36] Huang J, Li Y, Wan M, Yan Y, Feng F, Qu X, et al. Novel flat-plate photobioreactors for microalgae cultivation with special mixers to promote mixing along the light gradient. Bioresource Technology. 2014;**159**:8-16. Available from: http://www.sciencedirect.com/science/article/pii/S096085241400159X

[37] Rastogi RP, Pandey A, Larroche C, Madamwar D. Algal green energy – R&D and technological perspectives for biodiesel production. Renewable and Sustainable Energy Reviews. 2018;**82**(October 2017):2946-2969. DOI: 10.1016/j.rser.2017.10.038

[38] Barros AI, Gonçalves AL, Simões M, Pires JCM. Harvesting techniques applied to microalgae: A review. Renewable and Sustainable Energy Reviews. 2015;**41**:1489-1500. Available from: http://www.sciencedirect.com/science/article/pii/S1364032114008107

[39] Rashid N, Ur Rehman MS, Sadiq M, Mahmood T, Han J-I. Current status, issues and developments in microalgae derived biodiesel production. Renewable and Sustainable Energy Reviews. 2014;**40**:760-778 Available from: http://www.sciencedirect.com/science/article/pii/S1364032114005565

[40] Reis CER, De Souza Amaral M, Loures CCA, Da Rós PCM, Hu B, Filho HJI, et al. Microalgal feedstock for bioenergy: Opportunities and challenges. Biofuels in Brazil: Fundamental Aspects, Recent Developments, and Future Perspectives. 2014. p. 367-392. Springer International Publishing. Available from: https://doi.org/10.1007/978-3-319-05020-1_17

[41] Amaro HM, Guedes AC, Malcata FX. Advances and perspectives in using microalgae to produce biodiesel. Applied Energy.

2011;**88**(10):3402-3410. Available from: http://www.sciencedirect.com/science/article/pii/S0306261910005210

[42] Salam KA, Velasquez-Orta SB, Harvey AP. A sustainable integrated in situ transesterification of microalgae for biodiesel production and associated co-product-a review. Renewable and Sustainable Energy Reviews. 2016;**65**:1179-1198. Available from: http://www.sciencedirect.com/science/article/pii/S1364032116304075

[43] Vieira APDA, da Silva MAP, Langone MAP. Biodiesel production via esterification reactions catalyzed by lipase. Latin American Applied Research. 2006;**36**(4):283-288

[44] Leung DYC, Wu X, Leung MKH. A review on biodiesel production using catalyzed transesterification. Applied Energy. 2010;**87**(4):1083-1095. Available from: http://www.sciencedirect.com/science/article/pii/S0306261909004346

[45] Loures CCA, Amaral MS, da Rós PCM, Zorn SMFE, de Castro HF, Silva MB. Simultaneous esterification and transesterification of microbial oil from Chlorella minutissima by acid catalysis route: A comparison between homogeneous and heterogeneous catalysts. Fuel. 2018;**211**(September 2017):261-268

[46] Miao X, Wu Q. Biodiesel production from heterotrophic microalgal oil. Bioresource Technology. 2006;**97**(6):841-846. Available from: http://www.sciencedirect.com/science/article/pii/S0960852405002191

[47] Amaral MS, Loures CC, da Rós PCM, Machado SA, Reis CER, de Castro HF, et al. Evaluation of the cultivation conditions of marine microalgae chlorella sp. to be used as feedstock in ultrasound-assisted ethanolysis. Biofuel Research Journal. 2015;**2**(3):288-294

[48] Go AW, Sutanto S, Ong LK, Tran-Nguyen PL, Ismadji S, Ju Y-H. Developments in in-situ (trans) esterification for biodiesel production: A critical review. Renewable and Sustainable Energy Reviews. 2016;**60**:284-305. Available from: http://www.sciencedirect.com/science/article/pii/S1364032116001003

[49] Lewis T, Nichols PD, McMeekin TA. Evaluation of extraction methods for recovery of fatty acids from lipid-producing microheterotrophs. Journal of Microbiological Methods. 2000;**43**(2):107-116. Available from: http://www.sciencedirect.com/science/article/pii/S0167701200002177

[50] Skorupskaite V, Makareviciene V, Gumbyte M. Opportunities for simultaneous oil extraction and transesterification during biodiesel fuel production from microalgae: A review. Fuel Processing Technology. 2016;**150**:78-87. Available from: http://www.sciencedirect.com/science/article/pii/S0378382016301904

Chapter 21

Microalgae Growth under Mixotrophic Condition Using Agro-Industrial Waste: A Review

Izabel Pereira, Adriano Rangel, Bruna Chagas, Bruno de Moura, Stela Urbano, Roberto Sassi, Fabiana Camara and Cíntia Castro

Abstract

Microalgae has a great potential to produce biofuels and bioproduct but the cost is still too high mainly due to the biomass production. Mixotrophic cultivation has been pointed as microalgae cultivation mode for biomass/bioenergy production with lower cost and able to make remediation of organic waste. The proposals of this work was to make a review of microalgae growth under mixotrophic condition using agro-industrial waste. Agro-industrial by-products and wastes are of great interest as cultivation medium for microorganisms because of their low cost, renewable nature, and abundance. However biotechnological technologies are necessary to develop the production of microalgae on a large scale.

Keywords: microalgae, mixotrophic growth, biomass, agro-industrial waste, biorefinary

1. Introduction

Microalgae are a variety of autotrophic, prokaryotic or eukaryotic organisms, where their single-cell structure allows solar energy to be easily converted into chemical energy. This biochemical conversion is being used commercially to obtain the biomass, consequently, in the insertion in products with commercial application. The most used microalgae cultivation techniques are opens aerated lagoons and closed photobioreactors [1–4].

Due to the advantages that microalgae offer over many other species, researchers and entrepreneurs have shown great interest in the development of production processes for biofuels, functional foods and bio-products from different species. Compared to terrestrial crops, these microorganisms have photosynthetic efficiency, growth rate and higher biomass production, consequently mass cultivation for commercial microalgae production can be carried out efficiently [5]. In addition, the cultivation of microalgae does not require arable soil, and can be grown in saline, brackish and wastewater and in harsh conditions, not competing with the production of food that is currently a major challenge for the production of first and second biofuels generation [6]. Therefore, competition for arable land with other crops, especially for human consumption, is greatly reduced.

Although most microalgae grow exclusively through photosynthesis, some species are mixotrophic and use extracellular organic carbon when a light source is not available [7]. Microalgae can be a source of several important compounds, including hydrogen and hydrocarbons, pigments and dyes, food and feed, biopolymers, biofertilizers, insecticides, neutraceuticals (foods capable of providing health benefits) and pharmacological compounds, in addition to being a potential biomass for production of biofuels [8].

Although the production of microalgae does not directly compete with food production and can be grown in harsh conditions, economic viability does not yet exist in many of the processes of industrial interest. However, the improvement and mastery of technologies capable of making inserted industrial processes viable become essential. Despite of the microalgae have a wide potential for production and applications, there are many obstacles to the biodiversity of these algae, such as mastery of technologies for production, genetic improvement research of strains more resistant to pathogens and economic viability in large-scale production [9–10]. According to Georgianna and Mayfield [11], although promising, the success of inserting microalgae in the production of various products depends mainly on two important factors: high productivity and quality of biomass, as well as cost-effective production.

One of the viable solutions to reduce the costs of microalgae biomass production is to explore different forms of energy metabolism, highlighting the photoautotrophic, heterotrophic and mixotrophic for commercial production. Understanding these forms of metabolism allows the application of efficient crop strategies aimed at increasing the production of biomass and bioproducts on a large scale with cost optimization to couple the agroindustry waste treatment [7]. Microalgae are able to eliminate a variety of pollutants in wastewater mainly nitrogenated, phosphates and organic carbons [12].

Mixotrophic cultivation is a preferable microalgae growth mode for biomass production [13]. Compared to photoautotrophic and heterotrophic metabolism, mixotrophic cultures have been demonstrated many advantages, such as less risk of contamination, reduced cost and high biomass productivity. Even susceptible to contaminations, the use of photobioreactors minimizes this risk, but increases the cost of the process, which can be offset by the high biomass yield that can reach 5–15 g/L, being 3–30 times higher than those produced under autotrophic growth conditions [14, 15].

The use of waste for microalgae mixotrophic growth has been researched, mainly with the objective of expanding and diversifying in an alternative way the control and combating the inappropriate disposal of these in the respective industries, combined with the perspective of minimizing the operational costs of producing microalgae in large scale that are still considered high. The waste generated by the agribusiness has a high load of organic matter with high concentrations of Biochemical Oxygen Demand (BOD), Chemical Oxygen Demand (COD), ammonia, phosphates, suspended solids harmful to the environment, in addition to dissolved components such as sugars, fat and proteins originating from food, contributing to environmental pollution [16].

According to Patel et al. [7], research involving the mixotrophic cultivation of microalgae using organic matter as a source of carbon points to the production of high yields of biomass and biocomposites of industrial interest when compared to systems involving photoautotrophic and heterotrophic metabolisms. In this sense, recent studies have been carried out using agroindustrial waste to grow microalgae in a mixotrophic regime in order to minimize the cost of the biomass production process and treat the effluent adding value to the process, suggesting a microalgae biorefinery system [17].

Patel et al. [18] cultivated *C. protothecoides* UTEX-256 under mixotrophic conditions using dairy waste as source of carbon. The high CO2-emitting dairy industry obligated to treat waste and improve its carbon-footprints. In general, biochemical treatment was effective to remove respectively 99.7 and 91–100% of organic and inorganic pollutants and produce biomass and lipids fractions.

Xio-Bo Tan et al. [19] demonstrated that *Chlorella pyrenoidosa* (FACHB-9) cultivated under mixotrophic conditions using anaerobic digestate of sludge with an optimal addition of acidified starch wastewater improved biomass and lipids production by 0.5-fold (to 2.59 $g{\cdot}L^{-1}$) and 3.2-fold (87.3 $mg{\cdot}L^{-1}{\cdot}d^{-1}$), respectively. In addition, 62% of total organic carbon, 99% of ammonium and 95% of orthophosphate in mixed wastewater were effectively removed by microalgae.

Wang et al. [20] utilized glucose recovered from enzymatic hydrolysis of food waste as culture medium in mixotrophic cultivation of *Chlorella sp*. to obtain high levels of lipid and lutein. The algal biomass was 6.9 g L^{-1} with 1.8 g L^{-1} lipid and 63.0 mg L^{-1} lutein using hydrolysate with an initial glucose concentration of 20 g L − 1. Furthermore, lipid derived from microalgae biomass using food hydrolysate was at high quality in terms of biodiesel properties.

Due to the success of mixotrophic microalgae growth, the use of agro-industrial by-products stands out, adding value to production processes and reducing costs. The nutritional characteristics, availability and low cost of obtaining evidence the possibility of using the by-products in the cultivation of microalgae. This work reviews the mixotrophic cultivation system of microalgae using waste from agribusiness as a source of organic carbon, pointing out the benefits of this strategy as a solution to the environmental problems caused by these effluents, adding value to an industrial process for the production of biomass and biocompounds as biorefinery.

2. Microalgae

Microalgae is a generic term used to refer a widely diverse group of photosynthetic microorganisms [21]. There are several species of microalgae, which are found in aquatic environments of fresh water, brackish and saline [22]. Microalgae in general, have varying microscopic sizes, perform photosynthesis, use carbon dioxide as a nutrient source for growth, in addition to playing a fundamental role in ecosystems [23–25]. It is estimated that there are about 800 thousand species of microalgae, of which about 40 to 50 thousand are of scientific knowledge, which makes it an almost unexplored resource, demonstrating the great biodiversity of these algae [26–27]. In addition, most species are not yet known and very few are used for any purpose.

The basic composition of microalgae is based on carbohydrates, lipids, proteins, ash and nucleic acids, in addition to chlorophyll and other protective pigments and light capture that provide high photosynthetic capacity, allowing conversion of up to 10% of energy in biomass [28]. In conventional plants, this percentage is higher when compared to other conventional plants, whose conversion is limited to a maximum of 5% [29]. The predominant elements in the biomass of microalgae are carbon, nitrogen and phosphorus and some metals such as iron, cobalt, zinc is also found [28].

3. Application of microalgae

In recent years, several researches have been carried out seeking to develop technologies for the elaboration and diversification of products based on microalgae.

The growing expansion of these products is part of a wide range of utilities inserted in the most different commercial niches, expanding the possibilities of use and adding value to the market. According to Hu et al. [30], the global algae market is expected to be worth of about $ 1.1 billion by 2024.

Microalgae are inserted in a wide variety of species, distinguishing one from the other due to their biological structure. In this sense, these microorganisms offer potential possibilities for CO2 biofixation, remediation, effluent treatment, production of biofuels, high-value products including pharmaceuticals, food and neutraceutics [31].

According to Rizwan et al. [32], microalgae can be a source of antioxidant compounds, carotenoids, enzyme polymer, lipid, natural dye, polyunsaturated fatty acid, peptide, toxin and sterols, which are widely used in industry. In addition, they are used for the synthesis of antiviral, antimicrobial, antiviral, antibacterial and anticancer drugs [33].

Commercial microalgae cultivation systems are operated to produce mainly pigments and metabolites for nutritional supplements [34]. The algae that have technical and economic viability of production are Spirulina (Arthrospira) for supplements with a high protein content, Haematococcus as a source of astaxanthin and *Dunaliella salina* for the production of pro-vitamin A [31]. Spirulina represents 60% of all biomass produced on a large scale [35]. This species can be easily grown in tropical regions and is well adapted to extreme environments, being relatively less susceptible to contamination than other microalgae, making it the most favorable choice for large-scale production [36].

Spirulina consists mainly of proteins (50–70%), being widely used in human nutrition to combat malnutrition [37]. This species is rich in essential amino acids, beta-carotene, minerals, essential fatty acids, vitamins, polysaccharides, among others. Chlorella accumulate high concentrations of carotenoids (astaxanthin, lutein, β-carotene, violaxanthin and zeaxanthin), antioxidants, vitamins, polysaccharides, proteins, peptides and fatty acids [5, 38]. In addition to all the benefits mentioned, the bioactive compounds of microalgae can have a biological, immune, antiviral and anti-cancer properties, being highly active [39].

Global warming has been worrying environmentalists across the planet. Although there are different ways of capturing CO2, the biological method stands out as a potentially attractive alternative. The requirements for producing and obtaining biomass from microalgae are basically CO2 and a source of light, be it natural or artificial [2]. Carbon dioxide can be converted into organic matter by performing photosynthesis using sunlight as an energy source [40–42]. Microalgae are more efficient for fixing CO2 and have a higher productivity rate (ton/ha/year) when compared to terrestrial plants. In addition, CO_2 biofixation can be combined with other processes, such as the treatment of organic waste, being advantageous in terms of economic viability and environmental sustainability. Microalgae can also be grown in nutrient-rich organic effluents, salt and brackish water, reducing the use of fertile land and fresh drinking water [43].

Studies involving the mixotrophic cultivation of microalgae using industrial residues from agro-industry as a source of organic carbon have been carried out to minimize the cost of biomass production, treat the effluent and promote CO_2 biofixation [7]. In this sense, expanding the ways in which these residues are used, avoiding their incorrect disposal, minimizes the effects of environmental pollution and adds value to industrial processes, encouraging a cleaner and more sustainable bioeconomy.

Currently, fossil fuels represent the main source of energy in the world, but unsustainable and directly related to the pollution of air, land, water and climate change. The burning of fossil fuels consolidated to increase the atmospheric concentration of CO_2 being directly associated with global warming. Allied to this, the future oil scarcity is a major challenge for scientists, motivating a constant search for technologies capable of producing clean and sustainable fuels [44]. Among many biomasses, microalgae represent a promising source for the production of clean renewable energy, as they are capable of fixing CO_2 by performing photosynthesis with efficiency and productivity superior to that of conventional oilseeds and terrestrial plants used in the production of biodiesel and bioethanol. Among the available biomass sources, microalgae have been evaluated and investigated as generation third biomass, being researched to produce different types of biofuels, among which are biodiesel, bioethanol, bio-oil, char, hydrogen and synthesis gas [45]. Recent research involving the production of biofuels has been focused on third generation biomass, since the first and second raw materials are based on terrestrial cultures that compete with food production and can lead to food crises [46]. Algae biofuels are not yet obtained on a large scale due to the high cost of the process justifying the development of new technologies that can bring economic viability [47].

Bioremediation and biofuel production from waste resources by microalgae platform is mainly important to utilize abundantly available solar energy biofixing CO_2 and treat effluents through the mixotrophic growth of microalgae [7]. Algal bioremediation is a good strategy to produce biomass for biofuels production while remediating wastes, also improving carbon-footprint through carbon capturing and utilization technology.

A microalgae biorefinery enables to integrate fractionation and conversion processes to transform biomass into bioproducts such as food, feed, chemicals, and bio-energy as optimization of the use of the microalgae for reducing waste production, and maximizes process profit. After lipid transesterification for biodiesel, the residual biomass can be used to produce other biofuels such as methane, bio-oil and ethanol or biocompounds for food and pharmaceutical industry [48].

4. Cultivation systems of microalgae

There are currently four cultivation technologies in use for the production of commercial microalgae including open ponds and raceway ponds (open systems), photobioreactors and fermenters [49]. In open systems, microalgae are grown in open areas, including tanks, lakes, and ponds, deep channels, among others. In closed systems, crops are grown in transparent bioreactors, exposed to sunlight or artificial radiation for photosynthesis and fermenters.

4.1 Opens systems

Natural and artificial lakes and ponds, where most of the systems commonly used are large, shallow ponds and tanks, represent open pond systems. The main advantages of these systems are the ease of construction and operation when compared to photobioreactors and the possibility of operating hybrid processes involving the cultivation of algae associated with the treatment of wastewater. However, the disadvantages are inefficient light distribution, losses through evaporation, diffusion of CO_2 into the atmosphere, contamination and the requirement for large areas of land [50]. Open ponds are currently in use for wastewater treatment and production of *Dunaliella salina*, characterized as a hybrid process. These systems are used by Ognis Australia Pty Ltd. to produce β-carotene from *Dunaliella salina* in Hutt Lagoon and Whyalla. In terms of surface area used, these are among the largest algae production systems in the world.

Closed or artificial ponds (circular ponds and raceway ponds) are more efficient than open systems for producing microalgae, since control over the production environment is much better than open ponds or extensive ponds. The cost of raceway ponds is higher than that of open lagoon systems, but lower than that of photobioreactors. These systems are the most used due to their potential to produce large quantities of biomass for commercial application. Raceway pond ponds are commonly used to grow *Chlorella sp.*, *Spirulina platensis*, *Haematococcus sp.* and *Dunaliella salina*, with a biomass production rate of 60–100 mg of dry biomass/L/day [50]. Raceway ponds are used to produce Spirulina at Earthrise Nutraceuticals in the USA and Cyanotech Corp. in Hawaii [49].

4.2 Closed systems

Photobioreactors were designed to overcome the problems associated with open growth systems. It has been shown that cultivation of microalgae in these systems are capable of producing large amounts of biomass as they allow an effective control of process parameters, such as pH, temperature, CO_2 concentration, level of contamination, among others. However, photobioreactors are much more expensive than open ponds and raceway ponds. Commercial photobioreactor productions include the production of *H. pluvialis* in Israel and Hawaii and *C. vulgaris* in Germany. Production costs are very high, reaching $ 100/kg [49]. As a result, biofuel production based entirely on photobioreactors is generally considered unlikely to be commercially viable [6, 49, 50].

Closed fermenters are used for the production of heterotrophic algae, where sugars or other simple carbon sources are used for growth instead of CO_2 and light. Open fermenters similar to the fermenters used in the production of ethanol in industries are not suitable for the growth of algae, since these microorganisms have slow growth when compared to yeasts and bacteria. In the USA, India and China ω-3 fatty acids are produced from *Thraustochytrids* by heterotrophic fermentation through sugars and O_2. As it is a high value-added product, it is sold for 100 $ /kg, which justifies the high cost of the process [49].

5. Growth metabolism

Microalgae have different growth metabolisms, which characterizes their versatility. Cultivation conditions define the metabolic route for the production of biocompounds, including proteins, carbohydrates, pigments and fatty acids. Although the production of microalgae has traditionally been photoautotrophic, these microorganisms have different forms of energy metabolism including heterotrophic and mixotrophic, which use an organic carbon source for the growth and production of biomass. The understanding of these metabolisms allows the diversification of current growth systems aiming at increasing the production of biomass and certain specific metabolites.

5.1 Photoautotrophic

Photoautotrophic growth is the most common way to cultivate microalgae through photosynthesis. In these systems, high concentrations of CO2 are sequestered, but productivity is low when compared to heterotrophic and mixotrophic systems that provide high yields of biomass and secondary metabolites [51, 52]. Through photobioreactors is possible to obtain the maximum cell density of 40 g/L, while in outdoor open-pond or raceway-pond cultures, the cell concentration is usually lower than 10 g/L. This significantly increases the energy consumption of cell harvesting and the cost of biomass production [53]. Usually scale-up of microalgae cultivation in wastewater is fulfilled phototrophically, which may be hindered by inefficient illumination and the low biomass density, leading to poor removal of nutrients [54].

For photosynthesis, light is used as an energy source and CO2 is used as a carbon source [55]. The carbon source is essential for growth and the higher its concentration, the higher the productivity. Nitrogen sources in the culture medium are also essential for the synthesis of proteins, nucleic acids and other biocomposites necessary for cell growth and survival [56] and the concentration must be compatible with the amount of carbon in the medium [6]. The type of cultivation, the nutrients, the carbon source, the salinity of the medium, the irradiance and the temperature vary according to the chosen species and also considerably influence the success of the microalgae production.

5.2 Heterotrophic

Microalgae can grow in the absence of light in culture media assimilating organic carbon. Studies show that some species, including *Chlorella*, *C. protothecoides*, *C. vulgaris*, *C. zofingensis*, *C. minutissima*, *Tetraselmis* and *Neochloris* [57] are able to grow in both autotrophic and heterotrophic conditions [56]. According to Hosoglu et al. [58], microalgae of the *Chlorella Beyerinck* genus are those that have the greatest potential for large-scale heterotrophic production, with emphasis on the species *C. protothecoides* that has been widely studied.

In heterotrophic crops, microalgae acquire carbon and energy from organic sources via oxidative phosphorylation, consuming O2 and releasing CO2. According to Behrens [59], the cost of producing the kg of dry biomass produced in photoautotrophic conditions can be 5.5 times higher than in heterotrophic conditions. In heterotrophic systems there are no light limitation problems, since microalgae grow in the absence of light using organic carbon as a carbon and energy source reaching high concentrations of biomass reaching up to 100 g/L, which considerably facilitates the harvesting process [60]. In heterotrophic processes, 18% of the energy obtained can be converted into adenosine triphosphate (ATP) whereas in photoautotrophic cultures this percentage is only 10% [61]. However, despite the advantages, the cultivation of microalgae under heterotrophic conditions, due to the use of organic carbon, requires reactors and techniques of greater complexity and high cost to avoid contamination caused by other microorganisms.

5.3 Mixotrophic

Under mixotrophic conditions, microalgae grow both photoautotrophically and heterotrophically, being able to assimilate organic compounds as a carbon source and use inorganic carbon as an electron donor [62]. In this metabolism, the energy is captured through the catalysis of external organic compounds through respiration and the light energy is converted into chemical compounds via photosynthesis, being a promising solution in the processes of environmental remediation, being able to treat the effluents of the agribusiness and produce biomass rich in metabolites of industrial interest.

In photoautotrophic crops, self-shade caused by the high density of cells makes light penetration difficult, causing photoinhibition and may be the limiting factor to its propagation leading to low biomass yields. Under heterotrophic conditions, not all microalgae species are able to grow and the strict use of only organic substrates as a source of carbon and energy makes the process more prone to contamination.

Thus, an alternative to maximize production can be through mixotrophic cultivation. Therefore, the cells would multiply with autotrophic metabolism until reaching the maximum cell density, at which point a source of organic carbon is added to stimulate, also, heterotrophic growth. Thus, due to the high cell density in the medium, problems with contamination of microorganisms would be less likely to happen. For the success of this technique, the cultivated species must be able to grow in heterotrophic conditions without microbial contamination and the organic compound to be added, as well as its ideal concentration must be known.

Reducing use of light in mixotrophic processes decreases the demand for energy, which minimizes the operational cost in processes with artificial light. According to Perez-Garcia and Bashan [63] in these systems there is a better control capable of regulating the growth rate of the species, which minimizes the risk of contamination by photosynthetic microorganisms. Due to the significant heterotrophic contribution of the organic fraction, the reactor design does not require a maximized area to expose the microalgae to light to perform photosynthesis, decreasing the process cost.

According to Patel et al. [7], mixotrophic cultivation has advantages over photoautotrophic and heterotrophic metabolism. In this process, in addition to the higher growth rates reducing the microalgae growth cycle, there is also growth in the dark phase mediated by respiration, potentiating the production of biomass. There is also a comparatively prolonged exponential growth phase, great flexibility to change the metabolism from heterotrophic to photo-autotrophic and vice versa, prevention of photo-oxidative damage caused by O2 accumulated in closed photobioreactors and reduction in substrate uptake photoinhibition. It has been shown that microalgae cultivated under mixotrophic conditions can present, under controlled conditions, higher biomass productivity when compared to photoautotrophic and heterotrophic cultures [64].

6. Growth microalgae using waste from agro-industry

The advancement of agro-industry has generated a large amount and variety of waste causing serious environmental problems and is one of the sectors that generate more waste rich in organic matter. According to Dahiya et al. [65], approximately 1.3 billion tons of foods are wasted each year during its production, handling, storage, processing, distribution or consumption. The composition of food processing residues is extremely varied and depends on both the nature of the raw material and the production technique employed.

Due to the scarcity of available areas close to large urban centers for the disposal of industrial and urban waste, the vast majority of companies do not carry out treatment and/or correct disposal of this material, which contributes to the increase in environmental pollution. Pollution is due to high concentrations of organic matter and heavy metals causing contamination, eutrophication of water bodies, death of aquatic organisms and local vegetation, ecological imbalance and health problems for the population. In this sense, the use of agro-industrial waste as a source of organic carbon for the cultivation of microalgae is presented as an option for the bioremediation of these effluents added to the production of biomass rich in biocompounds with different applications.

A lot of studies have evaluated the mixotrophic growth of microalgae using glucose, glycerol and acetate as a source of organic carbon and observed that the biomass yields were higher, which could decrease the cost of the process. Liang et al. [66] investigated *C. vulgaris* strain under autotrophic, mixotrophic and heterotrophic growth conditions. Mixotrophic growing on glucose with light produced the highest lipid productivity compared with other growth modes. Garcia et al. [67] studied the mixotrophic growth of tricornutum UTEX-640 using acetate, lactic acid, glycine, glucose and glycerol. The best results were obtained using with urea, which resulted in maximum biomass and eicosapentaenoic acid productivities significantly higher than those obtained for the photoautotrophic control, which suggest the possibility of using mixotrophy for the mass production of microalgae. Cheirsilp & Torpee [68] cultivated *Chlorella sp.*, *Chlorella sp.*, *Nannochloropsis sp.* and *Cheatoceros sp.* under mixotrophic condition using glucose as source of organic carbon. They observed that the biomass and lipid production of all tested strains in mixotrophic culture were notably enhanced in comparison with photoautotrophic and heterotrophic cultures.

Due to the success of the microalgae mixotrophic cultivation method, recent studies have been evaluating the possibility of using residues from the agro-industry as source of organic matter as a strategy to treat the effluent, produce biomass and reduce the cost of the process.

Hu et al. [69] evaluated the mixotrophic cultivation of *Chlorella sp.* UMN271 utilizing swine manure as nutrient supplement for evaluate the nutrient removal efficiencies by alga. The results showed that addition of 0.1% (v/v) acetic, propionic and butyric acids, respectively, could promote algal growth, enhance nutrient removal efficiencies and improve total lipids productivities. They concluded that Chlorella sp. grown on acidogenically-digested manure could be used as a feedstock for high-quality biodiesel production.

Li et al. [70] investigated the effects of autotrophic and mixotrophic growth on cell growth and lipid productivity of green microalgae *Chodatella sp* and Piggery wastewater served as nutrient sources for mixotrophic growth. The specific growth rate, biomass production, and lipid productivity obtained with mixotrophic growth were until 5.6 times higher than those obtained with autotrophic growth. The mixotrophic cultivation simultaneously assimilated 99.7% ammonia nitrogen and 75.9% total phosphorus from piggery wastewater, which reduced the required nutrient for the culture of microalgae, thereby reducing the cost of biomass for diverse application.

León-Vaz et al. [71] cultivated *C. sorokiniana* microalgae under mixotrophic conditions utilized Oxidized wine waste lees among other agro-industrial wastes as carbon source. The fed-batch strategy and the medium optimization, with nutrient supplementation, have been found to be very effective in enhancing biomass and neutral lipid productivity, suggesting that this is a promising strategy for production of microalgal biomass. The algal biomass concentration was 11 g L-1 with a lipid content of 38% (w/w).

Bhatnagar et al. [72] evaluated mixotrophic growth of *Chlamydomonas globosa*, *Chlorella minutissima* and *Scenedesmus bijuga* under medium supplemented with different organic carbon substrates and wastewaters. The mixotrophic growth of these microalgae resulted in 3–10 times more biomass production relative to phototrophy. Poultry litter extract as growth medium recorded up to 180% more biomass growth compared to standard growth medium, while treated and untreated carpet industry wastewaters also supported higher biomass, with no significant effect of additional nitrogen supplementation.

Andrade & Coosta [73] determined the effects of molasses concentration and light levels on mixotrophic biomass production by *Spirulina platensis*. Molasses concentration was the main factor influencing maximum biomass concentration (Xmax) reached 2.94 g L− 1 and μmax 0.147 d − 1. Molasses, suggesting that this industrial by-product could be used as a low-cost supplement for the growth of Spirulina platensis, stimulated the production of biomass.

Melo et al. [74] evaluated the growth, nutrients and toxicity removal of *Chlorella vulgaris* cultivated under autotrophic and mixotrophic conditions using corn steep liquor, cheese whey and vinasse as source of organic matter. The results demonstrated that corn steep liquor toxicity was totally eliminated and cheese whey and vinasse toxicity were minimized by *C. vulgaris*. They demonstrated that the mixotrophic cultivation of *C. vulgaris* is able to increase cellular productivity and could be an alternative to remove the toxicity from agroindustrial by-products.

Hugo et al. [75] studied the growth of forty microalgae strains under mixotrophic conditions using sugarcane vinasse as source of organic matter. *Micractinium sp.* Embrapa|LBA32 and *C. biconvexa* Embrapa|LBA40 presented expressive growth in a light-dependent manner even in undiluted vinasse under non-axenic conditions. Microalgae strains presented higher biomass productivity in vinasse-based medium compared to autotrophic medium. This research showed the potential of using residues derived from ethanol plants to cultivate microalgae for the production of energy and bioproducts.

Mitra et al. [76] cultivated *Chlorella vulgaris* under mixotrophic/heterotrophic conditions using dry-grind ethanol thin stillage and soy whey as nutrient feedstock. The results showed the biomass yields from thin stillage, soy whey and modified basal medium after 4 days of incubation at mixotrophic conditions in the bioreactor were 9.8, 6.3 and 8.0 g.L− 1 with oil content at 43, 11, and 27% (w/w) respectively. This research highlights the potential of these agro-industrial co-products as microalgal growth media with consequent production of high-value microalgal oil and biomass.

Salati et al. [77] cultivated *Chlorella vulgaris* using cheese whey, white wine lees and glycerol as carbon sources under mixotrophic conditions. The mixotrophic biomass production was1.5–2 times higher than autotrophic growth. Furthermore, it gave much higher energy recovery efficiency, i.e. organic carbon energy efficiency of 32% and total energy efficiency of 8%, suggesting the potential for the culture of algae as a sustainable practice to recover efficiently waste-C and produce biomass.

Piasecka et al. [78] studied the growth of *Tetradesmus obliquus* by supplementation with beet molasses in photoheterotrophic and mixotrophic culture conditions. The highest protein content was obtained in the mixotrophic growth suggesting this metabolism promising for protein production.

Tsolcha et al. [79] evaluated a mixed cyanobacterial-mixotrophic algal population, dominated by the filamentous cyanobacterium *Leptolyngbya sp.* and the microalga *Ochromonas* under non-aseptic conditions for its efficiency to remove organic and inorganic compounds from second cheese whey, poplar sawdust, and grass hydrolysates. Nutrient removal rates, biomass productivity, and the maximum oil production rates were determined. The highest lipid production was achieved using the biologically treated dairy effluent (up to 14.8% oil in dry biomass corresponding to 124 mg L − 1), which also led to high nutrient removal rates (up to 94%). Lipids synthesized by the microbial consortium contained high percentages of saturated and mono-unsaturated fatty acids (up to 75% in total lipids) for all the substrates tested, which implies that the produced biomass may be harnessed as a source of biodiesel.

Grupta et al. [80] cultivated the *Chlorella* microalgae under mixotrophic conditions using a raw food-processing industrial wastewater. About 90% reduction in TOC and COD were obtained for all dilutions of wastewater. Over 60% of nitrate and 40% of phosphate were consumed by microalgae from concentrated raw wastewater. The degradation kinetics also suggested that the microalgae cultivation on a high COD wastewater is feasible and scalable.

Yeesang et al. [81] evaluated *B. braunii*, a microalgae rich in oil under mixotrophic cultivation using molasses, a cheap by-product from the sugar cane plant as a carbon source and under photoautotrophic cultivation using nitrate-rich wastewater supplemented with CO_2. The mixotrophic cultivation produced a high amount of biomass of 3.05 g L − 1 with a high lipid content of 36.9%. The photoautotrophic cultivation in nitrate-rich wastewater supplemented with 2.0% CO2 produced a biomass of 2.26 g L − 1 and a lipid content of 30.3%. They showed that these strategies could be promising ways for producing cheap lipid-rich microalgal biomass as biofuel feedstock and animal feeds.

Gélinas et al. [82] studied the mixotrophic growth and lipid production of *Chlorella consortium* using residual corn hydrolysate and corn silage juice as source of organic and compared to heterotrophic conditions. Maximum microalgal biomass of 0.8 g/L was obtained with 1 g/L of residual corn hydrolysate whatever the trophic strategy. Under mixotrophic conditions, the use of residual corn hydrolysate led to an increase of 21% and 22% in comparison with the biomass produced with glucose or silage juice, respectively. This increase varied between 11% and 28% under heterotrophic condition. They observed that at the end of the experiment, algae exposed to silage juice decreased significantly. Residual corn hydrolysate represented an interesting and efficient alternative as an organic carbon source. However, silage juice needs additional treatments to be implemented as a culture medium.

Nur et al. [83] studied palm oil mill effluent (POME), one of the wastewaters generated from palm oil mills, as source of organic carbon for mixotrophic microalgae growth. The aim of this research was to identify the growth of *Chlorella vulgaris* cultured in POME medium under mixotrophic conditions in relation to a variety of organic carbon sources added to the POME mixture. The research was conducted with 3 different carbon sources (D-glucose, crude glycerol and $NaHCO_3$) in 40% POME. They showed that *C. vulgaris* using D-glucose as carbon source gained a lipid productivity of 195 mg/l/d.

Manzoor et al. 2020 presented the growth of *Scenedesmus dimorphus* NT8c cultivated mixotrophically on sugarcane bagasse hydrolysate, a low-value agricultural by-product. Under mixotrophic conditions the *S. dimorphus* NT8c showed higher growth rates compared to photoautotrophic cultivation and the biomass productivity was 119.5 mg L d − 1, protein contents was 34.82% and fatty acid contents was 15.41%. They concluded that mixotrophically-cultivated microalgae are able to increase the biomass and lipid productivity. However, the concentrations of supplementation need to be studied because higher level of organic carbon can result in unfavorable levels of turbidity and bacterial growth, reducing microalgal biomass productivity.

7. Conclusions

Microalgal biomass represents a sustainable alternative to fossil consumption and bioproducts for food and pharmaceutical industry. Microalgae can growth under photoautotrophic, heterotrophic or mixotrophic modes where the latter two trophic modes require organic carbon to grow efficiently. Actually, researchers have highlighted the role of low cost-efficient agro-industrial by-products used as supplements in algal culture media. However, supplementation of organic carbon contributes significantly to a higher cost of microalgae production and this can compete with human and animal alimentation. Agro-industrial by-products and wastes are of great interest as cultivation medium for microorganisms because of their low cost, renewable nature, and abundance.

Faced with this scenario, biotechnological technologies are necessary to develop the production of microalgae on a large scale and expand the range of utilities that, in the short or long term, contribute to the improvement of industrial processes. In addition, mixotrophic microalgae growth is a great strategy to reduce environmental pollution generated by residues rich in organic matter and can reduce the cost of the industrial process. However, more investments, development and greater knowledge of the metabolism of these microalgae and their effectiveness in the generation of new bioproducts are increasingly necessary.

Acknowledgements

The authors gratefully acknowledge the Postgraduation Dean's Office of the Federal University of Rio Grande do Norte and Dean of Extension of Federal Institute of Rio Grande do Norte. The authors would like to thank Coordenação de Aperfeiçoamento de Pessoal de Nível Superior (CAPES) for the financial support.

Author details

Izabel Pereira[1], Adriano Rangel[1], Bruna Chagas[1*], Bruno de Moura[2], Stela Urbano[1], Roberto Sassi[3], Fabiana Camara[1] and Cíntia Castro[1]

1 Federal University of Rio Grande do Norte, Natal, Brazil

2 Federal Institute of Rio Grande do Norte, Natal, Brazil

3 Federal University of Paraíba, João Pessoa, Brazil

*Address all correspondence to: brunam.emerenciano@gmail.com

References

[1] Brennan L, Owende P. 2010. Biofuels from microalgae areview of technologies for production, processing, and extractions of biofuels and co-products. Renewable Sustainable Energy Review. 2010; 14:557-577.

[2] Chen CY, Yeh KL, Aisyah R, Lee DJ, Chang JS. Cultivation, photobioreactor design and harvesting of microalgae for biodiesel production: a critical review. Bioresource Technology. 2011;102: 71-81.

[3] Harun R, Singh M, Forde GM, Danquah MK. Bioprocess engineering of microalgae to produce a variety of consumer products. Renewable Sustainable Energy Review. 2010;14:1037-47.

[4] Singh A, Nigam PS, Murphy JD. Mechanism and challenges in commercialization of algal biofuels. Bioresource Technology. 2010;102: 26-34.

[5] Mata TM, Mendes AM, Caetano NS, Martins AA. Sustainability and economic evaluation of microalgae grown in brewery wastewater. Bioresource Technology. 2014;168:151-158.

[6] Milano J, Ong HC, Masjuki HH, Chong WT, Lam MK, Loh PK, Vellayan V. Microalgae biofuels as an alternative to fossil fuel for power generation. Renewable and Sustainable Energy Reviews. 2016;58:180-197.

[7] Patel AK, Choi YY, Sim SJ. Emerging prospects of mixotrophic microalgae: Way forward to sustainable bioprocess for environmental remediation and cost-effective biofuels. Bioresource Technology. 2020; 300:122-741.

[8] Tang YY, Khooa KS, Chewb KW, Taoc Y, Hod S-H, Showa PL. Potential utilization of bioproducts from microalgae for the quality enhancement of natural products. Bioresource Technology. 2020;304: 122997.

[9] Lam MK, Lee KT. Microalgae biofuels: a critical review of issues, problems and the way forward. Biotechnology Advances. 2012; 30:673e90.

[10] Chen H, Qiu T, Rong J, He C, Wang Q. Microalgal biofuel revisited: an informatics-based analysis of development date and future prospects. Applied Energy. 2015; 155:585-98.

[11] Georgianna DR, Mayfield SP. Exploiting diversity and synthetic biology for the production of algal biofuels. Nature. 2012;488:329-335.

[12] Salla ACV, Margarites AC, Seibel FI, Holz LC, Brião VB, Bertolin TE, Colla LM, Costa JAV. Increase in the carbohydrate content of the microalgae Spirulina in culture by nutrient starvation and the addition of residues of whey protein concentrate. Bioresource Technology. 2016;209: 133-141.

[13] Zhan J, Rong J, Wang Q. Mixotrophic cultivation, a preferable microalgae cultivation mode for biomass/bioenergy production, and bioremediation, advances and prospect, International Journal of Hydrogen Energy. 2017; 42(12):8505-8517.

[14] Ngo-mati ME, Pieme CA, Azabji-kenfack M, Moukette BM, Korosky E, Stefanini P, Ngogang JY, Mbofung CM. Impact of daily supplementation of S.platensis on the immune system of naïve HIV-1 patients in Cameroon: a 12-months single blind, randomized, multicenter trial. Nutrition Journal, 2015; 14:2-7.

[15] Pereira MIB, Chagas BME, Sassi R, Medeiros GF, Aguiar EM, Borba LHF,

Andrade Neto JC, Rangel AHN. Mixotrophic cultivation of Spirulina platensis in dairy wastewater: Effects on the production of biomass, biochemical composition and antioxidant capacity. PLoS ONE. 2019; 14(10): e0224294. https://doi org/10.1371/journal.pone.0224294.

[16] Bajan B, Mrówczyńska-Kamińska A. Carbon footprint and environmental performance of agribusiness production in selected countries around the world. Journal of Cleaner Production. 2020; 276:123389

[17] Sua M, D'Imporzanoa G, Veronesia D, Africa S, Adania F. Phaeodactylum tricornutum cultivation under mixotrophic conditions with glycerol supplied with ultrafiltered digestate: A simple biorefinery approach recovering C and N. Journal of biotechnology. 2020; 323:73-81.

[18] Patel AK, Joun J, Sim SJ. A sustainable mixotrophic microalgae cultivation from dairy wastes for carbon credit, bioremediation and lucrative biofuels. Bioresource Technology. 2020; 313:123681. https://doi.org/10.1016/j.biortech.2020.123681

[19] Tan, X-B, Yang L-B, Zhang WW, Zhao XC. Lipids production and nutrients recycling by microalgae mixotrophic culture in anaerobic digestate of sludge using wasted organics as carbon source. Bioresource Technology. 2020; 297:122379.

[20] Wang X, Zhang M-Man, Sun Z, Liu S-Fen, Qin Z-Hao, Mou J-Hua, Zhou Z-Gang, Lin CSK, Sustainable lipid and lutein production from Chlorella mixotrophic fermentation by food waste hydrolysate, Journal of Hazardous Materials. 2020); 400:123258.

[21] Sanchez-silva L, López-gonzález D, Garcia-minguillan AM, Valverde JL. Pyrolysis, combustion, and gasification characteristics of Nannochloro psisgaditana microalgae. Bioresource Technology. 2013;130:321-331.

[22] Scott SA, Davey MP, Dennis JS, Horst I, Howe CJ, Lea-Smith DJ, Smith AG. Biodiesel from algae: challenges and prospects. Current Opinion in Biotechnology. 2010; 21: 277-86.

[23] Suganya T, Varman M, Masjuki HHS. Macroalgae and microalgae as a potential source for commercial applications along with biofuels production: A biorefinery approach. Renewable and Sustainable Energy Reviews. 2016;55:909-941.

[24] Guedes AC, Amaro HM, MALCATA FX. Microalgae as sources of carotenoids. Marine Drugs. 2011a; 9:625-644.

[25] Guedes AC, Amaro HM, Malcata FX. Microalgae as sources of high added value compounds–a brief review of recent work. Biotechnology Progress. 2011b;27:597-613.

[26] Oncel SS. Microalgae for a macroenergy world. Renewable and Sustainable Energy Reviews. 2013;26:241-64.

[27] Hu Q, Sommerfeld M, Jarvis E, Ghirardi M, Posewitz M, Seibert M, Darzins A Microalgal triacylglycerols as feed stocks for biofuel production: perspectives and advances. Plant Journal. 2008;54:621-39.

[28] Grobbelaar JU. Algal biotechnology: real opportunities for Africa. South African Journal of Botany. 2004; 70(1):140-144.

[29] Parmar A, Singh NK, Pandey A, Gnansounou E, Madamwar D. Cyanobacteria and microalgae: a positive prospect for biofuels.

Bioresource Technology. 2011; 102:10163-10172.

[30] Hu J, Nagarajan D, Zhang Q, Chang JS, Lee DJ. (2018). Heterotrophic cultivation of microalgae for pigment production: a review. Biotechnology Advances. 2018; 36:54-67. doi: 10.1016/j.biotechadv.2017.09.009

[31] Neofotis P, Huang A, Sury K, Chang W, Joseph F, Gabr A, Twary S, Qiu W, Holguin O, Polle JEW. Characterization and classification of highly productive microalgae strains discovered for biofuel and bioproduct generation. Algal Research. 2016;15:164-178.

[32] Rizwan M, Mujtaba G, Memon SA, Lee K, Rashid N. Exploring the potential of microalgae for new biotechnology applications and beyond: a review. Renewable and Sustainable Energy Reviews. 2018;92:394-404.

[33] Lee Y-K. Commercial production of microalgae in the Asia-Pacific rim. Journal of Applied Phycology. 1997;9:403-11.

[34] Borowitzka MA. High-value products from microalgae – Their development and commercialisation. Journal of Applied Phycology. 2013;25(3):743-756. https://doi.org/10.1007/s10811-013-9983-9.

[35] Mitra M., Mishra S. (2019) A Biorefinery from Nannochloropsis spp. Utilizing Wastewater Resources. In: Gupta S., Bux F. (eds) Application of Microalgae in Wastewater Treatment. Springer, Cham.

[36] Morais Junior WG, Gorgicha M, Corrêa PS, Martins AA, Mata TM, Caetano NS. Microalgae for biotechnological applications: Cultivation, harvesting and biomass processing. Aquaculture. 2020; 528:735562.

[37] Habib MAB, Parvin M, Huntington TC, Hasan MR. A Review on Culture, Production and Use of Spirulina as Food for Humans and Feeds for Domestic Animals and Fish. FAO Fisheries and Aquaculture Circular. No. 1034 FAO, Rome (2008).

[38] Safi C, Ursu AV, Laroche C, Zebib B, Merah O, P.-Y. Pontalier P-Y, Vaca-Garcia C. Aqueous extraction of proteins from microalgae: effect of different cell disruption methods. Algal Research. 2014; 3:61-65.

[39] Montalvo GEB, Thomaz-Soccol V, Vandenberghe LPS, Carvalho JC, Faulds CB, Bertrand E, Prado MRM, Bonatto SJR, Soccola CR. Arthrospira maxima OF15 biomass cultivation at laboratory and pilot scale from sugarcane vinasse for potential biological new peptides production. Bioresource Technology. 2019; 273:103-113.

[40] Commault AS, Laer G, Novis P. Photosynthetic biocathode enhances the power output of a sediment-type microbial fuel cell. 2014. New Zealand Journal of Botany. 2014;52:48-59.

[41] Wu XY, Song TS, Zhu XJ, Wei P, Zhou C. 2013a.Construction and operation of microbial fuel cell with *Chlorella vulgaris* biocathode for electricity generation. Applied Biochemistry and Biotechnology. 2013;171:2082e2092.

[42] Wu Y, Guan K, Wang Z, Xu B, Zhao F. Isolation, identification and characterization of anelectrogenic microalgae strain. PLoSOne. 2013b;8:e 73442.

[43] Razzak SA, Mali A, Hossain MM, Delasa H. Biological CO2 fixation with production of microalgae in wastewater – A review. Renewable and Sustainable Energy Reviews. 2017;76:379-390.

[44] Alaswad A, Dassisti M, Prescott T, Olabi A G. Technologies and developmentes of third generation biofuel production. Renewable and Sustainable Energy Rviews. 2015;51:1446-1460.

[45] Lee OK, Seong DH, Lee CG, Lee EY. Sustainable production of liquid biofuels from renewable microalgae biomass. Journal of Industrial and Engineering Chemistry. 2015; 29:24-31.

[46] Singh A, Stig Olsen SI. A critical review of biochemical conversion, sustainability and life cycle assessment of algal biofuels. Applied Energy. 2011;88;3548-3555.

[47] Ganesan R, Manigandan S, Samuel MS, Shanmuganathan R, Brindhadevi K, Chi LTN, Duc PA, Pugazhendhi A. A Review on Prospective Production of Biofuel from Microalgae. A Review on Prospective Production of Biofuel from Microalgae. Biotechnology Reports. 2020; 27:e00509.

[48] Sakarika M, Kornaros M. Chlorella vulgaris as a green biofuel factory: comparison between biodiesel, biogas and combustible biomass production. Bioresource Technology. 2019;273:237-243.

[49] Darzins, A., Pienkos, P., Edye, L. 2010. Current status and potential for algal biofuels production. Report to IEA Bioenergy Task 39: T39-T2.

[50] Maity JP, Bundschuh J, Chen C-Y, Bhattacharya P. Microalgae for third generation biofuel production, mitigation of greenhouse gas emissions and wastewater treatment: present and future perspectives - A mini review. Energy. 2014;78:104-113.

[51] Verma R, Kumari KVKK, Srivastava A, Kumar A. Photoautotrophic, mixotrophic, and heterotrophic culture media optimization for enhanced microalgae production. Journal of Environmental Chemical Engineering. 2020; 8(5):104149.

[52] Yang S, Liu G, Meng, Y, Wang P, Zhou S, Shang H. Utilization of xylose as a carbon source for mixotrophic growth of *Scenedesmus obliquus*. Bioresource Technology, Essex Journal, 2014;172(8)180-185.

[53] Scaife MA, Nguyen GTDT, Rico J, Lambert D, Helliwell KE, Smith AG. Establishing *Chlamydomonas reinhardtii* as an industrial biotechnology host. Plant Journal.2015; 82:532-546.

[54] Murwanashyaka T, Shena L, Yanga Z, Chang J-S, Manirafashaa E, Ndikubwimanaf T, Chena C, Lua Y. Kinetic modelling of heterotrophic microalgae culture in wastewater: Storage molecule generation and pollutants mitigation. Biochemical Engineering Journal. 2020;157(15):107523.

[55] Chen C, Ma X, Liu K. Thermogravimetric analysis of microalgae combustion under different oxygen supply concentrations. Applied Energy. 2011;88(9):3189-3196.

[56] Hu J, Nagarajan D, Zhang Q, Chang JS, Lee DJ. Heterotrophic cultivation of microalgae for pigment production: a review. Biotechnology Advances. 2018;36:54-67.

[57] Sánchez E, Ojeda K, El-Halwagi M, Kafarov V. (2011). Biodiesel from microalgae oil production in two sequential esterification/ transesterification rea.tors: Pinch analysis of heat integration. Chemical Engineer Journal. 2011; 2011:176-177: 211-216.

[58] Isleten-Hosoglu M, Gultpe I, Elibol M. Optimization of carbon and nitrogen sources for biomass and lipid production by *Chlorella saccharophila* under heterotrophic conditions and

development of Nile red fluorescence based method for quantification of its neutral lipid content. Biochemical Engineering Journal, 2012; 61:11-19.

[59] Behrens PW, Kyle DJ. Microalgae as a source of fatty acids. Journal of Food Lipids. 1996; 3:259-272.

[60] Morales-Sánchez D, Martinez-Rodriguez OA, Kyndt J, Martinez A. Heterotrophic growth of microalgae: metabolic aspects. World Journal of Microbiology Biotechnology. 2015;31:1-9.

[61] Yang C, Hua Q, Shimizu K (2000) Energetics and carbon metabolism during growth of microalgal cells under photoautotrophic, mixotrophic and cyclic light-autotrophic/dark-heterotrophic conditions. Biochem Eng J 6:87-102. doi:10.1016/S1369-703X(00)00080-2

[62] Cecchin M, Benfatto S, Griggio F, Mori A, Cazzaniga S, Vitulo N, Delledonne M, Ballottari M. Molecular basis of autotrophic vs mixotrophic growth in *Chlorella sorokiniana*. Scientific Reports. 2018;8(1):6465.

[63] Perez-Garcia, O., De-Bashan, L. E., Hernandez, J. P., & Bashan, Y. (2010). Efficiency of growth and nutrient uptake from wastewater by heterotrophic, autotrophic, and mixotrophic cultivation of *Chlorella vulgaris* immobilized with Azospirillum brasilense. Journal of Phycology. 2010;46(4):800-812.

[64] Roostaei J, Zhang Y, Gopalakrishnan K, Ochocki AJ. Mixotrophic Microalgae Bioflm: A Novel Algae Cultivation Strategy for Improved Productivity and Costefciency of Biofuel Feedstock Production. Scientific Reports. 2018;8:12528.

[65] Dahiya S, Naresh KA, Sravan JS, Chatterjee S, Sarkar O, Mohan SV. Food waste biorefinery: Sustainable strategy for circular bioeconomy. Bioresource Technology. 2018; 248A:2-12.

[66] Liang YN, Sarkany N, Cui Y. Biomass and lipid productivities of *Chlorella vulgaris* under autotrophic, heterotrophic and mixotrophic growth conditions. Biotechnology Letters. 2009; 31:1043-1049

[67] Cerón Garcia MC, Mirón AS, Sevilla JMF, Grima EM, Camacho FG. 2005. Mixotrophic growth of the microalgae *Phaeodactylum tricornutum*. Influence of different nitrogen and organic carbon sources on productivity and biomass composition. Journal Process Biochemical. 2005;40(1): 297-305.

[68] B. Cheirsilp B, Torpee S. Enhanced growth and lipid production of microalgae under mixotrophic culture condition: effect of light intensity, glucose concentration and fed-batch cultivation. Bioresource Technology. 2012;110:510-516.

[69] Hu B, Min M, Zhou W, Li Y, Mohr M, Cheng Y, Lei H, Liu Y, Lin X, Chen P, Ruan R. Influence of exogenous CO 2 on biomass and lipid accumulation of microalgae Auxenochlorella protothecoides cultivated in concentrated municipal wastewater. Applied Biochemical Biotechnology. 2012;166:1661-1673.

[70] Lia Y-R, Tsaia W-T, Hsub YC, Xie M-Z, Chend J-J. Comparison of Autotrophic and Mixotrophic Cultivation of Green Microalgal for Biodiesel Production. Energy Procedia. 2014;52:371-376.

[71] León-Vaz A, León R, Díaz-Santos E, Vigara J, Raposo S. Using agro-industrial wastes for mixotrophic growth and lipids production by the green microalga *Chlorella sorokiniana*. New Biotechnology. 2019;51:31-38.

[72] Bhatnagar A, Chinnasamy S, Singh M, Das KC. Renewable biomass

production by mixotrophic algae in the presence of various carbon sources and wastewaters. Applied Energy. 2011; 88:3425-3431.

[73] Andrade MR, Costa JAV. Mixotrophic cultivation of microalga Spirulina platensis using molasses as organic substrate. Aquaculture, v. 264, p. 130-134, 2007.

[74] Melo RG, Andrade AF, Bezerra RP, Correia DS, Souza VC , Vidal ACB, Marques DAV, Ana Lúcia Porto ALF. Chlorella vulgaris mixotrophic growth enhanced biomass productivity. Chemosphere. 2018; 204:344e350.

[75] Hugo et al. 2017, Microalgae cultivation in sugarcane vinasse: Selection, growth and biochemical characterization. Bioresource Technology. 2017;228:-140.

[76] Mitra D, van Leeuwen JH, Lamsal B (2012) Heterotrophic/mixotrophic cultivation of oleaginous *Chlorella vulgaris* on industrial co-products. Algal Res 1:40-48

[77] Salati, S., D'Imporzano, G., Menin, B., Veronesi, D., Scaglia, B., Abbruscato, P., Mariani, P., Adani, F., 2017. Mixotrophic cultivation of Chlorella for local protein production using agro-food by-products. Bioresource Technology. 230, 82-89.

[78] Piasecka, A., Krzeminska, I. and Tys, J., 2017. Enrichment of Parachlorella kessleri biomass with bio-products: oil and protein by utilization of beet molasses. Journal of Applied Phycology, vol. 29, no. 4, pp. 1735-1743. http://dx.doi.org/10.1007/s10811-017-1081-y. PMid:28775655.

[79] Tsolcha ON, Tekerlekopoulou, A.G., Akratos, C.S. et al. A Leptolyngbya-based microbial consortium for agro-industrial wastewaters treatment and biodiesel production. Environ Sci Pollut Res 25, 17957-17966 (2018).

[80] Gupta, S., Pandey, R.A. & Pawar, S.B. Microalgal bioremediation of food-processing industrial wastewater under mixotrophic conditions: Kinetics and scale-up approach. Front. Chem. Sci. Eng. 10, 499-508 (2016).

[81] Gélinas M, Pham TTH, Boëns B, Adjallé K, Barnabé S. Residual corn crop hydrolysate and silage juice as alternative carbon sources in microalgae production. Algal Research. 2015;12:33-42.

[82] Nur MMA, Hadiyanto H. Enhancement of *Chlorella vulgaris* Biomass Cultivated in POME Medium as Biofuel Feedstock under Mixotrophic Conditions J. Eng. Technol. Sci., Vol. 47, No. 5, 2015, 487-497

[83] Manzoor, Maleeha, Ahmad, Qurat-ul-Ain, Aslam, Ambreen, Jabeen, Faiza, Rasul, Azhar, Schenk, Peer M., and Qazi, Javed I. (2019). Mixotrophic cultivation of *Scenedesmus dimorphus* in sugarcane bagasse hydrolysate. Environmental Progress and Sustainable Energy 39 (2) e13334.

Chapter 22

Magnetic Field Application to Increase Yield of Microalgal Biomass in Biofuel Production

Lucielen Oliveira Santos, Pedro Garcia Pereira Silva, Sharlene Silva Costa and Taiele Blumberg Machado

Abstract

Use of fuels from non-renewable sources has currently been considered unsustainable due to the exhaustion of supplies and environmental impacts caused by them. Climate change has concerned and triggered environmental policies that favor research on clean and renewable energy sources. Thus, production of third generation biofuels is a promising path in the biofuel industry. To yield this type of biofuels, microalgae should be highlighted because this raw material contains important biomolecules, such as carbohydrates and lipids. Technological approaches have been developed to improve microalgal cultivation under ecological conditions, such as light intensity, temperature, pH and concentrations of micro and macronutrients. Thus, magnetic field application to microalgal cultivation has become a viable alternative to obtain high yields of biomass concentration and accumulation of carbohydrates and lipids.

Keywords: magnetic effect, bioenergy, third generation biofuel, *Chlorella*, *Spirulina*, carbohydrates, lipids

1. Introduction

Due to exponential growth of the population and consequent increase in the demand for energy, the energy crisis has worsened globally. Various current energy sources are non-renewable fossil fuels, such as diesel, gasoline, coal, natural gas and oil [1]. Those that have been used in electricity, transportation and heating industries have the disadvantages of releasing toxic and harmful gases into the atmosphere and of polluting the environment [2]. Greenhouse Gas (GHG) levels in the environment have increased by a staggering 25% in the post-industrialization [1]. As a result, changes in climatic conditions have occurred and replacing fossil fuels by other sources that can benefit the environment has become an alternative [3].

Thus, biofuels derived from plant material have emerged as a promising solution to reduce supply of fossil fuels in recent decades [4]. Biofuels have advantages, such as the ability to reduce GHG emissions, continuous supply of raw material and ease of cultivation, harvesting, and transportation [5]. According to the Global Trends Report (GTR), 133.7 billion liters of ethanol and 47.4 billion liters of biodiesel were produced in 2019 worldwide; the three largest producers were the United States, Brazil and Indonesia. The increase was 7.8 billion liters, by comparison with 2018 [6].

Biofuels are classified into three generations, depending on their sources. The first generation is derived from plant sources; the second ones comes from agro-industrial waste and wood waste, while the third one comes from microalgae. The fact that first- and second-generation raw material are limited by competition with food production and arable land emphasizes the importance of using alternatives, such as microalgae, which are attractive options [7].

Microalgae, which are photosynthetic microorganisms with simple growth needs, adapt to a wide range of variations in the culture medium. Depending on the conditions, they can produce large amounts of compounds, such as carbohydrates and lipids that can be processed into bioethanol and biodiesel, respectively [8].

Magnetic Fields (MF) are capable of causing effects on biological systems and, therefore, application of this technique has attracted attention in biotechnology and bioenergy to increase production of biomolecules of interest [9]. MF correspond to the region in which magnetic forces act [10]. MF can be generated either by magnets composed of conductive material with characteristic magnetic intensity, such as ferrite and neodymium or by an electric current that results from straight conductors, circular loop, flat circular coil, electromagnets and solenoids [11]. Since magnetic forces can act differently on microorganisms due to distinct biochemical and physiological constitutions, their biological effects can be inhibitory, null or stimulating [9, 12].

Studies have reported that some changes that can affect production of compounds are electro activation of some enzymatic systems and metabolic routes [13], oxidative stress, changes in enzyme and protein activity, gene expression, electron and ion movements [14, 15], cell growth [16–19] and high activity of photosystem II [20]. However, different strains of microalgae, application time and MF intensity can give different responses. Therefore, previous studies of MF, applied at different exposure times and intensities during cultivation, should be investigated to evaluate their effects [21].

2. Biofuels

The world population has currently faced a major challenge, i. e., to associate economic development with sustainable practices [22]. The amount of fossil fuel consumed by the population and, consequently the number of environmental problems, such as excess of CO_2 in the atmosphere and global warming, increased significantly [1].

Biofuels are promising for the replacement of fossil fuels since they can reduce environmental impacts and meet the global demand for energy consumption [23]. The first generation biofuels, such as biodiesel is produced from oleaginous crops. It has currently been questioned due to the large amount of water it consumes, the use of agricultural land and its competition with food production [24].

Third generation bioethanol and biodiesel are biofuels that use microalgal biomass as raw material which has become an alternative for this generation of sustainable and renewable biofuels. Adequate cultivation conditions are necessary to obtain high biomass yields, desirable carbohydrate accumulation for bioethanol production [25, 26] and essential lipid levels for biodiesel production [27, 28].

Global interest in renewable energy sources, such as biofuel production has been continuously growing. Thus, microalgal biomass is an excellent alternative for bioethanol production, not only because it decreases the use of traditional energy sources, but also because of the large carbohydrate accumulation in biomass. Regarding the third-generation bioethanol production, three countries, i. e., the USA, Brazil and China, produced 14,806, 7093 and 813 million gallons, respectively, in 2015 [26, 29].

According to Costa and Morais [30] biomass production is significantly positive for biodiesel production. Estimated average annual productivity of microalgal biomass in a tropical country, such as Brazil is 1.53 kg m^3 day^{-1} with an average of 30% of extracted lipid and the biodiesel yield from microalga of 98.4 m^3 ha^{-1}. Microalgae have the ability to use nitrogen and phosphorus from culture medium in their photosynthetic process and synthesize lipids [31]. Thus, microalgae can produce 58.700 L ha^{-1} of algal oil and 121.104 L ha^{-1} of biodiesels. The Renewable Fuels Standard (RFS) estimated that microalga-based fuel production will obtain 36 billion gallons by 2022 [26].

3. Microalgae

Microalgae are photoautotrophic microorganisms that grow fast under relatively simple nutritional conditions. Therefore, they are considered promising organisms for biomass production due to their high-value biomolecules for commercial application [32, 33]. Due to the diversity of biomolecules, several studies have investigated the use of microalgae for biofuel production. Both genera *Chlorella* and *Spirulina* have great potential for this purpose, since they have high concentrations of composts of interest, such as carbohydrates and lipids [33, 34].

Microalgae can be cultivated in three forms: photoautotrophic, heterotrophic and mixotrophic cultivation. The mixotrophic one is a variant of the heterotrophic culture, where CO_2 and organic carbon are assimilated by the respiratory and photosynthetic metabolism with high growth rate and biomass productivity [35]. In this type of cultivation, an organic source of carbon, such as molasses, glycerol and glucose is added [36, 37]. The capacity for assimilating high concentrations of available carbon by microalgae tends to accumulate more carbohydrates and lipids [36], macromolecules of interest in biofuel production.

Production of metabolites by microalgae is determined by several factors, such as species, agitation, pH, nutrient composition, CO_2 concentration, light intensity and temperature [38, 39]. According to Khan et al. [26], light intensity and temperature are the main limiting factors in microalgal cultivation, since these physical stress factors directly influence biochemical processes, such as photosynthesis and biomass production yield.

4. MF application to increase biomass and carbohydrate production

New strategies of culture technologies with high yield of biomass concentration are necessary to enable biofuel production by microalgae [40]. Thus, MF application has been considered a new low-cost technological approach to stimulate cell growth and increased carbohydrate content in microalgal biomass. These outcomes may be achieved by the complex biochemical system in microalga cells, which may cause changes in their defense mechanism and activate proteins, some enzymatic systems and free radicals [21, 41].

Small et al. [42] evaluated the cultivation of *Chlorella kessleri* in a raceway bioreactor with Blue-Green Medium (BG-11) with static MF from 5 to 15 mT, generated by a water-cooled solenoid for 13 d. Cultivation with 10 mT had significant increase of 50% in biomass production while the carbohydrate content reached 42.2% at the end of cultivation. Bauer et al. [16] investigated the influence of 30 mT on *Chlorella kessleri* cultivation in BG-11 medium for 10 d. In relation to the control assay (without MF application), biomass concentration increased 23.5%; its carbohydrate content reached 21.4% with MF applied throughout cultivation (**Table 1**).

Microalga species	Intensity (mT)	Time of exposure (h d^{-1})	Increased biomass content (%)	Carbohydrate content (%)	References
Chlorella kessleri UTEX 398	10	0.3	50	42.2	[42]
Chlorella fusca LEB 111	60	24	21.4	31.4	[17]
Spirulina sp. LEB 18	30	24	40	30.3	[43]
Chlorella kessleri LEB 113	30	24	23.5	21.4	[16]
Spirulina platensis sp	30	6	30.4	12.8	[44]
Chlorella minutissima sp	30	24	30	60.5	[19]

Table 1.
Magnetic effect on biomass and carbohydrate contents of different species of microalga.

Deamici et al. [43] investigated physiological changes in *Spirulina* sp. cultivated in Zarrouk medium under the influence of 30 mT in different periods of MF application (24 h d^{-1} and 1 h d^{-1}) for 15 d. When the microalga was exposed to the permanent condition, biomass concentration increased 40% and reached the highest carbohydrate content of 30.3%, it was 133.2% higher than the one of the control. Shao et al. [44] evaluated enhancement of *Spirulina platensis* biomass with the application of 30 mT for 22 d. Different exposures times (3, 6 and 12 h d^{-1}) were evaluated and the highest biomass concentration was reached when MF were applied for 6 h d^{-1}, increasing 30.4% in relation to the control assay, with carbohydrate content of 12.8%.

5. MF application to increase biomass and lipid production

Cultivation strategies have been studied to increase biomass production and lipid synthesis by microalgae [23]. MF application affected the composition and production of biomass, fundamental parameters in biofuel production [21].

According to Albuquerque et al. [45], MF are capable of regulating metabolic pathways of microorganisms, gene expression and chemical reactions. The authors also commented that the influence of MF on cell metabolism and on the growth of biomass depends on the interaction between the intracellular and extracellular environment, such as the type of cell, characteristics of the culture medium and the existence of biomolecules which are susceptible to MF.

MF can affect the growth and metabolism of microorganisms positively and negatively, depending on its intensity, frequency, pulse shape, type of modulation and exposure time [9]. MF have been shown to be efficient to increase biomass and lipids, since their action can cause oxidative stress in cells of microorganisms, change energy levels and orientation of free radicals and affect enzymatic activity of cells [46–48].

Changes promoted by the MF are responses of the interaction between them and microorganisms, i. e., alteration in permeability of membranes and, consequently, in their cellular metabolism [49]. MF application is considered a low cost and promising tool to overcome some limitations of microalgae, such as lipid productivity [50]. **Table 2** shows the effects of MF on biomass concentration and lipid content.

Microalga species	Intensity and exposure time MF	Lipid content (%w w^{-1})	Biomass concentration (g L^{-1})	References
Chlorella pyrenoidosa FACHB-9	50 mT for 3 h d^{-1}	—	Increased 12.8%	[51]
Chlorella kessleri UTEX 398	10 mT for 0.3 h d^{-1}	Increased 47%	Increased 43%	[42]
Chlorella fusca LEB 111	60 mT for 24 h d^{-1}	Inhibition 23.2%	Increased 20.5%	[17]
Spirulina sp. LEB 18	60 mT for 1 h d^{-1}	Null effect	Increased 95.1%	[43]
Chlorella minutissima sp	30 mT for 24 h d^{-1}	—	Increased 30%	[19]
Chlorella pyrenoidosa FACHB-9	50 mT for 1 h d^{-1}	Increased 12.9%	Increased 8.2%	[52]
Chlorella homosphaera sp	15 mT for 1 h d^{-1}	Increased 22.4%	Inhibition 33%	[18]

Table 2.
Influence of the MF on the synthesis of biomass and lipids in different species of microalga.

Studies have shown that lipid content and productivity can be increased when there is an association between nitrogen reduction and MF application. Bauer et al. [16] identified that, when *Chlorella kessleri* was cultured in BG 11 medium and exposed to 60 mT for 1 h d^{-1}, there was an increase in biomass concentration of 15% and 13.7% in lipid synthesis by comparison with the control. Chu et al. [28] evaluated the influence of MF application on *Nannochloropsis oculate* culture with modified Walne's medium. The highest lipid productivity (30.9 mg L^{-1} d^{-1}) and lipid content (42.4%) were obtained when 20 mT was applied during 7 days of cultivation.

Nannochloropsis oculate was exposed to different intensities of MF (5, 10 and 15 mT) to evaluate the influence its biochemical composition and cell growth. Cultivation under influence of 10 mT increased biomass productivity (45%) and lipid productivity (57%) by comparison with the control [53]. Costa et al. [18] reported that the *Chlorella homosphaera* cultivated in the Bristol's Modified Medium (BMM) with 50% reduction in the nitrogen source associated with exposure to 30 mT and 60 mT for 1 h d^{-1}, increased lipid productivity in 108.4% (35 mg L^{-1} d^{-1}) and 135.1% (39.5 mg L^{-1} d^{-1}), respectively.

6. Conclusion

This chapter reported the potential of microalgal cultivation with MF application for biofuel production. The use of microalgae as raw material is an attractive alternative that can reduce the use of fossil sources and CO_2 emissions, and consequent pollution in the environment. Studies have suggested that MF application may be the most commercial production due to increased production of carbohydrates and lipids. For best results, in the case of every microalga species, parameters, such as MF intensity, exposure time, application period during cultivation and devices used to apply MF, should be evaluated. However, large-scale production of biofuels derived from microalgae has yet to be achieved if it is to be cost-effective.

Acknowledgements

This study was financed in part by the Coordenação de Aperfeiçoamento de Pessoal de Nível Superior - Brasil (CAPES) - Finance Code 001. The authors are thankful for the financial support provided by the National Counsel for Technological and Scientific Development – Brazil (CNPq).

Conflict of interest

The authors declare no conflict of interest.

Appendices and nomenclatures

MF	magnetic fields
GHG	greenhouse Gas
GTR	global trends report
RFS	renewable fuels standard
BG-11	blue-green medium
AO	Aiba and Ogawa medium
BMM	Bristol's modified medium

Author details

Lucielen Oliveira Santos*, Pedro Garcia Pereira Silva, Sharlene Silva Costa and Taiele Blumberg Machado
Laboratory of Biotechnology, School of Chemistry and Food, Federal University of Rio Grande, Rio Grande, RS, Brazil

*Address all correspondence to: santoslucielen@gmail.com

References

[1] Enamala MK, Enamala S, Chavali M, Donepudi J, Yadavalli R, Kolapalli B, Aradhyula TV, Velpuri J, Kuppam C. Production of biofuels from microalgae - A review on cultivation, harvesting, lipid extraction, and numerous applications of microalgae. Renew Sustain Energy Rev. 2018;94:49-68. DOI: 10.1016/j.rser.2018.05.012

[2] Trivedi J, Aila M, Bangwal DP, Kaul S, Garg MO. Algae based biorefinery - How to make sense? Renew Sustain Energy Rev. 2015;47:295-307. DOI: 10.1016/j.rser.2015.03.052

[3] Peng L, Fu D, Chu H, Wang Z, Qi H. Biofuel production from microalgae: A review. Environ Chem Lett. 2020;18:285-297. DOI: 10.1007/s10311-019-00939-0

[4] Suganya T, Varman M, Masjuki HH, Renganathan S. Macroalgae and microalgae as a potential source for commercial applications along with biofuels production: A biorefinery approach. Renew Sustain Energy Rev. 2016;55:909-941. DOI: 10.1016/j.rser.2015.11.026

[5] Jambo SA, Abdulla R, Mohd Azhar SH, Marbawi H, Gansau JA, Ravindra P. A review on third generation bioethanol feedstock. Renew Sustain Energy Rev. 2016;65:756-769. DOI: 10.1016/j.rser.2016.07.064

[6] REN21. Renewables 2020 Global Status Report. Paris: REN21 Secretariat, 2020.

[7] Bajpai P. Advances in Bioethanol. New Dehli: Springer, 2013. Epub ahead of print 2013. DOI: 10.1007/978-81-322-1584-4_1

[8] Brennan L, Owende P. Biofuels from microalgae-A review of technologies for production, processing, and extractions of biofuels and co-products. Renew Sustain Energy Rev. 2010;14:557-577. DOI: 10.1016/j.rser.2009.10.009

[9] Hunt RW, Zavalin A, Bhatnagar A, Chinnasamy S, Das KC. Electromagnetic biostimulation of living cultures for biotechnology, biofuel and bioenergy applications. Int J Mol Sci. 2009;10:4515-4558.

[10] Dini L, Abbro L. Bioeffects of moderate-intensity static magnetic fields on cell cultures. Micron. 2005;36:195-217. DOI: 10.1016/j.micron.2004.12.009

[11] Giancoli DC. Physics: principles with applications. 6th ed. New Jersey: Prentice Hall, Upper Saddle River, 2005.

[12] Tang H, Wang P, Wang H, Fang Z, Yang Q, Ni W, Sun X, Liu H, Wang L, Zhao G, Zheng Z. Effect of static magnetic field on morphology and growth metabolism of *Flavobacterium* sp. m1-14. Bioprocess Biosyst Eng. 2019. DOI: 10.1007/s00449-019-02186-7.

[13] Fologea D. Increase of *Saccharomyces cerevisiae* plating efficiency after treatment with bipolar electric pulses. 1998; 285-287.

[14] Sahebjamei H, Abdolmaleki P, Ghanati F. Effects of magnetic field on the antioxidant enzyme activities of suspension-cultured tobacco cells. Bioelectromagnetics. 2007; 28: 42-47. DOI: 10.1002/bem.20262

[15] Yang G, Wang J, Mei Y, Luan Z. Effect of magnetic field on protein and oxygen-production of *Chlorella vulgaris*. Math Phys Fish Sci. 2011;9:116-126.

[16] Bauer LM, Costa JAV, Rosa APC, Santos LO. Growth stimulation and synthesis of lipids, pigments and antioxidants with magnetic fields in *Chlorella kessleri* cultivations. Bioresour Technol. 2017;244:1425-1432. DOI: 10.1016/j.biortech.2017.06.036

[17] Deamici KM, Cardias BB, Costa JAV, Santos LO. Static magnetic fields in culture of *Chlorella fusca*: Bioeffects on growth and biomass composition. Process Biochem. 2016; 51: 912-916. DOI: 10.1016/j.procbio.2016.04.005

[18] Costa SS, Peres BP, Machado BR, Costa JAV, Santos LO. Increased lipid synthesis in the culture of *Chlorella homosphaera* with magnetic fields application. Bioresour Technol. 2020;315:123880. DOI: 10.1016/j.biortech.2020.123880

[19] Menestrino BC, Pintos THC, Sala L, Costa JAV, Santos LO. Application of static magnetic fields on the mixotrophic culture of *Chlorella minutissima* for carbohydrate production. Appl Biochem Biotechnol. 2020. DOI: 10.1007/s12010-020-03364-0

[20] Deamici KM, Santos LO, Costa JAV. Use of static magnetic fields to increase CO_2 biofixation by the microalga *Chlorella fusca*. Bioresour Technol. 2019;276:103-109. DOI: 10.1016/j.biortech.2018.12.080

[21] Santos LO, Deamici KM, Menestrino BC, Garda-Buffon J, Costa JAV. Magnetic treatment of microalgae for enhanced product formation. World J Microbiol Biotechnol. 2017;33:169. DOI: 10.1007/s11274-017-2332-4

[22] Chisti Y. Biodiesel from microalgae. Biotechnol Adv. 2007; 5:294-306. DOI: 10.1016/j.biotechadv.2007.02.001

[23] Gouveia L. SpringerBriefs in Microbiology. Microalgae as Feedstock for Biofuels. 2011.

[24] Gonzalez-Fernandez, Cristina Muñoz R. Microalgae-Based Biofuels and Bioproducts: From Feedstock Cultivation to End Products. Elsevier, 2017.

[25] Freitas BCB, Brächer EH, Morais EG, Atala DIP, Morais MG, Costa JAV. Cultivation of different microalgae with pentose as carbon source and the effects on the carbohydrate content. Environ Technol. 2019;40:1062-1070. DOI: 10.1080/09593330.2017.1417491

[26] Khan MI, Shin JH, Kim JD. The promising future of microalgae: Current status, challenges, and optimization of a sustainable and renewable industry for biofuels, feed, and other products. Microb Cell Fact. 2018;17:1-21. DOI: 10.1186/s12934-018-0879-x

[27] Menegazzo ML, Fonseca GG. Biomass recovery and lipid extraction processes for microalgae biofuels production: A review. Renew Sustain Energy Rev. 2019;107:87-107. DOI: 10.1016/j.rser.2019.01.064

[28] Chu FJ, Wan TJ, Pai TY, Lin HW, Liu SH, Huang CF. Use of magnetic fields and nitrate concentration to optimize the growth and lipid yield of *Nannochloropsis oculata*. J Environ Manage. 2020;253:109680. DOI: 10.1016/j.jenvman.2019.109680

[29] Hossain N, Zaini J, Indra Mahlia TM. Life cycle assessment, energy balance and sensitivity analysis of bioethanol production from microalgae in a tropical country. Renew Sustain Energy Rev. 2019;115:109371. DOI: 10.1016/j.rser.2019.109371

[30] Costa JAV, Morais MG. The role of biochemical engineering in the production of biofuels from microalgae. Bioresour Technol. 2011;102:2-9. DOI: 10.1016/j.biortech.2010.06.014

[31] Ríos SD, Torres CM, Torras C, Salvadó J, Mateo-Sanz JM, Jiménez L. Microalgae-based biodiesel: Economic analysis of downstream process realistic scenarios. Bioresour Technol. 2013;136:617-625. DOI: 10.1016/j.biortech.2013.03.046

[32] Parniakov O, Barba FJ, Grimi N, Marchal L, Jubeau S, Lebovka N,

Vorobiev E. Pulsed electric field assisted extraction of nutritionally valuable compounds from microalgae *Nannochloropsis* spp. using the binary mixture of organic solvents and water. Innov Food Sci Emerg Technol. 2015;27:79-85. DOI: 10.1016/j.ifset.2014.11.002

[33] Jaiswal KK, Banerjee I, Singh D, Sajwan P, Chhetri V. Ecological stress stimulus to improve microalgae biofuel generation: A review. Octa J Biosci. 2020; 8: 48-54.

[34] Rahman QM, Wang L, Zhang B, Xiu S, Shahbazi A. Green biorefinery of fresh cattail for microalgal culture and ethanol production. Bioresour Technol. 2015;185:436-440. DOI: 10.1016/j.biortech.2015.03.013

[35] Cheirsilp B, Torpee S. Enhanced growth and lipid production of microalgae under mixotrophic culture condition: Effect of light intensity, glucose concentration and fed-batch cultivation. Bioresour Technol. 2012;110:510-516. DOI: 10.1016/j.biortech.2012.01.125

[36] Margarites ACF, Costa JAV. Increment of carbohydrate concentration of *Chlorella minutissima* microalgae for bioethanol production. J Eng Res Appl. 2014;4:2248-2280.

[37] Wang J, Yang H, Wang F. Mixotrophic cultivation of microalgae for biodiesel production: Status and prospects. Appl Biochem Biotechnol. 2014;172:3307-3329. DOI: 10.1007/s12010-014-0729-1

[38] Ho SH, Ye X, Hasunuma T, Chang JS, Kondo A. Perspectives on engineering strategies for improving biofuel production from microalgae - A critical review. Biotechnol Adv. 2014;32:1448-1459. DOI: 10.1016/j.biotechadv.2014.09.002

[39] Magro FG, Decesaro A, Berticelli R, Colla LM. Bioethanol production using microalgae: A review. Semin Exact Technol Sci. 2016;37:159. DOI: 10.5433/1679-0375.2016v37n1p159

[40] Ganesan R, Manigandan S, Samuel MS, Shanmuganathan R, Brindhadevi K, Lan-Chi NT, Duc PA, Pugazhendhi A. A review on prospective production of biofuel from microalgae. Biotechnol Reports. 2020;27:e00509. DOI: 10.1016/j.btre.2020.e00509

[41] Zhang X, Yarema K, Xu A. Impact of Static Magnetic Field (SMF) on Microorganisms, Plants and Animals. In: Biological Effects of Static Magnetic Fields. 2017.

[42] Small DP, Hüner NPA, Wan W. Effect of static magnetic fields on the growth, photosynthesis and ultrastructure of *Chlorella kessleri* microalgae. Bioelectromagnetics. 2012;33:298-308. DOI: 10.1002/bem.20706

[43] Deamici KM, Costa JAV, Santos LO. Magnetic fields as triggers of microalga growth: evaluation of its effect on *Spirulina* sp. Bioresour Technol. 2016;220:62-67. DOI: 10.1016/j.biortech.2016.08.038

[44] Shao W, Ebaid R, Abomohra AEF, Shahen M. Enhancement of *Spirulina* biomass production and cadmium biosorption using combined static magnetic field. Bioresour Technol. 2018;265:163-169. DOI: 10.1016/j.biortech.2018.06.009

[45] Albuquerque WWC, Costa RMPB, Salazar e Fernandes T, Porto ALF. Evidences of the static magnetic field influence on cellular systems. Prog Biophys Mol Biol. 2016;121:16-28. DOI: 10.1016/j.pbiomolbio.2016.03.003

[46] Green LM, Miller AB, Agnew DA, Greenberg ML, Li J, Villeneuve PJ, Tibshirani R. Childhood leukemia and personal monitoring of residential

exposures to electric and magnetic fields in Ontario, Canada. Cancer Causes Control. 1999;10:233-243. DOI: 10.1023/A:1008919408855

[47] Repacholi MH, Greenebaum B. Interaction of static and extremely low frequency electric and magnetic fields with living systems: Health effects and research needs. Bioelectromagnetics. 1999;20:133-160. DOI: 10.1002/(SICI)1521-186X(1999)20:3

[48] Wang HY, Zeng XB, Guo SY, Li ZT. Effects of magnetic field on the antioxidant defense system of recirculation-cultured *Chlorella vulgaris*. Bioelectromagnetics. 2008;29:39-46. DOI: 10.1002/bem.20360

[49] David GF, Perez VH, Justo OR, Cubides DC, Cardona CA, Hristov J. Glycerol bioconversion in unconventional magnetically assisted bioreactor seeking whole cell biocatalyst (intracellular lipase) production. Chem Eng Res Des. 2016;111:243-252. DOI: 10.1016/j.cherd.2016.05.011

[50] Geada P, Rodrigues R, Loureiro L, Pereira R, Fernandes B, Teixeira JA, Vasconcelos V, Vicente AA. Electrotechnologies applied to microalgal biotechnology – Applications, techniques and future trends. Renew Sustain Energy Rev. 2018;94:656-668. DOI: 10.1016/j.rser.2018.06.059

[51] Han S, Jin W, Chen Y, Tu R, Abomohra AEF. Enhancement of lipid production of *Chlorella Pyrenoidosa* cultivated in municipal wastewater by magnetic treatment. Appl Biochem Biotechnol. 2016;180:1043-1055. DOI: 10.1007/s12010-016-2151-3

[52] eng X, Chen Y, Lv J, Han S, Tu R, Zhou X, Jin W, Ren N. Enhanced lipid production by *Chlorella pyrenoidosa* through magnetic field pretreatment of wastewater and treatment of microalgae-wastewater culture solution: Magnetic field treatment modes and conditions. Bioresour Technol. 2020;306:123102. DOI: 10.1016/j.biortech.2020.123102

[53] Oliveira M. Magnetic stimulation on the growth of the microalga *Nannochloropsis oculata*,. Electronic Thesis and Dissertation Repository. 2017. https://ir.lib.uwo.ca/etd/4597 (2017).

Chapter 23

Valorization of Lignocellulosic and Microalgae Biomass

Edward Kwaku Armah, Maggie Chetty, Jeremiah Adebisi Adedeji and Donald Tyoker Kukwa

Abstract

Lignocellulosic biomass has gained increasing recognition in the past decades for the production of value-added products (VAPs). Biomass feedstocks obtained from various sources, their composition, and pretreatment techniques employed for delignification into bioenergy production are discussed. The conversion processes of biomass into VAPs involve various methods. Notable among them are biochemical conversions; namely, anaerobic digestion and ethanol fermentation, and thermo-chemical conversions; namely, pyrolysis and gasification which are considered in this chapter. Microalgae can adapt to changes in the environment, producing biomass that serves as a precursor for a variety of biomolecules, such as proteins, which find their application in pharmaceutical, cosmetic, and biofuel industries. Suitable strains of freshwater microalgae biomass contain high levels of lipid which can be harnessed for bioenergy production. Hence, the advancement in the conversion of biomass into VAPs could help scientists and environmentalists for sustainable use of biomass in future developments.

Keywords: biomass, freshwater microalgae, lignocellulose, microalgae, value-added products

1. Introduction

Biomass resources are readily available globally as residual wastes derived from agricultural and industrial sources. Crop residues such as corn straw, wheat straw, and rice straw are classified as important and relatively abundant renewable biomass resources [1, 2]. With regards to the abundance of biomass resources, China still leads as one of the largest agricultural-based economies in the world, producing approximately about 216 million metric tons of corn straw per annum. For the aforementioned, more than half of that reported from China remain unutilized [3]. Lignocellulose arises from corn straw containing non-edible plant material, composed largely of cellulose, hemicellulose, and lignin. These three components comprise covalent cross-linkages between the polysaccharides (cellulose and hemicellulose) and lignin, making biomass a composite material [4]. The sources and compositions of lignocellulose play a very important role in predicting its potential as value added-products. The hemicellulose is present as the matrix that surrounds the cellulose skeleton, while lignin is present as an encrusting material and serves as a protective layer. However, biomass pretreatment is an essential tool for cellulose conversion processes as it changes the structure of cellulosic biomass to make

IntechOpen

cellulose more available to the enzymes that convert the carbohydrate polymers into fermentable sugars [5]. Other studies have reported that pretreatment of lignocellulosic biomass (LB) aids to overcome recalcitrance through the combination of chemical and structural changes to the lignin and carbohydrates. Some of the different methods of pretreatment include physical; namely, mechanical pretreatment, physicochemical; namely, steam explosion, chemical; namely, alkali and acidic pretreatment, and biological; namely, manure addition or mixed microorganisms [6, 7]. Nonetheless, these traditional methods of pretreatment are cost-intensive, as additional chemicals or energy are required [8]. Also, useful information for policymakers and researchers on lignin biorefinery is presented in this chapter.

2. Lignocellulose biomass (LB)

LB is a composite, based on intertwined biopolymers on a dry basis, consisting of 35–45% cellulose, 25–30% hemicellulose, and 25–30% lignin [9]. These are classified into four major proportions based on their source, namely, woody biomass, agricultural residues (for example, rice/wheat/barley straws, corn stover, sugarcane bagasse), energy crops (switchgrass, *Miscanthus* and short-rotation hardwood is specifically grown for biofuel production) and a group of cellulosic wastes (for example, municipal solid waste, pulp mill and lumber mill wastes) [10]. Cellulose and hemicellulose are broken down by enzymatic saccharification into simple sugars which are further digested by microorganisms through the anaerobic digestion process to produce bioenergy such as biogas [11]. Nonetheless, the application of LB for the net reduction of CO_2 emissions from the transport sector is considered environmentally benign [12]. As a result, pretreatment becomes very important to improve the digestibility of the LB [5, 13]. **Table 1** shows the various chemical compositions of sugarcane bagasse, a lignin-rich residue obtained from the sugar industry.

2.1 Cellulose

Cellulose is a linear polymer of β-D-glucopyranose units linked to each other by 1,4-glycosidic bonds. The linear cellulose chain has a very strong tendency to form intra and inter-molecular hydrogen bonds, which promotes the collection of parallel chains into basic microfibrils. Most wood species contain 40–45% cellulose based on oven-dry (OD) wood. Compression wood of softwoods contains less crystalline cellulose than non-compression wood [13]. The chemical structure of cellulose is shown in **Figure 1**.

Components (%)			References
Cellulose	Hemicellulose	Lignin	
47.0	27.0	23.0	[14]
38.8	26.0	32.4	[15]
45.5	27.0	21.1	[16]
38.4	23.2	25.0	[17]
45.0	25.8	19.1	[18]
39.5	25.6	30.4	[19]
43.6	17.2	22.0	[20]

Table 1.
Chemical composition of raw sugarcane bagasse (%w/w, dry basis).

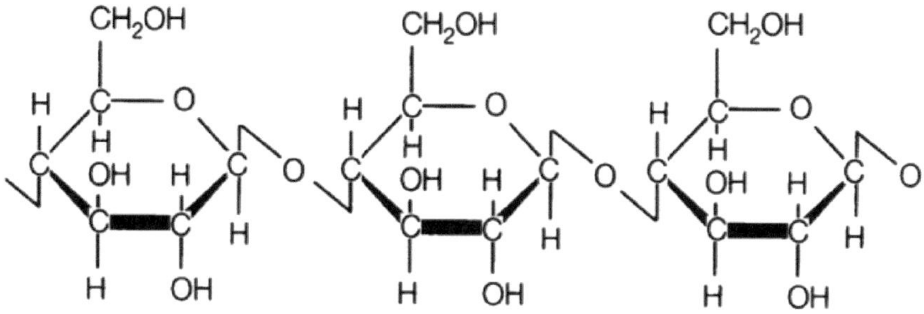

Figure 1.
Chemical structure of cellulose [21].

2.2 Hemicellulose

This is the second most abundant structural component of a typical plant cell wall after cellulose [21]. Cellulose microfibrils are thus linked together in a hydrogen bond with hemicellulose forming cellulose microfibrils (fibers). Hemicellulose has a random, amorphous structure and can be easily hydrolyzed by dilute acid or alkaline of various 5 and 6 carbon sugars, including arabion-xylans glucomannans, and galactans. Xylan is a family of polysaccharides most common to hemicellulose [22]. The schematic diagram is shown in **Figure 2**.

2.3 Lignin

Lignin is the third most abundant structural component in nature of a typical plant cell wall after cellulose and hemicellulose [14]. This amorphous heteropolymer consists of three different phenylpropane units, namely, p-coumaryl, coniferyl and sinapyl alcohol joined by different linkages (presented in **Figure 3**). Lignin was first discovered in 1813 by a Swiss botanist, A. P de Candolle who described it as fibrous, tasteless and insoluble in water and alcohol, but soluble in a weakly basic solution, thus, making it difficult for biodegradation [15]. It is a class of complex organic polymer forming structural materials for supporting tissues of vascular plants and offers impermeability and resistance to microbial attacks [16]. It strengthens stems and vascular tissue, allowing upward growth and permits

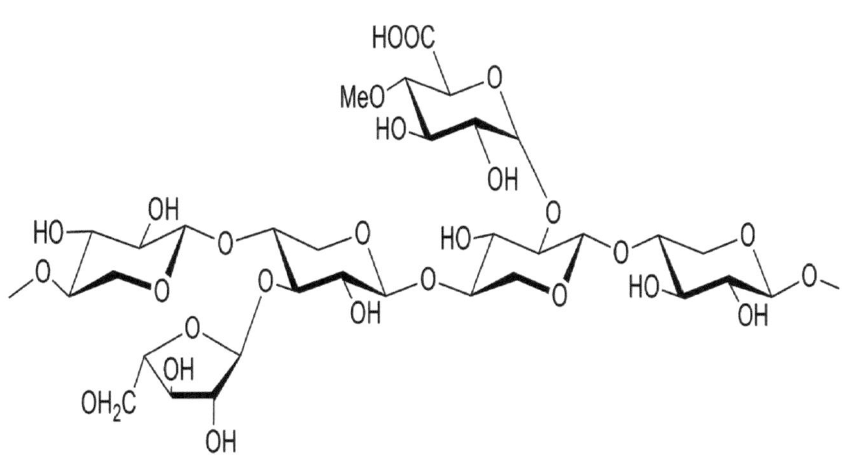

Figure 2.
Chemical structure of hemicellulose [14].

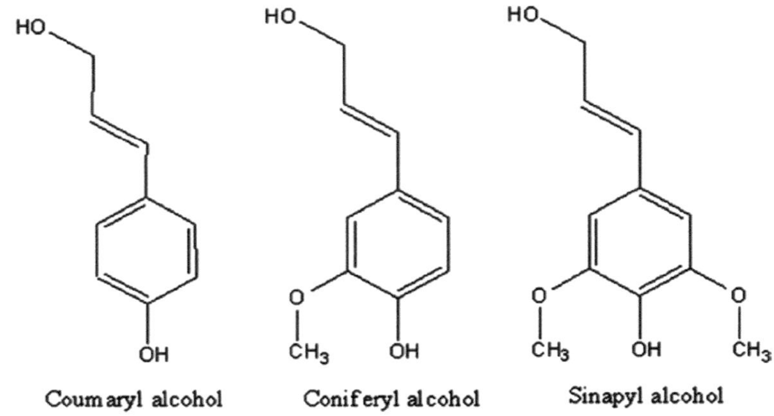

Figure 3.
Chemical structure of the main monolignols of lignin.

water and minerals to be conducted through the xylem under negative pressure without collapse of the tissue. In addition to mechanical support, lignin contributes to protective functions in plants by, for example, increasing resistance to biodegradation and environmental stresses, such as changes in humidity and water balance.

3. Lignocellulose pretreatment technologies

Application of biorefining to bagasse requires lignocellulose fractionation into cellulose, hemicelluloses and lignin [17]. This step involves pretreatment where a considerable part of hemicelluloses is solubilized. In this regard, the cellulose portion is activated initially towards enzymatic hydrolysis and subsequently, for ethanol/biogas production. The use of pretreatment in the conversion of LB for bioenergy production is to enhance the release of cellulose and disrupt lignin and hemicellulose [13]. The sole aim is to remove lignin and hemicellulose, thereby enhancing the cellulose crystallinity and porosity for easier accessibility of microbes to breakdown lignocellulosic feedstock [18]. Various feedstocks that have been employed in literature as pretreatments methods are presented in **Table 2**. Lignin is an amorphous and water-insoluble heteropolymer, and as stated earlier, it is composed of phenylpropane units (coniferyl, p-coumaryl and sinapyl alcohol) held together by different linkages as discussed earlier [19]. A simplified diagram of biomass pretreatment techniques showing the major components of lignocellulose is presented in this chapter (**Figure 4**). The fermentation of LB is difficult due to

Pre-treatment methods	Feedstocks	References
Hydrothermal	Sugarcane bagasse	[23]
Ultrasonic	Sugarcane bagasse	[24]
Ionic liquids	Water hyacinth	[25, 26]
Hydrothermal	Sugarcane bagasse	[14]
Alkali	Cattle dung	[27]
Thermochemical	Water hyacinth	[28]

Table 2.
Feedstocks and pretreatment methods.

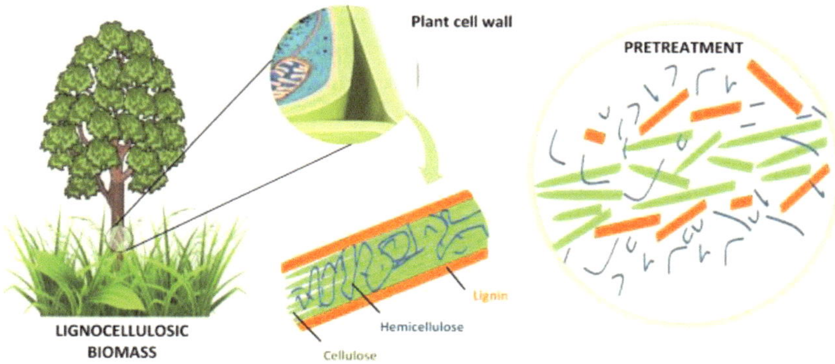

Figure 4.
Simplified diagram of biomass pretreatment technique showing the major components of lignocellulose (Accessed at: https://doi.org/10.1016/B978-0-12-802323-5.00001-3).

the high recalcitrant lignin and inadequate accessibility to sites for enzyme activity. Studies have shown that irrespective of its solubilization, the lignin content change is related to the solidification and re-deposition which is due to cooling after severe pre-treatment. Therefore, only re-allocation of lignin takes place, instead of lignin removal during pre-treatment at high temperatures and pressures [21]. Lignin inhibits hydrolysis by forming physical barriers and non-productive adsorption of cellulase enzymes. Thus, lignin restricts the enzymes from reaching the cellulose, thereby reducing the active enzyme sites for cellulose hydrolysis. Brandt et al. [20] observed that 80–90% of lignin was recovered from a solid fraction of hardwood through hydrothermal pre-treatment at 180–220°C. Therefore, as the severity of hydrothermal pre-treatment increases, the lignin content in the pre-treated solids also increases due to the simultaneous de- and re-polymerization reactions of lignin. Some pretreatment methods are summarized in the following sections.

3.1 Hydrothermolysis

During hydrothermolysis, the lignocellulosic changes that occur for bioenergy production were found to be an efficient method to disrupt lignin and hemicellulose and expose cellulose [20]. The authors [20] concluded that it is impossible for this pretreatment method to completely remove all the lignin present in a lignocellulosic feedstock. In the hydrothermolysis of sunflower oil cake for 1, 2, 4 and 6 h intervals at 25–200°C, it was observed that the cellulose solubilization rate was low (5%) while the hemicellulose content decreased from 13 to 6% at 200°C [29]. In the case of wheat straw at 200°C, cellulose crystallinity reduced as the cellulose hydrolysis rate was increased [30]. Lignin repolymerization occurred at 140°C -180°C for wood in 12–192 minutes with a removal of 75% of lignin [31]. Biogas production from sugarcane bagasse by hydrothermolysis was studied [25]. The authors [25] finding was that pretreatment by hydrothermolysis increased the biogas yield by approximately 15%.

3.2 Ionic liquid pretreatment

The search for a green solvent such as ionic liquids (IL) in the pretreatment of LB for biofuel production has gained increasing recognition for decades. ILs do solubilize complex biomass, thus providing industrial scale-up potential [23]. The unique abilities of ILs to selectively dissolve components of biomass or whole native biomass have been demonstrated [24]. Most ILs have been reported to be viscous in nature, requiring the use of co-solvents to enhance its fluidity and the recovery by a commonly employed aqueous biphasic system, or the use of acetone, sodium hydroxide or water [25]. Commonly used co-solvents are dimethyl sulfoxide (DMSO) and dimethylacetamide (DMAC). The application of ILs to LB in areas such as fractionation, cellulose composites preparation and its derivative and removal of pollutants is a new avenue for the efficient utilization of these solvents [26]. ILs have been found to be the most expensive research-grade solvents under investigation for the dissolution of biomass and provides further challenges with solvent recovery [20].

3.3 Acidic and alkaline pretreatment

Lignocellulosic pretreatment with acids at ambient temperatures are carried out to enhance hemicellulose solubilization, thereby, making cellulose accessible for enzyme degradation with a dilute or a strong acid [14]. In this process, solubilized hemicelluloses are exposed by hydrolytic reactions to produce monomers, furfural, and other volatile products under acidic conditions [27]. In this regard, solubilized lignin quickly condenses and precipitates into acidic conditions. Hemicellulose solubilization and lignin precipitation are therefore noticeable during strong acid pretreatment. A disadvantage of this method is the risk of the formation of inhibiting compounds [14]. However, the use of dilute acid pretreatment has gained numerous research interests over the use of concentrated acids [28]. This is due to the fact that concentrated acids are toxic, corrosive, hazardous, and require reactors that need expensive construction materials which are resistant to corrosion.

3.4 Biological pretreatment

The delignification of LB could also involve application of biological methods using enzymes or microorganisms. Wood degrading microbes including white, brown, soft rot fungi, and bacteria are used in biological applications [28]. Biodegradation releases the chemical components and opens up the structure of the LB which promotes enzyme action leading to further breakdown. The brown and soft rots have been reported to attack cellulose leading to lignin modifications, whilst the lignin components are degraded by the white rot fungi [28]. The biological pretreatment of wood chips with four different white-rot fungi for a period of 30 days was studied [3]. The glucose yield of the pretreated wood by *Trametes versicolor MrP 1* reached 45% by enzymatic hydrolysis while 35% solid was converted to glucose during fungi incubation. Some microbes that have been employed in the past decades include *Ceriporia lacerate, Sterum hirsutum, Polyporus brumalis* and *Phanerochaete chrysosporium.*

4. Microalgae-based systems for CO_2 sequestering and industrial biorefineries

Microalgae have the capacity to adapt to changes in the environment, producing biomass that serves as a precursor for a variety of biomolecules; such as proteins, pigments, vitamins, lipids, and carbohydrates, in addition to finding applications in pharmaceutical, cosmetic, food and biofuel industries [32]. In **Figure 5**, a process flow diagram for micro-algal system in a combined biofuel production system is presented. Microalgae have a promising physiological plasticity in that they have a wide range of pH which allows for a range of species that can convert biomass to high value applications [33]. Pollutants in wastewaters present themselves as nutrients to microalgae, thus providing application of microalgae technology in the

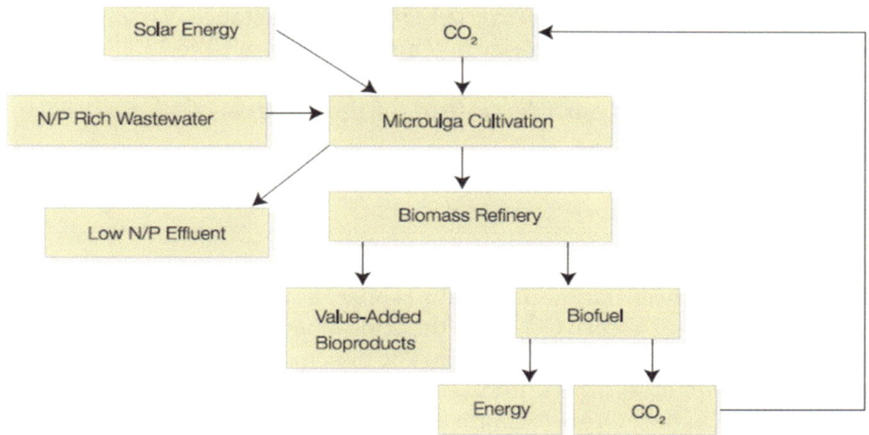

Figure 5.
Process flow for a micro-algal system for combined biofuels production, CO2 biomitigation, and N/P removal from wastewater [54].

wastewater treatment sector. These photosynthetic organisms grow under diverse luminous intensities and electromagnetic radiations to produce biomass of desired compositions. In addition, they sequester CO_2 from the environment and contribute to the global CO_2 balance, thus addressing the global warming phenomenon induced by emissions from fossil fuel combustion processes. Microalgae cultivation combined with metabolic techniques range from autotrophy and heterotrophy to mixotrophy, allowing the biomass a wide latitude for varied specific growth rate, productivity, and composition, which in turn can be enhanced by the hydrodynamics that are governed by the reactor configurations. Lipid recovery continues to be a significant bottleneck in biodiesel production due to high costs of harvesting the biomass in the first instance and the available lipid extraction techniques [34]. Microalgae growth is induced and sustained by factors such as (i) thermal energy, (ii) inorganic carbon supply, (iii) nutrient availability, (iv) luminous exposure, (v) organic carbon and (vi) water. Under photosynthesis protocol, these factors usually work in combination through different metabolic scenarios, which include autotrophy, heterotrophy and mixotrophy. The response to these factors and the nutritional programs depends on the microalgae species and strains. However, the quality of biomass produced from these photosynthetic metabolic scenarios depends on the hydrodynamic stress of the cultivation system [35].

4.1 Thermal energy

Most microalgae species are mesophilic in nature as they produce biomass in the temperature range of 15–35°C. However, some species are extremophiles, i.e. some strains are psychrophilic (*Chlamydomonas nivalis*, *Raphidonema* sp., *Mesotaenium berggrenii*, and *Chloromonas* sp. [CCCryo 020–99]) as they produce biomass under snowy conditions, while few other strains are thermophilic (*Phormidium sp.* and *Thermosynechococcus elongatus* BP-1 which are cyanobacteria; and *Desmodesmus sp.* F51, *Chlorella sorokiniana* UTEX 2805, *Desmodesmus sp*. F2 and F18 are green algae and; *Galdieria sulphuraria* 074G and *Galdieria sulphuraria* CCMEE 5587.1 are the red algae); they produce biomass at temperatures as high as 55–74°C. Depending on the species, microalgae at an optimum temperature with suitable nutrients media (nitrogen, phosphorus, and sulfur) and luminous exposure, produce biomass of varying properties and composition [36]. Temperature is a key variable that influences the composition of microalgae biomass. For instance, Varshney and co-workers [37] reported that an increase in temperature from 20 to 25°C doubled the lipid content of *Nannochloropsis oculata* (from 7.90 to 14.92%), whilst an increase from 25 to 30°C brought about a decrease of the lipid content of *Chlorella vulgaris* from 14.71 to 5.90%.

4.1.1 Inorganic carbon

Inorganic carbon that is accessible to microalgae is mostly CO_2 gas. This gas is available in dilute concentrations in the atmosphere at 0.035 mole percent (dry basis) [38]. Microalgae absorb CO_2 from the atmosphere to produce sugars by the physiochemical process of photosynthesis. The biological conversion of CO_2 results in products of the photosynthetic metabolism such as cells, oxygen biopolymers which are soluble in the culture medium and volatile organic compounds (VOC's). Zhao and Su [39] described photosynthesis as a two-stage process. The first stage is the light-dependent reaction which captures the energy of light for oxidative phosphorylation in the metabolic cycle that produces the energy-storage molecules, adenosine triphosphate (ATP) and nicotinamide adenine dinucleotide phosphate (NADPH) as shown in Eq. (1).

$$2H_2O + 2NADP^+ + 3ADP + 3P_i \xrightarrow{hv} 2NADPH + 2H^+ + 3ATP + O_2 \qquad (1)$$

Eq. (2) shows the second-stage reaction, which is carbon dioxide fixation; and it is not directly light-dependent. This photosynthetic dark-reaction captures and reduces carbon dioxide to carbohydrates and releases molecular oxygen [40].

$$CO_2 + 2H_2O \xrightarrow{photons} [CH_2O] + O_2 \qquad (2)$$

With a solubility of 0.1449 g CO_2/100 mL H_2O at 25°C and 101.325 kPa vapor pressure, carbon dioxide gas dissolves in surface water and slowly reacts with water to alter its chemistry as shown in Eqs. (3) and (4).

$$CO_2 + H_2O \rightleftharpoons H_2CO_3 \qquad (3)$$

$$H_2CO_3 \rightleftharpoons H^+ + HCO_3^- \qquad (4)$$

The release of H^+ ion in Eq. (4) causes the reduction in pH of the culture medium. However, the ability to thrive in a wide pH range has given microalgae the privilege to access nutrients from municipal and industrial wastewaters. In the presence of hydroxide ions, carbonate ions are also released as in Eq. (5).

$$OH^- + HCO_3^- \rightleftharpoons H_2O + CO_3^{2-} \qquad (5)$$

When the dynamic ionization equilibrium is attained, dissolved inorganic carbon (DIC) is available in the form of CO_2, HCO_3^-, CO_3^{2-}, and H_2CO_3. However, only CO_2 and HCO_3^- are accessible to microalgae and both species are utilized simultaneously to produce biomass [41]. The CO_2 conversion into biomass is increased only under conditions where the CO_2 mass loading rate is low. At a high CO_2 mass loading rate, the formation of VOCs is the main CO_2 biotransformation strategy [33].

4.1.2 Nutrients availability

Standard microalgae culture media have been developed and produced out of the need to produce desired products and are available in the market for freshwater microalgae growth management. Some of the media are the (i) Blue-Green medium, BG-11 (ii) Bold's Basal medium, BBM (iii) Bold's Basal medium modified, BBM-3 N (iv) CHU 13 and (v) Jaworski's Medium, JM. Both municipal and industrial wastewater have the basic nutrients common to all the artificial media designed for microalgae cultivation; and microalgae access these nutrients as nitrates and reactive phosphates from wastewaters to produce biomass and bioproducts [42, 43].

4.2 Metabolic flexibility of microalgae

Microalgae have three different metabolic pathways, namely, autotrophy, heterotrophy and mixotrophy. While all algae species are autotrophic, some stains have the ability to exhibit heterotrophy and mixotrophy; and any of the chosen photosynthetic metabolic depends on the microalgae species, and the quality of the biomass desired. Autotrophic metabolism utilizes inorganic carbon in the form of CO_2, gas and light energy. This mode of fixing CO_2 produces low density microalgal biomass. Heterotrophic metabolism takes advantage of the presence of organic carbon and utilizes it both as a source of carbon and energy. This is the dominant pathway during the night or dark phases. Some microalgae do metabolize mixotrophically. Under mixotophic mode, light energy is not the absolute growth limiting factor as organic carbon sources are also accessed and utilized for microalgal biomass production. Photoinhibition, a phenomenon that describes excessive light intensity thereby arresting photosynthetic metabolism, is overcome under the mixotrophic metabolic mode. Consequently, the growth rate is not interrupted and high density biomass is produced with recorded higher productivities when compared to autotrophic and heterotrophic metabolic scenarios [6].

4.2.1 Organic carbon

Organic carbon present in municipal and industrial wastewater are carbohydrates, fats, volatile fatty acids (VFAs), soaps, synthetic detergents, lignin, proteins and their decomposition products; as well as various natural and synthetic organic chemicals. Wastewater treatment and concomitant algal biofuel production has received increasing attention in recent years owing to its diverse environmental and economic benefits [44]. Organic carbon is accessible through monosodium glutamate wastewater, cheese whey permeates, sodium acetate, fruit peel, glucose, fructose, glycerol, etc. via mixotrophic microalgal growth mode. Tan and coworkers [45] reported that productivities of *C. vulgaris* cultured in wastewaters containing glucose and sodium acetate were 63.5 and 55.2 $mg\,L^{-1}\,day^{-1}$, respectively. This accounted for the leap of 2.61 and 2.27 times the productivities, respectively, achieved under autotrophic metabolic modes. Also, *Chlorella vulgaris* cultivated in sodium acetate and glucose wastewaters recorded productivities of lipid at 17.35 $mg\,L^{-1}\,day^{-1}$ and carbohydrate at 18.75 $mg\,L^{-1}\,day^{-1}$, respectively, indicating that sodium acetate and glucose wastewaters have the potential to boost microalgal lipid production, which in turn may serve as feedstock for the biorefinery [46].

4.2.2 Hydrodynamic stress

Biotransformation kinetics in microalgae are driven by two cultivation systems: (i) open cultivation systems, (OCS) and (ii) closed cultivation systems, (CCS). Open cultivation systems employ open ponds, tanks and raceway ponds while the closed cultivations system utilizes closed photobioreactors (PBR), such as bubble column reactors (BCR), airlift reactors (ALR), tubular reactors (TR), plastic bag (single-use) reactors (SUR), stirred tank reactors (STR), and plate reactors (PR). The OCS attracts minimal capital, operating cost, and lesser energy for culture mixing. However, OCS require large land-mass for scale-up operations as they are prone to contamination and adverse weather conditions wherein they suffer evaporation and temperature fluctuations [47]. The CCS on the other hand, are operated at highly controlled conditions and are more efficient in terms of quality. PBRs can be designed and optimized to cultivate a chosen microalgal strain; since they occupy minimal landmass with enhanced luminous exposure to the microalgae cells and encounter little or no contamination. However, PBRs do have bio-fouling issues, cleaning difficulties, benthic microalgal growth, and high build-up of dissolved oxygen (DO) leading to growth obstructions, and high capital cost [48, 49].

The microalgae culture mixing regimes may vary from one PBR to another, and since the purpose of mixing is to ensure adequate exposure of all the microalgae cells to the growth index, variables such as thermal energy, nutrients adequate illumination, gas exchange, and the quality of microalgal biomass churned from each PBR varies in terms of cell density and biomass composition [50].

PBRs using microalgae to treat wastewater and to produce biomolecules are based on five basic criteria: (i) full control of the reaction conditions, (ii) increased efficient use of light energy, (iii) an adequate mixing system, (iv) reduced hydrodynamic stress on the cells and (v) flexible scale-up operations [51].

4.2.3 Luminous exposure

Biomass productivity depends largely on the quantity and quality of light available to microalgae cells during exponential growth, especially in the autotrophic metabolic mode. Lighting has a great influence on the synthesis of co-products in microalgal biomass as the cells increase pigmentation. Large quantities of solar radiation storage are enhanced as biomass, which can be transformed into solid, liquid, or gaseous fuels [52]. On the other hand, the exposition of microalgae cells to excessive illumination can cause photoinhibition, a phenomenon which describes termination of photocatalytic activity in the presence of illumination. Both photoperiod and light intensity influence microalgal growth, pigment production, biomass, and lipid productivities. High biomass and lipid productivities have been reported for stepwise strategic light-intensity increases during mixotrophic cultivation of microalgae. Cheirsilp and Torpee [53] reported the influence of light intensity on the growth and lipid accumulation of marine *Chlorella sp. and Nannochloropsis sp.*; and observed that the growth of marine *Chlorella sp.* increased when the light intensity was increased from 2000 to 8000 lux. Increasing the light intensity to 10,000 lux registered a slight decrease in the lux indices, which could be due to photoinhibition.

5. Conversion of biomass to bioenergy production

Generation of bioenergy from biomass is achieved in various ways and may be classified into three main categories, namely, physio-chemical, bio-chemical, and thermo-chemical processes [54]. The following are common techniques utilized in the conversion of biomass into biofuels, i.e. mechanical extraction, transesterification, pyrolysis, anaerobic digestion, fermentation, gasification, liquefaction, and fuel cell systems as shown in **Figure 6**. This subsection gives an overview of the conversion process, the factors affecting each process and the main products derived.

5.1 Pre-treatment methods

Prior to the application of a specific technique of biofuel generation from biomass, various pre-treatment or pre-processing steps may be carried out to aid effective conversion. Two main pre-treatment methods broadly classified under physio-thermal and chemical methods are usually applied based on the lignocellulosic substrates as discussed in the latter sections on lignocellulosic and the related conversion techniques. Processes such as drying, sizing, crushing, powdering, pelletizing, torrefaction and heating are common physio-thermal methods for pre-treatment of lignocellulosic and algae biomass.

5.2 Physico-chemical means

This involves the mechanical extraction of oil from lignocellulosic and algae biomass, where the oil produced is further esterified to produce biodiesel. Biodiesel is

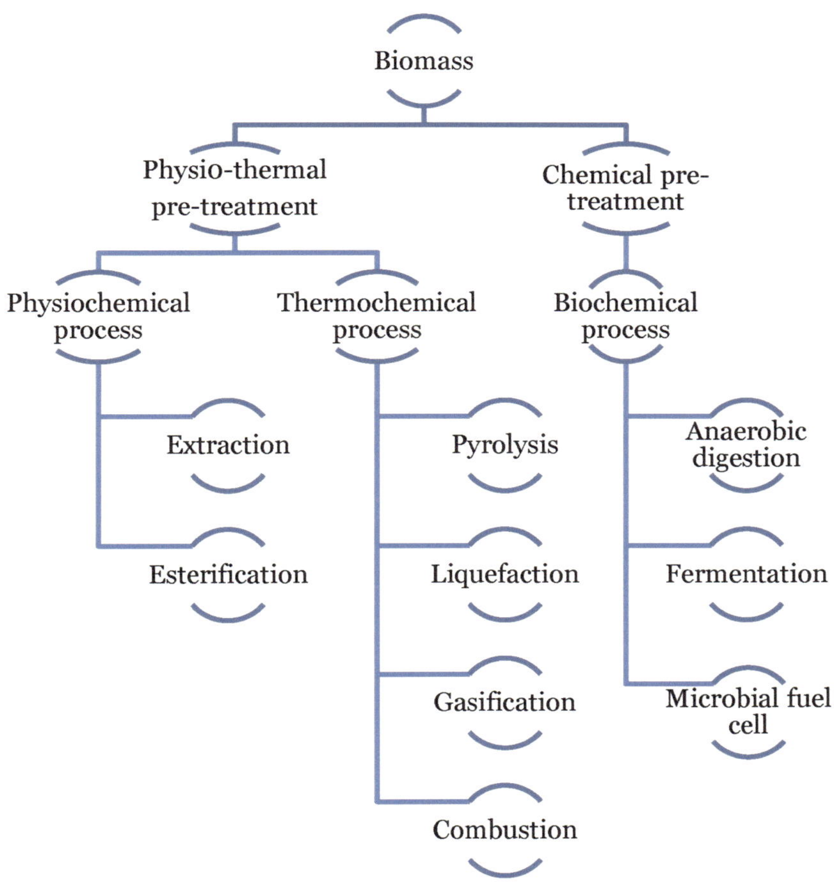

Figure 6.
Biomass conversion techniques.

usually blended with conventional diesel to be used as fuel for motor vehicles. The oil extracted from the biomass is highly viscous with polyunsaturated characteristics; therefore, transesterification processes which utilize either acids, bases, or enzymes as catalysts, convert the oil into fatty acid methyl esters (FAME) or fatty acid alkyl esters with glycerol as by-product. The esters produced from the transesterification process have lower viscosities and are comparable to conventional fuels [51].

5.3 Bio-chemical means

The biochemical conversion process utilizes the metabolic activity of microorganism for conversion of biomass into biofuel and by-products. Biomass conversion is environmentally friendly when compared to thermochemical methods where the residence time for the conversion process to be achieved is longer when compared to thermochemical conversion means. The main biochemical conversion methods are discussed below.

5.3.1 Anaerobic digestion

Anaerobic digestion (AD) is the microbial degradation of organic matter in the absence of oxygen to produce mainly biogas (biomethane and carbon dioxide). The conversion of organic matter into biogas is presented in **Figure 7**. It is a series of biochemical reactions where microorganisms anaerobically convert organic materials into products to be finally converted into biogas [2].

Microorganisms break down high molecular mass compounds such as polysaccharides, proteins, fats, cellulose, and hemicellulose into smaller molecular mass

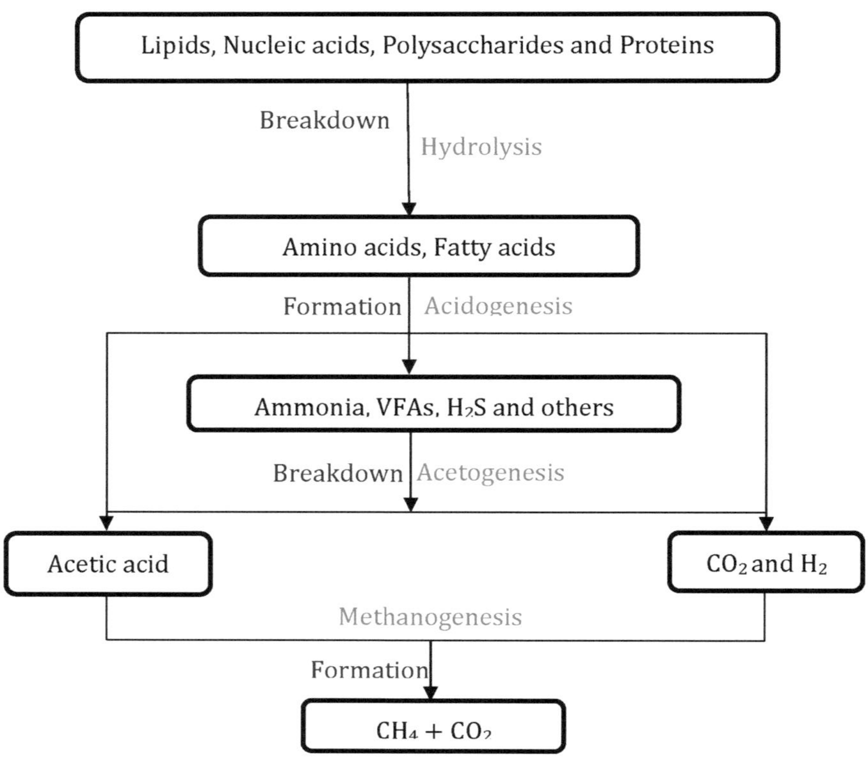

Figure 7.
Anaerobic digestion degradation process.

compounds which are later converted into biogas. The efficiency of AD processes is dependent on the components of the substrate and the activity of microorganisms. AD is a biochemical conversion process that is robust with proven reliable applications. AD has been widely applied in the treatment of organic waste streams, and its development for production of biogas goes far back as the 16th century [55]. The general equation to produce methane from organic matter is given as in Eq. (6).

$$C_nH_aO_b + \left(n - \frac{a}{4} - \frac{b}{2}\right)H_2O \rightarrow \left(\frac{n}{2} - \frac{a}{8} + \frac{b}{4}\right)CO_2 + \left(\frac{n}{2} + \frac{a}{8} - \frac{b}{4}\right)CH_4 \quad (6)$$

5.3.1.1 Stages of AD process

The four main stages involved in the AD of lignocellulosic biomass are hydrolysis, acetogenesis, acidogenesis and methanogenesis. Theses stages occur successively as the product from a preceding stage is utilized in the next stage. Hydrolysis is the first stage in the AD process, and it involves the breakdown of high molecular mass compounds into smaller ones. This stage is followed by acidogenesis, which is the conversion of the smaller molecular compounds into identifiable lower molecular compounds. The acetogenesis stage converts the volatile fatty acids (VFAs) from acidogenesis into acetate, CO_2 and H_2, where after these intermediate compounds are converted into methane and CO_2 with other by-products in trace amounts in methanogenesis stage, with the aid of methanogens. The stages of the AD process are affected by various parameters and the biomass chosen.

5.3.2 Fermentation

The conversion of biomass to simple sugars with subsequent transformation into alcohol and CO_2 with the aid of microorganisms, mainly, yeasts is known as fermentation. This process has been applied for centuries to produce ethanol from sugar crops. It has been used for conversion of lignocellulosic material and algae to ethanol. The by-product from this process after the distillation process are the non-fermentable products which are further directed for use as animal-feed, or as raw feed for the thermochemical conversion. Ethanol produced from the process can be used as fuel for motor vehicles, or as compliment for existing conventional fuels. It has been used as an additive for petrol to improve the octane rating and vehicle emissions reduction in countries such as Australia, Brazil, Sweden, and United States [20]. Anaerobic or dark fermentation of pre-treated lignocellulosic biomass is used in the production of biohydrogen, where the process is like the acidogenic stage in AD. The microorganisms used in this process are mainly hydrogen producing microbes such as *Thermoanaero bacterales, Clostridiaceae* and *Enterobacteriaceae.* This process is gaining interest as the combustion of hydrogen is free of any harmful emissions [56]. The challenge encountered is the low yield of H_2 generated, though there has been research carried out into the upgrading of CH_4 into H_2.

5.3.3 Microbial fuel cell (MFC)

MFCs employ the activity of micro-organisms to convert pre-treated biomass into bioelectricity that can be fed into an existing electricity grid. The process involves the oxidation of the substrates (cellulosic biomass) at the anode chamber of the cell by microorganisms, (electrode-reducing organisms) to electrons which are transferred to the cathode chamber through a conductive material. In the cathode region, the electrode-oxidizing organisms utilize the electrons for reduction of various compounds to other forms (such as CO_2 to acetate, nitrate to N_2 and O_2 to H_2O). It is an electrochemical reaction that utilizes microorganisms for catalysis; therefore, it is referred to as a bioelectrochemical process. The anode in an MFC is usually carbon based such as carbon cloth, felt, fiber, rod, and paper, while the cathode is either of the latter coated with platinum [57, 58]. **Figure 8** shows a schematic of the working principles of a microbial fuel cell. The catalyst used may differ based on the application of the fuel cell.

5.4 Thermo-chemical processes

Thermochemical conversion techniques involve the use of heat or chemicals for the conversion of biomass into fuel and heat (as in the case of combustion and gasification). The thermochemical means is sometimes preferred since it requires limited time, little or no pre-treatment, and generation of variable end-products as compared to biochemical means. The overview of the main thermochemical conversion techniques is summarized in the following section.

5.4.1 Pyrolysis

An irreversible chemical reaction impacted by heat in the absence of oxygen is known as pyrolysis. This method generally converts lignocellulosic biomass into solid, liquid, and gaseous fractions which are further processed into another product spectrum [59]. Pyrolysis is commonly adopted, since its end-product ranges from gaseous to solid fuels in varying percentages, when compared to other thermochemical biomass conversion methods. Pyrolysis can be a precursor to some other conversion processes such as gasification and combustion, as well as a succeeding step to some pre-treatment methods, such as torrefaction and degradative solvent extraction (DSE) [60]. Pyrolysis is a temperature, heating rate and time dependent process, and varying these conditions with addition of selected catalysts give a specified, desired end-product. Based on the specific conditions, pyrolysis is classified into slow, intermediate, and fast mode; and from which different percentages of solid, liquid, and gaseous product are derived. The classification of pyrolysis base on temperature and residence time is shown in **Table 3**, and it ranges from the low temperatures (~ 300°C) to high temperatures (approximately 900°C).

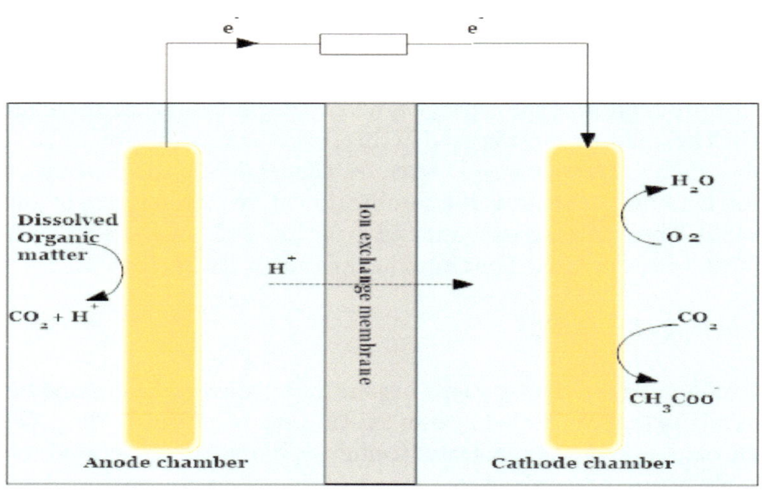

Figure 8.
Schematic of a microbial fuel cell.

Mode of pyrolysis	Heating rate	Temperature	Residence time	Product percent by weight (%wt)		
				Solid	Liquid	Gas
Slow		250–400°C	Long hours to days			
Carbonization	Low	~400°C*		35	30	35
Torrefaction Intermediate		~280°C	10–60 mins	80	3	20
	Medium	300–500°C	5–30 s	25	50	25
Fast	High	~500°C	< 2 s	12	75	13

**Approximate values.*

Table 3.
Mode of pyrolysis and product distribution [61, 62].

Slow pyrolysis involves lower heating rates between 0.1 and 2°C, and the products are mainly solids. Carbonization is a form of slow pyrolysis which is the old technique used in the production of charcoal (biochar) and the vapor produced during the process is usually not condensed, but rather used for heating. Torrefaction mainly aims at improving the energy density and biomass fuel properties, such as reduced weight and volume, which renders the fuel easy to transport and be crushed when needed. The intermediate pyrolysis process produces less solid and liquid with low viscosities. Fast pyrolysis is one of the biomass conversion technologies currently gaining prominence because of its ability to generate more liquid fuel (bio-oil) which could be easily upgraded to diverse valuable products, and transported [19, 61].

5.4.2 Gasification

Gasification is the conversion of biomass or carbonaceous feedstock into gaseous components at temperatures higher than 650°C. The gaseous component of the gasification process may vary depending on the operating temperature. At a lower temperatures below 1200°C, gas constituents may vary from CO, CO_2, CH_4 and H_2, which are collectively known as producer gas while at a higher temperatures, the constituents are CO and H_2, which are collectively known as synthesis gas or syngas [62]. Coal gasification is a well-known process in the generation of electricity which can also be used for biomass conversion into gas. The four steps involved in the gasification process are drying or heating, pyrolysis, gas–solid reaction, and gas phase reaction. These steps occur successively and may take a fraction of a minute depending on the reactor (gasifier) design [52]. Drying/heating is mainly adopted to remove the moisture content of biomass, thereby converting the biomass into dry mass. This is done to attain the required temperature for gasification and the desired products. This process is followed by pyrolysis. The gas–solid reaction step involves the reaction of the gas and solid (char) produced from the pyrolysis phase. The char which is carbon reacts with carbon dioxide, hydrogen, oxygen, and water (vapor) to form gaseous compounds as shown in the conversion Eqs. (7)–(10).

$$\text{Boudouard reaction}: C + CO_2 = 2CO \tag{7}$$

$$\text{Carbon} - \text{hydrogen reaction}: C + H_2 = CH_4 \tag{8}$$

$$\text{Carbon} - \text{oxygen reaction}: C + O_2 = CO_2 \tag{9}$$

$$\text{Carbon} - \text{water reaction}: C + H_2O = CO + H_2 \tag{10}$$

Thus, the gas phase reaction is shown in Eqs. (11)–(12).

$$CO + H_2O = CO_2 + H_2 \tag{11}$$

$$CH_4 + H_2O = CO + 3H_2 \tag{12}$$

5.4.3 Solvent liquefaction

This is the conversion of biomass into liquid or solubilized products at moderate temperatures (105–400°C) and pressure (2–20 MPa) with the aid of solvents. Solvents such as water, ethanol, phenol, tetralin, sulfuric acid, phosphoric acid, nitric acid, and other ionic solvents have been utilized. When water is used as the solvent in the liquefaction process, it is known as hydrothermal liquefaction [63]. For optimal efficiency of the process, the main parameter is the choice of solvent as the process requires the solvent to be in the liquid phase during the reaction. Solvents such as creosote have been utilized to achieve a bio-oil yield of 74 wt% as compared to water of 35 wt% yield or acetone of 10 to 60 wt% yield, for lignocellulosic biomass. This process has been mainly used for the processing of lignocellulosic biomass, algae, and other biomass feedstocks. Unlike other thermochemical conversion processes, this process does not need much residence time for intermediate drying/heating of the biomass as it could be used to process biomass with 15 to 80% moisture content.

5.4.4 Combustion

Combustion is the complete oxidation of carbon containing materials to CO_2 and H_2O in the presence of air (oxygen). For a complete combustion process, the four stages involved are heating or drying, pyrolysis, volatiles combustion (known as flaming) and char combustion (smoldering). These stages are similar to the stages in gasification and the only difference is that combustion requires excess air [64]. Combustion depends on the operating temperature, feedstock type, the particle size of the feedstock, the design of the reactor and atmospheric conditions. Other by-products of this process are nitrogen oxides, sulfur, ash, and particulate matter which are environmentally unfriendly [63].

6. Conclusion

The conversion of LB into value-added products is vital to meet the global demand for lignocellulosic products. The concept of biorefining arose since the potential of lignocellulosic and microalgae-based products were substituted for fossil fuel derived products, which accounted for increased usage of non-renewable fuels. The reduction in the demand on fossils, creation of opportunities in the job market and the provision of sustainable forms of energy has highlighted the role of biorefineries in tackling climate change. This chapter presented the insights into the various components of lignocellulose and microalgae, the pretreatment techniques adopted in the past decades for delignification and conversion into useful products, and the applications coupled with future prospects for valorization of biomass.

Author details

Edward Kwaku Armah*, Maggie Chetty, Jeremiah Adebisi Adedeji and Donald Tyoker Kukwa
Department of Chemical Engineering, Durban University of Technology, Durban, South Africa

*Address all correspondence to: edwardkarmah@gmail.com

References

[1] Amin FR, Khalid H, Zhang H, Rahman SU, Zhang R, Liu G, et al. Pretreatment methods of lignocellulosic biomass for anaerobic digestion. AMB Express. 2017;7:72

[2] Armah EK, Chetty M, Deenadayalu N. Biomethane potential of agricultural biomass with industrial wastewater for biogas production. Chemical Engineering Transactions. 2019;**76**: 1411-1416

[3] Chandel AK, Garlapati VK, Singh AK, Antunes FAF, da Silva SS. The path forward for lignocellulose biorefineries: Bottlenecks, solutions, and perspective on commercialization. Bioresource Technology. 2018;**264**:370-381

[4] Hassan SS, Williams GA, Jaiswal AK. Moving towards the second generation of lignocellulosic biorefineries in the EU: Drivers, challenges, and opportunities. Renewable and Sustainable Energy Reviews. 2019;**101**:590-599

[5] Bhatia SK, Joo H-S, Yang Y-H. Biowaste-to-bioenergy using biological methods–a mini-review. Energy Conversion and Management. 2018;**177**:640-660

[6] Wagner AO, Lackner N, Mutschlechner M, Prem EM, Markt R, Illmer P. Biological pretreatment strategies for second-generation lignocellulosic resources to enhance biogas production. Energies. 2018;**11**:1797

[7] Hernández-Beltrán JU, Lira H-D, Omar I, Cruz-Santos MM, Saucedo-Luevanos A, Hernández-Terán F, et al. Insight into pretreatment methods of lignocellulosic biomass to increase biogas yield: Current state, challenges, and opportunities. Applied Sciences. 2019;**9**:3721

[8] Meléndez-Hernández PA, Hernández-Beltrán JU, Hernández-Guzmán A, Morales-Rodríguez R, Torres-Guzmán JC, Hernández-Escoto H. Comparative of alkaline hydrogen peroxide pretreatment using NaOH and Ca (OH) 2 and their effects on enzymatic hydrolysis and fermentation steps. Biomass Conversion and Biorefinery. 2019:1-11

[9] Paul S, Dutta A. "challenges and opportunities of lignocellulosic biomass for anaerobic digestion," resources. Conservation and Recycling. 2018;**130**:164-174

[10] Kim JS, Lee Y, Kim TH. A review on alkaline pretreatment technology for bioconversion of lignocellulosic biomass. Bioresource Technology. 2016;**199**:42-48

[11] Sindhu R, Binod P, Pandey A, Madhavan A, Alphonsa JA, Vivek N, et al. Water hyacinth a potential source for value addition: An overview. Bioresource Technology. 2017;**230**:152-162

[12] Konda NM, Shi J, Singh S, Blanch HW, Simmons BA, Klein-Marcuschamer D. Understanding cost drivers and economic potential of two variants of ionic liquid pretreatment for cellulosic biofuel production. Biotechnology for Biofuels. 2014;**7**:86

[13] Gupta PK, Raghunath SS, Prasanna DV, Venkat P, Shree V, Chithananthan C, et al. An update on overview of cellulose, its structure and applications. In: Cellulose. Rijeka: IntechOpen; 2019

[14] Hendriks A, Zeeman G. Pretreatments to enhance the digestibility of lignocellulosic biomass. Bioresource Technology. Netherlands: Elsevier BV; 2009;**100**:10-18

[15] Fengel D, Wegener G. Wood: Chemistry, Ultrastructure, Reactions. Berlin: DeGruyter; 1984

[16] Martone P, Estevez J, Lu F, Ruel K, Denny M, Sommerville C, et al. Discovery of lignin in seaweed reveals convergent evolution of cell-wall architecture. Current Biology. 2009;**19**:169-175

[17] Rocha DGJ, Carlos M, Barbosa IS, Souto MAM, Macedo HB, Moraes ACD. Dilute mixed-acid pretreatment of sugarcane bagasse for ethanol production. Biomass and Bioenergy. 2011;**35**:663-670

[18] Fan S, Zhang P, Li F, Jin S, Wang S, Zhou S. A review of lignocellulose change during hydrothermal pretreatment for bioenergy production. Current Organic Chemistry. 2016;**20**:1-11

[19] De Wild PJ, Huijgen WJ, Gosselink RJ. "lignin pyrolysis for profitable lignocellulosic biorefineries," biofuels. Bioproducts and Biorefining. 2014;**8**:645-657

[20] Brandt A, Gräsvik J, Hallett JP, Welton T. Deconstruction of lignocellulosic biomass with ionic liquids. Green Chemistry. 2013;**15**:550-583

[21] Takara D, Khanal SK. Biomass pretreatment for biofuel production. In: Sustainable Bioenergy and Bioproducts. Springer; 2012. pp. 59-70

[22] Armah EK, Tetteh EK, Boamah BB. Overview of biogas production from different feedstocks. International Journal of Scientific and Research Publications. 2017;7:158-164

[23] Badgujar KC, Bhanage MB. Factors governing dissolution process of lignocellulosic biomass in ionic liquid: Current status, overview and challenges. Bioresource Technology. 2015;**178**:2-18

[24] Reddy P. A critical review of ionic liquids for the pretreatment of lignocellulosic biomass. South African Journal of Science. 2015;**111**:1-9

[25] Gao J, Chen L, Yan Z, Wang L. Effect of ionic liquid pretreatment on the composition, structure and biogas production of water hyacinth (*Eichhornia crassipes*). Bioresource Technology. 2013;**132**:361-364

[26] Zhu S. Perspective use of ionic liquids for the efficient utilization of lignocellulosic materials. Journal of Chemical Technology and Biotechnology. 2008;**83**:777-779

[27] Ramos LP. The chemistry involved in the steam treatment of lignocellulosic materials. Química Nova. 2003;**26**:863-871

[28] Zheng Y, Pan Z, Zhang R. Overview of biomass pretreatment for cellulosic ethanol production. International Journal of Agricultural and Biological Engineering. 2009;**2**:51-68

[29] Fernández-Cegrí V, De la Rubia MÁ, Raposo F, Borja R. Effect of hydrothermal pretreatment of sunflower oil cake on biomethane potential focusing on fibre composition. Bioresource Technology. 2012;**123**:424-429

[30] Ibbett R, Gaddipati S, Hill S, Tucker G. Structural reorganisation of cellulose fibrils in hydrothermally deconstructed lignocellulosic biomass and relationships with enzyme digestibility. Biotechnology for Biofuels. 2013;**6**:33

[31] Trajano HL, Engle NL, Foston M, Ragauskas AJ, Tschaplinski TJ, Wyman CE. The fate of lignin during hydrothermal pretreatment. Biotechnology for Biofuels. 2013;**6**:110

[32] He B, Hu Z, Ma L, Li H, Ai M, Han J, et al. Transcriptome analysis of different growth stages of *Aspergillus oryzae* reveals dynamic changes of distinct classes of genes during growth. BMC Microbiology. 2018;**18**:12

[33] Kumar KS, Dahms H-U, Won E-J, Lee J-S, Shin K-H. Microalgae–a promising tool for heavy metal remediation. Ecotoxicology and Environmental Safety. 2015;**113**:329-352

[34] Anastopoulos I, Kyzas GZ. Progress in batch biosorption of heavy metals onto algae. Journal of Molecular Liquids. 2015;**209**:77-86

[35] Zhou Q, Liao B, Lin L, Qiu W, Song Z. Adsorption of Cu (II) and Cd (II) from aqueous solutions by ferromanganese binary oxide–biochar composites. Science of the Total Environment. 2018;**615**:115-122

[36] Varshney P, Mikulic P, Vonshak A, Beardall J, Wangikar PP. Extremophilic micro-algae and their potential contribution in biotechnology. Bioresource Technology. 2015;**184**:363-372

[37] Rodrigues DB, Flores ÉM, Barin JS, Mercadante AZ, Jacob-Lopes E, Zepka LQ. Production of carotenoids from microalgae cultivated using agroindustrial wastes. Food Research International. 2014;**65**:144-148

[38] Veillette M, Chamoumi M, Nikiema J, Faucheux N, Heitz M. Production of biodiesel from microalgae. Advances in Chemical Engineering. 2012;**10**:245-260

[39] Zhao B, Su Y. Process effect of microalgal-carbon dioxide fixation and biomass production: A review. Renewable and Sustainable Energy Reviews. 2014;**31**:121-132

[40] WRAIGHT CA. Current attitudes in photosynthesis research. Photosynthesis: Energy Conversion by Plants and Bacteria. 2012;**1**:17-61

[41] Show K-Y, Lee D-J, Tay J-H, Lee T-M, Chang J-S. Microalgal drying and cell disruption–recent advances. Bioresource Technology. 2015;**184**:258-266

[42] Al-Hothaly KA, Mouradov A, Mansur AA, May BH, Ball AS, Adetutu EM. The effect of media on biomass and oil production in *Botryococcus braunii* strains Kossou-4 and Overjuyo-3. International Journal of Clean Coal and Energy. 2015;**4**:11

[43] Al-Hothaly KA, Adetutu EM, Taha M, Fabbri D, Lorenzetti C, Conti R, et al. Bio-harvesting and pyrolysis of the microalgae *Botryococcus braunii*. Bioresource Technology. 2015;**191**:117-123

[44] Banerjee S, Singh H, Das D, Atta A. Process optimization for enhanced biodiesel production by *Neochloris oleoabundans* UTEX 1185 with concomitant CO2 sequestration. Industrial & Engineering Chemistry Research. 2019;**58**:15760-15771

[45] Tan X-B, Meng J, Tang Z, Yang L-B, Zhang W-W. Optimization of algae mixotrophic culture for nutrients recycling and biomass/lipids production in anaerobically digested waste sludge by various organic acids addition. Chemosphere. 2020;**244**:125509

[46] Reijnders L. Lipid-based liquid biofuels from autotrophic microalgae: Energetic and environmental performance. Wiley Interdisciplinary Reviews: Energy and Environment. 2013;**2**:73-85

[47] Rohwer CM, Maciołek A, Dietrich S, Krüger M. Correlations and forces in sheared fluids with or without quenching. New Journal of Physics. 2019;**21**:073029

[48] Gross M, Rohwer CM, Dietrich S. Dynamics of the critical Casimir force for a conserved order parameter after a critical quench. Physical Review E. 2019;**100**:012114

[49] Junne S, Solymosi T, Oosterhuis N, Neubauer P. Cultivation of cells and microorganisms in wave-mixed disposable bag bioreactors at different scales. Chemie Ingenieur Technik. 2013;**85**:57-66

[50] Rinanti A. Biotechnology carbon capture and storage by microalgae to enhance CO2 removal efficiency in closed-system photobioreactor. Algae-Organisms for imminent biotechnology. 2016:133-156

[51] Deprá MC, Jacob-Lopes E, Zepka LQ. Life cycle assessment of biofuels from microalgae. In: Energy from Microalgae. Switzerland: Springer Nature; 2018. pp. 141-155

[52] Brown RC. Thermochemical Processing of Biomass: Conversion into Fuels, Chemicals and Power. Hoboken, New Jersey: John Wiley & Sons; 2019

[53] Cheirsilp B, Torpee S. Enhanced growth and lipid production of microalgae under mixotrophic culture condition: Effect of light intensity, glucose concentration and fed-batch cultivation. Bioresource Technology. 2012;**110**:510-516

[54] Mata-Alvarez J, Dosta J, Romero-Güiza M, Fonoll X, Peces M, Astals S. A critical review on anaerobic co-digestion achievements between 2010 and 2013. Renewable and Sustainable Energy Reviews. 2014;**36**:412-427

[55] Grando RL, de Souza Antune AM, Da Fonseca FV, Sánchez A, Barrena R, Font X. Technology overview of biogas production in anaerobic digestion plants: A European evaluation of research and development. Renewable and Sustainable Energy Reviews. 2017;**80**:44-53

[56] Bolatkhan K, Kossalbayev BD, Zayadan BK, Tomo T, Veziroglu TN, Allakhverdiev SI. Hydrogen production from phototrophic microorganisms: Reality and perspectives. International Journal of Hydrogen Energy. 2019;**44**:5799-5811

[57] Kim HJ, Park HS, Hyun MS, Chang IS, Kim M, Kim BH. A mediator-less microbial fuel cell using a metal reducing bacterium, *Shewanella putrefaciens*. Enzyme and Microbial Technology. 2002;**30**:145-152

[58] Liu H, Ramnarayanan R, Logan BE. Production of electricity during wastewater treatment using a single chamber microbial fuel cell. Environmental Science & Technology. 2004;**38**:2281-2285

[59] Cai J, Xu D, Dong Z, Yu X, Yang Y, Banks SW, et al. Processing thermogravimetric analysis data for isoconversional kinetic analysis of lignocellulosic biomass pyrolysis: Case study of corn stalk. Renewable and Sustainable Energy Reviews. 2018;**82**:2705-2715

[60] Dhyani V, Bhaskar T. A comprehensive review on the pyrolysis of lignocellulosic biomass. Renewable Energy. 2018;**129**:695-716

[61] Fazlollahi S, Maréchal F. Multi-objective, multi-period optimization of biomass conversion technologies using evolutionary algorithms and mixed integer linear programming (MILP). Applied Thermal Engineering. 2013;**50**:1504-1513

[62] Brown RC. Introduction to thermochemical processing of biomass into fuels, chemicals, and power. Thermochemical Processing of Biomass: Conversion into Fuels, Chemicals and Power. 2011:1-12

[63] Haverly MR, Schulz TC, Whitmer LE, Friend AJ, Funkhouser JM, Smith RG, et al. Continuous solvent liquefaction of biomass in a hydrocarbon solvent. Fuel. 2018;**211**:291-300

[64] Capareda SC. Biomass energy conversion. In: Nayeripour M, Kheshti M, editors. Sustainable Growth and Applications in Renewable Energy Sources. London, UK: IntechOpen; 2011. pp. 209-226

Chapter 24

Recent Advances in Algal Biomass Production

Meghna Rajvanshi and Richard Sayre

Abstract

The promise of algae to address the renewable energy and green-product production demands of the globe has yet to be realized. Over the past ten years, however, there has been a substantial investment and interest in realizing the potential of algae to meet these needs. Tremendous progress has been achieved. Ten years ago, the price of gasoline produced from algal biomass was 20-fold greater than it is today. Technoeconomic models indicate that algal biocrude produced in an optimized cultivation, harvesting, and biomass conversion facility can achieve economic parity with petroleum while reducing carbon-energy indices substantially relative to petroleum-based fuels. There is also an emerging recognition that algal carbon capture and sequestration as lipids may offer a viable alternative to direct atmospheric CO_2 capture and sequestration. We review recent advances in basic and applied algal biomass production from the perspectives of algal biology, cultivation, harvesting, energy conversion, and sustainability. The prognosis is encouraging but will require substantial integration and field testing of a variety of technology platforms to down select the most economical and sustainable systems to address the needs of the circular economy and atmospheric carbon mitigation.

Keywords: algae, biofuel, biomass, carbon sequestration, carbon index

1. Introduction

Over the last ten years (since 2010) there has been accelerated investment in research for the development of commercially viable algal biomass and coproduct production systems [1–5]. The challenge for algae biomass production systems has been that unlike crop biomass production systems having thousands of years of development history, algae until very recently were not the target of integrated research and development (R&D) strategies focused on efficient production of food, fuel, and coproducts [6]. Recent estimates indicate that there are globally more than 150,000 species of single cell and multicellular algae having polyphyletic origins, complex and diverse metabolic machinery, occupying vast environmental niches with immense ranges of biotic and abiotic stress tolerance, and having growth or biomass production rates that range over two magnitudes in yield compared to traditional agricultural production [7]. The challenge for the industry has been to identify the best algal production systems that are suitable for commercially viable industrial applications. Beginning with algal biology much effort has focused on identifying the best performing algal strains. The criteria for down-selecting the best performing strains have included, identifying algae with the greatest biomass production rates, optimizing algal growth media, CO_2 exchange and culture

IntechOpen

conditions, identifying algal strains that are the most resistant to pathogens and herbivory (minimizing pond crashes), and developing strains having enhanced performance characteristics through application of genetic engineering, breeding and genome editing tools [6]. Research and development for improved biomass production has also focused on developing enhanced cultivation, harvesting and biomass conversion technologies with the objective to achieve the lowest carbon emissions, recycle inorganic nutrients as efficiently as possible, minimize energy inputs at each stage in production, and integrate the algal biomass production systems into the existing energy infrastructure as seamlessly as possible.

In 2010, the US Department of Energy launched the largest government-funded integrated algal biomass, biofuels and bioproducts program carried out to date. The National Alliance for Advanced Biofuels and Bioproducts (NAABB) achieved notable advances in reducing the cost of producing biomass and making biofuels from microalgae. In three years NAABB developed and modeled a pathway to move the price point for producing a gallon gasoline equivalent (GGE) of fuel from microalgae from $150 to $8 a GGE [1–3]. More recently, the price point for a GGE produced from algal biomass has been reduced to < $5. Based on Reliance's demonstration scale studies, the technoeconomic modeling (TEM) for a 10 k barrels/ day (bpd) scale production of crude oil from microalgae was estimated to be at 100$/ barrel without any subsidy. The major factors contributing to the substantial cost reductions in producing fuel from algal biomass included, the discovery and development of more robust, high biomass producing algal strains for year-round consistent performance, identification of the best geographies to produce algal biomass, advance pond designs and improved culture mixing for effective light utilization, effective crop control methods that prevent pond crash and biomass loss, innovative harvesting techniques and effective water and nutrient recycling to maximize resource utilization. Also, advancements in biomass to biocrude conversion technologies including continuous flow hydrothermal liquefaction (HTL), the demonstration that algal biocrude coming from HTL could be used as a direct feedstock in existing oil refineries to produce fuels with performance characteristics similar to petroleum-based fuels, and the production of high value coproducts to offset the cost of producing fuels.

Stepping back, however, there remain many critical considerations that must be addressed if microalgal biomass is to be a commercial success in competition with other biomass sources in the world where the carbon energy index (g CO_2 emitted/ kJ energy produced) and the environmental impacts of any biomass production system must also be considered along with economics [6]. Beginning with first principles it is critical to identify what the thermodynamically most efficient biological mechanisms are for producing algal biomass that also have the highest carbon capture efficiency. Recent thermodynamic models suggest that the greatest energy efficiency for carbon capture and biomass production is achieved in algae that utilize light most efficiently and accumulate chemical energy in the form of carbohydrate polymers, e.g., starch rather than those that store oils [8, 9]. Additionally, algae with rapid division rates and/ or the ability to grow substantially in volume are likely to be greater biomass producers [10]. While most algal biofuel programs have focused on producing biomass from high lipid accumulating strains due to ease of conversion of lipids into biocrude it is becoming apparent that algae accumulating starch as a metabolic storage end product have the highest biomass production rates and thermodynamic efficiency [8–10]. While lipids have greater energy density and are more readily converted into fuels, starches have a greater chemical energy density per carbon per photon captured during photosynthesis [8]. One of the microalgal strains achieving the highest known biomass yields in cultivation is *Pseudoneochloris* which stores starch as an energy reserve, achieves high cell numbers at stationary phase of growth, and can increase its cellular volume as it grows by greater than 100-fold [8].

Cultivation systems are also a major cost factor in producing algal biomass. It is generally recognized that to produce low value algal biomass open pond production systems have the lowest capital and operating expenses and require less maintenance (to prevent fouling) than closed bioreactor cultivation systems. However, open pond systems require greater amounts of water to operate due to evaporation, have higher energy costs associated with concentrating more dilute cultures, and are more susceptible to contamination although biological contaminants in closed bioreactors may be more difficult to eradicate.

Regardless of the constraints and challenges mentioned above and the necessity to input higher capital investment in cultivation and downstream processing, production of microalgae biomass still stands out advantageous on many fronts in comparison to agriculture crops for food and fuel. Microalgae have high photosynthetic efficiency and short division time, making them highly suitable candidates for generating more biomass in less time. Growth rates of several microalgae have been reported to be 5–10 times higher than agriculture crops [9]. Moreover, microalgae can grow on low economic and ecological value lands and can utilize marine, brackish or fresh water for cultivation, depending on the species being used. CO_2 from industrial exhaust can be used for cultivation and nutrients from waste streams can be utilized for growth. Excess nutrients lost during harvesting process can be recycled back in the cultivation system, ensuring minimal wastage and maximum utilization [10, 11]. In contrast, agriculture depends on limited natural resources, like arable land and fresh water, with fresh water consumption being highest globally in agriculture. Over 80% of all water consumed globally is used for agricultural production. Agriculture also needs extensive application of fertilizers and pesticides to improve biomass productivity and prevent crop losses. However, nitrogen utilization is inefficient in crop plants, resulting in ~50% of nitrogen loss through leaching, soil erosion and gaseous evaporation [12]. Considering these facts, use of agricultural crops to meet growing biomass demands for food and energy will lead to land use change, environment pollution, loss of forest cover and biodiversity. Thus, from environment standpoint algal cultivation is much favored over traditional agriculture for feedstock production [12, 13].

Many microalgal species are good source of proteins, carbohydrates, lipids and other high value bioactive molecules, such as enzymes, pigments and vitamins. By altering the cultivation conditions or through metabolic engineering approaches, composition of algae can be manipulated to accumulate the specific biomolecule(s) of interest. Considering the higher growth rate and ability to accumulate high lipid content (≥30%), it is reported that microalgae can yield 58,700 L of oil/ ha as opposed to 172 L/ha for corn, 446 L/ha for soybean, 1892 L/ha for Jatropha and 5950 L/ha for oil palm [14]. Thus, the projected ability to produce oil from algae is ~10 times more compared to highest oil producing crop plant. Likewise, algae biomass can be a potential feedstock for bioethanol production because of its ability to accumulate starch even higher than 50% (w/w) of biomass under optimal conditions. Absence of lignin in algal cell wall makes its processing easier compared to lignocellulosic agricultural waste and woody biomass, where lignin removal is an additional step before processing for bioethanol production [13]. Moreover, lack of structural parts like leaves and roots in algae makes algal biomass more homogenous and might be less energy intensive to process compared to crop plants [13]. In an estimate, net energy from sugarcane ethanol and bagasse was 143 GJ/ha/year as opposed to 928 GJ/ha/year from microalgae, indicating microalgae to be significantly more efficient feedstock [15]. Protein is another commercially important component of algae biomass. Algae protein is comparable to other high-quality plant and animal protein sources, however, protein yield from algae happen to be between 4 and 15 tons/ha/year, which is significantly higher than 0.6–1.2 tons/ha/year, reported for soybean [16]. Clearly, microalgae supersede traditional agriculture on multiple aspects, however, biomass harvesting is an area which is well established in case of crop plants but highly energy intensive in case of algae due to its small size and low biomass density [10].

Regarding algal biomass harvesting systems the general objective has been to develop algae harvesting and concentrating systems that have parasitic energy consumption values of less than 10% of the total algal biomass energy content [6]. To reduce the costs of fuel production, recent efforts have focused on the direct conversion of harvested algal biomass into separate fuel and coproduct fractions in a continuous flow system while efficiently recycling water and nutrients. One of the more promising technology developments in this sector has been the development of two-stage HTL which allows for the separate recovery of coproducts and biocrude feedstock while recycling water and nutrients back to the pond thus avoiding the energy intensive step of drying the algal biomass before biomass to fuel conversion. The appropriate selection of what high value coproduct(s) to produce from algal biomass is critical for economic viability when coproduct production is coupled with fuel production. From this perspective the coproduct should have sufficient value based on biomass yields to be economically sustainable without saturating markets to the point of driving coproduct prices so low as to be economically untenable. As modeled by the US-DOE PACE algal biofuels consortium a fully integrated algal cultivation, harvesting, co-product and fuel production system with integrated water and nutrient recycling has the potential to recover over 60% of the energy content of the algae as biocrude while producing valuable coproducts that have a large global market demand (**Figure 1**).

Optimizing algal biomass production and carbon sequestration also has the potential to address the existential threat of global climate change associated with greenhouse gas emissions. Currently, biological carbon capture and sequestration (BCCS) is one of the more feasible means to remediate the earth's atmosphere. As a BCCS system, algae are particularly attractive not only for their high areal

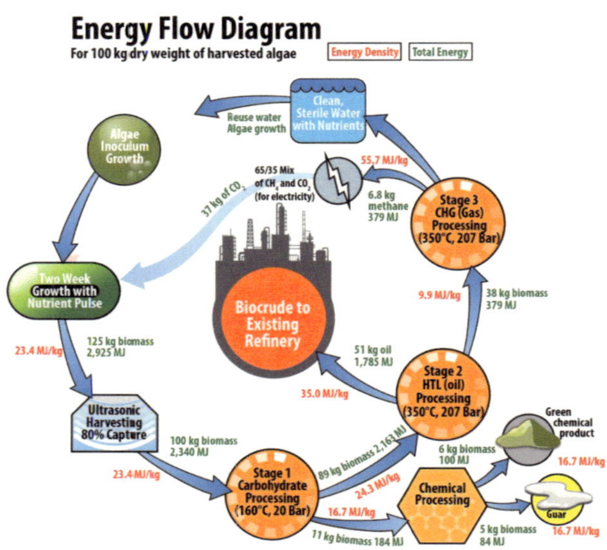

Figure 1.
PACE consortium working model for the integrated co-production of biofuels and co-products (green chemicals, polysaccharides (guar), and methane) from algae. Inorganic nutrients and wastewater are recycled. Algae are preloaded with nutrients (nutrient pulse) and grown in minimal media to reduce weedy species competition and continuously harvested at mid-log phase growth. HTL, hydrothermal liquefaction; CHG, catalytic gasification.

rates of carbon capture but also for their potential storage of carbon as lipids while recycling inorganic nutrients and water [17]. While not generally considered as a carbon sequestration material, lipids have several advantages over solid CO_2 as a carbon sequestration material [17]. Triacylglycerol ($C_{55}H_{98}O_6$) is 77% carbon by mass and has a density of 0.91 g/cm^3. In contrast, CO_2 is 27% carbon by mass and as a solid has a density of 1.96 g/ cm^3. Thus, lipids have a volumetric carbon density that is 32% greater than solid CO_2. Furthermore, being a liquid and not readily convertible to a gas, the ability of lipids to escape from deep geological sequestration is substantially less than CO_2 reducing potential long-term risk to aerobic organisms [17]. Overall, algae have great potential to address simultaneously fuel, food, green chemical, and environmental challenges.

In the following sections we will review recent advances in the sustainable production of algal biomass and coproducts for fuels and economic competitiveness with petroleum and non-algal coproduct production systems. Substantial achievements have been realized from an industry that has a truly short history compared to other biomass production systems.

2. Algal strains

Substantial efforts have focused on the identification of algal strains having maximum biomass yields under cultivation. Ideal biomass production strains must not only have fast growth rates but also must be robust and tolerate well abiotic (temperature, salinity, light) and biotic (pathogen, herbivore and weedy algae) stress conditions to minimize pond crashes and downtime in algal cultivation. There have been several large-scale algal surveys of wild algal species to identify those strains that perform well in cultivation [18]. In addition, screening systems for identifying strains with elevated performance characteristics in high light environments among others have led to some success in the identification of high performing algal strains [19]. Given that there as many as 150,000 species of algae have been identified and that limited resources have been available to screen algae for high biomass production, there remains a significant number of algae that remain to be assessed for biomass productivity in select environments [7]. In addition, substantial potential to improve algal productivity may also be achieved in traditional and molecular assisted breeding practices. Algae breeding efforts, except for laboratory strains such as *Chlamydomonas*, have been limited, however. This is because the means to induce gametogenesis to identify sexual mating types in most algae is not well understood. If the increased yield achieved through plant breeding are to serve as a prognosticator of the potential to enhance algal productivity it can be anticipated that algal breeding programs may enhance yields in the field by as much as ten-fold.

2.1 Modulating cultivation conditions to impact oil and carbohydrate yields

Given the fast rates of cell division and the absence of dedicated higher-order cellular structures including tissue and organs it is not unexpected that microalgae have an enhanced capability to metabolically remodel cellular functions under different growth conditions. Algae frequently live boom and busts cycles in the nutrient deserts of lakes and the open oceans. Thus, it is imperative that algae have flexible metabolic systems to survive in unpredictable and ever-changing environments and be unencumbered by programmed cell fates associated with the differentiation and organization of cells into higher order tissues and organs.

One of the manifestations of this metabolic flexibility is the ability to shift the biochemistry of the major cellular energy storage products from low energy density carbohydrates to high energy density hydrocarbons including triacylglycerol (TAG) and/ or polyterpenoids [20]. The metabolic shift from carbohydrate to hydrocarbon accumulation is typically induced by nutrient depravation. Upon shifting from a nitrogen-, sulfur- and/ or micronutrient-rich condition to a nutrient poor condition many algae will facultatively shift the metabolism of energy storage product accumulation from carbohydrates (starch) to hydrocarbons [21–25]. Hydrocarbons have more than 60% the energy density per fixed carbon of carbohydrates. Importantly, the facultative shift to hydrocarbon production allows algae to continue to generate and utilize reducing energy generated by the photosynthetic apparatus. Significantly, the accumulation of triacylglcerols may not only involve de novo synthesis but the remodeling of existing chloroplast membrane lipids into more fully reduced TAGs [26–29]. Given the desirability of hydrocarbons as a feedstock for biocrude production the ability to shift metabolism from carbohydrate to hydrocarbon production has been exploited to produce hydrocarbon rich biofuel feedstocks. The challenges with this strategy (nutrient deprivation) for facultative hydrocarbon production is that it can also lead to reduced rates of cell division and overall biomass accumulation. In a comprehensive empirical analysis of the impact of nitrogen deprivation on cell division rates, TAG accumulation, lipid remodeling, biomass accumulation and total caloric or biochemical energy accumulation in the green alga, *Chlorella sorokiniana*, it was demonstrated that upon shifting algae to a nitrogen-free growth medium there was a substantial increase in TAG accumulation and a redistribution of total cellular fatty acid profile to more energy dense saturated fatty acids [30]. Under the two-week nitrogen deprivation period employed in this study there was no statistically significant reduction in the rates of cell division or biomass (dry weight) accumulation. However, during the nutrient deprivation period the total chemical energy accumulated in biomass increased by greater than 60% associated with a 20-fold increase in TAG content. It is perhaps surprising that the two-week nitrogen deprivation period did not impair cell division and biomass accumulation suggesting that the alga had the capability to sequester nutrients and/ or catabolize and remodel existing nitrogen rich (proteins) molecules [30]. Not all algal species, however, exhibit similar responses to nutrient deprivation. For many algal species growth rates and biomass accumulation are substantially impaired during nutrient deprived growth conditions [31–33].

An additional practical application of nutrient deprivation for oil production is that growth in nutrient depleted media may reduce competition from weedy algal species [34]. This observation has led to the application of nutrient pulse technology to simultaneously induce oil accumulation during nutrient stress and inhibit the growth of weedy algal species. Under ideal growth conditions the limiting nutrients are withheld until there is an impairment in growth. At this transition point a pulse of the limiting nutrient is added to the growth media to support continued high growth rates [34]. Overall, the ability to induce oil production if managed well can lead to sustained high growth rates while enhancing the energy density of the biomass and the increased accumulation of biofuel feedstocks such as TAGs.

3. Genetic enhancement

Given the aforementioned challenges to breed wild algal strains for improved yield performance traits and the fact that substantial progress has been made in algal genomics and the development of robust genetic transformation systems substantial research efforts have focused on engineering microalgae with improved biomass performance traits. Most algal genetic engineering efforts have focused on the manipulation of metabolic pathways for increased biomass and coproduct production. The production and accumulation of biomass can be broadly divided into four phases known as source (push), sink (pull), storage (accumulate) and turnover (metabolism). Providing an over-riding template on this simplistic model of biomass accumulation is the genetic and developmental control of cell size and cell division or replication rates. Source strength is effectively the primary photosynthetic processes associated with light conversion into chemical energy and the fixation of carbon dioxide into storage products. Sink strength refers to the impact of downstream metabolic processes on biomass accumulation including metabolic feedback control of carbon flux from photosynthesis to production of carbon storage products. The carbon storage products must also be compartmentalized in the cell to support night-time respiration and biomass accumulation. In algae, starch is first primary carbon storage product and is stored in plastids. Algae may also accumulate high energy density hydrocarbons including triacylglycerols or oils. Oil is stored in specialized droplets packaged by outer membranes having surface displayed amphipathic proteins or oil droplet proteins. The extent of accumulation of these storage compartments can be regulated at the level of gene expression and thus is the subject of genetic manipulation impacting overall product yields. However, algal cell division rates and control of cell volume are among the more important determinants of algal biomass production. While many single celled algae have fixed cell volumes that determine the timing of cytokinesis some single celled algae are capable of over 100-fold increases in cell volume as they grow while having variable rates of cell division [35, 36].

In the following paragraphs we focus on progress that has been made at the molecular level to engineer or breed algae with enhanced source and sink strength, increased storage product accumulation, and accelerated cell division rates leading to enhanced yields. As is evident from the success achieved to date two- to five-fold increases in the rate of biomass production and yields are feasible.

3.1 Alterations in source strength

The efficiency of solar energy conversion into chemical energy stored in biomass by plants and algae ranges from 3 to 5% of available solar energy. Theoretically, efficiencies as high as 11% for conversion of solar energy into the chemical energy in biomass can be achieved utilizing just the photosynthetically active radiation (400–700 nm) in the solar spectrum. Maximum efficiencies of energy conversion as high as 30% can be achieved using just red light (~650–700 nm) which is most efficiently harvested by the photosynthetic pigments [8, 37, 38]. Thus, it is conceivable that 2- to 4- fold increases in biomass yields are feasible through improvements in photosynthetic efficiency. It has long been recognized that the greatest potential for increasing photosynthetic efficiency is through enhanced light use efficiency by the photosynthetic apparatus (**Figure 2**) [39–41]. During photosynthesis, light saturates in all plants and algae at approximately one quarter of full sunlight intensity [38, 41]. Thus 75% of the energy captured by the photosynthetic pigments does no productive work leading to biomass production. Since the excess energy captured by the photosynthetic pigments does not drive electron transfer and carbon fixation processes it must dissipate through non-productive energy emission and/ or energy conversion pathways (heat, fluorescence, production of reactive oxygen species (ROS)) some of which (ROS) can lead to substantial damage to the photosynthetic apparatus further reducing biomass yields [42].

One approach to deal with the challenge of excess light absorption by the photosynthetic apparatus has been to reduce the optical cross section of the light-harvesting antenna complex to better couple the rate of light capture with rate-limiting electron transfer processes, i.e., plastohydroquinone oxidation by the

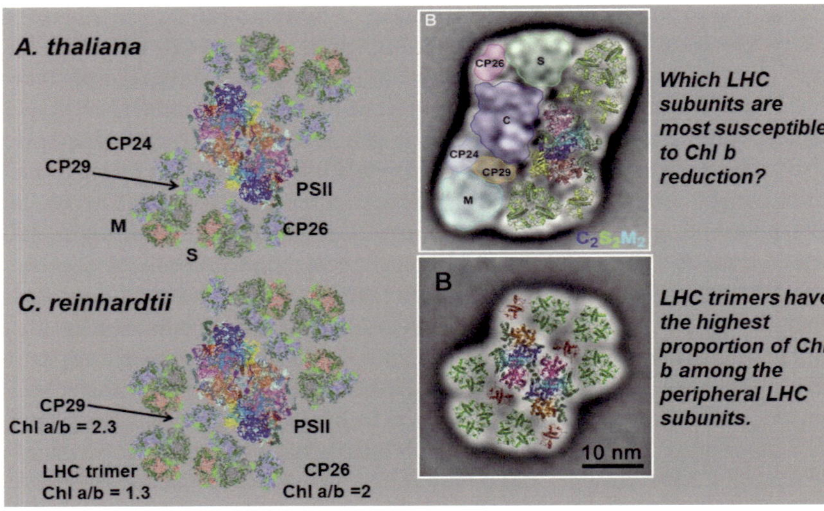

Figure 2.
Organization of the peripheral light harvesting antenna complexes adjacent to the dimeric photosystem II reaction center in plants and green algae. Chlamydomonas transgenics having chlorophyll a/b ratios of 5 have lost the equivalent of one peripheral light-harvesting complex II trimer (LHC trimer). Figure modified from Dr. Jun Minagawa.

cytochrome b6f complex and the development of an electron transport limiting trans-thylakoidal pH gradient [43, 44]. Various strategies have been developed to reduce the size of the light harvesting complex ranging from reducing the expression of the light harvesting complex proteins to targeted reductions in specific light harvesting pigment content often resulting in pleiotropic effects that indirectly affect photosynthetic efficiencies both negatively and positively [41, 45, 46]. Through the analysis of algae having a range in reduction in the light harvesting antenna size it has been empirically determined that the loss of approximately one third of the light harvesting apparatus (LHC2) results in maximum increases in photosynthetic efficiency of 20–30% and increases in biomass yield (40% greater) in both plants and green algae grown under outdoor cultivation conditions (**Figure 2**) [41, 45, 46]. A range in reductions of light harvesting antenna size were achieved by differential expression of the chlorophyllide a oxygenase gene (CAO) which produces chlorophyll b (Chl *b*). Chl *b* is present only in the light harvesting antenna complex proteins and not the photosynthetic reaction center. Since Chl *b* stabilizes the Chl *a/b* binding proteins, its reduction results in a corresponding loss in light harvesting antenna pigment-protein complexes. Significantly, a Chl *a/ b* ratio of 5 has been demonstrated both in plants and green algae to be optimal for achieving the greatest photosynthetic efficiency for plants and algae having altered light harvesting antenna sizes when grown at full sunlight intensity. Lesser or greater reductions in pigment (Chl *b*) content result in less than optimal photosynthetic performance due to indirect effects of Chl *b* reductions on the abundance of select light harvesting pigment-protein complexes, alterations in membrane architecture, reductions in energy transfer processes between the two photosystems, and increased susceptibility to photoinhibition [47, 48].

In nature, however, light intensities vary substantially over the course of the day, with depth in the canopy architecture or algal pond, and seasonally [48]. Theoretically, a light-harvesting apparatus that could be continuously adjusted in size to respond positively to differing light regimes would facilitate greater light use efficiency in dynamic light environments [47]. Recently, Negi et al. (2020) described a strategy for the continuous (daily) adjustment of the light-harvesting antenna size in response to light intensity shifts in the green alga *Chlamydomonas reinhardtii* [47]. This dynamic antenna size regulation system is based on light regulated post transcriptional control of CAO activity. Protochlorophyllide a oxygenase (CAO) catalyzes the synthesis of Chl *b* which is found only in the peripheral, nuclear-encoded light-harvesting pigment-protein complexes. Light intensity-dependent regulation of the light-harvesting complex size was achieved using as a host a CAO minus mutant which had been engineered to express a gene fusion product between the 5′ light regulated element (LRE) and the CAO gene [46]. A light regulated translational repressor, NAB1, binds to the LRE element and at high light represses translation of the modified CAO transcript reducing Chl *b* synthesis and decreasing the light harvesting antenna size. In low light such as occurs in dense cultures CAO translational repression by the NAB1 protein is reduced resulting in increased Chl *b* levels and increased light harvesting antenna size. Significantly, when the LRE-CAO transgenics were grown as monocultures under conditions mimicking those of a commercial production pond the transgenics had biomass yields that were more than two-fold higher than their wild-type parental strains. These are the greatest increases in biomass yield observed to date for algae engineered for improved photosynthetic efficiency.

Significantly, additional enhancements in photosynthetic rate are feasible in algae with optimized light harvesting antenna sizes. When the LRE-CAO transgenics were exposed to elevated bicarbonate concentrations there was an additional 20% increase in photosynthetic rates indicating that improvements in downstream carbon fixation processes could further enhance photosynthetic efficiency and biomass yield [46]. Obviously, elevated chloroplast CO_2 concentrations could potentially suppress RubisCO oxygenase activity and photorespiration [49].

In addition to targeting single gene traits to enhance biomass productivity, engineering strategies based on altering the expression of master growth regulatory genes in algae has proven fruitful for increasing biomass yields. In *Chlamydomonas reinhardtii*, the blue light photoreceptor phototropin (Phot) plays a vital role in progression of the sexual life cycle [50, 51], the control of the eye spot size and light sensitivity and in the control of blue-light mediated changes in the expression of genes involved in the synthesis of chlorophylls, carotenoids, chlorophyll binding proteins [52]. Thus, it was anticipated that Phot expression could potentially play a role in regulating photosynthesis and biomass productivity. Negi et al., tested this hypothesis as well as identified downstream genes in the Phot regulatory pathway that were known to be master regulators of carbohydrate metabolism in plants including analogues of the *Arabidopsis* KIN10 and KIN11 genes [53].

Based on a comparison of the photosynthetic attributes of two independent Phot mutants to their independent parental strains Negi et al., [50] demonstrated that the Chl a/b ratios were significantly greater in Phot mutants (2.9) than in wild type (2.0) grown at low light indicative of a smaller light harvesting antenna size in Phot mutants. When grown at high light intensities there was a further reduction in Chl a/b ratio (3.4) in Phot mutants indicating an ability to reduce the size of the light harvesting antenna grown resulting in increased light use efficiency [50]. The net result was that for Phot mutants photosynthetic rates were light-saturated at intensities 3-fold greater than for wild-type cells resulting in substantially accelerated cell division rates and biomass accumulation. RNAseq experiments indicated that these increases in productivity in Phot mutants were associated alterations in the patterns of expression for genes encoding enzymes involved photosynthesis, carbon metabolism, and those controlling cell division rates. Phot mutants had a 2- to 5-fold increase in the expression levels of multiple rate-limiting enzymes including; the Rieske Fe-S protein, ribulose-1,5-bisphosphate carboxylase/oxygenase, sedoheptulose 1,7 bisphosphatase glyceraldehyde-3- phosphate dehydrogenase, carbonic anhydrase, ADP glucose pyrophosphorylase, starch synthase, and genes involved in respiration and fatty acid biosynthesis. Additionally, genes involved in cell cycle control including; NIMA (never in mitosis), NEK2, NEK6 (NIMA related kinases), RCC1 (regulator of chromosome condensation, cyclin and cyclin-dependent kinases (CDK): Cyclin-dependent kinases, and MAT3 a homolog of retinoblastoma protein (MAT3/RB) were upregulated 2–15-fold in Phot mutants relative to their parental wild-type strains. The net result of this global alteration in gene expression was a two-fold increase in biomass productivity in Phot mutants relative to wild type [50].

Additional improvements in photosynthetic efficiencies have also been achieved by reducing apparent rate limitations in the Calvin–Benson–Bassham cycle (CBBC). Previous studies have demonstrated that the CBBC enzymes, fructose 1,6-bisphosphate aldolase (aldolase), sedoheptulose1,7-bisphosphatase (SBPase), and transketolase (TK), have the highest metabolic flux control coefficient values (maximum 0.55, 0.75, and 1.0, respectively) of any CBBC enzymes and thus have been targets for metabolic engineering to enhance carbon flux and accumulation in engineered plants and algae [54, 55]. Overexpression of the cyanobacterial dual functional fructose 1,6–/sedoheptulose 1,7-bisphosphatase (FBP/SBPase) and/ or plant SBPase was shown to significantly increase photosynthetic rates and growth in transgenic plants or algae [55, 56]. Similar to plants, mutagenesis studies in algae have demonstrated that hexokinase globally regulates genes involved in photosynthesis and hydrocarbon production and similar to Phot mutants can be manipulated to control biomass accumulation [57]. Thus, substantial gains in biomass productivity are feasible through targeted manipulations in both the light reactions and dark (CBBC) reactions of photosynthesis.

3.2 Alterations in carbon sink strength

Given the primary role of starch metabolism as carbon reserve and an intermediate in the production of hydrocarbons it is not unanticipated that alterations in starch metabolism may impact hydrocarbon and biomass yields [58]. For example, *Chlamydomonas* sta6 [ADP-glucose pyrophosphorylase] and sta7–10 [isoamylase] mutants having reduced capacity to synthesize starch had substantial increases in lipid accumulation during nitrogen deprivation relative to the wild-type controls but suppressed total biomass accumulation [58]. In addition, suppression of starch metabolism has been shown to impair upstream CBBC activity resulting in the dissipation of excess photosynthetically produced electrons through non-productive reduction of oxygen [54]. These results point to the central role of starch metabolism and accumulation in overall cellular homeostasis and biomass accumulation in algae and its impact on the thermodynamic efficiency of light energy conversion into chemical energy (biomass) [8, 58]. It has been estimated that carbohydrate metabolism can accounts for as great as 20% reductions in thermodynamic efficiency of photosynthesis [39]. These efficiencies can be further reduced by partitioning carbon into hydrocarbon storage products instead of starch. This is due to the central role of pyruvate (3C) metabolism in hydrocarbon (lipids, terpenes, and waxes) production. The production of acetyl CoA (2C) via the decarboxylation of pyruvate for hydrocarbon production results in the loss of 1/3 of the previously fixed carbon. In contrast, starch production from photosynthetically derived sugars has no associated decarboxylation steps. Hydrocarbons, however, have nearly twice the energy density of carbohydrates due to their more reduced state. Modeling studies indicate that the production of carbohydrates using solar photons is potentially 10–20% more efficient for solar energy conversion than hydrocarbon production [8]. Furthermore, the kinetics of lipid production are substantially slower than starch synthesis. Thus, algae that primarily store starch may accumulate biomass faster than algae that store hydrocarbons as energy reserves. The ecological downside of starch storage, however, is that starch has high volumetric density (1.56 g/cm^3) while lipids have a density of 0.91 g/cm^3 or less than that of water. Thus, algae that store starch must invest energy in motility devices and associated energy expenditures to avoid sinking to depths where light availability may be limiting for photosynthetic growth. It might then be predicted that algae that store starch, e.g., *Chlamydomonas*, predominantly inhabit soil environments that provide physical support whereas lipid accumulating algae, e.g., *Nanochloropsis*, tend to occupy aquatic environments where they are less dense or near the density of water and can remain at levels in the water column where light is not limiting for photosynthesis. To date, the relative energetic costs needed to support motility in starch accumulators versus lipid accumulators remains to be assessed.

3.3 Product storage and metabolism

Following the metabolic engineering paradigm for increasing product yield, i.e., push, pull, sequester and block storage product turnover, less attention has been directed towards the metabolic engineering of storage and product turnover in microalgae. As stated previously, energy reserves in algae fall into two classes, carbohydrates, and lipids. The genetic manipulation of starch accumulation in algae has received much attention. The chloroplast is the site of starch synthesis and storage in plants and algae. In contrast to plant cells, however, microalgae typically have only a single chloroplast per cell since chloroplast division must be synchronized with cell division to ensure that each progeny has a chloroplast [59]. Thus, there is no differentiation of plastids in single-celled microalgae into specialized starch storing amyloplasts as occurs in plants. As a result, increasing starch storage sites is not a viable strategy for increasing starch accumulation. Starch accumulation in a plastid can be genetically manipulated, however. Structurally, starch is composed of two types of glucose polymers, amylose and amylopectin, that differ in their degree of branching. The glucose density of starch granules and their size is controlled by the levels of starch branching and debranching enzyme activities. Genetic manipulations of enzymes controlling starch branching has been shown to substantially impact biomass production [58].

Enhanced lipid storage in microalgae has been achieved by over-expression of enzymes implicated in fatty acid and TAG biosynthesis [60–63], or by repression of lipid catabolism [62, 63]. Additionally, genetic manipulations to decrease starch accumulation also leads to substantial increases in storage lipid accumulation per cell. A *C. reinhardtii* mutant blocked in starch accumulation nearly doubled the amount of lipids accumulated under nitrogen deprivation relative to the control strain, indicating that TAG can act as an alternate sink for excess carbon and photosynthetic reducing equivalents [62]. High energy dense hydrocarbons are primarily stored as TAGs in microalgae and contained in membrane bound lipid droplets. Lipid droplet size and numbers are regulated in part by the production of lipid droplet proteins which are present in the membranes surrounding lipid droplets. Reductions in the expression of major lipid droplet proteins using RNA silencing techniques has been shown to significantly decrease the size of lipid droplets [63]. However, genetic manipulations to increase TAG accumulation by enhancing lipid droplet protein production to our knowledge has not been reported to date. Overall, genetic manipulation of genes controlling select aspects of source, sink, storage, metabolism, and cell growth rates have all proven to enhance biomass yields. Integration of multiple aspects of carbon metabolism, storage and growth leading to enhanced biomass yields have been achieved by alterations in mastery regulatory genes. But much remains to be characterized to achieve maximum thermodynamic efficiency for conversion of photons to the chemical energy of biomass.

4. Cultivation

Cultivation is a vital starting point in algae biomass production and hence choice of production site, strain and cultivation system are very crucial in attaining high biomass productivity. In addition, seasonal influence, crop losses, harvesting processes and nutrient and water recycling are some of the primary governing factors influencing biomass yield and production economics (**Figure 3**). The following section will cover the recent advances in some of the key areas mentioned above.

4.1 Criteria for siting production facilities

First and critical aspect in establishing successful algae cultivation facility is selection of suitable cultivation site. Site selection is quite a complex task and involves considerable attention on terrain, land costs, sunlight availability, seasonal temperatures, proximity to CO_2 and water sources, well-connected transport system, power supply etc. Economical, non-arable flat land with constructible soil is needed for raceway pond installation. Availability of adequate acreage is also an important criterion, as algal cultivation facility should be of scale where production of algae meets economics [64]. Another very important aspect in algal cultivation is availability of enough sunlight. Therefore, it is important to select a geographic location, which is less prone to seasonal variations, receives less rainfall and is climatically suitable to the strain being cultivated. For example, low altitude regions having warm climate and average solar radiation availability for 250 h/month are considered as good sites climatically [65]. CO_2 is regarded as free of cost, but its transportation can add substantial cost to the algae production if the CO_2 generation facility is far from the cultivation site [66]. Water availability is another important criterion. Proximity to sea in case of marine microalgae cultivation and assessment of water scarcity footprint in the region in case of fresh water algae cultivation is essential while selecting a site [67]. Various site selection models that consider parameters, such as soil properties, water availability, growth rate, infrastructure proximity etc. have been reported for identification of a suitable site for algae cultivation [68, 69]. These models can serve as useful tools for algae production site selection.

4.2 Seasonal challenges for biomass yield

Microalgal outdoor cultivation is subjected to diurnal and seasonal variations in temperature, solar irradiance, photoperiod and humidity, which in turn affect physiological responses and biomass yield. For instance, light is essential for photosynthesis, but excess light leads to photoinhibition, oxidative stress, damage of proteins

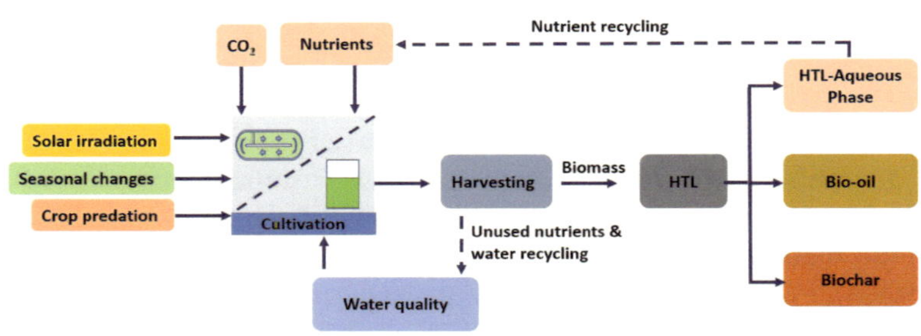

Figure 3.
Factors affecting algae biomass production.

involved in electron transfer and in turn affects CO_2 fixation in photosynthesis and biomass yield. Similarly, low light also reduces photosynthetic efficiency and thus biomass yield [70]. O_2 buildup in culture, which increases from morning till noon also can inhibit growth if O_2 concentration is more than 20 mg/L [71]. Temperature is another important factor, which is affected by light intensity, photoperiod and season. Optimal growth temperature for majority of algal strains lies between 20 and 25°C. However, temperatures above 35°C increases photorespiration, affects nutrient availability, increases the concentration of NH_3 in the medium, decreases CO_2 solubility and increases evaporation losses leading to salinity variations [71, 72]. The impact of these environmental variations is significant on biomass productivity but there are very few reports on quantification of effects of seasonal variations on algae biomass productivities in large scale production systems. For instance, growth performance of *Scenedesmus obtusiusculus* was studied in airlift extended loop photobioreactor operated in outdoor conditions. Yearlong study revealed that biomass productivity was maximum during spring (0.29 g/L/d) where irradiance (2035 μmol/m^2/s) and temperature (11–47°C) were highest, followed by autumn (0.22 g/L/d), summer (0.21 g/L/d) and winter (0.19 g/L/d). However, biomass productivity was much higher (0.97 g/L/d)under optimum laboratory conditions [73]. In another study, *Scenedesmus* sp. cultivated in outdoor pilot scale raceway ponds for waste water treatment resulted in biomass productivities, which ranged from 4 ± 0 g/m^2/d in December when average temperature was 13°C and irradiance was 300 ± 157 w/m^2 and 17 ± 1 g/m^2/d in July when average temperature was 23°C and irradiance was 468 ± 292 w/m^2 [74]. In another study, five microalgal sp. namely, *Chlorella vulgaris*, *Botryococcus braunii*, *Chlamydomonas reinhardtii*, *Euglena gracilis* and *Nannochloropsis oculata* were evaluated in open bioreactors with 30 L capacity in green house conditions. Experiments were conducted from March to April, June to July and Oct to Nov to evaluate growth response of these micro-algae with seasonal variations. *C. vulgaris, B. braunii* resulted in highest growth during month of March and April when average temperature was 28.5°C and irradiance was 15.9 MJ/m^2/d. Whereas, *C. reinhardtii* and *N. oculata* grew best in the month of June when average temperature was 36.1°C and irradiance was 6.6 MJ/m^2/d. while, *E. gracilis* grew comparably during March and June months. In general growth response was low for all five microalgae tested during the month of Oct and Nov, when both temperature and irradiance were low [75].

It is clear from these studies that seasonal variations play significant role in microalgal biomass production and it is important to note that the effect of the environmental changes is strain specific. As complete control of abiotic factors is not possible at large scale outdoor cultivation complete, careful strains selection and adoption of right cultivation practices ensuring effective light and nutrient utilization can help in tackling the seasonal variability to some extent.

4.3 Recycling nutrients

Optimal microalgal growth relies on continuous and adequate supply of nutrients (nitrogen, phosphorous, carbon, potassium, trace elements and water) and sunlight. Nutrient input can be in the form of fertilizer and waste-water streams. Nutrient supply in the form of fertilizers can incur significant cost to the cultivation and is also a competition to fertilizer for agriculture [65]. Therefore, it is important to minimize nutrient losses during cultivation. One way is through stoichiometrically balanced nutrient management to minimize nutrient losses during cultivation [76] and other ways are by recycling of spent medium (water recycle) and nutrient recycling post biomass conversion process.

4.3.1 Water recycling

During growth not, all the nutrients are used completely, and these unused nutrients will be lost if the water is not recycled post harvesting. Water recycling is important not just for nutrient recycling but also from an economics perspective. Water reuse reduces the need to acquire new water for cultivation, thus reducing the water foot print for cultivation and lowering energy usage in pumping water from source to site [77]. There is a finite possibility that water recycling can affect subsequent growth performance of the algae if the recycled water quality does not meet required standards. Primary factors influencing recycled water quality can be increased salinity of the water, use of chemical based harvesting system, accumulation of extracellular metabolites (protein, carbohydrate, fatty acids, nitrogen rich small organic molecules, cell wall debris and other particulate matter) which may be directly inhibit algal growth or increase the dissolved organic carbon (DOC) leading to increased bacterial load and gradual accumulation of toxic metabolites [77–79]. However, multiple studies, both at small and large scale have successfully demonstrated recycling of water without negatively affecting algal growth. Recycled water obtained after electro-flocculant, bio-flocculant, nanno-chitosan, filtration, and centrifugation based harvesting methods had shown no negative effect on the growth of tested algal species [80–84]. Flocculation-based methods have been predicted to be better for water recycling than other methods because they do not lyse the cells and help in reducing dissolved organic matter during harvesting [78]. Farooq et al. (2015) compared chemical flocculation ($FeCl_3$ or alum) of *Chlorella vulgaris* against centrifugation and showed that recycled media obtained after centrifugation or flocculation with $FeCl_3$ had positive effect on growth and lipid productivity. However, recycled medium obtained through treatment with alum even in low dose (<5 ppm) inhibited the growth of *C. vulgaris* due to the toxic effect of residual Al in the recycled water [85]. Similar results were obtained in case of *Scenedesmus* sp., where growth was affected in recycled medium, when alum (1 mM) was used to harvest the cells [84]. Likewise, in another study strain dependent growth inhibition was observed due to accumulated DOC in recycled water. Growth of *Navicula* sp. and *Chlorella* sp. were comparable to fresh medium, while growth of *Staurosira* sp. was completely inhibited in reused medium [79]. It is important to note that stage at which the culture is harvested also affects DOC concentration. Water recycled from exponentially growing cells was found to be more supportive of growth than cells in late log phase or stationary phase, conditions that lead to the maximum accumulation of growth inhibitory substances secreted by algae. As DOC accumulation is more during late log and stationary phases due to the release of secondary metabolites into extracellular space, it is better to avoid recycling water from cultures harvested from these phases [78]. Pretreatment of water before recycling can be considered to improve water quality for long term cultivation with recycled medium. Filtration, high speed centrifugation and sterilization methods have been studied for pretreatment, but their commercial scale application is questionable [77]. In one study, activated carbon was used to process recycled medium to remove humic and fulvic acid like growth inhibitors. This step moderately improved growth of *Nannochloropsis oceanica* in recycled water [86]. Recently, advanced oxidation process has been evaluated for pretreatment of recycled water. It was observed that UV/peroxydisulfate and UV/H_2O_2 processes are quite effective in addressing organic matter load in the water. Oxidation method could degrade and converts inhibitory substances into nutrient source for algal growth. This method helps in utilization of DOC in recycled water rather than its removal [87].

Thus, recycling of spent medium is commercially viable and practically feasible option, which not only helps in saving loss of unused nutrients but also reduces the overall nutrient input.

4.3.2 Nutrient recycling from HTL aqueous phase

Hydrothermal liquification (HTL) is a potential technology to convert wet algal biomass into bio-oil with biochar and aqueous phase (AP) as byproducts. AP is substantial portion because high moisture containing (~10–20% algal slurry) biomass is used as feedstock in HTL [88]. AP is nutritionally rich, containing organic carbon as short chain organic acids, like acetic and propionic acid, nitrogen as NH_4^+, nitrate and other nitrogen containing compounds, phosphorous as orthrophosphates and other macro and micro nutrients [89]. This makes AP a potential nutrients source for microalgae when recycled back into cultivation, which are otherwise lost. It is also reported that even harmful algal blooms are also good feedstock for HTL and AP produced is promising nutrient source for microalgae cultivation [90]. AP also has growth inhibitory compounds like phenols, amides, pyrazines, indole, metal ions like Ni etc., which either must be removed or diluted to the extent that they are no more growth inhibitory [89, 91]. Composition of AP is quite variable and depends on algal feedstock used for HTL, processing parameters, biomass loading and use of AP separation method from bio-oil. For instance, high protein content in feedstock leads to higher organic carbon and nitrogen content in AP [92]. Likewise, increasing resident time in HTL process also has shown to result in increased total nitrogen in the AP. Since, the concentration of nutrients and toxic compounds is often high in AP, substantial dilution of AP is needed to bring concentration of nutrients in the usable range and dilute growth inhibitory toxic elements. There are multiple studies reported where AP is used as sole nutrient source for algal cultivation or a supplement with systematic heavy dilutions made either with water or combination of water and standard nutrient medium. Outcome of these studies is quite variable and was dependent on AP composition and strain being used for cultivation. When AP was used as sole nutrient source, growth of the tested algae was relatively compromised. For instance, AP obtained from *Spirulina* HTL was used as sole nutrient source for cultivating *Chlorella minutissima,* where AP consisted ~16,200 mg/L N and 795 mg/L P along with other nutrients. Biomass productivity obtained was 0.035 g/L/d at 0.2% AP (500X dilution), which was significantly less than BG11 control, having 0.07 g/L/d productivity [91]. Likewise, APs obtained from HTL of *Chlorella vulgaris, Scenedesmus dimorphous* or *Spirulina platensis* as feedstocks were also evaluated as sole nutrient source at various dilutions to grow these stains. Growth of *Chlorella* and *Scenedesmus* was less in comparison to standard medium even at 400X dilution, however, *Spirulina* showed comparable growth in AP and standard medium [93]. Alba et al. (2013), presented comparative account of AP diluted with water versus standard medium for cultivation of *Desmodesmus* sp. A substantial reduction in growth was observed when AP was diluted with water, however, when mixture of water and AP was enriched with standard medium, growth comparative to standard medium was observed. This study clearly indicates that it is not just N and P content that is important for growth but balancing AP in such a way that other macro and micro nutrients are also not limiting is essential for successful use of AP for cultivation [94]. Similar results were obtained in other studies, where AP diluent was enriched with desired nutrients [95–100]. Interestingly, Lopez Barreiro et al. (2015) observed that growth in AP diluted with standard medium was strain dependent. *Nannochloropsis gaditana* and *Chlorella vulgaris* could grow well in AP diluted with standard medium, however, *Phaeodactylum tricornutum* and *Scenedesmus almeriensis* showed poor performance [98]. Apart from deficiency of essential nutrients in AP, other factors which have been reported for inhibited growth are presence of phenolic compounds [91], high Ni concentration [93], NH_3 toxicity [92, 101], limitation of carbon availability and generation of toxic metabolites [102]. HTL technology is evolving to address these issues. In direct HTL at temperature 300°C or above, protein converts into pyrazines, pyrroles and amines, whereas, polysaccharides convert into cyclic ketones and phenols [103]. These non-fuel components lower bio-oil quality, in the process polysaccharides are lost and toxic metabolites are generated and accumulated in AP. To improve quality of bio-oil and prevent loss of polysaccharides, sequential HTL (SEQHTL) is developed, where AP is recovered in first stage of HTL operated at lower temperature (~160°C) [104]. Polysaccharides constitute major portion in the AP from SEQHTL in contrary to AP from direct HTL, where N and P dominate. In nutrient reuse experiments using AP from SEQHTL, it was shown that *Chlorella sorokiniana and Chlorella vulgaris* could utilize 77% and 64% of hydrolyzed polysaccharides, respectively, however, *Galdieria sulphuraria* could not use the polysaccharides from AP, suggesting again that the utilization of nutrients from AP of HTL is strain dependent [88]. Apart from altering HTL conditions and dilution of AP, other ways to reduce toxicity of AP is through removal of toxic substance by absorbents like activated charcoal, zeolite and ion exchange resins. In recent study it was shown that AP treated with ion-exchange resin, Dowex 50WX8 supported the growth of *Chlorella vulgaris* at 100X dilution similar to control medium and better than activated charcoal treated AP [105].

Thus, outcome of multiple studies suggests that for successful utilization of HTL-AP for algal cultivation, selection of right strain is crucial, which can grow mixotrophically and can utilize N as NH_4^+. Appropriate dilution of AP or treatment with absorbents to reduce toxic metabolites load and supplementation with limiting nutrients are also essential for overcoming growth inhibition in AP.

4.4 Pond crashes and mitigation

Large scale algae cultivation ponds and photobioreactors are usually prone to contamination by unwanted foreign organisms due to nonsterile cultivation conditions. Moreover, suboptimal cultivation conditions (light, temperature, nutrients), poor culture mixing, old and sick cells, allow predators and contaminants overtake and crash the culture [106]. Common contaminants in algae cultivation include, grazers (ciliates, rotifers, flagellates, crustaceans, amoeba), pathogens (bacteria and virus) and parasites (fungi, vampyrellids). Multiple studies have reported culture crash due to these organisms. For instance, chytrid contamination in *Haematococcus pluvialis* [107], *Poterioochromonas* sp. (flagellate) [108] and *Euplotes* sp. (ciliate) [109] contamination in *Chlorella,* pleomorphic bacterial (FD111) contamination in *Nannochloropsis* [110], *Colpoda steinii* (ciliate) contamination in *Synechocystis* sp. [111], *Amoeboaphelidium protococcarum* (amoeba) contamination in *Scenedesmus* sp. [112] etc. are some of the studies where contamination resulted in collapse of the culture at mass scales. Since culture crash results in substantial biomass loss, a scalable, environmentally friendly and economical crop control measures are crucial.

Various chemical and physical methods are available for crop protection; however, selection of a method at large scale depends on its activity against predators, non-toxicity towards algae of interest, scalability and cost effectiveness. In case of chemical methods, availability, stability of the chemical and its environmental toxicity should also be considered [109]. Various chemicals belonging to antimicrobials, fungicides, herbicides, oxidants, pesticides, natural compounds, antiparasitic, antifeeding categories have been evaluated to control predators in algae cultivation. Majority of chemicals tested at lab scale are not suitable for large scale operation because of environmental toxicity or they are very expensive for use in algae cultivation. However, copper has been successfully used to selectively control rotifer- *Brachionus calyciflorus* at 1.5 ppm concentration in open pond cultivation of *Chlorella kessleri* [113]. Similarly, sodium hypochlorite (NaOCl) at a dosage of 0.45 to 0.6 mg Cl/L with dosing frequency of every two hours also inhibited predation by *B. calyciflorus* while no growth inhibition was observed in *C. kessleri* [114]. Use of NaOCl might be practically more feasible in open ponds as chlorine dissipates rapidly, leaving no long-lasting residual effects. Moreover, it is effective at lower dosage in comparison to commonly used insecticides Fenitrothion (6.7 mg/L) and Chlorpyrifos (12 mg/L) for controlling *Brachionus calyciflorus* [115, 116]. Recently, Karuppasamy et al. (2018), have screened around 100 chemicals and out of these 21 were effective against *Euplotes* sp. and *Oxyrrhis* sp., and did not have noticeable detrimental effect on *Chlorella vulgaris.* Further, considering cost, availability, stability and effectiveness, benzalkonium chloride (a quaternary amine) at a concentration of 2 mg/L was evaluated and recommended for preventing pond crash [109]. Apart from chemical control, temporary alteration of cultivation conditions has also been reported to be effective in pond crash mitigation. For instance, limitation of P in the medium does not affect algae severely but affects growth of zooplanktons. Slowest zooplankton growth was observed under high light/P ratio [117]. Flynn et al. (2017) also reported through predictive modeling that low level of P stress can be strategically applied to create suboptimal conditions to zooplankton growth without causing detrimental conditions for algal growth [118]. Another potential strategy to control certain type of predators is use of high level of CO_2 in the culture medium. Ma et al. (2017) demonstrated that CO_2 purging temporarily lowered *C. sorokiniana* GT-1 culture pH to 6–6.5 and helped in controlling *Poterioochromonas malhamensis* by lowering its intracellular pH and resulting in cell death. This strategy can be implemented for controlling *P. malhamensis* and other protozoans in large scale cultivation [119]. In addition, CO_2 asphyxiation was found to be effective in causing acute mortality of all zooplankton species in $t < 10$ min [120]. *Poterioochromonas* sp. contamination could also be controlled through cultivation at high pH (>pH 11) as reported in *Synechocystis* sp. PCC 6803 cultures [121].

Apart from chemical methods, there are multiple physical methods, which have been developed for grazer control in algae cultivation. Hydrodynamic cavitation (HC), ultrasonication, foam flotation, pulse electric field, filtration and electromagnetic stratagem are some of the technologies used for crop protection. HC is considered as simple and economical method to kill zooplanktons in waste-water treatment. Kim et al. (2017) have extended this technology in controlling rotifers in algal cultivation. This method could successfully control 99% rotifers in four passes with little effect on *Nannochloropsis* [122]. Likewise, flagellate *Poterioochromonas* sp., a deleterious contaminant in *Chlorella* mass cultivation was disrupted using ultrasonication. This method was tested at 60 L scale and has potential to be used at mass scale. Ultrasonication was also shown to be effective in controlling fungi, amoeba, and ciliates [108]. Electrocution is another technology, which was successfully tested outdoors in 1 and 20 m^2 ponds. Here, 5–10 mA current was applied through graphite rods for 6 h or more to control ciliates and dinoflagellates, however, algae growth was not affected [123]. Pulse electrophoresis is another technology which has been used to effectively control rotifers in tubular PBR. Technology however, can be used for freshwater algal cultures [124]. Umar et al. (2018) evaluated foam flotation, a physiochemical method to remove ciliates *Tetrahymena pyriformis* from *C. vulgaris* culture grown in PBR. Addition of SDS at 40 mg/L concentration lysed ciliates without affecting algal cells [125].

It is clear from the above description that there are multiple methods available to control the crop loss. However, not all methods are equally effective in controlling all types of predators. Therefore, careful selection of a chemical or physical method based on algae and its intended use is needed to prevent the pond crashes or to control the predators without affecting the algal growth.

4.5 Harvesting efficiencies and energy targets

Harvesting and dewatering of microalgae is a very challenging process due to their small cell size (<20 μm), low biomass concentration (0.2–1 g/L in ponds and 2–9 g/L in PBRs) [126], density comparable to water (1.08–1.13 g/mL) and negative charge on algal cells, keeping cells in suspension due to repulsive forces [127]. Common harvesting technologies of microalgae include flocculation, centrifugation, sedimentation, filtration and flotation. These methods can be used individually or in combination to improve the effectiveness and economics of harvesting. For example, flocculation can be combined with sedimentation or dissolved air flotation (DAF), DAF can be combined with filtration or centrifugation. First stage of algae harvesting is generally called primary harvesting process, which concentrates cells up to 2–7% and the second stage is called secondary harvesting or dewatering. It uses primary harvested biomass as feed and further concentrates it up to 15–25% [128]. Fasaei et al. (2018) have discussed 28 combinations of primary and secondary harvesting and recommended filtration followed by centrifugation or flocculation followed by membrane filtration and a finishing step with spiral plate technology or centrifugation as economically attractive solutions. Further, when initial biomass concentration and separation techniques are considered, the estimated operational costs and energy consumption for various harvesting methods were estimated to be in the range of 0.1–2 €/kg and 0.1–5 kWh/kg, respectively. Based on these estimates, harvesting cost was projected to be between 3 and 15% of the production cost, which is significantly lower than the earlier estimate of 20–30%, reported in other studies [129, 130].

Flocculation is most common primary harvesting technique, where cell aggregation is achieved through charge neutralization by cationic flocculants, polymers and metal salts like ferric chloride, alum, aluminum sulfate and ferric sulfate [128]. The flocks formed in association with chemicals are either allowed to settle under gravity in a settling tank or floated by attaching micro-bubbles to their surface using a DAF. Energy consumption range for this process as reviewed by Mo et al. (2015) is 0.1–14.8 kWh/m^3 [131]. Chemical flocculation has resulted in variable outcome as harvesting efficiency of flocculation is dependent on the flocculent dosage, pH of the culture medium, surface charge and salinity. Under optimal conditions, greater than 90% harvesting efficiency was achieved in many studies, for instance, flocculation of *Chlorella sorokiniana, Chaetoceros muelleri, Chlorella vulgaris* and *Scenedesmus costatum* with chitosan [132–134]. Likewise, *Chlorococcum* sp. and *Dunaliella tertiolecta* were harvested with more than 90% harvesting efficiency using $Al_2(SO_4)_3$ or $Fe_2(SO_4)_3$ as flocculants [135]. However, chemical flocculation in large scale algae production may not be economically viable because of high cost of chemical and high dosage requirement. Also, accumulation of residual flocculant in the harvested water and with microalgae might affect the downstream process and may pose environmental concerns [136].

Filtration is another promising harvesting method, which can give 100% biomass recovery and clean biomass, as the process is devoid of chemical input. However, low flux, frequent membrane fouling and high cost of filtration process are key bottlenecks in the large-scale operations. To improve filtration performance and reduce membrane fouling, filtration process has been clubbed with accessory technologies, like aeration [137], vibration [138], use of electro membrane [139] and rotating disk [140]. Bilad et al. (2012) used submerged microfiltration equipped with vibrator for harvesting *Chlorella vulgaris* and *Phaeodactylum tricornutum* and reported energy consumption of 0.27 kWh/m^3 (0.64 kWh/kg) and 0.25 kWh/m^3 (0.98 kWh/kg), respectively [138]. Corresponding energy for electro-coagulation flocculation process is reported to be 1.3–9.5 kWh/m^3 for the same species, which was substantially high [141]. Recently, pilot scale ultra-filtration membrane trial clubbed with air assisted backwashing technology has been successfully used to harvest *Scenedesmus acuminatus*. The culture was concentrated from 0.5 g/L to 136 g/L with 93% biomass recovery. The energy consumption reported was 0.59 kWh/kg dry biomass [142]. Though filtration is less energy intensive [130], further improvements in filtration technology is required and can also be achieved by using membranes with advanced hydrophilic material and introducing negative surface charges [126].

Centrifugation is another physical method of harvesting, but the harvesting efficiency is less than filtration and highly depends on the gravitational forced applied. Centrifugation is also energy intensive, difficult to scaleup, requires high maintenance and considered expensive for low value products like oil. Using centrifugation as sole harvesting method is not recommended as energy consumption and cost of harvesting is significantly higher compared to a process, where centrifugation is used as secondary harvesting method. In a study where Evodos spiral plate centrifuge was solely used to harvest 10,000 L of *Chlorella* culture, energy consumption was 55 kWh/m^3, as opposed to 5.5 kWh/m^3, when centrifugation was used as secondary harvesting step [128]. Other common centrifuge types are disc stack and decanter. Disc stack is the most common industrial centrifuge with reported energy consumption ~1 kWh/m^3. However, energy consumption was further reduced to 50% by design changes, like modifying flow paths of rotor, reduction of aerodynamic losses by air removal outside rotor and use of direct drive instead of belt or gear drive [143]. In case of decanter centrifuge, energy consumption ranged between 1.3–8 kWh/m^3. In another study by National Renewable Energy Laboratory (NREL), energy consumption in concentrating microalgae from 13–20% using centrifuge was estimated to be 1.3 kWh/m^3, with a dewatering efficiency of 97% [144].

In conclusion, it is clear from above description that significant developments are made in harvesting technology but none of the techniques seems to be economical and efficient enough. Combination of two to three technologies have been proposed to give economically viable solution but still significant optimization and innovation is necessary in current technologies and there is substantial scope for development of new, cheaper and more efficient harvesting technologies.

5. Commercial scale up

High cost of biomass production and subsequent extraction processes have limited the progress of upscaling of microalgae for commercial fuel and other value-added products. The technoeconomic analyses reported thus far have a wide variation in the cost estimates, primarily due to non-existence of standardized cost assumptions across different geographic locations. For example, in a study conducted in the US, production of microalgal biomass is estimated at $4.92/kg with current technology status [145]. In another study conducted in Europe, production cost was estimated to be €4.95, 4.16, and 5.96/kg of biomass from open ponds, horizontal tubular and flat panel photobioreactors, respectively [146]. Even the biomass production cost drops down to $0.5/ kg, still scaling-up of microalgae for standalone production of biofuel is economically infeasible due to swift competition with fossil fuel [145]. Hence, cost reduction and integration of additional revenue generation steps could help in successful scale-up.

While microalgae are primarily sought-after for biodiesel production through utilization of lipids, valorization of other components through a biorefinery approach, as proposed in many studies might enhance the chances of commercialization. Microalgae are traditionally utilized for food and feed, cosmetics, nutraceutical and pharmaceutical applications because of the presence of high content of protein, carbohydrate, pigments, antioxidants, ω-3 fatty acids and other industrially important chemicals. Extraction of these compounds as co or byproducts can improve the overall process economics [147, 148]. In microalgal biorefinery, valorization of different components of microalgae biomass is achieved through a series of unit operations for extraction, purification and biomass conversion [149]. Based on the type of the primary product being extracted, biorefineries can be classified as energy driven or material driven biorefinery. In energy driven biorefinery, oil for biofuel is extracted first, and the de-oiled biomass is used for extraction of value-added products or in a bioconversion processes like fermentation, anaerobic digestion, pyrolysis, hydrothermal liquification (HTL) etc. The best possible sequence of extraction of compounds for valorization of biomass can be evaluated through cost effectiveness assessment (CEA), which is the ratio of total outcomes from a biorefinery to the total cost of producing products [150]. Also, for successful biorefinery scheme, the net energy ratio (NER) assessment is important. It is the ratio of energy output over energy input and should be greater than unity. Higher the values of CEA and NER are, higher would be the feasibility of that biorefinery scheme [148, 150].

Several microalgae biorefineries have been proposed and tested in the literature but their implementation at large scale is still far from reality. **Table 1** summarizes some of the recent biorefinery approaches reported in the literature and **Figure 4** represents various possible biorefinery approaches. Razon and Tan (2011) evaluated a biorefinery for production of biodiesel and biogas from *Haematococcus pluvialis* and *Nannochloropsis* [161]. The NER was less than one for both the cases indicating negative energy balance, even when best performance estimates were taken for unit operations. However, economics of the system can be improved if cultivation is integrated with waste-water plant, thus eliminating the need for chemical fertilizers. Also, wet extraction should be followed thus saving on drying cost [161]. Similarly, Andersson et al. (2014) evaluated the biodiesel and biogas production through integration of cultivation with waste-water treatment plant, flue gas as carbon source and excess heat from industrial cluster. Production of biodiesel and biogas in biorefinery scheme resulted in net positive outcome compared to biogas alone [162]. In another biorefinery approach, Ansari et al. (2015) evaluated lipid extracted algae (LEA) for its use as protein or reducing sugar source and observed comparable yields of these products from whole algae and LEA as source material. Also, oven drying over sun drying and microwave assisted lipid extraction resulted in highest lipid yield compared to other methods tested [154]. In another study first protein was recovered from *Botryococcus braunii* under alkaline conditions, followed by lipid extraction for biodiesel and finally spent biomass was used for bio-oil production through pyrolysis. This biorefinery process resulted in 10% protein recovery, 2% lipid recovery and 33% bio-oil recovery. Bio-oil being obtained in this scheme was at neutral pH and hence non-corrosive for combustion engines but bio-oil recovery from spent biomass was less than that of whole algae. Due to poor recovery of lipids, their extraction step can be omitted and scheme can be simplified [153]. Recently, in *Chlorella vulgaris* biorefinery, where protein extraction was integrated with pyrolysis, extraction under alkaline condition (12 pH) at

Microalgae	Biorefinery products	Remarks	Reference
Nannochloropsis sp.	Biodiesel; Carotenoids; Bio hydrogen	CO_2 super critical fluid extraction plus ethanol (20 wt.%) could extract 45% (dry weight basis) of lipids and recover 70% of the pigments Dark fermentation of left-over biomass by *E. aerogenes* yielded maximum 60.6 mL H_2/g Dry biomass of alga.	[151]
Dunaliella tertiolecta	Biodiesel; Bioethanol	Enzymatic saccharification of de-oiled biomass followed by fermentation resulted in the yield of 0.14 g ethanol/g residual biomass equivalent of 82% of the theoretical fermentation yield.	[152]
Botryococcus braunii	Protein; lipid; bio-oil	Protein extraction followed by bio-oil recommended. Neutral pH was found in bio-oil from microalgal biomass.	[153]
Scenedesmus obliquus	Lipid; Protein/ Reducing sugars	Microwave assisted extraction from oven dried samples provided highest lipid yield. Protein and reducing sugar yield comparable in lipid extracted algae vs. whole algae. Sun drying resulted in poor outcome.	[154]
Nannochloropsis sp.	Lipid; fuel gases; Nitrogen as NH_4^+	75% recovery of energy in SCWG process and 100% recovery of N from lipid extracted hydrochar.	[155]
Spirogyra sp.	Carotenoids; Biohydrogen	Electrocoagulation and solar drying reduced the energy requirements by 90% for harvesting and dewatering. 0.12 g/100 g dry biomass of total pigments with 56% free astaxanthin, 16% beta-carotene & 5% of lutein and canthaxanthin. Fermentation of residual biomass produced hydrogen yield of 47 mL/g d.w. Carotenoid extraction with acetone is expensive and hydrogen yields have to improve by increasing sugar content in the biomass through altered cultivation practices.	[156]
Scenedesmus acutus	Bioethanol; biodiesel	Whole algal slurry after acid pretreatment is directly used for ethanol fermentation. No losses of fermentable sugars in the solids, which are otherwise separated from the sugar rich supernatant. $0.95/ GGE cost reduction in biofuel production.	[157]
Phaeodactylum tricornutum	Pigments; Fatty acids	Green processes: pressurized liquid extraction (PLE) and microwave-assisted solvent extraction (MAE) were evaluated for extraction of bioactive compounds. Optimum extraction conditions were 50°C, 100% EtOH, 20 min for PLE, while optimum conditions for MAE were 30°C, 100% EtOH and 2 min. Higher recovery of fucoxanthin enriched with EPA were obtained with PLE method.	[158]

Microalgae	Biorefinery products	Remarks	Reference
Chlorella vulgaris	Protein; bio-oil	Hydrolysis with sonication under alkaline conditions yielded high protein recoveries. Scheme is economically feasible if extracted protein is used for food application. Profit is 1.51 $/Kg of microalgae biomass	[159]
Chlorella vulgaris	Lutein; Protein	Pulse electric field treatment enhanced the lutein (2.2 ± 0.1-fold) and chlorophyll yields (5.2 ± 3.4-fold) compared to non-treated cells single-stage ethanol extraction process. Protein extraction cost estimated to be US$4.16/kg of protein with 50% extraction yield and 57% purity. Further improvement in yield and purity is needed to make this biorefinery economically viable.	[160]

All the studies mentioned are conducted at lab scale.

Table 1.
Experimental demonstration of microalgal biorefinery approaches.

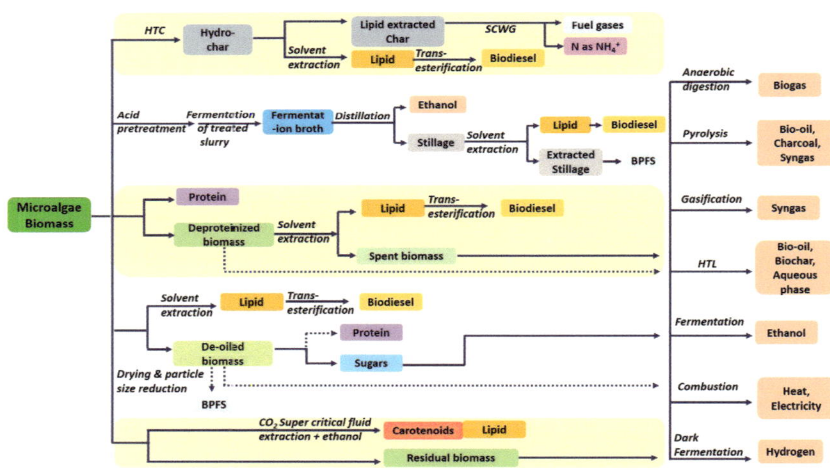

Figure 4.
Possible microalgal biorefinery approaches (dotted line: Alternate route, HTL: Hydrothermal liquification, HTC: Hydrothermal carbonization, SCWG: Super critical water gasification, BPFS: Bioplastic feed stock, stillage: Fermentation broth after removal of ethanol, extracted stillage: Broth after extraction of ethanol and lipids, spent biomass: Deproteinized and de-oiled biomass).

50°C for 90 min followed by sonication resulted in 80% protein recovery. Bio-oil obtained from protein extracted biomass was better in quality and comparable in quantity with whole algal biomass extraction. Based on technoeconomic analysis, it was proposed that the extracted protein if used for food application then the profit can increase by 1.51 USD/ kg of microalgae biomass [159]. Lu and Savage (2015) processed *Nannochloropsis* slurry through hydrothermal carbonization (HTC) and resultant hydrochar was further used for lipid extraction. Lipid extracted residual char was converted into fuel gases through a process called super critical water gasification and almost complete recovery of N as NH_4^+ was achieved, which can be used for nutrient recycling. This scheme is attractive as multiple products can be extracted efficiently and use of HTC process in the beginning eliminates the energy intensive step of drying for lipid extraction [155]. Dong et al. (2016) tested another biorefinery approach, where *Scenedesmus acutus* slurry was subjected to fermentation after acid pretreatment at 155°C for 15 min. Ethanol was recovered through distillation of fermentation broth and lipids were recovered from stillage by solvent extraction. Advantage of this approach was that monomeric sugar was fully utilized in the fermentation process as sugar-rich liquor was not separated from solid residues post pretreatment. Also, using whole cell algal slurry (post pretreatment) for fermentation resulted in microalgal biofuel cost reduction by $0.95/GGE [157]. In another study, production of bioplastic feed stock (BPFS) and biofuel were integrated in algal biorefinery. Open raceway pond (ORP) cultivation followed by utilization of dried biomass as BPFS was found to be the most economical with minimum selling price (MSP) estimated to be $970/ ton. Other scenarios, were, lipid extraction or fractionation prior to use of biomass as BPFS. These biorefineries with an estimate of MSP of $1370 and $1460/ ton of lipid extracted or fractionated biomass, respectively, were proposed to be competitive if cultivation cost is reduced [163].

Though biorefinery concept gives greater product and economic flexibility, the technologies needed for processing of residual streams of microalgae biomass are still in nascent stages of development and hence many biorefinery models are faced with technoeconomic hurdles. Cultivation, harvesting and drying are highly cost and energy intensive steps and needs substantial innovations and advancements to improve economics. Economics of downstream processing steps, which include cell disruption, extraction, purification and biomass conversion are not thoroughly assessed and reported, moreover, technology for multiproduct extraction is neither fully mature nor evaluated at large scale [164]. The economic analyses reported on biorefineries thus far are mostly based on small scale studies and limited knowledge on end-to-end biorefinery trials at large scale, affects the reliability of economic analysis [162]. Other significant challenges in successful implementation of microalgae biorefineries are; consistent availability of algal biomass, variation in microalgal composition based on cultivation conditions and strain specificity. Therefore, adequate control of cultivation parameters and selection of appropriate strain is important. When biorefinery products are intended for food industry, then the production process from cultivation to final product should adhere to regulations set by regulatory agencies in respective geographic locations [149]. Product stability is another key challenge and must be ensured throughout the storage period.

In conclusion, though microalgae are an excellent feedstock for implementation of biorefinery approaches, a concerted effort is still needed to make the production process economically viable and environmentally sustainable.

6. Conclusions

6.1 Will algal biomass production ever be economically viable?

Though microalgae technologies have evolved tremendously in the past decade and have shown greater promise as renewable feedstocks for food, fuel and other high value products, their commercial scale production is still in its infancy. Companies like Sapphire Energy, Aurora Biofuels, Solazyme, and Algenol started with the aim of producing biofuel from algae at a large scale but could not sustain their operations due to economic infeasibility. Some companies have stopped the operations, while others changed their focus to produce algae for food or other non-fuel products. Considering the technoeconomic analysis of fuel production from microalgae, production of algae for food, nutraceutical, cosmetics etc. has higher chances of success, as these products provide lot of opportunities to innovate and higher value of these compounds in comparison to fuel can fetch higher returns on investments. However, it must be noted that the market for these products is either substantially small or still in early stages of evolution. Moreover, availability of several cheap alternatives and lack of awareness among people about algae products are also critical stumbling blocks in market acceptability of algal products.

To bring microalgae production into mainstream both cost and market awareness must be improved. Integrated biorefinery approaches, discussed in detail in previous section, can be a viable option in this direction if technological and financial challenges are overcome. For that, focused research in both fundamental and applied areas to bridge the gap between lab to field translatability is imperative. Understanding biology for high biomass production and tweaking production strains through mutation, genetic and metabolic engineering approaches to increase the efficiency of accumulating desirable products and building the capability to withstand biotic and abiotic stresses would be a step towards success of commercial scale algal biomass production. In parallel, optimization of unit operations in cultivation, harvesting and downstream processing by improving their efficiency, lowering cost and finally integrating biological and engineering systems to ultimately develop economically viable end-to-end process is crucial for success. Lastly, government support in terms of well-defined policy, setting clear renewable energy targets, funding and subsidies on environmentally sustainable technologies would be a strong push in making algal biomass production at commercial scale a reality.

6.2 Next generation systems

It is clear from the discussion above that substantial improvements are needed in multiple processes of algal biomass production. Next generation systems should focus on improving pond design and better hydrodynamics, which can enhance fluid mixing and minimize dead zones resulting in improved biomass productivity, reduction in contaminant growth and pond crashes. Pond design should also support improved light and dark cycle leading to better light utilization, thus enhancing biomass productivity. Cost reduction through innovative low-cost pond lining is another important focus area for next generation systems. Development of efficient and inexpensive CO_2 delivery systems, where CO_2 wastage can also be minimized is an area of active research and such novel delivery methods should be part of next generation systems. Harvesting incurs significant cost to the algal biomass production, hence, combining two or more harvesting strategies and identifying coagulation, flocculation and dewatering chemical recipes that also can work effectively under saline conditions for microscopic algae will add in improving economics of biomass production. Strain modification and developing robust strains should also be the focus area of next generation systems. One example is propiconazole resistant *Chlorella* strain developed through mutagenesis, also harbors trait of high temperature tolerance. These two traits make the strain apt for cultivation in outdoor conditions [165].

6.3 Biological carbon capture and sequestration

There is growing recognition that the greatest existential threat facing the planet is anthropomorphic climate change. There is growing evidence that reductions in carbon emissions may not be sufficient to push global temperatures beyond a tipping point that would lead to an inhabitable planet for much of life as we know it today. Perhaps the greatest irony is that the geological sequestration of microalgal biocrudes may be one of the most efficient and sustainable means to sequester atmospheric carbon [35, 166, 167]. Instead of extracting non-renewable petroleum (ancient algal biomass) from the earth it may become necessary to sequester atmospheric carbon by returning algal biocrude to the earth perhaps through the same pumps and wells that were used to extract petroleum. Carbon capture by algae is sustainable given efficient recycling of water and nutrients. The major concern is public inertia to mitigate carbon and economics. The costs associated with algal biocrude or carbon sequestration may be attractive. The economics of algal biocrude sequestration can be offset in part by the co-production of high volume/ low value animal feeds (proteins and carbohydrates) and the production of high value commodities minimizing the need for governmental financial support of atmospheric carbon mitigation technologies. To date, an algal BCCS system linked with food and valuable coproduct production has not been modeled for carbon capture efficiency and costs. The challenge for the next generation of algal scientists and economists is to consider whether algal BCCS is a workable solution to mitigate atmospheric carbon and address the looming specter of climate change.

Acknowledgements

Richard Sayre acknowledges financial support from the New Mexico Consortium to write this article. Meghna Rajvanshi acknowledges Dr. Santanu Dasgupta and Dr. Ajit Sapre from Reliance Industries Ltd., India for providing the opportunity and support in the preparation of this chapter.

Author details

Meghna Rajvanshi[1] and Richard Sayre[2*]

1 Reliance Industries Limited, Mumbai, India

2 New Mexico Consortium, Los Alamos, USA

*Address all correspondence to: rsayre@newmexicoconsortium.org

References

[1] Unkefer CJ, Sayre RT, Magnuson JK, Anderson DB, Baxter I, Blaby IK, et al. Review of the algal biology program within the National Alliance for advanced biofuels and bioproducts. Algal Research. 2017;**22**:187-215

[2] Lammers PJ, Huesemann M, Boeing W, Anderson DB, Arnold RG, Bai X, et al. Review of the cultivation program within the National Alliance for advanced biofuels and bioproducts. Algal Research. 2017;**22**:166-186

[3] Marrone BL, Lacey RE, Anderson DB, Bonner J, Coons J, Dale T, et al. Review of the harvesting and extraction program within the National Alliance for advanced biofuels and bioproducts. Algal Research. 2018;**33**:470-485

[4] Benedetti M, Vecchi V, Barera S, Dall'Osto L. Biomass from microalgae: The potential of domestication towards sustainable biofactories. Microbial Cell Factories. 2018;**17**(1):173

[5] Fabris M, Abbriano RM, Pernice M, Sutherland DL, Commault AS, Hall CC, et al. Emerging Technologies in Algal Biotechnology: Toward the establishment of a sustainable, algae-based bioeconomy. Frontiers in Plant Science. 2020;**11**:279

[6] Barry A, Wolfe A, English C, Ruddick C, Lambert D. 2016 National Algal Biofuels Technology Review; USDOE Office of Energy Efficiency and Renewable Energy (EERE), Bioenergy Technologies Office (EE-3B), 2016 DOE/EE-1409; Other: 7421 United States 10.2172/1259407 Other: 7421 EE-LIBRARY English.

[7] AlgaeBase is a global algal database of taxonomic, nomenclatural and distributional information. https://www.algaebase.org/.

[8] Subramanian S, Barry AN, Pieris S, Sayre RT. Comparative energetics and kinetics of autotrophic lipid and starch metabolism in chlorophytic microalgae: Implications for biomass and biofuel production. Biotechnology for Biofuels. 2013;**6**(1):150

[9] Zullaikah S, Utomo AT, Yasmin M, Ong LK, Ju YH. 9. Ecofuel conversion technology of inedible lipid feedstocks to renewable fuel. In: Azad K, editor. Advances in Eco-Fuels for a Sustainable Environment: Woodhead Publishing. 2019. pp. 237-276

[10] Pienkos PT, Darzins A. The promise and challenges of microalgal-derived biofuels. Biofuels, Bioproducts and Biorefining. 2009;**3**(4):431-440

[11] Rösch C, Roßmann M, Weickert S. Microalgae for integrated food and fuel production. GCB Bioenergy. 2019;**11**(1):326-334

[12] Campbell BM, Beare DJ, Bennett EM, Hall-Spencer JM, Ingram JSI, Jaramillo F, et al. Agriculture production as a major driver of the earth system exceeding planetary boundaries. Ecology and Society. 2017;**22**:4

[13] Van Iersel S. Algae-Based Biofuels. A Review of Challenges and Opportunities for Developing Countries. Rome (Italy): FAO; Global Bioenergy Partnership GBEP, c/o Environment, Climate Change and Bioenergy Division, Food and Agriculture Organization of the United Nations FAO, Viale delle Terme di Caracalla, 00153 Rome (Italy); Ecofys, Utrecht (Netherlands); 2009. Medium: X; Size: 60 pages p

[14] Chisti Y. Biodiesel from microalgae. Biotechnology advances. 2007;**25**(3):294-306

[15] Chisti Y. Response to Reijnders: Do biofuels from microalgae beat biofuels

from terrestrial plants? Trends in Biotechnology. 2008;**26**(7):351-352

[16] MMv K, Bikker P, IMvd M, CMCvd P-S, Vereijken JM. Cultivation, Processing and Nutritional Aspects for Pigs and Poultry of European Protein Sources as Alternatives for Imported Soybean Products. Lelystad: Wageningen UR Livestock Research; 2013. pp. 1570-8616

[17] Sayre R. Microalgae: The potential for carbon capture. Bioscience. 2010;**60**(9):722-727

[18] Neofotis P, Huang A, Sury K, Chang W, Joseph F, Gabr A, et al. Characterization and classification of highly productive microalgae strains discovered for biofuel and bioproduct generation. Algal Research. 2016;**15**:164-178

[19] Weissman JC, Likhogrud M, Thomas DC, Fang W, Karns DAJ, Chung JW, et al. High-light selection produces a fast-growing *Picochlorum celeri*. Algal Research. 2018;**36**:17-28

[20] Breuer G, Lamers PP, Martens DE, Draaisma RB, Wijffels RH. The impact of nitrogen starvation on the dynamics of triacylglycerol accumulation in nine microalgae strains. Bioresource Technology. 2012;**124**:217-226

[21] Cakmak T, Angun P, Demiray YE, Ozkan AD, Elibol Z, Tekinay T. Differential effects of nitrogen and sulfur deprivation on growth and biodiesel feedstock production of *Chlamydomonas reinhardtii*. Biotechnology and Bioengineering. 2012;**109**(8):1947-1957

[22] Gong Y, Guo X, Wan X, Liang Z, Jiang M. Triacylglycerol accumulation and change in fatty acid content of four marine oleaginous microalgae under nutrient limitation and at different culture ages. Journal of Basic Microbiology. 2013;**53**(1):29-36

[23] Lv J-M, Cheng L-H, Xu X-H, Zhang L, Chen H-L. Enhanced lipid production of *Chlorella vulgaris* by adjustment of cultivation conditions. Bioresource Technology. 2010;**101**(17):6797-6804

[24] Rodolfi L, Chini Zittelli G, Bassi N, Padovani G, Biondi N, Bonini G, et al. Microalgae for oil: Strain selection, induction of lipid synthesis and outdoor mass cultivation in a low-cost photobioreactor. Biotechnology and Bioengineering. 2009;**102**(1):100-112

[25] Siaut M, Cuiné S, Cagnon C, Fessler B, Nguyen M, Carrier P, et al. Oil accumulation in the model green alga *Chlamydomonas reinhardtii*: Characterization, variability between common laboratory strains and relationship with starch reserves. BMC Biotechnology. 2011;**11**(1):7

[26] Benning C. Mechanisms of lipid transport involved in organelle biogenesis in plant cells. Annual Review of Cell and Developmental Biology. 2009;**25**(1):71-91

[27] Fan J, Andre C, Xu C. A chloroplast pathway for the de novo biosynthesis of triacylglycerol in *Chlamydomonas reinhardtii*. FEBS Letters. 2011;**585**(12): 1985-1991 Epub 2011/05/18

[28] Martin GJO, Hill DRA, Olmstead ILD, Bergamin A, Shears MJ, Dias DA, et al. Lipid profile remodeling in response to nitrogen deprivation in the microalgae *Chlorella* sp. (Trebouxiophyceae) and Nannochloropsis sp. (Eustigmatophyceae). PLoS ONE. 2014;**9**(8):e103389-e

[29] Lourenço SO, Barbarino E, Marquez UML, Aidar E. Distribution of intracellular nitrogen in marine microalgae: Basis for the calculation of specific nitrogen-to-protein conversion factors. Journal of Phycology. 1998;**34**(5):798-811

[30] Negi S, Barry AN, Friedland N, Sudasinghe N, Subramanian S, Pieris S, et al. Impact of nitrogen limitation on biomass, photosynthesis, and lipid accumulation in *Chlorella sorokiniana*. Journal of Applied Phycology. 2016;**28**(2):803-812

[31] Spolaore P, Joannis-Cassan C, Duran E, Isambert A. Commercial applications of microalgae. Journal of Bioscience and Bioengineering. 2006;**101**(2):87-96

[32] Huerlimann R, de Nys R, Heimann K. Growth, lipid content, productivity, and fatty acid composition of tropical microalgae for scale-up production. Biotechnology and Bioengineering. 2010;**107**(2):245-257

[33] Sharma KK, Schuhmann H, Schenk PM. High lipid induction in microalgae for biodiesel production. Energies. 2012;**5**(5):1532-1553

[34] Shi T-Q, Wang L-R, Zhang Z-X, Sun X-M, Huang H. Stresses as first-line tools for enhancing lipid and carotenoid production in microalgae. Frontiers in bioengineering and biotechnology. 2020;**8**:610

[35] Aketo T, Hashizume R, Yabu Y, Hoshikawa Y, Nojima D, Maeda Y, et al. Characterization of a novel marine unicellular alga, *Pseudoneochloris* sp. strain NKY372003 as a high carbohydrate producer. Journal of Bioscience and Bioengineering. 2020;**129**(6):687-692

[36] Matsuo T, Iida T, Ohmura A, Gururaj M, Kato D, Mutoh R, et al. The role of ROC75 as a daytime component of the circadian oscillator in *Chlamydomonas reinhardtii*. PLoS Genet. 2020;**16**(6):e1008814-e

[37] Okada K, Fujiwara S, Tsuzuki M. Energy conservation in photosynthetic microorganisms. The Journal of General and Applied Microbiology. 2020;**66**(2):59-65

[38] Blankenship RE, Tiede DM, Barber J, Brudvig GW, Fleming G, Ghirardi M, et al. Comparing photosynthetic and photovoltaic efficiencies and recognizing the potential for improvement. Science. 2011;**332**(6031):805-809

[39] Zhu X-G, Song Q, Ort DR. Elements of a dynamic systems model of canopy photosynthesis. Current Opinion in Plant Biology. 2012;**15**(3):237-244

[40] Ort DR, Merchant SS, Alric J, Barkan A, Blankenship RE, Bock R, et al. Redesigning photosynthesis to sustainably meet global food and bioenergy demand. Proceedings of the National Academy of Sciences. 2015;**112**(28):8529-8536

[41] Perrine Z, Negi S, Sayre RT. Optimization of photosynthetic light energy utilization by microalgae. Algal Research. 2012;**1**(2):134-142

[42] Demmig-Adams B, Garab G, Iii W, Govindjee. Non-photochemical quenching and energy dissipation in plants, algae and cyanobacteria 2014.

[43] Kiirats O, Cruz J, Edwards G, Kramer D. Feedback limitation of photosynthesis at high CO_2 acts by modulating the activity of the chloroplast ATP synthase. Functional Plant Biology. 2009;**36**(10-1):893-901

[44] Baker NR, Harbinson J, Kramer DM. Determining the limitations and regulation of photosynthetic energy transduction in leaves. Plant, Cell & Environment. 2007;**30**(9):1107-1125

[45] Friedland N, Negi S, Vinogradova-Shah T, Wu G, Ma L, Flynn S, et al. Fine-tuning the photosynthetic light harvesting apparatus for improved photosynthetic efficiency and biomass yield. Scientific Reports. 2019;**9**(1):13028

[46] Negi S, Perrine Z, Friedland N, Kumar A, Tokutsu R, Minagawa J, et al. Light regulation of light-harvesting antenna size substantially enhances photosynthetic efficiency and biomass yield in green algae†. The plant journal. 2020;**103**(2):584-603

[47] Wu G, Ma L, Sayre RT, Lee C-H. Identification of the optimal light harvesting antenna size for high-light stress mitigation in plants. Frontiers in Plant Science. 2020;11(505).

[48] Mirkovic T, Ostroumov EE, Anna JM, van Grondelle R, Govindjee, Scholes GD. Light absorption and energy transfer in the antenna complexes of photosynthetic organisms. Chemical Reviews. 2017;**117**(2):249-293

[49] Wang Y, Duanmu D, Spalding MH. Carbon dioxide concentrating mechanism in *Chlamydomonas reinhardtii*: Inorganic carbon transport and CO2 recapture. Photosynthesis Research. 2011;**109**(1-3):115-122

[50] Negi S, Sayre RT, Starkenburg SR. Improved Productivity and Bioproduct Formation in Phototropin Knock/out Mutants in Microalgae. USA Patent WO 2016/197/136 A2. 2016.

[51] Huang K, Beck CF. Phototropin is the blue-light receptor that controls multiple steps in the sexual life cycle of the green alga *Chlamydomonas reinhardtii*. Proceedings of the National Academy of Sciences of the United States of America. 2003;**100**(10):6269-6274 Epub 2003/04/25

[52] Im CS, Eberhard S, Huang K, Beck CF, Grossman AR. Phototropin involvement in the expression of genes encoding chlorophyll and carotenoid biosynthesis enzymes and LHC apoproteins in *Chlamydomonas reinhardtii*. The Plant Journal. 2006;**48**(1):1-16

[53] Baena-González E, Rolland F, Thevelein JM, Sheen J. A central integrator of transcription networks in plant stress and energy signalling. Nature. 2007;**448**(7156):938-942

[54] Saroussi S, Karns DAJ, Thomas DC, Bloszies C, Fiehn O, Posewitz MC, et al. Alternative outlets for sustaining photosynthetic electron transport during dark-to-light transitions. Proceedings of the National Academy of Sciences. 2019;**116**(23):11518-11527

[55] Banerjee S, Banerjee S, Ghosh AK, Das D. Maneuvering the genetic and metabolic pathway for improving biofuel production in algae: Present status and future prospective. Renewable and Sustainable Energy Reviews. 2020;**133**:110155

[56] Yang B, Liu J, Ma X, Guo B, Liu B, Wu T, et al. Genetic engineering of the Calvin cycle toward enhanced photosynthetic CO_2 fixation in microalgae. Biotechnology for Biofuels. 2017;**10**(1):229

[57] Roth MS, Westcott DJ, Iwai M, Niyogi KK. Hexokinase is necessary for glucose-mediated photosynthesis repression and lipid accumulation in a green alga. Commun Biol. 2019;**2**(347):019-0577

[58] Work VH, Radakovits R, Jinkerson RE, Meuser JE, Elliott LG, Vinyard DJ, et al. Increased lipid accumulation in the *Chlamydomonas reinhardtii* sta7-10 starchless isoamylase mutant and increased carbohydrate synthesis in complemented strains. Eukaryotic Cell. 2010;**9**(8):1251-1261 Epub 2010/06/18

[59] de Vries J, Gould SB. The monoplastidic bottleneck in algae and plant evolution. Journal of Cell Science. 2018;**131**(2):203414

[60] Leyland B, Boussiba S, Khozin-Goldberg I. A review of diatom lipid droplets. Biology. 2020;**9**:38

[61] Wang X, Wei W, Li N-J, Yuan W, Ding Y, Yang W-D, et al. Heterogeneous expression of human PNPLA3 triggers algal lipid accumulation and lipid droplet enlargement. Algal Research. 2018;**31**:276-281

[62] Ischebeck T, Krawczyk HE, Mullen RT, Dyer JM, Chapman KD. Lipid droplets in plants and algae: Distribution, formation, turnover and function. Seminars in Cell & Developmental Biology. 2020

[63] Moellering ER, Benning C. RNA interference silencing of a major lipid droplet protein affects lipid droplet size in *Chlamydomonas reinhardtii*. Eukaryotic Cell. 2010;**9**(1):97-106

[64] US-DOE. Algae Cultivation for Carbon Capture and Utilization Workshop Summary Report. Orlando, Florida: 2017.

[65] Slade R, Bauen A. Micro-algae cultivation for biofuels: Cost, energy balance, environmental impacts and future prospects. Biomass and Bioenergy. 2013;**53**:29-38

[66] Bhujade R, Chidambaram M, Kumar A, Sapre A. Algae to economically viable low-carbon-footprint oil. Annual Review of Chemical and Biomolecular Engineering. 2017;**8**:335-357

[67] Xu H, Lee U, Coleman AM, Wigmosta MS, Wang M. Assessment of algal biofuel resource potential in the United States with consideration of regional water stress. Algal Research. 2019;**37**:30-39

[68] Sun N, Skaggs RL, Wigmosta MS, Coleman AM, Huesemann MH, Edmundson SJ. Growth modeling to evaluate alternative cultivation strategies to enhance national microalgal biomass production. Algal Research. 2020;**49**:101939

[69] Venteris ER, McBride RC, Coleman AM, Skaggs RL, Wigmosta MS. Siting algae cultivation facilities for biofuel production in the United States: Trade-offs between growth rate, site constructability, water availability, and infrastructure. Environmental Science & Technology. 2014;**48**(6):3559-3566

[70] Janssen M. Chapter 4 - microalgal photosynthesis and growth in mass culture. In: Legrand J, editor. Advances in Chemical Engineering. Academic Press; 2016. pp. 185-256

[71] Posadas E, Alcántara C, García-Encina PA, Gouveia L, Guieysse B, Norvill Z, et al. 3 - microalgae cultivation in wastewater. In: Gonzalez-Fernandez C, editor. Muñoz R, editors. Microalgae-Based Biofuels and Bioproducts: Woodhead Publishing; 2017. pp. 67-91

[72] Emerson K, Russo RC, Lund RE, Thurston RV. Aqueous ammonia equilibrium calculations: Effect of pH and temperature. Journal of the Fisheries Research Board of Canada. 1975;**32**(12):2379-2383

[73] Morales M, Sánchez L, Revah S. The impact of environmental factors on carbon dioxide fixation by microalgae. FEMS Microbiology Letters. 2017;**365**(3)

[74] Posadas E, MdM M, Gomez C, Acién FG, Muñoz R. Influence of pH and CO_2 source on the performance of microalgae-based secondary domestic wastewater treatment in outdoors pilot raceways. Chemical Engineering Journal. 2015;**265**:239-248

[75] Metsoviti MN, Papapolymerou G, Karapanagiotidis IT, Katsoulas N. Comparison of growth rate and nutrient content of five microalgae species cultivated in greenhouses. Plants (Basel). 2019;**8**(8):279

[76] Rajvanshi M, Gautam K, Manjre S, Krishna Kumar GR, Kumar C, Govindachary S, et al. Stoichiometrically balanced nutrient management using a newly designed nutrient medium for large scale cultivation of *Cyanobacterium aponinum*. Journal of Applied Phycology. 2019;**31**(5):2779-2789

[77] Farooq W, Suh WI, Park MS, Yang J-W. Water use and its recycling in microalgae cultivation for biofuel application. Bioresource Technology. 2015;**184**:73-81

[78] Loftus SE, Johnson ZI. Cross-study analysis of factors affecting algae cultivation in recycled medium for biofuel production. Algal Research. 2017;**24**:154-166

[79] Loftus SE, Johnson ZI. Reused cultivation water accumulates dissolved organic carbon and uniquely influences different marine microalgae. Frontiers in bioengineering and biotechnology. 2019;**7**(101)

[80] Fon Sing S, Isdepsky A, Borowitzka MA, Lewis DM. Pilot-scale continuous recycling of growth medium for the mass culture of a halotolerant *Tetraselmis* sp. in raceway ponds under increasing salinity: A novel protocol for commercial microalgal biomass production. Bioresource Technology. 2014;**161**:47-54

[81] Erkelens M, Ball AS, Lewis DM. The influences of the recycle process on the bacterial community in a pilot scale microalgae raceway pond. Bioresource Technology. 2014;**157**:364-367

[82] Hadj-Romdhane F, Jaouen P, Pruvost J, Grizeau D, Van Vooren G, Bourseau P. Development and validation of a minimal growth medium for recycling *Chlorella vulgaris* culture. Bioresource Technology. 2012;**123**:366-374

[83] Farid MS, Shariati A, Badakhshan A, Anvaripour B. Using nano-chitosan for harvesting microalga *Nannochloropsis* sp. Bioresource Technology. 2013;**131**:555-559

[84] Kim D-G, La H-J, Ahn C-Y, Park Y-H, Oh H-M. Harvest of *Scenedesmus* sp. with bioflocculant and reuse of culture medium for subsequent high-density cultures. Bioresource Technology. 2011;**102**(3):3163-3168

[85] Farooq W, Moon M, Ryu B-G, Suh WI, Shrivastav A, Park MS, et al. Effect of harvesting methods on the reusability of water for cultivation of *Chlorella vulgaris*, its lipid productivity and biodiesel quality. Algal Research. 2015;**8**:1-7

[86] Zhang X, Lu Z, Wang Y, Wensel P, Sommerfeld M, Hu Q. Recycling *Nannochloropsis oceanica* culture media and growth inhibitors characterization. Algal Research. 2016;**20**:282-290

[87] Wang W, Sha J, Lu Z, Shao S, Sun P, Hu Q, et al. Implementation of UV-based advanced oxidation processes in algal medium recycling. Science of The Total Environment. 2018;**634**:243-250

[88] Chen L, Zhu T, Fernandez JSM, Chen S, Li D. Recycling nutrients from a sequential hydrothermal liquefaction process for microalgae culture. Algal Research. 2017;**27**:311-317

[89] Leng L, Li J, Wen Z, Zhou W. Use of microalgae to recycle nutrients in aqueous phase derived from hydrothermal liquefaction process. Bioresource Technology. 2018;**256**:529-542

[90] Kumar V, Kumar S, Chauhan PK, Verma M, Bahuguna V, Joshi HC, et al. Low-temperature catalyst based hydrothermal liquefaction of harmful macroalgal blooms, and aqueous phase nutrient recycling by microalgae. Scientific Reports. 2019;**9**(1):11384

[91] Jena U, Vaidyanathan N, Chinnasamy S, Das KC. Evaluation of microalgae cultivation using recovered aqueous co-product from thermochemical liquefaction of algal biomass. Bioresource Technology. 2011;**102**(3):3380-3387

[92] Li Z, Haifeng L, Zhang Y, Shanshan M, Li B, Zhidan L, et al. Microalgae production for bioenergy from post hydrothermal liquefaction wastewater: Effects of strain, nutrients concentration and inoculum size on microalgae culture. International Journal of Agricultural and Biological Engineering. 2017;**10**:194-204

[93] Biller P, Ross AB, Skill SC, Lea-Langton A, Balasundaram B, Hall C, et al. Nutrient recycling of aqueous phase for microalgae cultivation from the hydrothermal liquefaction process. Algal Research. 2012;**1**(1):70-76

[94] Garcia Alba L, Torri C, Fabbri D, Kersten SRA, Brilman DWF. Microalgae growth on the aqueous phase from hydrothermal liquefaction of the same microalgae. Chemical Engineering Journal. 2013;**228**:214-223

[95] Bagnoud-Velásquez M, Schmid-Staiger U, Peng G, Vogel F, Ludwig C. First developments towards closing the nutrient cycle in a biofuel production process. Algal Research. 2015;**8**:76-82

[96] Hognon C, Delrue F, Texier J, Grateau M, Thiery S, Miller H, et al. Comparison of pyrolysis and hydrothermal liquefaction of *Chlamydomonas reinhardtii*. Growth studies on the recovered hydrothermal aqueous phase. Biomass and Bioenergy. 2015;**73**:23-31

[97] Yao C, Wu P, Pan Y, Lu H, Chi L, Meng Y, et al. Evaluation of the integrated hydrothermal carbonization-algal cultivation process for enhanced nitrogen utilization in *Arthrospira platensis* production. Bioresource Technology. 2016;**216**:381-390

[98] López Barreiro D, Bauer M, Hornung U, Posten C, Kruse A, Prins W. Cultivation of microalgae with recovered nutrients after hydrothermal liquefaction. Algal Research. 2015;**9**:99-106

[99] Godwin CM, Hietala DC, Lashaway AR, Narwani A, Savage PE, Cardinale BJ. Algal polycultures enhance coproduct recycling from hydrothermal liquefaction. Bioresource Technology. 2017;**224**:630-638

[100] Edmundson S, Huesemann M, Kruk R, Lemmon T, Billing J, Schmidt A, et al. Phosphorus and nitrogen recycle following algal bio-crude production via continuous hydrothermal liquefaction. Algal Research. 2017;**26**

[101] Talbot C, Garcia-Moscoso J, Drake H, Stuart BJ, Kumar S. Cultivation of microalgae using flash hydrolysis nutrient recycle. Algal Research. 2016;**18**:191-197

[102] Selvaratnam T, Reddy H, Muppaneni T, Holguin FO, Nirmalakhandan N, Lammers PJ, et al. Optimizing energy yields from nutrient recycling using sequential hydrothermal liquefaction with *Galdieria sulphuraria*. Algal Research. 2015;**12**:74-79

[103] Yang W, Li X, Li Z, Tong C, Feng L. Understanding low-lipid algae hydrothermal liquefaction characteristics and pathways through hydrothermal liquefaction of algal major components: Crude polysaccharides, crude proteins and their binary mixtures. Bioresource Technology. 2015;**196**:99-108

[104] Martinez-Fernandez JS, Chen S. Sequential hydrothermal liquefaction characterization and nutrient recovery assessment. Algal Research. 2017;**25**:274-284

[105] Chen PH, Venegas Jimenez JL, Rowland SM, Quinn JC, Laurens LML.

Nutrient recycle from algae hydrothermal liquefaction aqueous phase through a novel selective remediation approach. Algal Research. 2020;**46**:101776

[106] Borowitzka MA, Vonshak A. Scaling up microalgal cultures to commercial scale. European Journal of Phycology. 2017;**52**(4):407-418

[107] Han D, Li Y, Hu Q. Biology and Commercial Aspects of *Haematococcus pluvialis*. In: Richmond A, Hu Q, editors. Handbook of Microalgal Culture. 2013. pp. 388-405

[108] Wang Y, Gong Y, Dai L, Sommerfeld M, Zhang C, Hu Q. Identification of harmful protozoa in outdoor cultivation of *Chlorella* and the use of ultrasonication to control contamination. Algal Research. 2018;**31**:298-310

[109] Karuppasamy S, Musale AS, Soni B, Bhadra B, Gujarathi N, Sundaram M, et al. Integrated grazer management mediated by chemicals for sustainable cultivation of algae in open ponds. Algal Research. 2018;**35**:439-448

[110] Lee PA, Martinez KJL, Letcher PM, Corcoran AA, Ryan RA. A novel predatory bacterium infecting the eukaryotic alga *Nannochloropsis*. Algal Research. 2018;**32**:314-320

[111] Troschl C, Fritz I, Sodnikar K, Drosg B. Contaminations in mass cultivation of cyanobacteria: Highly resilient *Colpoda steinii* leads to rapid crash of *Synechocystis* sp. cultures and is inhibited by partially anoxic conditions. Algal Research. 2017;**28**:229-234

[112] Letcher PM, Lopez S, Schmieder R, Lee PA, Behnke C, Powell MJ, et al. Characterization of *Amoeboaphelidium protococcarum*, an algal parasite new to the *Cryptomycota* isolated from an outdoor algal pond used for the production of biofuel. PLoS One. 2013;**8**(2):e56232

[113] Pradeep V, Van Ginkel SW, Park S, Igou T, Yi C, Fu H, et al. Use of copper to selectively inhibit *Brachionus calyciflorus* (predator) growth in *Chlorella kessleri* (prey) mass cultures for algae biodiesel production. International journal of molecular sciences. 2015;**16**(9):20674-20684

[114] Park S, Van Ginkel SW, Pradeep P, Igou T, Yi C, Snell T, et al. The selective use of hypochlorite to prevent pond crashes for algae-biofuel production. Water Environment Research. 2016;**88**(1):70-78

[115] Snell TW, Moffat BD, Janssen C, Persoone G. Acute toxicity tests using rotifers: IV. Effect of cyst age, temperature, and salinity on the sensitivity of *Brachionus calyciflorus*. Ecotox Environ Safe. 1991;**21**(3):308-317

[116] Ferrando MD, Sancho E, Andreu-Moliner E. Chronic toxicity of Fenitrothion to an algae (*Nannochloris oculata*), a rotifer (*Brachionus calyciflorus*), and the Cladoceran (*Daphnia magna*). Ecotox Environ Safe. 1996;**35**(2):112-120

[117] Hessen DO, Færøvig PJ, Andersen T. Light, nutrients, and P:C ratios in algae: Grazer performance related to food quality and quantity. Ecology. 2002;**83**(7):1886-1898

[118] Flynn KJ, Kenny P, Mitra A. Minimising losses to predation during microalgae cultivation. Journal of Applied Phycology. 2017;**29**:1829-1840

[119] Ma M, Yuan D, He Y, Park M, Gong Y, Hu Q. Effective control of *Poterioochromonas malhamensis* in pilot-scale culture of *Chlorella sorokiniana* GT-1 by maintaining CO_2-mediated low culture pH. Algal Research. 2017;**26**:436-444

[120] Montemezzani V, Duggan IC, Hogg ID, Craggs RJ. Screening of potential zooplankton control

technologies for wastewater treatment high rate algal ponds. Algal Research. 2017;**22**:1-13

[121] Touloupakis E, Cicchi B, Benavides AMS, Torzillo G. Effect of high pH on growth of *Synechocystis* sp. PCC 6803 cultures and their contamination by golden algae (*Poterioochromonas* sp.). Applied Microbiology and Biotechnology. 2016;**100**(3):1333-1341

[122] Kim D, Kim EK, Koh HG, Kim K, Han J-I, Chang YK. Selective removal of rotifers in microalgae cultivation using hydrodynamic cavitation. Algal Research. 2017;**28**:24-29

[123] Venkata Subhash G, Rangappa M, Raninga S, Prasad V, Dasgupta S. Raja Krishna Kumar G. electromagnetic stratagem to control predator population in algal open pond cultivation. Algal Research. 2019;**37**:133-137

[124] Rego D, Redondo LM, Geraldes V, Costa L, Navalho J, Pereira MT. Control of predators in industrial scale microalgae cultures with pulsed electric fields. Bioelectrochemistry. 2015;**103**:60-64

[125] Umar A, Caldwell GS, Lee JGM. Foam flotation can remove and eradicate ciliates contaminating algae culture systems. Algal Research. 2018;**29**:337-342

[126] Sinha A, Pandharbale A, Patel JR. Current issues with a sustainable approach in the industrial-scale operations of microalgae. In: Gayen K, Bhowmick TK, Maity SK, editors. Sustainable downstream processing of microalgae for industrial application. Florida, USA: CRC Press (Taylor and Francis Group); 2019. pp. 231-291

[127] Park JBK, Craggs RJ, Shilton AN. Wastewater treatment high rate algal ponds for biofuel production. Bioresource Technology. 2011;**102**(1):35-42

[128] Sharma KK, Garg S, Li Y, Malekizadeh A, Schenk PM. Critical analysis of current microalgae dewatering techniques. Biofuels. 2013;**4**(4):397-407

[129] Fasaei F, Bitter JH, Slegers PM, van Boxtel AJB. Techno-economic evaluation of microalgae harvesting and dewatering systems. Algal Research. 2018;**31**:347-362

[130] Molina Grima E, Belarbi EH, Acién Fernández FG, Robles Medina A, Chisti Y. Recovery of microalgal biomass and metabolites: Process options and economics. Biotechnology advances. 2003;**20**(7):491-515

[131] Mo W, Soh L, Werber JR, Elimelech M, Zimmerman JB. Application of membrane dewatering for algal biofuel. Algal Research. 2015;**11**:1-12

[132] Xu Y, Purton S, Baganz F. Chitosan flocculation to aid the harvesting of the microalga *Chlorella sorokiniana*. Bioresource Technology. 2013;**129**:296-301

[133] Chang Y-R, Lee D-J. Coagulation–membrane filtration of *Chlorella vulgaris* at different growth phases. Drying Technology. 2012;**30**(11-12):1317-1322

[134] Chen G, Zhao L, Qi Y, Cui Y-L. Chitosan and Its derivatives applied in harvesting microalgae for biodiesel production: An outlook. Journal of Nanomaterials. 2014;**2014**:217537

[135] Eldridge RJ, Hill DRA, Gladman BR. A comparative study of the coagulation behaviour of marine microalgae. Journal of Applied Phycology. 2012;**24**(6):1667-1679

[136] Branyikova I, Prochazkova G, Potocar T, Jezkova Z, Branyik T. Harvesting of microalgae by flocculation. Fermentation. 2018;**4**(4):93

[137] Castaing JB, Massé A, Séchet V, Sabiri NE, Pontié M, Haure J, et al. Immersed hollow fibres microfiltration (MF) for removing undesirable micro-algae and protecting semi-closed aquaculture basins. Desalination. 2011;**276**(1):386-396

[138] Bilad MR, Vandamme D, Foubert I, Muylaert K, Vankelecom IFJ. Harvesting microalgal biomass using submerged microfiltration membranes. Bioresource Technology. 2012;**111**:343-352

[139] Kim D-Y, Hwang T, Oh Y-K, Han J-I. Harvesting *Chlorella* sp. KR-1 using cross-flow electro-filtration. Algal Research. 2014;**6**:170-174

[140] Hwang K-J, Wu S-E. Disk structure on the performance of a rotating-disk dynamic filter: A case study on microalgae microfiltration. Chemical Engineering Research and Design. 2015;**94**:44-51

[141] Vandamme D, Pontes S, Goiris K, Foubert I, Pinoy L, Muylaert K. Evaluation of electro-coagulation-flocculation for harvesting marine and freshwater microalgae. Biotechnology and Bioengineering. 2011;**108**

[142] Wang L, Pan B, Gao Y, Li C, Ye J, Yang L, et al. Efficient membrane microalgal harvesting: Pilot-scale performance and techno-economic analysis. Journal of Cleaner Production. 2019;**218**:83-95

[143] Szepessy S, Thorwid P. Low energy consumption of high-speed centrifuges. Chemical Engineering & Technology. 2018;**41**(12):2375-2384

[144] Davis R, Markham J, Kinchin C, Grundl N, Tan ECD, Humbird D. Process design and economics for the production of algal biomass: Algal biomass production in open pond systems and processing through dewatering for downstream conversion. Golden, CO (United States): National Renewable Energy Lab. (NREL); 2016. NREL/TP-5100-64772 United States 10.2172/1239893 NREL English

[145] Kumar D, Singh B. Algal biorefinery: An integrated approach for sustainable biodiesel production. Biomass and Bioenergy. 2019;**131**:105398

[146] Norsker N-H, Barbosa MJ, Vermuë MH, Wijffels RH. Microalgal production — A close look at the economics. Biotechnology advances. 2011;**29**(1):24-27

[147] Koyande AK, Show P-L, Guo R, Tang B, Ogino C, Chang J-S. Bio-processing of algal bio-refinery: A review on current advances and future perspectives. Bioengineered. 2019;**10**(1):574-592

[148] Zhu L. Biorefinery as a promising approach to promote microalgae industry: An innovative framework. Renewable and Sustainable Energy Reviews. 2015;**41**:1376-1384

[149] Chandra R, Iqbal HMN, Vishal G, Lee H-S, Nagra S. Algal biorefinery: A sustainable approach to valorize algal-based biomass towards multiple product recovery. Bioresource Technology. 2019;**278**:346-359

[150] Roux J-M, Lamotte H, Achard J-L. An overview of microalgae lipid extraction in a biorefinery framework. Energy Procedia. 2017;**112**: 680-688

[151] Nobre BP, Villalobos F, Barragan BE, Oliveira A, Batista AP, Marques P, et al. A biorefinery from *Nannochloropsis* sp. microalga–extraction of oils and pigments. Production of biohydrogen from the leftover biomass. Bioresource Technology. 2013;**135**:128-136

[152] Lee OK, Kim AL, Seong DH, Lee CG, Jung YT, Lee JW, et al. Chemo-enzymatic saccharification

and bioethanol fermentation of lipid-extracted residual biomass of the microalga, *Dunaliella tertiolecta*. Bioresour Technol. 2013;**132**:197-201

[153] Muñoz R, Navia R, Ciudad G, Tessini C, Jeison D, Mella R, et al. Preliminary biorefinery process proposal for protein and biofuels recovery from microalgae. Fuel. 2015;**150**:425-433

[154] Ansari FA, Shriwastav A, Gupta SK, Rawat I, Guldhe A, Bux F. Lipid extracted algae as a source for protein and reduced sugar: A step closer to the biorefinery. Bioresource Technology. 2015;**179**:559-564

[155] Lu Y, Savage PE. Supercritical water gasification of lipid-extracted hydrochar to recover energy and nutrients. The Journal of Supercritical Fluids. 2015;**99**:88-94

[156] Pacheco R, Ferreira AF, Pinto T, Nobre BP, Loureiro D, Moura P, et al. The production of pigments & hydrogen through a *Spirogyra* sp. biorefinery. Energy Conversion and Management. 2015;**89**:789-797

[157] Dong T, Knoshaug EP, Davis R, Laurens LML, Van Wychen S, Pienkos PT, et al. Combined algal processing: A novel integrated biorefinery process to produce algal biofuels and bioproducts. Algal Research. 2016;**19**:316-323

[158] Gilbert-López B, Barranco A, Herrero M, Cifuentes A, Ibáñez E. Development of new green processes for the recovery of bioactives from *Phaeodactylum tricornutum*. Food Research International. 2017;**99**:1056-1065

[159] Phusunti N, Cheirsilp B. Integrated protein extraction with bio-oil production for microalgal biorefinery. Algal Research. 2020;**48**:101918

[160] Soto-Sierra L, Kulkarni S, Woodard SL, Nikolov ZL. Processing of permeabilized *Chlorella vulgaris* biomass into lutein and protein-rich products. Journal of Applied Phycology. 2020;**32**(3):1697-1707

[161] Razon LF, Tan RR. Net energy analysis of the production of biodiesel and biogas from the microalgae: *Haematococcus pluvialis* and *Nannochloropsis*. Applied Energy. 2011;**88**(10):3507-3514

[162] Andersson V, Broberg Viklund S, Hackl R, Karlsson M, Berntsson T. Algae-based biofuel production as part of an industrial cluster. Biomass and Bioenergy. 2014;**71**:113-124

[163] Beckstrom BD, Wilson MH, Crocker M, Quinn JC. Bioplastic feedstock production from microalgae with fuel co-products: A techno-economic and life cycle impact assessment. Algal Research. 2020;**46**:101769

[164] Dixon C, Wilken LR. Green microalgae biomolecule separations and recovery. Bioresources and Bioprocessing. 2018;**5**(1):14

[165] Kumar C, Soni N, Soni BR, Das G, Dasgupta S, Reliance Industries Limited, India, Assignee. Propiconazole Resistant Mutants of *Chlorella* Species. United States of America Patent US 10,694,752 B2. 2020.

[166] Krause-Jensen D, Lavery P, Serrano O, Marbà N, Masque P, Duarte CM. Sequestration of macroalgal carbon: The elephant in the blue carbon room. Biology Letters. 2018;**14**(6):20180236

[167] Packer M. Algal capture of carbon dioxide; biomass generation as a tool for greenhouse gas mitigation with reference to New Zealand energy strategy and policy. Energy Policy. 2009;**37**(9):3428-3437

Section 6

Biotechnological Products

Chapter 25

Use of Olive Mill Wastewaters as Bio-Insecticides for the Control of *Potosia Opaca* in Date Palm (*Phoenix dactylifera L.*)

Abdelilah Meddich, Abderrahim Boutasknit, Mohamed Anli, Meriame Ait Ahmed, Abdelilah El Abbassi, Hanane Boutaj, Mohamed Ait-El-Mokhtar and Ali Boumezzough

Abstract

The date palm is one of the most economically important perennial plants of the North Africa and in Morocco, where it is extensively cultivated for food and many other commercial purposes. Palm trees are threatened by many pests such as *Potosia opaca* newly identified in Morocco, especially in Marrakesh and Errachidia regions. In addition, olive mill wastewaters (OMW) are an environmental problem in olive oil producing countries such as Morocco. Generally, these effluents are drained into ecosystems without any pre-treatment. To reduce their negative impact and to get benefits in particular from their high phenolic content, OMW were used as bio-insecticides in crude form. The results showed that crude OMW were effective to control this pest causing a weight loss similar to Cordus insecticide (17% *vs.* 15%) and mortality almost similar to Kemaban insecticide. OMW's biocide potential was related principally to their high phenolic content. Based on HPLC analysis, ten phenolic molecules were identified, including two which were revealed as the major monomeric phenolic compounds in OMW, 0.248 g/L of hydroxytyrosol and 0.201 g/L of tyrosol. In this chapter, the potential use of OMW as bio-insecticides for the control of *P. opaca* in date palm is discussed.

Keywords: olive mill wastewaters, *Potosia opaca*, date palm trees, insecticidal activity, biological control

1. Introduction

The date palm trees have many important socio-economic and ecological roles in oases ecosystems [1]. In North Africa and in Morocco, the oases are facing several constraints related to urbanization, drought, salinity, desertification, poor soils in organic matter and nutrients, genetic erosion, aging, diseases like Bayoud palm caused by *Fusarium oxysporum* fsp *albidinis* [2–4] and pests attacks [5, 6]. Palm trees are strongly threatened by the red weevil caused by *Rhynchophorus ferrugineus,* which causes huge economic losses [7]. The red weevil causes economy loss, resulting in millions of dollars each year, related to agricultural production or costs related to pest control [8]. In the Gulf countries and the Middle East, US$ 8 million is spent every year to cut down contaminated trees [8, 9]. In Spain, red palm weevil has appeared since 1999 and damaged almost 20,000 palms of *Phoenix dactylifera* [10]. In the North of Morocco and more precisely in Tangier, the number of *Phoenix canariensis* prospected during 2009–2016 is 244,393 [11]. The number of *P. canariens is* infested with *R. ferrugineus* was 904, which 896 were incinerated [11], whereas no *P. dactylifera* has been infested with *R. ferrugineus*. In Morocco, *Potosia opaca* var. cardui Gyllenhal has been observed for the first time by Meddich and Boumezzough [12]. Indeed, in Marrakesh and Errachidia regions, it attacks *P. dactylifera* L. and *P. canariensis* by consuming their wood, which causes faster degradation. Thus, to remedy the damage caused by *P. opaca*, most farmers were forced to use synthetic pesticides. However, the intensive use of these pesticides are generally effective in protecting crops [13], but they are toxic to wildlife and to organisms from different levels of the ecosystems [14–17]. Over time, the permanent use of insecticides may be accompanied by the development of resistant strains in some treated species. Biocontrol strategies for pests need to be investigated and developed to provide an ecological substitute or alternative approach to the conventional methods. Some sub-products such as olive oil mill waste waters (OMW) are currently used to control pests, which is essential for crop protection [18, 19]. Most of the OMW phenolic compounds derived from olive polyphenols have many other biological properties [20, 21], as well as biocide activities [22] and phytotoxic effects [23]. Due to their particular characteristics, these effluents are a serious problem for the Mediterranean region, which annually produce around 30 million m^3 of OMW with a damaging effect on the environment [23] and accounts for approximately 95% of olive oil production in the world [19]. In addition, different physicochemical methods have been proposed to treat OMW, including natural and forced evaporation [24], electro-coagulation [25], oxidation by ozone and Fenton reagent [26] as well as their agricultural spreading [21], which is an alternative among the suggested solutions. However, the agronomic application of OMW is limited by the doses to be applied and the risk of polyphenols accumulation in the soil after consecutive applications [21–27]. In parallel with researches made on the treatment of OMW, many studies have been carried out aiming the recovery of OMW phenolic compounds. Recent studies tried to take its advantage from the antimicrobial and phytotoxic properties by using it as biopesticide for crops protection [28–30] or as insecticides to control *P. opaca* larvae [31]. This contribution summarized the quality of palm health status, OMW characteristics and its application as insecticides to control *P. opaca* larvae in date palm, especially in *P. dactylifera* L.

2. Materials and methods

2.1 Study area and plant material

This study was conducted during the period of 2014–2016 in the oasis of Marrakesh located in the central region of Morocco and the oasis of Errachidia situated in the southeastern part of the country. The majority of their territory presents arid climate, hot summer and cold winter. Palms (*Phoenix*) constitute one of the important botanical families, and include some of the world's most important economic plants. In North Africa regions, dates production provides jobs for estimated around 50 million people [32]. It plays by now an undeniable role in maintaining human populations in arid regions where natural resources are limited and living conditions are difficult [32, 33]. Like in Zagora, Errachidia, Ouarzazate which are located in the Draa-Tafilalet region. It is built on terraces, crowned by an old Glaoui kasbah of a hill and surrounded by a major oasis of palm grove. This includes in particular the biosphere reserve of the oases of south-eastern Morocco [34], which forms an agglomeration of ksours (small castles). This area is located in an

Figure 1.
Pictures of P. dactylifera *(A) and* P. canariensis *(B) in Marrakesh city.*

environment covered by vegetation and many dunes were transformed into regs (vast stone expanses) where the vestiges of khetaras (old local irrigation system) are still visible. As Marrakesh city, the area is under a semi-arid climate regime characterized by relatively cold and humid winters and hot and dry summers with a large diurnal temperature range [35]. This study was performed on *P. dactylifera* and *P. canariensis*. 60-year-old trees with a diameter of 70–90 cm were selected for larvae sampling (**Figure 1**). Analyzed leaves were 4–5 m long with 80–100 segments on each side of the spine. No specific permissions were required for these locations and activities. The field studies did not involve endangered or protected species. The pathogen presence was visually confirmed. No plants were available at very early stage of infection.

2.2 Phytopathological analysis and fungi isolation

The present study used augers to perform localized sampling and perform microbiological isolates. To deepen the diagnosis of *P. canariensis*, we made samplings of rachis, leaflets palms, dry and green leaf bases at the crown. We have carried out cultures and incubation of extracts of rachis (1 cm) and palm leaves puny on selective and non-selective media.

2.3 Sampling techniques and prospecting the crown of palm

During the exploration of the palm crown, using a scaffold (**Figure 2A**), a number of larvae (white grubs) sluggish and arched with strong mandibles were harvested at the base of green and/or dead rachis. Similarly, insect larvae were removed from the base of green and dried palms for laboratory breeding in incubators with controlled conditions of temperature, humidity and photoperiod. Dead rachis (**Figure 2B** and **C**) were brought back to the laboratory to explore them further and also put them in terrariums to ensure the follow-up and development of the larvae trapped inside the hope of having imagos (adult forms). Identification of larvae was performed according to the key proposed by Mico and Galante [36].

2.4 Breeding of larva and nymphs

The collected larvae were immediately placed into breeding boxes; transparent, with holes on the sides and on the top of the boxes to ensure oxygenation and avoid asphyxiation. The holes on the boxes are closed with a fabric scrim (muslin type). Rectangular boxes were used for rearing larvae harvested from *P. canariensis* and

Figure 2.
Exploration of the P. canariensis *palm crown using a scaffold (A); base of leaves (B and C).*

P. dactylifera. The breeding substrate was composed of a mixture of untreated natural soil and debris of dead wood, rotted wood and sawdust. Care was taken not to import diseases on bringing boxes of dry dung in breeding substrate in order to increase its acidity, which disadvantages the development of diseases. The breeding substrate was constantly renewed as soon as the feces of larvae appear in large quantities on the surface and more debris and wood in the rearing environment was observed. This breeding operation continued in incubators refrigerated and illuminated with controlled temperature and humidity. However, larvae are lucifugous (escape behavior of light); the optimum temperature is between 25 and 30°C. The nymphal hulls were placed in boxes with slightly damp peat. The duration of pupation varies according to the temperature supported by the larvae and also the male or female sex. The infected nymphal hulls were removed as soon as possible from the breeding environment to avoid pathological contamination of the rest of the cocoons.

2.5 Chemical properties of palm wastes

Date palm wastes were used as food for *P. opaca* larvae. The sampling of these wastes was carried out in April 2016 in Marrakesh area (Morocco). **Table 1** summarizes the main chemical composition of date palm wastes.

2.6 Main chemical composition of OMW

Sampling of OMW was carried out in a semi-modern three-phase olive mill installed in Marrakesh (Morocco) and the samples were conserved at 4°C. The determination of the volatile matter (VM) was performed by differentiating between the dry matter (DM) obtained by evaporation at 105°C and the ash residue obtained from calcination at 550°C over a two-hour period.

2.7 Phenolic compounds analysis

2.7.1 Phenols extraction from OMW

OMW total phenolic compounds were obtained by liquid–liquid extraction according to the method described by El Abbassi et al. [37]. HCl (2 M) was added to OMW

Parameters	Mean ± SD
Ph	7.03 ± 0.17
TOC (%)	40.80 ± 2.47
NTK (%)	1.06 ± 0.08
C/N	38.60 ± 4.88
Ashes (%)	29.80 ± 1.58
NH_4^+ (mg/g)	738.00 ± 30.1
NO_3^- (mg/g)	0.70 ± 0.08
Available phosphorus (µg/g)	9.00 ± 0.80
NH_4^+/NO_3^- (×1000)	1.05 ± 0.75

TOC, total organic carbon; TKN, total Kjeldahl nitrogen.

Table 1.
Chemical characteristics of date palm waste.

samples (5 mL) to adjust pH to2.0.OMW were defatting using n-hexane and two extractions were performed with ethyl acetate. The aqueous ethyl extracts were dried at 40°C under reduced pressure via a rotary evaporator and then recovered in methanol (5 ml).

2.7.2 Total phenolic content

Estimation of the total phenol content was determined by the Folin–Ciocalteu calorimetric method [38] where gallic acid was used as the standard. Therefore, it was measured as gallic acid equivalent (GAE) and expressed as g of GAE/L of OMW.

2.7.3 OMW phenolic compounds identification

HPLC analysis was conducted at the Center for Analysis and Characterization (Cadi Ayyad University, Marrakesh, Morocco) with C18 column (Eurospher II 100–5, 250 x 4.6 mm) in gradient system (eluting solution A = acetonitrile; eluting solution B = *o*-phosphoric acid/water (pH = 2.6), 5/95 v/v). A volume of 10 µl was injected at a flow rate of 1 mL/min and pressure of 117 bar. The characterization of phenolic compounds was carried out using their UV–Vis diode-array detector at a spectrum of 280 nm and their identification was performed by comparing their retention time (RT) with standards. Then these compounds were quantified through the calibration curve of the corresponding standards. The results obtained were expressed in g/L.

2.8 Larvae cultures

Sampling of larvae of *P. opaca* var. cardui Gyllenhal (**Figure 3**) was conducted according to section sampling techniques. The larvae were reared in round boxes (8 cm x 5 cm: diameter x height) containing a mixture of palm waste (150 g). The larvae culture was maintained in darkness at an optimal temperature between 25 and 30°C inside an incubator. The experiments were conducted in the same conditions as those for the cultures.

2.9 Spray toxicity bioassay

The insecticidal activity of crude OMW and the two commercial insecticides (used as positive controls, Cordus and Kemaban 48 EC) to control *P. opaca* larvae was assessed by a spray toxicity bioassay conducted using 5 g of palm compost in plastic boxes. Cordus and Kemaban are composed of 50 and 48% chlorpyrifos-ethyl, respectively. Chlorpyrifos is an insecticide, acaricide and organophosphate miticide used mainly against insect pests. Concentrations of 0.5 and 1 μL/mL of Cordus and Kemaban were dissolved in distilled water. Many preliminary tests have been performed to select the doses to be used for positive controls. For each solution and crude OMW, a volume of 5 mL was sprayed on the surface of the palm compost every 24 hours for 6 days. Based on dry matter, the dose of OMW was calculated to be 94.86 g/L (≈95 mg/mL) (**Table 2**). The cumulative dosage of OMW polyphenols applied over a six days treatment was calculated. For each treatment, six larvae were placed in each box using ten replicates. Weight loss and mortality of positive controls and OMW were recorded 8 hours per day every 2 hours. Negative control was treated with distilled water. The larva was considered dead, when no movement was recorded when shaking. Observations were made on all treated larvae until their death.

2.10. Statistical analysis

All results were analyzed statistically with the CO-STAT software (Statistical Software, New Style Anova). The study includes an analysis of variance followed by the Newman and Keuls test at the 5% threshold.

Figure 3.
Potosia opaca *larvae isolated from* P. canariensis *and* P. dactylifera *palms.*

Parameters	Mean ± SD
pH	4.70 ± 0.11
EC (mS/cm)	23.50 ± 0.50
TOC (g/L)	26.23 ± 1.40
DM (g/L)	94.86 ± 1.66
TSS (g/L)	21.79 ± 0.50
Ash (g/L)	11.35 ± 0.67
TPC (g GAE/L)	8.38 ± 0.14
Residual Oil (g/L)	2.20 ± 0.30

EC, electrical conductivity; TOC, total organic carbon; DM, dry matter; TSS, total suspended solids; TPC, total phenolic content.

Table 2.
Physicochemical characteristics of crude OMW.

Probit analysis [39] was conducted to estimate lethal times (LT_{50} and LT_{90}) with their 95% confidence interval by SPSS 20.0 Statistical software; LT values were considered significantly different when their respective 95% confidence interval did not overlap.

3. Results

3.1 Microbiological analysis of crown palm

In laboratory, the microbiological analysis showed fungal formations (blackish, whitish and greenish spots) observed on the Petri dishes that contain the extracts [12]. After observation on microscope, it has been found as saprophytic fungi in particular the Dimaties, with blackish spots; those take advantage of the damage already noted and the genus *Fusarium* with a non-virulent strain developed in whitish and pinkish spots with some conidia. The genus *Trichoderma* wide spread, in greenish spots and is often a key fungus in symbiosis and in biological control.

3.2 Sampling techniques and prospecting the crown of palm

During the exploration of the palm crown (*P. canariensis and P. dactylifera*), a number of soft arched larvae (grubs) with strong mandibles were harvested at the base of the green spine and / or dead. The first diagnosis of these larvae showed that they have a powerful mandibles, form "melolonthoides," sub-cylindrical strongly arched, whitish, with head, stigma and brownish legs; the head was always perpendicular to the body axis, with blackish posterior end; related to the Scarabaeidae beetles larvae (3 pairs of legs, antennae with 3 segments). Maximum width of head capsule: 4.6–4.9 mm. These larvae were either rhizophagous (melolonthoids), saprophytophagous or saproxylophagous (Cetoniidae). Initial research has shown that in the forms of phytophagous and saproxylophagous larvae, egg-laying can include about 100 eggs placed directly in the soil or in the wood, sometimes with the aid of an auger [40]. According to field investigations, Meddich and Boumezzough [12] observed the theft of a number of scarabaeidae beetles from the Cetoniidae family, but linking the larvae harvested at the level of the palm crown and the adults captured on the flight seems to be an illusory thing especially since all the larvae of Coleoptera belonging to the family Scarabaeidae were similar and that the systematic identification at the generic and specific level requires a rearing of the larvae under appropriate conditions of temperature and humidity for obtaining the adult. The ultimate stage of development of this Scarabaeidae and whose characteristics were indispensable for the identification of specimens. The evolution of the larvae is carried out at the base of the palms and rachis weakened and attacked by saprophytic fungi, which can lead to the appearance of rot and the dieback of the weakened palm [12]. Their presence at the top of the palm can also be explained by the fact that this beetle found an ideal biotope for the proliferation of larvae. Probably the adult manages to lay his eggs at the base of the weakened rachis and continues its development process while damaging the foliar bases as well as the heart of the palm. The larvae of this insect developing in the crown of the stem can infect the whole leaf mass and cause the crown tilt, which can be fatal to the palm. It should be noted that important attacks are observed at the palms base, at the crown; Attacks marked by the formation of galleries with a rejection of sawdust and dejections of white grubs (**Figure 4**) [12].

3.3 Breeding of larva and nymphs

After one month of larval rearing, the majority of stage III (L3) larvae are transformed into nymphs (the last stage before adult release) [12]. At the end of the last stage of development, the larva becomes more yellowish (accumulation of adipose tissue to the detriment of the stercoral volume of the rectal sac). The premymphal phase lasts a few days during which the elderly larva (L3) no longer feeds and migrates to the bottom of the substratum to build a pupal cocoon, which it generally chooses to make against a support (**Figure 5**).

Pupation occurs in the dorsal position since the larvae move on the back, which distinguishes them from Oryctes larvae (Rhinoceros). In general, its cycle development from the egg to the imago known to be set on one year [40]. However, its metabolic activity is closely related to the ambient temperature, the low temperatures slow down this passage, which may last 2 years [40]. The oases, Moroccan as already said are relatively warm, which could privilege this passage. The nymph of brown-orange color has appendages entirely free and folded down on its ventral surface. During this critical phase of development, the insect does not feed, its mobility is very limited and it is very dependent on the conditions of the environment (temperature, humidity and predation). During this passive stage of development, the nymph gradually acquires a darker color. This pigmentation is perfected in the days before the molt. On the occasion of this final metamorphosis, the insect takes a ventral position, in order to facilitate the deployment of wings and elytra. Its tissues harden progressively in the presence of oxygen from the air. After gaining greater rigidity, the adult perforates its cocoon and migrates to the surface of the substrate in order to begin its phase of aerial life. The adult that has just hatched is sometimes still a little soft and often presents colors less sustained and clearer than its older congeners (**Figure 6**) [12].

Examination of adults under binocular loupe and using a dichotomous key and reference collections from the laboratory led us to the species of Coleoptera Scarabaeidae: *Potosia opaca* var. Cardui Gyllenhal. This species varies greatly in size (from 14 to 24 mm), its morphology with the sides of the pronotum, which can

Figure 4.
*Larvae found in the palm crown (*P. canariensis*) (A) and (B) larvae, (C) leaf base.*

Figure 5.
Observed cocoons (A) and nymphs (B).

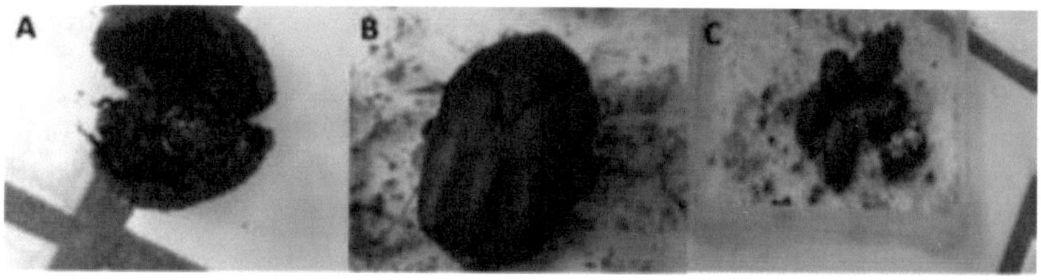

Figure 6.
Adult observed in its damaged cocoon (A), (B) and (C) free adults.

be weakly indented before the posterior angles, or not indented. The, the general coloring on the topis black, passes more rarely to the black green (typical form) and even to bronze and green with coppery metallic reflection. On the underside, the color is sometimes black, sometimes bluish, or sometimes greenish or white.

Systematic
Hexapoda (Insecta)
Coleoptera,
Scarabaeidae
Cetoniini
Potosia opaca Fabricius.

3.4 Olive mill wastewater characteristics

3.4.1 Physicochemical characteristics of crude OMW

Table 2 shows the physicochemical characteristics of the OMW according to Boutaj et al. [31]. In addition, these effluents have an acidic pH of 4.7, a high electrical conductivity of 23.5 mS/cm, a residual oil of 2.2 g/L, a high polyphenol content of 8.38 g GAE/L of crude OMW, and an average dry matter content of 94.86 g/L.

3.4.2 Identification and quantification of OMW phenolic compounds

As described by Boutaj et al. [31], OMW present a high phenol content. **Table 3** summarizes the qualification and quantification of these principal phenolic compounds identified by HPLC analysis. Based on comparisons of their retention times and their UV spectra with standards analyzed under the same conditions, 10 free compounds were provisionally identified and quantified in crude OMW (**Table 3**). HPLC analysis revealed that the two main monomeric phenolic compounds in OMW were hydroxytyrosol (0.248 g/L) and tyrosol (0.201 g/L).

3.5 Olive mill wastewater as bio-insecticides to control *P. OPACA*

3.5.1 Weight loss of treated larvae

The effect of OMW spray on the larvae was significantly important compared to the control larvae sprayed with distilled water (**Figure 7**). Over time, larvae treated with OMW showed a significant weight loss from 2.38 to 2.02 g after 216 h. In contrast, the negative control was increased from 2.38 to 2.45 g after 168 h, and then decreased slightly from 2.41 to 2.39 g from 192 to 456 h. In comparison with the crude OMW and the negative control, the two positive controls (Cordus and Kemaban) showed a significant difference (**Figure 7**). Indeed, Cordus is a very effective insecticide resulting from the combination of two active substances, whose weight loss was similar to that of OMW for the two doses applied. The greatest weight loss over the first three days compared to the other treatments was achieved at a dose of 1 μL/mL, which resulted from the slow decrease in weight. Then, a more or less similar weight loss was observed for both doses. Kemaban insecticide is formed by a single active substance and resulted in significant weight loss at both doses compared to control and other treatments (Cordus and OMW) (**Figure 7**). Significant weight loss was seen during the first 4 days at a dose of 1 μL/mL of Cordus compared to 0.5 μL/mL of Kemaban. Thereafter, weight loss was found to be almost similar until 174 h, and then became slight and stable at a dose of 1 μL/mL until 224 h. The stability was the result of larval death.

3.5.2 Mortality rate of treated larvae

Mortality due to crude OMW was 33, 67 and 100% at 216, 224 and 456 h, respectively (**Table 4**). In the same way, Kemaban exhibited 33% mortality at 192 and 174 h with application of doses of 0.5 and 1 μL/mL, respectively. The mortality

Peak	Retention time (min)	Area (mAU)	Concentration (g/L)	Compounds
1	3.950	654.660	0.146	Gallic acid
2	9.967	3191.702	0.248	Hydroxytyrosol
3	13.317	2005.142	0.201	Tyrosol
4	14.450	53.766	0.122	Hydroxybenzoic acid
5	15.167	186.037	0.128	4-dihydroxybenzoic acid
6	16.117	102.685	0.124	Vanillic acid
7	17.317	42.461	0.122	Caffeic acid
8	22.250	53.637	0.122	Coumaric acid
9	30.933	39.469	0.122	Oleuropein
10	44.417	674.747	0.147	Quercetin

Table 3.
OMW phenolic compounds determined by HPLC.

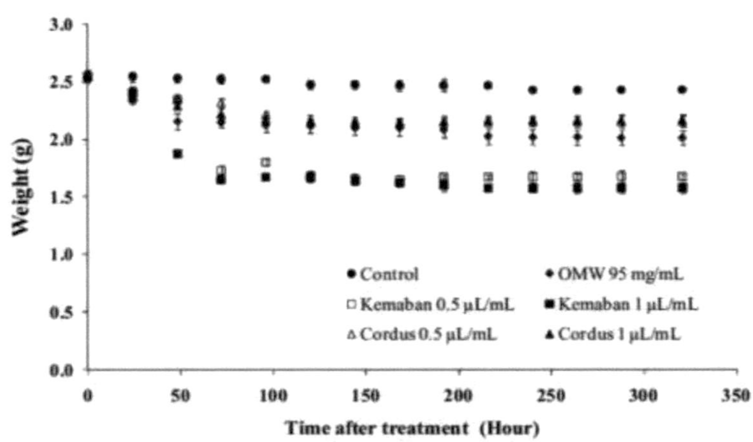

Figure 7.
Weight loss of P. opaca *larvae treated with crude OMW and different doses of commercial insecticides.*

	Time after treatment (Hour)										
Treatment	**0**	**144**	**146**	**174**	**192**	**198**	**216**	**218**	**224**	**288**	**456**
Control	0.00	0.00	0.00	0.00	0.00	0.00	0.00	0.00	0.00	0.00	0.00
Crude OMW											
95 mg/mL	0.00	0.00	0.00	0.00	0.00	0.00	33.33	33.33	66.67	66.67	100
Cordus											
0.5 μL/mL	0.00	0.00	33.33	66.67	66.67	66.67	100	100	100	100	100
1 μL/mL	0.00	33.33	33.33	66.67	66.67	100	100	100	100	100	100
Kemaban											
0.5 μL/mL	0.00	0.00	0.00	0.00	33.33	33.33	33.33	66.67	100	100	100
1 μL/mL	0.00	0.00	0.00	33.33	33.33	66.67	66.67	100	100	100	100

Table 4.
Percentage of dead Potosia opaca larvae for each treatment.

Treatments	LT_{50}(h) (95% CL)[a]	LT_{90} (h) (95% CL)[a]	Slope ± SE[b]	χ^2	Df[c]
Crude OMW					
95 mg/mL	245.39 (224.25–326.42)	323.86 (281.69–568.58)	0.01 ± 0.01	2.17	4
Cordus					
0.5 μL/mL	172.85 (128.56–201.58)	211.00 (188.72–383.17)	0.02 ± 0.01	1.98	4
1 μL/mL	160.02 (66 13–184.92)	199.23 (177.22–430.20)	0.02 ± 0.01	1.30	4
Kemaban					
0.5 μL/mL	208.01 (173.03–249.35)	233.91 (217.58–623.46)	0.04 ± 0.24	2.52	4
1 μL/mL	197.53 (164.23–237.69)	228.65 (209.02–507.80)	0.02 ± 0.02	2.07	4

[a]95% lower and upper confidence limits are shown in parenthesis.
[b]SE, standard error.
[c]Df, degree of freedom.

Table 5.
LT_{50} and LT_{90} values of OMW and two positive controls applied by using spray toxicity bioassay to control Potosia opaca larvae.

rate was 100% when larvae were treated with Kemaban at 218 and 216 h for 0.5 and 1 μL/mL, respectively. Otherwise, mortality of larvae treated with Cordus was significantly higher compared to other treatments. After 144 and 146 h, the effect began with doses of 1 and 0.5 μL/mL, respectively. However, larvae treated with OMW, Kemaban and Cordus started to die for the first time from day 9, 8 and 6 respectively. In contrast, the two commercial insecticides caused 100% mortality within 8 (Cordus) and 9 (Kemaban) days for all doses tested, compared to OMW which showed 33 and 100% mortality after 9 and 19 days, respectively.

3.5.3 Spray toxicity bioassay

Boutaj et al. [31] reported that OMW showed insecticidal activity to control *P. opaca* larvae with LT_{50} and LT_{90} values of 245 h and 324 h, respectively (**Table 5**). Furthermore, a positive correlation was found between treatments and the duration of exposure (Letal time to kill 50% and 90% of the population). The highest efficiency was recorded for Cordus insecticide with LT50 of 160 h and 173 h for 1 and 0.5 μL/mL doses, respectively and LT90 of 199 h and 211 h for 1 and 0.5 μL/mL doses, respectively. Median lethal times (LT_{50} and LT_{90}) generally decrease when insecticide concentrations increase. The least effectiveness was observed in *P. opaca* larvae for crude OMW. However, the results were close to Kemaban at a dose of 0.5 μL/mL.

4. Discussion

The palm represents the symbolism of life in the arid and semi-arid area and being one of the oldest domesticated trees with multifold social and economical status [41, 42]. The *P. dactylifera* is a former species, which constitutes the pivot of the oasis agriculture in the south of Morocco. Out of an overall area estimated at 84500 ha in 1948, the Moroccan palm groves in 1994 covered an area of 44,450 ha occupied by a total of 4.42 million palm trees [9]. This population is currently estimated at 5.12 million palms on an area of 48,000 ha. The importance of the palm by province showed that the provinces of Ouarzazate (1,873,000 palm trees), Errachidia (1,250,000), Tata (Bani) (800,000), Marrakesh (799,000), Tiznit (139,140), Guelmime (138,000) and Figuig (125,500) were the most important and thus constitute the largest phoenicultural areas (quoted by Sedra (2003) and updated by Meddich [43]. Meddich and Boumezzough [12] note that the prospected and infected *P. canariensis* palms were cultivated and located in the North-East palm grove of Marrakesh. This will allow exchanges of attacks between the two palm species studied. For the oasis ecosystem, the problem will cause a lot of damage and a serious socio-economical problem. Moroccan palms suffer from invasions by larvae and adults of *Rhinoceros Borer*; this is the case in Tunisia, the Middle East and Iran [33]. In Morocco, no *R. Borer* larvae or adults were found; also, the colonization of the wounds by the saprophytic fungi can lead to the appearance of rot and the dieback of the weakened palm. According to INRA (Institut National de Recherche Agronomique) reports and their research axis developed on *P. opaca* in oases newly identified in Morocco by Meddich and Boumezzough [12]. Tauzin [40] indicated the presence of this species in Anti-Atlas (Ifni, Tiznit), middle and southern of Morocco. The presence of *P. opaca* in the crown of the *P. canariensis* and *P. dactylifera* palm which can be explained only by the fact that adults have found an ideal biotope for egg laying and larval development. Meddich and Boumezzough [12] showed that *P. opaca* occurred in decaying wood of and *P. dactylifera*, also, where they consumed the wood and promote more rapid decay and laid their eggs in the hollows of branches. The finding of Meddich and Boumezzough [12] was supported by Mico and Galante [36]. Besides, adults of the Cetoniidae fly above the vault of the trees (including the palm tree) and feed on nectar plants and fruit trees and probably the inflorescences of the palm trees. In this way, it can be assumed that this species has undergone mutations by changing biotope and passing from saprophagous larvae (dead organic matter, compost) to saproxylophagous larvae (dead woods, rachis and dead and / or alive palms of the *P. canariensis* and *P. dactylifera*). As finding by Meddich and Boumezzough [12], *P. opaca* larvae was found of all studied sites. The phytophagous species (Scarabaeidae) were active at night. In the broad sense, saproxylophages cause damage by attacking either roots or leaves. Larvae (white worms) were generally the most harmful to palms. However, Meddich and Boumezzough [12] conclude that the degradation of the Moroccan palm grove may be linked to the attack of *P. opaca* larvae. They observed that the frequency of palms prospected and infected with *P. opaca* remains higher in the Marrakesh palm grove than in the south of Morocco. Date palm grove of Marrakesh is more confronted with anthropogenic constraints such as urban extension and air pollution by dust exceeding the tolerance threshold [43, 44]. Suspended particle concentrations in certain areas of Marrakesh are slightly above the limit value for health protection in Morocco [44]. This may induce the creation of biotopes necessary for the development of *P. opaca* larvae in this city.

Many researches carried out in other countries have highlighted the danger of the red palm weevil for both species *P. dactylifera* and *P. canariensis*. In the Arab countries, the efforts deployed to control *R. ferrugineus* were based mainly on modified cultural practices, the application of traditional insecticides and traps that uses pheromones to lure *R. ferrugineus* [38]. Moreover, the control of *P. opaca* insect pests involves spreading chemicals (in the form of toxic pellets) as uniformly as possible in areas where the larvae are causing damage. The elimination of diseased palms (attacked bylarvae) can reduce the spread of the pest and consequently limit the damage.

For preventive treatments of the aerial parts and the crown of the palm:

- Use Imidacloprid (200 g/L) by spraying at the crown (stipe, palm and heart) 2 to 3 times every 3 months. It is a systemic insecticide for the selective control of scarabaeidae beetles.

- Spray acetamiprid 100 g/L at 2–3 times / year.

- Also, use entomo-pathogenic nematodes with a dose of 180 million nematodes per 100 L of water, which act by reducing the size of the larvae and consequently affect the number of female adults capable of laying the eggs.

- Trapping control using pheromones. The use of such specific substances for sexual confusion may be a means.

Efforts now focus on the development of integrated pest management methods based on biological control and pheromone traps rather than on conventional insecticides [45]. Since it is an internal tissue borer, *R. ferrugineus* is difficult to control in the early stage of attack [38–46]. Initial efforts to control red palm weevil in the Kingdom of Saudi Arabia using chemical insecticides were failed [47]. An integrated pest management strategy, developed in India, has successfully suppressed the pest in the date plantations in the Kingdom of Saudi Arabia [38]. The strategy is modeled on the lines of tackling the pest on coconut. The pheromone traps has been used successfully to monitor and mass attract the pest, and it could be considered as the core of in any integrated pest management [48–50].

There is a great danger from chemicals such as insecticides and fungicides in human's food and animal use. Recognizing this real danger, farmers and consumers turned their efforts to environmental and eco-friendly practices by using as well as consuming biological and healthy products. The polyphenols are natural molecules present in OMW from the olive fruit which could be an alternative and an asset for pest control. However, the amount of OMW polyphenols may vary based on multitude factors, such as climatic conditions, olive variety and fruit ripening stage as well as the harvest period [51–53]. The OMW phenolic compounds content, can be considerably affected by the technological processes used for olive oil extraction [54]. In this context, the phenolic compounds content of OMW which, presented potential insecticidal activity has been assessed and investigated by Boutaj et al. [31] in view to develop new valorization strategies. Additionally, an application of a hydroxytyrosol-rich OMW extract by spraying it against olive psyllid (*Euphyllura olivine*), in a drip-irrigated olive orchard for evaluating the insecticidal activity of OMW, was carried out in 2008 and 2009 [29]. The extract from OMW had a strong insecticidal activity control this insect when the applied concentration was 2 g/L. In addition, the authors observed a significant biocide effect depending on OMW phenolic extracts concentration on *E. olivina* larvae as well as adults. Indeed, OMW showed similar toxicity to the Kemaban insecticide at 0.5 μL/mL dose. Nevertheless, it is clear that the obtained results were attributed to the chemical molecules that contain the two commercial insecticides. Cordus presents two active molecules namely chlorpyriphos ethyl and cypermethrin. As for Kemaban contains a single active molecule which, is chlorpyriphos ethyl. These molecules act on the spread of nerve impulses along the axon (cypermethrin action) and inhibit the acetylcholine esterase by blocking the transmission of the nerve flux (chlorpyriphos ethyl action) [55, 56]. The main mechanisms which explain the OMW's biocide effect on invasive species in general including insects are not clarified. It has been suggested that the transmission of the nerve flux may be blocked by the high phenolic compounds content in OMW [57, 58]. A significant inhibition of acetylcholine esterase activity in a marine mollusk (*Mytilusgallo provincialis*) has been reported by Danellakis et al. [57] after exposition to OMW. While, Campani et al. [58] reported that the inhibition of acetylcholine esterase may be attributed to the potential presence in OMW of organophosphates and carbamates, two pesticides which, are strong inhibitors of acetylcholine esterase activity and commonly used to treat the olive fruit fly (*Bactroceraoleae*). However, the authors did not dismiss the inhibition of acetylcholine esterase possible which, could be explained, by the phenolic compounds as well as metals and ammonia contained in OMW. Thus, the used OMW crude showed a toxic effect on *P. opaca* larvae. Danellakis et al. [57] noted that the toxicity is provided by the phenolic compounds and trace metals contained in OMW. Furthermore, Barbera et al. [59] showed that during their growth cycle, no phytotoxic effects were observed when OMW were applied on crops. This is related to plant phenological stage, the application modalities and the applied doses. The absence of harmful residues is the main advantage of OMW application to control plant pests as well as pathogens. Therefore, we can assume that there will be no need for a preharvest interval on crops, after the application of OMW.

5. Conclusions and future directives

In this chapter, we present a new eco-friendly approach to control the spreading of *P. opaca* which started in Morocco. Microbiological analysis show the presence of saprophytic fungi and genus *Fusarium* with a non-virulent strain. On the other hand, the two insecticides used separately and crude OMW are toxic on *P. opaca* var. Cardui Gyllenhal larvae. These results are promising and suggest the possibility of using OMW due to their high content of phenolic compounds as a means of biological control to overcome environmental problems caused by synthetic pesticides. The OMW and their phenolic extract compounds could be used in agricultural systems. Moreover, focused field researches (each plant-pathogen system) could be carried out to understand and evaluate the effects of OMW on specific *in situ* pest problems. Based on the main findings, it is clear that OMW may contribute to improve the date palm protection to control *P. opaca* and could be used as bio-insecticides. Nevertheless, OMW could be used safely as a challenge to control plant pest without affecting negatively the soil and plants. Besides, the use of OMW combined with other pest bio-control practical methods which could be a sustainable approach to minimize the potential risks. In this context and to understand the beneficial effects of OMW, more investigations could be required to assess the feasibility of OMW application in bio-controlled systems at large-scale, and determining the limitations and advantages on the long term. Further, research works are needed to test, besides crude OMW, pretreated OMW (ultra-filtered or heated) and its phenolic extracts as biodegradable pesticides.

Author details

Abdelilah Meddich[1*], Abderrahim Boutasknit[1], Mohamed Anli[1], Meriame Ait Ahmed[1], Abdelilah El Abbassi[1], Hanane Boutaj[1], Mohamed Ait-El-Mokhtar[1,2] and Ali Boumezzough[3]

1 Laboratory of Agro-Food, Biotechnologies and Valorization of Vegetal Bioresources, Faculty of Semlalia Sciences, Cadi Ayyad University, Marrakesh, Morocco

2 Department of Biology, Faculty of Sciences and Techniques Mohammedia, Hassan II University, Morocco

3 Ex-Full Professor University, Morocco

*Address all correspondence to: a.meddich@uca.ma and aboumezzough@gmail.com

References

[1] Sedghiani S, Raboudi F, Bouktila D, Makni H, Makni M. A practical molecular diagnostic tool of the date moth *Ectomyelois ceratoniae* (Lepidoptera: Pyralidae) in Tunisia. Journal of the Entomological Research Society. 2017;**19**(1):81-90

[2] Awad MA. Water spray as a potential thinning agent for date palm flowers (*Phoenix dactylifera* L.) c.v. "Lulu.". Scientia Horticulturae (Amsterdam). 2006;**111**(1):44-48

[3] Meddich A, Oihabi A, Jaiti F, Bourzik W, Hafidi M. Role of arbuscular mycorrhizal fungi on vascular wilt and drought tolerance in date palm (*Phoenix dactylifera* L.). Canadian Journal of Botany. 2015;**93**(6):369-377

[4] Meddich A, Jaiti F, Bourzik W, El Asli A, Hafid M. Use of mycorrhizal fungi as a strategy for improving the drought tolerance in date palm (*Phoenix dactylifera* L). Scientia Horticulturae (Amsterdam). 2015;**192**:468-474

[5] Howard F, Moore W, Giblin-Davis DRM, Abad RG. The Animal Class Insecta and the Plant Family Palmae. New York: Insects on Palms. Cabi Publishing; 2001. pp. 1-32

[6] El-Shafie HAF. Review: List of arthropod pests and their natural enemies identified worldwide on date palm, *Phoenix dactylifera* L. Agriculture and Biology Journal of North America [Internet]. 2012 Dec;**3**(13):516-524. [Cited: 26 July 2018]. Available from: http://www.scihub.org/ABJNA/PDF/2012/12/ABJNA-3-12-516-524.pdf

[7] Mahmoud EA, Gabarty A. Impact of gamma radiation on male proboscis of *Rhynchophorous ferrugineus* (Olivier, 1790) (Coleoptera: Curculionidae). Journal of the Entomological Research Society. 2017;**19**(2):53-65

[8] FAO. Palmiers: Un plan d'action international contre le charançon rouge. APS. 2017

[9] FOA. Données statistiques de production végétale. site FAOSTAT Database. 2003

[10] Faleiro JR. A review of the issues and management of the red palm weevil *Rhynchophorus ferrugineus* (Coleoptera: Rhynchophoridae) in coconut and date palm during the last one hundred years. International Journal of Tropical Insect Science. 2006;**26**(3):135-154

[11] Ben AA. Presentation of the status of infestations by the red weevil in Morocco. Notions on the biology of the pest. A strategy for the control of the disease. In: Workshop on the Technique of Endotherapy Applied in the Frame Work of the Strategy against the Red Palm. 2017

[12] Joseph SV, Taylor AG. Effect of insecticide-coated seeds on *Protaphorura fimata* (Collembola: Poduromorpha: Onychiuridae) feeding damage. Journal of Entomological Science [Internet]. 2017;**52**(4):463-467. Oct 8 [Cited: 26 July 2018]. Available from: http://www.bioone.org/doi/10.18474/JES17-46.1

[13] Herrero-Hernández E, Andrades MS, Álvarez-Martín A, Pose-Juan E, Rodríguez-Cruz MS, Sánchez-Martín MJ. Occurrence of pesticides and some of their degradation products in waters in a Spanish wine region. Journal of Hydrology. 2013 Apr 12;**486**:234-245

[14] Palma P, Köck-Schulmeyer M, Alvarenga P, Ledo L, Barbosa IR, López de Alda M, et al. Risk assessment of pesticides detected in surface water of the Alqueva reservoir (Guadiana basin, southern of Portugal). Science of the Total Environment [Internet]. 2014 Aug 1;**488-489**:208-219. [Cited: 25 July 2018].

Available from: http://www.ncbi.nlm.nih.gov/pubmed/24836129

[15] Ieromina O, Peijnenburg WJGM, De Snoo GR, Vijver MG. Population responses of *Daphnia magna*, *Chydorus sphaericus* and *Asellus aquaticus* in pesticide contaminated ditches around bulb fields. Environmental Pollution. 2014;**192**:196-203

[16] Jovana M, Tanja M, Mirjana S. Effects of three pesticides on the earthworm Eisenia fetida (Savigny 1826) under laboratory conditions: Assessment of mortality, biomass and growth inhibition. European Journal of Soil Biology [Internet]. 2014 May 1;**62**: 127-131. [Cited: 26 July 2018]. Available from: https://www.sciencedirect.com/science/article/pii/S1164556314000387

[17] Haouache N, Boughdadi A. Utilisation des margines contre Aphis pomi (De Geer, 1773) (Homoptera, Aphididae). In: Neuvième Congrès de l'Association Marocaine de Protection des Plantes; 2014. pp. 1-15

[18] El-Abbassi A, Saadaoui N, Kiai H, Raiti J, Hafidi A. Potential applications of olive mill wastewater as biopesticide for crops protection. Science of the Total Environment [Internet]. 2017 Jan;**576**:10-21. [Cited: 05 November 2016]. Available from: http://linkinghub.elsevier.com/retrieve/pii/S0048969716321982

[19] Shi X, Tang F, Zhou X, Bu X. In vitro inhibition of polyphenol oxidase activity by insecticides and allelochemicals in *Clostera anastomosis* (Lepidoptera: Notodontidae) larvae and poplar trees. Journal of Entomological Science [Internet]. 2017 Jul 14;**52**(3): 239-247. [Cited: 25 July 2018]. Available from: http://www.bioone.org/doi/10.18474/JES16-38.1

[20] Belaqziz M, El-Abbassi A, Lakhal EKEK, Agrafioti E. Galanakis CMCM. Agronomic application of olive mill wastewater: Effects on maize production and soil properties. Journal of Environmental Management [Internet]. 2016 Apr;**171**:158-165. [Cited: 13 June 2016]. Available from: http://linkinghub.elsevier.com/retrieve/pii/S0301479716300470

[21] Mishra RK, Bohra A, Kamaal N, Kumar K, Gandhi K, GK S, et al. Utilization of biopesticides as sustainable solutions for management of pests in legume crops: achievements and prospects. Egyptian Journal of Biological Pest Control [Internet]. 2018 Dec 30;**28**(1):3. [Cited: 25 July 2018]. Available from: https://ejbpc.springeropen.com/articles/10.1186/s41938-017-0004-1

[22] El Hassani FZ, Zinedine A, Amraoui MB, Errachidi F, Alaoui SM, Aissam H, et al. Characterization of the harmful effect of olive mill wastewater on spearmint. Journal of Hazardous Materials [Internet]. 2009 Oct 30;**170**(2-3):779-785. [Cited: 26 July 2018]. Available from: https://www.sciencedirect.com/science/article/pii/S0304389409007717?via%3Dihub

[23] Masi F, Bresciani R, Munz G, Lubello C. Evaporation-condensation of olive mill wastewater: Evaluation of condensate treatability through SBR and constructed wetlands. Ecological Engineering. 2015;**80**:156-161

[24] Hanafi F, Mountadar M, Assobhei O. Combined electrocoagulation and fungal processes for the treatment of olive mill wastewater. Fourteenth International Water Technology Conference. 2010;**14**:269-281

[25] Lafi WK, Shannak B, Al-Shannag M, Al-Anber Z, Al-Hasan M. Treatment of olive mill wastewater by combined advanced oxidation and biodegradation. Separation and Purification Technology [Internet]. 2009 Dec 10;**70**(2):141-146. [Cited: 26 July 2018]. Available from:

https://www.sciencedirect.com/science/article/pii/S1383586609003979

[26] Mechri B, Ben MF, Baham M, Ben ES, Hammami M. Change in soil properties and the soil microbial community following land spreading of olive mill wastewater affects olive trees key physiological parameters and the abundance of arbuscular mycorrhizal fungi. Soil Biology and Biochemistry [Internet]. 2008 Jan;**40**(1):152-161. [Cited: 09 April 2013]. Available from: http://linkinghub.elsevier.com/retrieve/pii/S003807170700315X

[27] Yangui T, Sayadi S, Gargoubi A, Dhouib A. Fungicidal effect of hydroxytyrosol-rich preparations from olive mill wastewater against *Verticillium dahliae*. Crop Protection [Internet]. 2010 Oct;**29**(10):1208-1213. [Cited: 09 April 2013]. Available from: http://linkinghub.elsevier.com/retrieve/pii/S0261219410001110

[28] Larif M, Zarrouk A, Soulaymani A, Elmidaoui A. New innovation in order to recover the polyphenols of olive mill wastewater extracts for use as a biopesticide against the *Euphyllura olivina* and *Aphis citricola*. Research on Chemical Intermedates [Internet]. 2013;**39**(9):4303-4313. Available from: http://link.springer.com/10.1007/s11164-012-0947-5

[29] Lykas C, Vagelas I, Gougoulias N. Effect of olive mill wastewater on growth and bulb production of tulip plants infected by bulb diseases. Spanish Journal of Agricultural Research [Internet]. 2014;**12**(1):233. Available from: http://revistas.inia.es/index.php/sjar/article/view/4662

[30] Boutaj H, Boutasknit A, Anli M, Ait Ahmed M, El Abbassi A, Meddich A. Insecticidal effect of olive mill wastewaters on *Potosia opaca* (Coleoptera: Scarabaeidae) larva. Waste and Biomass Valorization. 2019;**11**(7):3397-3405

[31] EL Hadrami I, EL Hadrami A. Breeding date palm. In: Breeding Plantation Tree Crops: Tropical Species. New York, NY: Springer; 2009. pp. 191-216

[32] Ehsine M, Belkadhi MS, Chaieb M. Seasonal and nocturnal activities of the Rhinoceros Borer (Coleoptera: Scarabaeidae) in the North Saharan Oases ecosystems. Journal of Insect Science [Internet]. 2014;**14**(1):256. [Cited 26 July 2018]. Available from: http://www.ncbi.nlm.nih.gov/pubmed/25527574

[33] Tahraoui A, El-Hilaly J, Israili ZH, Lyoussi B. Ethnopharmacological survey of plants used in the traditional treatment of hypertension and diabetes in South-Eastern Morocco (Errachidia province). Journal of Ethnopharmacology. 2007;**110**(1):105-117

[34] Mico E, Galante E. Larval morphology and biology of four Netocia and Potosia species (Coleoptera: Scarabaeoidea: Cetoniidae: Cetoniinae). European Journal of Entomology. 2003;**100**(1):131-142

[35] El-Abbassi A, Kiai H, Hafidi A. Phenolic profile and antioxidant activities of olive mill wastewater. Food Chemistry [Internet]. 2012 May;**132**(1):406-412. [Cited: 13 March 2013]. Available from: http://linkinghub.elsevier.com/retrieve/pii/S0308814611015925

[36] Finney DJ. Statisical logic in the monitoring of reactions to therapeutic drugs. Methods Inf Med [Internet]. 1971 Oct;**10**(4):237-245. [Cited: 26 July 2018]. Available from: http://www.ncbi.nlm.nih.gov/pubmed/5124593

[37] Meddich A, Boumezzough A. First detection of *Potosia opaca* larva attacks on *Phoenix dactylifera* and *Phoenix canariensis* in Morocco: Focus on pests control strategies and soil quality of

prospected palm groves. Journal of Entomology and Zoology Studies. 2017;**5**(4):984-991

[38] Tauzin P. Chorologie de Protaetia (Potosia) opaca en France (Coleoptera Cetoniinae Cetoniini). Cetoniimania. 2007;**1**(1856):19-48

[39] Zohary D, Hopf M. Domestication of Plants in the Old World: The Origin and Spread of Cultivated Plants in West Asia, Europe and the Nile Valley. Oxford, UK: Oxford University Press; 2000

[40] Chao CCT, Krueger RR. The date palm (*Phoenix dactylifera* L.): Overview of biology, uses, and cultivation. HortScience. 2007;**42**(5):1077-1082

[41] Meddich A. Program for the safeguarding and development of the Marrakech palm grove. In: National workshop on the palm grove and the gardens of Marrakech state of progress and assessment, organized by the Marrakech Palm Grove Observatory Congress palace, Marrakech, Morocco. 2014

[42] OREDD. Rapport sur l'Etat de l'Environnement de la Région de Marrakech. Observatoire Régional del'Environnement et de Développement durable. Ministère Marocain de l'Energie, des Mines, de l'Eau etde l'Environnement. 2013

[43] Abraham VA, Shuaibi MAA, Faleiro JR, Abozuhairah RA, Vidyasagar PSPV. An integrated management approach for red palm Weevil *Rhynchophorus ferrugineus* Oliv. A key pest of date palm in the Middle East. Journal of Agriculture and Marine Sciences [JAMS] [Internet]. 1998 Jan 1;**3**(1):77. [Cited: 28 July 2018] Available from: https://journals.squ.edu.om/index.php/jams/article/view/522

[44] Murphy S, Briscoe B. The red palm weevil as an alien invasive: Biology and the prospects for biological control as a component of IPM. Biocontrol News and Information. 1999;**20**(1):35-46

[45] Ferry M, Gomez S. The red palm weevil in the Mediterranean area. Palms. 2002;**46**(4):172-178

[46] Bokhari UG, Abozuhairah RA. Diagnostic tests for redpalm weevil. *Rhynchophorus ferrugineus* infested date palm trees. Arab Gulf Journal of Scientific Research. 1992;**10**:93-104

[47] Vidhyasagar PSPV, Haji M, Abozuhairah RA, ElMohanna O, Al SA. Impact of mass pheromonetrapping on red palm weevil : Adult population andinfestation level in date palm gardens of Saudi Arabia. Plant Kuala Lumpur. 2000;**76**:347-355

[48] Faleiro JR, Ashok Kumar J, Rangnekar PA. Spatial distribution of red palm weevil *Rhynchophorus ferrugineus* Oliv. (Coleoptera: Curculionidae) in coconut plantations. Crop Protection. 2002;**21**(2):171-176

[49] Al-Saoud AH. Effect of ethyl acetate and trap colour on weevil captures in red palm weevil *Rhynchophorus ferrugineus* (Coleoptera: Curculionidae) pheromone traps. International Journal of Tropical Insect Science. 2013;**33**(3):202-206

[50] Yousfi K, Cert RM, García JM. Changes in quality and phenolic compounds of virgin olive oils during objectively described fruit maturation. European Food Research and Technology [Internet]. 2006 May 28;**223**(1):117-124. [Cited: 24 July 2018]. Available from: http://link.springer.com/10.1007/s00217-005-0160-5

[51] Ben Ahmed C, Ben Rouina B, Boukhris M. Effects of water deficit on olive trees cv. Chemlali under field conditions in arid region in Tunisia. Scientia Horticulturae (Amsterdam)

[Internet]. 2007 Jul 20;**113**(3):267-277. [Cited: 28 July 2018]. Available from: https://www.sciencedirect.com/science/article/pii/S0304423807001306

[52] Gómez-Rico A, Fregapane G, Salvador MD. Effect of cultivar and ripening on minor components in Spanish olive fruits and their corresponding virgin olive oils. Food Research International [Internet]. 2008 Jan 1;**41**(4):433-440. [Cited: 26 July 2018]. Available from: https://www.sciencedirect.com/science/article/pii/S0963996908000409

[53] Butinar B, Bučar-Miklavčič M, Lipnik-Štangelj M. Antioxidants in virgin olive oils produced from two olive cultivars of Slovene Istria. Annales, Series Historia Naturalis [Internet]. 2006;**2**(16):201-208. [Cited: 28 July 2018]. Available from: https://repozitorij.upr.si/IzpisGradiva.php?id=5269&lang=eng

[54] McCarthy MM. Stretching the truth-why hippocampal neurons are so vulnerable following traumatic brain injury. Experimental Neurology. 2003;**1**(148):40-43

[55] Tian Y-T, Liu Z-W, Yao Y, Zhang T, Yang Z. Effects of alpha- and theta-cypermethrin insecticide on transient outward potassium current in rat hippocampal CA3 neurons. Pesticide Biochemistry and Physiology [Internet]. 2008 Jan 1;**90**(1):1-7. [Cited: 24 July 2018]. Available from: https://www.sciencedirect.com/science/article/abs/pii/S0048357507001058

[56] Danellakis D, Ntaikou I, Kornaros M. Dailianis S. Olive oil mill wastewater toxicity in the marine environment: Alterations of stress indices in tissues of mussel *Mytilus galloprovincialis*. Aquatic Toxicology [Internet]. 2011 Jan;**101**(2):358-366. [Cited: 27 June 2016]. Available from: https://linkinghub.elsevier.com/retrieve/pii/S0166445X10004327

[57] Campani T, Caliani I, Pozzuoli C, Romi M, Fossi MC, Casini S. Assessment of toxicological effects of raw and bioremediated olive mill waste in the earthworm Eisenia fetida: A biomarker approach for sustainable agriculture. Applied Soil Ecology [Internet]. 2013 Mar 1;**119**:43-53. [Cited: 28 July 2018]. Available from: https://www.sciencedirect.com/science/article/pii/S0929139317300707

[58] Barbera AC, Maucieri C, Cavallaro V, Ioppolo A, Spagna G. Effects of spreading olive mill wastewater on soil properties and crops, a review. Agricultural Water Management [Internet]. 2013 Mar 1;**119**: 43-53. [Cited: 28 July 2018]. Available from: https://www.sciencedirect.com/science/article/pii/S0378377412003344

[59] Barbera AC, Maucieri C, Ioppolo A, Milani M, Cavallaro V. Effects of olive mill wastewater physico-chemical treatments on polyphenol abatement and Italian ryegrass (*Lolium multiflorum* Lam.) germinability. Water research. 2014;**52**:275-281

Chapter 26

Fungal Biomass Load and *Aspergillus flavus* in a Controlled Environment

Alfred Mitema and Naser Aliye Feto

Abstract

Fungal biomass quantification is critical in understanding the interactions between the pathogen and susceptibility or resistance of the host plant as well as identifying competition between individual fungal spp. in disease progression. In the present chapter, two maize lines grown in different climatic regions of Kenya were infected with an aflatoxigenic *A. flavus* isolate (KSM014) and fungal colonization of the maize plant tissues was monitored by measuring fungal biomass load after 14 days in a controlled environment. The objective of the study was to determine whether the maize line colonized was a factor in increasing or limiting the growth of an aflatoxigenic strain of *Aspergillus flavus*.

Keywords: fungi, *Aspergillus flavus*, aflatoxins, maize lines

1. Introduction

Fungal biomass quantification is critical in understanding the interactions between the pathogen and susceptibility or resistance of the host plant in identification and competition in individual species of fungi in diseases progression [1].

The quantification and detection of biomass of fungi in plant host tissues have been conducted using polymerase chain reaction methods [2, 3]. Some authors, Sanzani et al. [2] showed that, the sensitivity of quantitative polymerase chain reaction (qPCR) can be applied to measure infections at very low volumes, thus, corresponding to the quantity of the pathogen that might be present during the latent or time of and or at asymptomatic infections. qPCR also enables the evaluation of stages of infection in plant tissues and the quantification of a fungal pathogen throughout the entire disease cycle [2].

Coninck et al. [1] developed a qPCR assay for detection and quantification of *Cercospora beticola* fungi in leaves of sugar beet. Moreover, Waalwijk and co-workers [4], using a gene involved in fumonisin biosynthesis developed a qPCR assay to quantify and detect fumonisin producing *Fusarium verticillioides* strains from maize obtained from South African subsistence farmers. These results were then compared to the fungal DNA content and with the fumonisin levels of the respective *F. verticillioides* strain. A qPCR assay was also developed by Nicolaisen et al. [5] using Ef1α for quantification and detection of 11 *Fusarium* spp. isolated from field materials associated with wheat and maize. Similarly, Korsman and co-workers [6], demonstrated the use of qPCR assay for detection and quantification of Gray leaf spot disease in maize leaves using cytochrome P450 reductase gene. These studies demonstrated the potential use of qPCR for detection and quantification of fungal pathogens and for probable selection of resistant plant cultivars in breeding lines. This also helps in understanding the processes involved in infection in a host-pathogen system and providing information on the bioecology [7, 8].

Mayer et al. [9] and Jurado et al. [10] used single copy mycotoxin biosynthetic genes to develop PCR assays for detecting mycotoxigenic fungi. Assay sensitivity increased when ITS1 and ITS2 spacer regions were included as, these regions have sufficient variability to enable discrimination of closely related species in the genus *Aspergillus* [11]. Subsequently, these regions have been successfully used for detection and identification of aflatoxigenic *Aspergillus* spp. [12, 13].

The objective of this study was to develop a sensitive, specific qPCR assay for quantifying *A. flavus* biomass in infected maize tissues. The assay was used to measure, the sensitivity of two dry land African maize lines grown in Kenya KDV1 and GAF4, when infected with an aflatoxigenic isolate (KSM014). Similar studies have been done with other fungal species, but this study is the first where the biomass of *A. flavus* from infected maize was detected and quantified with qPCR. This approach also could be used to discriminate between inbred maize lines that are sensitive or resistant to specific *A. flavus* strains and to help understand the mechanism of the maize defense response to *A. flavus*.

2. Materials and methods

2.1 Cultures of fungi

The aflatoxigenic *A. flavus* KSM014 isolate was cultivated and maintained as described previously [14] and thereafter stored as spore suspension in 15% glycerol for short term storage at -20°C or for long term storage at -80°C prior to DNA/RNA extraction.

2.2 Maize cultivars

GAF4 maize lines and KDV1 varieties were obtained from Kenya Agricultural and Livestock Research Organization (KALRO), Kenya. The selection of the varieties are focused mainly on their drought tolerance and the agro-ecological zones in which they were grown. Striga tolerant variety (GAF4) is produced by KALRO Kibos, Kisumu County. GAF4 is cultivated in Homa Bay, Kisumu and Busia counties [15]. The Kenya Dryland Varieties 1 is an open pollinated hybrid recommended for medium to low altitude areas. KDV1 is drought tolerant, matures early and produce flowers after germination between 45 and 52 days. It is mainly cultivated in Homa Bay and Makueni regions (http://drylandseed.com).

2.3 Media preparation and reagents

Phytagel, Nicotinic acid, Glycine, Thiamine hydrochloride, Murashige and Skoog medium (MS), Potassium hydroxide, Pyridoxine hydrochloride and Myo-inositol were from Sigma-Aldrich (USA). MS vitamins; 5 g myo-inositol, 500 mg Thiamine-HCl, 500 mg pyridoxine-HCl, 250 mg nicotinic acid and 100 mg glycine were filter sterilized after preparation in distilled water and thereafter stored at -20°C according to the instructions of the manufacturer's (Sigma-Aldrich, USA). The modified MS media was briefly prepared, 2.15 g MS salt was dissolved in sterile H_2O, thereafter, 10 ml MS vit. added and pH 5.7 adjusted using 1 M KOH and volume further adjusted to 1 l using sterile H_2O. 5 g of phytagel was added to MS media and heated in microwave to dissolve the salts. Fifty milliliters of the media was dispensed into tissue culture bottles, autoclaved and thereafter, cooled in the level 2 biosafety cabinet for approx. 1 h prior to inoculations as previously described [14].

2.4 Seed sterilization and *Aspergillus flavus* infection

The seeds were sterilized in a biosafety cabinet, level 2 [Contained Air Solutions (CAS) BioMAT2, UK]. Twenty milliliters of 95–100% ethanol was used for sterilization of viable seeds for 1 min and briefly shaken for 15 s. The alcohol was replaced with 20 ml of sodium hypochlorite (2.5%). After 15 min of reaction at room temperature, the mixture was shaken for 30 s and thereafter, the liquid discarded. 30 ml of sterile H_2O was used 5× to wash the seeds with intermittent shaking after every wash. 50 ml of sterile H_2O was added and left to stand for 1 hr at rmt. The H_2O was replaced with 20 ml of 2% Tween 20 and shaken for 30 s. Conidia suspensions adjusted to 1×10^6 conidia ml^{-1} using a hemocytometer was used to inoculate the seeds. The seeds in the tubes were para filmed after sealing and kept for 30 min in a shaking incubator at 30°C. Controls were treated with sterile H_2O instead of spores of conidia and thereafter, incubated following the same conditions. The seeds were left to dry in Petri dishes after inoculations overlaid with filter paper overnight (Whatman No. 1). The seeds were germinated in a plant growth chamber, Conviron (Winnipeg, Manitoba, Canada) set at 28°C after subsequent inoculations onto tissue culture bottles. The germination and growth were observed for a 14-day period, tissues of the plant (roots and shoots) were separately harvested, flash frozen in liquid nitrogen prior to DNA/RNA extraction and stored at −80°C.

2.5 The extraction of gDNA from maize tissues and *Aspergillus flavus*

A 100 mg of each of the following samples: control healthy maize tissues, infected and *A. flavus* KSM014 mycelia was used to extract DNA following the method of Möller et al. [16] with some modification. Briefly, 100 mM Tris pH 8.0, 2% SDS, modified TES buffer, 2% (w/v) polyvinylpyrrolidone (PVP) and 10 mM EDTA was prepared. 5 μl RNase (10 mg/ml) and 450 μl of TES buffer was added to microtube (2 ml) containing the tissues and thereafter, homogenized by vortexing for 15 min or with a microtube pestle. Twenty microlitres of Proteinase K (1 μg/μl) was added and vortexed for 1 min, thereafter, incubated for 1 h at 60°C. Seventy microliters of 10% CTAB (0.1 vol.) and 160 μl of 5 M NaCl (0.3 vol. was added and incubated at 65°C for 10 min). Seven hundred and fifty microliters of chloroform/isoamyl-alcohol (24:1) was added, vortexed for 5 min, incubated on ice for 30 min and centrifuged at 14,000 rpm for 10 min. The aqueous phase was transferred cautiously onto a new microtube (2 ml) and 300–350 μl isopropanol (0.55 vol.) added and left to stand for 30 min at RT after mixing gently for 30 s. The mixture was centrifuged for 10 min at 14,000 rpm. Supernatant was cast-off, the pellets rinsed with chilled 700 μl of 70% ethanol twice and centrifuged again at 14,000 rpm for 2 min after mixing gently. The pellets were air dried and dissolved in 40 μl TE buffer (10 mM Tris-Cl pH 8, 1 mM EDTA pH 8) or nuclease free water after discarding the ethanol. The integrity of DNA was assessed on a 1% agarose/EtBr gel and the concentration quantified on a Nano-Drop™ 1000 spectrophotometer (Nano Drop Technologies, USA). DNA was diluted to 10 ng/μl for further analysis.

2.6 Designing of primers

Sets of 3 primers (**Table 1**); elongation factor 1 alpha (*Ef1a*), *β*-tubulin, and membrane protein (*MEP*) were used in the current research. *MEP* and *Ef1a* were

Primer name	Forward primer (5′-3′)	Reverse primer (5′-3′)	Product size (bp)	Ta	Reference
Membrane Protein (*MEP*)	TGTACTCGGCAATGCTCTTG	TTTGATGCTCCAGGCTTACC	203	$64\,^{0}$ C	Manoli et al. [17]
Elongation Factor 1 alpha (*EF1α*)	CGTTTCTGCCCTCTCCCA	TGCTTGACACGTGACGATGA	102	$62\,^{0}$ C	Nicolaisen et al. [5]
β-TubM	TCTTCATGGTTGGCTTCGCT	CTTGGGTCGAACATCTGCT	118	$62\,^{0}$ C	Mitema et al. [18]

Table 1.
Specific primers used in the current study.

obtained from Dr. Shane Murray (pers. comm) whereas, *β*-tubulin was designed in Primer3 ver. 4.0 programme [19]. Secondary structure formation was evaluated in DNAMAN software ver. 6.0 (Lynnon LLC., USA) and further verified in OligoAnalyzer Tool (Integrated DNA Technologies). The melt curve and PCR analysis were used to identify both non-specific and specific amplification.

2.7 PCR amplification

Conventional polymerase chain reaction amplification was carried out in volumes of 25 μl and consisted of 0.5 μl of 10 μM dNTPs (Bioline), 10× reaction buffer with $MgCl_2$, 1 μl of 10 μM forward and reverse primers, 0.2 μl Kapa Taq, 1 μl of 10 ng DNA template, and sterile H_20. Protocol performed and followed for cycling conditions were: 1 cycle for 5 min at 94°C followed by 35× (for 30 s at 94°C, for 45 s at 60°C, for 90 s at 72°C). Elongation step was achieved at 72°C for 7 min and finally at 4°C for 1 min. The products of PCR were assessed on 2% agarose/EtBr gel in TAE1 X buffer (Tris–acetate 40 mM and EDTA 1.0 mM). Fermentas (100 bp DNA ladder) was used as a molecular size marker.

2.8 Standard curves and fungal quantification

Ten-fold serial dilutions of pooled 10 ng gDNA extracts from *A. flavus* and control plants were used to create standard curves. The threshold cycle (Ct) values for each dilution were plotted against the logarithm of the starting quantity of the template. The amplification efficiencies were created from the std. curve slopes according to the methods [13, 20]. Additionally, linear regression curves were drawn, and the qPCR efficiency was calculated as: $E = 10\left(\frac{-1}{Slope}\right)$.

The quantity of targeted DNA in an unknown sample was inferred from the respective std. curves.

Ten nanograms of DNA isolated from infected and healthy maize roots and shoots respectively were used to assess primer specificity. For the exclusion of false negative results, template DNA samples from fungi were assessed for polymerase chain reaction amplification with primer pairs *EF1a* and *β*-Tub. DNA extracted from pure fungal cultures (*A. flavus*) and control plant tissues were pooled, diluted to 10 ng/μl and used to evaluate the quantity of fungal DNA template in the infected plant tissue. The final fungal DNA template concentrations were 1, 5×10^{-1}, 2.5×10^{-1}, 1.25×10^{-1}, 6.25×10^{-2}, 3.125×10^{-2} ng/μl. These dilutions were used to estimate the detection limits of the *EF1a* and β-Tub primer pair in the infected plant tissues. Serial dilutions of extracted DNA from healthy maize tissue were prepared to gauge the detection limits of the *MEP*. For normalization and quantification of gene between different samples, the amount of fungal DNA as calculated by the *Ct* value for *β*-Tub and/or *EF1a* was divided by the amount of maize DNA as calculated by the *Ct* values for MEP. Rotor Gene 6000 2 plex HRM (Corbett Life Science Research, Australia) was used to assess the profiles of gene expression. Kapa *SYBR* Fast Kit, Master mix (Kapa BioSystems, South Africa) containing DNA polymerase, reaction buffers, dNTPs and 3 mM $MgCl_2$ were used for each polymerase chain reaction. Final concentrations of 10 μM gene specific primers (0.4 μl reverse and 0.2 μl forward), 1× Kapa SYBR green and 1 μl gDNA template were prepared to 20 μl total volume using nuclease free H_2O. Primer sets of specific genes (**Table 1**) were used in separate reactions which were performed in triplicate.

The quality and integrity of the isolated DNA, samples from infected and control tissues of the maize, and saprophytic fungi were subjected to polymerase chain reaction analysis with the reference genes under the following amplification conditions: for 10 min at 95°C; 35 cycles for 3 s at 95°C, for 20 s at 64°C, for 1 s at 72°C for *MEP* and at annealing temperature 62°C for both *Ef1a* and β-tubulin.

2.9 Statistical analysis

The statistical analysis was performed as previously described [14].

3. Results and discussion

3.1 Gene specificity and qPCR assays

To our knowledge, a qPCR assay for the detection and quantification of *A. flavus* biomass using extracted fungal DNA from control or infected maize tissues has not been previously reported. Since this is the first report, our discussion will be in comparison with reports for *Fusarium* spp. and related fungi where this assay is more commonly used.

In this study, the qPCR assay was developed to specifically detect and quantify *A. flavus* gDNA in maize tissues. Primers were designed, and their specificity was confirmed by testing against control and infected tissues (**Figure 1**). The fungal biomass in the co-infected shoots differed from the fungal biomass in the roots according to 1-way ANOVA analysis and TMCT test ($P < 0.05$).

Amplification of the MEP gene (203 bp) was used to detect maize DNA, while amplification of *β*-tubulin (118 bp) and *Ef1a* (102 bp) were used to detect *A. flavus* DNA (**Table 1**; **Figure 1**). The specificity of the primer pairs was determined by conventional PCR (**Figure 1**) after *A. flavus* KSM014 infection of GAF4 and KDV1 maize lines. *A. flavus* DNA extracted from infected maize plant tissues, for both lines, gave an amplification product for both *β*-tubulin (118 bp) and *Ef1a* (102 bp) (**Figure 1**). However, there was amplification product for *Ef1a* than there was for to *β*-tubulin (**Figure 1**), especially in the roots. The *MEP* gene (203 bp) was amplified in both control and infected maize plants for both lines (**Figure 1**). *MEP* amplification was plant specific and *β*-Tub and *Ef1a* were fungal specific. Based on these results, *β*-Tub is a better marker for detecting *A. flavus* in infected maize tissues than was *Ef1a* (**Figure 1**) and was used for fungal biomass determination.

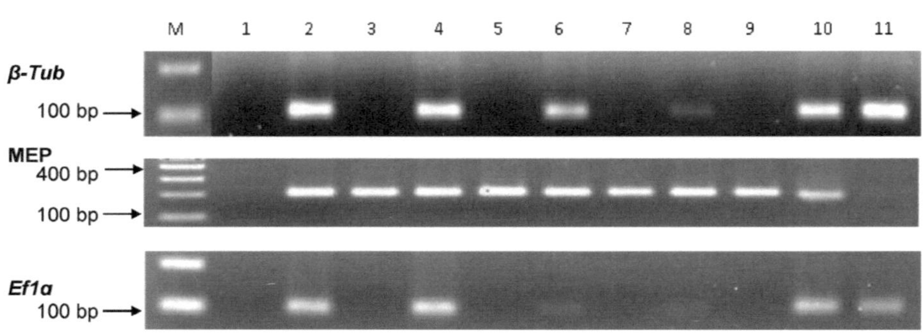

Figure 1.
*Gel images of the quantitative polymerase chain reaction amplicon sizes for maize maker gene (*MEP*) and* A. flavus *maker genes (*Ef1α*, β-tub) assessed on 2% agarose/EtBr gel run at 80 v for 45 min. M. 100 bp ladder; 1. NTC; 2. Pooled samples (maize gDNA and pure fungal gDNA); 3. GAF4 (control roots); 4. GAF4 (infected roots) 5. GAF4 (control shoots); 6. GAF4 (infected shoots); 7. KDV1 (control shoots); 8. KDV1 (infected shoots); 9. KDV1 (control roots); 10. KDV1 (infected roots); 11. KSM014 (positive control).*

3.2 Colonization of plant tissues by *Aspergillus flavus*

Aspergillus flavus KSM014 infection of both maize lines resulted in changes in maize phenotype with the KDV1 showing more severe symptoms that GAF4 (**Figure 2**). After 3–14 days post infection, the infected kernels for both maize lines showed stunted growth compared to control kernels (**Figure 2**). Additionally, the shoots and roots exhibited minimal growth with the *A. flavus* fungi colonizing the kernels and this could possibly explain the reason for stunted growth or germination. The phenotypic observations suggest that KDV1 maize line grown in Makeuni is more susceptible to fungal infection (*A. flavus*), whereas GAF4, grown in Kisumu and Homa bay appeared more resistant to the infection (**Figure 2**).

The observed phenotypic characteristics were further supported by the detection and quantification of fungal biomass load in gDNA extracted from infected and control plant tissues as revealed by the qPCR assay (**Figure 3**).

Insignificant difference was observed in biomass of fungi between infected plant tissues for the GAF4 and the control maize line (**Figure 3a**). In contrast, significant differences in biomass of fungi for the KDV1 maize line was exhibited upon infection ($p < 0.05$) for both the shoot and root tissue (**Figure 3b**). Fungal gDNA level was observed to be lower in the infected GAF4 maize line tissues compared to KDV1 suggesting that GAF4 was more resistant to *A. flavus* KSM014 infection than KDV1 (**Figure 3**).

The fungal biomass of *Alternaria dauci* was observed to be equivalent in two carrot cultivars between 1 and 15 days of post-inoculation, whereas it was found to be four-fold higher in the more susceptible cultivar between 21 and 25 days post-inoculation [21]. This suggests that fungal pathogens may colonize both susceptible and resistant cultivars in a similar manner during the first stages of the interaction, but fungal development is subsequently restricted in the partially resistant cultivar due to putative plant defense mechanisms [21].

It must be noted that we measured fungal biomass 14 days after infection when symptoms of the infection was phenotypically visible. However, other fungal biomass studies have shown that specific fungi could be identified even before the development of the symptoms. The presence of *Colletotrichum acutatum* by qPCR in

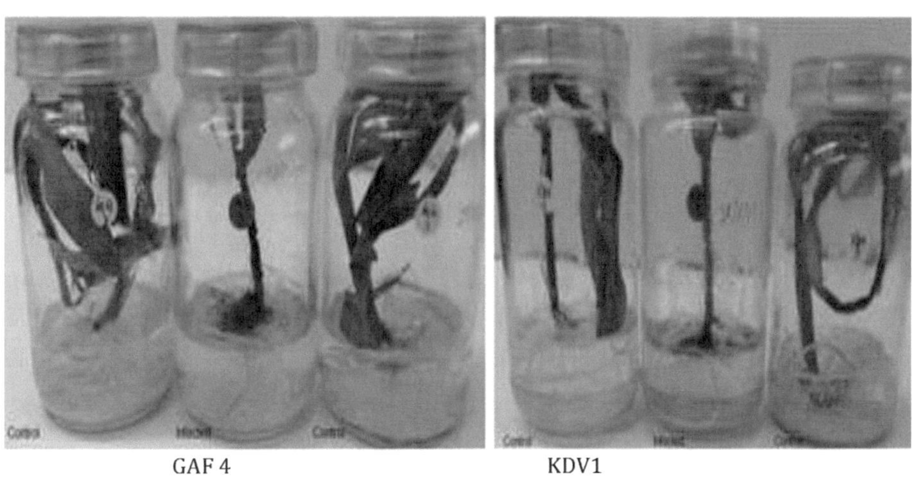

Figure 2.
The GAF4 and KDV1 maize lines after 14 days of growth with and without Aspergillus flavus *KSM014 infection. The red sticker shows infected maize plants while the white stickers are the control, uninfected maize plants.*

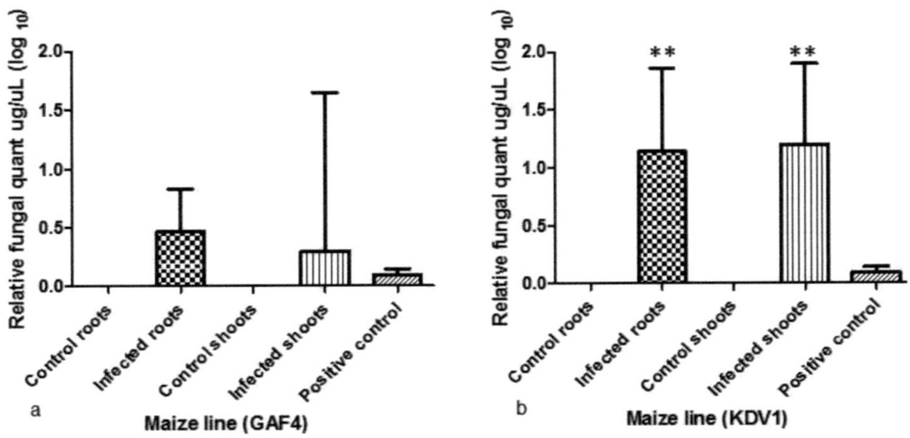

Figure 3.
Quantitative polymerase chain reaction analysis indicating fungal load of A. flavus *KSM014 in the shoot and root tissues of KDV1 and GAF4 maize* lines respectively. Biomass of fungi was measured in infected and non-infected *(control) GAF4 (a) and KDV1 (b) maize lines after 14 days where the A. flavus β-tub gene was used for quantification of fungi against the maize* MEP *gene. Tukey's multiple comparison test and one-way ANOVA ($P < 0.05$), was done where the asterisks indicate significance and the error bars shows standard mean deviation.*

strawberry leaves was detected by Debode et al. [22] 2 h post-inoculation whereas the initial symptoms of the disease appeared only after 96 h. Similarly, *Fusarium langsethiae* gDNA was accurately measured by Divon and Razzaghian [23] in oats independently from symptoms of the disease. These findings show the specificity and efficiency of the qPCR assay for the detection and quantification of fungal pathogens upon infection at early stages, before symptomatic appearances.

GAF4 is a *Striga* spp. resistant maize line cultivated in Kisumu, Kibos, Homa Bay and some parts of Nandi, while KDV1 is an open pollinated maize variety cultivated in Makueni and the neighboring counties. The observation that KDV1 maize line as more susceptible to aflatoxigenic *A. flavus* (KSM014) infection could be one of the contributing factors to why Makueni and the neighboring regions are more prone to frequent aflatoxicosis outbreak and high levels of aflatoxin contamination of the maize used for consumption.

The current study relates to the previous findings on Makueni maize samples [18] where they screened the strains of *A. flavus* isolated from maize kernels obtained from Makueni region on CAM media and found that there was significant variation in production of blue (toxigenic) and green (atoxigenic) fluorescence by most isolates. Seventy eight percent of the isolates from Makueni were observed to produce high amounts of aflatoxin AFB1, AFB2, the most potent carcinogen compared to other regions under study [18]. Additionally, studies conducted by Probst et al. [24] in eastern Kenya, revealed a similar result where they performed culture-based methods to monitor and describe the population structures of aflatoxigenic fungi and its closely associated strains on maize kernels. Moreover, a related study by Lewis et al. [25] and Klich [26] observed that in sub-Saharan Africa, products from subsistence farmers may reach the final consumer without the appropriated monitoring, resulting in critical risks for human health.

Moreover, the current study developed a qPCR assay using *A. flavus* gDNA and the *β*-tubulin gene for the quantification of *A. flavus* in maize tissue. Due to its high sensitivity and specificity, qPCR has been incorporated in official protocols of the European Plant Protection Organization (http://archives.eppo.org/index.htm) for the production, certification and assessment of healthy plant materials [27, 28]. This could therefore, in future, provide a screening strategy for finding African maize cultivars that are resistant to *A. flavus* infection or as an assessment of healthy maize plants. Zhao et al. [29] developed a qPCR assay for the detection of *Magnaporthe poae* resistant *Poa pratensis* (Kentucky bluegrass turf), which typically needed 3 weeks to detect using conventional culture-based methods. Further, Montes-Borrego et al. [30] demonstrated that fungal presence can be detected earlier, enabling the selection of resistant plants even when samples are indistinguishable based on visual assessment. Lastly, the early detection of latent infections of rust on leaves of cereals was used to estimate infection levels before the appearance of the disease [2].

The genomic DNA extracted from the co-infected shoots of both maize lines showed varied concentrations of fungal biomass load compared to the roots according to analysis using 1-way ANOVA and TMCT test ($p < 0.05$). The quantification of *Verticillium dahliae* gDNA in different tomato cultivars also revealed the concentration of pathogen DNA in plant tissues increased and decreased in susceptible and resistant cultivars, respectively [31]. Similarly, significant differences were found in the amount of *F. oxysporum* DNA in roots of different chickpea cultivars [32], while the detection of *Phomopsis sclerotioides* in pumpkin, melon, cucumber and watermelon showed that infection and rate of disease development of this polyphagous pathogen may vary according to the host [33]. In general, Vandemark and Barker [34], concluded that low levels of pathogen DNA in resistant plants is indicative of a mechanism that inhibits pathogen growth, whereas, the presence of a relatively high amount of pathogen DNA in asymptomatic plants indicates a resistance mechanism based on tolerance rather than on true resistance.

4. Conclusion

The study demonstrated that KDV1 maize line was more susceptible to *A. flavus* infection when compared to GAF4. This also implies that a possible reason for the frequent cases of aflatoxicosis in Makeuni county is the fact that the KDV1 maize line is grown in that region is more susceptible to *A. flavus* infection.

The *β*-tubulin gene is a potential marker for quantification of the *A. flavus* biomass load in maize plants compared to *Ef1a*. The *MEP* gene for maize gDNA was also found to be plant specific by the absence of cross-reaction with fungal gDNA. The specificity of the qPCR assay for *A. flavus* biomass quantification makes it a useful tool in other areas such as screening of *A. flavus* resistant maize lines for breeding, determining possible asymptomatic infection and in plant-pathogen interaction studies.

The next chapter will focus on in vitro biocontrol approach in aflatoxin mitigation and bio-analytical approaches to detect and quantify aflatoxins. The aim is to determine whether biocontrol can minimize aflatoxin production and to find important metabolites that are produced by specific *A. flavus* isolates.

Acknowledgements

The work was supported by the University Science, Humanities and Engineering Partnerships in Africa (USHEPiA) Fund and South African Bio-Design Initiative (SABDI) grant number 420/01 SABDI 16/1021. Also, the authors acknowledge the University of Nairobi, Kenya and Vaal University of Technology, Vanderbijlpark, South Africa for providing laboratory space and funding.

Authors contribution

Data curation, Alfred Mitema; Formal analysis, Alfred Mitema and Naser Aliye Feto; Funding acquisition, Nasesr Aliye Feto; Investigation, Alfred Mitema; Methodology, Alfred Mitema and Naser Aliye Feto; Resources, Alfred Mitema and Naser Aliye Feto; Validation, Alfred Mitema; Writing—original draft, Alfred Mitema; Writing—review & editing, Alfred Mitema and Naser Aliye Feto.

Conflicts of interest

The authors declare no conflict of interest. The authors are responsible for the content and writing of the paper.

Author details

Alfred Mitema[1,2]* and Naser Aliye Feto[1]

1 Department of Biotechnology, Vaal University of Technology, Vanderbijlpark, South Africa

2 School of Biological Sciences, University of Nairobi, Nairobi, Kenya

*Address all correspondence to: alfmite@yahoo.com

References

[1] Coninck BMA De, Amand O, Delaure SL, Lucas S, Hias N, Weyens G, et al. The use of digital image analysis and real-time PCR fine-tunes bioassays for quantification of Cercospora leaf spot disease in sugar beet breeding. Plant Pathology 2012;**61**:76-84

[2] Sanzani SM, Li Destri Nicosia MG, Faedda R, Cacciola SO, Schena L. Use of quantitative PCR detection methods to study biocontrol agents and phytopathogenic fungi and oomycetes in environmental samples. Journal of Phytopathology. 2014;**162**(1):1-13

[3] Mackay IM. Real-Time PCR in Microbiology From Diagnosis to Characterization. 2007

[4] Waalwijk C, Koch SH, Ncube E, Allwood J, Flett B, de Vries I, et al. Quantitative detection of *Fusarium* spp. and its correlation with fumonisin content in maize from south African subsistence farmers. World Mycotoxin Journal. 2008;**1**(February):39-47

[5] Nicolaisen M, Kærgaard L, Lazzaro I, Henrik N, Fejer A. Real-time PCR for quantification of eleven individual *Fusarium* species in cereals. Journal of Microbiological Methods. 2009;**76**:234-240

[6] Korsman J, Meisel B, Kloppers FJ, Crampton BG, Berger DK. Quantitative phenotyping of grey leaf spot disease in maize using real-time PCR. European Journal of Plant Pathology. 2012;**133**:461-471

[7] Demontis MA, Cacciola SO, Orrù M, Balmas V, Chessa V, Maserti BE, et al. Development of real-time PCR systems based on SYBR® Green I and TaqMan® technologies for specific quantitative detection of *Phoma tracheiphila* in infected citrus. European Journal of Plant Pathology. 2008;**120**:339-351

[8] Covarelli L, Beccari G, Steed A, Nicholson P. Colonization of soft wheat following infection of the stem base by *Fusarium culmorum* and translocation of deoxynivalenol to the head. Plant Pathology. 2012;**61**:1121-1129

[9] Mayer Z, Bagnara A, Färber P, Geisen R. Quantification of the copy number of nor-1, a gene of the aflatoxin biosynthetic pathway by real-time PCR , and its correlation to the cfu of *Aspergillus flavus* in foods. International Journal of Food Microbiology. 2003;**82**:143-151

[10] Jurado M, Vazquez C, Marın S, Sanchis V, Jaen MTG. PCR-based strategy to detect contamination with mycotoxigenic *Fusarium* species in maize. Systematic and Applied Microbiology. 2006;**29**:681-689

[11] Edwards SG, Callaghan JO, Dobson ADW. PCR-based detection and quantification of mycotoxigenic fungi. Mycological Research. 2002;**106**(9):1005-1025

[12] González-Salgado A, Patiñeo B, Gil-Serna J, Vázquez C, González-Jaén MT. Specific detection of *Aspergillus carbonarius* by SYBR® green and TaqMan® quantitative PCR assays based on the multicopy ITS2 region of the rRNA gene. FEMS Microbiology Letters. 2009;**295**(1):57-66

[13] Sardiñas N, Vázquez C, Gil-Serna J, González-Jaén MT, Patiño B. Specific detection and quantification of *Aspergillus flavus* and *Aspergillus parasiticus* in wheat flour by SYBR® Green quantitative PCR. International Journal of Food Microbiology. 2011;**145**:121-125. DOI: 10.1016/j.ijfoodmicro.2010.11.041

[14] Mitema A, Okoth S, Rafudeen MS. The development of a qPCR assay to measure *Aspergillus flavus* biomass

in maize and the use of a biocontrol strategy to limit aflatoxin production. Toxins (Basel). 2019;**11**(179):1-18

[15] Sunda W, Ochuodho J, Ngode L, Okalebo JR, Othieno CO, Nekesa AO, et al. Development of integrated Striga management package to improve maize production in Western Kenya. In: Third RUFORUM Biennial Meeting; 2012. pp. 375-381

[16] Möller EM, Bahnweg G, Sandermann H, Geiger HH. A simple and efficient protocol for isolation of high molecular weight DNA from filamentous fungi, fruit bodies, and infected plant tissues. Nucleic Acids Research. 1992;**20**(22):6115-6116

[17] Manoli A, Sturaro A, Trevisan S, Quaggiotti S, Nonis A. Evaluation of candidate reference genes for qPCR in maize. Journal of Plant Physiology. 2012;**169**(8):807-815. DOI: 10.1016/j.jplph.2012.01.019

[18] Mitema A, Okoth S, Rafudeen MS. Vegetative compatibility and phenotypic characterization as a means of determining genetic diversity of *Aspergillus flavus* isolates. Fungal Biology. 2018;**122**(4):203-213. DOI: 10.1016/j.funbio.2017.11.005

[19] Untergasser A, Cutcutache I, Koressaar T, Ye J, Faircloth BC, Remm M, et al. Primer3-new capabilities and interfaces. Nucleic Acids Research. 2012;**40**(15):1-12

[20] Kubista M, José Manuel A, Bengtsson M, Forootan A, Jonak J, et al. The real-time polymerase chain reaction. Molecular Aspects of Medicine. 2006;**27**:95-125

[21] Boedo C, Le Clerc V, Briard M, Simoneau P, Chevalier M, Georgeault S, et al. Impact of carrot resistance on development of the Alternaria leaf blight pathogen (*Alternaria dauci*). European Journal of Plant Pathology. 2008;**121**:55-66

[22] Debode J, Van Hemelrijck W, Baeyen S, Creemers P, Heungens K, Maes M. Quantitative detection and monitoring of *Colletotrichum acutatum* in strawberry leaves using real-time PCR. Plant Pathology. 2009;**58**:504-514

[23] Divon HH, Razzaghian J. *Fusarium langsethiae* (Torp and Nirenberg), investigation of alternative infection routes in oats. European Journal of Plant Pathology. 2012;**132**:147-161

[24] Probst C, Schulthess F, Cotty PJ. Impact of *Aspergillus* section *Flavi* community structure on the development of lethal levels of aflatoxins in Kenyan maize (*Zea mays*). Journal of Applied Microbiology. 2010;**108**(2):600-610. Available from: http://www.ncbi.nlm.nih.gov/pubmed/19674186

[25] Lewis LMO, Njapau H, Schurz-Rogers H, George Luber SK, Nyamongo J, Backer L, et al. Aflatoxin contamination of commercial maize products during an outbreak of acute aflatoxicosis in Eastern and Central Kenya. Environmental Health Perspectives. 2005;**113**(12):1763-1767 Available from: http://www.ehponline.org/ambra-doi-resolver/10.1289/ehp.7998

[26] Klich MA. *Aspergillus flavus:* The major producer of aflatoxin. Molecular Plant Pathology. 2007;**8**(6):713-722. DOI: 10.1111/j.1364-3703.2007.00436.x

[27] Blanco-Meneses M, Ristaino JB, Pathology P, Carolina N, Rica C, Carolina N. Detection and quantification of *Peronospora tabacina* using a real-time polymerase chain reaction assay. Plant Disease. 2011;**95**(6):673-682

[28] Boutigny AL, Beukes I, Small I, Zühlke S, Spiteller M, Van Rensburg BJ, et al. Quantitative detection of *Fusarium* pathogens and their mycotoxins in South African maize. Plant Pathology. 2012;**61**(3):522-531

[29] Zhao C, Waalwijk C, De Wit PJGM, Tang D, Van Der Lee T. RNA-Seq analysis reveals new gene models and alternative splicing in the fungal pathogen *Fusarium graminearum*. BMC Genomics. 2013;**14**(21):1-16

[30] Montes-borrego M, Protection C, Agriculture S, Viso CC, Sevilla C, Jiménez-díaz RM, et al. Real-time PCR quantification of *Peronospora arborescens*, the opium poppy downy mildew pathogen, seed stocks and symptomless infected plants. Plant Disease. 2011;**February**:143-152

[31] Gayoso C, Ilarduya OM, Federico P, Merino F. Assessment of real-time PCR as a method for determining the presence of *Verticillium dahliae* in different Solanaceae cultivars. European Journal of Plant Pathology. 2007:199-209

[32] Jiménez-Fernández D, Montes-Borrego M, Navas-Cortés JA, Jiménez-Díaz RM, Landa BB. Identification and quantification of *Fusarium oxysporum* in planta and soil by means of an improved specific and quantitative PCR assay. Applied Soil Ecology. 2010;**46**(3):372-382

[33] Shishido M, Kubota I. Comparison of PCR assays for detection and quantification of *Phomopsis sclerotioides* in plant and soil. Journal of General Plant Pathology. 2013;**79**:18-27

[34] Vandemark GJ, Barker BM. Quantifying *Phytophthora medicaginis* in susceptible and resistant alfalfa with a real-time fluorescent PCR assay. Journal of Phytopathology. 2003;**151**:577-583

Chapter 27

Role of Decomposers in Agricultural Waste Management

Nusrat Iqbal, Amrish Agrawal, Saurabh Dubey and Jitender Kumar

Abstract

In this chapter, agricultural waste residue management by bio-organisms is discussed along with different types of decomposition processes. Tons of agricultural wastes are produced every year. These agricultural wastes create major environmental problems without effective means of management methods. There are many technologies being used for the decomposition, which mainly include anaerobic decomposition, compositing, fermentation, etc. All these decomposition processes depend upon the different soil-inhabiting microbes. These microbes are the key components of agri-residue decomposition process. Every step of decomposition requires different microbes. Various sets of catalytical enzymes are involved for the catabolic procedures of organic matter. By successive catabolic reactions, all the organic matters are mineralized into soil essential constituents, which will be the most effective sources of macro- and micronutrients for the soil fertility. Working efficiency of these microbes depends upon different parameters like moisture, temperature, pH, etc. The vitality and efficiency of microbes can be enhanced by using various inert carriers. If the efficiency of these soil microbes enhances by various factors, then the rate of decomposition could be enhanced to handle this ever-increasing problem of agriculture residue in near future.

Keywords: agri-waste, pollution, decomposition, mineralization, microbes

1. Introduction

Agricultural waste is produced by various agricultural operations, and it contains sugarcane baggage, paddy and wheat straw and husk, waste of vegetables, food products, jute fibers, crop stalks, etc. About 998 million tons of agri-residue waste is produced every year [1]. The abundance of agricultural waste production causes a lot of environmental pollution and generates many environmental contamination problems. During certain periods of time, characteristics of waste materials have changed and cause harmful and toxic effects toward human beings. Approximately, 2 tons per day agri-waste is produced in rural areas. Besides this, cow houses and sugar industry produced an average 20 million tons of waste, which is a rich source of nutrients and manures. Conventionally, farmers burnt up or left the agricultural waste in the field, but this causes lots of air pollution and soil pollution. Various techniques have been utilized for agri-residue disposal.

Agricultural waste mainly consists of crop residues which have lots of organic carbon content and a supply of plant nutrients. Crop residues' retention after harvesting reduces soil erosion [2]. Combine harvester machines contributes 75% residue after harvesting but due to high silica content animals are not likely to feed these residues. Then farmers start to burn the residues, but burning emits 8.77 Mt. of CO, 141.15 Mt. CO2, 0.23 Mt. of NO, and 0.12 Mt. of NH3 [3], which causes air pollution and loss of organic content of approximately 80–90% N, 25% of P, 20% of K, and 50% of Singh et al. [4]. So, the rice straw management is a challenging task in rice-producing regions. So, there is a need for an effective waste disposal technology for converting this waste into some valuable form. Physical, chemical, and biological decompositions break the lingo-cellulose bonds in crop residues and result in the enhancing of the nutrient content of soil [5]. Biological decomposition is the main and efficient decomposition method in which bacterial and fungal spores speed up the decomposition of waste under aerobic and anaerobic conditions. Microbial decomposition enhances nutrient content by nitrogen fixing, phosposorous solublization, and cellulose decomposition of decomposed final product [6]. There are a variety of bio-decomposers such as bacteria, fungi, protozoa, etc. and they are capable to degrade cellulose by depolymerizing cellulases which hydrolyze ligno-celluloses. Most commonly known bio-decomposers are fungi which include *Humicola*, *Trichoderma*, and *Penicillium aspergillus* [7]. The market sale value of soil microbes are increasing nowadays [8]. Indian government is working for food self-sufficiency and environmental sustainability. Due to the high market value, the production of soil microbial-based decomposer product would be expected to increase in coming years. Researchers have been able to identify and isolate multiple types of bio-decomposers, but still no formulation is available for the efficient use of these microbes. Researchers are working to improve the efficiency and storage of multiple types of microbes in an effective formulation product in one package which would have a high commercial value.

2. Production of agricultural waste internationally

About 998 million tons of agri-residue waste are produced every year [1]. Eighty percent of organic waste can be converted into organic manure with the rate of 5.27 kg/day/1000 kg wt basis [9]. Development of agricultural waste management methods for sustainable and eco-friendly approach is required for effective disposal of agri-waste [10]. Agro-waste may cause several health and environments issues if not disposed properly so, it requires a very safe method for disposal [11]. Many studies and research work have been focused on development of agro-waste based bio-fertilizers. Bio-fertilizers basically contain living organisms and enrich the soil with different soil nutrient and minerals which is very essential for the growth and development of plant [12]. Microbial inoculants in industries were being used in Malasiya in late 1940s [13].

Eighty percent of organic waste can converted into organic manure with the rate of 5.27 kg/day/1000 kg wt basis by sustainable and efficient waste disposable methods [9, 10]. Agro-waste may cause several health and environments issues if not disposed properly so it requires a very safe method for disposal (Sud et al., 2008).

Many studies and research works have been focused on development of agro-waste based bio-fertilizers. The bio-fertilizer obtained from the natural resources is a sustainable source of fertility inducer for the small farmers [14]. Bio-fertilizers basically contain living organisms and enrich the soil with different soil nutrients and minerals which are very essential for the growth and development of plant [12].

Bio-fertilizers are derivatives of agricultural residues such as straw, corn stalks, etc. that are decomposed by bio-decomposers and are also known as recycled

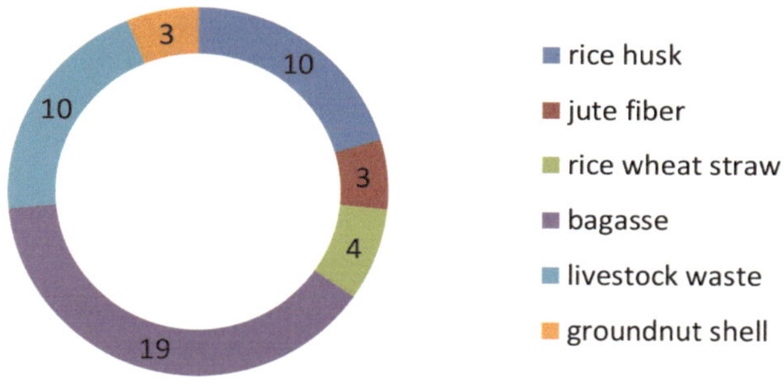

Figure 1.
Agriculture waste percentage from different sources.

products. The leafy vegetables, fruits, and cereal crops residues are an efficient source of basic nutrients such as nitrogen, phosphorus, and potassium as well as secondary and micronutrients such as calcium, boron, magnesium, and manganese. The bio-decomposer enhances the soil fertility and soil texture (**Figure 1**).

3. Types of agricultural waste

There are three major types of agricultural waste (**Figure 2**):

1. **Livestock manure**: The agricultural land is generally devoted to three farm animals: cattle, pig, and poultry, and lands used for the farms are about 15.3, 0.10, and 1.3 million hectares, respectively [15]. Around 120 million tons of manure are produced from these livestock sectors per year.

2. **Postharvest agricultural waste:** These wastes are in the category of primary agricultural residue. It mainly includes straw, husk, and stalks of planted crop, which are left after harvest. These can be used in fodder and rest is decomposed by different waste decomposers or burnt. These residues are rich in cellulosic fibers [16].

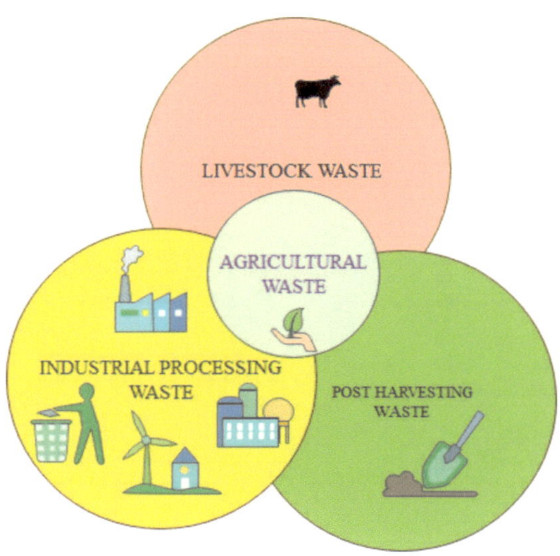

Figure 2.
Types of agricultural waste.

3. **Agro-industrial residues**: These are in the category of secondary agricultural residuals produced after processing of agricultural crop into various bio-product forms. These include husk, molasses, bagasse, peels, hulls, husk, etc.

4. Waste management processes

There are different methods of solid waste management which are well known. Among the various methods, bio-process-based are the most suitable for managing the waste in a safe way and they generate the nontoxic degraded foams.

1. **Landfills:** Landfill is the most primitive and commonly used waste disposal method (**Figure 3**). In this method, waste organic materials are simply collected and thrown in depth or pits. There are two main processes involved in an efficient landfilling method: 1. Collection of waste in a very small and confine place. 2. Compaction of collected waste by compactors or bulldozers.

There are five phases involved in this method:

i. Phase I: Adjustment of waste with O_2 supply

The waste compacted in this phase with void spacing for oxygen exchange (O_2). Initially, O_2 level is high due to activity of microbes for aerobic decomposition process but O_2 level gradually decreased due to high microbial density.

ii. Phase II: Transition from aerobic to anaerobic

Due to high microbial density, O_2 level decreases and CO_2 level increases. The anaerobic condition occurs in the layers and primary electron acceptors are sulfates and nitrates instead of O_2.

iii. Phase III: Generation of organic acids by acidification

This phase is also known as acidification phase, and in this, volatile fatty acids (VFA) are decomposed to acetic acid (CH3COOH), CO2, and H2. The H2 formation phase occurred in the last and activates the H2 oxidizing bacteria. The acidogenic bacteria increase the decomposition rate of waste.

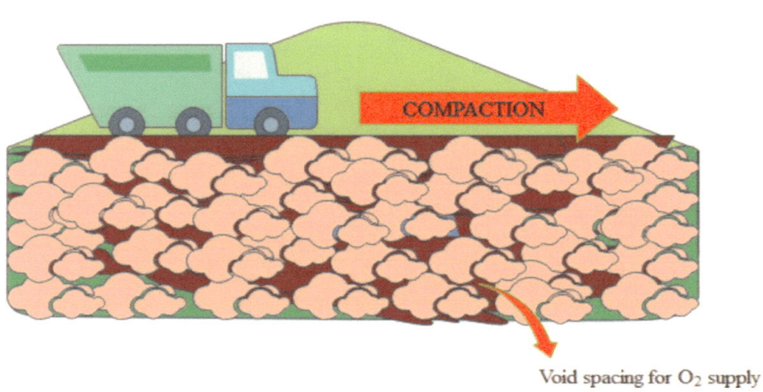

Figure 3.
Landfill mechanism of waste decomposition.

iv. Phase IV: Methonogenesis

The organic acids products, viz., acetic, propionic, and butyric acids are converted into CH4 and CO2 as intermediate products by methoanogenic microorganisms. In this phase, pH changes from acidic to neutral. This phase is the longest phase of decomposition.

v. Phase V: Humification

All the microbial activity slows down in this phase and CH4 production also completely disappears. All the remaining organic materials are converted into gas phase by the oxidative processes and converted into humic acid compounds.

2. **Anaerobic decomposition:** Anaerobic decomposition is gaining attention in various energy sources to fulfill the demand of increasing population in a safer and economical way [17]. There are around 267 plants that have been distributed all over the world. In anaerobic decomposition, microorganisms decompose the solid waste into biodegradable and nontoxic forms in the absence of oxygen (1). There are two types of bacteria which play a significant role in anaerobic decomposition reactions, which include hydrolytic bacteria and the acidogenic bacteria. These two bacteria belong to two kingdoms, the archea and the bacteria [18].

There are four successive stages of anaerobic decomposition which are: hydrolysis, acidogenesis, acetogenesis, and methanogenesis. These four stages are carrying out by the interaction of various microbes [19].

a. Hydrolysis

The decomposition pathways start with the hydrolysis of the waste agricultural biomass, which is rich in complex polymeric constituents. The hydrolytical cleavage broke the polymeric interconnecting bonds by the hydrolysis process. Agricultural waste is rich in organic polymers such as cellulose, hemicelluloses, lignin, etc. [20]. Hydrolysis breakdown is an electrochemical process driven by the hydrolytic bacteria by cellulolytic enzymes, lipase enzymes, etc. Hydrolysis cleavage converts these insoluble long chain polymers into soluble simpler derivatives like ammonium and other organic constituents [21].

Kinetically, hydrolysis reactions are the first-order reactions in which complex biodegradable waste material is hydrolyzed at constant temperature and pH [22].

$$\mathrm{(C6H10O5)n + n\ H2O \rightarrow n\ C6H12O6 + n\ H2}$$

$$\frac{\mathrm{dXdegr}}{\mathrm{dt}} = -Kh.\mathrm{Xdegr} \tag{1}$$

where Xdegr = Amt. of Decomposed material (kg/m^3)
t = time interval (days)
kh = hydrolysis constant in first order reaction.

Hydrolysis reactions in anaerobic decomposition can be considered as rate-determining steps, which depend upon the ration of hydrolytic product to microorganisms, pH 5–7, and 30–35°C [23]. The hydrolysis reactions make the agricultural postharvesting biodegradable waste residue into e simpler forms and make it accessible for acidogenic bacteria for further breakup.

b. Acidogenesis

The hydrolysis soluble products of biodegradable material are easily diffused in to the cell membranes of acidogenic microorganisms. Acidogenic reactions are associated with obligate and facultative microbes, viz., *Micrococcus*, *Peptococcus*, *Streptococcus*, *Desulfomonas*, and ***Escherichia coli***. The acidogenic microorganisms further convert the hydrolytic products into CO_2, H_2, and various organic acids which are known as volatile fatty acids (VFA) like acetates, propionate and butyrate along with this small amount of ethanol and lactate [24, 25]. The rate of reaction of VFA production is more in lower pH conditions, that is, pH 5 and ethanol production occurs in pH 4 [26]. The final acidic product can be used as a liquid and dry fertilizer [27]. VFA also creates the precursors for the final stage of methanogenesis. Chemical reactions that occur in acidogenesis are:

$$C6H12O6 \leftrightarrow 2\ CH3CH2OH + 2\ CO2$$

$$C6H12O6 + 2\ H2 \leftrightarrow 2\ CH3CH2COOH + 2\ H2O$$

$$C6H12O6 \rightarrow 3\ CH3COOH$$

c. Acetogenesis

The acidogenesis products like acetates and other organic acids further convert into hydrogen gas by dehydrogenation reactions. The hydrogen produced in this stage is the main substrate for methanogenic microorganisms. The reactions involved in this stage are [28, 29]:

$$CH3CH2COO - +3\ H2O \leftrightarrow CH3COO - +H + HCO3 - +3\ H2$$

$$C6H12O6 + 2\ H2O \leftrightarrow 2\ CH3COOH + 2\ CO2 + 4\ H2$$

$$CH3CH2OH + 2\ H2O \leftrightarrow CH3COO - +3\ H2 + H+$$

The hydrogen produced in this step causes an inhibitory effect toward acetogenic microorganisms and the produced hydrogen is available for the consumption of methanogens. So, microorganisms of acetogenisis and methanogenisis work symbiotically in association.

d. Methanogenesis

In the final stage of anaerobic digestion, all the intermediates convert into H2, CH4, CO2, and CH3COOH [30]. The microbial species mainly include *Methanobrevibacter ruminantium*, *M. bryantic* and *M. thermoautotrophicum*, *Methanogenium cariaci and M. marinsnigri*, etc. [31]. Methane-producing bacteria can be divided into two groups, namely, acetophilic and hydrogenophilic.

The reaction equations in the methanogenic stage are as follows [32]:

$$CH3COOH \rightarrow CH4 + CO2$$

$$CO2 + 4\ H2 \rightarrow CH4 + 2\ H2O$$

$$2\ CH3CH2OH + CO2 \rightarrow CH4 + 2\ CH3COOH$$

3. **Compositing:** Compositing is defined as the process of decomposition of complex waste organic matter into the simpler molecular chains. These smaller molecular compounds further decompose and form soil humus [33].

Compositing process is also mediated by different microorganisms in aerobic environment so it is also known as aerobic decomposition. The most common microorganisms are: bacteria, fungi, algae, and protozoans [34].

Compositing process can be explained by this equation:

$$\text{Organic matter} + O_2 \rightarrow \text{Compost} + CO_2 + H_2O + NO_3^{-} + SO_4^{2-} + \text{Heat}$$

Compositing process occurs in two phases which include: 1. degradation and 2. maturation.

1. Degradation: In this phase simpler organic matter are degraded by aerobic microorganisms. This phase is for several weeks to few months and occurs with high speed. This step needs temperature with full aeration for cooling the support also termed as thermophillic phase. Various phytotoxins also release as decomposition proceeds [35]. After degradation of simpler organic matter then decomposition of complex organic matter starts. At the end of this phase, most of the microorganisms die due to nonavailability of sufficient food. In addition to this, thermophillic condition changed to mesophillic and temperature drops from 50-55C to 25 C. According to several authors [33, 34, 36], this stage occurred for longer period of time.

2. Maturation: During the mesophillic phase or maturation phase, actinomyces appear and start degrading compost material of first phase into finer particles. Lots of invertebrates like earthworms, ticks, and centipedes further disintegrate by chemical and biological transformations (**Figure 4**). These transformations are termed as humification which is endorsed by the oxidative polymerization of phenolic compounds acquired by degradation of organic matter in the first phase.

3. Solid-state fermentation (SSF) process is another bioprocess for solid waste disposal in absence of free water [37]. Microbial growth is very important for

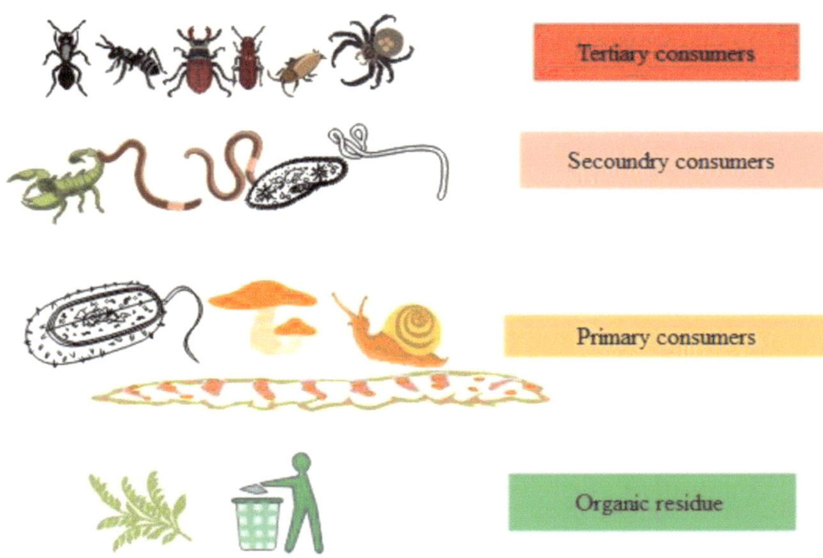

Figure 4.
Hierarchy of bio-decomposers in compositing process.

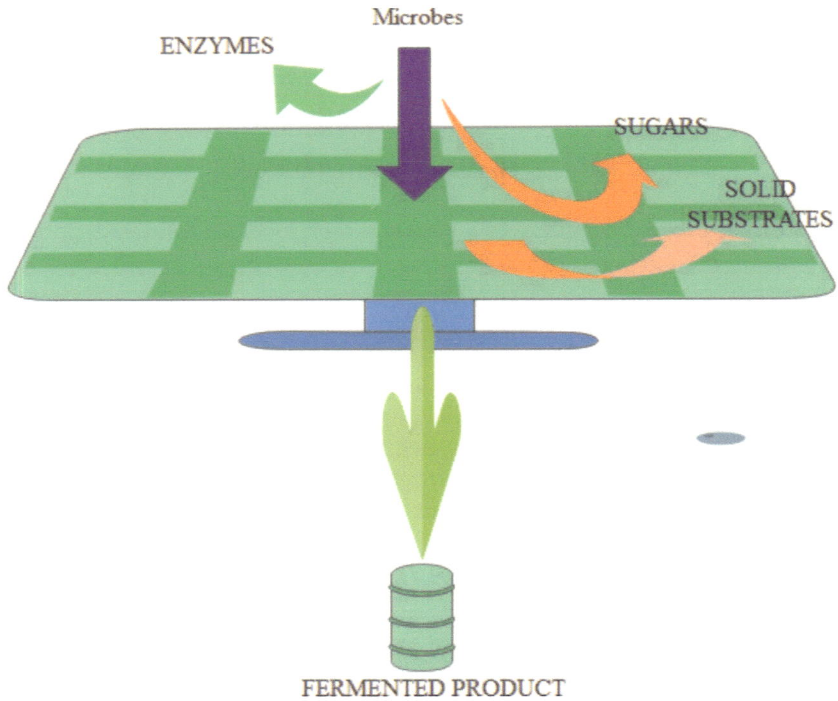

Figure 5.
Solid fermented product.

SSF which are very much affected by various environmental factors like moisture content, temperature, Ph, etc. [38]. Moisture is very crucial factor in hindrance of microbial enzymatic activity. The main bio-products of SSF are antibiotics, are organic acids, aromas, biopesticides, bio-fuel, biosurfactants, and bioplastics (**Figure** 5).

5. Classification of decomposer biota

There are three main biota involved in agricultural waste management, which include:

1. Microbes involved in soil organic matter formation.

2. Microbes involved in mineralization process.

3. Microbes involved in channelization of energy involved in decomposition process.

There are vast variety of decomposers like bacteria, protozaons, and some larger vertebrates. These are classified on the basis of size as (**Figures 6, 7**):

1. Microflora and fauna: Bacteria, fungi, actinomycetes, and protozoans.

2. Mesofauna: Acarina, collembola, etc.

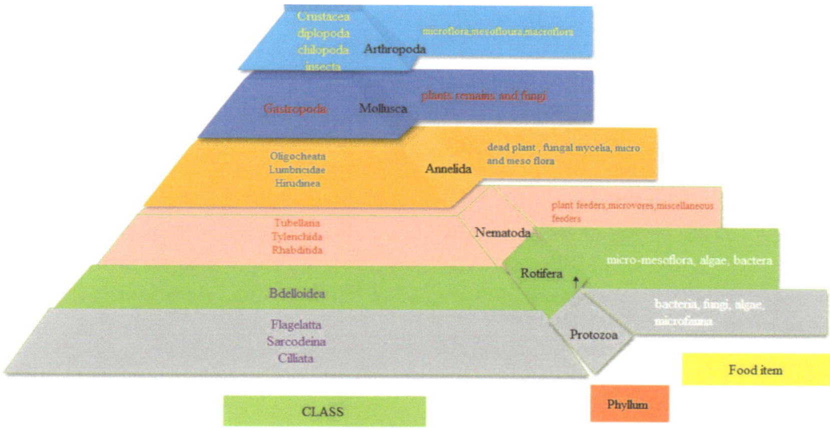

Figure 6.
Taxonomical biodeversity of waste decomposers.

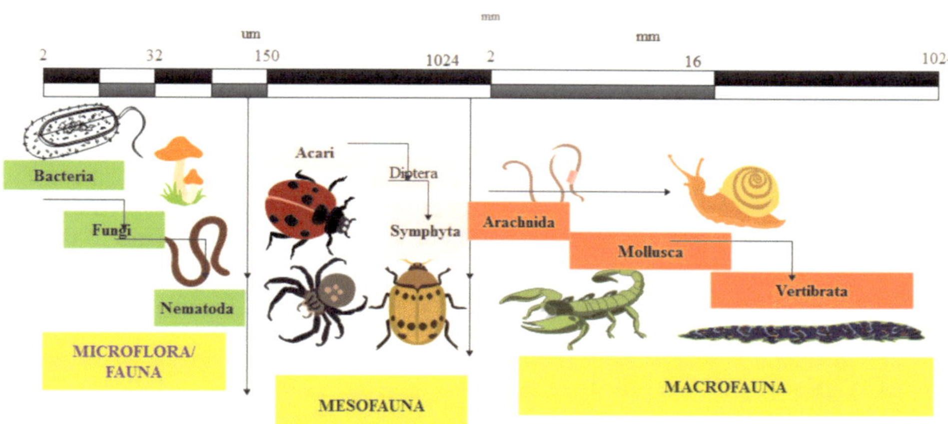

Figure 7.
Classification of bio-decomposers on the basis of size.

3. Macrofauna, including lumbricids, enchytraeids, millipedes, gastropods, and arthropods.

These decomposers further classified on the bases of role in decomposition process primary and secondary (**Figure 4**).

1. **Primary decomposers:** These microbes have the potential to decompose the compounds present in organic matter. These microbes have the requisite enzymes like pectinase, ligase, cellulose, etc. to break the complex agricultural residues complex components like cellulose, hemicelluloses, lignin, etc.

2. **Secondary decomposers**: These decomposers play a significant role in the fragmentation and comminution of decomposing agricultural waste. These microbes do not require enzymes for decomposition. These decomposers constituted earthworms, collembola, mites, enchytraeids, etc. and the rate of decomposition gets slower without secondary decomposers [39, 40].

Major decomposers of soil are bacteria, fungi, nematodes, micro-arthropods, and earthworms [41, 42].

1. **Bacteria**: Bacteria are the most effective decomposers due to their high degree of versatility. On the basis of mobility, there are two types of bacteria: mobile and immobile, due to the presence of whip-like structure flagella. They are further classified by the number of flagella; uniflagillated or multiflagillated. The average size of bacteria is in the range of 2 mm. The bio-decomposer bacterial species are the most important and initiator of decomposition process. There are about 5000 bacteria of estimated 109 bacterial cells present in one gram of organic soil. The cellulose-producing bacteria are the potent bacterial cell lines in hydrolytical cleavage of agricultural residue in decomposition process which includes *bacillus* spp., *pseudomonas*, *cellulomonas*, etc.

2. **Fungi:** Fungi constitute a very large decomposer biomass [43]. The organic matter decomposition is mainly regulated by fungi [44, 45]. The mobility of fungus is prompted by the mycelia. There are approximately 100–120 fungal species present in natural soil without any contamination. Among all the species, mycorrhizal fungi is the most destructive bio-decomposer of any types of agri-residue. The degradation rate is very high by these fungal species. These fungal species mainly include Aspergillus, sclerotium, white-rot fungi, Trichoderma, etc. These fungal species produce enzymes for the degradation of cellulose and lignin during decomposition process [46].

3. **Nematodes:** Nematodes are the free living organisms. These work in a dual way, decomposition as well as recycling of nutrients in soil. Nematodes do not feed directly the soil organic matter but feed bacteria, fungi, protozaons, etc. On the basis of feeding habits, nematodes are categorized in four types-bacterivores (feed bacteria), fungivores (feed fungus), predators (protoazoans, rotifers, etc.) and omnivores (bacteria, fungi, algae, protozoans, etc.). Decomposition process is accelerated in the presence of nematodes. The minerals and other nutrients return back to the soil by the nematodes' feeding cycle and are easily absorbed by the plants.

4. **Micro-arthropods:** Micro-arthropods play significant role in decomposition of litter standing crop. Mineralization of N, P, and K from the litter is enhanced by 23% due the feeding activities of micro-arthropods [47]. The soil micro-arthropods include chelicerates (spiders, mites, and pseudoscorpions), crustaceans (small aquatic forms), myriapods (millipedes, centipedes), springtails, and other insects.

5. **Mechanism of decomposition:** In organic matter decomposition two main process are cellulolytic and lignocellulolytic process, which is governed by the several bio-organisms [48]. These microorganisms mainly decompose organic waste by enzymatic reactions which involves breakdown of cellulose, hemicelluloses, pectin and poly aromatic inputs such as lignin, humus and phenolic acids [45]. These microflora accelerate the decomposition in composting process by the action of cellulolytic and lignocellulolytic enzymatic activity [49]. The cellulolytic microbes include mesophillic bacterial species that degraded proteins, amino acids, peptide bonds, etc. Fungi and bacteria are the main microbes which play significant role in release of hydrolytic and oxidative enzymes. These cellulolytic and lignocellulolytic activities transform the organic waste to humus substances [50, 51]. Therefore, many studies have revealed that these enzymatic activities results into high quality of composts.

6. Different phases of decomposition process along with enzymatic activity

6.1 PHASE I: Hydrolysis

Polysaccharides/lipids/proteins ⟶ Mono-sacharides/glycerol and fatty acid/amino acid.

Hydrolytic enzymes

Name of enzymes	Hydrolytic product
Cellulase; endo-1,4-beta-d-glucanase	Cellulose, lichenin, and cereal beta-d-glucans
Cellulose 1,4-beta-cellobiosidase (nonreducing end)	Cellulose and cellotetraose
Beta-glucosidase	Nonreducing beta-d-glucosyl residues
Endo-1,4-beta-xylanase	Xylans
Xylan 1,4-beta-xylosidase	D-xylose
Mannan endo-1,4-beta-mannosidase	Mannans, galactomannans, and glucomannans
Beta-mannosidase	Terminal, nonreducing beta-d-mannose residues in beta-d-mannosides
Alpha-galactosidase	Hydrolysis of terminal galactose oligosaccharides, galactomannans, and galactolipids
Alpha-glucuronidase	Alpha-d-glucuronoside
Peptidases	Acting on peptide bonds

6.2 Phase II: Acidogenic

$$\text{Sugars/amino acids/peptides} \xrightarrow{\text{Fermentation}} \text{Propionate, butyrate}$$

Enzymes involved in acidogenic pathways

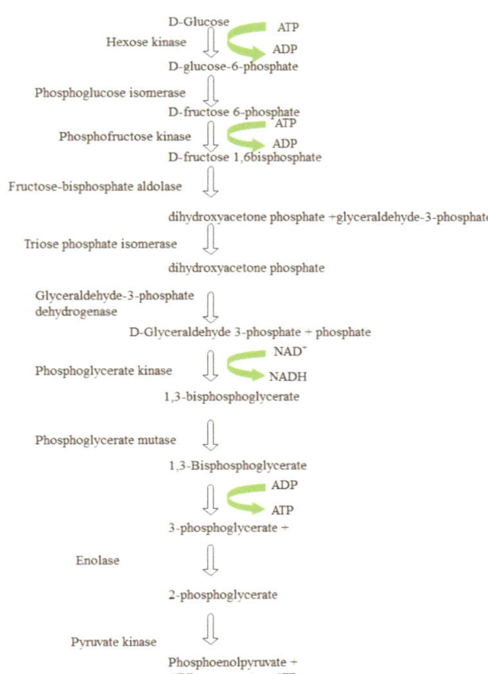

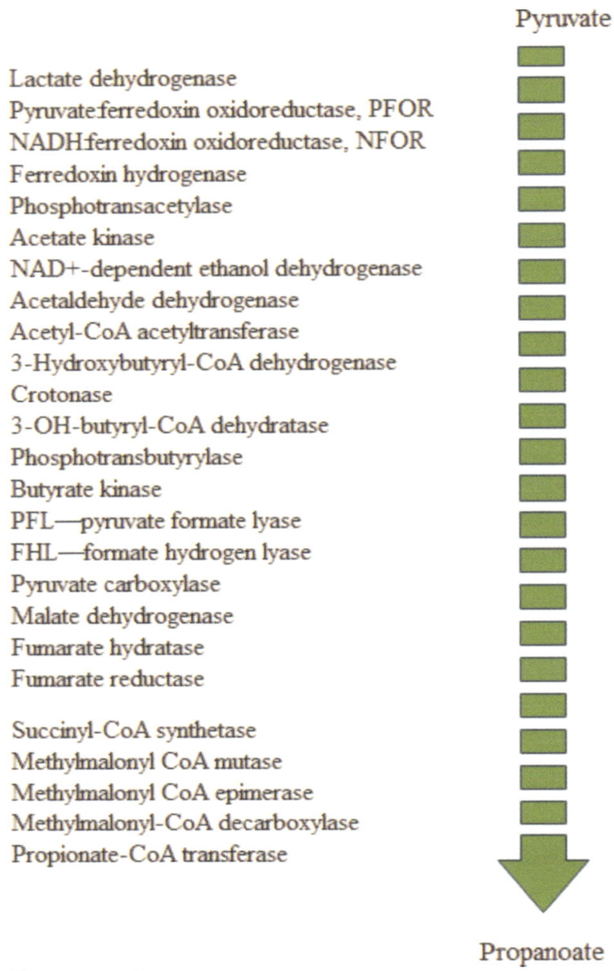

Amino acids fermentations

$$RCH(NH_4^+)COO^- + H_2O \rightarrow RCOCOO^- + NH_4^+ + H_2$$

$$RCOCOO^- + H_2O \xrightarrow{\text{Oxidative Decarboxylation}} RCOO^- + CO_2 + H_2$$

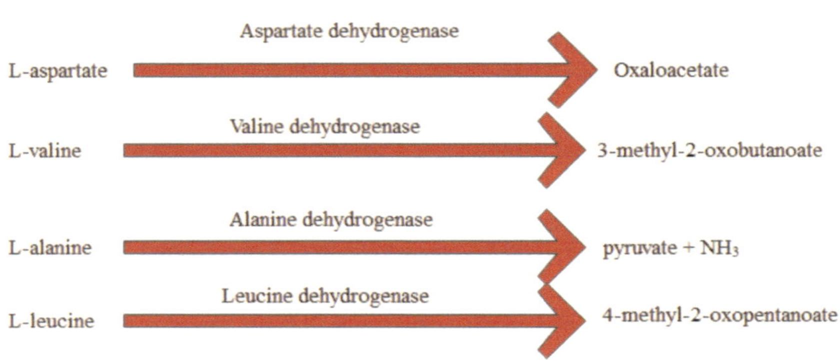

6.3 Phase III-Acetogenesis

Enzymes involved in acetogenesis reaction pathways
Citric acid cycle

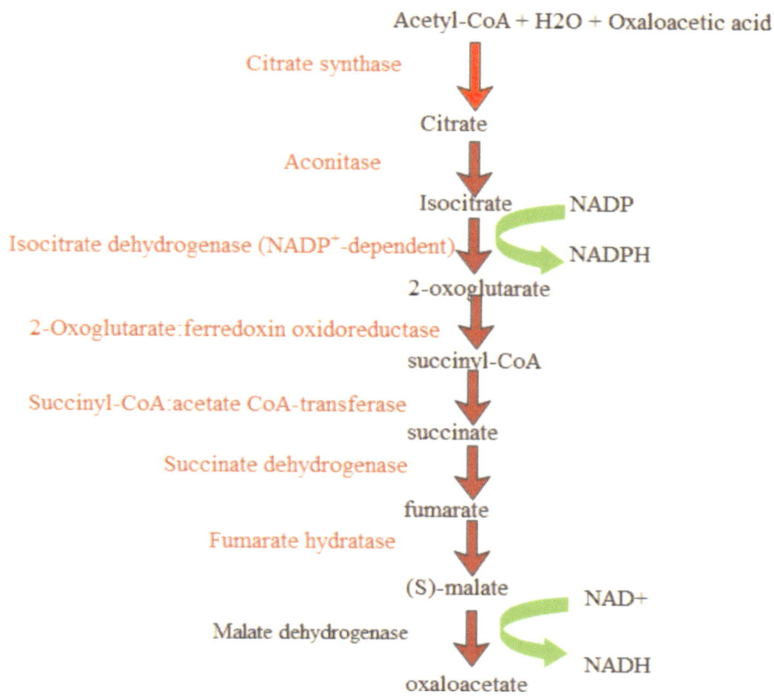

7. Process for improving the vitality of bio-decomposer microbes

In 1998, Kale's finding was that if solid by-products of agricultural waste are dumped as landfilling, a traditional disposal method of agricultural waste disposal, it results in the contamination of groundwater, soil, as well as food resource materials [52, 53]. The contamination is due to the presence of fibrous substances in agri-residues which have very high BOD and COD for decomposition or oxidation process.

7.1 Addition of inert materials to enhance the performance of bio-organisms

Various scientists offered some safe inert ingredients for managing this agro waste along with microorganisms which is environmentally friendly, economical and accepted by the farmers [56]. A compositing technology is then come into application in which waste is processed to remove bio-toxic compounds by anaerobic decomposition method. Previous studies revealed that vermin-compositing was an appropriate technology for converting agri-waste into a valuable product [54]. In vermin-compositing, earthworms and associated micro-flora convert waste into soil nutrients which boost up the soil fertility [55]. For decomposition of hazardous substance like pesticides, fertilizers, etc., mix with some bulky inert ingredients like saw dust and cow dung in 1:1 ratio for vermin-compositing trails for maximum decomposition by earthworms [56]. Textile sludge mixing with biogas plant slurry in combination of 2:3 ratio further increases decomposition [57]. Vermin-compositing of waste by beverage industry in combination with cow dung enhances the nutrient end product and earthworm performance in 1:1 ratio [58]. So, all these studies revealed that vermin-compositing performance is enhanced in presence of some bulky inert materials in suitable proportion.

7.2 Addition of some moisture-absorbing inert materials

Moisture is very crucial factor in hindrance of microbial enzymatic activity in solid-state fermentation decomposition method. So, some moisture-absorbing inert ingredient is very important for maintaining the decomposition activity.

7.3 Methods to increase the rate of survival of microorganisms

The success rate of microorganisms depends upon the survival of bio-decomposers in soil [59]. Formulation of bio-inoculants extends the period of storage and maximizes the decomposition rate in field and produce high quality of inoculants [60]. US Patent no. 7097830 reveals synergistic bio-inoculant *Trichoderma harzianum* with Bacillus strains in combinations with carrier which results into promotion of plant growth. In formulated product, various types of bio-inoculants can be combined which will provide economical and effective crop growing systems. Along with this formulated product would have longer shelf life while maintaining the easy usability and handling of agriculturally important microbial bio-inoculant [61].

As waste decomposer is very valuable for the Indian farmers and the value can be further increases if it is formulated into formulations like wettable powder and suspension concentrate. In formulated products their viability, shelf life and efficiency would be further enhanced by adding carriers, stabilizers, wetting and dispersing agents.

8. Challenges of agricultural waste decomposition process

1. **Nonuniform compositing:** Incomplete or immature composts cause many deleterious effects to soil, which in turn affects plant growth and ecosystem function [50]. Similar study also explained that application of immature compost for nitrogen fixation will release the toxic gases or compete for o2 in rhizosphere and results hindered growth of plant [62]. Mulec et al. also reported that immature compost cause phyto-toxicity due to presence of heavy metals and ammonia. Immature compost causes competition for oxygen between plant and microbes and results into toxic gases NO2 and H2S [50].

2. **Emission of gases:** Many studies have reviewed the gas emission problem. The rate of gas emission during compositing is the qualitative parameter of final compost. High emissions of these gases cause environmental pollution and other deleterious effect in ecosystem [63]. Most common odorous gaseous by-products are N2O, NH3, CH4, etc. [64].

3. **Leaching of compost:** Compost is rich in organic carbon content, which is very beneficial to plant growth. In addition to this beneficial aspect, leachate generated from compost is problematic to environment and causes groundwater contamination [65, 66]. Therefore, the compost cannot be stored in a same place for longer period of time [67, 68].

9. Future prospective of agricultural waste management

1. **Farmer-friendly**: Integrated approach of farm production along with compost formation by using agricultural waste will be beneficial for the farm

productivity and soil health. By following this approach, farmers need not depend on costly fertilizers and manure.

2. **Technological advancement for quick composting:** Compositing process is a time-consuming process, but due to some technological advancement, the compositing process's duration could be shortened. According to Lim et al. [69], addition of polyethylene glycol and jaggery speeds up the compositing process and results in a superior quality compost. These additives are cost-effective. More research studies should be done in these additives for finding more cost-effective and efficient additives.

3. **Use of genetically modified stains of innoculum:** Genetically modified stains of microbial inoculums will be used to accelerate the bioconversion of organic matter. These modified stains are more efficient than the natural innoculum. These innoculum will produce uniform compositing in an efficient way.

4. **Government agencies support:** Government supports the compositing process by providing loans and grants for compositing facilities for farmers. These steps of government will ensure sustainability and create job opportunities and uplift the rural areas. By this approach, government encourages organic farming by use of composts as soil conditioner and soil fertilizers [70].

5. **Installation of bioreactors to reduce the toxic gas emissions:** Bioreactors can be used to reduce the emission of toxic volatile organic constitutes. This strategy would manage the toxic gases emissions and save the environment [71, 72].

6. **Utilization of heat release in other energy sources:** Heat released during the compost formation can be utilized as a reliable energy source [73]. The energy generated by these heat harvests would be a sustainable and renewable source of energy in near future. In coming years, compost formation will not only decompose the residues but will also be a source of many other transformations.

Author details

Nusrat Iqbal*, Amrish Agrawal, Saurabh Dubey and Jitender Kumar
Institute of Pesticide Formulation Technology, Gurugram, Haryana, India

*Address all correspondence to: nusratsiddiqa20@gmail.com

References

[1] Agamuthu P, Khidzir KM, Hamid FS. Drivers of sustainable waste management in Asia. 2009;**27**(7):625-633

[2] Singh B, Rengel Z. The role of crop residues in improving soil fertility. In: Marschner P., Rengel Z. (eds) nutrient cycling in terrestrial ecosystems. Soil Biology. 2007;**10**(3):102-134. DOI: 10.1007/978-3-540-68027-7_7

[3] Butchaiah G, Sébastien B, Christoph M, Savitri G. Air pollutant emissions from rice straw open field burning in India, Thailand and the Philippines. Environmental Pollution. 2009;**157**(5): 1554-1155. DOI: 10.1016/j.envpol.2009.01.004

[4] Singh R, Babu N, Kumar R, Srivastava P, Singh P, Raghubanshi AS. Multifaceted application of crop residue biochar as a tool for sustainable agriculture: an ecological perspective. Ecological Engineering. 2015;**77**:324-347

[5] Kamara AH, Kamar S, Kamara MS. Effect of rice straw biochar on soil quality and the early growth and biomass yield of two rice varieties. Journal of Agricultural Sciences. 2015;**6**: 798-806

[6] Manna MC, Hazra JN, Ganguly NB, Sinha TK. Enrichment of compost by bioinoculants and mineral amendments. Journal of the Indian Society of Soil Science. 1997;**45**(6):831-833

[7] Gautam SP, Bundela PS, Pandey AK, Jain RK. Biodegradation and recycling of urban solid waste. American Journal of Environmental Sciences. 2009;**5**:45-60

[8] Sandhu HS, Wratten SD, Cullen R. Organic agriculture and ecosystem services. Environmental Science & Policy. 2010;**13**:1-7

[9] Overcash MR, Humenik FJ, Miner JR, editors. Livestock Waste Management. Vol. 1. Boca Raton: CRC Press; 1983

[10] Hai HT, Tuyet NTA. Benefits of the 3R Approach for Agricultural Waste Management (AWM) in Vietnam. Hayama, Japan: IGES; 2010

[11] Sud D, Mahajan G, Kaur MP. Agricultural waste material as potential adsorbent for sequestering heavy metal ions from aqueous solutions – A review. Bioresource Technology. 2008;**99**: 6017-6027

[12] Ngampimol H, Viyada K. The study of shelf life for liquid biofertilizer from vegetable waste. AU JT. 2008;**11**(4): 204-208

[13] Rahim KA. Biofertilizers in Malaysian agriculture: perception, demand and promotion. 2002 FNCA Joint Workshop on Mutation and Biofertilizer. 2002. p. 2023

[14] Ladha JK, Reddy PM. Nitrogen fixation in rice systems: state of knowledge and future prospects. Plant and Soil. 2000;**252**:151-167

[15] Agreste (2008) Enquête Aviculture Available online at: http://agreste.agriculture.gouv.fr/enquetes/pratiques-d-elevage/enquete-aviculture-2008/.

[16] Clancy-Hepturn M. Agricultural residues: a promising alternative to virgin wood fiber. Issue in Resources Conservation, Briefing Series. 1998;**1**

[17] Asam ZUZ, Poulsen TG, Nzami A-S, Raxique R, Kiely G, Murphy JD. How can we improve biomethane production per unit of feedstock in biogas plant. Applied Energy. 2011;**88**(3):2013-2018

[18] Dugba PN, Zhang R. Treatment of diary waste water with two stage

anaerobic sequencing batch reactor systems—Thermophilic versus Mesophilic operations. Bioresource Technology. 1999;**68**:225-233. DOI: 10.1016/S0960-8524(98)00156-4

[19] Verma S. Anaerobic Digestion of Biodegradable Organics in Municipal Solid Wastes. New York, NY, USA: Columbia University; 2002. p. 2002

[20] Gujer W, Zehnder AJB. Conversion processes in anaerobic digestion. Water Science and Technology. 1983;**15**(2):127-167. DOI: 10.2166/wst.1983.0164

[21] Santerre MT, Smith KR. Measures of appropriateness. The resource requirements of anaerobic digestion (biogas) system. World Development. 1982;**10**:239-261. DOI: 10.1016/0305-750X(82)90013-4

[22] Paslotathis SG, Girardo-Gomez E. Kinetics of Anaerobic Treatment A Critical Review. Critical Reviews in Environmental Control. 1991;**21**:411-490

[23] Luo K, Yang Q, Li X, Yang G, Liu Y, Wang D, et al. Hydrolysis kinetics in anaerobic digestion of waste activated sludge enhanced by α-amylase. Biochemical Engineering Journal. 2012. 1990;**62**:17-21. DOI: 10.1016/j.bej.2011.12.009.

[24] Bergman EN. Energy contributions of volatile fatty acids from the gastrointestinal tract in various species. Physiological Reviews. 1990;**70**(2):567-590

[25] Van Lier JB, Mahmoud N, Zeeman G. Biological Wastewater Treatment: Principles, Modelling and Design. Vol. 13. No. 4. London, UK: International Water Association, Anaerobic Wastewater Treatment; 2008. pp. 401-442

[26] Bajpai P. Basics of anaerobic digestion process. In: Anaerobic Technology in Pulp and Paper Industry. Berlin/Heidelberg, Germany: Springer; 2017. pp. 7-12

[27] Yamada K, Xu H. Properties and applications of an organic fertilizer inoculated with effective microorganisms. Journal of Crop Production. 2001;**3**(2):255-268. DOI: 10.1300/J144v03n01_21

[28] Van Haandel A, van der Lubbe J. Handbook Biological Wastewater 2007. 2018. Available from: http://www.wastewaterhandbook.com/documents/sludge_treatment/831_anaerobic_digestion_theory.pdf [Accessed: May 18, 2018]

[29] Yadvika S, Sreekrishnan TR, Kohli S, Rana V. Enhancement of biogas production from solid substrates using different techniques—A review. Bioresource Technology. 2004;**95**(5):1-10

[30] Ferry JG. The chemical biology of methanogenesis. Planetary and Space Science. 2010;**58**:1775-1783. DOI: 10.1016/j.pss.2010.08.014

[31] Kosaric N, Blaszczyk R. Industrial effluent processing. In: Lederberg J, editor. Encyclopedia of Microbiology. New York, NY, USA: Academic Press Inc; 1992. pp. 473-491

[32] Zupanˇciˇc GD, Grilc V. Anaerobic treatment and biogas production from organic waste. In: Kumar S, editor. Management of Organic Waste. Vol. 13. No. 2. London, UK: Intechopen; 2012. pp. 1-2

[33] Sequi P. The role of composting in sustainable agriculture. In: de Bertoldi M, Sequi P, Lemmes B, Papi T, editors. The Science of Composting. Dordrecht: Springer; 1996. DOI: 10.1007/978-94-009-1569-5_3

[34] Tuomela M, Vikman M, Hatakka A, Itävaara M. Biodegradation of lignin in a

compost environment: a review. Bioresource Technology. 2000;**72**:169-183

[35] Beck-Friis B. Formation and emission of N_2O and CH_4 from compost heaps of organic household waster. Environmental Monitoring and Assessment. 2000;**62**(3):317-331

[36] Wu JQ, Zhao Y, Qi HS, Zhao XY, Yang TX, Du YQ, et al. Identifying the key factors that affect the formation of humic substance during different materials composting. Bioresource Technology. 2017;**244**:1193-1196

[37] Mienda B, Idi A, Umar A. Microbiological features of solid state fermentation and its applications - An overview. Research in Biotechnology. 2011;**2**:14-45. Available from: http://updatepublishing.com/journal/index.php/rib/article/view/2381

[38] Pérez-Guerra N, Torrado-Agrasar A, López-Macias C, Pastrana L. Main characteristics and applications of solid substrate fermentation. Electronic Journal of Environmental, Agricultural and Food Chemistry. 2003;**2**:3-85

[39] Kurcheva GF. Role of invertebrates in the decomposition of oak litter. Soviet Soil Science. 1960;**4**:360-365

[40] Edwards C A, Reichle D E, Crossley DA. The role of soil invertebrates in turnover of organic matter and nutrients analysis of temperate forest ecosystems. 2015;**4**:147-172

[41] Woods LE. Active organic matter distribution in the surface 15 cm of undisturbed and cultivated soil. Biology and Fertility of Soils. 1989;**8**: 271-278

[42] Michael Beare H, Robert W, Parmelee Paul F, Hendrix Weixin C, Cheng C, David D, et al. Microbial and Faunal Interactions and Effects on Litter Nitrogen and Decomposition in Agroecosystems1992. DOI: 10.2307/2937317

[43] Hawksworth DL. The fungal dimension of biodiversity: magnitude, significance, and conservation. Mycological Research. 1991;**95**:641-655. DOI: 10.1016/S0953-7562(09)80810-1

[44] Dighton J, Boddy L. Role of fungi in nitrogen, phosphorus and sulphur cycling in temperate forest. Ecosystems. 1989

[45] Wainwright M. An introduction to fungal biotechnology. In: The Role of Fungi in Litter Decomposition and Nutrient Cycling. Mycology series. USA: John Wiley & Sons; 1992

[46] Rashmi A, Kalpna B, Harsh N. Temperature stress and redox homeostasis in agricultural crops. Frontiers in Environmental Science. 2015;**3**:11-14

[47] Seastedt TR. The role of microarthropods in decomposition and mineralization processes. Annual Review of Entomology. 2003;**29**:25-46. DOI: 10.1146/annurev.en.29.010184.000325

[48] Gupta P, Samant K, Sahu A. Isolation of cellulose-degrading bacteria and determination of their cellulolytic potential Int. Journal of Microbiology. 2012;**12**:13-25

[49] Gaurav K, Loganathan K, Kokati V, Bhaskara R. Antimicrobial activity of latex of calotropis gigantea against pathogenic microorganisms - An in vitro study. Pharmacology. 2010;**3**:155-163

[50] López-González JA, Suárez-Estrella F, Vargas García MC, López MJ, Jurado MM, Moreno J. Dynamics of bacterial microbiota during lignocellulosic waste composting: Studies upon its structure. Functionality and Biodiversity Bioresource Technology. 2015;**175**:406-416. DOI: 10.1016/j.biortech.2014.10.123

[51] Wu JQ, Zhao Y, Qi HS, Zhao XY, Yang TX, Du YQ, et al. Identifying the key factors that affect the formation of humic substance during different materials composting. Bioresource Technology. 2017;**244**:1193-1196

[52] Suthar S. Vermistabilization of municipal sewage sludge amended with sugarcane trash using epigeic *Eisenia fetida* (Oligochaeta). Journal of Hazardous Materials. 2009;**163**:199-206

[53] Suthar S, Singh S. Comparison of some novel polyculture and monoculture vermicomposting reactors to decompose organic wastes. Ecological Engineering. 2008;**33**:210-219

[54] Kale RD. Earthworms: nature's gift for utilization of organic wastes. 1998

[55] Suthar S. Pilot-scale vermireactors for sewage sludge stabilization and metal remediation process: comparison with small-scale vermireactors. Ecological Engineering. 2010;**36**:703-712

[56] Suthar S. Production of vermifertilizer from gaur gum industrial wastes by using composting earthworm *Perionyx sansibaricus* (Perrier). Environmentalist. 2007;**27**:329-335

[57] Garg VK. Livestock excreta management through vermicomposting using an epigeic earthworm Eisenia foetida. Environmentalist. 2006;**26**(4): 269-276

[58] Singh RP, Singh P, Araujo ASF, Ibrahim MH, Sulaiman O. Management of urban solid waste: vermicomposting a sustainable option. Resources, Conservation and Recycling. 2011;**55**: 719-729

[59] Souza R, Ambrosini A, Passaglia L. Plant growth-promoting bacteria as inoculants in agricultural soils PMID: 26537605, PMCID: PMC4763327. 2015. DOI: 10.1590/S1415-47573842015005

[60] JHG S, Rask HM. Inoculant production and formulation. Field Crops Research. 2000;**65**:249-258. DOI: 10.1016/S0378-4290(99)00090-8

[61] Jurez M, Barbara P, Andreas W, Heribert I, Ingrid H. Franke-Whittle. Co-composting of biowaste and wood ash, influence on a microbially driven-process. Waste Management. 2015;**46**: 155-164

[62] Guo R, Li G, Jiang T, Schuchardt F, Chen T, Zhao Y, et al. Effect of aeration rate, C/N ratio and moisture content on the stability and maturity of 474 compost. Bioresource Technology. 2012; **112**:171-178

[63] Nasini L, De Luca G, Anna R. Gas emissions during olive mill waste composting under static pile conditions. International Biodeterioration & Biodegradation. 2016;**107**:70-76

[64] Jiang J, Liu X, Huang Y, Huang H. Inoculation with nitrogen turnover bacterial agent appropriately increasing nitrogen and promoting maturity in pig manure composting. Waste Management. 2015;**39**:78-85

[65] Chatterjee A, Sarkar C, Adak A, Mukherjee U, Ghosh S, Raha S. Ambient air quality during diwali festival over Kolkata – A mega-city in India. Aerosol and Air Quality Research. 2013;**13**:1133-1144. DOI: 10.4209/aaqr.2012.03.0062

[66] Tyrrel SF, Seymour JA, Harris. Bioremediation of leachate from a green waste composting facility using waste-derived filter. Bioresource Technology. 2008;**99**:7657-7664

[67] Trujillo ME, Kroppenstedt RM, Schumann P, Carro L, Martinez-Molina E. *Micromonospora coriariae* sp. nov., isolated from root nodules of *Coriaria myrtifolia*. International Journal of Systematic and Evolutionary Microbiology. 2006;**56**:2381-2385

[68] Maleki A, Zazouli MA, Izanloo H, Rezaee R. Composting plant leachate treatment by coagulation-flocculation process. American-Eurasian J Agric. & Environ. Sci. 2009;**5**:638-643

[69] Lim CH, Mohammed IY, Abdalla Abakr Y, Kazi FK, Yusup S, Lam HL. Novel input-output prediction approach for biomass pyrolysis. Journal of Cleaner Production. 2016;**136**:51-61

[70] Platt B, Nora G. Why compost? Biocycle. 2014;**19**:12-34

[71] Godbout S, Hamelin L, Lemay SP, Pelletier F, Zegan D, Belzile M, et al. Reduction of odour emissions from swine buildings: comparison of three reduction techniques. In: XVIIth World Congress of the International Commission of Agricultural Engineering (CIGR) Hosted by the Canadian Society for Bioengineering (CSBE/SCGAB) Québec City, Canada June 13-17, 2010

[72] Matthieu G. Biodegradation in animal manure management. IntechOpen: Biodegradation-Engineering and Technology; 2013

[73] Irvine G, Lamont ER, Antizar-Ladislao B. Energy from Waste: Reuse of Compost Heat as a Source of Renewable Energy, Bioprocess Development for Biofuels and Bioproducts. Volume 2010: Article ID 627930. Available from: https://doi.org/10.1155/2010/627930

Section 7

Thermochemical Processes

Chapter 28

Gasification of Biomass

Thanh Phong Mai and Dinh Quan Nguyen

Abstract

Gasification is an indirect combustion of solid and liquid biomass by converting them to combustive syngas. Gasification is an alternative process for the traditional combustion, in which the emission of dust and toxic gases can be minimized. In this chapter, a comparison of these two biomass-to-heat conversion processes applied on biomass is presented in term of environmental impacts and technological benefits with a hope to provide readers a basic view of choices. Gasification is classified as in term of gasification agents, non-catalytic and catalytic process, and plasma assisted process. Popular types of gasification equipment, aka gasifiers, are introduced with working principles, through which the advantages and weakness of technology are briefly discussed.

Keywords: gasification, gasifier, biomass, indirect combustion, direct combustion, syngas, plasma, updraft, downdraft, cross-draft, bridging, tar, charcoal, hydrogen, steam, carbon dioxide, flue gas

1. Introduction

For the last decades, the demand for renewable energy has been increasing intensively due to the crude-oil crisis and the alert of global warming. Among the alternatives for fossil fuels to generate heat, biomass is an abundant neutral carbon source, of which its conversion to heat does not break the balance of the atmosphere's air contents [1]. Combustion of biomass has been the most direct and simple process to produce energy. However, the traditional combustion of biomass, such as wood, charcoal, straw, husks, etc., often leads to the emission of smoke, dust, fumes, volatile compounds and toxic gases due to incomplete reactions and fine particles dragged out of the system by the flue gas [2]. Although several combustion methods were invented to increase efficiency and reduce emission of pollutants, such as fixed bed rocket type, and fluidized bed technology, the direct combustion of solid fuels is still one of the major causes of the industrial air pollutant in the world [3].

In contrast, gasification of biomass can minimize the emission of pollutants. Syngas produced from gasification of biomass can be optionally purified before being combusted. Ultimately, the combustion of gaseous fuels inherently has higher efficiency than that of solid matters. That is because the oxidation of a solid object in oxygen/air is gradually happening from its outer surface into the inner layers, which can be described as a heterogeneous process, while a combustive gas like syngas can be burned at a very high mass transfer rate in a homogeneous process. A comparison is presented in **Table 1**.

The gasification phenomenon of carbonaceous materials was possibly observed in the human history as very early as the invention of fire. Gasification was found as the ignition and combustion of smoke released from smoldering coal, wood, straw,

	Combustion of syngas from gasification of biomass	Direct combustion of solid biomass
Type of reactions	Homogeneous	Heterogeneous
Uniformity	Very high	None
Process nature	Simple	Complex
Mass transfer rate	Almost instant	Slow, depending on the solid surface – oxygen/air contact

Table 1.
Combustion of syngas vs. combustion of solid biomass.

Figure 1.
Gasification of oil-extracted cashew nut shell at Laboratory of Biofuel and Biomass Research, Ho chi Minh City University of Technology (HCMUT).

grass, or other organic substances in the lack of oxygen. In 1792, the first industrial gasification system to generate electricity was reported [4]. Gasification is a thermal decomposition process of solid or liquid substances to syngas in the presence of gasification agents through a series of chemical reactions mentioned in the following sections. This technology can help converting variable low-energy-density fuels to combustive gases. It attracts significant interests in both academic and industrial fields. **Figure 1** shows a very strong flame torch produced by gasification of oil-extracted cashew nut shell.

Gasification is an advanced technology to convert biomass to syngas fuel under different atmospheres (oxygen/air, steam, H_2, CO_2). The product syngas can also be used as precursors to synthesize valuable chemicals via Fischer-Tropsch (F-T) reactions [5]. **Table 2** highlights some key differences between gasification and direct combustion of biomass.

2. Biomass gasification reactions

The combustion of a solid fuel is a thermal and oxidation decomposition with the involvement of oxygen in air. Generally, for biomass, it can be simply expressed as:

$$C_aH_bO_cN_dS_e + O_2/\text{air} \rightarrow CO_2 + H_2O + N_xO_y + SO_z + \text{Heat} \qquad (1)$$

This process can be observed with two visual phenomena: first, thermal decomposition on the outer surface of the solid phase to release volatile and combustive components, which join thermal reactions in the gas phase secondly, as the formation of flames [6]. Differing from direct combustion, gasification limits the process at the first step to produce syngas. Conventionally, oxygen/air is used as gasification agent in this case. However, other gasification agents also can be employed to enhance the conversion efficiency as presented followings.

	Gasification of biomass	Direct combustion of biomass
Input feedstock	Low-energy-density and wet biomass is still feasible	The biomass fuel must have acceptable moisture content and relatively flammable to guarantee a sustainable operation.
Output flame	Smokeless, free of dust and toxic gases if the syngas is purified.	Smoky and dusty with fly ash.
Impact to the heat exchangers' surface	Minimized	Silica fume, dusty aerosol, and corrosive gases can shorten the lifetime of equipment.
Applicability for internal combustion engines	Yes	No
Equipment design complexity	Complex	Simple
Heat receiver arrangement	Mobile	Fixed to the burner
Side product	Char, ash (solids), tar, bio-oil, wood vinegar (liquids)	Ash

Table 2.
A brief comparison between biomass gasification and combustion.

In this context, to simplify the theory, biomass can be formulated with its main general composition $C_aH_bO_c$ due to the much lower contents of other elements, such as N, S, P, and halogens. The involvement of inorganic compounds is not considered.

2.1 Oxygen/air as gasification agent

The thermal decomposition of biomass in insufficient presence of oxygen/air, known as incomplete combustion, is the most conventional gasification. Logically, the whole process can be described below as rearranged from theory [7].

Drying: firstly, once entering the reactor, the biomass is dried due to heat.

Combustion: secondly, a part of the solid biomass, which was ignited and in contact with locally excess oxygen/air, is combusted to generate heat as the energy source for later reactions to occur.

$$C_aH_bO_c\ (\text{biomass}) + O_2 \rightarrow CO_2 + H_2O + Q\ (\text{kJ/mol}) \tag{2}$$

Pyrolysis: heat from the combustion zone is transferred via radiation, conduction, and convective hot streams to the surrounding biomass where oxygen/air is not sufficient or absent. Due to the heat, pyrolysis occurs to form CO_2, CO, CH_4, C_2H_4, H_2O, char (C), and other organic solids and liquids as primary tar (2).

Reduction: after the above two steps, hot reactants react in situ with the biomass and with each other via a series of reactions.

$$\text{Water gas reaction}: C + H_2O + 118.5\ \text{kJ/mol} \rightarrow CO + H_2 \tag{3}$$

$$\text{Methanation reaction}: C + 2H_2 \rightarrow CH_4 + 87.5\ \text{kJ/mol} \tag{4}$$

$$\text{Steam reforming methanation}: CH_4 + H_2O + 206\ \text{kJ/mol} \rightarrow CO + 3H_2 \tag{5}$$

$$\text{Water gas shift reaction}: CO + H_2O \rightarrow CO_2 + H_2 + 40.9\ \text{kJ/mol} \tag{6}$$

$$\text{Boudouard reaction}: C + CO_2 + 159.9\ \text{kJ/mol} \rightarrow 2CO \tag{7}$$

$$\text{Reverse water gas shift reaction}: CO_2 + H_2 + 41.2\ \text{kJ/mol} \rightarrow H_2O + CO \tag{8}$$

The main weakness of gasification by oxygen/air is due to a large portion of inert nitrogen in the agent (79–80%), which makes the resulted syngas diluted. It can be roughly estimated that syngas from this type of gasification mainly contains around 30–60% of nitrogen and 10–15% of CO_2 since its heating value is typically between 4 and 6 MJ/m^3 (for comparison, HHV of H_2 = 12.76 MJ/m^3, CO = 12.63 MJ/m^3, CH_4 39.76 MJ/m^3 and CH_4 is commonly much less than CO and H_2) [7–9]. Low quality syngas is the main disadvantage of this technique for applications which require high temperature and steady operation, such as internal combustion engine, metallurgy, and melting glass industries.

Air-based gasification processes are sensitive and complex, which are influenced by a number of factors, such as biomass composition and particle geometry, gasification agent composition and flow rate, equipment design, etc. Among these, the ratio of actual air-fuel ratio to the stoichiometric air-fuel ratio (ER) is used as a parameter to calculate and to simulate the process [10].

$$ER = \frac{Stoichiometric\ Air\ (Nm^3)}{Actual\ Air\ Supplied\ (Nm^3)} \text{ and ER} < 1.0 \tag{9}$$

Gasification ER is theoretically usually from 0.19 to 0.43, and a range of 0.25–0.29 was studied to be considered as the optimum ER in gasification of some popular biomass [11].

2.2 Oxygen-enriched air

To obtain more concentrated syngas, nitrogen must be limited from the gasification agent in air-based systems while sufficient oxygen is still guaranteed for combustion to generate heat [12]. This method does not change the nature of the gasification process since nitrogen is an inert gas not involved in the reactions. Several techniques were introduced to remove nitrogen, thus increase oxygen content in the input air stream, such as pressure swing adsorption (PSA) [13], temperature swing adsorption [14], carbon membranes [15], etc. Oxygen concentration in studies on gasification with oxygen- enriched air is found limited by less than 50%, and no study on 100% oxygen gasification, possibly because of a high risk of explosion [16–18].

Figure 2 shows the visual change in an air-based syngas flame (wood pellet as feedstock) when oxygen concentration in the gasifying agent increased from that of normal air to 30%. With normal air, the syngas flame is thinner with smoke, while oxygen-enriched air makes the flame stronger, thicker, and less smoke. The flame temperature was measured as 874 and 933°C, respectively.

a) b) c)

Figure 2.
Experimental gasification of wood pellet (a) showing the flame of syngas when using (b) normal air (21% vol. as O_2) and (c) oxygen-enriched air (30% vol. as O_2)

2.3 Saturated and superheated steam as gasification agent

Water gas (3) and water gas shift (6) reactions are the reasons steam can be introduced to oxygen/air gasification or wet biomass is accepted, of which moisture is more tolerated than that in direct combustion. Higher generation yields of H_2 and CO are obtained so the final syngas mixture gets higher heating value. However, these two reactions are endothermic while the vaporization enthalpy of water has a large value (at atmospheric pressure that is 40.65 kJ/mol) so saturated steam or water can make the pyrolysis zone lose heat, drop temperature, leading to lower conversion yield. Lower quantity becomes a contrast to higher quality of syngas formation in this case. Subsequently, the process even gets faded if sufficient heat is not guaranteed. To achieve both quantity and quality of syngas, heat should be redeemed by using superheated steam instead of saturated steam or water in wet biomass so that the gasification temperature is maintained above 750–800°C [19].

The ratio of steam to carbon content of the biomass fuel (SCR) is used as a crucial operating parameter in biomass gasification with steam feeding [20]:

$$SCR = \frac{Steam\ mass\ flow\ rate\ \left(\frac{kg}{s}\right)}{Carbon\ feed\ rate\ \left(\frac{kg}{s}\right)} \quad (10)$$

Steam flow rate (kg/s) to biomass (kg/s) ratio (S/B) is also used like SCR [21]. Steam feeding makes the ratio of hydrogen to carbon in the whole reaction mixture increase, which was found to yield more H_2, and increase the heating value of the syngas, while tar content decreases significantly [22]. This technique is positively meaningful in biomass gasification because it does not only increase the quality of the syngas but also reduce tar-clogging problems to sustain the process.

2.4 Other gasification agents: H_2 and CO_2

Not many studies on gasification by hydrogen and carbon dioxide were found although these two agents are reactants in methanation (4) and Boudouard (7) reactions.

Methanation reaction can be increased when more H_2 exists in the reaction zone of a gasifier. Since methanation is exothermic, hydrogen can be mixed with air in air-based gasification or can be used as the only gasification agents without slagging problems in the gasifiers like conventional oxygen/air gasification. Pure hydrogen gasification is expected to be able to run at lower temperature and milder conditions because less heat is generated from methanation reaction ($\Delta H = -87.5$ kJ/mol) than from combustion step in air-based gasification [23], which may lead to the absence of oils and tars [24]. However, catalysts are needed because the reaction rates are very low [25]. Otherwise, hydrogen gasification should be carried out in high H_2 pressure, which rises several safety concerns.

CO_2 is a Boudouard reactant, as well as it can react with H_2 in the mixture via reverse water gas shift reaction. Hot flue gas is a popular product in industry, which includes steam, CO_2, and heat from direct combustion of fuel, thus can be considered as a gasification agent [26]. This technique is available if a combustion process is combined with gasification because air-based gasification already has its combustion zone. CO_2 utilization and enhancement of CO formation can be the purposes of CO_2-gasification [27].

2.5 Catalysts

The reactions in gasification can proceed with higher yields and less energy input if appropriate catalysts are employed. Catalysts can facilitate the process by reducing slagging problems, by which in severe cases, gasifiers need to be shut down for maintenance. Together with slagging of low-melting-point inorganic compounds, tar and soot formation also interrupts the operation because matters can be vaporized at high temperature, then condense at cooler zones and clog the systems. Catalysis helps limit the formation of such undesired side-products or decompose them to workable substances by cracking reactions. The mechanism of tar catalytic cracking can be assumed as follows [28]:

- Organic and hydrocarbon compounds are dissociated from the biomass and absorbed on the catalytic sites.

- Catalytic dehydrogenation reactions happen.

- Water is hydroxylated to OH radicals, which oxidize the hydrocarbon fragments.

- Syngas, CH_4, and lighter hydrocarbons are formed then.

In contrast, catalytic gasification has some disadvantages, such as material costs and fading catalyst performance over reaction time. Theoretically, catalysts can be recovered after the process. But in fact, they are easily poisoned and contaminated by variable products, which are formed from the complex interactions in gasification.

Alkali metal salts seem to be the earliest catalysts to be examined for gasification [29]. Alkali elements were studied to catalyze gasification of char and biomass, and they were proved to reduce the formation of tar and soot [30, 31]. The employment of catalysts is preferred for entrained-flow gasifiers, which will be discussed later [32].

Natural minerals, precious metal and synthetic catalysts are also studied for their application in biomass gasification, as well as coal and syngas conversion [33–35].

2.6 Plasma gasification

Plasma, which can be produced by an electric arc discharged to a gas, is a very hot and highly ionized gaseous mixture. The initial gas interacts with the electric arc to become dissociated into electrons and ions at temperatures often exceeding thousands of Celsius degree. When biomass and a non-oxidizing gasifying agent are fed into a plasma reactor, the gasification can proceed at high temperatures without combustion to generate heat as in conventional process. Therefore, plasma gasification can convert organic substances to syngas that preserve all its chemical and heat energy, while converts inorganic mineral ash to inert vitrified glass or slag. As a result, contamination and dilution of syngas are minimized and the process control is easy to yield expected syngas composition [36, 37].

Microwave was also used to generate plasma in plasma gasification [38]. However, microwave plasma system is not easy to scale up for industrial purposes like electric arc type.

With the principle of supplying intensive heat for endothermic reactions, plasma gasification was used to produce hydrogen with steam injection as discussed in Section 2.3 [20]. Carbon dioxide gasification was studied with a various biomass feedstock to show input plasma energy was lowest while syngas formation yield was highest [39]. Experimental results showed that steam or catalysts added to plasma gasification can significantly reduce the formation of tars [40].

3. Gasifiers

Gasification is a complicated process, which is influenced by many factors, among which equipment design plays a very important role. Popular types of gasifiers are listed and briefly discussed as bellows.

3.1 Fixed-bed gasifiers

There are three ways of arrangement for biomass and gasifying agents entering to react with each other in the reactors: updraft, downdraft, and cross draft as illustrated in **Figure 2a–c**.

- Updraft gasifiers (**Figure 3a**): in this type of reactor, biomass is fed downward from the top and gasifying agents is fed upward from the bottom in a counter flow arrangement. Ash is collected at the bottom of the equipment with air-lock design. The biggest weakness of updraft gasifiers is the accumulation of tar, moisture, and soot on the top of the reactors, which becomes hard clogging blocks inside the equipment. **Figure 4** is the actual photo of a very thick and hard layer of tar and soot attached to the inner wall on the top of an updraft biomass gasification reactor (the photos were taken at the Laboratory of Biofuel and Biomass Research, Ho Chi Minh City University of Technology, HCMUT). This counter flow process also makes syngas from updraft gasifiers carries much contamination. In contrast, the operation of updraft gasifiers is the easiest among the three types of fix-bed gasifiers above. Its design is also simple and available for multi-feed stock purpose.

- Downdraft gasifiers (**Figure 3b**): a narrow throat at the combustion zone is the typical design of this type of equipment. Since syngas is obtained at the bottom of the reactor, biomass and gasifying agents move in a co-current direction and get in contact for combustion at the device throttle. The flow rate of the

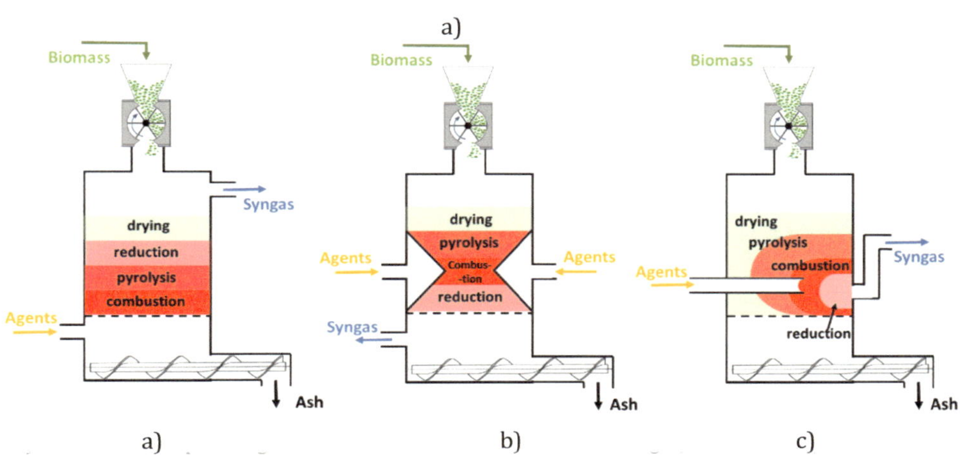

Figure 3.
Fix-bed gasifier types. (a) Updraft gasifier. (b) Downdraft gasifier. (c) Crossdraft gasifier.

a) b)

Figure 4.
(a) an updraft gasifier converting rice husk to syngas, (b) the inside wall of the top opening is clogged with a thick layer of condensed tar and soot.

gasifying agent gets maximum at this position due to the decreasing cross-sectional area of the orifice. As a result of this structure, the combustion increased sharply at the throttle while the amount of feeding agents is still. Downdraft gasifiers have higher conversion yield than that of their updraft models [41]. Syngas from downdraft gasifiers have much less tar and incomplete decomposed substances because they have to pass the combustion zone before exit with the syngas. However, downdraft gasifiers cannot be scaled up easily due to difficulties in controlling the movement of solid fuels through the throttle. Another difficulty in designing and fabricating downdraft gasifiers is "bridging problems" for feedstock with low densities [42]. The downward flow of the solid fuel is dictated by gravity while the pyrolysis zone is right above the narrow throat. The melting and adhesivity of lignin in biomass, as well as the local condensation of volatile substances, also facilitate the formation of stiff domes above the device throat, blocking the coming feedstock. It was observed that a rice husk downdraft gasifier kept stop working within some minutes of operation due to this problem and it was not an easy job to remove the bridging dome of "melting" rice husk inside the equipment (**Figure 5**).

- Crossdraft gasifiers (**Figure 3c**): as an intermediate between downdraft and updraft design, crossdraft gasifiers has the simplest design when biomass is fed from the top, gasifying agent from the rear side, and syngas is withdrawn from the other rear side of the reactor. Thanks to this arrangement, the pyrolysis zone is separated from reduction zone, where syngas is obtained, and between them is the combustion zone to reduce tar and soot. Bridging problem is not a concern in this case, and scaling up is feasible.

3.2 Fluidized bed gasifiers

Fluidization is an advance technique for solid fuel combustion. It is also applied for gasification. Inert materials (sand, dolomite, crushed stone, etc.) are employed to hold fluidization. The gasifying agents enter the reactor from the bottom upward to the top at velocities of 1–3 m/s through the biomass + inert material bed. Gasification reactions occur inside the bed then the resulted gases drag the particle before going up like "bubbling". This technique provides the mixture a uniformity for heat

Figure 5.
Fixing a downdraft gasifier after a bridging problem happened.

exchange. Cyclones are installed to collected solid particles and return them to the reactors. With high gasification efficiency, fluidized bed gasifiers are known for tar and char reduction [43].

The operating temperature of fluidized-bed gasifiers is limited to the melting point of the inert medium. The gasifying agents also play a role as fluidization fluids so the input flow rate must be high enough. Therefore, gasification agents in fluidized bed gasifiers are usually rather than only oxygen/air, which need to be at a limited mass ratio to the biomass [44, 45].

3.3 Entrained flow gasifiers

Entrained flow gasifiers are applied for biomass with small particle sizes so that the specific contact area with gasifying agents is large enough for suitable reaction rate. Simply described as illustrated in **Figure 6a**, the solid and the gas agents are fed co-currently to the reactor in the same downward direction. The agent surrounds the solid particles and react to convert the biomass to syngas. At the end of the falling routine to the bottom of the reactor of the feed, only ash and slag are expected to be remained solid collected in cyclone systems while syngas is passing through. The operation is carried out at high temperature and in high pressure. The

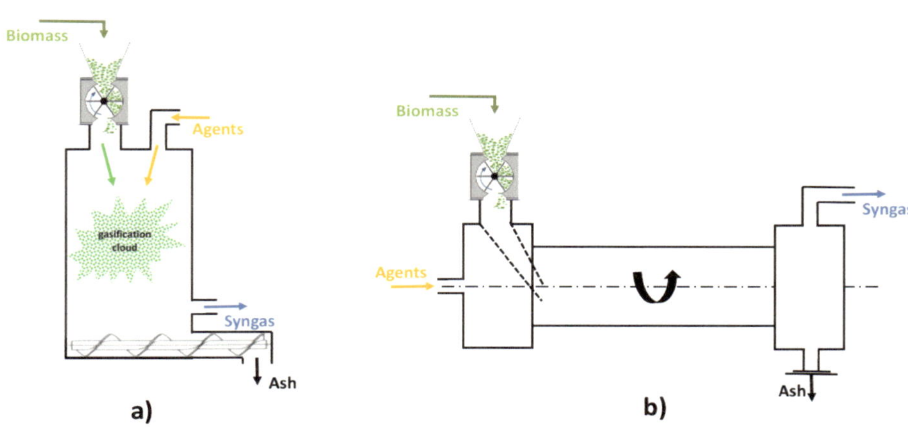

Figure 6.
(a) Entrained flow gasifier, (b) rotary drum gasifier.

extremely turbulent flow of the aerosol mixture causes rapid conversion and allows high throughput [46].

3.4 Rotary drum gasifiers

To reach uniformity of the biomass during gasification without combustion (using gasifying agents rather than oxygen/air), mechanical mixing can be applied as rotary kiln type reactor (**Figure 6b**). In this rotating cylinder, biomass is well mixed in contact with gasifying agents. Differing from fluidized bed and entrained flow equipment, the gasifying agents' flow rates can be at any value in rotary drum gasifiers.

4. Conclusion

Gasification is a big subject in biomass and chemical engineering. Among the renewable technologies converting biomass to fuels and energy with environmental preservation concern, gasification is superior over combustion with variable feasible application. Gasification process includes many reactions, which make it complex and sensitive to many factors. The diversity in the thermochemistry of gasification gives researchers and engineers a big space for creativity in R&D. This context introduced some brief theory and technical discussion on gasification technology with a humble hope to contribute to that vision.

Acknowledgements

This research was funded by Vietnam National University Ho Chi Minh City (VNU-HCM) under grant number B2018-20-02. We acknowledge the support of time and facilities from Ho Chi Minh University of Technology (HCMUT), VNU-HCMUT for this study. We also acknowledge the technical support and consultancy from Tin Thanh Group for Laboratory of Biofuel and Biomass Research in the last 5 years of this study.

Author details

Thanh Phong Mai and Dinh Quan Nguyen*
Laboratory of Biofuel and Biomass Research, Ho Chi Minh City University of Technology (HCMUT), Ho Chi Minh City, Vietnam

*Address all correspondence to: ndquan@hcmut.edu.vn

References

[1] Tran TTA, Le TKP, Mai TP, et al. Bioethanol Production from Lignocellulosic Biomass. In: Yun Y (ed) *Alcohol Fuels - Current Technologies and Future Prospect*. London, UK: IntechOpen, p. 1.

[2] Cormier SA, Lomnicki S, Backes W, et al. Origin and Health Impacts of Emissions of Toxic By-Products and Fine Particles from Combustion and Thermal Treatment of Hazardous Wastes and Materials. *Environ Health Perspect* 2006; 114: 810–817.

[3] Nussbaumer T, Hustad JE. Overview of Biomass Combustion. In: *Developments in Thermochemical Biomass Conversion*. Dordrecht: Springer Netherlands, pp. 1229–1243.

[4] Farzad S, Mandegari MA, Görgens JF. A critical review on biomass gasification, co-gasification, and their environmental assessments. *Biofuel Res J* 2016; 3: 483–495.

[5] Santos RG dos, Alencar AC. Biomass-derived syngas production via gasification process and its catalytic conversion into fuels by Fischer Tropsch synthesis: A review. *Int J Hydrogen Energy* 2020; 45: 18114–18132.

[6] Caillat S, Vakkilainen E. Large-scale biomass combustion plants: an overview. In: *Biomass Combustion Science, Technology and Engineering*. Elsevier, pp. 189–224.

[7] Ciferno JP, Marano JJ. Benchmarking biomass gasification technologies for fuels, chemicals and hydrogen production. *US Dep Energy Natl Energy* 2002; 58.

[8] Waldheim, L. Nilsson T. Heating value of gases from biomass gasification. *IEA Bioenergy Agreement, Task 20 - Therm Gasif Biomass* 2001; 61.

[9] Tran PNU, Vu LVK, Nguyen DQ, et al. Energy balance of small-scale biorefinery system. *Environ Sci* 2013; 26: 489–496.

[10] Rukshan J, Souman R. Numerical and Experimental Investigation of Equivalence Ratio (ER) and Feedstock Particle Size on Birchwood Gasification. *Energies* 2017; 10: 1232.

[11] James AM, Yuan W, Boyette MD, et al. The effect of air flow rate and biomass type on the performance of an updraft biomass gasifier. *BioResources* 2015; 10: 3615–3624.

[12] Cao Y, Wang Q, Du J, et al. Oxygen-enriched air gasification of biomass materials for high-quality syngas production. *Energy Convers Manag* 2019; 199: 111628.

[13] Reiss G. Oxygen Enrichment of Air With Molecular Sieve Zeolites Using The Psa/Vsa Technique. pp. 607–614.

[14] Eriksson T, Kiros Y. Temperature swing adsorption device for oxygen-enriched air. *J Clean Prod* 2014; 76: 174–179.

[15] Haider S, Lindbråthen A, Lie JA, et al. Carbon membranes for oxygen enriched air – Part II: Techno-economic analysis. *Sep Purif Technol* 2018; 205: 251–262.

[16] Liu L, Huang Y, Cao J, et al. Experimental study of biomass gasification with oxygen-enriched air in fluidized bed gasifier. *Sci Total Environ* 2018; 626: 423–433.

[17] Liu C, Huang Y, Niu M, et al. Influences of equivalence ratio, oxygen concentration and fluidization velocity on the characteristics of oxygen-enriched gasification products from biomass in a pilot-scale fluidized bed. *Int J Hydrogen Energy* 2018; 43: 14214–14225.

[18] Wang Z, He T, Qin J, et al. Gasification of biomass with oxygen-enriched air in a pilot scale two-stage gasifier. *Fuel* 2015; 150: 386–393.

[19] Kumar A, Jones DD, Hanna MA. Thermochemical biomass gasification: A review of the current status of the technology. *Energies* 2009; 2: 556–581.

[20] Favas J, Monteiro E, Rouboa A. Hydrogen production using plasma gasification with steam injection. *Int J Hydrogen Energy* 2017; 42: 10997–11005.

[21] Maschio G, Lucchesi A, Stoppato G. Production of syngas from biomass. *Bioresour Technol* 1994; 48: 119–126.

[22] Narváez I, Orío A, Aznar MP, et al. Biomass Gasification with Air in an Atmospheric Bubbling Fluidized Bed. Effect of Six Operational Variables on the Quality of the Produced Raw Gas. *Ind Eng Chem Res* 1996; 35: 2110–2120.

[23] Sheth AC, Sastry C, Yeboah YD, et al. Catalytic gasification of coal using eutectic salts: reaction kinetics for hydrogasification using binary and ternary eutectic catalysts. *Fuel* 2004; 83: 557–572.

[24] LEE S, LEE J, KIM J, et al. Hydrogasification characteristics of bituminous coals in an entrained-flow hydrogasifier. *Fuel* 2006; 85: 803– 806.

[25] Klinghoffer NB, Castaldi MJ, Nzihou A. Catalyst Properties and Catalytic Performance of Char from Biomass Gasification. *Ind Eng Chem Res* 2012; 51: 13113–13122.

[26] Maj I, Kudlek E, Ostrowski P, et al. Investigation of sewage sludge gasification with use of flue gas as a gasifying agent. *E3S Web Conf* 2017; 14: 02001.

[27] Sadhwani N, Adhikari S, Eden MR, et al. Aspen plus simulation to predict steady state performance of biomass-CO 2 gasification in a fluidized bed gasifier. *Biofuels, Bioprod Biorefining* 2018; 12: 379–389.

[28] Ren J, Cao J-P, Zhao X-Y, et al. Recent advances in syngas production from biomass catalytic gasification: A critical review on reactors, catalysts, catalytic mechanisms and mathematical models. *Renew Sustain Energy Rev* 2019; 116: 109426.

[29] Suzuki T, Ohme H, Watanabe Y. Mechanisms of Alkaline-Earth Metals Catalyzed CO2 Gasification of Carbon. *Energy & Fuels* 1994; 8: 649–658.

[30] Postma RS, Kersten SRA, van Rossum G. Potassium-Salt-Catalyzed Tar Reduction during Pyrolysis Oil Gasification. *Ind Eng Chem Res* 2016; 55: 7226–7230.

[31] Furusjö E, Ma C, Ji X, et al. Alkali enhanced biomass gasification with in situ S capture and novel syngas cleaning. Part 1: Gasifier performance. *Energy* 2018; 157: 96–105.

[32] Umeki K, Häggström G, Bach-Oller A, et al. Reduction of Tar and Soot Formation from Entrained-Flow Gasification of Woody Biomass by Alkali Impregnation. *Energy & Fuels* 2017; 31: 5104–5110.

[33] Yang Z, Cao J-P, Ren X-Y, et al. Preparation of hierarchical HZSM-5 based sulfated zirconium solid acid catalyst for catalytic upgrading of pyrolysis vapors from lignite pyrolysis. *Fuel* 2019; 237: 1079–1085.

[34] Park SY, Oh G, Kim K, et al. Deactivation characteristics of Ni and Ru catalysts in tar steam reforming. *Renew Energy* 2017; 105: 76–83.

[35] Zhao X-Y, Ren J, Cao J-P, et al. Catalytic Reforming of Volatiles from Biomass Pyrolysis for Hydrogen-Rich Gas Production over Limonite Ore. *Energy & Fuels* 2017; 31: 4054–4060.

[36] Hlina M, Hrabovsky M, Kavka T, et al. Production of high quality syngas from argon/water plasma gasification of biomass and waste. *Waste Manag* 2014; 34: 63–66.

[37] Rutberg PG, Bratsev AN, Kuznetsov VA, et al. On efficiency of plasma gasification of wood residues. *Biomass and Bioenergy* 2011; 35: 495–504.

[38] Delikonstantis E, Sturm G, Stankiewicz AI, et al. Biomass gasification in microwave plasma: An experimental feasibility study with a side stream from a fermentation reactor. *Chem Eng Process - Process Intensif* 2019; 141: 107538.

[39] Kuo P-C, Illathukandy B, Wu W, et al. Plasma gasification performances of various raw and torrefied biomass materials using different gasifying agents. *Bioresour Technol* 2020; 314: 123740.

[40] Mei D, Wang Y, Liu S, et al. Plasma reforming of biomass gasification tars using mixed naphthalene and toluene as model compounds. *Energy Convers Manag* 2019; 195: 409–419.

[41] Muthu Dinesh Kumar R, Anand R. Production of biofuel from biomass downdraft gasification and its applications. In: *Advanced Biofuels*. Elsevier, pp. 129–151.

[42] Guangul FM, Sulaiman SA. Mitigation of Bridging Problem in Biomass Gasification by a Novel Approach. *Asian J Sci Res* 2013; 6: 331–338.

[43] Gómez-Barea A, Ollero P, Leckner B. Optimization of char and tar conversion in fluidized bed biomass gasifiers. *Fuel* 2013; 103: 42–52.

[44] Udomsirichakorn J, Basu P, Salam PA, et al. Effect of CaO on tar reforming to hydrogen-enriched gas with in-process CO2 capture in a bubbling fluidized bed biomass steam gasifier. *Int J Hydrogen Energy* 2013; 38: 14495–14504.

[45] Kirnbauer F, Wilk V, Kitzler H, et al. The positive effects of bed material coating on tar reduction in a dual fluidized bed gasifier. *Fuel* 2012; 95: 553–562.

[46] Wu Y, Smith PJ, Zhang J, et al. Effects of Turbulent Mixing and Controlling Mechanisms in an Entrained Flow Coal Gasifier. *Energy & Fuels* 2010; 24: 1170–1175.

Chapter 29

Numerical and Experimental Analysis of Thermochemical Treatment for the Liquefaction of Lemon Bagasse in a Jacketed Vessel

Brenno S. Leite, Daniel J.O. Ferreira, Sibele A.F. Leite and Vanessa F.C. Lins

Abstract

In this work, it was investigated the time evolution of thermal profile inside a liquefaction vessel and how the temperature and time of reaction influenced liquefaction yield. Liquefaction was performed in two different ways: (1) Experimental Analysis; (2) Numerical 3-D model, using Computational Fluid Dynamics (CFD). Liquefaction was performed using lemon bagasse samples, glycerol and sulphuric acid, as catalyst. Temperature and liquefaction Yield (LY) were measured for different time of reaction (30, 60 and 90 minutes). From experimental data, LY were higher than 70 wt% for 90 minutes reaction. The increase in the temperature inside the reactor occurred due to the conduction and natural convection phenomena. Although the jacketed vessel was fed with steam at 125°C, working conditions allowed the heating of the mixture to less than 100°C. CFD thermal profile was in accordance with experimental data. They showed it was necessary 60 minutes to achieve a steady state of heating in the mixture inside this liquefaction vessel. From CFD transient simulations, it was observed some oscillations and detachment from experimental data, which may be due to changes in fluids properties along the process. Despite this consideration CFD could satisfactory analyse heat transfer in this liquefaction process.

Keywords: biomass, polyol, CFD, liquefaction yield, glycerol

1. Introduction

The use of biomass and agricultural waste as an alternative to fossil products has been widely investigated in the last decades. Several processes have been developed to provide more efficient use of those renewable sources and generate the so-called "green products", with properties and performance comparable to petroleum derivative products [1]. Polyol is one of these potential green products, since it is an essential polymer in the polyurethane compounds industry. It may produce foams, elastomers, inks, paints and others, which are currently obtained from petroleum derivatives.

A "green polyol" can be obtained from biomass liquefaction process [2]. This process consists of the endothermic reaction between biomass, solvent and catalytic agent, followed by an exothermic recombination of molecules, during a specific time interval. Biomass liquefaction is usually carried out in a jacketed reactor, that can be a stirred tank or a single vessel, without mechanical stirring [3].

The liquefaction yield depends on the technology used to modify the structure of the lignocellulosic biomass and the raw material. Because of the importance of the temperature on liquefaction process, some works performed an investigation of liquefaction yield and polyol properties as a function of temperature and time reaction [2, 4–6].

According to earlier literature, a wide range of temperature have been used on liquefaction processes. Dimitriadis and Bezergianni [7] reported a temperature range between 200–450°C (473.15–723.15 K) for the hydrothermal liquefaction, which varied according to the solvent, the biomass and the process used. Rafiqul [8] set autoclave temperature around 350–450°C (623.15–723.15 K) to perform co-liquefaction of bituminous coal with bagasse. To liquefy rice straw, using glycerol as solvent, Cao [9] used autoclave, with thermopar probes, set in the temperature between 220–300°C (493.15–573.15 K). Ye [2] evaluated the liquefaction process, of bamboo shoot with two types of glycerol, using heat and stirring. In this study, temperature varied around 110–150°C (383.15–423.15 K) with the best liquefaction yield at 150°C. Also, there was no significant difference on liquefaction yield, after 80 minutes of reaction (2). Li [10] investigated the liquefaction of wheat straw using alcohol/water mixed solvent. They observed better results for liquefaction yield at 270°C (543.15 K), using a residence time lower than 120 minutes.

Despite the significant importance of the temperature and time on the liquefaction process, it is still necessary to set a method to help researches choose these operational parameters. As shown in this review, it has been seen considerable variation among temperatures values and its range. Therefore, a suitable analysis of the heat transfer would be an interesting tool for industrial and academic applications to understand temperature and time reaction, in the liquefaction process.

Hence, in this work, the time evolution of thermal profile inside the liquefaction vessel and how temperature and time of reaction influenced liquefaction yield were investigated. Based on the above considerations, the analysis of liquefaction was performed in two different ways:

1. Experimental Analysis: the liquefaction of lemon bagasse with crude glycerol, to produce polyol, was conducted in a heated jacket reactor, supplied by steam, at constant values of temperature and pressure.

2. Numerical 3-D model: Computational Fluid Dynamics (CFD) was used considering two computational domains. The first domain considered the fluid flow of superheated steam inside the heated jacket at steady state flow rate. The second domain represented natural convection of glycerol and lemon bagasse mixture (polyol) considering transient fluid motion.

The experimental and numerical results were compared in order to validate the CFD simulation. It is expected that the simulation could be a helpful tool to evaluate velocity and temperature profiles for transient and steady state operations in further liquefaction experiments, despite geometries and scales used. In this way, it may guide researches choose temperatures and time reaction, in order to have better liquefaction yield with lower energy consumption.

2. Materials and methods

2.1 Materials

Lemon bagasse samples were collected between May and October of 2016 in the southeast region of Brazil (19° 53′ 12″ - S; 44° 25′ 56″ - W). The samples were dried at 105°C using an oven-dry, until a constant weight was achieved. Then, the biomass was cut in a knife mill to get fibbers of 0.5 mm length. The crude glycerol, used as liquefaction solvent, was kindly provided by Petrobrás (Usina Darcy Ribeiro - Montes Claros - MG, Brazil). Sulphuric acid (Synth) was used as catalyst for this reaction.

2.2 Experimental analysis

The experiments were performed in a jacket vessel. The reactor was supplied with a mixture of lemon bagasse (biomass), crude glycerol (solvent) and a sulphuric acid solution (95 wt%), as catalyst [11]. The steam used to feed the reactor was produced in an autoclave (water steam at 125°C - 398.15 K). The reaction was performed using three length of time: 30, 60 and 90 minutes. The experimental parameters: temperature, pressure, time and catalyst, have significant impact on the liquefaction process [5, 12]. Hence, in the present work, length of time reaction was investigated for three different solvent/biomass ratio. The solvent/biomass ratio conditions and their respective obtained Polyol are shown in **Table 1**.

From the reaction, it was produced a mixture of polyol and residues. Each resulting mixture was filtered to separate the two components: polyol and solid residues. The solid residues were dehydrated at 75°C (348.15 K), for 72 hours, to calculate the liquefaction yield (LY), obtained in weight percentage (wt%) as shown in Eq. (1) [8, 12]:

$$LY = \left(\frac{\text{Biomass weight} - \text{Residue weight}}{\text{Biomass weight}} \right) \cdot 100 \tag{1}$$

Where the biomass weight is the lemon bagasse weight (g) and the residue weight is the insoluble lemon weight, after the liquefaction process (g). The higher the LY, the higher is the polyol production.

2.3 Analysis of heat transfer

The temperatures in the centre of liquefaction vessel were recorded by a thermopar probe Digital Thermometer model K-type-1-channel brand Thermocouple Thermometer. It was done for 90 minutes, in time intervals of 5 minutes,

Variables	Mixture para obtenção do poliol		
	Polyol 1	Polyol 2	Polyol 3
Ratio solvent/biomass (wt%)	2:1	3.5:1	5:1
Lemon Bagasse (g)	115.00	78.00	58.00
Glycerol (g)	230.00	273.00	290.00
Suphuric Acid (g)	7.00	7.00	7.00

Table 1.
Liquefaction variables used to obtain the polyols.

considering the three solvent/biomass ratios presented (Polyol 1, 2 and 3). Temperature probes were located inside the reactor: one at the centre of the axis and another at the centre of the bottom. Data obtained was used to compare with CFD results.

The temperature data collections were performed at intervals of the $\Delta t = 5$ minutes, considering the initial temperature Ti = 25°C (298.15 K). Temperature variations (ΔT) and final temperature (Tf) were measured for each experimental run according to their respective intervals of time (0-30 minutes, 30-60 minutes, 60-90 minutes).

2.4 Reactor configurations

The jacketed reactor used in the present work is made of stainless steel and its geometry is showed in **Figure 1**.

The dimensions presented in **Figure 1** are: (a) jacket internal radius: 56,5 mm, (b) jacket external radius: 76,5 mm, (c) diameter of the input and output for steam in the jacketed: 20,0 mm, (d) thickness of the heat transfer surface: 2,0 mm, (e) total height of the reactor: 220 mm, (f) internal height of the reactor: 200 mm and (g) diameter of the steam input: 25,4 mm.

2.5 Numerical 3-D CFD model from

The Computational Fluid Dynamics (CFD) is a numerical technique used to solve fluid flow problems described by transport equations for mass, energy and momentum conservations employing the Finite Volume Methods (FVM) in three dimensional geometries. In the present CFD simulations, it was assumed that the internal shell thickness, between the steam and polyol domains, is negligible compared to dimensions of the heated jacket and liquefaction. The thermal resistance of the liquefaction vessel shell can be assumed negligible due to its high conductivity of the stainless steel compared to steam and glycerol mixture. It was also assumed constant temperature at the top surface of the vessel in the polyol domain and external adiabatic walls in the jacket domain. Each case was composed by two different simulations:

i. a steady state analysis for steam at 125°C flow passing through the jacket domain;

ii. a transient analysis of the natural convection of the polyol mixture inside the reactor vessel domain, heated by external walls which heat transfer is function of the wall heat transfer coefficient imported from jacket domain results.

The geometry of the reactor domain must fit perfectly inside the internal cylindrical hole of the jacket domain. Therefore, the reactor domain and the internal cylinder of the heat jacket have 200 mm height and 113 mm diameter. The computational mesh used for the liquefaction vessel domain for jacket domain were composed by 2,504,216 structured cells and 1,0763,346 non-structured cells, respectively.

Since it was not expected that temperature variations inside liquefaction vessel can perturb the thermal profile of the jacket domain, it was assumed one-way "communication" between the two computational domains. The results from jacket domain were used as boundary condition of wall heat transfer coefficient for polyol domain simulations. In the jacket domain the fluid employed was Water Vapor at 100 °C, available

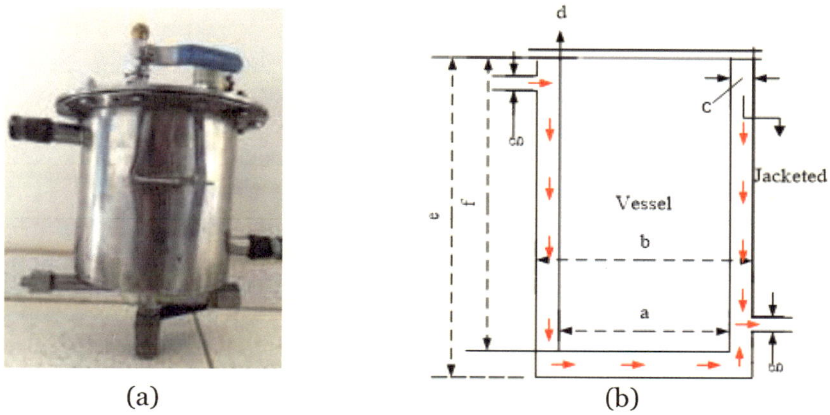

Figure 1.
The reactor model: (a) Jacketed reactor; (b) the geometry of reactor (values expressed in mm), red arrows represent water vapor flux.

Domain jacket			Domain reactor		
Fluid	**Water vapor at 100°C**	**Fluids**	**Polyol 1; 2 and 3**		
Simul. type	Steady state	Simul. type	Transient 90 min time-step: 0.1 s max. iter. per t-s: 40 convergence per t-s: 10-4		
Turbulence	SST (5%)	Turbulence	Laminar, SST (5 %)		
Heat transfer	Thermal energy	Heat transfer	Thermal energy		
		Buoyancy	g [m/s^2] = (0.00; -9.81; 0.00)		
Boundary conditions			**Boundary conditions**		
Inlet	0.01 kg/s; 125°C	External wall	Imported wall heat transfer coeff. profile with ext. 125°C		
Outlet	0.01 kg/s	bottom walls			
Internal walls	Fixed temp. 25°C	upper wall	Fixed temp. 25°C		
External walls	Adiabatic	Initial temp.	25°C		
Fluid physical properties					
Parameter	**Glycerol**	**Lemon Bagasse**	**Polyol 1**	**Polyol 2**	**Polyol 3**
ρ [kg/m^3]	1260.0	1300.0	1273.2	1268.8	1266.8
ν [kg/(m.s)]	1.4800	1.5000	1.4866	1.4866	1.4866
k[1] [W/m.K]	0.29	0.061	0.2102	0.2352	0.2465
cp[2] [J/kg.K]	2428.0	2825.52	2559.2	2515.5	2495.6

[1]*[13].*
[2]*[14].*

Table 2.
Physical properties, configurations, simulation parameters and boundary conditions employed in the superheated steam flow and natural convection simulations of jacket and polyol domains.

in the library of Ansys CFX software. In order to analyse the heat transfer inside the reaction vessel, it was assumed ideal mixture of glycerol and lemon bagasse for the following physical properties: density, viscosity, thermal conductivity and specific heat capacity. The transient natural convection simulation for biomass liquefaction used single-phase fluids.

The physical properties were estimated by data for the three solvent/biomass rates: 2:1, 3.5:1, 5:1, whose cases are denominated Polyol 1, Polyol 2 and Polyol 3, respectively. In the reactor domain, the flow was assumed laminar due to the motion of the glycerol mixture (promoted by natural convection), which is characterized by low values of Ra and Nu inside the liquefaction vessel. In the jacket domain, it was considered turbulent flow represented by the SST (Shear Stress Transport) model. It is a RANS (Reynolds Averaged Navier Stokes) model that considers isotropic velocities fluctuations in the fluid flow. Despite of the low Re values, which is an evidence of laminar flow, it was expected the formation of three dimensional natural convection cells that evaluates in time, which are characteristic of turbulent flows. Therefore, for the polyol domain simulations, it was performed two runs of CFD simulations for each case, considering laminar and turbulent (SST model) flow, respectively. The simulation time was 90 minutes with time-step of 0.1 seconds and 40 iterations for each time-step. The fluid physical properties such as specific mass (ρ), cinematic viscosity (υ), thermal conductivity (k) and calorific capacity (c_p) were obtained considering the ideal mixture for its components (glycerol and lemon bagasse properties). The boundary conditions used in the simulations of the jacket and polyol domains are showed in **Table 2**.

3. Results and discussion

3.1 Liquefaction yield (LY) and analysis of heat transfer

The experimental results obtained from the three solvent/biomass (s/b) mass ratios are shown in **Table 3**. The experiments started at 25°C and the lateral and the bottom of the reactor walls were at 125°C (due to the vapor steam in the heated jacket). The temperature variation ΔT, in **Table 3**, expresses the average heating rate, for 30 minutes time interval, from time 0(zero) until 90 minutes, for each experiment.

According to data in **Table 3**, the liquefaction yields were influenced by the time of reaction, which is coherent with previous works about other types of biomass [5, 15]. It was observed yields higher than 70 wt% in the three solvent/biomass ratios, for 90 minutes reaction, which are satisfactory results compared to experimental results in literature data [12, 16–18].

Although the highest heating rate (ΔT) was observed in the first 30 minutes (58.6 ± 1.38°C) and followed by the 30-60 minutes interval (9.6 ± 1.83°C), 90 minutes reaction was important to guarantee biomass conversion. It was observed a decrease of the solid residue with time reaction due to the gradual conversion of solid matters into polyol (reduction of 57.91 to 44.46 to 13.48 wt%).

The largest temperature variation occurred in the first 30 minutes due to the high initial gradient temperature between the vessel wall (125.15°C at t = 0 s) and the bulk mixture (25.15 K at t = 0 s). Therefore, at the beginning of liquefaction, it was expected the higher heat transfer rate.

It was also observed that, the higher the amount of glycerol, faster it achieves the final temperature. This behaviour was coherent since the greater the mass of glycerol used (liquid fraction), the higher the thermal conductivity of the mixture (k) and higher the heat transfer rate [19].

Since the steam employed in the liquefaction comes from an autoclave, it was observed a continuous rising of the solution temperature, until it reaches a thermal steady state equilibrium. However, the exchange of thermal energy was limited by

Sample	Time (min)	Temperature (K)	LY (wt%)	ΔT (°C)
Polyol 1	30	356.95	48.63	58.40
	60	366.55	40.41	9.60
	90	369.25	86.52	2.70
Polyol 2	30	357.25	42.09	58.60
	60	367.65	55.54	10.40
	90	369.15	77.62	1.50
Polyol 3	30	360.85	34.27	61.60
	60	367.75	65.79	6.90
	90	370.95	88.37	3.20
	30	361.45	—	62.80
Glycerol	60	367.05	—	5.60
	90	369.15	—	2.10

Table 3.
Experimental results for liquefaction yield (LY) and heating rate (ΔT), according to time reaction and temperatures measured at the vessel.

the heat transfer surface and the combination of the internal thermal resistances, which can be observed through polyol temperature that does not exceed 100°C, due to the physical characteristic of solvent and mixture. This behaviour is coherent with observations of Choi and Mills [20], which showed that apart from the thermal resistance, the tendency of temperature stabilization and its delay may be due to supply energy to promote the endothermic liquefaction reaction.

The knowledge about the evolution of the temperature distribution along the time inside the reactor can be used to estimate the amount of the energy required for the process. The analysis of heat transfer from jacket into the reaction vessel is shown in the following section.

3.2 Computational Fluid Dynamics (CFD) analysis

Despite the superheated vapor flows inside the heated jacket was a steady state simulation, the thermal profiles of polyol mixture were obtained from transient simulation. Therefore, the wall heat flux obtained from jacket domain was used as boundary conditions for polyol domain transient simulations. **Figure 2** shows the simulated results for wall heat flux of jacket domain considering internal surface temperature at 25°C.

According to the profiles presented in **Figure 2**, the wall heat flux from heated jacket domain to reactor vessel domain was not uniform or symmetrical and it was necessary to simulate three-dimensional profiles of fluid velocities and temperature inside the reaction vessel. It was expected that the liquefaction process is affected by temperature, since most of the cases of biomass conversion is well defined by Arrhenius equation [21]. Hence, the evolution in time of the thermal profile behaviour should be related with the evolution in time of the liquefaction yield inside the reaction vessel. However, it worth mentioning that the heated jacket simulation was steady state and the thermal profile in the reactor vessel domain is representative for 90 minutes of operation obtained in a transient simulation. According to the obtained results showed below, this is time enough to the system achieves a natural convection behaviour, able to present a symmetrical

Figure 2.
Wall heat flux profile (W m^{-2}) in the internal surface of the jacket domain visualized by the opposite sides in a semi-isometric view.

profile and significant contribution of conduction mechanism rather than convection in the net heat transfer.

Additionally, in the boundary conditions specifications, it was imported the profile for heat transfer coefficient with external temperature of 125°C. When the temperature near the walls achieves higher values next to 125°C, the effect of the heat flux profile as boundary condition becomes less influent than the specification of fixed temperature and the temperature profile tends to become uniform. Since the temperature gradient drives the heat transfer, it was expected higher heat transfer rates at the beginning of the process, enhanced by convection of momentum and energy, followed by a soft decreasing of its contribution until to establish the temperature profile as the temperature gradient also decreases. Therefore, the heat flux was more influent in the initial time-steps due to the high values of temperature gradient. **Figure 3** presents the evolution of temperature profile inside the liquefaction vessel. It can be seen that the temperature profile becomes stable between 45 and 60 min of heating, considering the reaction vessel without mechanical agitation and the mixtures Polyol 1, 2 and 3. These results are in accordance with those results from experimental analysis, present in **Table 3**.

According to the temperature profiles of **Figure 3**, there was a tendency for occurrence of a natural convection structure after 20 minutes of heating. It also can be observed that high temperatures appear first in the superior part and there was a slow heating from above to below. However, there was a centreline at the vessel axis where the temperatures remain significantly lower than in rest of the domain, due to the downward stream produced by the natural convection cell. After the natural convection structure has been established, it was possible to see that there were two heat fluxes inside the liquefaction vessel. The first was the main (primary) flux with vertical rising streams next to the walls and a descendant stream passing by the reactor axis. The secondary flux, slower than main flux, transports heat predominantly by conduction from the top to the bottom and from axis to walls in radial direction.

The thermal profile for a vertical plane in the two computational domains, vessel and jacket, can be visualized in **Figure 4**. It considered the simulation of Polyol 1, where the thermal profile shown represents the results for steady state jacket domain and for reactor vessel domain at the final time-step (90 min). A simplified

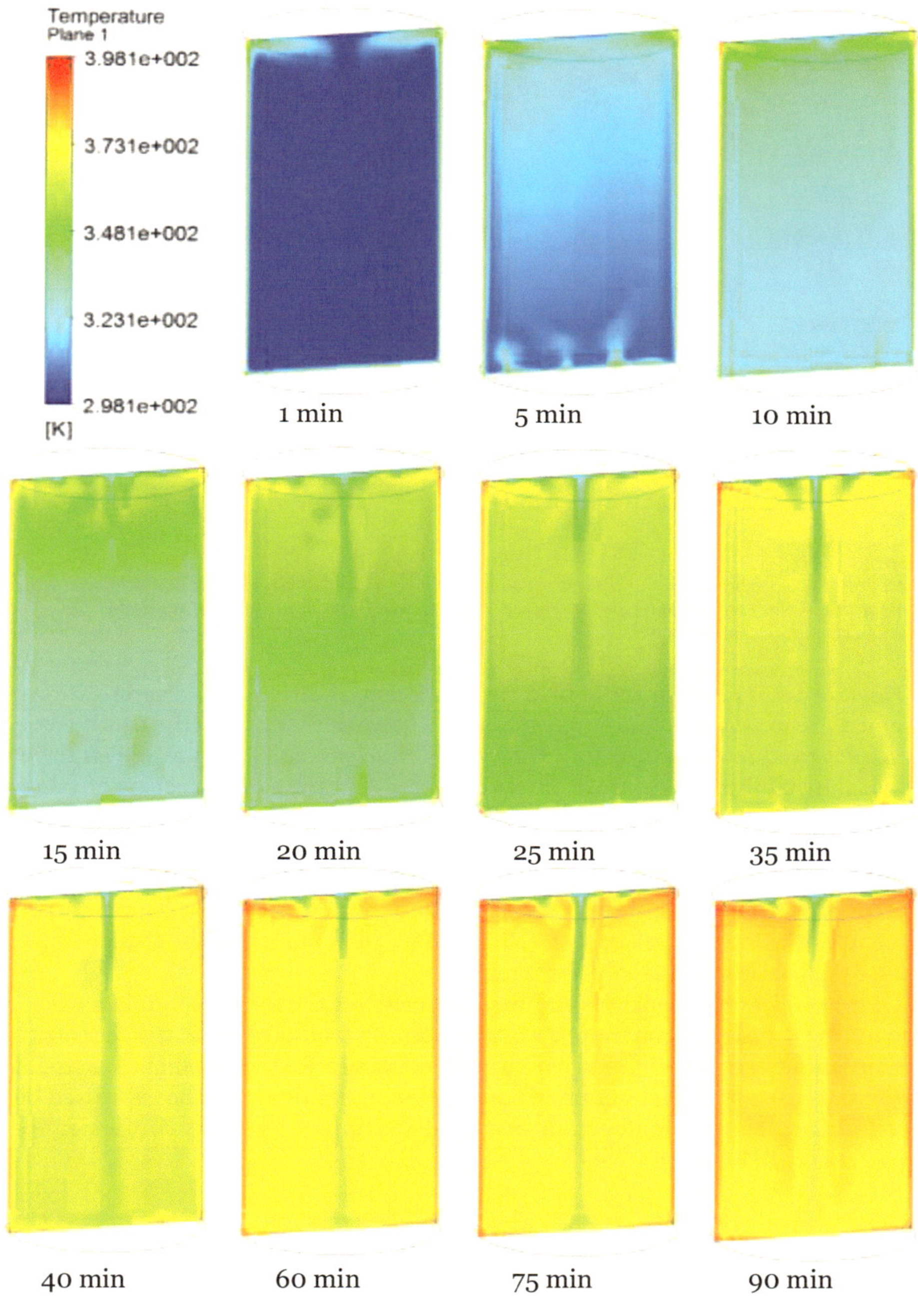

Figure 3.
Time evolution of thermal profile in vertical plane for the glycerol mixture inside the liquefaction vessel.

scheme with the two main heat fluxes, represented by continuous and dashed lines, after the formation of natural heat convection cell is also presented in **Figure 4**.

Despite of the non-uniform profile of wall heat transfer coefficient showed in **Figure 2**, the temperature profile inside liquefaction vessel looked like symmetrical after natural convection has been formed, according to **Figures 3** and **4**. This indicates that, after few minutes of heating, the temperature inside liquefaction vessel becomes more affected by flow pattern than by heat transfer from heated walls.

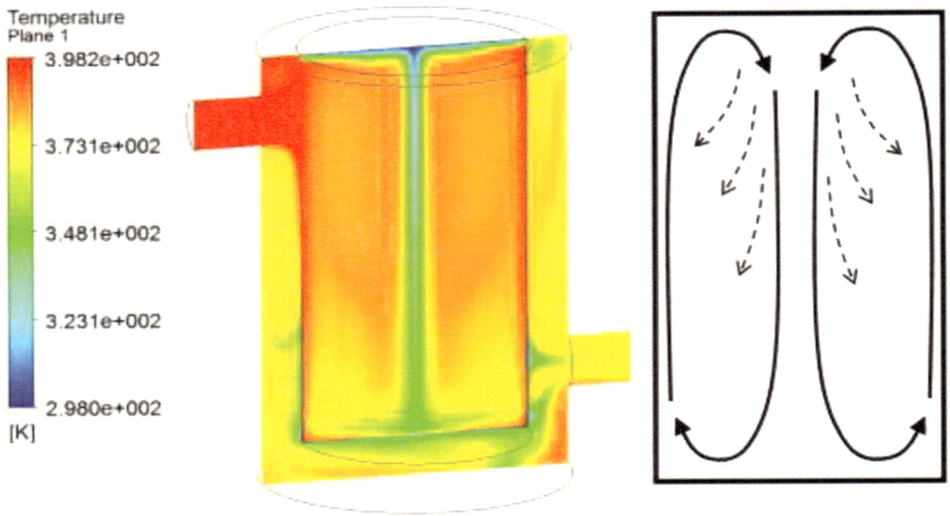

Figure 4.
Thermal profile in vertical plane for the two simulated domains showed simultaneously and the simplified scheme of the primary (continuous) and secondary (dashed) heat fluxes inside reaction vessel after the formation of natural convection cell.

However, it is hard for the hybrid (upwind/central difference) discretization scheme to represent a not imposed symmetry (zero derivative) in the central axis for velocity and temperature [22]. Hence, CFD solver results can presents transient and local instabilities. Therefore, it was expected that the eddy structure of falling cold flow in the axis domain oscillates significantly when the natural convection cell was formed. Since the liquefaction is influenced by temperature, it is convenient to evaluate the evolution of the temperature over the time. Considering that heat transfer occurs through out the walls, it can be expected the lowest temperatures at the centre of reactor vessel axis (0.0, 0.2, 0.0 m), this point was the chosen one to analyse the heating of the polyol mixture.

Despite of the low velocities and high viscosity for the mixture biomass with solvent, the transient temperature values assuming laminar flow at the centre of domain present large oscillation over the time. Using a RANS turbulence model, the vector and scalar variables obtained from CFD simulations could be expressed by a time average according to Reynolds average definitions for velocity components u_i (Eq. (2)) and T (Eq. (3)):

$$u_{i,turb} = U_i = \bar{u}_i = \frac{1}{\Delta t}\int_0^{\Delta t} u_i(t)dt \quad (2)$$

$$T_{turb} = \bar{T} = \frac{1}{\Delta t}\int_0^{\Delta t} T(t)dt \quad (3)$$

Because every value calculated under the turbulent RANS is time averaged, results from models, such as standard k-ε or SST, should present less oscillation than those observed from laminar regime [22]. However, to compare the simulated results with experimental data, those oscillations are still large and should be attenuated. In order to represent the simulated results with less variation, it was employed a moving average in the simulated temperatures (Eq. (4)).

$$T_{mov\,ave,j} = \frac{1}{5}\sum_{j-2}^{j+2} T_{turb,j} \quad (4)$$

Where the subscript j refers to a specific time-step and each $T_{mov\ ave,j}$ is the average temperature value of the five values around of the its determinate time j. The comparison between the experimental data and simulated results, expressed by the moving average temperatures, is presented in **Figure 5**.

According to samples Polyol 1 and 2, there are two different periods of time to be analysed: before 25 minutes and after 25 minutes of heating. In the first period, the experimental temperatures are lower than simulated results and in the second period the experimental values are higher than simulation temperatures. However, this behaviour, that presents two different periods, is not observed for Polyol 3 and the simulated results presented temperatures below the experimental data. Polyol 3 behaves similar to crude glycerol, since it has higher solvent/biomass ratio (**Table 3**).

Due to the liquefaction of solid biomass, the liquid mass increases. This transformation occurs mainly above 30 minutes. From this time, it can be observed that the behaviour of Polyol 1 and 2, started being similar to Polyol 3 (and crude glycerol). Although the polyols have higher density and viscosity compared to crude glycerol [23], the heat changes the fluid properties, which may favour the heat transfer by conduction and natural convection.

Changes in the fluid properties may also be the main reason of the difference between experimental results and CFD simulation, since the model used did not consider dynamic density and viscosity values.

At last, from **Figure 5**, it can be observed the tendency of a flat plateau formation. It has a little variation in temperature between 60 and 90 minutes of heating, which may be an evidence of the stabilization of the system or a period that all supplied energy is consumed to promote the endothermic liquefaction reaction.

In a nutshell, CFD simulations is suitable to represent the experimental data. Although it overestimates the temperatures before 25 minutes of heating for the Polyol 1 and Polyol 2 and underestimates it after 25 minutes of operation, the

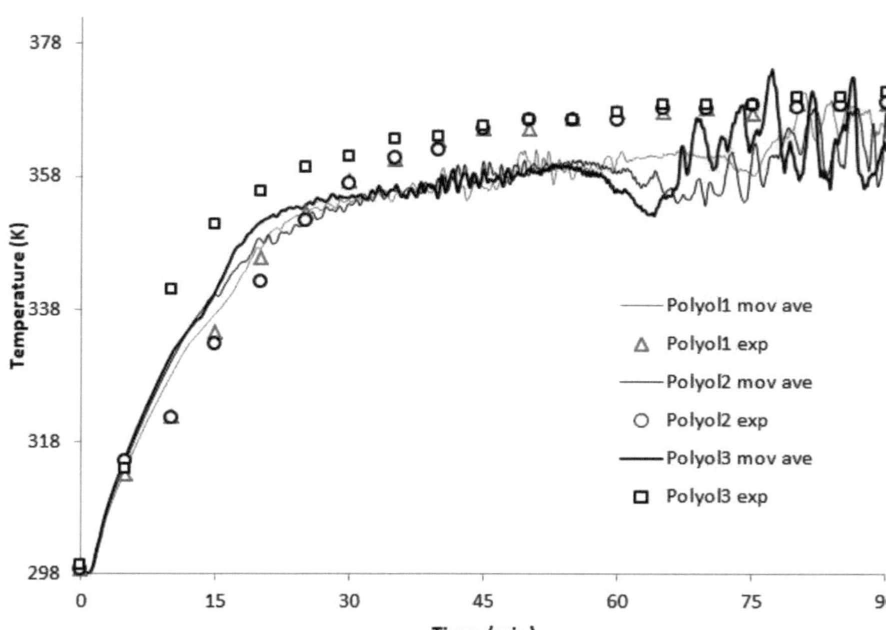

Figure 5.
Evolution in time of temperature: comparison between experimental data (exp) and simulation results treated by moving average (mov ave), for the samples Polyol 1, 2 and 3.

simulation underestimates the temperatures inside the reaction vessel for Polyol 3 during all the time of experiment. It is worth mentioning that several endothermic and exothermic reactions occur during the experiments but the CFD simulations does not consider any chemical reaction neither changes in fluids properties.

4. Conclusion

From the CFD thermal profile, in vertical plane, it was necessary 60 minutes to achieve a steady state of heating in the mixture inside this liquefaction vessel. The model is accordance with experimental data, which shows the heating hate (ΔT) was significant lower in the last stage (60 to 90 minutes). Although, the highest ΔT was observed in the first steps, the 90 minutes reaction was important to guarantee biomass conversion into polyols (LY greater than 77 wt%).

The increase in the temperature of the mixture inside the reactor occurred due to the conduction and natural convection phenomena. These phenomena of heat transfer were favoured by the modification in the proprieties of the fluids, due to the heating of them and due to the reduction of the biomass and formation of the polyol. However, from the experimental data and CFD simulation it was observed that mixture temperature did not exceeded 100°C. The temperature inside the reactor was limited to the thermal conductivity of the system and its reagents.

In conclusion, Computational Fluid Dynamics transient simulations, even presenting some oscillations, can be a satisfactory way to analyse heat transfer in liquefaction process, using crude glycerol as solvent.

Acknowledgements

The authors thanks Universidade Federal de Viçosa (UFV—Campus Florestal), Federal University of Minas Gerais (UFMG), Fundação de Amparo à Pesquisa de Minas Gerais –FAPEMIG and Conselho Nacional de Desenvolvimento Científico e Tecnológico – CNPQ.

Author contributions

B.S. Leite and S.A.F. Leite conducted the liquefaction process; D.J.O. Ferreira conducted the CFD analysis; V.F.C. Lins analysed the results and reviewed the paper; all authors had approved the final version.

Conflict of interest

The authors declare no conflict of interest.

Author details

Brenno S. Leite[1], Daniel J.O. Ferreira[2], Sibele A.F. Leite[1] and Vanessa F.C. Lins[3]*

1 Institute of Science and Technology, Federal University of Vicosa (UFV—Campus Florestal), Florestal, Brazil

2 Department of Chemical Engineering, Federal University of São Paulo (UNIFESP), Diadema, Brazil

3 Institute of Science and Technology, University Federal of Vicosa (UFV—Campus Florestal), Belo Horizonte, Brazil

*Address all correspondence to: vlins@deq.ufmg.br

References

[1] Hassan EM, Shukry N. Polyhydric alcohol liquefaction of some lignocellulosic agricultural residues. Ind Crops Prod. 2008 Jan;**27**(1):33-38

[2] Ye L, Zhang J, Zhao J, Tu S. Liquefaction of bamboo shoot shell for the production of polyols. Bioresour Technol. 2014;**153**:147-153

[3] Nakano T. Mechanism of Thermoplasticity for Chemically-Modified Wood. Vol. 48, Holzforschung - International Journal of the Biology, Chemistry, Physics and Technology of Wood. 1994. p. 318.

[4] Lee S-H, Yoshioka M, Shiraishi N. Liquefaction of corn bran (CB) in the presence of alcohols and preparation of polyurethane foam from its liquefied polyol. J Appl Polym Sci. 2000 Oct;**78**(2):319-325

[5] Briones R, Serrano L, Llano-Ponte R, Labidi J. Polyols obtained from solvolysis liquefaction of biodiesel production solid residues. Chem Eng J. 2011;**175**(1):169-175

[6] Lee JH, Lee JH, Kim DK, Park CH, Yu JH, Lee EY. Crude glycerol-mediated liquefaction of empty fruit bunches saccharification residues for preparation of biopolyurethane. J Ind Eng Chem. 2016;**34**:157-164

[7] Dimitriadis A, Bezergianni S. Hydrothermal liquefaction of various biomass and waste feedstocks for biocrude production: A state of the art review. Renew Sustain Energy Rev. 2017;**68**(May 2016):113-125

[8] Rafiqul I, Lugang B, Yan Y, Li T. Study on co-liquefaction of coal and bagasse by Factorial Experiment Design Method. Fuel Process Technol. 2000;**68**(1):3-12

[9] Cao L, Zhang C, Hao S, Luo G, Zhang S, Chen J. Effect of glycerol as co-solvent on yields of bio-oil from rice straw through hydrothermal liquefaction. Bioresour Technol. 2016;**220**:471-478

[10] Li H, Feng S, Yuan Z, Wei Q, Xu CC. Highly efficient liquefaction of wheat straw for the production of bio-polyols and bio-based polyurethane foams. Ind Crops Prod. 2017 Dec;**109**:426-433

[11] Leite SAF, Leite BS, Carriço CS, Rosa YG, Figueiredo MTD. Biopolyol and Foam Production From Lemon Bagasse Liquefaction. 2018;**65**:376-381

[12] Carriço CS, Fraga T, Pasa VMD. Production and characterization of polyurethane foams from a simple mixture of castor oil, crude glycerol and untreated lignin as bio-based polyols. Eur Polym J. 2016;**85**:53-61

[13] Lamb J. Influence of water on the thermal properties of foods. Chem Ind. 1976;**24**:1046

[14] Renaud T. Mesure et modélisation de la conductivité et de la diffusivité thermiques des produits alimentaires congelés et non congelés. 1990.

[15] Juhaida MF, Paridah MT, Mohd. Hilmi M, Sarani Z, Jalaluddin H, Mohamad Zaki AR. Liquefaction of kenaf (*Hibiscus cannabinus* L.) core for wood laminating adhesive. Bioresour Technol. 2010 Feb;101(4):1355-60.

[16] Hejna A, Kirpluks M, Kosmela P, Cabulis U, Haponiuk J, Piszczyk Ł. The influence of crude glycerol and castor oil-based polyol on the structure and performance of rigid polyurethane-polyisocyanurate foams. Ind Crops Prod. 2017 Jan;**95**:113-125

[17] Hu S, Li Y. Polyols and polyurethane foams from base-catalyzed liquefaction of lignocellulosic biomass by crude glycerol: Effects of crude glycerol

impurities. Ind Crops Prod [Internet]. 2014 Jun 1 [cited 2018 Jun 26];57:188-94. Available from: https://www.sciencedirect.com/science/article/pii/S0926669014001666

[18] Kairytė A, Vėjelis S. Evaluation of forming mixture composition impact on properties of water blown rigid polyurethane (PUR) foam from rapeseed oil polyol. Ind Crops Prod. 2015 Apr;**66**:210-215

[19] Smith JM, Van Ness HC, Abbott MM, Swihart MT. Introduction to Chemical Engineering Thermodynamics. 8th ed. Education M-HH, editor. 2017. 768 p.

[20] Choi H, Mills AF. Heat and mass transfer in metal hydride beds for heat pump applications. Int J Heat Mass Transf. 1990 Jun;33(6):1281-8.

[21] Syed AH. CFD modeling for direct liquefaction of biomass in hydrothermal media [Internet]. Norwegian University of Science and Technology; 2013. Available from: https://core.ac.uk/reader/52098590

[22] Versteeg HK, Malalasekera W. Computational Fluid Dynamics – The Finitie Volume Method. 2nd ed. Pearson; 2007. 503 p.

[23] Leite BS, Ferreira DJO, Leite SAF, Bastos FS, Lins VFC, Castro BT. Investigation of Agricultural Biomass Residues in Liquefaction Process. 2019;10(11).

Section 8

Kinetic Models and Tools for Biomass Measurement

Chapter 30

Investigation of Nonisothermal Combustion Kinetics of Isolated Lignocellulosic Biomass: A Case Study of Cellulose from Date Palm Biomass Waste

Emmanuel Galiwango and Ali H. Al-Marzouqi

Abstract

The efficient and high yielding acid-base and Organosolv methods were studied for cellulose isolation from date palm lignocellulose waste biomass and thereafter analyzed for nonisothermal kinetic and thermodynamic parameter determination using model-free methods. The structural and chemical characterization of the isolated celluloses revealed structures and functional groups characteristics of cellulose. Thermal decomposition analysis revealed one major peak with average mass loss of 72.51 ± 0.7% and 55.82 ± 1.1% for the acid-base and Organosolv method, respectively. This occurred in the temperature region between 250 and 350°C associated with cellulose degradation and contrasted with the three peaks detected in the original biomass. The kinetic and thermodynamic results revealed a strong relationship between the average activation energy and average change in enthalpy with a difference of 5.23 and 147.07 kJmol^{-1} for Organosolv and acid-base methods, respectively. The Gibbs's free energy results revealed that Organosolv cellulose pyrolysis would reach equilibrium faster in KAS, Starink and FWO models with average ΔG values of 115.80 ± 36.62, 115.89 ± 36.65, and 119.45 ± 37.98 kJmol^{-1}, respectively. The acid-base method for FWO model gave negative entropy values. The Malek method revealed the acid-base and Organoslv cellulose pyrolysis mechanism as $(g(\alpha) = [-\ln(1-\alpha)]^{\frac{1}{4}})$ and $(g(\alpha) = [-\ln(1-\alpha)]^{\frac{1}{3}})$, characterized by random nucleation and growth, respectively.

Keywords: lignocellulose biomass, nonisothermal kinetics, isolation methods, characterization

1. Introduction

Lignocellulosic biomass is the most abundant, renewable, and one of the cheapest carbon neutral raw materials in the biosphere that can be used to produce sustainable products such as biofuels, using different technologies [1]. The lignocellulosic biomass consists of mainly cellulose carbohydrate polymer, hemicellulose, and the aromatic component, lignin [2]. Lignocellulose biomass can store up to 47

$MJkg^{-1}$ more energy than lithium ion batteries (0.8 $MJkg^{-1}$) [3]. Lignocellulose biomass is considered a potential candidate to sustainable green alternative source of energy and chemicals due to its high energy density, volatile matter content, and global widespread [3]. The release of volatile matter and other contents in biomass has been extensively studied using pyrolysis technology. Pyrolysis involves the conversion of biomass into bio-oil, gases (volatile matter) and biochar, in the absence of oxygen [4, 5]. The technique is robust and essential in providing vital knowledge of kinetics of devolatilization of any biomass prior to further processing via different conversion technologies. In addition, pyrolysis is effective in reducing the bulky biomass into uniform, energy dense, and easily transportable fuel [6]. Despite always being the first stage in most combustion or gasification process, there are no accurate and enough data on the kinetics and reaction mechanisms of different lignocellulosic biomass [7]. This is attributed to the complexity and the varying physico-chemical properties in different lignocellulosic biomass [7]. In addition, there may be many reactions occurring from the extremely complex pyrolysis process of the lignocellulose biomass [8]. Therefore, developing accurate kinetic models to account for all reactions taking place remains a challenge [6]. Isolation of the complex lignocellulosic biomass into individual fractions and characterization of the individual fractions can provide a better understanding of the combustion kinetics and reaction mechanism. Different biomass isolation/extraction techniques such as liquid-liquid, liquid-solid, acid-base, ultrasound, and microwave-assisted extractions, among others, have been reported before [9]. The choice of the method depends on the biomass type and its fraction to be isolated [9]. Hence, each procedure affects the sample's product yields and physical, chemical, kinetic, and thermodynamic properties differently. Despite the studies on the yields and operating parameters such as solvent and time [9], less or no information is available regarding the kinetic and thermodynamic parametric studies for the combustion of the isolated lignocellulose fractions to assess the difference in the extraction processes. Different general kinetic models on lignocellulose biomass have been suggested [10, 11].

Date palm waste constitutes about 500,000 metric tons per year from ca. 44 million date palm trees found in the United Arab Emirates where this research was conducted. The aims of this research are to isolate cellulose from date palm lignocellulose complex using low concentration acid-base solutions and Organosolv techniques and to model nonisothermal combustion kinetics using model-free methods and finally to predict the most probable mechanistic reaction mechanism of the isolated celluloses. Using thermal-gravimetric technique at different heating rates, kinetic and thermodynamic parameters were calculated using model-free methods, namely Kissinger-Akahira-Sunose (KAS), Flynn-Wall-Ozawa (FWO), and Starink models. The FWO model-free method compensates the experimental measurement errors. However, the KAS and Starink methods depend on the choice of good constant degree of conversion from the derivative mass loss function to provide precision of the kinetic data [12]. In addition, application of different model-free methods involves wide conversion range that allows for study of change in mechanism during a reaction and reduces mass transfer limitations by using multiple heating rates [13].

2. Isolation techniques and nonisothermal kinetic studies

The rachis part of adult date palm waste (DPW) (10–15 years old) was supplied by the UAE University farm, Al foah, Al Ain. The samples were ground to 180-micron particle size to reduce the effects of heat and mass transfer limitations. All the solvents (ACS grade) and reagents were supplied by Sigma Aldrich and were used with no further purification. Prior to cellulose isolation, 10 g biomass was valorized with benzene/ethanol (2, 1 v/v) for 48 h using Soxhlet extraction to reduce extractives such as waxes and resins surrounding the lignocellulose complex.

2.1 Acid-base and Organosolv cellulose isolation methods

For acid-base isolation; DPW (5 g) extractive-free sample of particle size 180 μm was placed in a 250-mL beaker and leached with 200 mL of 0.1 M HCl while heating at 100°C for 2 h under stirring at 150 rpm. After vacuum filtration, the cellulose and lignin-rich residue was washed with 20 mL of deionized water to remove any residual hemicellulose and then air dried overnight. The hemicellulose was solubilized by HCl and heating due to its labile nature making it easy to dissolve out of the lignocellulose complex. The cellulose- and lignin-rich residue was further treated with 200 mL of 0.1 M NaOH while heating at 100°C for 2 h under constant stirring at 150 rpm. After subsequent vacuum filtration of the mixture, the cellulose-rich residue was washed with 20 mL of 0.1 M NaOH to remove any residual lignin. The isolated cellulose was air dried under laboratory conditions overnight prior to characterization.

For Organosolv isolation, DPW were isolated using methanol/water solvents as reported in literature, with some modifications [14]. The 6.7 g sample of particle size 180 μm was placed in high pressure/temperature reactor vessel (Parr 4848, U.S.A). A mixture of 84 mL sulfuric acid (0.045 N), 13.4 mL formaldehyde (37 wt.%), and 84 mL methanol was added to the reactor vessel containing the sample. The reactor was sealed and purged with nitrogen gas (6–10 bars), and the reaction was performed for 1 h at 160°C under constant stirring at 700 rpm. The product mixture was vacuum filtered after cooling to room temperature. The cellulose-rich residues were air dried overnight prior to characterization, and the yield was determined by a gravimetric analysis technique. The ultimate analysis was conducted, and the results are recorded in **Table 1**.

2.2 Fourier transform infrared (FTIR) spectroscopy

The IRTrace-100 FTIR spectrophotometer (Shimadzu, Kyoto, Japan) was used for the FTIR analysis. The extracted celluloses were analyzed to investigate the difference in the functional groups after extraction. The spectral results were recorded within a range of 500–4000 cm^{-1} wavelength using 4 cm^{-1} spectral resolution and 34 scans. **Figure 1** shows several major absorption bands and the difference between the samples. DPW sample before isolation showed typical lignocellulose strong band absorption bonds. For instance, the bands at 1037 cm^{-1}

Proximate and ultimate analyses	Date palm lignocellulose	Organosolv cellulose	Acid-base cellulose
Moisture content (wt.%)	6.72 ± 0.4	7.08 ± 0.4	8.72 ± 0.4
Volatile matter (wt.%)	78.62 ± 0.04	65.22 ± 0.02	66.92 ± 0.01
Ash content (wt.%)	6.12 ± 0.1	7.36 ± 0.04	7.24 ± 0.01
Fixed carbon (wt.%)	5.40 ± 0.01	4.40 ± 0.10	4.80 ± 0.14
HHV ($MJkg^{-1}$)	17.28	15.46	15.18
Cellulose yield (wt.%)	—	43.15 ± 2.40	64.15 ± 2.40

Table 1.
Physicochemical analysis of biomass (dried basis).

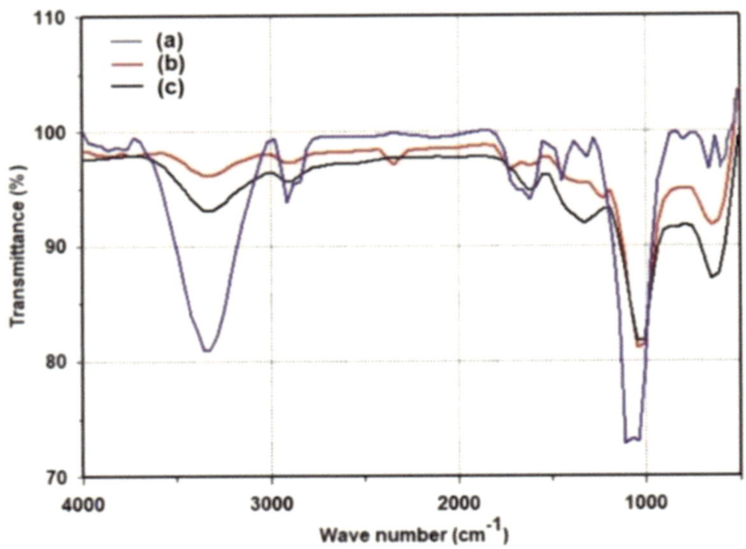

Figure 1.
FTIR spectrum of (a) lignocellulose DPW, (b) Organosolv cellulose and (c) acid-base cellulose.

indicating C—O, C=C and C—C bond stretching, between 2840 and 2926 cm^{-1} indicating C—H stretching and 3200–3474 cm^{-1} for O—H stretching, were assigned to cellulose, hemicellulose and lignin, respectively. Similar results were reported for other biomass in the literature [15]. However, there were noticeable changes in the functional groups of celluloses from the same biomass with reduction in the peak intensity, an indication of component(s) removal (plausibly lignin and hemicellulose). For instance, the would-be lignin and hemicellulose band intensity at 1037 cm^{-1} greatly reduced an indication of component(s) removal. The C—H bond stretching in the region of 2840 assigned to lignin in DPW was absent in isolated cellulose samples. In addition, the reduced peak intensity between 845 and 1156 cm^{-1} associated with the C—O—C asymmetrical stretching and glycosidic bond, a characteristic of cellulose, was observed. Furthermore, the decrease in OH vibration strength around 3200–3474 cm^{-1} indicates a reduction in some of the OH-containing compounds which are phenolics from lignin. It is worth to note that both extraction methods showed similar functional groups except that the Organosolv cellulose had C—H bond assigned to lignin in the region around 2326 and 2363 cm^{-1} which was absent for the acid-base cellulose samples. The FTIR results showed the effectiveness of the cellulose isolation methods from DPW biomass complex. The samples were further characterized for their morphological differences using SEM imaging technique.

2.3 Scanning electron microscopy (SEM)

The structural morphologies of the isolated cellulose were analyzed using the scanning electron microscope (JEOL Neoscope JCM-5000, Tokyo Japan). The samples were Au/C coated using vacuum spatter while clamped on the sample holder. The images were captured on spot size of 40 using 10 kV. The SEM results in **Figure 2** show a difference in the structural morphologies between the cellulose samples from the two methods. **Figure 2(a)** shows the original DPW with ring-like structures (see the arrow point) plausible to be the cellulose chiral nematic ordering, surrounded by irregular shaped structures which could be assumed to be lignin

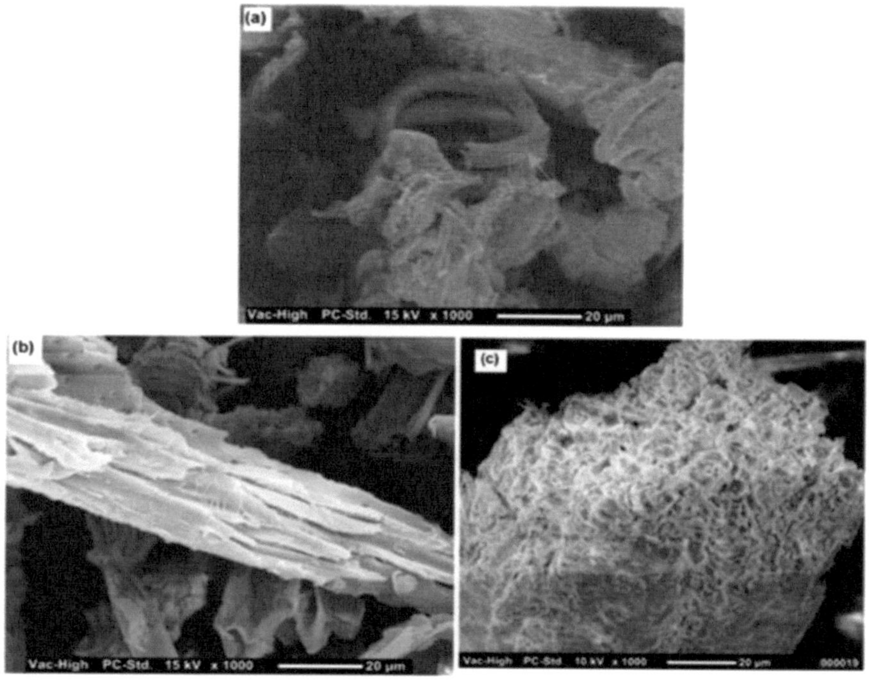

Figure 2.
SEM images of the (a) original rachis, (b) acid-base cellulose and (c). Organosolv cellulose captured at magnification X1000.

and hemicellulose. Acid-base cellulose in **Figure 2(b)** showed porous surface similar to those reported for cellulose from the teak wood [16]. However, Organosolv cellulose (**Figure 2(c)**) showed an aggregate of cellulose block structure with uneven polished surface.

2.4 Thermogravimetric analysis (TGA)

The combustion characteristics of isolated celluloses were studied using thermogravimetric analysis (TGA). The analysis was done on a TGA (Q500, TA instrument). Samples of 6 mg (±1.0) were first equilibrated at 25°C for 5 min and then heated at specific heating rates of 10, 15, 20, and 25°C/min to 900°C. The process was performed under constant nitrogen environment flowing at 20 mL/min. As the thermal decomposition progressed, the change in weight was recorded continuously as a function of temperature and time. **Figure 3** shows the isoconversion versus temperature at different heating rates for the isolated celluloses from DPW. The conversion curves for acid-base (colored) and Organosolv (black) methods below 300 and 340°C, respectively, showed similar thermal decomposition patterns at all heating rates. There was a slight shift toward higher temperature side with increasing heating rates, possibly due to the increasing thermal energy in the system [17]. However, at higher temperatures, the conversion pattern changed for both methods, possibly due to the change in the degradation chemistry of components under pyrolysis. It is worth to note that Organosolv cellulose showed better thermal stability than the acid-base cellulose. **Figure 4** shows the differential thermogravimetric (DTG) results against temperature at different heating rates for the DPW and the isolated celluloses. The results showed a typical thermal degradation of lignocellulose biomass. The curves of both samples moved downward as the heating

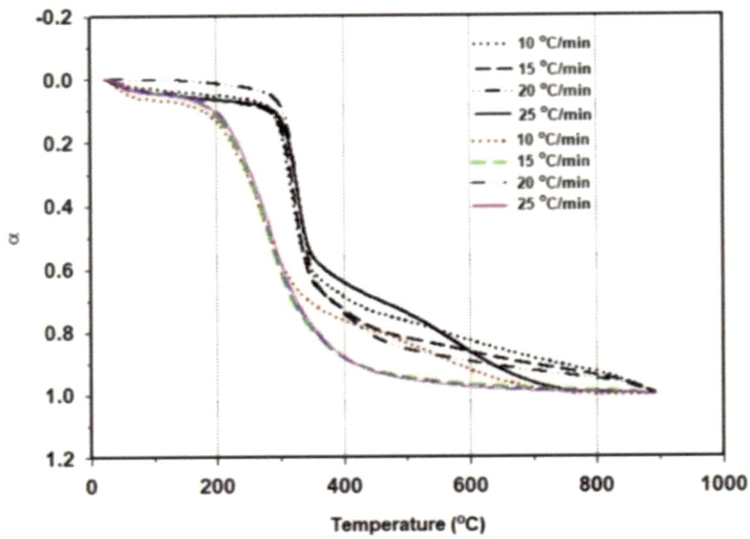

Figure 3.
The relationship of conversion against temperature for acid-base cellulose (colored) and Organosolv cellulose (black).

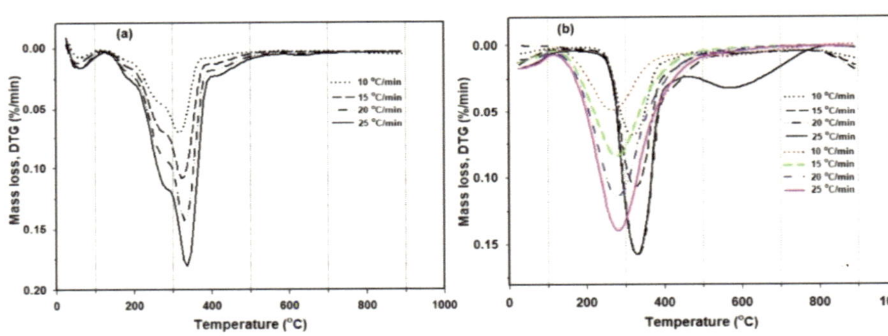

Figure 4.
The relationship of DTG against temperature for (a) date palm waste and (b) isolated celluloses (acid-base, colored; Organosolv, black).

rate increased, owing to the shorter reaction time at increasing temperature, a phenomenon known as thermal hysteresis. However, **Figure 4(a)** showed a noticeable difference between the peak mass loss patterns compared to **Figure 4(b)** for the two cellulose methods, which suggests a difference in the degradation chemistry. **Table 2** shows the temperature ranges that define the major stages of mass loss in response to increasing temperature for isolated celluloses from both methods, as given in **Table 3**. Stage I started from minimum temperature, T_{min} to T_1, the total average celluloses mass loss for acid-base and Organosolv were 9.46 ± 0.1 and $5.28 \pm 0.1\%$, respectively. This was attributed to the inherent moisture and water molecules embedded in the intercellular and intracellular void spaces of the celluloses. Stage II, from T_1 to T_3 for both methods, there was only one major clear peak (**Figure 4(b)**) and the average mass loss in this region was 72.51 ± 0.7 and $55.82 \pm 1.1\%$, for acid-base and Organosolv celluloses, respectively. The weight loss in this stage is associated with pyrolysis of mainly cellulose and to a lesser extent hemicellulose [18]. Compared to the three peaks observed in **Figure 4(a)** for the original DPW, this clearly shows that both methods were effective for cellulose

Heating rate (°C min^{-1})	T_{min} (°C)	T_1 (°C)	T_2 (°C)	T_3 (°C)	T_{max} (°C)
		Acid-base cellulose			
10	30	192	266	402	900
15		197	273	406	
20		199	279	411	
25		201	381	415	
		Organosolv cellulose			
10	30	220	318	422	900
15		226	324	435	
20		227	328	441	
25		228	330	450	

Table 2.
Characteristic temperatures associated with mass loss during pyrolysis of cellulose.

Stages	Temperature	Heating rate (°C min^{-1})			
		10	15	20	25
		Acid-base cellulose			
Stage I, WL%	T_{min}–T_1	10.06	9.17	9.33	9.26
Stage II, WL %	T_1–T_3	71.59	72.76	73.16	72.52
Stage III, WL %	T_3–T_{max}	10.18	8.88	7.30	8.35
Final residue at 900-100 °C (%)		7.99	9.19	11.21	9.87
		Organosolv cellulose			
Stage I, WL %	T_{min}–T_1	4.35	5.24	5.72	5.79
Stage II, WL %	T_1–T_3	54.42	56.16	57.16	55.55
Stage III, WL %	T_3–T_{max}	35.57	33.30	32.24	33.69
Final residue at 900-100 °C (%)		5.66	5.30	4.88	4.97

Table 3.
Mass loss (%) during different stages of cellulose pyrolysis.

isolation from the complex lignocellulose matrix of DPW. Stage III had total average mass loss of 8.68 $\pm$ 1.2 and 33.08 $\pm$ 0.8%, for acid-base and Organosolv methods, respectively. This represented combustion of the carbonaceous and some part of char oxidation [19]. In addition, the higher mass loss for Organosolv cellulose was plausibly due to residual lignin. Moreover, the FTIR results showed some lignin functional groups for this method. The last stage was associated with charring process and ash formation. The average total mass loss for acid-base and Organosolv methods in this stage were 9.57 $\pm$ 1.3 and 5.20 $\pm$ 0.4%. TGA analysis data was used for kinetic modeling using the model-free methods.

2.5 Nonisothermal kinetic analysis

The TGA data were used to calculate the nonisothermal kinetic and thermodynamic parameters using model-free equations of Flynn-Wall-Ozawa (FWO), Kissinger-Akahila-Sunose (KAS), and Starink reported in the literature [20].

$$FWO\ model:\ \ln(\beta) = \ln\left(\frac{AE_\alpha}{g(\alpha)R}\right) - 5.331 - 1.052\left(\frac{-E}{RT}\right) \tag{1}$$

$$KAS\ model:\ \ln\left(\frac{\beta_i}{T_{\alpha,i^2}}\right) = \ln\left(\frac{AR}{E_\alpha}\right) - \left(\frac{E_\alpha}{RT_\alpha}\right) + \ln\left(\frac{df(\alpha)}{d\alpha}\right) \tag{2}$$

$$Starink\ model:\ \ln\left(\frac{\beta_i}{T_{\alpha,i^{1.92}}}\right) = \ln\left(\frac{AR^{0.92}}{g(\alpha)E_\alpha^{0.92}}\right) - 1.0008\left(\frac{E_\alpha}{RT_\alpha}\right) - 0.312 \tag{3}$$

where $T_{\alpha,i}$ is the time to reach a given extent of conversion at temperature T_i. At α, the value of E_α is determined from the slope of the plot $\ln(\beta)$, $\ln(\beta/T_{\alpha,i^2})$, and ln $(\beta/T_{\alpha,i^{1.92}})$ versus $1,000/T_{\alpha,i}$.

$$\alpha = \frac{\mathrm{m_1 - m_t}}{\mathrm{m_1 - m_\infty}} \tag{4}$$

where $\mathrm{m_1}$ is the initial biomass weight, $\mathrm{m_t}$is the change in weight at a particular time during the experiment, and $\mathrm{m_\infty}$ is the residual weight after time of the experiment.

The choice for these model-free methods is because no previous knowledge about reaction mechanism is required to determine the reaction activation energy [21]. The preexponential factor (A) and thermodynamic parameters [enthalpy (ΔH), entropy (ΔS), and Gibb's free energy (ΔG)] were calculated using equations in literature [22].

$$A = \frac{\left[\beta.\exp\left(\frac{E_\alpha}{RT_m}\right)\right]}{\left[RT_m^2\right]} \tag{5}$$

$$\Delta H = E_\alpha - RT \tag{6}$$

$$\Delta G = E_\alpha + RT_m \ln\left(\frac{K_B T_m}{hA}\right) \tag{7}$$

$$\Delta S = \frac{\Delta H - \Delta G}{T_m} \tag{8}$$

where β is the heating rate, E_α is the activation energy, T_m is the maximum peak temperature, K_B is the boltzman constant, and h is the plank constant.

The activation energies for both sample methods were calculated using the three models, namely FWO, KAS, and Starink. These model-free methods avoid the shortcomings during model fitting and kinetic compensation effects. The FWO model-free method compensates the experimental measurement errors. However, the KAS and Starink methods depend on the good constant degree of conversion from the derivative mass loss function to provide precision of the kinetic data [12]. Therefore, application of different model-free methods involves a wide conversion range that allows for the study of change in mechanism during a reaction and reduces mass transfer limitations by using multiple heating rates [13]. **Figure 5** shows the relationship of activation energy and enthalpy from the three model-free methods for acid-base and Organosolv celluloses. Results showed little or no difference between E_α and ΔH. This closeness in E_α and ΔH values signifies the formation of activation complex and little extra energy might be required to achieve product formation [23]. Organosolv cellulose E_α and ΔH values were higher than acid-base cellulose especially at higher temperatures ($\alpha > 0.6$). This was possibly due to a

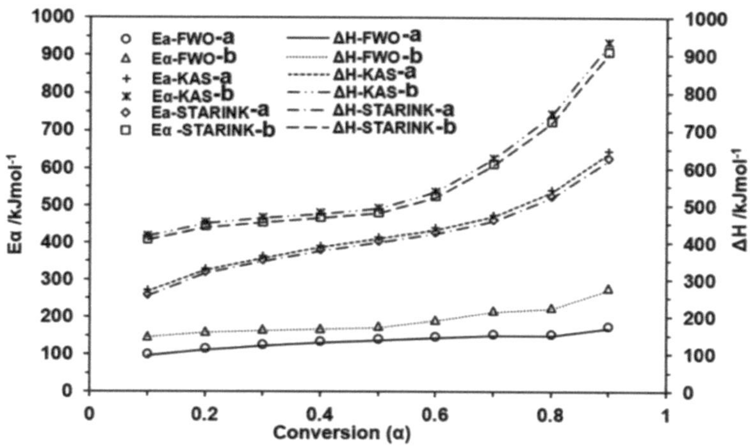

Figure 5.
Activation energy and enthalpies of (a) acid-base cellulose and (b) Organosolv by three model-free methods.

difference in the cellulose structure between the two methods. In addition, the residual lignin fractions detected in Organosolv method could also have resulted in the increased energy of activation, E_α, and bond dissociation, ΔH, needed to overcome the carbon number distribution from other components other than cellulose. It was noted that the ΔH values for all samples were positive, an indication of energy consumed during pyrolysis process, and are used to release various volatile and biochar products. Furthermore, the calculated solid-state process parameters were different due to the fundamental differences in the model-free methods [12, 13]. The first difference arises from the slope, S, of straight lines which is directly proportional to the activation energy, that is, FWO, KAS, and Starink slope, $S = -\frac{1.052E}{RT_\alpha}, S = -\frac{E}{RT_\alpha}$ *and* $S = -\frac{1.0008E}{RT_\alpha}$, respectively. The second difference is in the time to reach the extent of conversion (T_{α,i^n}) at a given temperature, which is different across different models (for (T_{α,i^n}) term, n = 0, 2, and 1.92 for FWO, KAS, and Starink, respectively).

Tables 4–6 show other thermodynamic parameters from the three model-free methods for the acid-base and the Organosolv cellulose samples, respectively. The ΔG values for Organosolv cellulose for all model-free methods were lower than those of acid-base cellulose samples. Gibb's free energy gives the measure of how favorable a reaction is to reach chemical equilibrium [24]. In context of the first and second laws of thermodynamics, the sample with higher values of ΔG (acid-base cellulose), the further its reaction is from equilibrium and the further its reaction must shift to reach equilibrium. However, the entropy, ΔS values were lower for the acid-base celluloses for all model-free methods, with negative entropy values for the FWO model. This implies that the degree of disorder of initial reactants was higher than that of the products formed by bond dissociations [22]. In addition, it was already discussed previously that the heat input during the thermal decomposition was for bond dissociation of the reactants. In the context of reaction energy, the acid-base cellulose sample required lower activation energy and enthalpy to form products than Organosolv cellulose samples. On the other hand, the preexponential factor of Organosolv cellulose was ca. two times higher than that of acid-base cellulose. This was plausibly because the activation energy had a similar trend as already discussed above. The preexponential factor and activation energy both influence chemical kinetics and reaction dynamics in pyrolysis of biomass involving complex heterogeneous reactions [25]. The R^2 of all model-free parameters was above 0.98, signifying accuracy of the models.

	FWO[a]			FWO[b]		
α	Log A (s^{-1})	ΔG ($kJmol^{-1}$)	ΔS ($Jmol^{-1}$)	Log A (s^{-1})	ΔG ($kJmol^{-1}$)	ΔS ($Jmol^{-1}$)
0.1	10.17	151.46	−0.12	27.20	84.18	0.21
0.2	10.96	164.16	−0.11	27.24	91.96	0.21
0.3	11.34	171.76	−0.10	27.26	94.86	0.21
0.4	11.79	178.07	−0.09	27.15	97.26	0.21
0.5	12.13	183.36	−0.08	27.17	100.17	0.21
0.6	12.25	189.25	−0.08	27.53	109.59	0.21
0.7	12.17	198.03	−0.08	25.07	136.91	0.17
0.8	11.15	216.34	−0.10	21.64	167.53	0.10
0.9	11.28	243.04	−0.10	23.20	192.64	0.13
Av	**11.47 ±0.68**	**188.39 ±27.89**	**−0.10 ±20.01**	**25.94 ± 2.16**	**119.45 ±37.98**	**0.18 ±0.04**

R^2 were above 0.98.
[a]Acid-base.
[b]Organosolv.

Table 4.
The kinetic and thermodynamic parameter values of celluloses determined by FWO model.

	KAS[a]			KAS[b]		
α	Log A (s^{-1})	ΔG ($kJmol^{-1}$)	ΔS ($Jmol^{-1}$)	Log A (s^{-1})	ΔG ($kJmol^{-1}$)	ΔS ($Jmol^{-1}$)
0.1	31.09	138.51	0.28	76.86	81.66	1.16
0.2	34.78	149.86	0.35	76.54	89.24	1.15
0.3	36.56	156.67	0.38	76.62	92.06	1.16
0.4	38.05	162.34	0.41	76.69	94.38	1.16
0.5	39.18	167.12	0.43	76.76	97.21	1.16
0.6	40.29	172.40	0.46	76.72	106.42	1.16
0.7	41.65	180.30	0.48	72.62	132.86	1.08
0.8	43.89	196.73	0.52	71.44	162.07	1.05
0.9	46.97	220.82	0.58	78.19	186.28	1.18
Av	**39.16 ± 4.78**	**171.64 ± 25.07**	**0.43 ± 0.09**	**75.83 ± 2.23**	**115.80 ± 36.62**	**1.14 ± 0.04**

R^2 were above 0.98.
[a]Acid-base
[b]Organosolv.

Table 5.
The kinetic and thermodynamic parameter values of celluloses determined by KAS model.

2.6 Reaction model determination

Malek method which is the commonly used approach to determine probable reaction mechanism involving heterogeneous reaction was used [26]. The Malek method is described by the following equation.

$$Z(\alpha) = f(\alpha)g(\alpha) = \left(\frac{d\alpha}{dt}\right)_{\alpha} T_{\alpha}^{2}\left[\frac{\Pi(x)}{\beta T_{\alpha}}\right] \quad (9)$$

	STARINK[a]			STARINK[b]		
α	Log A (s^{-1})	ΔG ($kJmol^{-1}$)	ΔS ($Jmol^{-1}$)	Log A (s^{-1})	ΔG ($kJmol^{-1}$)	ΔS ($Jmol^{-1}$)
0.1	29.65	138.69	0.25	74.89	81.73	1.12
0.2	33.82	149.97	0.33	74.76	89.30	1.12
0.3	35.63	156.78	0.37	74.59	92.13	1.12
0.4	36.99	162.47	0.39	74.71	94.46	1.12
0.5	38.15	167.24	0.41	74.80	97.28	1.12
0.6	39.22	172.53	0.43	74.83	106.50	1.12
0.7	40.55	180.43	0.46	70.70	132.96	1.04
0.8	42.60	196.89	0.50	69.46	162.20	1.01
0.9	45.58	221.00	0.55	76.04	186.43	1.14
Av	**38.02 ± 4.75**	**171.78 ± 25.05**	**0.41 ± 0.09**	**73.86 ± 2.21**	**115.89 ± 36.65**	**1.10 ± 0.04**

R^2 were above 0.98.
[a]Acid-base
[b]Organosolv.

Table 6.
The kinetic and thermodynamic parameter values of celluloses determined by STARINK model.

where $\left(\frac{d\alpha}{dt}\right)_\alpha$ is rate of reaction at a given conversion, α, and heating rate, β, $\Pi(x)$ approximates the temperature integral profile and $x = E_\alpha/RT_\alpha$. The x values used were in a range of 5–20 and the temperature approximation $\Pi(x)$ function is defined by the following equation [27].

$$\Pi(x) = \frac{x^3 + 18x^3 + 88x + 96}{x^4 + 20x^3 + 120x^3 + 240x + 120} \tag{10}$$

The theoretical z(α) plots versus α depend on $f(\alpha)$ and g(α) functions. However, the experimental z(α) values can be obtained by using a specific heating rate for a specific value of $\frac{d\alpha}{dt}$, E_α and T_α. The experimental z(α) master plots as a function of α are compared with known theoretical model functions [28]. The best fit between the experimental z(α) master plots and theoretical model functions describes the probable biomass reaction mechanism. **Figures 6** and 7 show the experimental z(α) master plots and fitted model plots of acid-base and Organosolv, as determined by model-free methods at different heating rates, respectively.

The experimental and the fitted z(α) master plots of acid-base cellulose showed a normal distribution behavioral curve trend for all model-free methods at investigated heating rates. However, Organosolv cellulose showed a sigmoid curve skewed more to the left hand side. The correlation coefficient of acid-base method ranged between 0.9789 and 0.9884, while that of Organosolv ranged between 0.9525 and 0.9795, signifying the accuracy in the reported data. It is worth to note that both methods had best fit at 15°C/min. The data were fit with polynomial curves of n = 3 and n = 4 for Organosolv and acid-base celluloses, respectively, implying third and fourth dimension growth as described by general Avrami-Erofeev model of multidimensional nuclei and random growth reaction mechanism $\left(g(\alpha) = [-\ln(1-\alpha)]^{\frac{1}{n}}\right)$. This type of mechanism is often as a result of hydration, adsorption, dissolution, and defects on the crystallite within particle size of the

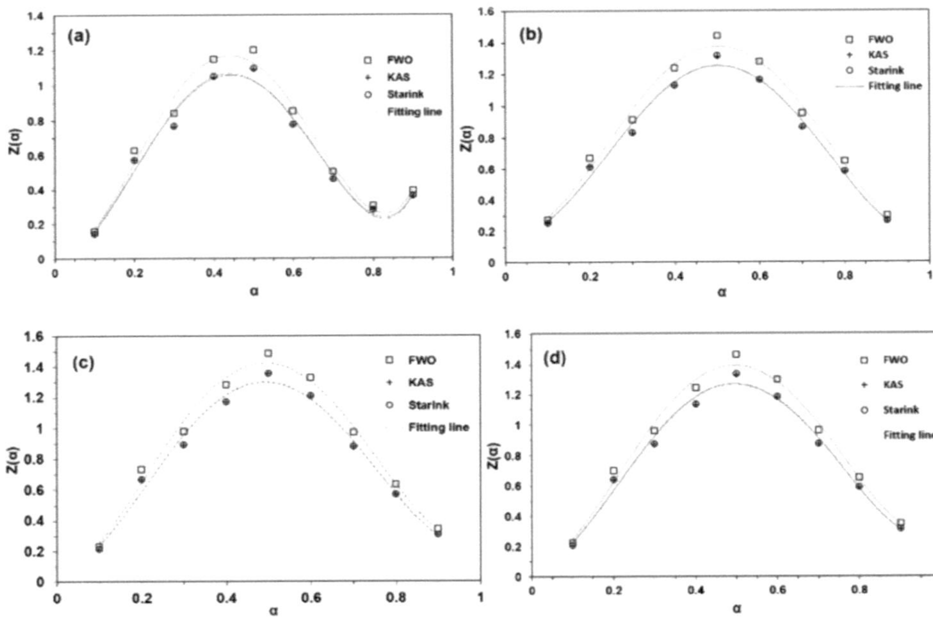

Figure 6.
Experimental and theoretical $Z(\alpha)$ master plots for pyrolysis of acid-base cellulose at (a) $10^{\circ}Cmin^{-1}$, (b) $15^{\circ}Cmin^{-1}$, (c) $20^{\circ}Cmin^{-1}$ and (d) $25^{\circ}Cmin^{-1}$.

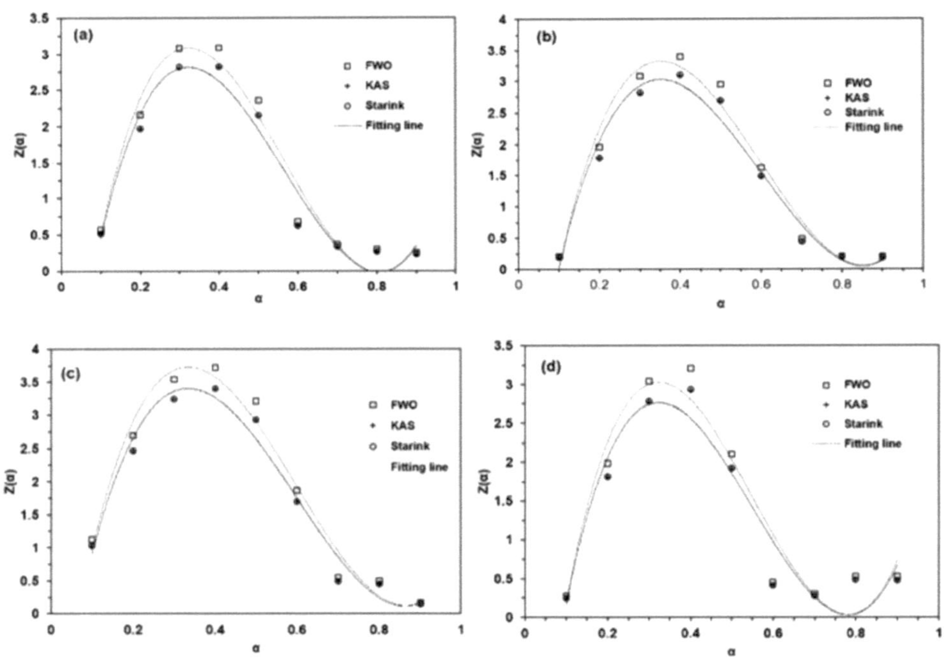

Figure 7.
Experimental and theoretical $Z(\alpha)$ master plots for pyrolysis of Organosolv cellulose at (a) $10^{\circ}Cmin^{-1}$, (b) $15^{\circ}Cmin^{-1}$, (c) $20^{\circ}Cmin^{-1}$ and (d) $25^{\circ}Cmin^{-1}$.

sample that can cause thermodynamic inhibition leading to varying activation energies [29]. Therefore, random nucleation and growth is the most probable reaction mechanism for the pyrolysis of celluloses isolated from DPW by acid-base and Organosolv methods.

3. Conclusions

The low cost and high yield acid-base and Organosolv methods were assessed for isolation of cellulose from date palm lignocellulose waste biomass. The structural, chemical, and morphological characterizations of the isolated celluloses were studied. The nonisothermal combustion studies were investigated using three different model-free methods. The reaction mechanism was studied using Malek method.

- The SEM images revealed chiral nematic orderings structures distinctive of cellulose. The change in FTIR peak intensity and the difference in the vibrational bond stretching among the isolated celluloses and between original biomass signified component removal from the lignocellulose complex. The TGA results from both methods showed one major decomposition peak assigned to cellulose in contrast to original biomass with three peaks. The results further revealed a possible difference in the degradation chemistry at higher temperature where isoconversion was higher than 0.6.

- The FWO model for the acid-base method gave the lowest activation energy (99.77–173.76 $kJmol^{-1}$) and the Organosolv method by KAS model gave the highest activation energy (419.63–934.49 $kJmol^{-1}$). There was a strong relationship between activation energy and enthalpy, and the positive enthalpy values confirmed that endothermic reaction took place during the pyrolysis of the cellulose samples. The Gibbs's free energy, ΔG, results revealed that Organosolv cellulose pyrolysis reaction would easily reach equilibrium, much easier in a trend of KAS > Starink> FWO models. The measure for disorder was less favorable for the acid-base method with negative entropy values in the FWO model-free method.

- The reaction mechanism by Malek method was described by Avrami-Erofeev model mechanism ($g(\alpha) = [-\ln(1-\alpha)]^{\frac{1}{4}}$) for the acid-base method and ($g(\alpha) = [-\ln(1-\alpha)]^{\frac{1}{3}}$) for the Organoslv method.

- The results of this study confirm the existence of multistep mechanism occurring in solid-state reactions due to variations in activation energy with the heating rates. The study provides important data information and a robust approach to understanding the cellulose pyrolysis structures and mechanisms by different isolation methods across a broad range of temperature and different heating rates.

Acknowledgements

This work was financially supported by the Emirates Centre for Energy and Environment Research, UAEU (31R107).

Conflict of interest

The authors declare no conflict of interest.

Author details

Emmanuel Galiwango and Ali H. Al-Marzouqi*
United Arab Emirates University, Al Ain, UAE

*Address all correspondence to: hassana@uaeu.ac.ae

References

[1] Liu C, Wang H, Karim AM, Sun J, Wang Y. Catalytic fast pyrolysis of lignocellulosic biomass. Chemical Society Reviews. 2014;**43**(22):7594-7623

[2] Wu W, Mei Y, Zhang L, Liu R, Cai J. Effective activation energies of lignocellulosic biomass pyrolysis. Energy & Fuels. 2014;**28**(6):3916-3923

[3] Liao JC, Mi L, Pontrelli S, Luo S. Fuelling the future: Microbial engineering for the production of sustainable biofuels. Nature Reviews. Microbiology. 2016;**14**(5):288

[4] Bridgwater AV. Review of fast pyrolysis of biomass and product upgrading. Biomass and Bioenergy. 2012;**38**:68-94

[5] Mohan D, Pittman CU Jr, Steele PH. Pyrolysis of wood/biomass for bio-oil: A critical review. Energy & Fuels. 2006; **20**(3):848-889

[6] Huang L, Ding T, Liu R, Cai J. Prediction of concentration profiles and theoretical yields in lignocellulosic biomass pyrolysis. Journal of Thermal Analysis and Calorimetry. 2015;**120**(2): 1473-1482

[7] Müller-Hagedorn M, Bockhorn H, Krebs L, Müller U. A comparative kinetic study on the pyrolysis of three different wood species. Journal of Analytical and Applied Pyrolysis. 2003; **68**:231-249

[8] Ibbett R, Gaddipati S, Davies S, Hill S, Tucker G. The mechanisms of hydrothermal deconstruction of lignocellulose: New insights from thermal–analytical and complementary studies. Bioresource Technology. 2011; **102**(19):9272-9278

[9] Segneanu A-E, Cziple F, Vlazan P, Sfirloaga P, Grozescu I, Gherman VD. Biomass extraction methods. In: Biomass Now-Sustainable Growth and Use. IntechOpen; 2013. pp. 390-399. DOI: 10.5772/55338

[10] Koufopanos CA, Papayannakos N, Maschio G, Lucchesi A. Modelling of the pyrolysis of biomass particles. Studies on kinetics, thermal and heat transfer effects. The Canadian journal of chemical engineering. 1991;**69**(4): 907-915

[11] Wu W, Cai J, Liu R. Isoconversional kinetic analysis of distributed activation energy model processes for pyrolysis of solid fuels. Industrial and Engineering Chemistry Research. 2013;**52**(40): 14376-14383

[12] Gordina NE, Prokof'ev VY, Hmylova OE, Kul'pina YN. Effect of ultrasound on the thermal behavior of the mixtures for the LTA zeolite synthesis based on metakaolin. Journal of Thermal Analysis and Calorimetry. 2017;**129**(3):1415-1427

[13] Guida MY, Bouaik H, ElMouden L, Moubarik A, Aboulkas A, Elharfi K, et al. Utilization of starink approach and avrami theory to evaluate the kinetic parameters of the pyrolysis of olive mill solid waste and olive mill wastewater. Journal of Advanced Chemical Engineering. 2016;**6**:1-8

[14] Luo H, Abu-Omar MM. Lignin extraction and catalytic upgrading from genetically modified poplar. Green Chemistry. 2018;**20**(3):745-753

[15] Xu F, Yu J, Tesso T, Dowell F, Wang D. Qualitative and quantitative analysis of lignocellulosic biomass using infrared techniques: A mini-review. Applied Energy. 2013;**104**:801-809

[16] Rohmawati B, Sya'idah FAN, Alighiri D, Eden WT. Synthesis of bioplastic-based renewable cellulose

acetate from teak wood (*Tectona grandis*) biowaste using glycerol-chitosan plasticizer. Oriental Journal of Chemistry. 2018;**34**(4):1810

[17] Wang G, Li W, Li B, Chen H. TG study on pyrolysis of biomass and its three components under syngas. Fuel. 2008;**87**(4–5):552-558

[18] Mohammed MAA, Salmiaton A, Azlina WW, Amran MSM. Gasification of oil palm empty fruit bunches: A characterization and kinetic study. Bioresource Technology. 2012;**110**: 628-636

[19] Lapuerta MN, Hernández JJ, Rodriguez JN. Kinetics of devolatilisation of forestry wastes from thermogravimetric analysis. Biomass and Bioenergy. 2004;**27**(4):385-391

[20] Hu Y, Wang Z, Cheng X, Ma C. Non-isothermal TGA study on the combustion reaction kinetics and mechanism of low-rank coal char. RSC Advances. 2018;**8**(41):22909-22916

[21] Shuping Z, Yulong W, Mingde Y, Chun L, Junmao T. Pyrolysis characteristics and kinetics of the marine microalgae Dunaliella tertiolecta using thermogravimetric analyzer. Bioresource Technology. 2010;**101**(1): 359-365

[22] Kim YS, Kim YS, Kim SH. Investigation of thermodynamic parameters in the thermal decomposition of plastic waste−waste lube oil compounds. Environmental Science & Technology. 2010;**44**(13): 5313-5317

[23] Vlaev LT, Georgieva VG, Genieva SD. Products and kinetics of non-isothermal decomposition of vanadium (IV) oxide compounds. Journal of Thermal Analysis and Calorimetry. 2007;**88**(3):805-812

[24] Coker AK. Ludwig's Applied Process Design for Chemical and Petrochemical Plants. Burlington, VT, USA: Elsevier, Gulf Professional Publishing; 2014

[25] White JE, Catallo WJ, Legendre BL. Biomass pyrolysis kinetics: A comparative critical review with relevant agricultural residue case studies. Journal of Analytical and Applied Pyrolysis. 2011;**91**(1):1-33

[26] Málek J. The kinetic analysis of non-isothermal data. Thermochimica Acta. 1992;**200**:257-269

[27] Senum GI, Yang R. Rational approximations of the integral of the Arrhenius function. Journal of Thermal Analysis and Calorimetry. 1977;**11**(3): 445-447

[28] Stanko M, Stommel M. Kinetic prediction of fast curing polyurethane resins by model-free isoconversional methods. Polymers. 2018;**10**(7):698

[29] Burnham AK, Weese RK, Weeks BL. A distributed activation energy model of thermodynamically inhibited nucleation and growth reactions and its application to the $\beta-\delta$ phase transition of HMX. The Journal of Physical Chemistry B. 2004;**108**(50): 19432-19441

Chapter 31

Economics, Sustainability, and Reaction Kinetics of Biomass Torrefaction

Temitope Olumide Olugbade

Abstract

Biomass torrefaction is capable of significantly improving the quality and properties of solid biofuels. It is often referred to as complex reactions involving the decomposition of lignin, cellulose, and hemicellulose as well as moisture evaporation due to several reactions involved. To evaluate the efficiency of the torrefaction process as well as the reactor performance, considering the economics of biomass torrefaction including the total production cost and capital investment, production capacity, feedstock input, feedstock type, pre-treatment, procurement and transportation costs is of high importance. In this Chapter, the economics of torrefaction process will be discussed. In addition, ways to ensure competitiveness of torrefaction technology will be explained provided factors including the use of plant with larger capacity, integrated system features such as pelletization, and moisture content of the feedstock, are properly considered. Thereafter, the concept of sustainability of biomass torrefaction in relation with the environmental factor (sustainable forest management), social factor (revitalization of rural areas), and economic factor (fossil fuels dependence and renewable energy consumption) will be presented.

Keywords: biomass, fuels, torrefaction, renewable energy, lignocellulosic, lignin, briquettes

1. Introduction

Biomass has been widely recognized as an important source of renewable energy due to its inherent properties such as availability, abundant supply, carbon stability, organic nature, etc. Unlike non-renewable resources such as oil and coal, biomass is a renewable natural resource and organic material mostly derived from animals and plants for the production of fuels at local and commercial scales, which is the dream of many biofuel producers and energy experts over the years. Interest in biomass as a renewable resource is increasing with time thanks to its ability to be burned directly for heat or converted to renewable fuels via several thermal decomposition methods including torrefaction, gasification, hydrotreating, carbonization, and pyrolysis [1–4].

It is important to know that biomass which can be crops, wood, landfill gas, alcohol fuels, and garbage contribute the largest percentage of energy used in many sectors including the electric power, commercial, residential, commercial, transportation, and industrial. For instance, wastes derived from biomass and woods are used to produce electricity in electric power sector, renewable natural gas derived from municipal solid waste (MSW) are consumed and sold in commercial sector whereas the wood pellets and firewood are mostly consumed in residential sector. In addition, the plant and animal-based biomass are used to generate liquid biofuels including biomass-based diesel and ethanol which finds major application in most transportation and industrial sectors.

Biomass can be converted to energy via several methods such as biological conversion for the production of fuels (gaseous and liquid), chemical conversion for producing liquid fuels, thermochemical conversion for the production of liquid, gaseous, and solid fuels; and direct combustion to generate heat. In biological conversion process, biomass can be converted into renewable natural gas or biogas [5] through anaerobic digestion method or ethanol through fermentation process. Meanwhile, in chemical conversion process, greases, animal fats, and plant-based biomass such as vegetable oils can be converted into fatty acid methyl esters (FAME) which are mostly utilized for the production of biodiesel through trans-esterification method. On the other hand, through thermal decomposition methods including torrefaction, gasification and pyrolysis, biomass can be thermochemically converted to produce bio-oil (hydrotreating), hydrogen, methane, renewable diesel (pyrolysis), or synthesis gas and carbon monoxide (CO) (gasification) [1, 2, 4]. Furthermore, biomass can be converted to energy in direct combustion method such as the generation of electricity in steam turbines, industrial process heat, and heating buildings.

Generally, biomass sources for energy include but not limited to human sewage and animal manure, biogenic materials in MSW (wood wastes, yard, food, wool products, cotton, paper), waste materials and agricultural crops (woody plants, switchgrass, sugar cane, soybeans, corn, etc.), and wood processing wastes (paper mills, sawdust, wood chips, wood pellets, firewood) [6–11]. For easy storage and transportation, biomass can be interestingly made or densified into briquettes for producing fuels and biogas [5]. Briquettes are the solid biofuels with compact shape and size for producing renewable energy which can be made with binders [6, 8–11] or without the use of binders [7]. Biomass has been densified into solid briquettes in the past using the same or different materials including corncob, rice bran [8–10], mesocarp fiber (MF), fruit fresh bunches (FFB), palm kernel shell (PKS) [8–11], bagasse, tea waste, cotton stalk, sugarcane bagasse (SCB), empty fruit bunches (EFB), etc. Most of the performance related problems such as low yield and energy content are linked to the effectiveness of the binder materials, type and compositions [6].

As a form of thermal decomposition processes of converting biomass to energy, torrefaction is presently receiving wide attentions for producing high grade solid biofuels with greater energy density/high heating value (HHV). Torrefaction process is capable of significantly improving the quality and properties of solid biofuels. It improves the physicochemical properties of biomass for a long-term storage. In addition, torrefied biomass can serve as a good replacement for coal in the generation of heat and electricity. The three common types of torrefaction types are the dry torrefaction (DT), wet torrefaction (WT), and ionic-liquid assisted torrefaction (ILA) [6]. The DT is usually carried out in a gas-phase environment, while the WT is often performed under pressure of a liquid-phase environment. Meanwhile, the ILA is a combination of a typical torrefaction and a pretreatment process, which is aimed at improving the reaction rate. The ionic-liquid assisted torrefaction (ILA) usually involves the use of ionic liquids which is often referred to as green solvents due to their ability to dissolve lignocellulosic biomass under normal conditions due to their special properties such as recyclability and high thermal stability [4].

However, an in-depth understanding on the economics and sustainability of the torrefaction process is lacking. With the aim of giving new insight into further study, a comprehensive overview on the reactor design for commercialization purposes, reaction kinetics and mechanism, economics, as well as the sustainability of biomass torrefaction is presented in the next section of this chapter.

2. Biomass torrefaction: A general overview

Biomass torrefaction is the process of producing high-quality and attractive solid biofuels from several sources of ordinary agro residues or woody biomass, with the sole aim of improving biomass properties and performance for gasification [1–3] and combustion applications via thermal decomposition at temperature ranging from 200–300°C [12] under atmospheric pressure. Through torrefaction, a coal-like material can be generated from biomass with superior fuel properties and quality when compared with the parent materials. Torrefaction is a mild pyrolysis process where biomass is thermally treated in a controlled environment (in a non-fluidized bed reactor or fluidized bed) with low or no traces of air or oxygen [13] resulting in the production of torrefied biomass which is water resistant, brittle and stable with less energy intensive and easy grindability. During torrefaction, drying of biomass and partial devolatilization occurs leading to mass reduction without losing or decreasing the energy content. Heating biomass at typical temperatures between 200°C and 300°C often lead to the evaporation of moisture or unbound water (H_2O) through thermo-condensation process (at temperature above 160°C) and the removal of volatiles (low-calorific parts), resulting in the decomposition of hemicellulose in the biomass hence the transformation of biomass from a low-quality fuel into an excellent high-quality fuel. In a bid to improve the biochemical, chemical, and physical properties of biomass, the basic principle behind the biomass torrefaction process can thus be summarized as the removal of volatiles via several decomposition reactions. With torrefaction, there is no biological activity, hydrophobicity and higher durability can be obtained, excellent grindability and higher bulk density can be achieved. In addition, more homogenous product and a fuel comparable to coal can be produced with higher calorific value as compared with original feedstock.

It should be noted that the torrefaction processing parameters such as the residence time and torrefaction temperature have significant effects on the overall properties and performance of torrefied biomass. In other words, there is a direct relationship between the torrefaction processing parameters and the physicochemical properties of torrefied biomass. For instance, high torrefaction temperature and short residence time tend to optimize the material flow via the torrefaction reactors thereby producing a cost-efficient torrefied biomass on a large scale. By increasing the torrefaction temperature, the fixed carbon and ash contents in biomass can be markedly increased with a decrease in volatile contents. This can lead to a decrease in atomic ratios of oxygen-carbon (O/C) and hydrogen-Carbon (H/C), as well as decreasing the oxygen content resulting in improved calorific value, which ultimately enhance the overall fuel features and performance of the biomass products. Moreover, the acid content in biomass materials can be significantly reduced with increasing the torrefaction temperature. The decrease in the acidity of the biomass say bio-oil for example [14] can be attributed to the fact that the acetic acid solely originates from the deacetylation reaction and decomposition of hemicellulose component of the biomass. In addition, increasing the torrefaction temperature can reduce the moisture content of the biomass, hence improving the quality of the biomass. By this, it can be said that torrefaction can increase the carbon yield of aromatic hydrocarbon and decrease the number of compounds containing oxygen in the biomass such as furans, sugars, and acids [15]. Furthermore, an increase in the residence time and torrefaction temperature can improve the hydrophobicity and the calorific value and lignin contents can be increased by decreasing the contents of hemicellulose and cellulose [16]. While studying the influence of torrefaction temperature on the production composition, physicochemical properties, structure, and yield of bio-oil [14], an increase in torrefaction temperature reportedly reduce the crystallinity index as a result of recrystallization and degradation, increase the pore volume and the residual carbon contents, decrease the pyrolysis peak temperature and increases the carbon-carbon contents.

The residence time during torrefaction process can also directly influences the biomass properties including the energy density, surface area, and grindability. For instance, increasing the residence time tends to increase the energy density and grindability, leading to the production of biomass with fine and small particle sizes thereby resulting in larger surface area. In addition, due to the disintegration of the fiber structure of biomass, increasing the residence time can increase the fixed carbon and volatile matter compositions. In other words, the higher the residence time, the better the properties and compositions of biomass materials. Heating value can also be increased with an increase in residence time and the disintegration of fiber structure continues with increasing the residence time, thereby enhancing the grindability and improving the energy density of the biomass products. Meanwhile, grindability is an important property in biomass torrefaction which is the resistance of biomass materials to be ground.

In terms of composition, biomass can be broadly categorized as non-lignocellulosic (mostly rich in fatty acids and protein) with animal manure and fat, and sewage sludge as the main components; or lignocellulosic with hemicellulose, cellulose, and lignin as the major components, together with small amounts of minerals, proteins, and pectins [17–21]. As indicated in **Figure 1**, torrefaction process involve several stages ranging from the initial to the solids cooling [22–25].

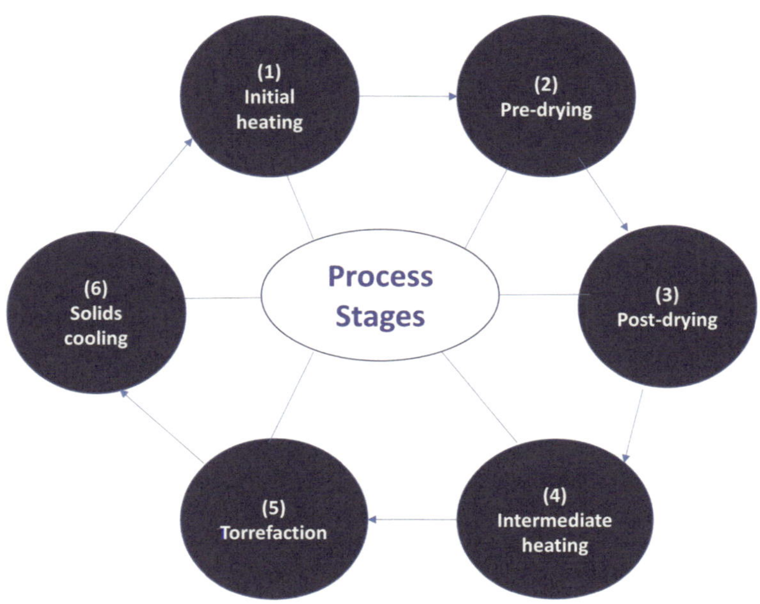

Figure 1.
The main stages in the total torrefaction process.

Figure 2 presents the mechanism of the thermal treatment during dry torrefaction, from decarboxylation to aromatization chemical reactions. During the thermal conversion in dry torrefaction, an improvement in the fungal resistance [26–34], dimensional stability, absorption, water vapor, and durability can be achieved. As illustrated in **Figure 3**, the binderless briquetting technology can be summarized into several theorems including capillary, adhesion molecule, bituminous, humic acid, colloid, and denser water.

As compared with the original biomass, torrefied biomass offers many advantages including homogeneity, increased density in briquettes, reduced grinding

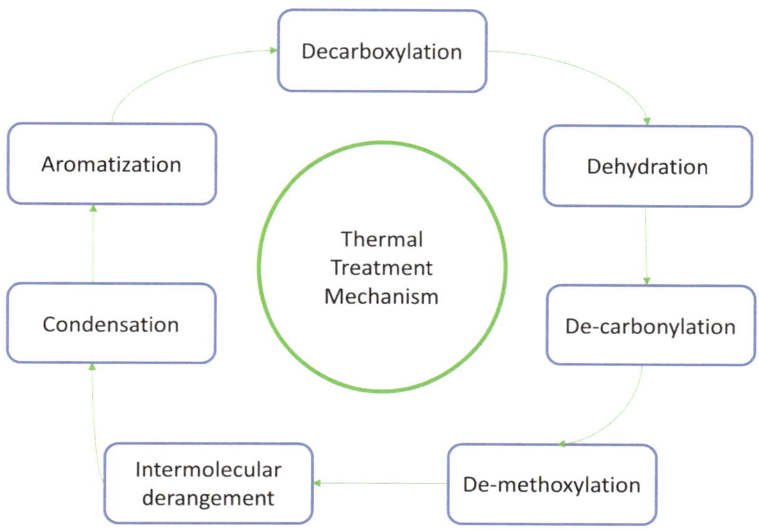

Figure 2.
Mechanism of the thermal treatment during dry torrefaction.

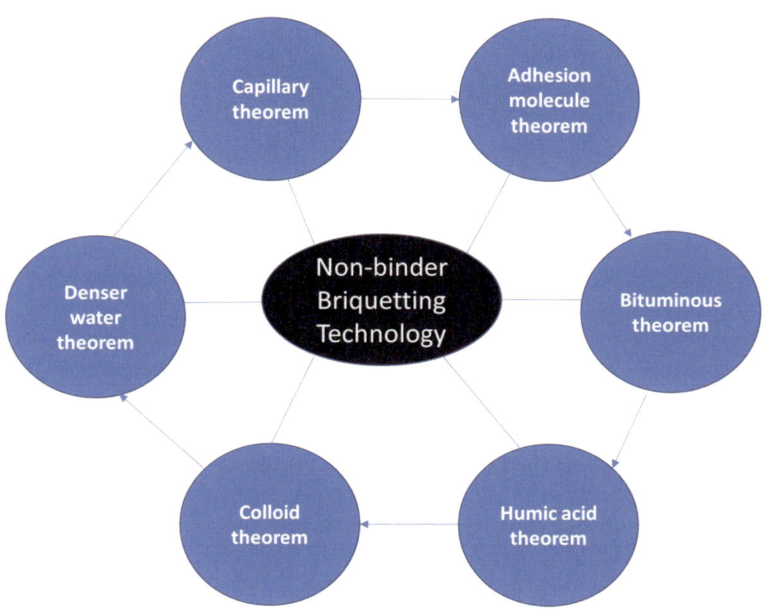

Figure 3.
Binderless briquetting technology.

energy, hydrophobicity, increased energy density, and lower moisture content. As a source of a power source, with torrefied biomass, low net carbon energy source and renewable fuel for baseload power can be achieved.

3. Economics of biomass torrefaction

While ensuring the conversion of biomass into renewable energy for the benefits of mankind, it is necessary to consider the economics aspect of biomass torrefaction. That is, the costs of generating high-grade fuel should be affordable. The economic features of the torrefaction process covers the total production cost, total capital investment, production capacity, feedstock input, feedstock type, procurement costs, transportation costs, pre-treatment, etc. [4].

For effective economy analysis, economic optimization of the system, including the extended fuel supply to application system, and integrated process between torrefaction and gasification as well as torrefaction and gasification should be properly looked into. The cost of the torrefaction process which is higher than that of coal at times can be greatly reduced by improving the empirical cumulation and torrefaction plant equipment size and as well as the utilization of carbon credits market. The total costs during the torrefaction process is often influenced by the important sensitivity parameters including the torrefaction plant CAPEX and torrefaction mass yield, drying technology, biomass moisture content, logistics equipment, biomass premium, and the quantity of the available biomass [2, 3]. Meanwhile, the depreciation, biomass delivery costs, energy consumption, labor, capital expenditure, biomass delivered costs are the common influencing factors for production with CAPEX.

Biomass torrefaction is widely regarded as a breakthrough technology accounting for the world largest renewable energy which reduces the investment for co-firing application as well as decreasing the storage and handling costs thereby making the process economically viable and serving as a potential way of replacing coal in power plants.

4. Sustainability of biomass torrefaction

Biomass torrefaction involves the conversion of biomass into a coal-like material with improved fuel properties as compared with the original biomass. It is presently a vital tool for the sustainable development in many developed and developing countries with the aim of supporting large scale utilization of bioenergy through the reduction of carbon dioxide (CO_2) at source and other emissions which may be harmful to the community. That is, the production of solid sustainable energy can be obtained through biomass torrefaction. It is important to note that biomass torrefaction can pose threats to humanity since some harmful toxins and greenhouse gases are usually generated and released into the atmosphere during biomass combustion resulting into environmental problem. Carbon dioxide (CO_2) remain the greenhouse gas with the largest volume and percentage released into the atmosphere when compared to other greenhouse gases. Biomass as a renewable energy which has some similar features with fossil fuels still have some environmental challenges. It can cause deforestation and it is not entirely clean.

Sustainability concept of biomass torrefaction can be properly broadly categorized into three factors; (1) economic factor, (2) environmental factor, and (3) social factor [4]. Generally speaking, the economic factor is majorly related to the fossil fuels dependence and renewable energy consumption whereas the environmental factor is associated with the sustainable forest management. On the other hand, the social factor is linked with the regeneration of rural areas and more jobs.

5. Reaction kinetics of biomass torrefaction

To a large extent, torrefaction process can influence the overall kinetics and mechanism [35] as well as the reactivity, reaction behavior, and thermal conversion performance of biomass which is often estimated through thermogravimetric analysis (TGA). Through TGA, the reaction kinetics parameters including the mechanism function, pre-exponential factor, and activation energy [36, 37], as well as the thermal features including temperature at peak value, reaction period, peak value, and thermogravimetric (TG) data can be determined. In addition, biomass torrefaction is often regarded as a complex mechanism due to the several reactions involved including the decomposition of the common biomass components including lignin, cellulose, and hemicellulose as well as moisture evaporation.

The decomposition of the biomass components depends to some extent on the temperature at which the torrefaction process is carried out. For instance, at the temperature range of 160–900°C, lignin slowly decomposes, cellulose decomposes at 315–400°C, while the hemicellulose decomposes at the lowest temperature ranging from 220–315°C [38]. For instance, while the torrefaction of hemicellulose leads to more devolatilization and carbonization, the depolymerization of cellulose plays a vital role in the decomposition mechanism. The study on the kinetics of biomass torrefaction is very important because it represents the torrefaction reaction thereby predicting the optimum thermal degradation conditions which can ultimately enhance the process control for continuous torrefaction reactor [39, 40]. Meanwhile, at the temperature range of 230–300°C [41], the kinetics of torrefaction reactions can be best described by a two-step mechanism; cellulose and hemicellulose decomposition.

6. Conclusions

Biomass upgrading for the production of high-grade solid fuels with greater energy density, excellent grindability, and enhanced durability can be achieved by subjecting the original raw biomass to thermal mild pretreatment process under inert atmosphere without the presence of air or oxygen. This process is commonly referred to as torrefaction. The major aim is to improve the chemical, thermal, and physical properties of biomass for a long-term storage through the elimination of oxygen, reduction of moisture content and change of chemical compositions. After torrefaction, some challenges pertaining to technological applications of biomass such as difficulty to obtain a small particle size and high oxygen–carbon ratio, can be properly addressed.

Upon the mild thermal pyrolysis of raw biomass in oxygen-free or N_2 atmosphere at moderate temperatures over a period of time, the biomass fiber structure tends to break down which makes the biomass easy to grind, hence an enhanced energy density. By this process, the properties of raw biomass including low calorific value, grindability, hydrogen-carbon and, hygroscopicity, can be greatly improved. At first, torrefaction process can reduce the weight of the biomass to about 30%, but the final solid biofuel produced can retain about 90% of the original biomass energy content.

As compared to the original raw biomass, torrefied biomass can serve as a good replacement for coal in the generation of heat and electricity, as well as input for gasification, densification, and iron making processes, with many positive attributes, like grinding and burning like coal, lower ash and sulfur content, lower transport and shipping costs, lower feedstock costs, and the ability to produce non-intermittent renewable energy. Hence, further studies to understand the mechanism behind the torrefaction process in producing more uniform biomass products and the influence of torrefaction process parameters on the biomass feedstock upgrading is necessary to open the market for the mass production of high-grade solid biofuels with enhanced energy density and hydrophobicity for a long-term storage.

Attributed to the several reactions involved, biomass torrefaction is sometimes referred to as the complex reactions including the decomposition of the common biomass components including lignin, cellulose, and hemicellulose as well as moisture evaporation. The economic, environmental, and social factors are the three major concepts of sustainability as regards the biomass torrefaction. While the economic factor is majorly related to the renewable energy consumption, the environmental factor focused more on the sustainable forest management while the regeneration of rural areas and more jobs is related to social factor. The economics of biomass torrefaction including the total production cost, total capital investment, production capacity, feedstock input, feedstock type, pre-treatment, and procurement costs, transportation costs are necessary to evaluate the efficiency of the torrefaction process as well as the reactor performance.

Nomenclature

MSW	municipal solid waste
FAME	fatty acid methyl esters
CO	carbon monoxide
CO_2	carbon dioxide
O_2	oxygen
H_2O	water
HHV	high heating value
PKS	palm kernel shell
MF	mesocarp fiber
FFB	fruit fresh bunches
EFB	empty fruit bunches
DT	dry torrefaction
WT	wet torrefaction
ILA	ionic-liquid assisted torrefaction
TGA	thermogravimetric analysis
SCB	sugarcane bagasse
CAPEX	capital expenditures
O/C	oxygen–carbon
H/C	hydrogen-carbon
TG	thermogravimetric

Author details

Temitope Olumide Olugbade
Department of Industrial and Production Engineering, Federal University of Technology, P.M.B. 704, Akure, Ondo State, Nigeria

*Address all correspondence to: tkolugbade@futa.edu.ng

References

[1] Xin S, Mi T, Liu X, Huang F. Effect of torrefaction on the pyrolysis characteristics of high moisture herbaceous residues. Energy. 2018;152:586-593

[2] Bergman PCA, Kiel JHA. Torrefaction for Biomass Upgrading. In Proceedings of the 14th European Biomass Conference, Paris, France. 2005;17-21

[3] Chen Q, Zhou JS, Liu BJ, Mei QF, Luo ZY. Influence of torrefaction pretreatment on biomass gasification technology. Chin. Sci. Bull. 2011;56:1449-1456

[4] Olugbade TO, Ojo OT. Biomass torrefaction for the production of high-grade solid biofuels: a review. Bioenerg. Res. DOI: 10.1007/s12155-020-10138-3

[5] Akinnuli BO, Olugbade TO. (2014) Development and performance evaluation of piggery and water hyacinth waste digester for biogas production. Int J Engineering and Innovative Tech. 2014;3:271-276

[6] Olugbade T, Ojo O, Mohammed T. Influence of binders on combustion properties of biomass briquettes: A recent review. Bioenergy Research. 2019;12:241-259

[7] Olugbade TO, Ojo OT. Binderless briquetting technology for lignite briquettes: a review. Energy, Ecology and Environment. DOI: 10.1007/s40974-020-00165-3

[8] Olugbade TO, Mohammed TI. Fuel developed from rice bran briquettes and palm kernel shells. Int. Journal of Energy Engineering. 2015;5:9-15

[9] Mohammed TI, Olugbade TO. Characterization of briquettes from rice bran and palm kernel shell. Int. Journal of Material Science Innovations.2015;3:60-67

[10] Mohammed TI, Olugbade TO. Burning rate of briquettes produced from rice bran and palm kernel shells. Int. Journal of Material Science Innovations.2015;3:68-73

[11] Oke PK, Olugbade TO, Olaiya NG. Analysis of the effect of varying palm kernel particle sizes on the calorific value of palm kernel briquette. British Journal of Applied Science and Technology. 2016;14:1-5

[12] Pestaño LDB, José WI. Production of solid fuel by torrefaction using coconut leaves as renewable biomass. International Journal of Renewable Energy Development. 2016;5(3):187-197

[13] Nunes LJR, Matias JCO, Catalão JPS. A review on torrefied biomass pellets as a sustainable alternative to coal in power generation. Renew. Sustain. Energy Rev. 2014;40:153-160

[14] Zheng Y, Tao L, Yang X, Huang Y, Liu C, Gu J, Zheng Z. Effect of the torrefaction temperature on the structural properties and pyrolysis behavior of biomass. *BioRes.* 2017;12(2):3425-3447

[15] Chen D, Li Y, Deng M, Wang J, Chen M, Yan B, Yuan Q. Effect of torrefaction pretreatment and catalytic pyrolysis on the pyrolysis poly-generation of pine wood. *Bioresource Technol.* 2016;214:615-622. DOI: 10.1016/j.biortech.2016.04.058

[16] Li MF, Chen LX, Li X, Chen C Z, Lai YC, Xiao X, Wu YY. Evaluation of the structure and fuel properties of lignocelluloses through carbon dioxide torrefaction. *Energ. Convers. Manage.* 2016;119:463-472. DOI: 10.1016/j.enconman.2016.04.064

[17] Hossain N. Characterization of novel moss biomass, *Bryum dichotomum* Hedw as solid fuel feedstock. Bioenergy Res. 2020;13:50-60

[18] Pathomrotsakun J, Nakason K, Kraithong W, Khemthong P, Panyapinyopol B, Pavasat P. Fuel properties of biochar from torrefaction of ground coffee residue: effect of process temperature, time, and sweeping gas. Biomass Convers Biorefin. DOI: 10.1007/s13399-020-00632-1

[19] Grycova B, Pryszcz A, Krzack S, Lestinsky P. Torrefaction of biomass pellets using the thermogravimetric analyser. Biomass Convers Biorefin. DOI: 10.1007/s13399-020-00621-4

[20] Manatura K. Inert torrefaction of sugarcane bagasse to improve its fuel properties. Case Stud Therm Eng. 2020;19:100623

[21] Azevedo SG, Sequeira T, Santos M, Mendes L. Biomass related sustainability: a review of the literature and interpretive structural modeling. Energy. 2019;171:1107-1125

[22] Mochizuki Y, Ma J, Kubota Y, Uebo K, Tsubouchi N. Production of high-strength and low-gasification reactivity coke from low-grade carbonaceous materials by vapor deposition of tar. Fuel Process Technol. 2020;203:106384

[23] Puig-Arnavat M, Bruno JC. Alberto Coronas, Review and analysis of biomass gasification models. Renew Sust Energ Rev. 2010;14:2841-2851

[24] Taba LE, Irfan MF, Daud WAMW. The effect of temperature on various parameters in coal, biomass and CO-gasification: a review. Renew Sust Energ Rev. 2012;16:5584-5596

[25] Pirraglia A, Gonzalez R, Saloni D, Denig J. Technical and economic assessment for the production of torrefied lignocellulosic biomass pellets in the US. Energy Convers Manag. 2013;66:153-164

[26] Dubey MK, Pang S, Walker J. Changes in chemistry, color, dimensional stability and fungal resistance of Pinusradiata D. Don wood with oil heat treatment. Holzforschung. 2012;66:49-57

[27] Pfriem A, Zauer M, Wagenfuhr A. Alteration of the unsteady sorption behaviour of maple (*Acer pseudoplatanus* L.) and spruce (Piceaabies (L.) Karst.) due to thermal modification. Holzforschung 2010;64(2):235-241

[28] Ates S, Akyildiz MH, Ozdemir H. Effects of heat treatment on calabrian pine (*Pinus brutia* ten.) wood. Bioresources 2009;4(3):1032-1043

[29] Borrega M, Karenlampi PP. Mechanical behavior of heat-treated spruce (Piceaabies) wood at constant moisture content and ambient humidity. Holz Roh Werkst 2008;66:63-69.

[30] Hakkou M, Petrissans M, Gerardin P, Zoulalian A. Investigations of the reasons for fungal durability of heat-treated beech wood. Polym Degrad Stabil 2006;91:393-7

[31] Tjeerdsma BF, Militz H. Chemical changes in hydrothermal treated wood: FTIR analysis of combined hydrothermal and dry heat-treated wood. Holz Roh Werkst. 2005;63:102-11

[32] Kamdem DP, Pizzi A, Jermannaud A. Durability of heat-treated wood. Holz Roh Werkst. 2002;60:1-6

[33] Pelaez-Samaniego MR, Yadama V, Lowell E, Espinoza-Herrera R. A review of wood thermal pretreatments to improve wood composite properties. Wood Sci Technol. 2013;47(6):1285-319

[34] Funke A, Ziegler F. Hydrothermal carbonization of biomass: a summary and discussion of chemical mechanisms for process engineering. Biofuels, Bioprod Biorefin. 2010;4:160-177

[35] Vyazovkin S, Burnham AK, Criado JM, Pérez-Maqueda LA, Popescu C, SbirrazzuoliN. ICTAC kinetics committee recommendations for performing kinetic computations on thermal analysis data. Thermochim Acta. 2011;520:1-19.

[36] Xu Y, Zhang Y, Zhang G, Guo Y, Zhang J, Li G. Pyrolysis characteristics and kinetics of two Chinese low-rank coals. J Therm Anal Calorim. 2015;122:975-84.

[37] Xu Y, Zhang Y, Zhang G, Guo Y. Low temperature pyrolysates distribution and kinetics of Zhaotong lignite. Energy Convers Manage. 2016;114:11-19

[38] Chen WH, Peng J, Bi XT. A state-of-art review of biomass torrefaction, densification and applications, RenewSust Energ Rev. 2015;44:847-866

[39] Prins MJ, Ptasinski KJ, Janssen FJJG. Torrefaction of wood, Part 1. Weight loss kinetics J. Anal. Appl. Pyrolysis. 2006;77:28-34

[40] Onsree T, Tippayawong N, Williams T, McCullough K, Barrow E, Pogaku R, Lauterbach J. Torrefaction of pelletized corn residues with wet flue gas. Bioresour Technol. 2019;285:1213302

[41] Li SX, Chen CZ, Li MF, Xiao X. Torrefaction of corncob to produce charcoal under nitrogen and carbon dioxide atmospheres. Bioresour Technol. 2018;249:348-353

Chapter 32

Soft Sensors for Biomass Monitoring during Low Cost Cellulase Production

Chitra Murugan

Abstract

Low cost cellulase production has become a major challenge in recent years. The major hurdle in the production of biofuel and other products from biomass is the lack of efficient economically feasible cellulase. This can be achieved by proper monitoring and control of bioprocess. In order to implement any control scheme, the accurate representation of the system in the form of a model is necessary. There are many challenges associated with modeling the fermentation process such as inherent nonlinear dynamic behavior, complexity of process due to co-existence of viable and nonviable cells, presence of solid substrates, etc. Toward the achievement of this goal, researchers have been developing new techniques that can be used to monitor the process online and at-line. These newer techniques have paved the way for designing better control strategies that can be integrated with quality by design (QbD) and process analytic technology (PAT).

Keywords: biomass, lignocellulosic substrates, hybrid model, NARX, LSTM

1. Introduction

Monitoring and control of biomass plays a vital role during fermentation process [1]. Low cost Production of ethanol/cellulase from lignocellulosic substances has been a topic of research for the past few years as it uses the carbon source from industrial or agricultural waste [2, 3]. These processes mostly involve fungi such as *Aspergillus, Trichoderma*. One of the main challenges during production of low cost cellulase is the estimation of biomass in the presence of insoluble solid substrates. The conventional methods such as dry weight filtration, optical density become unusable during fungal biomass estimation [4]. Methods such as monitoring DNA concentration, Image analysis were widely used. However, for the control of fermentation process, continuous monitoring of biomass is essential.

2. Biomass estimation techniques

The first principles model based on mass or energy balance equations is widely used in industries. Unstructured models such as monod model captures the process dynamics effectively only during log phase and on the other side, structured models have more parameters and are difficult to use [5]. The kinetic parameters in first principles model such as specific growth rate, biomass yield are found using laborious experimentation methods [6].

New process monitoring approaches use several online sensors to determine the concentration of viable biomass that are found useful during control of fermentation process. A simple on-line method for fungal biomass estimation based on agitation rate has been evolved for DO stat cultures. The estimator is developed based on changes in dissolved oxygen concentration in the initial transient time and yield change. The estimation parameters are found using agitation rate at 20% of DO concentration [7].

Dielectric spectroscopy has been widely used for monitoring biomass during submerged fermentation. The working of capacitance probe is based on cell membrane polarization. When the microbial cells are placed in ionic solution and are subjected to alternating electric field, they will act as capacitor due to restriction of ion movement by cell membrane. The on-line capacitance value represents viable biomass as dead cells do not polarize [8]. The Biomass Monitor TM, model 214 M (Aber Instruments, Aberystwyth, UK), dual frequency version (0.2–1.0 MHz, and approximately 9.5 MHz) is the commonly used capacitance probe.

Another method for estimation of viable biomass is the **radio-frequency (RF) impedance spectroscopy [9]**. This method provides useful information on the live cell concentration both in fixed as well as in dual frequency mode. To identify important changes in the process or to control the biomass at a constant level, the determination of on-line live cell concentration can be useful. **Electrochemical impedance spectroscopy (EIS)** is another method used to monitor biomass during fermentation [10]. The increase in biomass during cultivation is proportional to the increase in the double layer capacitance (Cdl), determined at frequencies below 1 kHz. A good correlation of Cdl with cell density is found and in order to get an appropriate verification of this method, different state-of-the-art biomass measurements are performed and compared. Since measurements in this frequency range are largely determined by the double layer region between the electrode and media, rather minor interferences with process parameters (aeration, stirring) are to be expected. It is shown that impedance spectroscopy at low frequencies is a powerful tool for cultivation monitoring. Though these dielectric spectroscopy, impedance spectroscopy techniques have been reported in literature, these methods require costlier instrumentation.

In recent years, soft sensors are widely used for the estimation of biomass. **Soft sensors** estimate the unknown state variable by using some other measured variables that influences the unknown state [11]. The data-driven methods widely used for the soft-sensor modeling are support vector machine, multiple least square support vector machine, neural network, deep learning, fuzzy logic and probabilistic latent variable models.

Artificial Neural Networks (ANN) based soft sensor have the capability to learn nonlinearity of the process using experimental plant data and thus can be used to estimate the state of bioprocess such as biomass concentration [12, 13]. The rapid development of algorithms and information technology is the major motivation behind the broad application of ANNs in research and development [14]. Currently, ANNs are employed in the prediction of various outcomes including process control, medicine, forensic science, biotechnology, weather forecasting, finance and investment and food science. However, it is noteworthy to state that the use of ANNs in biofuel production is currently in the early phases of its development. Generally, microbial fermentations exhibit non-linear relationships which could pose several problems during bioprocess modeling and optimization. The application of robust models such as ANN helps to capture this nonlinear behavior, and thus provides a model that links the process inputs to the corresponding output

Input parameters	Output parameters	ANN Type	ANN structure	R^2 value	References
pH	Biomass concentration	—	1–2-1	—	[3]
S_{in} (glucose)	Biomass concentration	HNN	1–3-1	—	[24]
S_{in} (glucose), sodium nitrate concentration, yeast extract concentration	Lipid Productivity, biomass concentration	FFBPNN	3–10-1	0.99	[25]

S_{in}, initial substrate concentration; R^2, coefficient of determination.

Table 1.
Application of ANN in bioprocess industries.

parameters. On comparison with other empirical models, neural networks are relatively less sensitive to noise and hence can be applied to process control systems with higher level of uncertainty [15]. During batch/fed-batch, ANN can be effectively used for estimation of biomass or product, optimization of fed-batch run and online control of bioprocess systems [16–18]. ANNs are suitable for many applications such as nonlinear filtering, prediction of output using input are widely used in the modeling of dynamic systems [19]. Several ANNs such as feed forward back-propagation neural network (FFBPNN) [20], Hopfield [21], radial basis function (RBF) networks [22], recurrent [23] and hybrid neural network (HNN) [24] found extensive application in bioprocess industries based on their functions. The BPNN with supervised learning has been reported in biofuel process modeling as shown in **Table 1**.

3. Biomass modeling methods

The Food and Drug Administration (FDA) of United States has initiated the online monitoring and closed loop control of a bioprocess via Process Analytical Technology (PAT) initiative. PAT highlights the concept of process understanding in order to deliver high quality products and this can be achieved by the design of accurate bioprocess models. The mathematical models are in general classified as black box, white box and gray box models.

3.1 Black box models

Black box models are input/output models that do not require *a priori* knowledge about the process and describe the system based on experimental data. An example of input/output model that has an output y which depends on past and present inputs is given in Eq. (1) as follows:

$$y(t_x) = \frac{B(q)}{A(q)} u(t_x) \tag{1}$$

where q is the backward shift operator for the polynomials $A(q)$ and $B(q)$ as given in Eq. (2), Eq. (3) and Eq. (4)

$$q^{-j}(y(t_i)) = y(t_{i-j}) \tag{2}$$

$$A(q) = 1 + a_1 q^{-1} + a_2 q^{-2} + \ldots + a_n q^{-n} \tag{3}$$

$$B(q) = b_0 + b_1 q^{-1} + b_2 q^{-2} + \ldots + b_m q^{-m} \quad (4)$$

The input output experimental data is used to determine the values of variables a, b and the order of the polynomials n and m. Artificial Neural Network (ANN) models are another type of black box models that has wide application in bioprocess technology due to their ability to represent non-linear functions.

3.2 White box models

White box models are mechanistic models that are been used widely used incorporates the available process knowledge in the form of first principle model equations. As an example, the first principles model widely used for fungal fermentation during cellulase production are represented in Eq. (5), Eq. (6), Eq. (7) as follows [26].

$$\frac{dX}{dt} = \mu_m \frac{SX}{K_s + S} - k_d X \quad (5)$$

$$\frac{dE_t}{dt} = \frac{k_1 X}{1 + \frac{S}{K_i}} - k_2 E_t \quad (6)$$

$$\frac{dS}{dt} = -\frac{1}{Y_{X/S}} \frac{dX}{dt} - m_s X \quad (7)$$

where X represents the biomass concentration (g/l), S represents the substrate concentration (g/l), μ_m is the maximum specific growth rate (1/h), K_s is the substrate saturation constant (g/l), k_d is the cell death constant (1/h), k_1, k_2 are rate constants for cellulase synthesis (IU/ml h) and cellulase decay (1/h), K_i is the substrate inhibition coefficient(g/l), E_t is the total cellulase activity, $Y_{X/S}$ is the stoichiometric biomass yield coefficient (g/g) and m_s is the specific maintenance coefficient (1/h).

3.3 Hybrid models

Gray box models also known as hybrid models are considered as an effective tool for model identification. These models combine *a priori* knowledge of the process and black box representations. The black-box model can be ANN, Fuzzy, NARX, Neuro-Fuzzy etc. It provides the flexibility to develop model based on both process data and available knowledge about the process [27, 28]. Hybrid models provide higher estimation accuracy, interpretability and extrapolation. In particular, during fungal fermentation using lignocellulosic substrates, the kinetic parameters μ_m and $Y_{X/S}$ are generally assumed to be constant throughout the batch/fed-batch process. However, there might be a slight change in their values and accurate estimation of these kinetic parameters such as aids greatly in process control and product enhancement. The multivariate interactions during the process operation can be found using statistical design of experiments (DoE). The hybrid models are commonly represented in two configurations as parallel and cascade as shown in **Figure 1**.

The parallel configuration **Figure 1(a)** uses complete first principles model and the error between outputs of this model and real-time process are modeled. The serial configuration **Figure 1(b)** is used when there is less number of unknown parameters. This is the most frequently used hybrid model structure. The choice of hybrid configuration strongly depends on the First principles model structure.

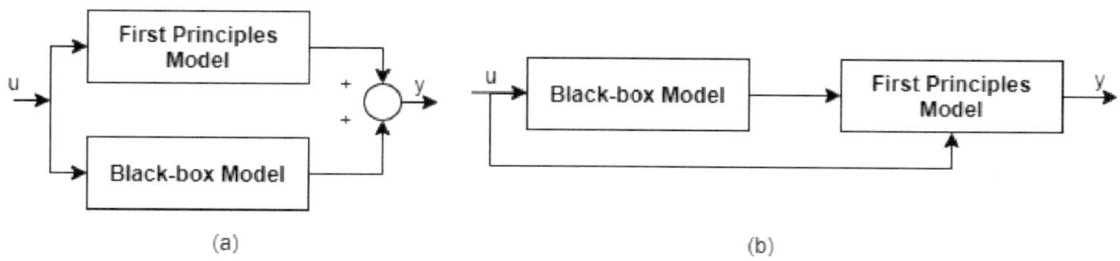

Figure 1.
Configurations of a hybrid model. (a) Parallel form (b) Serial form.

When the First principles model is not accurate, Parallel configuration is a better choice as parallel hybrid model can compensate for the First principles model mismatch [29]. When the First principles model is accurate, serial configuration seems to be a better choice as it offers better extrapolation [30].

3.3.1 Hybrid fuzzy models

A non-linear process such as bioreactor can be modeled using Fuzzy logic via Rule base analysis. Initially a model structure is chosen wherein the number of inputs and number of fuzzy sets per variable are found, then estimation of fuzzy membership function parameters with regard to its shape, position is carried out and finally the rule base mapping from fuzzy sets to functions are carried out. The fuzzy logic system draws decision with the help of Fuzzy Inference System which uses "If ..Then" rules along with OR, AND connectors. For example, if specific growth rate is high, biomass concentration is high; else if specific growth rate is low, biomass concentration is low. Fuzzy models render transparency and therefore have the advantage of interpretability when compared to other data- driven methods. The popular fuzzy models in use are the Takagi-Sugeno and the Mamdani type. Takagi-Sugeno type fuzzy models are suitable for modeling the non-linearity of the process and can be represented as many linear models in parallel, wherein the sub-models are chosen based on some specified rule. These are generally well suited to perform mathematical analysis and can be applied in Multiple Input Single Output (MISO) systems. Also, in Takagi-Sugeno model, output membership function are either constant or linear. The Mamdani type fuzzy model has a fuzzy logic set as the output of each rule. These are well suited to perform with manual input and can be applied in Multiple Input Single Output (MISO) systems and Multiple Input Multiple Output (MIMO) systems. The output membership function is present and the output is not continuous. Moreover, the Mamdani type fuzzy model is more accurate than Takagi-Sugeno, but it requires estimation of huge number of parameters.

Hybrid Fuzzy models comprises both first principles model equations and Fuzzy models [31]. The parameter estimation of hybrid fuzzy models are commonly done by Kalman filter. Fuzzy models are identified from experimental process data commonly by Fuzzy Clustering Method (FCM). Improvements in accuracy and performance of fuzzy models can be achieved by implementation of FCM based on Artificial Bee Colony (FCMABC), FCM based on Chicken Swarm Oprimization (FCMCSO), etc.

A case study to exhibit the Hybrid fuzzy modeling approach, the low cost cellulase production process is considered where the objective is to find the product concentration (cellulase) based on the interactions between the biomass and substrate.

The following assumptions are made during the batch/fed-batch operation of the bioreactor. The reactor is completely mixed and the feed flow rate (F) is known. Measurements for biomass concentration (X), Substrate concentration (S), product concentration (P) and volume (V) are available. The sampling interval is 30 minutes for these measurements.

The first principles model consists of 4 state equations for biomass concentration X (g/l), Substrate concentration S (g/l), Product concentration P (g/l) and volume of the bioreactor V (l) as defined in Eq. (8–11).

$$\frac{dx}{dt} = \mu X - \frac{F}{V}X \tag{8}$$

$$\frac{ds}{dt} = -q_s X + \frac{F}{V}(S_{in} - S) \tag{9}$$

$$\frac{dP}{dt} = q_p X - P\left(\frac{F}{V} + K\right) \tag{10}$$

$$\frac{dV}{dt} = F \tag{11}$$

where μ is the specific growth rate (h^{-1}), F represents substrate feed rate, S_{in} is the substrate concentration in the feed, q_s is the substrate consumption rate (h^{-1}), q_p is the product formation rate (h^{-1}), K is the product decay constant (h^{-1}). No direct measurements were made for these kinetic parameter rates. Therefore, a fuzzy model structure is represented as in Eq. (12–14).

$$q_s = \frac{\mu_m S}{Y_{x/s}(K_s + S)} + \frac{q_p}{Y_{p/s}} + m_s X \tag{12}$$

$$\mu = f_{fuzzy}(S, X) \tag{13}$$

$$q_p = f_{fuzzy}(S, X) \tag{14}$$

where μ_m, $Y_{x/s}$, $Y_{p/s}$, K_s, m_s are constants.

The kinetic parameters μ and q_p depends on X and S and their values are unknown. Hence, fuzzy models are developed to estimate their values. An extended Kalman filter is designed to obtain the estimated values of parameters. The filter is tuned by fixing the process noise covariance matrix Q as [0.001,0.001, 0.001, 0.054,0.003] and measurement error covariance matrix R as [0.1, 0.05, 0.03, 0.1]. The filter performance is evaluated by stability border criterion λ and significance level, L_α (**Table 2**). Smaller values of these two criterions represent good tuning of the Kalman filter.

From the table, it is observed that the filter is tuned properly. In this work, the fuzzy sub-model identification for specific growth rate and product formation rate are done with fuzzy clustering. The basic idea is to form clusters (similar groups) with the available experimental data. Each cluster exhibit an independent rule in the rule

λ	0.32
L_α for X	5
L_α for S	3
L_α for P	7

Table 2.
Results of Kalman filter.

DOI: http://dx.doi.org/10.5772/intechopen.96027

base. The advantage of using fuzzy clustering method is that the experimental data is focused and from that, the fuzzy model with independent rules is developed. The fuzzy models for kinetic parameters are represented in **Figure 2** and the hybrid model output in comparison with experimental data is illustrated in **Figure 3**. The optimization of fuzzy model parameters will improve the performance of hybrid model.

3.3.2 Hybrid ANN models

Hybrid ANN models are combination of first principles and ANN wherein the ANNs are used to estimate the kinetic parameters (black box models) [32]. The hybrid model shown in **Figure 4** is a combination of neural network estimator with the Mass Balance equations (Mathematical model). The neural network estimator is capable of estimating the process parameters from the real time measurements and these kinetic parameters (μ and $Y_{s/x}$) are updated in the mass balance equations to give the value of the state variables in the next time instant.

In general, the kinetic parameters are determined offline from experimental data. Due to the ability of neural networks to learn and model non-linear relationships, the parameter values can be estimated after proper training. Neural network with varying number of hidden neurons has been trained and MSE between the actual data and estimated data are calculated. Network with less MSE has been selected to find optimal hidden neurons. In this case study, a neural network structure comprising of two layer feed-forward network with sigmoid hidden neuron and linear output neuron is used. The state variables X_t $X(t)$ and $S(t)$ S(t) are the inputs and the parameters $\hat{\mu}$ and $\hat{Y}_{S/X}$ are the outputs of the neural network. The parameters are found using the Eq. (15) and Eq. (16) given below for every time instant.

$$\mu = \mu_{max} \frac{S(t)}{kS + S(t)} \tag{15}$$

where S(t) is the substrate concentration, k_S – saturation constant which is found experimentally to be 0.01 g/L.

$$\hat{Y}_{S/X} = \frac{S(t) - S(t-1)}{X(t) - X(t-1)}$$

$$Y_{S/X} = \frac{S_t - S_{t-1}}{X_t - X_{t-1}} \tag{16}$$

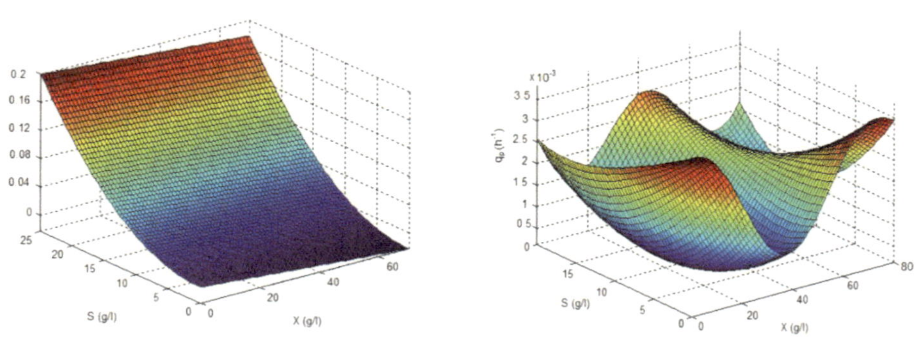

(a) (b)

Figure 2.
Fuzzy model for kinetic parameters μ *and* Yp/s.

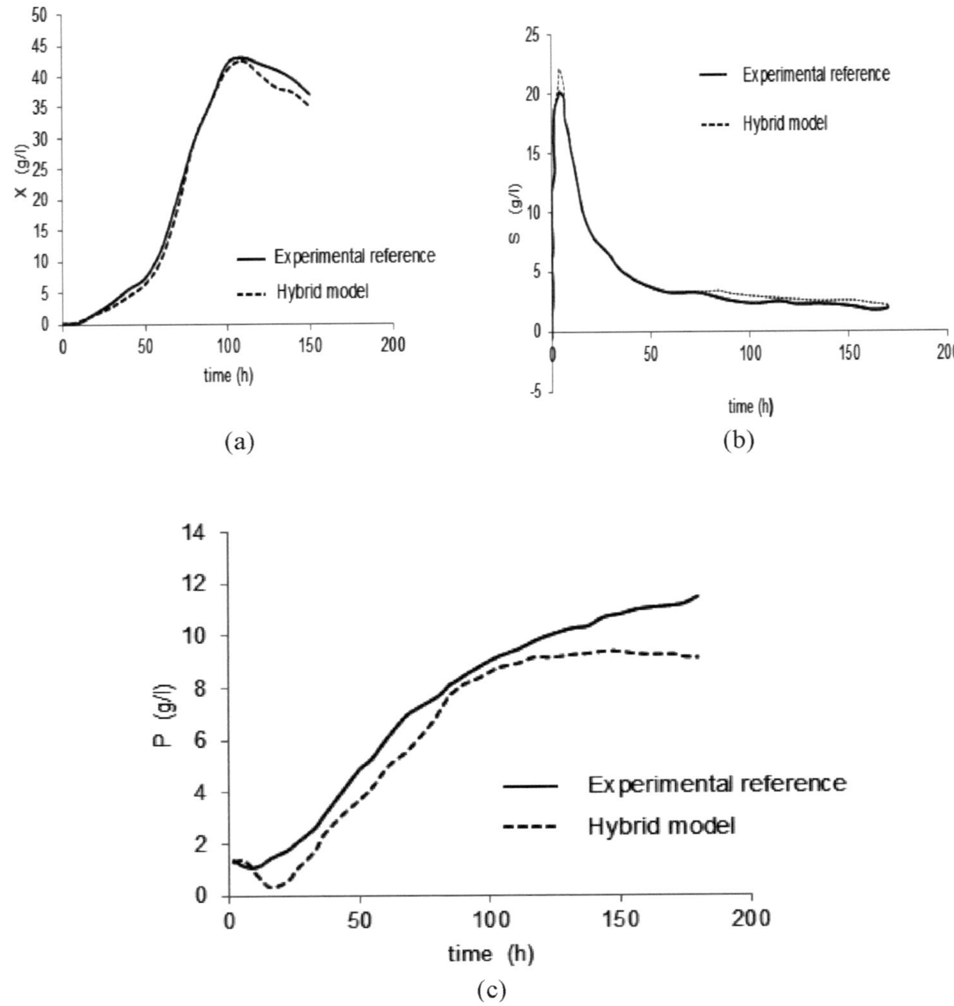

Figure 3.
Performance of hybrid model for biomass(a), substrate(b) and product (c) concentrations.

Where $S(t)$ and $S(t-1)$ are the present and past substrate concentrations, $X(t)$ X_t and X_{t-1} $X(t-1)$ are the present and past biomass concentrations respectively. For generalization, the experimental data is divided for training, testing and validation in the ratio 70:15:15. The training of neural network has been carried out in MATLAB. Levenberg–Marquardt algorithm is used with single hidden layer. Sigmoid and Linear activation functions are used for hidden and output layers respectively. Mean Square Error is the performance evaluation criterion and accordingly the number of hidden neurons is chosen to be 15. The number of iterations is fixed at 1000.

The response from the hybrid model obtained for training input is shown in **Figure 5(a)** and test input is shown in **Figure 5(b)**.

The process parameters μ and $Y_{X/S}$ change with respect to time and the corresponding state variable measurements. The time varying natures of the parameters are shown in **Figure 6**.

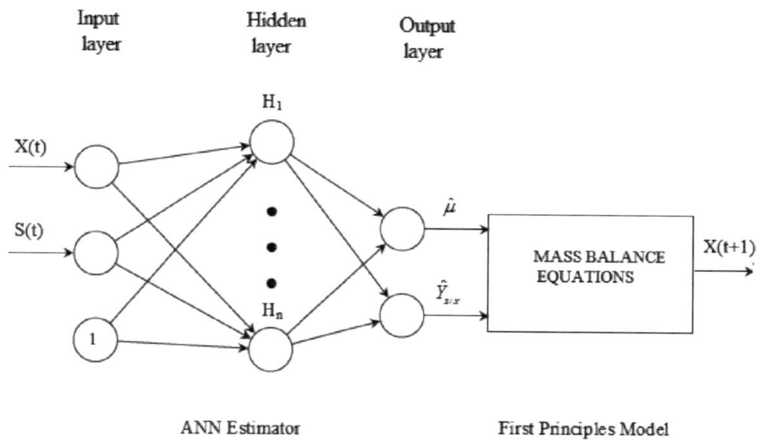

Figure 4.
Hybrid model structure.

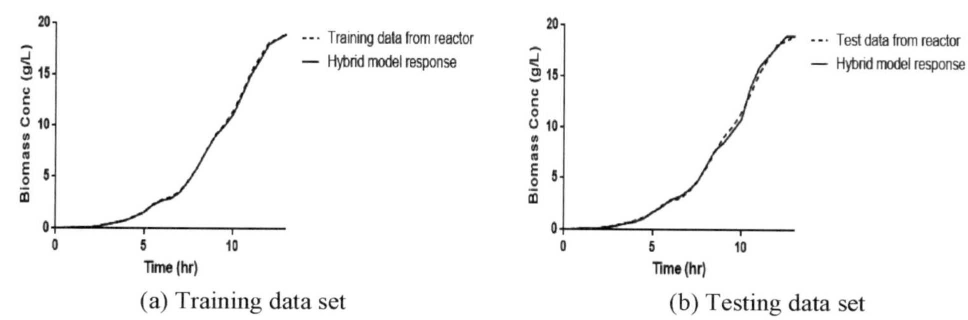

Figure 5.
Validation of hybrid model with training and testing data set.

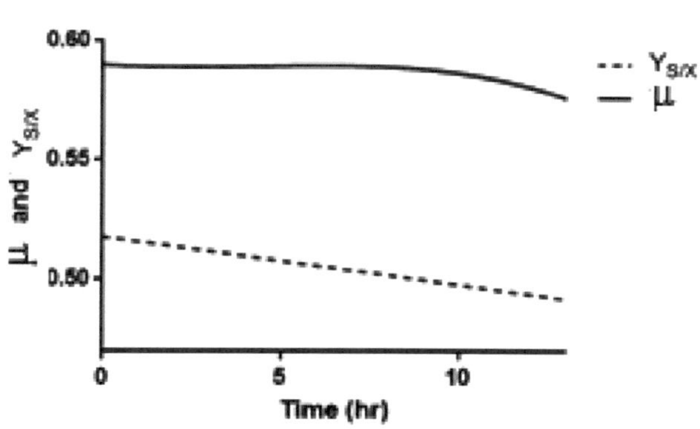

Figure 6.
Variation of kinetic parameters μ and YX/S with respect to time.

3.4 NARX models

Non-linear regressive models with exogenous input (NARX) are found to be effective for non-linear system identification, as they have good predictive capability. To predict the system behavior without a deep mathematical knowledge [33, 34], model identification with input–output measurements is generally used. For closer approximations of actual process, a NARX model is commonly employed [35]. The NARX is a recurrent dynamic network, with feedback connections enclosing several layers of the network. The defining equation for the NARX model is given in Eq. (17).

$$y(t) = f(y(t-1), y(t-2), y(t-3), \ldots \ldots y(t-ny), u(t-1) \ldots u(t-nu)) + e(t) \tag{17}$$

where $y(t)$ and $u(t)$ are output and input signal, 'f' is a nonlinear function, ny and nu are the output and input delays of nonlinear model and $e(t)$ is the error term. The next value of the dependent output signal is regressed on previous values of the output signal and previous values of an independent (exogenous) input signal. The successful estimation of process state by soft sensor greatly depends on input output data set. The input variable should be chosen such that it has a direct or indirect relation with the estimation variable. The microbial cell metabolism is influenced by pH value, agitation speed and substrate concentration inside the bioreactor vessel, hence it is directly related to the biomass concentration [36].

As a case study [37], a NARX model is developed for estimation of biomass concentration using the dataset of pH, agitation speed and substrate concentration values starting from the time of inoculation till the end of fed batch process. The experimental data is divided into training, testing and validation in the ratio 70:15:15. The inputs for the NARX model are present values of pH, substrate concentration (S), agitation speed and previous sampling instant biomass concentration, X(k-1) and the output is the estimated biomass concentration, X(k) as shown in **Figure 7**.

To obtain best performance from NARX model, two hidden layers are used and the numbers of hidden neurons in each layer are chosen based on Mean Square Error (MSE). ANN parameters used in the NARX model development are listed in **Table 3**.

Experimental validation of NARX Model is done with the help of capacitance probe. The annular dielectric probe when inserted into the bioreactor, gives a capacitance value that can be directly related to the concentration of biomass. The probe is useful particularly during fungal biomass cultivation due to the nonexistence of accurate offline measurements [38]. As the probe measures only the viable cells excluding the dead cells and insoluble substrates, it is an optimal choice for process validation. The probe generates a dielectric spectrum at 2 different frequencies given, based on cell size, morphology etc. [39]. In this case study, the

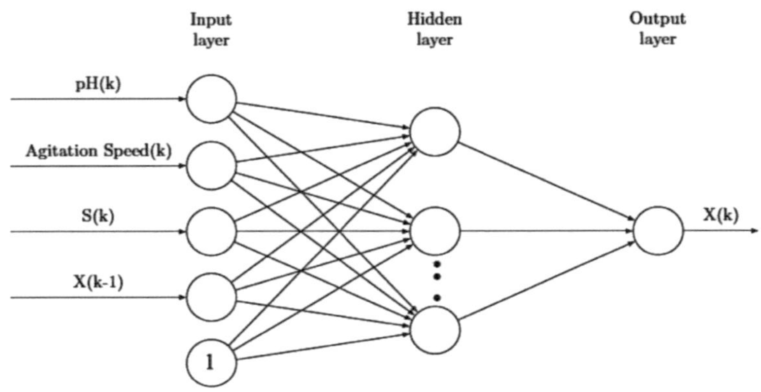

Figure 7.
NARX model for estimation of biomass concentration.

Parameters	Value
Architecture	Dynamic neural network (NARX)
Training Algorithm	Levenberg–Marquardt algorithm
Number of iterations	1000
Number of hidden layer	2
Number of hidden neurons in first layer	18
Number of hidden neurons in second layer	10
Bias	1
Activation function (hidden layer)	Sigmoid
Activation function (Output)	Linear
Performance Evaluation	Mean Square Error

Table 3.
ANN parameters for NARX model development.

capacitance probe is utilized in microbial mode and the biomass is measured at frequencies 0.6 MHz and 15 MHz. The 15 MHz reading is used as a form of auto zero and subtracted from the 0.6 MHz. Therefore, the capacitance of background matter is automatically subtracted from the signal. A resolution of 0.1 pF/cm on the instrument typically represents 106 Cells/ml, or 0.5 grams per liter. The dynamic NARX network are trained with different sets of input–output data. The response of NARX network to test inputs is shown in **Figure 8**.

It is inferred from **Figure 8**, that the trained dynamic NARX network can be used in the place of biosensor. The error response for single NARX network is shown in **Figure 9**.

The performance of NARX network is analyzed based on the performance criteria, Root Mean Square Error (RMSE) and coefficient of determination (R^2). RMSE is calculated by the Eq. (18).

$$RMSE = \sqrt{\sum_{i=1}^{n} \frac{(yi - ti)^2}{n}} \tag{18}$$

where N is the length of data, *yi is* the predict value and *ti is* the target value (Rafsanjani *et al.* 2016). The RMSE and R^2 values of NARX network are obtained as 0.01 and 0.8789 respectively and the correlation graph is shown in **Figure 10**.

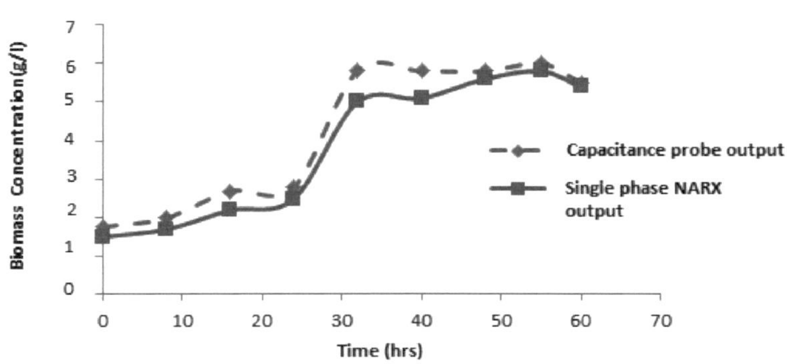

Figure 8.
Comparison of NARX model output with experimental biomass concentration data.

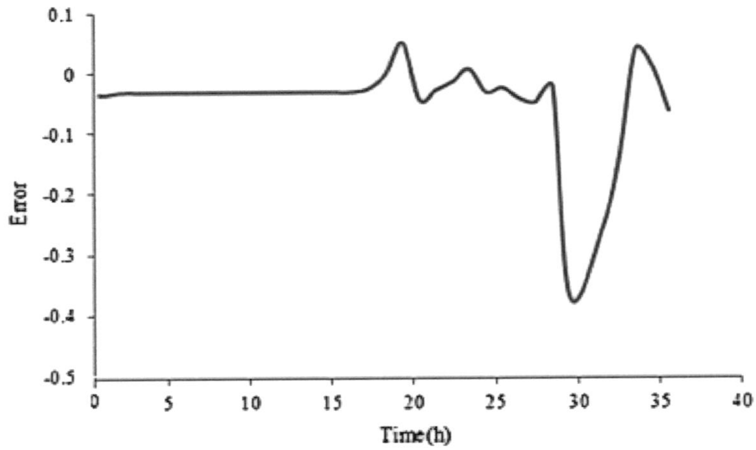

Figure 9.
Error plot for the NARX model.

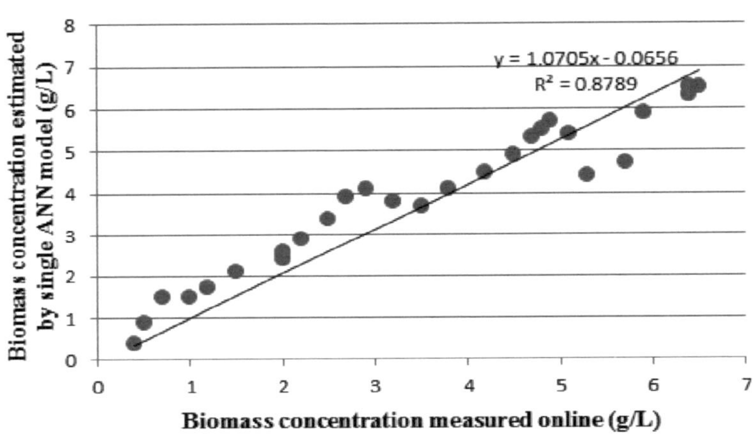

Figure 10.
Correlation graph between biomass concentrations monitored with real-time capacitance probe and estimated by the NARX model.

It is observed that the NARX model has a low value of RMSE, a very high value of R^2 and good correlation to real time probe data, which confirms that this dynamic neural network soft sensor performs well in the estimation of biomass concentration.

3.5 LSTM models

Recurrent Neural Networks (RNNs) are a type of deep networks that are structured to capture the temporal dependencies of the process effectively [40]. Long Short-Term Memory (LSTM) networks are a type of recurrent neural network capable of learning order dependence in sequence prediction problems. The LSTM network was invented with the goal of addressing the vanishing gradients problem. The key insight in the LSTM design was to incorporate nonlinear, data-dependent controls into the RNN cell, which can be trained to ensure that the gradient of the objective function with respect to the state signal does not vanish [41] and hence LSTMs are well suited for classification and prediction problems. The LSTM model

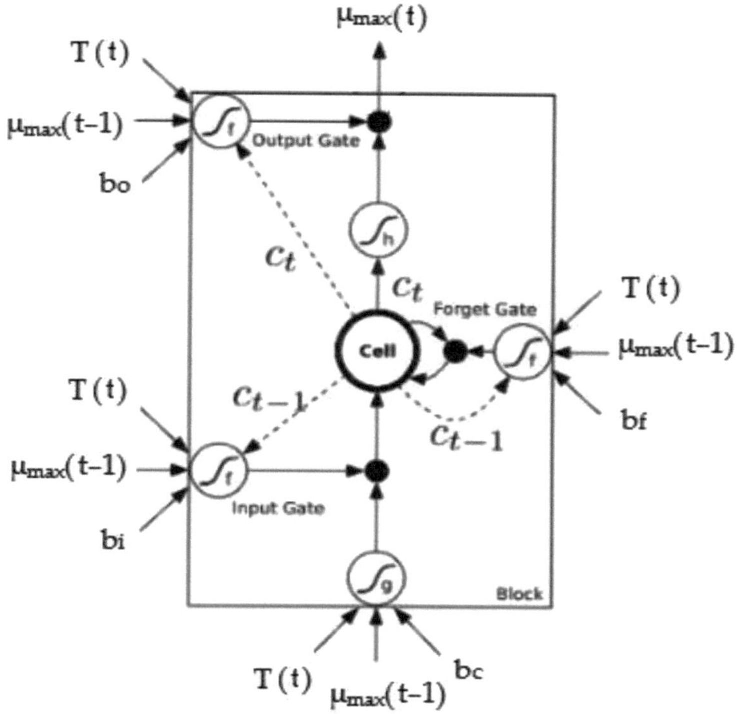

Figure 11.
LSTM model developed for maximum specific growth rate estimation.

developed for the estimation of maximum specific growth rate is shown in **Figure 11**. The input gate, forget gate and output gate equations are given by Eq. (19), (20) and (21) respectively.

The new memory cell and final memory cell equations are given as Eq. (22) and Eq. (23). The hidden state equation is represented in Eq. (24).

$$i^{(t)} = \sigma\left(W^{(i)}x^{(t)} + U^{(i)}h^{(t-1)}\right) \tag{19}$$

$$f^{(t)} = \sigma\left(W^{(f)}x^{(t)} + U^{(f)}h^{(t-1)}\right) \tag{20}$$

$$o^{(t)} = \sigma\left(W^{(o)}x^{(t)} + U^{(o)}h^{(t-1)}\right) \tag{21}$$

$$\bar{c}^{(t)} = \tanh\left(W^{(c)}x^{(t)} + U^{(c)}h^{(t-1)}\right) \tag{22}$$

$$c^{(t)} = f^{(t)}\bar{c}^{(t-1)} + i^{(t)}\bar{c}^{(t)} \tag{23}$$

$$h^{(t)} = o^{(t)} \tanh\left(c^{(t)}\right) \tag{24}$$

where,

$x^{(t)}$	Input word
$i^{(t)}$	Input state
$f^{(t)}$	forget state
$\bar{c}^{(t-1)}$, $\bar{c}^{(t)}$, $c^{(t)}$	Past, new and final memory
$h^{(t-1)}$,$h^{(t)}$	Previous and current hidden state

$o^{(t)}$	Output state
$W^{(i)}, W^{(f)}, W^{(c)}, W^{(o)}$	Weight vectors connecting previous and current hidden layers
$U^{(f)}, U^{(c)}, U^{(o)}$	Vectors connecting inputs to the current hidden layer
σ	Sigmoid activation function

Similar to the development of NARX model, the modeling of LSTM network also includes data collection, parameter determination, training, testing, validation. The modeling can be done either in MATLAB or using python coding. As a case Study, consider a LSTM network in which two hidden layers are chosen and the number of neurons in each hidden layer is varied till the MSE reaches minimum value. The parameters chosen to frame the LSTM network are listed in **Table 4**. The training to testing ratio is chosen as 67:33. The predicted values of maximum specific growth rate calculated from the LSTM model are presented in **Figure 12**.

The performance of LSTM model is evaluated by the statistical measures RMSE, R^2 and Accuracy factor (A_f). The A_f averages the distance between every point and the line of equivalence as a measure of finding the closeness between predicted and observed values. The RMSE, R^2, A_f values of the LSTM model are found to be 0.011, 0.994 and 1.024 respectively. The RMSE and A_f values are minimum which suggests that the LSTM predictive model fit well with the experimental data.

Parameter	Choice	
Batch size	50	
No. of hidden layers	2	
No. of hidden neurons	Hidden layer 1	Hidden layer 2
	15	47
Activation function	Sigmoid	
Optimizer	Adam	
Learning rate	0.01	
No. of epochs	800	

Table 4.
Parameters for LSTM model development.

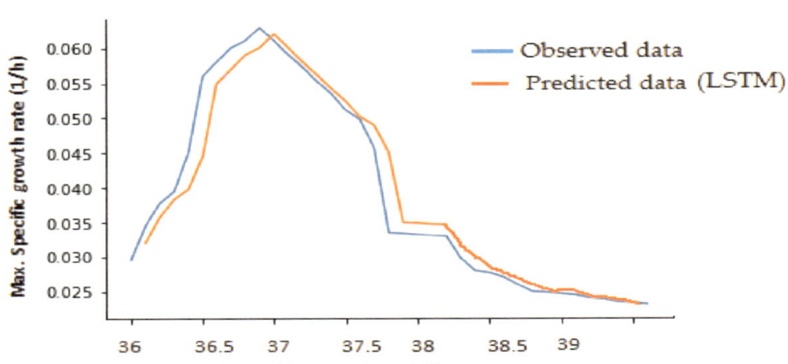

Figure 12.
Comparison of LSTM predicted maximum specific growth rate data with experimental data.

4. Conclusions

Several modeling techniques that will aid in the monitoring and estimation of fungal biomass in the presence of lignocellulosic substrates during fed-batch fermentation are discussed in this chapter. Moreover, the bioprocess models are validated with experimental data as discussed in case studies. The use of these soft sensors in industries with accompanying control system will improve the cellulase concentration yield.

Acknowledgements

The author acknowledge Anna University and Department of Science and Technology for their funding and support.

Conflict of interest

The author declares no conflict of interest.

Author details

Chitra Murugan
Department of Instrumentation Engineering, Madras Institute of Technology (MIT) Campus, Anna University, Chennai, India

*Address all correspondence to: chitramurugan.g@gmail.com

References

[1] Neves AA, Pereira DA, Vieira LM, Menezes JC. Real time monitoring biomass concentration in Streptomyces clavuligerus cultivations with industrial media using a capacitance probe. Journal of biotechnology. 2000 Nov 17; 84(1):45–52.

[2] Ellilä S, Fonseca L, Uchima C, Cota J, Goldman GH, Saloheimo M, Sacon V, Siika-Aho M. Development of a low-cost cellulase production process using Trichoderma reesei for Brazilian biorefineries. Biotechnology for biofuels. 2017 Dec 1;10(1):30.

[3] Sarkar N, Ghosh SK, Bannerjee S, Aikat K. Bioethanol production from agricultural wastes: an overview. Renewable energy. 2012 Jan 1;37(1):19–27.

[4] Lynd LR, Weimer PJ, Van Zyl WH, Pretorius IS. Microbial cellulose utilization: fundamentals and biotechnology. Microbiology and molecular biology reviews. 2002 Sep 1; 66(3):506–77.

[5] Lin J, Lee SM, Lee HJ, Koo YM. Modeling of typical microbial cell growth in batch culture. Biotechnology and Bioprocess Engineering. 2000 Oct 1; 5(5):382–5.

[6] Zelić B, Vasić-Rački Đ, Wandrey C, Takors R. Modeling of the pyruvate production with *Escherichia coli* in a fed-batch bioreactor. Bioprocess and Biosystems Engineering. 2004 Jul 1;26 (4):249–58.

[7] Jeong-Geol N, Kim HH, Chang YK. On-line estimation of cell growth from agitation speed in DO-stat culture of a filamentous microorganism, Agaricus blazei. Biotechnology and Bioprocess Engineering. 2005 Dec 1;10(6):571.

[8] Norsyahida A, Rahmah N, Ahmad RM. Effects of feeding and induction strategy on the production of BmR1 antigen in recombinant *E. coli*. Letters in applied microbiology. 2009 Nov;49(5):544–50.

[9] Holland T, Blessing D, Hellwig S, Sack M. The in-line measurement of plant cell biomass using radio frequency impedance spectroscopy as a component of process analytical technology. Biotechnology journal. 2013 Oct;8(10):1231–40.

[10] Slouka C, Wurm DJ, Brunauer G, Welzl-Wachter A, Spadiut O, Fleig J, Herwig C. A novel application for low frequency electrochemical impedance spectroscopy as an online process monitoring tool for viable cell concentrations. Sensors. 2016 Nov;16 (11):1900.

[11] Bogaerts P, Hanus R. On-line state estimation of bioprocesses with full horizon observers. Mathematics and computers in simulation. 2001 Jun 11;56 (4–5):425–41.

[12] James S, Legge R, Budman H. Comparative study of black-box and hybrid estimation methods in fed-batch fermentation. Journal of process control. 2002 Jan 1;12(1):113–21.

[13] Won H, Yoon-Keun C. An artificial neural network for biomass estimation from automatic pH control signal. Biotechnology and Bioprocess Engineering. 2006 Aug 1;11(4):351–6.

[14] Veloso AC, Rocha I, Ferreira EC. Monitoring of fed-batch *E. coli* fermentations with software sensors. Bioprocess and biosystems engineering. 2009 Apr 1;32(3):381–8.

[15] Zelić B, Bolf N, Vasić-Rački Đ. Modeling of the pyruvate production with *Escherichia coli*: comparison of mechanistic and neural networks-based models. Bioprocess and biosystems engineering. 2006 Jun 1;29(1):39–47.

[16] Garcíaa-Gimeno RM, Hervás-Martíanez C, Barco-Alcalá E, Zurera-Cosano G, Sanz-Tapia E. An artificial neural network approach to *Escherichia coli* O157: H7 growth estimation. Journal of Food Science. 2003 Mar;68(2): 639–45.

[17] Vlassides S, Ferrier JG, Block DE. Using historical data for bioprocess optimization: modeling wine characteristics using artificial neural networks and archived process information. Biotechnology and Bioengineering. 2001 Apr 5;73(1):55–68.

[18] Haider MA, Pakshirajan K, Singh A, Chaudhry S. Artificial neural network-genetic algorithm approach to optimize media constituents for enhancing lipase production by a soil microorganism. Applied biochemistry and biotechnology. 2008 Mar 1;144(3):225–35.

[19] Menezes Jr JM, Barreto GA. A new look at nonlinear time series prediction with NARX recurrent neural network. In2006 Ninth Brazilian Symposium on Neural Networks (SBRN'06) 2006 Oct 23 (pp. 160–165). IEEE.

[20] Whiteman JK, Kana EG. Comparative assessment of the artificial neural network and response surface modelling efficiencies for biohydrogen production on sugar cane molasses. BioEnergy Research. 2014 Mar 1;7(1): 295–305.

[21] Kmet T, Kmetova M. Adaptive critic design and Hopfield neural network based simulation of time delayed photosynthetic production and prey–predator model. Information Sciences. 2015 Feb 10;294:586–99.

[22] Dewasme L. Neural network-based software sensors for the estimation of key components in brewery wastewater anaerobic digester: an experimental validation. Water Science and Technology. 2019 Nov 15;80(10): 1975–85.

[23] Huang Y, Kangas LJ, Rasco BA. Applications of artificial neural networks (ANNs) in food science. Critical reviews in food science and nutrition. 2007 Apr 1;47(2):113–26.

[24] Wu Z, Shi X. Optimization for high-density cultivation of heterotrophic Chlorella based on a hybrid neural network model. Letters in applied microbiology. 2007 Jan;44(1):13–8.

[25] Mohamed MS, Tan JS, Mohamad R, Mokhtar MN, Ariff AB. Comparative analyses of response surface methodology and artificial neural network on medium optimization for Tetraselmis sp. FTC209 grown under mixotrophic condition. The Scientific World Journal. 2013 Oct;2013.

[26] Ma L, Li C, Yang Z, Jia W, Zhang D, Chen S. Kinetic studies on batch cultivation of Trichoderma reesei and application to enhance cellulase production by fed-batch fermentation. Journal of biotechnology. 2013 Jul 20; 166(4):192–7.

[27] Oliveira R. Combining first principles modelling and artificial neural networks: a general framework. Computers & Chemical Engineering. 2004 May 15;28(5):755–66.

[28] Peres J, Oliveira R, De Azevedo SF. Knowledge based modular networks for process modelling and control. Computers & Chemical Engineering. 2001 May 1;25(4–6):783–91.

[29] Bhutani N, Rangaiah GP, Ray AK. First-principles, data-based, and hybrid modeling and optimization of an industrial hydrocracking unit. Industrial & engineering chemistry research. 2006 Nov 8;45(23):7807–16.

[30] Corazza FC, Calsavara LP, Moraes FF, Zanin GM, Neitzel I. Determination of inhibition in the enzymatic hydrolysis of cellobiose using hybrid neural modeling. Brazilian

Journal of Chemical Engineering. 2005 Mar;22(1):19–29.

[31] Van Lith PF, Betlem HB, Roffel B. Hybrid fuzzy-first principles modelling using fuzzy clustering, genetic algorithms and neuro-fuzzy methods: a comparison. IFAC Proceedings Volumes. 2000 Jun 1;33(10):455–60.

[32] Valente E, Rocha I, Rocha M. Modelling fed-batch fermentation processes: an approach based on artificial neural networks. In2nd International Workshop on Practical Applications of Computational Biology and Bioinformatics (IWPACBB 2008) 2009 (pp. 30–39). Springer, Berlin, Heidelberg.

[33] Linder J, Enqvist M. Identification of systems with unknown inputs using indirect input measurements. International Journal of Control. 2017 Apr 3;90(4):729–45.

[34] Ang J, Ingalls B, McMillen D. Probing the input–output behavior of biochemical and genetic systems: system identification methods from control theory. InMethods in enzymology 2011 Jan 1 (Vol. 487, pp. 279–317). Academic Press.

[35] Chetouani Y. Non-linear modeling of a reactor-exchanger by using NARX neural networks. InProceedings of European Congress of Chemical Engineering (ECCE-6) Copenhagen 2007 Sep 16 (Vol. 1620).

[36] Won H, Yoon-Keun C. An artificial neural network for biomass estimation from automatic pH control signal. Biotechnology and Bioprocess Engineering. 2006 Aug 1;11(4):351–6.

[37] Murugan C, Natarajan P. Estimation of fungal biomass using multiphase artificial neural network based dynamic soft sensor. Journal of microbiological methods. 2019 Apr 1;159:5–11.

[38] Carvell J, Lee M, Bhat AR. Recent developments in scaling down and using single use probes for measuring the live cell concentration by dielectric spectroscopy. InBMC proceedings 2015 Dec 1 (Vol. 9, No. S9, p. P46). BioMed Central.

[39] Kiviharju K, Salonen K, Moilanen U, Eerikäinen T. Biomass measurement online: the performance of in situ measurements and software sensors. Journal of industrial microbiology & biotechnology. 2008 Jul 1;35(7):657–65.

[40] Wong WC, Chee E, Li J, Wang X. Recurrent neural network-based model predictive control for continuous pharmaceutical manufacturing. Mathematics. 2018 Nov;6(11):242.

[41] Sherstinsky A. Fundamentals of recurrent neural network (rnn) and long short-term memory (lstm) network. Physica D: Nonlinear Phenomena. 2020 Mar 1;404:132306.